Electronics

Electromechanics

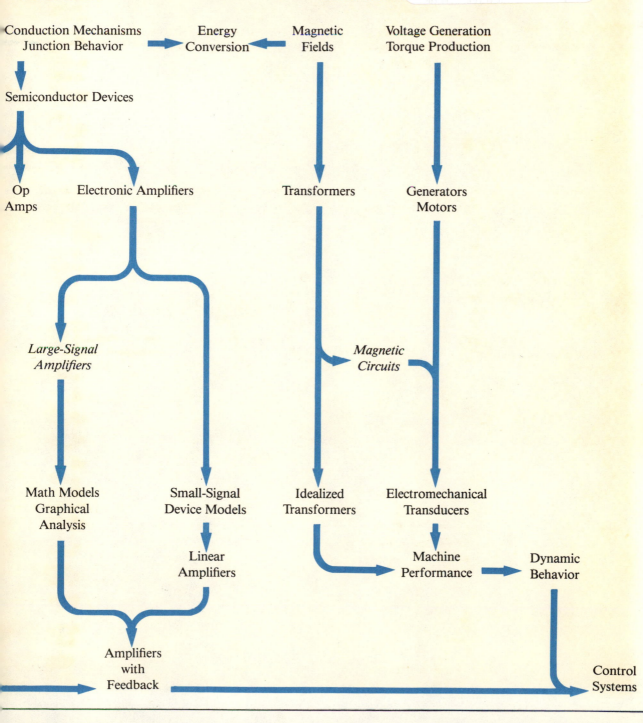

Conduction Mechanisms
Junction Behavior

→ Energy
Conversion ←

Magnetic
Fields

Voltage Generation
Torque Production

Semiconductor Devices

Op
Amps

Electronic Amplifiers

Transformers

Generators
Motors

*Large-Signal
Amplifiers*

*Magnetic
Circuits*

Math Models
Graphical
Analysis

Small-Signal
Device Models

Idealized
Transformers

Electromechanical
Transducers

Linear
Amplifiers

Machine
Performance

Dynamic
Behavior

Amplifiers
with
Feedback

Control
Systems

Circuits, Devices, and Systems

FOURTH EDITION

Also by Ralph J. Smith

Engineering as a Career, Fourth Edition, 1983

Electronics: Circuits and Devices, Second Edition, 1980

Circuits, Devices, and Systems

A First Course in Electrical Engineering

FOURTH EDITION

Ralph J. Smith

Professor of Electrical Engineering, Stanford University

JOHN WILEY & SONS
New York • Chichester • Brisbane • Toronto

Cover photograph by Chuck O'Rear

Library of Congress Cataloging in Publication Data
Smith, Ralph Judson.
  Circuits, devices, and systems.

  Includes index.
  1. Electrical engineering.   I. Title.
TK145.S616   1984   621.3   83-16874
ISBN 0-471-87496-5

Printed in the United States of America

10 9 8 7 6 5 4

To the students whose desire to learn and willingness to work challenged me to do my best

Preface

This fourth edition of *Circuits, Devices, and Systems* reflects changes in the rapidly advancing field of electrical engineering, and it incorporates many helpful suggestions from teachers and students who used the previous edition. The basic objectives and pedagogical approaches remain the same; the aims of the revision have been to provide a more modern emphasis in applications, a more effective arrangement of material, and an improved set of examples and assignments.

The major change is the modernization of the electronics material to place proper emphasis on operational amplifiers, digital devices, and microprocessors. The microprocessor is more than just a new device with a host of diverse applications. Its computing power, small size, and low cost have completely changed our way of thinking. It has altered our philosophy of design; to solve an information processing problem we now take a mass-produced "brain" with a flexible "memory" and "instruct" it to solve our problem. Fortunately, the microprocessor is an ideal subject for study; it integrates our knowledge of registers, counters, memories, and logic units. Mastery requires no difficult-to-understand concepts or hard-to-acquire mathematics. Students who understand microprocessors have gained a general understanding of digital systems as well as a powerful tool for solving engineering problems. Teachers have the satisfaction of knowing that they have made a significant contribution to their students' education.

OBJECTIVES

The purpose of this book is to provide a working knowledge of the fundamental principles of electrical engineering and practice in applying abstract concepts to real problems. The approach is to employ modern techniques and interesting applications in developing mastery of important principles and powerful methods. The ultimate goal is to develop in students the ability to predict the behavior of common circuits and systems and to design simple subsystems by specifying the critical device parameters.

It is assumed that students using this book have completed basic physics courses in mechanics, electricity, and magnetism; however, the essential physical concepts are reviewed as they are introduced. It is also assumed that students have a working knowledge of differential and integral calculus, and the discussions and assignments are designed to use the mathematical skills that most sophomore engineering students possess. Prior knowledge of differential equations is not required; the analytical technique employed here provides a good introduction to the formulation and solution of first- and second-order linear differential equations in engineering situations.

For this first course in electrical engineering, topics and techniques have been selected to provide the basis for further work in circuits, electronics, digital systems, electromechanics, or control systems. For electrical engineers who will later study all these in detail, a primary objective of this book is to provide a clear picture of the relationships among these subject areas. Majors in other fields should gain a good grasp of the fundamentals that will help them select courses for subsequent study and provide a solid background for such study.

CONTENTS

Part I, *Circuits*, is designed to provide a solid introduction to network analysis and applications; it has been rearranged to improve teachability. Chapters 1 and 2 lay the foundations of electrical quantities and circuit principles. Chapter 3, Signal Processing Circuits, starts with a description of signals and waveforms; then ideal amplifiers and diodes are introduced to illustrate interesting applications of simple circuit concepts. Chapters 4, 5, and 6 provide a careful development of natural, forced, and complete response using the powerful impedance concept. Chapters 7 and 8 treat important topics in ac circuits and general network analysis. Chapter 9 introduces systems concepts such as feedback and transfer functions and provides an appropriate termination to Part I.

The first three chapters of Part II, *Electronics*, develop a quantitative understanding of electron behavior, semiconductor physics, and device characteristics. Electron motion is described in Chapter 10, and the cathode-ray oscilloscope is introduced here for use in accompanying lab work. Chapter 11 focuses on the semiconductor diode; conduction, doping, and junction phenomena are used in deriving the external characteristics and a simple mathematical model. Chapter 12 focuses on transistors and thyristors and their integrated circuit forms; explanations of physical phenomena lead to quantitative descriptions of external characteristics.

The next three chapters provide an introduction to digital electronics. Chapter 13 uses simple models of diodes and transistors to explain the operation of electronic switches, logic gates, and flip-flops. In Chapter 14 the emphasis is on binary representation, the analysis and synthesis of logic circuits, and their application in registers, counters, and memories. Chapter 15 provides a first-level treatment of the microprocessor: basic concepts, computer architecture, programming, and the application of practical microprocessors to information processing.

The next four chapters provide an introduction to linear electronics. Chapter 16 describes practical operational amplifiers and shows how inexpensive IC op amps can be used in the design of practical amplifiers, buffers, integrators, converters, regulators, oscillators, or analog computers. Chapter 17 treats amplifiers in general and large-signal discrete-transistor amplifiers in particular; bias design and efficiency calculations for audiofrequency amplifiers are included. In Chapter 18, small-signal models of varying levels of sophistication are derived for field-effect and bipolar junction transistors. The models are applied to the analysis and design of small-signal amplifiers in Chapter 19, and the virtues of positive and negative feedback are explored.

Part III, *Electromechanics,* starts with a brief look at energy conversion phenomena and then focuses on devices employing magnetic coupling. The magnetic field and circuit concepts established in Chapter 20 are applied to transformers in Chapter 21, and simple circuit models are derived for predicting transformer performance. Basic principles of electromechanics are derived in Chapter 22 and applied to direct-current machines (Chapter 23) and alternating-current machines (Chapter 24). The emphasis here is on predicting the steady-state performance of conventional machines, but dynamic behavior is discussed in connection with automatic control systems (Chapter 25).

APPROACHES

The *Concept Flow Chart* on the front endpaper indicates how unity has been achieved in treating diverse topics. To meet the needs and interests of engineering students, a quantitative physical explanation accompanies the description of each new device. For example, an explanation of conduction in semiconductors and potential barrier effects accompanies the description of diodes and transistors, and basic principles of electromechanics are used in deriving expressions for rotating machine performance.

To provide coherence over a broad range of topics, a few basic tools are used repeatedly. For example, the same modeling process is employed for transistors, transformers, and rotating machines. The use of simplifying assumptions to reduce a complicated system with inherent nonlinearities to a linear, relatively simple form is emphasized. Other frequently used techniques are: the pole-zero concept, the impedance function, the transfer function, the load-line approach, and block diagrams.

The emphasis throughout is on principles; however, the topics selected are those of practical importance in modern electrical engineering. The illustrative examples are designed to appeal to students with various interests. Although a wide range of topics is included, this is more than just a survey. By limiting the discussions to the more basic aspects of each major topic, there is time to treat these aspects with sufficient rigor so that the results are meaningful, quantitative, and useful. (Students report that their experience with microprocessors makes them more employable.)

As in the previous editions, I have tried to help the student to learn and the instructor to teach. Again the assignments at the end of each chapter are grouped into three categories. The *Review Questions* are primarily for the student's own use in testing his or her understanding of the concepts and terminology introduced in that chapter. The *Exercises* vary in difficulty but, in general, are straightforward applications of the new principles to specific questions. The *Problems* are more involved and may require extending a concept, making simplifying assumptions, or putting ideas together in a design. Answers to Selected Exercises are included in the *Appendix*.

The *Instructor's Manual,* available from the publisher to teachers adopting the book for class use, contains complete solutions to the *Exercises* and the *Problems*. Also included are suggested lecture and reading schedules for courses of different length and emphasis, along with a brief description of each chapter from the teacher's viewpoint. Because laboratory work is such a valuable complement to the text discussions, an appropriate laboratory course is described and sample experiment instructions are included. Here at Stanford, we offer to all engineering students a three-quarter course in circuits, electronics, and electromechanics; electronics is *not* prerequisite to electromechanics. I welcome correspondence on any aspect of the book or the manual.

ACKNOWLEDGMENTS

It is a pleasure to acknowledge the assistance of many persons in this revision. Stanford colleagues Robert Helliwell, Larry Manning, and Umran Inan have given me valuable suggestions for improvement. Kurt Jaggers of Rolm Corporation and Celeste Baranski of Grid Systems made significant contributions to the material on digital systems and microprocessors. The revised manuscript was read critically by Professor Herbert Hacker Jr. of Duke University; his incisive comments have been very helpful. Solutions to the Exercises and Problems were prepared by Anne-Marie Fauchet, Aziz Inan, and Mongkol Dejnakarintra; their careful work on this enormous task is greatly appreciated.

Production of the book was supervised by Mary Forkner of Publication Alternatives. I benefitted from her professional competence, her sensitivity to an author's needs, and her unfailing good humor. I am grateful to Merrill Floyd and Gene Davenport of John Wiley & Sons for making her talents available on this complex production job.

For his enthusiastic support of my writing and teaching, I thank Department Chairman Robert L. White. For her skillful typing from very rough notes, I thank Yvonne Oden. Last, but by no means least, I am grateful to my wife Louise for her careful collating, patient proofreading, and cheerful performance of all the details of preparing another revision.

Stanford, California   Ralph J. Smith

Contents

P A R T **III** Electromechanics

PART I

Circuits

CHAPTER 1

Electrical Quantities

Introduction

Definitions and Laws

Circuit Elements

Energy and information are the hallmarks of civilization, and electrical engineers influence social change by improving energy conversion and information processing. In transformation of the power of falling water to electrical form in a remote hydroelectric plant, transmission at high voltage to urban load centers, and utilization in lights, motors, and industrial processes, *energy* is the significant quantity. In an observation satellite converting sensor output to digital form, performing data reduction according to instructions, and sending the results back to earth for display on your personal computer, energy is merely the means for processing *information*.

In general, electrical engineers are responsible for optimizing the generation, storage, transmission, control, and use of electrical energy or information. But all engineers employ electrical and electronic devices and systems, and the effective application of components and concepts depends on mastery of the fundamentals of electrical engineering. Furthermore, techniques of device modeling and system analysis originally developed to handle purely electrical problems now find application in all branches of engineering.

INTRODUCTION

Observations over the past two centuries, some random and some the result of careful experiment, have been organized, interpreted, and made available to us in the form of a relatively few basic principles. Ingenious methods have been developed for applying these principles to the analysis of existing devices and systems and in the creation of new and improved designs. The new devices and systems are being applied in an ever increasing diversity of products that affect the quality of life. The combination of fundamental principles, general methods, and interesting applications constitute the subject matter of this book.

Forces and Fields

Electric charges are defined by the forces they exert on one another; experimentally, the forces are found to depend on the magnitudes of the charges, their relative positions, and their velocities. Forces due to the position of charges are called *electric forces,* and those due to the velocity of charges are called *magnetic forces*. All electric and magnetic phenomena of interest in electrical engineering can be explained in terms of the forces between charges.

In a television picture tube, the electrodes are designed to provide accelerating and deflecting *electric fields* to control the path of electrons in forming the picture. In a computer memory bank, minute magnetic "bubbles" become magnetized by pulses of current and then "remember" the digital information in the form of *magnetic fields*. In a motor, the combination of steel armature and copper winding is shaped to create an intense magnetic field through which the conductors pass as they develop torque. Fields are characteristically distributed throughout a region and must be defined in terms of two or three dimensions.

Circuits

In contrast to fields, the behavior of a *circuit* can be completely described in terms of a single dimension, the position along the path constituting the circuit. In an electric circuit, the variables of interest are the voltage and current at various points along the circuit. In circuits in which voltages and currents are constant (not changing with time), the currents are limited by resistances. In the case of a battery being charged by a generator, 10 meters of copper wire may provide a certain resistance to limit the current flow. The same effect could be obtained by 10 centimeters of resistance wire, or by 1 centimeter of resistance carbon, or by 1 millimeter of semiconductor.

When the dimensions of a component are unimportant and the total effect can be considered to be concentrated at a point or "lumped," the component can be represented by a *lumped parameter*. In contrast, the behavior of 100 meters of copper wire as an antenna is dependent on its dimensions and the way in which voltage and current are distributed along it; an antenna must be represented by *distributed parameters*. In this book, we are concerned with lumped parameter circuits only.

Devices

Circuits are important in guiding energy within *devices* and also to and from devices that are combined into *systems*. Electrical devices perform such functions as generation, amplification, modulation, and detection of signals. For example, at an AM radio broadcasting station a modulator changes the amplitude of the transmitted wave in accordance with a musical note to produce amplitude modulation. In a microcomputer, logic gates, counters, and registers are organized to receive, process, and store digital signals. A *transducer* is a device that converts energy or information from one form to another; a microphone is a transducer that converts the acoustical energy in an input sound wave into the electrical energy of an output current. A cathode-ray tube converts the information stored in a solid-state memory to a graphic display.

Systems

Systems incorporate circuits and devices to accomplish desired results. A communication system includes a microphone transducer, an oscillator to provide a high-frequency carrier for efficient radiation, a modulator to superimpose the sound signal on the carrier, an antenna to radiate the electromagnetic wave into space, a receiving antenna, a detector for separating the desired signal from the carrier, various amplifiers and power supplies, and a loudspeaker to transduce electrical current into a replica of the original acoustic signal. A space-vehicle guidance system includes a transducer to convert a desired heading into an electrical signal, an error detector to compare the actual heading with the desired heading, an amplifier to magnify the difference, an actuator to energize the precise control jets, a sensor to determine the actual heading, and a feedback loop to permit the necessary comparison.

Models

Circuits are important for another reason: frequently it is advantageous to represent a device or an entire system by a *circuit model*. Assume that the 10 meters of copper wire mentioned previously is wound into the form of a multiturn coil and that a voltage of variable frequency is applied. If the ratio of applied voltage V to resulting current I is measured as a function of frequency, the observations will be as shown in Fig. 1.1. Over region A, the coil can be represented by a single lumped parameter (resistance R); in other words, the results obtained from the circuit of Fig. 1.1a are approximately the same as the results obtained from the actual coil of wire. Similarly, the behavior of the coil in region B can be represented by the circuit in Fig. 1.1b which

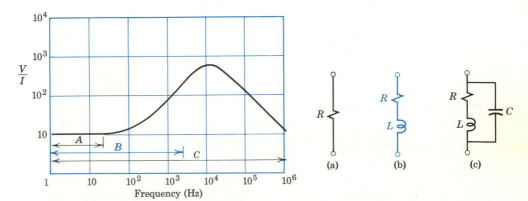

Figure 1.1   Coil characteristics and circuit models.

contains another lumped parameter (inductance L). To represent the coil over a wide range of frequencies, an additional parameter (capacitance C) is necessary, and the circuit model of Fig. 1.1c is used.

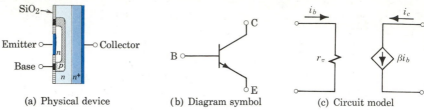

(a) Physical device   (b) Diagram symbol   (c) Circuit model

Figure 1.2   Transistor representation.

The technique of representing, approximately, a complicated physical device by a relatively simple model is an important part of electrical engineering. In this book we use circuit models to represent such devices as transistors and synchronous motors. One advantage of such a model is that it is amenable to mathematical analysis. Representing a transistor by a circuit model (Fig. 1.2) permits the use of well-known circuit laws in predicting the behavior of the actual transistor in an amplifier circuit.

A Preview

This book has three parts. The first, Circuits, is devoted to developing the set of basic principles and powerful techniques needed to predict the behavior of a wide variety of circuits. The design of simple circuits that perform interesting and useful functions is included to illustrate the application of circuit concepts. The final chapter of Part I provides an introduction to systems engineering.

In the second part, Electronics, we study the physical principles of electron motion and see how these principles are employed in diodes, transistors, and integrated circuits. Then we examine applications of basic elements in digital devices, microprocessors, operational amplifiers, and other types of amplifiers.

The last part, Electromechanics, starts with a look at energy sources and magnetic fields. Then we study the fundamentals of electromechanical energy conversion and see their application in practical devices with emphasis on rotating machines.

Although this book is primarily concerned with electrical phenomena, many of the principles, techniques, and approaches discussed are of basic importance and they find application throughout engineering.

DEFINITIONS AND LAWS

In beginning the study of circuits, we must first define the important circuit quantities and adopt a standard set of units, symbols, and abbreviations. Much of this material is a review of basic physics, but it deserves careful attention because it constitutes the "language" in which ideas are presented, concepts formed, and conclusions stated.

Next we take a new look at three laws based on early experiments conducted on resistors, inductors, and capacitors. Using these experimental results and modeling techniques, we invent idealized circuit components with highly desirable characteristics and then see how they behave in circuits and how they transform energy.

The International System of Units

In engineering, we must be able to describe physical phenomena quantitatively in terms that will mean the same to everyone. We need a standard set of units consistent among themselves and reproducible any place in the world. In electrical engineering, we use the SI (System International) system in which the *meter* is the unit of length, the *kilogram* the unit of mass, and the *second* the unit of time. Another basic quantity is temperature, which in the SI system is measured in *kelvins*. To define electrical quantities, an additional unit is needed; taking the *ampere* as the unit of electric current satisfies this requirement. The *candela* is needed to define illumination quantities. All quantities encountered in this book can be defined in terms of the six units displayed in Table 1-1. When data are specified in other units, they are first converted to SI units and then substituted in the applicable equations. Three conversions frequently needed are: 1 meter = 39.37 inches, 1 kilogram = 2.205 pounds (mass), and 1 newton = 0.2248 pound (force). Other useful conversion factors are given in the Appendix.

Table 1-1   Basic Quantities

Quantity	Symbol	Unit	Abbreviation
Length	l	meter	m
Mass	m	kilogram	kg
Time	t	second	s
Temperature	τ	kelvin	K
Current	i	ampere	A
Luminous intensity	I	candela	cd

Definitions

For quantitative work in circuits, we need to define the quantities displayed in Table 1-2 on page 8. These are probably familiar from your previous study, but a brief review here may be helpful.

Force. A force of 1 newton is required to cause a mass of 1 kilogram to change its velocity at a rate of 1 meter per second per second. In this text we are concerned primarily with electric and magnetic forces.

Energy. An object requiring a force of 1 newton to hold it against the force of gravity (i.e., an object "weighing" 1 newton) receives 1 joule of potential energy when it is raised 1 meter. A mass of 1 kilogram moving with a velocity of 1 meter per second possesses $\frac{1}{2}$ joule of kinetic energy.

Table 1-2 Important Derived Quantities

Quantity	Symbol	Definition	Unit	Abbreviation	(Alternate)
Force	f	push or pull	newton	N	$(kg \cdot m/s^2)$
Energy	w	ability to do work	joule	J	$(N \cdot m)$
Power	p	energy/unit of time	watt	W	(J/s)
Charge	q	quantity of electricity	coulomb	C	$(A \cdot s)$
Current	i	rate of flow of charge	ampere	A	(C/s)
Voltage	v	energy/unit charge	volt	V	(W/A)
Electric field strength	ε	force/unit charge	volt/meter	V/m	(N/C)
Magnetic flux density	B	force/unit charge momentum	tesla	T	(Wb/m^2)
Magnetic flux	ϕ	integral of magnetic flux density	weber	Wb	$(T \cdot m^2)$

Power. Power measures the rate at which energy is transformed. The transformation of 1 joule of energy in 1 second represents an average power of 1 watt. In general, instantaneous power p and average power P are defined by

$$p = \frac{dw}{dt} \quad \text{and} \quad P = \frac{W}{T} \tag{1-1}$$

Charge. The quantity of electricity is electric charge, a concept useful in explaining physical phenomena. Charge is said to be "conservative" in that it can be neither created nor destroyed. It is said to be "quantized" because the charge on 1 electron (1.602×10^{-19} C) is the smallest amount of charge that can exist. The coulomb can be defined as the charge on 6.24×10^{18} electrons, or as the charge experiencing a force of 1 newton in an electric field of 1 volt per meter, or as the charge transferred in 1 second by a current of 1 ampere.

Current. Electric field effects are due to the presence of charges; magnetic field effects are due to the motion of charges. The current through an area A is defined by the electric charge passing through the area per unit of time. In general, the charges may be positive and negative, moving through the area in both directions. The current is the *net* rate of flow of *positive* charges, a scalar quantity. In the specific case of positive charges moving to the right and negative charges to the left, the net effect of both actions is positive charge moving to the right; the instantaneous current to the right is given by the equation

$$i = \frac{dq}{dt} = +\frac{dq^+}{dt} + \frac{dq^-}{dt} \tag{1-2}$$

In a neon light, for example, positive ions moving to the right and negative electrons moving to the left contribute to the current flowing to the right. In a current of 1 ampere, charge is being transferred at the rate of 1 coulomb per second.

Voltage. The energy-transfer capability of a flow of electric charge is determined by the electric potential difference or voltage through which the charge moves. A charge of 1 coulomb receives or delivers an energy of 1 joule in moving through a voltage of 1 volt or, in general, instantaneous voltage is defined by

$$v = \frac{dw}{dq} \tag{1-3}$$

The function of an energy source such as an automobile battery is to add energy to the current; a 12-V battery adds twice as much energy per unit charge as a 6-V battery.

Electric Field Strength. The "field" is a convenient concept in calculating electric and magnetic forces. Around a charge we visualize a region of influence called an "electric field." The electric field strength $\mathcal{E}$, a vector, is defined by the magnitude and direction of the force **f** on a unit positive charge in the field. In vector notation the defining equation is

$$\mathbf{f} = q\mathcal{E} \tag{1-4}$$

where magnitude $\mathcal{E}$ could be measured in newtons per coulomb. However, bearing in mind the definitions of energy and voltage, we note that

$$\frac{\text{force}}{\text{charge}} = \frac{\text{force} \times \text{distance}}{\text{charge} \times \text{distance}} = \frac{\text{energy}}{\text{charge} \times \text{distance}} = \frac{\text{voltage}}{\text{distance}}$$

and electric field strength in newtons per coulomb is just equal and opposite to the *voltage gradient*[†] or

$$\mathcal{E} = -\frac{dv}{dl} \qquad \text{in volts per meter} \tag{1-5}$$

Magnetic Flux Density. Around a moving charge or current we visualize a region of influence called a "magnetic field." In a bar magnet the current consists of spinning electrons in the atoms of iron; the effect of this current on the spinning electrons of an unmagnetized piece of iron results in the familiar force of attraction. The intensity of the magnetic effect is determined by the magnetic flux density **B**, a vector defined by the magnitude and direction of the force **f** exerted on a charge q moving in the field with velocity **u**. In vector notation the defining equation is

$$\mathbf{f} = q\mathbf{u} \times \mathbf{B} \tag{1-6}$$

A force of 1 newton is experienced by a charge of 1 coulomb moving with a velocity of 1 meter per second normal to a magnetic flux density of 1 tesla.

Magnetic Flux. Historically, magnetic fields were first described in terms of *lines of force* or *flux*. The flux lines (so called because of their similarity to flow lines in a moving fluid) are convenient abstractions that can be visualized in the familiar iron-

---

[†]In the vicinity of a radio transmitter, the radiated field may have a strength of a millivolt per meter; in other words, one meter of antenna may develop a millivolt of signal voltage that can be fed into a receiver for amplification.

filing patterns. Magnet flux ϕ (phi) in webers is a total quantity obtained by integrating magnetic flux density over area **A**. The defining equation is

$$\phi = \int \mathbf{B} \cdot d\mathbf{A} \tag{1-7}$$

Because of this background, magnetic flux density is frequently considered as the derived unit and expressed in webers per square meter. In this text, however, we consider magnitude B in teslas as the primary unit.

Electrical Power and Energy

A common problem in electric circuits is to predict the power and energy transformations in terms of the expected currents and voltages. Since, by definition, $v = dw/dq$ and $i = dq/dt$, instantaneous power is

$$p = \frac{dw}{dt} = \frac{dw}{dq}\frac{dq}{dt} = vi \tag{1-8}$$

Therefore, total energy is

$$w = \int p \; dt = \int vi \; dt \tag{1-9}$$

EXAMPLE 1

The "electron gun" of a cathode-ray tube provides a beam of high-velocity electrons.

(a) If the electrons are accelerated through a potential difference of 20,000 V over a distance of 4 cm (Fig. 1.3), calculate the average field strength.

(b) Calculate the power supplied to a beam of 50 million billion electrons per second.

(a) By definition, $\mathcal{E} = -dv/dl$ or

$$\mathcal{E}_{av} = \frac{\Delta v}{\Delta l} = \frac{20,000}{0.04} = 5 \times 10^5 \text{ V/m}$$

(b) By definition, $i = dq/dt$ or

$$i = \frac{\text{charge}}{\text{electron}} \times \frac{\text{electrons}}{\text{second}}$$

$$= 1.6 \times 10^{-19} \times 50 \times 10^6 \times 10^9 = 0.008 \text{ A}$$

By Eq. 1-8, the power is

$$p = vi = 2 \times 10^4 \times 8 \times 10^{-3} = 160 \text{ W}$$

Note: In calculations, enter all values in SI units. Only in the results are units shown.

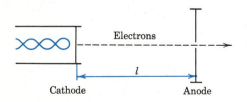

Cathode   Anode

Figure 1.3   Current and power.

Experimental Laws

In contrast to the foregoing arbitrary definitions, three laws have been formulated (about 200 years ago) from experimentally observed facts. Our present understanding of electrical phenomena is much more sophisticated and these laws are readily derived from basic theory. At this stage in our treatment of circuits, however, it is preferable to consider these relationships as they would be revealed in experiments on real devices in any laboratory. From observations on the behavior of real devices we can derive idealized models of circuit elements and rules governing the behavior of such elements when combined in simple or complicated circuits.

Resistance

Consider an experiment (Fig. 1.4a) in which a generator is used to supply a varying current i to a copper rod. The current and resulting voltage v are observed on

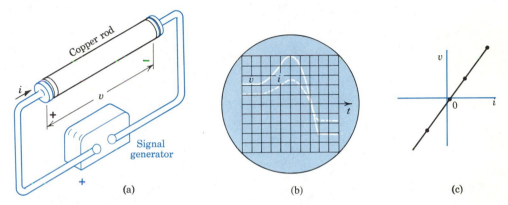

Figure 1.4   Experimental determination of resistor characteristics.

an oscilloscope, and voltage is plotted as a function of current (Fig. 1.4c). If all other factors (such as temperature) are held constant, the voltage is observed to be approximately proportional to the current. The experimental relation for this *resistor* can be expressed by the equation

$$v \cong Ri \tag{1-10}$$

where R is a constant of proportionality. Where v is in volts and i in amperes, Eq. 1-10 defines the *resistance R* in *ohms* (abbreviated Ω) and is called Ohm's law in honor of Georg Ohm, the German physicist, whose original experiments led to this simple relation. The same relation can be expressed by the equation

$$i \cong Gv \tag{1-11}$$

which defines the *conductance G* in siemens (S) in honor of the German inventor Werner von Siemens.[†]

We now know that a metallic conductor such as copper contains many relatively free electrons. The application of a voltage creates an electric field that tends to accelerate these *conduction* electrons, and the resulting motion is superimposed on the random thermal motion of the electrons at, say, room temperature. Electrons are accelerated by the field, collide with copper atoms, and give up their energy. They are accelerated again, gaining energy from the electric field, collide again, and give up

---

[†]In the United States, conductance is frequently designated in mhos ($\mho$).

their energy. Superimposed on the random motion due to thermal energy, there is an average net directed motion or *drift* due to the applied electric field. The speed of drift is found to be directly proportional to the applied electric field. In a given conducting element, therefore, the rate of flow of electric charge or the current is directly proportional to the electric field, which, in turn, is directly proportional to the applied voltage. In this way, Ohm's law can be derived directly from a consideration of the conduction mechanism.

Capacitance

Now consider an experiment (Fig. 1.5a) in which a copper rod is cut and reshaped into two flat plates separated by air as an insulator. Available, but not shown, are a

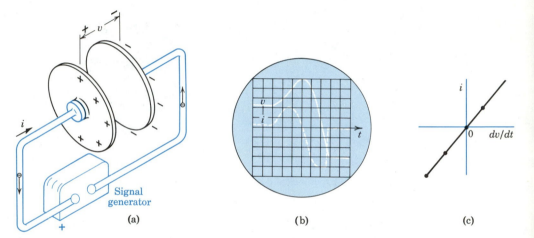

Figure 1.5   Experimental determination of capacitor characteristics.

voltmeter, an ammeter, and an electroscope to measure voltage, current, and charge. When a voltage is applied as shown, it is observed that positive charge appears on the left-hand plate and negative on the right. If the generator is disconnected, the charge persists. Such a device that stores charge is called a *capacitor*.

As long as the voltage is constant, the charge is maintained and no current flows. If the voltage is changing with time (Fig. 1.5b), the current is observed to be approximately proportional to the *rate of change* of voltage. The experimental relation can be expressed by the equation

$$i \cong C\frac{dv}{dt} \tag{1-12}$$

where C is a constant of proportionality called *capacitance* and measured in *farads* (abbreviated F) in honor of the ingenious English experimenter, Michael Faraday.

Equation 1-12 can be obtained in another way. If the charge on the capacitor is measured accurately, it is observed to be approximately proportional to the voltage applied or

$$q \cong Cv \tag{1-13}$$

where C has the same value as determined previously. Equation 1-12 can be derived from this relation since $i = dq/dt$ by definition.

Inductance

If, instead, the copper rod is drawn out and formed into the configuration of Fig. 1.6a, an entirely different result is obtained. Now the generator is used to supply current to a multiturn coil. It is observed that only a small voltage is required to maintain an appreciable steady current, but a relatively large voltage is required to produce a rapidly changing current. As indicated in Fig. 1.6c, the voltage is observed to be approximately proportional to the *rate of change* of current. The experimental relation can be expressed by the equation

$$v \cong L\frac{di}{dt} \qquad (1\text{-}14)$$

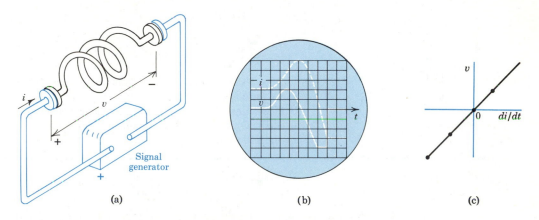

(a)   (b)   (c)

Figure 1.6 Experimental determination of inductor characteristics.

where L is a constant of proportionality called *inductance*[†] and measured in *henrys* (abbreviated H) in honor of the American inventor and experimenter, Joseph Henry.

This important relation can be derived by using our knowledge of magnetic fields. We know that charge in motion produces a magnetic field and the multiturn coil is an efficient configuration for concentrating the effect of all the moving charges in the conductor. In the region in and around the coil, the magnetic flux density is high for a given current. But we also know (and this was Henry's great discovery) that a changing magnetic field *induces* a voltage in any coil that it links and the induced voltage is directly proportional to the rate of change of the magnetic field. If the magnetic field is due to the current in the coil itself, the induced voltage always tends to oppose the change in current that produces it. The voltage required of the generator is determined primarily by this voltage of *self-induction* and is approximately proportional to the time rate of change of magnetic flux density. For the coil of Fig. 1.6, magnetic flux density is proportional to current and Eq. 1-14 is confirmed.

[†]More properly called "self-inductance"; *mutual inductance M* is defined by the equation $v_{12} = M\, di_2/dt$ where v_{12} is the voltage induced in coil 1 due to changing current i_2 in a second coil.

EXAMPLE 2

A current varies as a function of time as shown by the colored line in Fig. 1.7.

(a) Determine and plot the voltage produced by this current in a 30-Ω resistor.

(b) Determine and plot the voltage produced by this current in a coil of 0.5-Ω resistance and 4-H inductance.

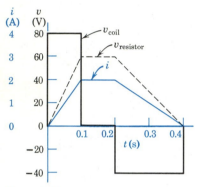

Figure 1.7   Current-voltage relations in resistor and inductor.

(a) Assuming that this is a linear resistor whose behavior is defined by $v = Ri$, the voltage is directly proportional to the current. For $i = 2$ A, $v = 30 \times 2 = 60$ V; the voltage as a function of time is as shown by the dashed line.

(b) For the maximum current of 2 A, the voltage due to the coil resistance is

$$v_R = Ri = 0.5 \times 2 = 1 \text{ V}$$

The voltage due to the coil inductance is

$$0 < t < 0.1 \quad v_L = L\frac{di}{dt} = 4\frac{2}{0.1} = 80 \text{ V}$$

$$0.1 < t < 0.2 \quad v_L = L\frac{di}{dt} = 4\frac{0}{0.1} = 0$$

$$0.2 < t < 0.4 \quad v_L = L\frac{di}{dt} = 4\frac{-2}{0.2} = -40 \text{ V}$$

Later we shall learn how to combine these two effects. For many purposes, however, satisfactory results can be obtained by neglecting v_R where v_L is large and considering v_R where v_L is small. On this basis the voltage across the coil is as shown by the solid line.

Decimal Notation

To specify the *value* of a measurable quantity, we must state the *unit* and also a *number*. We shall use the SI units, but the range of numbers encountered in electrical engineering practice is so great that a special notation has been adopted for con-

Table 1-3   Standard Decimal Prefixes

Multiplier	Prefix	Abbreviation	Pronunciation
10^{12}	tera	T	tĕr′ ȧ
10^{9}	giga	G	jĭ′ gȧ
10^{6}	mega	M	mĕg′ ȧ
10^{3}	kilo	k	kĭl′ ŏ
10^{2}	hecto	h	hĕk′ tŏ
10^{1}	deka	da	dĕk′ ȧ
10^{-1}	deci	d	dĕs′ ĭ
10^{-2}	centi	c	sĕn′ tĭ
10^{-3}	milli	m	mĭl′ ĭ
10^{-6}	micro	μ	mī′ krŏ
10^{-9}	nano	n	năn′ ŏ
10^{-12}	pico	p	pē′ cŏ
10^{-15}	femto	f	fĕm′ tŏ
10^{-18}	atto	a	ăt′ tŏ

venience. For example, the power of a public utility system may be millions of watts, whereas the power received from a satellite transmitter may be millionths of a watt. The engineer describes the former in *mega*watts and the latter in *micro*watts.

The notation is based on the decimal system that uses powers of 10. The standard prefixes in Table 1-3 are widely used to designate multiples and submultiples of the fundamental units. Thus 20 MW is read "20 megawatts" and is equal to 20×10^6 or 20 million watts; and 20 μW is read "20 microwatts" and is equal to 20×10^{-6} or 20 millionths of a watt. In Example 3, note that all numerical quantities are first converted to SI units before being substituted in the applicable equations.

EXAMPLE 3

A current varies as a function of time as shown in Fig. 1.8. Predict and plot the voltage produced by this current flowing in an initially uncharged 1-μF capacitor.

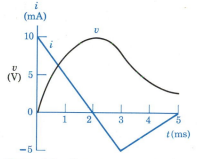

Figure 1.8   Current-voltage relations in a capacitor.

For an ideal capacitor, $i = C \, dv/dt$.

$$\therefore v = \frac{1}{C} \int_0^t i \, dt + V_0$$

For $0 < t < 3$ ms, $V_0 = 0$ and

$$i = 10 \times 10^{-3} - (15 \times 10^{-3}/3 \times 10^{-3})t$$
$$= 0.01 - 5t$$

$$\therefore v = \frac{1}{10^{-6}} \int_0^t (0.01 - 5t) \, dt + 0$$

$$= 10^6(0.01t - 2.5t^2), \text{ a parabola}$$

At $t = 2$ ms, for example,

$$v = 10^6(0.01 \times 2 \times 10^{-3} - 2.5 \times 4 \times 10^{-6})$$
$$= 20 - 10 = 10 \text{ V}$$

A similar calculation for $3 < t < 5$ ms (the voltage at 3 ms is 7.5 V) yields the curve shown.

CIRCUIT ELEMENTS

An electric circuit consists of a set of interconnected components. Components carrying the same current are said to be connected in *series*. A simple series circuit might include a battery, a switch, a lamp, and the connecting wires. Components subjected to the same voltage are said to be connected in *parallel*. The various lights and appliances in a residence constitute a parallel circuit. The terms *circuit* and *network* are sometimes used interchangeably, but usually network implies a more complicated interconnection such as the communication network provided by the telephone company. There is also an implication of generality when we refer to network analysis or network theorems.

Circuit Components

From our standpoint, the distinguishing feature of a circuit component is that its behavior is described in terms of a voltage-current relation at the terminals of the component. This relation, called the "*v-i* characteristic," may be obtained analytically by using field theory in which the geometry of the associated electric or magnetic field is important, or it may be obtained experimentally by using measurements at the terminals. Once the *v-i* characteristic is known, the behavior of the component in combination with other components can be determined by the powerful methods of *circuit theory,* the central topic of this part of the book.

Using hypothetical experiments and applying elementary field concepts, we obtained *v-i* characteristics for a resistor, a capacitor, and an inductor. In writing mathematical expressions for the *v-i* characteristics (Eqs. 1-10, 1-12, and 1-14) we neglected the distributed nature of the fields involved and we assumed linearity. But we know that no real physical device is exactly linear. For example, the resistance of the incandescent lamp of Fig. 1.9a increases rapidly with current, and the inductance of the coil of Fig. 1.9b is greatly affected by the value of average current.[†] Therefore we shall invent three ideal circuit components, or *models,* that *are* linear.

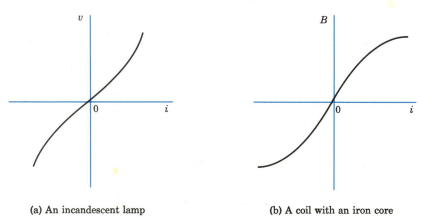

(a) An incandescent lamp   (b) A coil with an iron core

Figure 1.9   Characteristics of real circuit components.

There are good reasons for doing this. Some real components can, within limits, be represented by such lumped linear models in many problems. Other real components can be represented accurately by combinations of two or more ideal components. Also, by using the mathematical methods applicable to linear models we can derive general results of great value in predicting the behavior of complicated devices and systems of devices.

Circuit Element Definitions

The three linear models are shown symbolically in Fig. 1.10 along with their defining equations and a set of alternative equations.[‡] As you may recall, a real coil

---

[†]Since v is proportional to dB/dt and, in general, B is a nonlinear function of i.

[‡]The plus sign above the "v" indicates that the upper terminal is the positive reference direction of potential. The arrow under the "i" indicates that flow to the right is the positive reference direction of current. These conventions are discussed in detail in Chapter 2.

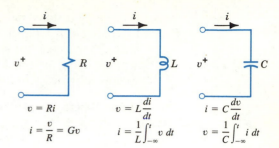

Figure 1.10 Circuit elements.

$$v = Ri$$
$$i = \frac{v}{R} = Gv$$

$$v = L\frac{di}{dt}$$
$$i = \frac{1}{L}\int_{-\infty}^{t} v\, dt$$

$$i = C\frac{dv}{dt}$$
$$v = \frac{1}{C}\int_{-\infty}^{t} i\, dt$$

or inductor exhibits the property of inductance; but every real inductor also exhibits the property of resistance. The corresponding linear model is called an "ideal inductor" or, since it exhibits only a single property, it is called an "inductance element," or simply an "inductance." The three linear models—resistance, inductance, and capacitance—are referred to as *circuit elements;* with these elements as building blocks, an infinite number of circuits can be devised.

Energy Storage in Linear Elements

Valuable insights into the behavior of real circuit components can be obtained by considering the energy transformations that occur in the corresponding linear models. Recalling that energy $w = \int vi\, dt$, we see that, once the v-i characteristic is defined, the energy storage or dissipation property is determined.

Inductance. Where $v = L\, di/dt$ and $i = 0$ at $t = 0$,

$$w_L = \int_0^T L\frac{di}{dt} i\, dt = \int_0^I Li\, di = \tfrac{1}{2}LI^2 \tag{1-15}$$

The total energy input to an inductance is directly proportional to the square of the final current. The constant of proportionality is $L/2$; in fact, this expression for energy could have been used to define inductance.

Inductance is a measure of the ability of a device to store energy in the form of moving charge or in the form of a magnetic field.

The equation reveals that the energy is stored rather than dissipated. If the current is increased from zero to some finite value and then decreased to zero, the upper limit of the integration becomes zero and the net energy input is zero; the energy input was stored in the field and then returned to the circuit.

Capacitance. Where $i = C\, dv/dt$ and $v = 0$ at $t = 0$,

$$w_C = \int_0^T vC\frac{dv}{dt} dt = \int_0^V Cv\, dv = \tfrac{1}{2}CV^2 \tag{1-16}$$

Can *you* interpret this equation? In words, the total energy input to a capacitance is directly proportional to the square of the final voltage. The constant of proportionality is $C/2$.

Capacitance is a measure of the ability of a device to store energy in the form of separated charge or in the form of an electric field.

The energy is stored rather than dissipated; if the final voltage is zero, the energy stored in the field is returned to the circuit.

These expressions for stored energy remind us of the expressions for kinetic energy of a moving mass and potential energy of a stretched spring. In such mechanical systems, displacement $dx = u\ dt$ and $w = \int f\ dx = \int fu\ dt$.

Mass. Where $f = M\ du/dt$ and $u = 0$ at $t = 0$,

$$w_M = \int_0^T M\frac{du}{dt}u\ dt = \int_0^U Mu\ du = \tfrac{1}{2}MU^2 \tag{1-17}$$

As expected, this equation indicates that the mass M of a body is a measure of its ability to store kinetic energy.

Compliance. Where $x = Kf$ and $f = 0$ at $t = 0$,

$$w_K = \int_0^F f\ d(Kf) = \int_0^F Kf\ df = \tfrac{1}{2}KF^2 \tag{1-18}$$

In a similar way, the spring compliance K is a measure of the spring's ability to store potential energy.

Energy Dissipation in Linear Elements

Each of these elements (and there are others in thermal and chemical systems) has the ability to store energy and then return it to the circuit or system. When a similar analysis is made of an electrical resistance, the results are quite different.

Resistance. Where $v = Ri$ and $i = 0$ at $t = 0$,

$$w_R = \int_0^T Ri\ i\ dt = \int_0^T Ri^2\ dt \tag{1-19}$$

To evaluate the total energy supplied, we must know current i as a function of time t. For the special case where the current is constant or $i = I$,

$$w_R = \int_0^T RI^2\ dt = RI^2T \tag{1-20}$$

In general, for finite values of i (either positive or negative) and finite values of t, the energy supplied to the resistor is finite and positive. There is no possibility of controlling the current in such a way as to return any energy to the circuit; the energy has been *dissipated*. In a real resistor, the dissipated energy appears in the form of heat; in describing the behavior of the linear model we say that the energy has been dissipated in an *irreversible transformation,* irreversible because there is no way of heating an ordinary resistor and obtaining electric energy. (Some extraordinary resistors are described in Chapter 20.)

The *rate* of dissipation of energy or power is a useful characteristic of resistive elements. By definition, $p = dw/dt$; therefore,

$$p_R = \frac{dw_R}{dt} = Ri^2 \tag{1-21}$$

The power dissipation in a resistance is directly proportional to the square of the current. The constant of proportionality is R and this expression for power is frequently used to define resistance.

Resistance is a measure of the ability of a device to dissipate power irreversibly.

Friction. At this point an alert student would wonder if there is in mechanical systems a corresponding dissipative element. There is and it is called *frictional resistance*. The shock absorber is a friction device designed to dissipate the energy received by an automobile when it goes over a bump in the road. The shock absorber, or "dash pot," contains a viscous fluid in a piston-cylinder arrangement. Relative motion of the body and chassis forces the fluid through a small hole and the kinetic energy is converted to heat in an irreversible process.

Over limited ranges, the friction force developed is approximately proportional to velocity or

$$f = Du \tag{1-22}$$

where D is the frictional resistance in units of newtons/(meters per second) or newton·seconds/meter. In the linear model of a dash pot where $f = Du$ and $u = 0$ at $t = 0$,

$$w_D = \int_0^X f \, dx = \int_0^T Du \, u \, dt = \int_0^T Du^2 \, dt \tag{1-23}$$

and the power dissipation in friction is

$$p_D = \frac{dw}{dt} = Du^2 \tag{1-24}$$

Continuity of Stored Energy

It is common knowledge that a massive body tends to oppose rapid changes in velocity; we say that such a body has *inertia*. In solving problems in dynamics we take advantage of the fact that the velocity of a mass cannot be changed instantaneously; for example, the velocity just after a force is applied must be the same as that just before the force is applied. Similar conclusions can be drawn about electric circuit quantities by applying the fundamental principle that in any physical system the energy must be a continuous function of time.

Since power is the time rate of change of energy ($p = dw/dt$), an instantaneous change in energy would require an infinite power. But the existence of an infinite power is contrary to our concept of a physical system, and we require that the energy stored in any element of a system, real or ideal, be a continuous function of time. Recalling that the energy of a moving mass is $\frac{1}{2}Mu^2$, we conclude that the velocity u cannot change instantaneously.

Following the same reasoning, we note that the energy stored in an inductance is $\frac{1}{2}Li^2$ and, therefore,

The current in an inductance cannot change instantaneously.

Since the energy stored in a capacitance is $\frac{1}{2}Cv^2$,

The voltage across a capacitance cannot change instantaneously.

These concepts are particularly useful in predicting the behavior of a circuit just after an abrupt change, such as closing a switch. Note that in considering energy changes there is no limitation on the rapidity with which inductance voltage or capacitance current can change. Also, because resistance does not store energy, there is no limitation on the rapidity with which resistance voltage and current changes can occur.

EXAMPLE 4

The ignition coil of an automobile can be represented by a series combination of inductance L and resistance R (Fig. 1.11). A source of voltage V, the battery, is applied through contacts or "points" represented by switch S.

(a) With the switch open, the current i is zero. What is the current just after the switch is closed?

(b) After a finite time the current has reached a value of 1 A and the switch is suddenly opened. What is the current just after the switch is opened?

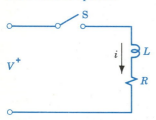

Figure 1.11   Ignition coil.

(a) Since the current in the inductance cannot change instantaneously, the current just after closing the switch must be zero (but starting to increase rapidly for proper operation of the ignition system).

(b) The current just after the switch is opened is still 1 A because it takes time to dissipate the energy stored in the inductance ($\frac{1}{2}Li^2$). Some of the energy goes to ionize the air between the switch contacts and an arc is formed; the points are "burned."

In practice, a "condenser" (a capacitor) is connected across the switch to minimize arcing by absorbing energy from the coil. The role of the condenser will become more clear in Chapter 4.

Energy Sources and Reversible Transformations

Energy can be stored in inductance or capacitance and dissipated in resistance. These are called *passive* circuit elements in contrast to *active* elements that can serve as sources of electrical energy. The many different forms of electrical "generators" include the rotating dynamo, the chemical storage battery, the solar cell, the fuel cell, and the thermocouple. In each of these real devices the electrical energy is "generated" by the conversion of some other form of energy. The conversion represents a *reversible transformation* in that the conversion process can go either way, in contrast to the situation in a resistor. To illustrate: in a dynamo operating as a "generator," mechanical energy is converted into electrical energy; but electrical input to a typical dynamo will cause it to operate as a "motor" producing mechanical energy.

Although every generator supplies both voltage and current ($w = \int vi\, dt$), it is desirable to distinguish between two general classes of devices. If the voltage output is relatively independent of the circuit to which it is connected (as in the battery of Fig. 1.12a), the device is called a "voltage generator." If over wide ranges the output

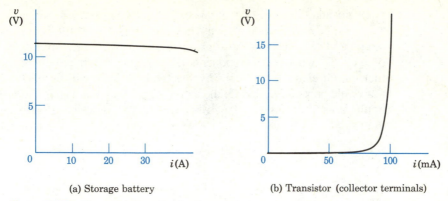

(a) Storage battery   (b) Transistor (collector terminals)

Figure 1.12   Characteristics of real voltage and current generators.

current tends to be independent of the connected circuit (as in the transistor of Fig. 1.12b), it is treated as a "current generator."

An ideal voltage generator in which the output voltage is completely independent of the current is called an "ideal voltage source," or simply a "voltage source." The output voltage is specified, usually as a function of time, and it is unaffected by changes in the circuit to which it is connected. At low currents, an automobile battery supplies a nearly constant voltage and may be represented by a 12-V voltage source. At high currents, however, the output voltage is appreciably less than 12 V.[†] The difference is due to the fact that the energy-conversion process is not perfect; part of the chemical energy is converted to heat in an irreversible process. This result suggests the possibility of representing a real electrical generator by a linear model consisting of a pure voltage source and a resistance, the resistance accounting for the irreversible energy transformations.

The circuit symbol and the defining equation for a voltage source are shown in Fig. 1.13a. The corresponding ideal current generator or "current source" is shown in Fig. 1.13b. Note that here the output current is specified, usually as a function of time, and it is independent of the voltage across the source. As we shall see, a combination of a current source and a resistance also may be used to represent a real generator in predicting the behavior of real circuits. There is a finite limit to the power supplied by such a combination, whereas an ideal voltage or current source can supply, or absorb, an unlimited amount of power.

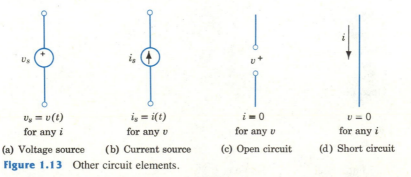

$v_s = v(t)$ $i_s = i(t)$ $i = 0$ $v = 0$
for any i for any v for any v for any i

(a) Voltage source   (b) Current source   (c) Open circuit   (d) Short circuit

Figure 1.13   Other circuit elements.

[†]How is this fact evidenced when an automobile is started with its headlights on?

Two additional circuit elements need to be defined before we begin circuit analysis. The *open circuit* shown symbolically in Fig. 1.13c is characterized by finite voltage and zero current; the *short circuit* is characterized by finite current and zero voltage. (To what real components do open and short circuits correspond?) While other circuit elements are conceivable (and, in fact, will be desirable later on), the passive elements R, L, and C, the active sources $v(t)$ and $i(t)$, and the open and short circuits constitute a set that will be sufficient to describe a large number of useful circuits.

SUMMARY

■ The important electrical quantities, their definitions, and their SI units are summarized in Table 1-2 on p. 8.
In electrical terms, instantaneous power and energy are expressed by

$$p = \frac{dw}{dt} = vi \quad \text{and} \quad w = \int p \, dt = \int vi \, dt$$

■ The approximate behavior of real circuit components is used to define the five passive and two active circuit elements displayed in Table 1-4.
Voltage and current sources represent reversible energy transformations.

Table 1-4   Electrical Circuit Elements

Element	Unit	Symbol	Characteristic
Resistance (Conductance)	ohm (siemen)	$i \quad R$ $+v$ (G)	$v = Ri$ $(i = Gv)$
Inductance	henry	$i \quad L$ $+v$	$v = L\frac{di}{dt} \quad i = \frac{1}{L}\int_{-\infty}^{t} v \, dt$
Capacitance	farad	$i \quad C$ $+v$	$i = C\frac{dv}{dt} \quad v = \frac{1}{C}\int_{-\infty}^{t} i \, dt$
Short circuit		i	$v = 0$ for any i
Open circuit		$+v$	$i = 0$ for any v
Voltage source	volt	v_s	$v = v_s$ for any i
Current source	ampere	i_s	$i = i_s$ for any v

■ Resistance is a measure of the ability of a device to dissipate energy irreversibly ($p_R = Ri^2$).

Inductance is a measure of the ability of a device to store energy in a magnetic field ($w_L = \frac{1}{2}Li^2$).

Capacitance is a measure of the ability of a device to store energy in an electric field ($w_C = \frac{1}{2}Cv^2$).

■ Energy stored in an element must be a continuous function of time.

The current in an inductance cannot change instantaneously.

The voltage on a capacitance cannot change instantaneously.

REVIEW QUESTIONS

(These questions are primarily for use by the student in checking his or her familiarity with the ideas in the text.)

1. Without referring to the text, list with their symbols and units the six quantities basic to the SI system.
2. Without referring to the text, define "force," "energy," and "power," and give their symbols and units.
3. Without referring to the text, define "current," "voltage," "electric field strength," and "magnetic flux density," and give their symbols and units.
4. Some automobile manufacturers are considering going to an 18-V battery. What specific advantages would this have?
5. Explain the minus sign in Eq. 1-5.
6. Given two points a and b in a field of known strength $\mathcal{E}$, how could voltage v_{ab} be calculated?
7. Explain the difference between an "electric signal" that moves along a wire at nearly the speed of light and an "electric charge" that drifts along at only a fraction of 1 m/s.
8. Define "resistance," "inductance," and "capacitance" in terms of voltage-current characteristics and also in terms of energy transformation characteristics.

9. Draw a curve representing current as a function of time and sketch on the same graph curves of the corresponding voltage across inductance and capacitance.
10. Draw a curve representing voltage as a function of time and sketch on the same graph curves of the corresponding current in an inductance and a capacitance.
11. How can a current flow "through" a capacitor containing an insulator?
12. Once the effect of an electric field has been represented by a circuit element, is the geometry of the field significant?
13. List three physical components not mentioned in the text that have the ability to store energy.
14. The concept of continuity of stored energy places what restrictions on the circuit behavior of inductances and capacitances?
15. Give three examples of reversible transformations involving electrical energy.
16. Explain, in terms of source characteristics, the dimming of the house lights when the refrigerator motor starts.
17. Distinguish between an "inductor" and an "inductance"; give an example of each.

EXERCISES

(These exercises are intended to be straightforward applications of the concepts of this chapter.)

1. A woman weighing 110 lb runs up a 12-ft (vertical distance) flight of stairs in 3 s.

   (a) Express in newtons the force such a woman would exert on a bathroom scale and in kilograms the "reading" of the scale.

   (b) Express in joules and in kilowatt-hours the net work done in climbing the stairs.

   (c) Express in watts the average power expended in climbing the stairs.
2. A glass tube with a cross-sectional area of 10^{-2} m^2 contains an ionized gas. The densities are 10^{15} positive ions/m^3 and 10^{11} free electrons/m^3. Under the influence of an applied volt-

age, the positive ions are moving with an average axial velocity of 5×10^3 m/s. At that same point the axial velocity of the electrons is 2000 times as great. Calculate the electric current.

3. A laser printer produces circular dots 125 μm in diameter. The exposure required is 1.5 μJ/cm^2. Calculate the dot writing rate for a 5-mW helium-neon laser. If a typical letter requires 10 dots, how long would it take to print a page of text in this book?

4. A voltage of 2000 V exists across the 1-cm insulating space between two parallel conducting plates. An electron of mass $m = 9.1 \times 10^{-31}$ kg is introduced into the space.
   (a) Calculate the electric field strength in the space.
   (b) Calculate the force on the electron and the resulting acceleration.

5. In a picture-tube deflection system, electrons moving at a velocity of 39,000 km/s enter a magnetic field of flux density 0.2 T.
   (a) Predict the force exerted on the electrons.
   (b) If flux is from left to right on the screen, what is the direction of the resulting deflection? (See Fig. 10.3 for the definition of **u** × **B**.)

6. A 12-V automobile light is rated at 30 W.
   (a) What total charge flows through the filament in 1 min?
   (b) How many electrons does this represent?
   (c) What is the cost of 24 hours' operation at 6 cents per kilowatt-hour?
   (d) What is the resistance of the hot filament?
   (e) How many 100-μF capacitors would be required to store at 12 V the charge for 1 minute's operation?

7. Ammeter AM and voltmeter VM, connected as shown in Fig. 1.14, measure instantaneous current and voltage. An ammeter reads "upscale" or positive when current flows into the meter at the + terminal. Is power flowing *into* or *out of* device D when:

(a) AM reads positive and VM reads negative?
(b) AM reads negative and VM reads positive?
(c) AM reads negative and VM reads negative?

8. In a TV picture tube, the accelerating voltage is 25 kV.
   (a) Express in joules the energy gained by an electron moving through this potential.
   (b) If the power delivered to the fluorescent screen by the high-velocity electrons is to be 2 W, calculate the necessary beam current.

9. A $\frac{1}{4}$-hp dc motor under rated load draws 2 A at 120 V. Calculate the input power and the rated efficiency (output/input) in percent.

10. Predict the total energy available from a 9-V radio battery rated at 0.6 A·h and the total operating time at a current of 0.018 A.

11. In Fig. 1.15, the conductance of the resistor is 0.02 S. For $i = 2.5$ A, what voltage appears across the resistor?

12. In Fig. 1.15, current $i(t) = I_m \sin \omega t$.
    (a) Write an expression for the voltage across the resistor of R ohms.
    (b) Write an expression for the voltage across the inductor of L henrys.

13. The current in Fig. 1.15 is $i(t) = 5t$ A where t is in seconds. For time $t = 5$ ms, predict the voltages across the 120-Ω resistor and across the 3-H inductor.

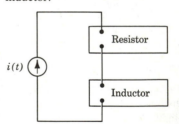

Figure 1.15

14. The voltage in Fig. 1.16 is $v(t) = 2000t$ V where t is in seconds. For time $t = 5$ ms, predict the current through the 4-Ω resistor and through the 300-μF capacitor.

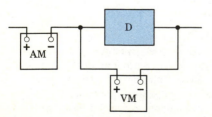

Figure 1.14

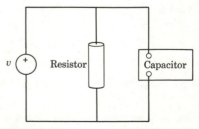

Figure 1.16

15. The current shown in Fig. 1.17 flows through a 5-H inductor and an initially uncharged 2-μF capacitor. Plot to scale graphs of v_L and v_C.

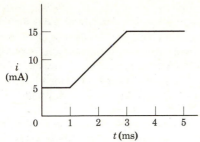

Figure 1.17

16. The voltage shown in Fig. 1.18 is applied to a 2-μF capacitor and a 0.5-H inductor (no initial current). Plot to scale graphs of i_C and i_L.

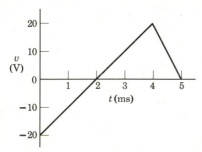

Figure 1.18

17. (a) Calculate the energy stored in an inductance $L = 10$ H at time $t = 2$ ms by the waveform of Fig. 1.17.
(b) Calculate the energy stored in a capacitance $C = 100$ μF at time $t = 3$ ms by the waveform of Fig. 1.18.
(c) Explain the difficulty in performing calculation (a) on Fig. 1.18 and calculation (b) on Fig. 1.17.

18. Plot as a function of time the power to the inductor of Exercise 15 and the energy stored in the inductor.

19. Plot as a function of time the power to the capacitor of Exercise 16 and the energy stored in the capacitor.

20. The current source in the circuit of Fig. 1.15 provides an output defined by $i = 5t$ A where t is in seconds. For time $t = 2$ s, calculate:
(a) The power dissipated in the 15-Ω resistor.
(b) The power supplied to the 10-H inductor.
(c) The power delivered by the source.
(d) The total energy dissipated in the resistor.
(e) The total energy stored in the inductor.

21. The voltage source in the circuit of Fig. 1.16 provides an output defined by $v = 200t$ V where t is in seconds. For time $t = 0.5$ s, calculate:
(a) The power dissipated in the 200-Ω resistor.
(b) The power supplied to the 1000-μF capacitor.
(c) The power delivered by the source.
(d) The total energy dissipated in the resistor.
(e) The total energy stored in the capacitor.

22. In Fig. 1.19, capacitor C is "charged up" by closing switch S_1. After equilibrium is reached, switch S_2 is closed.
(a) Just *before* closing S_2, what is the voltage across C? The current in inductor L?
(b) Just *after* closing S_2, what is the voltage across C? The current i_L? The current i_2?

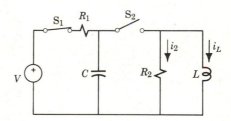

Figure 1.19

PROBLEMS

(These problems require extending the basic concepts of this chapter to new situations.)

1. A "Q-switching" carbon dioxide laser capable of producing 50 W of continuous-wave power will produce 50 kW of pulsed power in bursts 150 ns long at a rate of 400 bursts per second. Calculate the average power output in the pulsed mode and the energy per burst.

2. Show that the following mathematical models describe a resistance, capacitance, or inductance.

(a) $v = kq$
(b) $v = k \, dq/dt$
(c) $v = k \, d^2q/dt^2$

3. Would it be physically possible to have a passive circuit component for which the voltage was proportional to the second derivative of current? If a current $i = I_m \sin t$ flowed in such a component, what power would be absorbed?

4. As we shall see (in Chapter 3), the circuit shown in Fig. 1.20 can be used to smooth out or "filter" variations in input current i so that a fairly constant voltage appears across load resistor R. For good filtering, the capacitor should be able to store 10 times as much energy as is dissipated in one cycle by R. If $R = 10 \text{ k}\Omega$ and variations in i occur 60 times per s, specify the value of C.

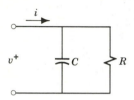

Figure 1.20

5. Laboratory tests on a coil yield the following data: a change in current from $+10$ mA to -10 mA in 50 μs results in an average voltage of 2 V; a steady current of 10 mA requires a power input of 0.2 W. Devise a linear circuit model to represent the coil.

6. A fuse protects a circuit by melting when the current becomes excessive. The resistance of the fuse is given by $R = R_R(1 + cT)$ where R_R is the resistance at room temperature, c is the temperature coefficient, and T is the rise above room temperature. The rise in temperature is proportional to the heat generated in the fuse or $T = kP$. Derive an expression for R in terms of current I and evaluate the current at which the fuse "blows."

7. Devise a linear model incorporating a current source to represent the solar battery whose output characteristics are shown in Fig. 1.21. Specify numerical values valid for currents up to 40 mA.

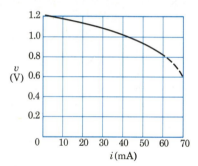

Figure 1.21

2

Circuit Principles

Circuit Laws

Network Theorems

Nonlinear Networks

The emphasis so far has been on the behavior of individual components and the ideal circuit elements that serve as models. Now we are ready to consider combinations of elements into circuits, using two well-known experimental laws. Our first objective is to learn how to formulate and solve circuit equations.

Circuits of considerable complexity or generality are called *networks* and circuit principles capable of general application are called *network theorems*. Some of these theorems are useful in reducing complicated networks to simple ones. Other theorems enable us to draw general conclusions about network behavior. This chapter focuses on networks consisting of resistances and steady (dc) sources; later the theorems will be extended to include networks containing other elements and other sources.

We start with *linear* networks, i.e., combinations of components that can be represented by the ideal circuit elements R, L, and C, and ideal energy sources V and I. However, most electronic devices and many practical components are nonlinear. Therefore, we also consider how linear methods can be used in analyzing some nonlinear networks, and we learn some new techniques applicable to nonlinear devices.

CIRCUIT LAWS

The foundations of network theory were laid about 150 years ago by Gustav Kirchhoff, a German university professor, whose careful experiments resulted in the laws that bear his name. In the following discussion, a *branch* is part of a circuit with two terminals to which connections can be made, a *node* is the point where two or more branches come together, and a *loop* is a closed path formed by connecting branches.

Kirchhoff's Current Law

To repeat Kirchhoff's experiments in the laboratory, we could arrange to measure the currents in a number of conductors or "leads" soldered together (Fig. 2.1a). In every case, we would find that the sum of the currents flowing into the common point (a node) at any instant is just equal to the sum of the currents flowing out.

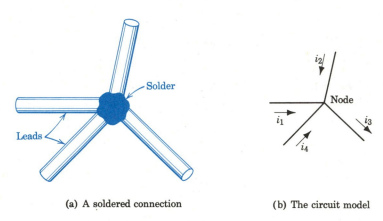

(a) A soldered connection   (b) The circuit model

Figure 2.1   Kirchhoff's current experiment.

In Fig. 2.1b, a circuit model is used to represent an actual connection. The arrows define the *reference* direction for positive current where current is defined by the motion of positive charges. The quantity "i_1" specifies the *magnitude* of the current and its algebraic *sign* with respect to the reference direction. If i_1 has a value of "+5 A," the effect is as if positive charge is moving toward the node at the rate of 5 C/s. If $i_1 = +5$ A flows in a metallic conductor in which charge is transported in the form of negative electrons, the electrons are actually moving away from the node but the effect is the same. If i_2 has a value of "−3 A," positive charge is in effect moving away from the node. In a practical situation the direction of current flow is easy to determine; if an ammeter reads "upscale," current is flowing into the meter terminal marked + and through the meter to the terminal marked −.

By Kirchhoff's current law, the algebraic sum of the currents into a node at any instant is zero.

It is sometimes convenient to abbreviate this statement, and write "$\Sigma i = 0$," where the Greek letter sigma stands for "summation." As applied to Fig. 2.1b,

$$\Sigma i = 0 = i_1 + i_2 - i_3 + i_4 \tag{2-1}$$

Obviously some of the currents may be negative.

EXAMPLE 1

In Fig. 2.2, if $i_1 = +5$ A, $i_2 = -3$ A, and $i_4 = +2$ A, what is the value of i_3?

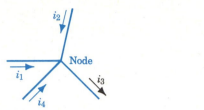

Figure 2.2   Current calculation.

By Kirchhoff's current law,

$$\Sigma i = 0 = i_1 + i_2 - i_3 + i_4$$

$$\Sigma i = 0 = +5 - 3 - i_3 + 2$$

or

$$i_3 = +5 - 3 + 2 = +4 \text{ A}$$

Note: We could just as well say that the algebraic sum of the currents *leaving* a node is zero.

Kirchhoff's Voltage Law

The current law was originally formulated on the basis of experimental data. The same result can be obtained from the principle of conservation of charge and the definition of current. Kirchhoff's voltage law also was based on experiment, but the same result can be obtained from the principle of conservation of energy and the definition of voltage.

To repeat Kirchhoff's observations about voltages, we could set up an electrical circuit and arrange to measure the voltages across a number of components that form a closed path. Only a portion of the circuit is shown in Fig. 2.3, but the combination

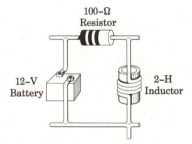

Figure 2.3   A voltage experiment.

of a battery, a resistor, an inductor, and the associated leads forms the desired closed path. In every case, we would find that the voltages around the loop, when properly combined, add up to zero.

The circuit model of Fig. 2.4 is more convenient to work with. There the voltages are labeled and the + signs define the *reference* direction for positive voltage or potential difference. The quantity v_S specifies the *magnitude* of the source voltage and its algebraic *sign* with respect to the reference. If v_S has a value of $+12$ V, the voltage of node b with respect to node a is positive. Since by definition, voltage is energy per unit charge, a positive charge of 1 C moving from node a to b gains 12 J of energy from the voltage source. If v_R has a value of $+5$ V, a positive charge of 1 C moving from b to c loses 5 J of energy; this energy is removed from the circuit and dissipated in the resistance R. If v_L has a value of -7 V, node d is at a higher voltage than c and a positive charge of 1 C moving from c to d *gains* 7 J of energy. A current flowing from c to d could receive energy previously stored in the magnetic field of inductance L.

By Kirchhoff's voltage law, the algebraic sum of the voltages around a loop at any instant is zero.

As an abbreviation, we may write "$\Sigma v = 0$." As applied to Fig. 2.4,

$$\Sigma v = 0 = v_{ba} + v_{cb} + v_{dc} + v_{ad} \qquad (2\text{-}2)$$

where v_{ba} means "the voltage of b with respect to a"; if v_{ba} is positive, terminal b is at a higher potential than terminal a. In applying the law, the loop should be traversed in one continuous direction, starting at one point and returning to the same point. Starting from node a in this case,

$$\Sigma v = 0 = v_S - v_R - v_L + 0 \qquad (2\text{-}3)$$

A minus sign is affixed to the v_R term because the plus sign on the diagram indicates that node b is nominally at a higher potential than c or $v_{cb} = -v_R$.

EXAMPLE 2

In Fig. 2.4, if $v_S = +12$ V and $v_R = +5$ V, what is the value of v_L?

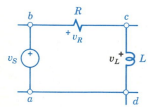

Figure 2.4 Voltage calculation.

By Kirchhoff's voltage law,

$$\Sigma v = 0 = v_S - v_R - v_L + 0$$

$$\Sigma v = 0 = +12 - 5 - v_L + 0$$

or

$$v_L = +12 - 5 = +7 \text{ V}$$

The loop can be traversed in either direction, starting at any point. Going counterclockwise, starting at node c,

$$\Sigma v = 0 = \ + v_R - v_S + 0 + v_L$$

which agrees with Eq. 2-3.

Application of Kirchhoff's Laws

To illustrate the application of Kirchhoff's laws in solving electric circuits, consider the circuit shown in Fig. 2.5. In this case, the source voltages and the resistances are given and the element currents and voltages are to be determined.

The first step is to label the unknown currents arbitrarily; as drawn, the arrow to the right indicates the *reference direction* of i_1. If the value of i_1 is calculated and found to be positive, current i_1 actually flows to the right and an ammeter inserted between node a and the 2-Ω resistor with the $+$ terminal at a would read *upscale* or positive.

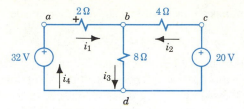

Figure 2.5 Application of Kirchhoff's laws.

If the value of i_1 is found to be negative, current i_1 actually flows to the left (the ammeter inserted as previously would read *downscale* or negative).

Applying Kirchhoff's current law to node a,

$$\Sigma i_a = 0 = +i_4 - i_1$$

or

$$i_4 = i_1$$

From this we conclude that the current is the same at every point in a series circuit. Since $i_4 \equiv i_1$, it is not another unknown and will be considered no further. Note that i_2 is the current in the 20-V source as well as in the 4-Ω resistance.

The next step is to define the unknown element voltages in terms of the arbitrarily assumed currents. In flowing from a to b, the positive charges constituting current i_1 lose energy in the 2-Ω resistance; this loss in energy indicates that the potential of a is higher than that of b, or $v_{ab} = Ri_{ab} = +2i_1$. The left-hand terminal of the 2-Ω resistance is then marked with a small $+$ to indicate the polarity of the element voltage *in terms of the assumed current direction*. Following the same analysis, the upper end of the 8-Ω resistance and the righthand end of the 4-Ω resistance could be marked $+$ in accordance with the following *element equations:*

$$v_{ab} = +2i_1 \qquad \text{or} \qquad v_{ba} = -2i_1$$

$$v_{bd} = +8i_3 \qquad \text{or} \qquad v_{db} = -8i_3 \tag{2-4}$$

$$v_{cb} = +4i_2 \qquad \text{or} \qquad v_{bc} = -4i_2$$

Additional relations are obtained in the form of *connection equations*. Applying Kirchhoff's current law to node b,

$$\Sigma i_b = 0 = +i_1 + i_2 - i_3 \tag{2-5}$$

Since we are summing the currents *into* node b, i_3 is shown with a minus sign. Applying the current law to node d,

$$\Sigma i_d = 0 = -i_1 - i_2 + i_3$$

Note that this equation is *not independent;* it contributes no new information.

Applying the voltage law to the left-hand loop *abda*, and considering each element in turn,

$$\Sigma v = 0 = v_{ba} + v_{db} + v_{ad} \tag{2-6}$$

For the right-hand loop *bcdb*,

$$\Sigma v = 0 = v_{cb} + v_{dc} + v_{bd} \tag{2-7}$$

Kirchhoff's voltage law applies to any closed path; for the outside loop *abcda*,

$$\Sigma v = 0 = v_{ba} + v_{cb} + v_{dc} + v_{ad}$$

But this equation is just the sum of Eqs. 2-6 and 2-7 and no new information is obtained. Although this equation is not independent, it can be valuable in checking.

It is always possible to write as many independent equations as there are un-knowns. For a circuit with six unknowns (three voltages and three currents), we have written six equations (three element and three connection equations). Other equations

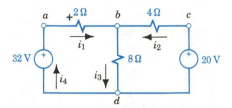

Figure 2.6   Writing circuit equations.

could be written but they would contribute no new information. If the currents are of primary interest, the unknown voltages may be eliminated by substituting Eqs. 2-4 into Eqs. 2-6 and 2-7, which become

$$\Sigma v_{abda} = 0 = -2i_1 - 8i_3 + 32$$

$$\Sigma v_{bcdb} = 0 = +4i_2 - 20 + 8i_3$$

When these are rewritten along with Eq. 2-5, we have for Fig. 2.6

$$i_1 + i_2 - i_3 = 0 \tag{2-8}$$

$$+2i_1 \qquad + 8i_3 = 32 \tag{2-9}$$

$$+ 4i_2 + 8i_3 = 20 \tag{2-10}$$

To evaluate the unknowns, these three equations are to be solved simultaneously. Some commonly employed methods are illustrated in the solution of this problem.

Solution by Determinants

The method of determinants is valuable because it is systematic and general; it can be used to solve complicated problems or to prove general theorems. The first step is to write the equations in the *standard form* of Eqs. 2-8, 2-9, and 2-10 with the constant terms on the right and the corresponding current terms aligned on the left.

A determinant is an array of numbers or symbols; the array of the coefficients of the current terms is a *third-order* determinant

$$D = \begin{vmatrix} 1 & 1 & -1 \\ 2 & 0 & 8 \\ 0 & 4 & 8 \end{vmatrix} \tag{2-11}$$

The value of a determinant of second order is

$$\begin{vmatrix} a_1 & a_2 \\ b_1 & b_2 \end{vmatrix} = a_1 b_2 - a_2 b_1 \tag{2-12}$$

The value of a determinant of third order is

$$\begin{vmatrix} a_1 & a_2 & a_3 \\ b_1 & b_2 & b_3 \\ c_1 & c_2 & c_3 \end{vmatrix} = +a_1 b_2 c_3 + a_2 b_3 c_1 + a_3 b_1 c_2 - a_1 b_3 c_2 - a_2 b_1 c_3 - a_3 b_2 c_1 \tag{2-13}$$

A simple rule for second- and third-order determinants is to take the products "from upper left to lower right" and subtract the products "from upper right to lower left." Determinants of higher order can be evaluated by the process called "expansion by minors."[†]

Repeating the first two columns for clarity and using the rule of Eq. 2-13,

$$D = \begin{vmatrix} 1 & 1 & -1 \\ 2 & 0 & 8 \\ 0 & 4 & 8 \end{vmatrix} \begin{matrix} 1 & 1 \\ 2 & 0 \\ 0 & 4 \end{matrix} = \begin{aligned} &+(1)(0)(8) + (1)(8)(0) + (-1)(2)(4) \\ &-(1)(8)(4) - (1)(2)(8) - (-1)(0)(0) \end{aligned}$$

$$= 0 + 0 - 8 - 32 - 16 - 0 = -56$$

Replacing the coefficients of i_3 with the constant terms on the right-hand side of the equations in the standard form yields a new determinant:

$$D_3 = \begin{vmatrix} 1 & 1 & 0 \\ 2 & 0 & 32 \\ 0 & 4 & 20 \end{vmatrix} = 0 + 0 + 0 - 128 - 40 - 0 = -168 \tag{2-14}$$

By Cramer's rule for the solution of equations by determinants,

$$i_n = \frac{D_n}{D} \quad \text{or} \quad i_3 = \frac{D_3}{D} = \frac{-168}{-56} = 3 \text{ A}$$

Currents i_1 and i_2 could be obtained in a similar manner.

Substitution Method

When the number of unknowns is not large, the method of *substitution* is sometimes more convenient than using determinants. In this method, one of the equations is solved for one of the unknowns and the result is substituted in the other equations, thus eliminating one unknown. Solving Eq. 2-8 for i_3 yields

$$i_3 = i_1 + i_2$$

Substituting in Eqs. 2-9 and 2-10 yields

$$2i_1 + 8(i_1 + i_2) = 32$$

$$4i_2 + 8(i_1 + i_2) = 20$$

[†]See Gere and Weaver: *Matrix Algebra for Engineers,* Brooks/Cole, Monterey, CA, 1983, or any college algebra textbook.

which can be rewritten as

$$(2 + 8)i_1 + \qquad 8i_2 = 32 \qquad (2\text{-}15)$$

$$8i_1 + (8 + 4)i_2 = 20 \qquad (2\text{-}16)$$

Solving Eq. 2-15 for i_1 yields

$$i_1 = \frac{32 - 8i_2}{10}$$

Substituting in Eq. 2-16 yields

$$8\frac{32 - 8i_2}{10} + 12i_2 = 20 \qquad \text{or} \qquad i_2 = \frac{-56}{56} = -1 \text{ A}$$

and the voltage across the 4-Ω resistance is

$$v_{cb} = 4i_2 = 4(-1) = -4 \text{ V}$$

The interpretation of the minus sign is that in reality node b is at a higher potential than c, and positive current flows to the right in the 4-Ω resistance.

Loop-Current Method

The reduction in the number of unknowns and in the number of simultaneous equations achieved by substitution can be obtained automatically by an ingenious approach to circuit analysis. A *loop current* I_1 is assumed to circulate around loop *abda* in Fig. 2.7, and another loop current I_2 is assumed to circulate around loop *cbdc*. By inspection, the branch current i_1 is equal to loop current I_1, and i_2 is equal to I_2; but branch current i_3 is the sum of loop currents I_1 and I_2.

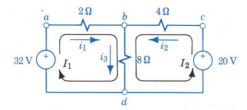

Figure 2.7 Loop currents.

By applying Kirchhoff's voltage law to the two loops we obtain

$$\Sigma v_{abda} = 0 = -2I_1 - 8(I_1 + I_2) + 32$$

$$\Sigma v_{cbdc} = 0 = -4I_2 - 8(I_1 + I_2) + 20$$

which can be rewritten as

$$(2 + 8)I_1 \qquad +8I_2 = 32 \qquad (2\text{-}17)$$

$$8I_1 + (8 + 4)I_2 = 20 \qquad (2\text{-}18)$$

These are similar to Eqs. 2-15 and 2-16, but the physical interpretation is different. Equation 2-17 says: "The source voltage in loop 1 is equal to the sum of two voltage drops. The first, $(2 + 8)I_1$, is the product of loop current I_1 and the sum of all the resistances in loop 1. The second, $8I_2$, is the product of loop current I_2 and the sum of all resistances that are common to loops 1 and 2." The *self-resistance* of loop 1 is $2 + 8 = 10$ Ω; the *mutual resistance* between the two loops is 8 Ω. A similar statement could be made by using Eq. 2-18. To obtain I_1, multiply Eq. 2-17 by 3 and Eq. 2-18 by (-2) and add the resulting equations. Then

$$30I_1 + 24I_2 = 96$$
$$-16I_1 - 24I_2 = -40$$
$$\overline{14I_1 = 56} \quad \text{or} \quad I_1 = 4 \text{ A} = i_1$$

Checking

In solving complicated circuits there are many opportunities for mistakes and a reliable check is essential. This is particularly true when the solution is obtained in a mechanical way, using determinants. A new equation such as that obtained by writing Kirchhoff's voltage law around the outside loop of Fig. 2.7 provides a good check. With a little experience you will be able to write (by inspection, without first writing the element equations)

$$\Sigma v = 0 = -2i_1 + 4i_2 - 20 + 32$$

Substituting $i_1 = 4$ A and $i_2 = -1$ A,

$$\Sigma v = 0 = -2(4) + 4(-1) - 20 + 32 = 0$$

and the values of i_1 and i_2 are checked.

EXAMPLE 3

Given the circuit of Fig. 2.8, calculate the current in the 10-Ω resistance using loop currents.

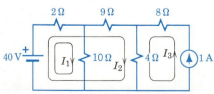

Figure 2.8 Loop-current analysis.

1. Assume loop currents so that the desired current is I_1 by definition and $I_3 = 1$ A as given.
2. Applying Kirchhoff's voltage law to two loops,

$$\Sigma v = 0 = 40 - 2(I_1 + I_2) - 10I_1$$

$$\Sigma v = 0 = 40 - 2(I_1 + I_2) - 9I_2 - 4(I_2 + 1)$$

In terms of self and mutual resistance,

$$(10 + 2)I_1 + 2I_2 = 40 \qquad (2\text{-}19)$$

$$2I_1 + (2 + 9 + 4)I_2 + 4 \times 1 = 40 \qquad (2\text{-}20)$$

From Eq. 2-20, $I_2 = (36 - 2I_1)/15$.

Substituting into Eq. 2-19,

$$12I_1 + (72 - 4I_1)/15 = 40$$

or

$$I_1 = \frac{600 - 72}{180 - 4} = \frac{528}{176} = 3 \text{ A}$$

Node-Voltage Method

The wise selection of loop currents can significantly reduce the number of simultaneous equations to be solved in a given problem. In the preceding example, the number of equations was reduced from three to two by using loop currents. Would it be possible to get all the important information in a single equation with a single unknown? In this particular case, the answer is "yes" if we choose as the unknown the voltage of node b with respect to a properly chosen reference. In some practical devices many components are connected to a metal "chassis," which in turn is "grounded" to the earth. Such a ground, often shown as a common lead at the bottom of a circuit diagram, is a convenient reference. In this problem the greatest simplification will result if we choose node d; now the voltage of any node is understood to be with respect to node d, i.e., the voltage of node d is zero. (See Fig. 2.9.)

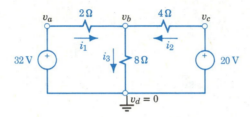

Figure 2.9 Node voltages.

Next, apply Kirchhoff's current law to each independent node. Here there is only one independent node (since the voltages of nodes a and c are fixed by the voltage sources), so the sum of the currents into node b is

$$\Sigma i_b = 0 = i_1 + i_2 - i_3 \tag{2-21}$$

The current i_1 to node b from node a is equal to the difference in potential between nodes a and b divided by the resistance between a and b, or

$$i_1 = \frac{v_a - v_b}{R_{ab}} \tag{2-22}$$

Similarly,

$$i_2 = \frac{v_c - v_b}{R_{cb}} \quad \text{and} \quad i_3 = \frac{v_b - 0}{R_{bd}} \tag{2-23}$$

where all voltages are in reference to node d. In this particular problem, v_a is given as $+32$ V, and v_c is given as $+20$ V.

Finally, bearing in mind Eq. 2-21 and the concepts represented by Eqs. 2-22 and 2-23, we write one equation with a single unknown,

$$\Sigma i_b = 0 = \frac{32 - v_b}{2} + \frac{20 - v_b}{4} - \frac{v_b - 0}{8} \tag{2-24}$$

Multiplying through by 8,

$$128 - 4v_b + 40 - 2v_b - v_b = 0$$

Therefore, the unknown voltage at independent node b is

$$v_b = \frac{168}{7} = 24 \text{ V}$$

By inspection of Fig. 2.9 we can write

$$i_3 = \frac{24}{8} = 3 \text{ A}, \qquad i_1 = \frac{32 - 24}{2} = 4 \text{ A}, \qquad \text{and} \qquad i_2 = \frac{20 - 24}{4} = -1 \text{ A}$$

which agree with the results previously obtained. As illustrated in Example 4, the node-voltage approach can be applied to more complicated circuits.

EXAMPLE 4

Given the circuit of Fig. 2.10, calculate the current in the 10-Ω resistance using node voltages.

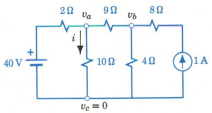

Figure 2.10   Node-voltage analysis.

1. Choose node c (one end of the branch of interest) as the reference node.
2. Apply Kirchhoff's current law to independent nodes a and b.

$$\Sigma i_a = 0 = \frac{40 - v_a}{2} + \frac{v_b - v_a}{9} - \frac{v_a - 0}{10}$$

$$\Sigma i_b = 0 = \frac{v_a - v_b}{9} + 1 - \frac{v_b - 0}{4}$$

Multiplying through by 90 and 36, respectively,

$$1800 - 45v_a + 10v_b - 10v_a - 9v_a = 0$$

$$4v_a - 4v_b + 36 - 9v_b = 0$$

Collecting terms,

$$64v_a - 10v_b = 1800$$

$$4v_a - 13v_b = -36$$

3. Solving by determinants,

$$v_a = \frac{D_a}{D} = \frac{\begin{vmatrix} 1800 & -10 \\ -36 & -13 \end{vmatrix}}{\begin{vmatrix} 64 & -10 \\ 4 & -13 \end{vmatrix}} = \frac{-23{,}400 - 360}{-832 + 40}$$

$$\therefore v_a = 30 \text{ V} \qquad \text{and} \qquad i_{10} = v_a/10 = 3 \text{ A}$$

Formulation of Equations

Many books are devoted to electrical network analysis at intermediate and advanced levels; these books give a thorough treatment of a variety of methods for solving networks of great complexity and generality. Although many significant problems can be solved by using the three approaches outlined in this chapter, the discussion here is intended to be introductory and illustrative rather than comprehensive.

At this point you should be familiar with electrical quantities (their definitions, units, and symbols), you should understand the distinction between real circuit components and idealized circuit elements, you should know the voltage-current characteristics of active and passive circuit elements, and you should be able to apply Kirchhoff's laws to obtain circuit equations. In engineering, the solution of equations is usually less demanding than the formulation, and it is proper to place primary emphasis on the formulation of equations.

If voltages and currents are unchanging with time, as in the preceding example, the behavior of a circuit is determined by resistance alone. A more general problem is illustrated in Fig. 2.11, which resembles the tuning circuit found in every transistor

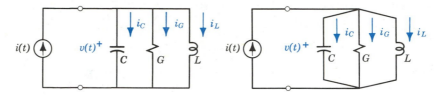

Figure 2.11 A parallel *CGL* circuit drawn in two ways.

radio. Following the procedure used previously, we write element equations and then connection equations. Assuming a voltage $v(t)$ with the reference polarity indicated by the $+$ sign, the corresponding current directions are as shown. Assuming no initial current in the inductance ($i_L = 0$ at $t = 0$), the element equations are

$$i_C = C\frac{dv}{dt}, \qquad i_G = Gv, \qquad i_L = \frac{1}{L}\int_0^t v\, dt + 0$$

Apply Kirchhoff's current law to the upper node to obtain the connection equation,

$$\Sigma i = 0 = i(t) - i_C - i_G - i_L$$

Substituting and rearranging,

$$C\frac{dv}{dt} + Gv + \frac{1}{L}\int_0^t v\, dt = i(t) \tag{2-25}$$

If the element values and $i(t)$ are known, the solution of this *integro-differential* equation consists in finding $v(t)$. In the general case, this can be quite difficult (see Chapter 6).

As another illustration, consider the dynamics problem of Fig. 2.12a. A mass restrained by a spring slides on a friction surface under the action of an applied force. The similarity between an electric circuit and this *mechanical circuit* is made clear by the application of d'Alembert's principle. This principle, which changes the dynamics problem into a statics problem, states that the resultant of the external forces acting on a body and the inertia force equals zero. In other words, if the inertia force (a

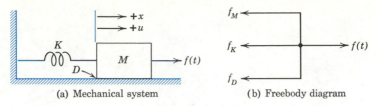

(a) Mechanical system (b) Freebody diagram

Figure 2.12 A dynamics problem.

reaction) is considered in the same manner as external forces, the sum of the forces acting equals zero, or $\Sigma f = 0$.

Again, we write "element" equations and "connection" equations. For positive displacement to the right, the force f_K of a linear spring is to the left and proportional to the displacement. The friction force f_D may be assumed to be proportional to velocity and in such a direction as to oppose motion. The inertia force f_M is proportional to the mass and in such a direction as to oppose acceleration. Assuming no initial displacement and no initial velocity, the element equations are

$$f_K = \frac{1}{K}x = \frac{1}{K}\int_0^t u\,dt + 0, \qquad f_D = Du, \qquad f_m = Ma = M\frac{du}{dt}$$

directed as shown in Fig. 2.12b. The connection equation, obtained by applying d'Alembert's principle, is

$$\Sigma f = 0 = f(t) - f_M - f_D - f_K$$

Substituting and rearranging,

$$M\frac{du}{dt} + Du + \frac{1}{K}\int_0^t u\,dt = f(t) \tag{2-26}$$

If the element values and $f(t)$ are known, the solution of this integro-differential equation consists in finding $u(t)$.

Two conclusions can be drawn from these illustrations. First, there is a basic procedure in formulating equations that is applicable to a variety of electrical circuits and to problems in other fields as well. Second, for even simple circuits the equations are formidable if the variables are functions of time. If practical problems are to be solved, it appears that special mathematical tools and techniques are necessary.

NETWORK THEOREMS

A common problem in circuit analysis is to determine the response of a two-terminal circuit to an input signal. In Fig. 2.13a there is one "port" at which excitation is applied and response is measured. If the desired response is measured across R_2 as in Fig. 2.13b, this circuit becomes a *three-terminal circuit* (one terminal is common to input and output) or a *two-port circuit*, or simply a *two-port*. Transistors, transformers, and transmission lines are two-ports; typically there are an input port and a separate output port. The two-input diode **OR** gate of Fig. 2.13c is a *three-port*.

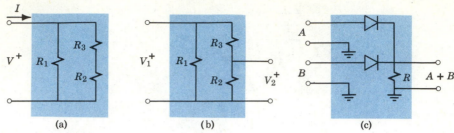

Figure 2.13 One-port, two-port, and multiport networks.

Equivalence

First, we consider how complicated two-terminal circuits can be reduced to simpler circuits that are *equivalent*. By definition:

Two one-ports are equivalent if they present the same v-i characteristics.

Two one-ports consisting of various resistances connected in series and parallel are equivalent if they have the same input resistance or the same input conductance. To calculate the response I to an input V in Fig. 2.13a, what single resistance could replace the three resistances?

Network Reduction

Replacing a complicated network with a simpler equivalent is advantageous in network analysis. Applying Kirchhoff's voltage law to the circuit of Fig. 2.14a, ("closing the loop" by going through V),

$$V = V_1 + V_2 = IR_1 + IR_2 = I(R_1 + R_2) = IR_{\text{EQ}}$$

In general, for n resistances in series the equivalent resistance is

$$R_{\text{EQ}} = R_1 + R_2 + \cdots + R_n \tag{2-27}$$

Applying Kirchhoff's current law to the circuit of Fig. 2.14b,

$$I = I_1 + I_2 = VG_1 + VG_2 = V(G_1 + G_2) = VG_{\text{EQ}}$$

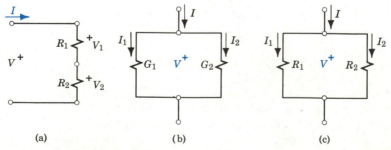

Figure 2.14 Resistances in series and parallel.

In general, for n conductances in parallel the equivalent conductance is

$$G_{EQ} = G_1 + G_2 + \cdots + G_n \tag{2-28}$$

For two resistances in parallel (Fig. 2.14c), a convenient relation is

$$R_{EQ} = \frac{1}{G_{EQ}} = \frac{1}{G_1 + G_2} = \frac{1}{(1/R_1) + (1/R_2)} = \frac{R_1 R_2}{R_1 + R_2} \tag{2-29}$$

The corresponding relationship for two conductances in series is

$$G_{EQ} = \frac{1}{(1/G_1) + (1/G_2)} = \frac{G_1 G_2}{G_1 + G_2} \tag{2-30}$$

By following a similar approach, expressions can be derived for series and parallel combinations of inductances and capacitances (see Problems 3 and 4).

For n inductances in series,

$$L_{EQ} = L_1 + L_2 + \cdots + L_n \tag{2-31}$$

For n capacitances in parallel,

$$C_{EQ} = C_1 + C_2 + \cdots + C_n \tag{2-32}$$

Other useful tools in network analysis include the *voltage divider* and the *current divider*. Referring again to Figs. 2.14a and b, we see that

$$V_2 = \frac{R_2}{R_1 + R_2} V = \frac{R_2}{R_{EQ}} V \quad \text{and} \quad I_2 = \frac{G_2}{G_1 + G_2} I = \frac{G_2}{G_{EQ}} I \tag{2-33}$$

Another useful form of the current-divider equation can be obtained by using $R = 1/G$. Then

$$I_2 = \frac{G_2}{G_1 + G_2} I = \frac{1/R_2}{(1/R_1) + (1/R_2)} I \cdot \frac{R_1 R_2}{R_1 R_2} = \frac{R_1}{R_1 + R_2} I \tag{2-34}$$

When a particular voltage or current in a circuit is to be determined, using these relations is preferable to writing and solving loop or node equations. (See Example 5, p. 42.)

Thévenin's Theorem

The one-port networks discussed so far have been *passive;* they absorb energy from the source. In contrast, *active* networks include energy sources. A valuable method for representing active networks by simpler equivalents is described in the following statement of *Thévenin's theorem:*

Insofar as a load is concerned, any one-port network of resistance elements and energy sources can be replaced by a series combination of an ideal voltage source V_T and a resistance R_T where V_T is the open-circuit voltage of the one-port and R_T is the ratio of the open-circuit voltage to the short-circuit current.

EXAMPLE 5

In the circuit of Fig. 2.15a, determine the current in the 12-Ω resistance.

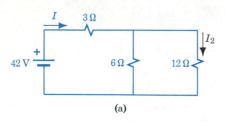

(a)

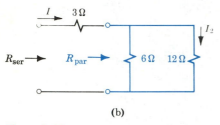

(b)

Figure 2.15 Current-divider application.

For resistances in parallel as in Fig. 2.15b,

$$R_{\text{par}} = \frac{R_1 R_2}{R_1 + R_2} = \frac{6 \times 12}{6 + 12}$$

For resistances in series,

$$R_{\text{ser}} = R + R_{\text{par}} = 3 + \frac{6 \times 12}{6 + 12}$$

and

$$I = \frac{V}{R_{\text{ser}}}$$

Using the current-divider principle, Eq. 2-34,

$$I_2 = \frac{R_1}{R_1 + R_2} I = \frac{R_1}{R_1 + R_2} \cdot \frac{V}{R_{\text{ser}}}$$

$$= \frac{6}{6 + 12} \cdot \frac{42}{3 + (6 \times 12)/(6 + 12)} = \frac{1}{3} \cdot \frac{42}{3 + 4} = 2 \text{ A}$$

By this method I_2 is obtained directly without writing simultaneous equations.

If the two networks of Fig. 2.16 are to be equivalent for all values of load resistance, they must be equivalent for extreme values such as $R_L = \infty$ and $R_L = 0$. The value $R_L = \infty$ corresponds to the open-circuit condition; by comparison of the two networks, the open-circuit voltage V_{OC} of the original network is equal to V_T of the equivalent. The value $R_L = 0$ corresponds to the short-circuit condition; by comparison of the two networks, the short-circuit current I_{SC} of the original network is equal to V_T/R_T of the equivalent. Therefore, the equations

$$V_T = V_{\text{OC}} \quad \text{and} \quad R_T = \frac{V_T}{I_{\text{SC}}} = \frac{V_{\text{OC}}}{I_{\text{SC}}} \tag{2-35}$$

define the components of the Thévenin equivalent network.

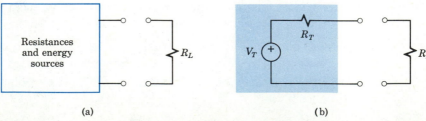

(a)   (b)

Figure 2.16 A one-port network and the Thévenin equivalent.

Alternatively, R_T is the "resistance seen by looking in" at the terminals with all independent[†] energy sources "removed." Voltage sources must be removed by short-circuiting, and current sources must be removed by open-circuiting. By way of explanation, the resistance "seen" by the network in Fig. 2.16b is R_L; the resistance "looking in" at the terminals of the Thévenin equivalent when V_T is short-circuited is R_T.

EXAMPLE 6

Using Thévenin's theorem, determine the current in the 3-Ω resistance of Fig. 2.17a.

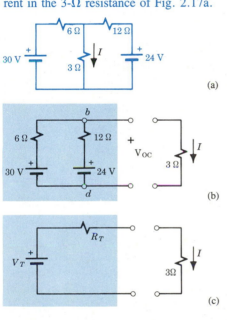

(a)

(b)

(c)

(d)

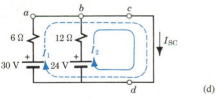

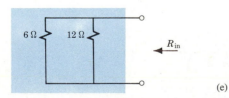

(e)

Figure 2.17 Thévenin's theorem.

The resistance of interest is isolated as a load (Fig. 2.17b) and the Thévenin equivalent is drawn (Fig. 2.17c). Once the components of the equivalent circuit are determined, I can be easily calculated.

Under open-circuit conditions (Fig. 2.17b), the node-voltage method yields

$$\Sigma i_b = 0 = \frac{v_b - 30}{6} + \frac{v_b - 24}{12}$$

or

$$2v_b - 60 + v_b - 24 = 0$$

and

$$3v_b = 84 \qquad \text{or} \qquad v_b = 28 \text{ V} = V_{\text{OC}}$$

Under short-circuit conditions (Fig. 2.17d),

$$\Sigma V_{dacd} = 0 = 30 - 6I_1 = 0 \therefore I_1 = 5 \text{ A}$$

$$\Sigma V_{dbcd} = 0 = 24 - 12I_2 = 0 \therefore I_2 = 2 \text{ A}$$

and

$$I_1 + I_2 = 5 + 2 = 7 \text{ A} = I_{\text{SC}}$$

Hence, by Eq. 2-35,

$$R_T = \frac{V_{\text{OC}}}{I_{\text{SC}}} = \frac{28}{7} = 4 \text{ }\Omega$$

Alternatively, the resistance looking into Fig. 2.17e is

$$R_T = R_{\text{in}} = \frac{6 \times 12}{6 + 12} = 4 \text{ }\Omega$$

Therefore (Fig. 2-17c),

$$I = \frac{V_T}{R_T + 3} = \frac{28}{4 + 3} = 4 \text{ A}$$

and the current in the 3-Ω resistance is 4 A downward.

[†]All energy sources considered up to this point have been *independent* of voltages and currents in other parts of the circuit. *Controlled* sources, for example those used in representing transistors (Chapter 18), must not be removed in applying Thévenin's theorem.

Thévenin's theorem is particularly useful when the load is to take on a series of values or when a general analysis is being performed with literal numbers. We shall use it in analyzing such devices as transistor biasing circuits and feedback amplifiers. No general proof of the theorem is given here, but Example 6 is a demonstration of its validity. (As a check on the result, solve for I by the loop-current method.)

The significance of the restriction "insofar as a load is concerned" is revealed if we attempt to use the Thévenin equivalent to determine internal behavior in Example 6. The power developed internally in the Thévenin equivalent is $P_T = V_T I = 28 \times 4 = 112$ W. In the original circuit, $I_6 = 3$ A and $I_{12} = 1$ A, and the total power supplied is $P = 30 \times 3 + 24 \times 1 = 114$ W. The Thévenin circuit is "equivalent" only in a restricted sense.

Norton's Theorem

Our experience with networks leads us to expect that there is a corollary to Thévenin's theorem. An alternative proposition is described in the following statement called *Norton's theorem*:

> *Insofar as a load is concerned, any one-port network of resistance elements and energy sources can be replaced by a parallel combination of an ideal current source I_N and a conductance G_N where I_N is the short-circuit current of the one-port and G_N is the ratio of the short-circuit current to the open-circuit voltage.*

If the two networks of Fig. 2.18 are to be equivalent for all values of load conductance, they must be equivalent for extreme values such as $G_L = \infty$ and $G_L = 0$. The value $G_L = \infty$ corresponds to the short-circuit condition; by comparison of the two networks, the short-circuit current I_{SC} of the original is equal to I_N of the equivalent. The value $G_L = 0$ corresponds to the open-circuit condition; by comparison of the two networks, the open-circuit voltage V_{OC} of the original network is equal to I_N/G_N of the equivalent. Therefore,

$$I_N = I_{SC} \qquad \text{and} \qquad G_N = \frac{I_N}{V_{OC}} = \frac{I_{SC}}{V_{OC}} \tag{2-36}$$

define the components of the Norton equivalent network. *Note: $G_N = 1/R_T$.*

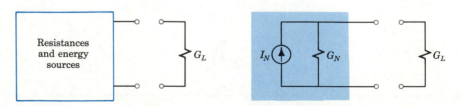

Figure 2.18   A one-port network and the Norton equivalent.

E X A M P L E 7

Use Norton's theorem to determine the current in the 3-Ω resistance of Example 6.

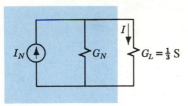

Figure 2.19   Application of Norton's theorem.

From the solution of Example 6,

$$V_{OC} = 28 \text{ V} \quad \text{and} \quad I_{SC} = 7 \text{ A}$$

By Eqs. 2-36, the parameters in Fig. 2.19 are

$$I_N = I_{SC} = 7 \text{ A}$$

and

$$G_N = \frac{I_{SC}}{V_{OC}} = \frac{7}{28} = 0.25 \text{ S}$$

By the current-divider principle,

$$I = \frac{G_L}{G_L + G_N} I_N = \frac{0.333}{0.333 + 0.25} \times 7 = 4 \text{ A}$$

As a by-product of the development of Thévenin's and Norton's theorems, we have obtained another general principle:

Any voltage source V with its associated series resistance R can be replaced by an equivalent current source I with an associated parallel conductance G where I = V/R and G = 1/R.

This principle of "source transformation" can be very useful in network analysis; for example, it permits the analyst to choose either an ideal voltage source or an ideal current source in representing a real energy source.

Superposition Theorem

The concept of *linearity* is illustrated in Fig. 2.20; an element of force ΔF produces the same element of deflection Δy at any point in the linear region. In general, where effect y is a function f of cause F, or $y = f(F)$, function f is *linear* if

$$f(F + \Delta F) = f(F) + f(\Delta F) = y + \Delta y \tag{2-37}$$

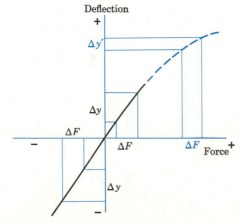

Figure 2.20   The concept of linearity.

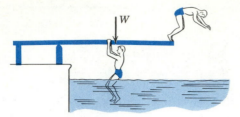

Figure 2.21   A linear beam.

In the nonlinear region, however, an equal element of force ΔF produces a different element of deflection $\Delta y'$.

In any situation where effect is directly proportional to cause, it is permissible to consider several causes individually and then combine the resulting individual effects to find the total effect. It is common practice to calculate the deflection of beams in this way. If a cantilever beam, such as the springboard of Fig. 2.21, carries three loads, the total deflection is the sum of the separate deflections due to the weight of the board itself, the swimmer hanging below, and the diver pushing on the tip. The general principle is called the *superposition theorem* and may be stated as follows:

> *If cause and effect are linearly related, the total effect of several causes acting simultaneously is equal to the sum of the effects of the individual causes acting one at a time.*

In electrical circuits, the causes are excitation voltages and currents and the effects are response voltages and currents. A typical case is the circuit of Fig. 2.22a. If the base of the current source is chosen as the reference node, the potential of node a is $I_2 R_2$ and the sum of the currents into a is

$$\frac{V - I_2 R_2}{R_1} + I - I_2 = 0$$

Then

$$V - I_2 R_2 + I R_1 - I_2 R_1 = 0$$

or

$$I_2 = \frac{V + I R_1}{R_1 + R_2} = \frac{V}{R_1 + R_2} + \frac{R_1}{R_1 + R_2} I \tag{2-38}$$

Equation 2-38 indicates that the current I_2 consists of two parts, one due to V and the other due to I. If the current source is removed (by open-circuiting) as in Fig. 2.22b, a component of current $I_V = V/(R_1 + R_2)$ flows in R_2. If the voltage source is

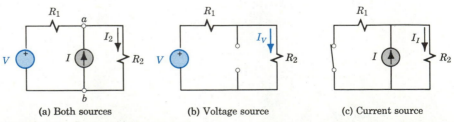

(a) Both sources   (b) Voltage source   (c) Current source

Figure 2.22   Superposition components.

removed (by short-circuiting) as in Fig. 2.22c, the current divides and a component $I_I = IR_1/(R_1 + R_2)$ flows in R_2. As indicated by the superposition theorem, I_2 is equal to $I_V + I_I$, the sum of the effects produced by the two causes separately. Note that ideal voltage sources are removed by short-circuiting and ideal current sources by open-circuiting. Real energy sources always have internal resistances that must remain in the circuit (see Example 8).

EXAMPLE 8

A load of 6 Ω is fed by a parallel combination of two batteries. Battery A has an open-circuit voltage of 42 V and an internal resistance of 12 Ω; battery B has an open-circuit voltage of 35 V and an internal resistance of 3 Ω. Determine the current I supplied by battery B, and calculate the power dissipated internally by the battery.

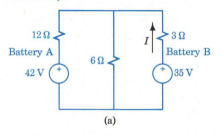

(a)

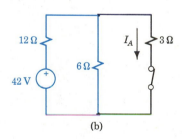

(b)

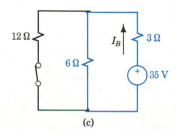

(c)

Figure 2.23 Application of the superposition principle.

If the batteries are represented by linear models, the circuit is as shown in Fig. 2.23a. With the 35-V source removed (Fig. 2.23b), the component of current due to battery A can be written in one step as

$$I_A = \frac{6}{6 + 3}\frac{42}{\left(12 + \frac{3 \times 6}{3 + 6}\right)} = \frac{2}{3}\left(\frac{42}{12 + 2}\right) = 2\ \text{A downward}$$

if the principles of current division (Eq. 2-34) and resistance combination (Eqs. 2-27 and 2-29) are applied simultaneously.

If the 42-V source is removed (the internal resistance of the battery remains), the component of current due to battery B (Fig. 2.23c) is

$$I_B = \frac{35}{3 + \frac{6 \times 12}{6 + 12}} = \frac{35}{3 + 4} = 5\ \text{A upward}$$

Then, by superposition,

$$I = -I_A + I_B = -2 + 5 = 3\ \text{A upward}$$

To calculate the power dissipated in the 3-Ω resistance, we might note that with the 35-V source removed, the power is

$$P_A = I_A^2 R = 2^2 \times 3 = 12\ \text{W}$$

With the 42-V source removed, the power is

$$P_B = I_B^2 R = 5^2 \times 3 = 75\ \text{W}$$

But, the actual power dissipated is *not* 12 + 75 = 87 W. Why not? What principle would be violated in such a calculation?

The answer is that power is not linearly related to voltage or current and therefore superposition cannot be applied in power calculations. Superposition is applicable only to linear effects such as the current response. Having found the current by superposition, we can calculate the power in the 3-Ω resistance as

$$P = I^2 R = (-I_A + I_B)^2 R = 3^2 \times 3 = 27\ \text{W}$$

NONLINEAR NETWORKS

In the preceding discussions we assumed linearity, but real devices are never strictly linear. In some cases, nonlinearity can be disregarded; linear approximation yields results that predict the behavior of the real devices within acceptable limits. In other cases, nonlinearity is annoying and special steps must be taken to avoid or eliminate its effect; later we shall learn how to use feedback to minimize the distortion introduced in nonlinear electronic amplifiers. Sometimes nonlinearity is desirable or even essential; the distortion that is annoying in an amplifier is necessary in the harmonic generator for obtaining output signals at frequencies that are multiples of the input signal.

Nonlinear Elements

A dissipative element for which voltage is not proportional to current is a *nonlinear resistor*. An ordinary incandescent lamp has a characteristic similar to that in Fig. 2.24a; the "resistance" of the filament increases with temperature and, therefore, under steady-state conditions, with current.

As we shall see, most electronic devices are inherently nonlinear. Because of its v-i characteristic (Fig. 2.24b), the semiconductor *diode* is useful in discriminating between positive and negative voltages. If a sinusoidal voltage is applied to such a diode, the resulting current has a large dc component and the diode is functioning as a *rectifier*.

To increase the energy-storage capability of an inductor, an iron core with favorable magnetic properties is used. (These properties are discussed in Chapter 21.) In such an inductor, the magnetic flux ϕ is not proportional to current (Fig. 2.24c); at large values of current, a given increment of current produces only a small increment of flux. One interpretation is that "inductance" is not constant. This nonlinearity is troublesome in a power transformer, but it may be useful in a control system.

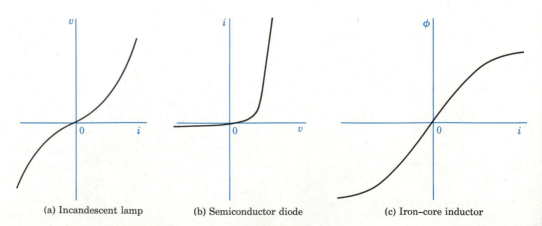

(a) Incandescent lamp   (b) Semiconductor diode   (c) Iron–core inductor

Figure 2.24 Characteristics of nonlinear elements.

Methods of Analysis

The method of analysis employed depends on the nature of the problem, the form of the data, and the computation aids available. The modern digital computer and analog computer permit the ready solution of problems that were hopelessly time consuming a few years ago, but here we are concerned only with simple series and parallel circuits.

Analytical Solution. If an analytical expression for the v-i characteristic can be obtained from physical principles or from experimental data, some simple problems can be solved algebraically. A useful model is the power series

$$i = a_0 + a_1v + a_2v^2 + a_3v^3 + \cdots \qquad (2\text{-}39)$$

The first two terms provide a linear approximation. The first three terms give satisfactory results in many practical nonlinear problems, although more terms may be necessary in some cases.

EXAMPLE 9

A voltage $v = v_m \cos \omega t$ is applied to the semiconductor diode of Fig. 2.25. Determine the nature of the resulting current.

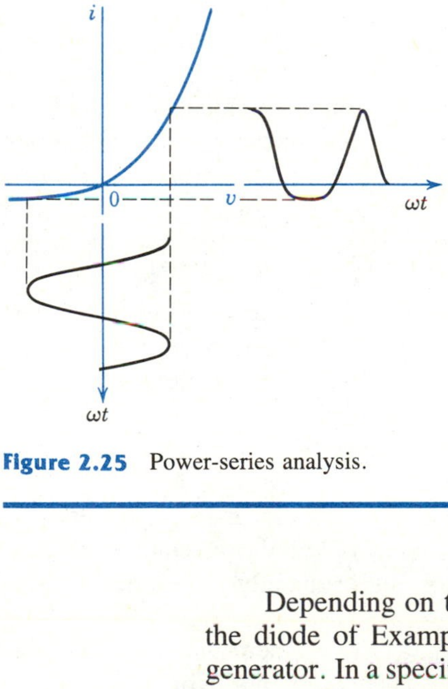

Figure 2.25   Power-series analysis.

Assuming that the v-i characteristic can be represented by the first three terms of the power series of Eq. 2-39, the current is

$$i = a_0 + a_1v + a_2v^2$$

But we note that for $v = 0$, $i = 0$; therefore, $a_0 = 0$. For the given voltage,

$$i = a_1V_m \cos \omega t + a_2V_m^2 \cos^2 \omega t$$

or

$$i = a_1V_m \cos \omega t + \frac{a_2V_m^2}{2}(1 + \cos 2\omega t)$$

$$= \frac{a_2V_m^2}{2} + a_1V_m \cos \omega t + \frac{a_2V_m^2}{2} \cos 2\omega t \qquad (2\text{-}40)$$

Equation 2-40 indicates that the resulting current consists of a steady or dc component, a "fundamental" component of the same frequency as the applied voltage, and a "second-harmonic" component at the new frequency 2ω.

Depending on the relative magnitudes of a_1 and a_2 and the associated circuitry, the diode of Example 9 could be used as a rectifier, an amplifier, or a harmonic generator. In a specific problem, the coefficients in the power series can be determined by choosing a number of points on the v-i characteristic equal to the number of terms to be included, substituting the coordinates of each point in Eq. 2-39, and solving the resulting equations simultaneously.

Piecewise Linearization. Where approximate results are satisfactory, a convenient approach is to represent the actual characteristic by a series of linear "pieces." The characteristic of the iron-core inductor of Fig. 2.26a can be divided into two regions, each approximated by a straight line. Below the "knee" of the curve, the behavior is adequately described by $\phi = k_1 i$; above the knee, a different expression must be used. Usually a trial solution will indicate in which region operation is occurring, and the problem is reduced to one in linear analysis. (This approach is used in developing models for electronic devices in Chapter 11.)

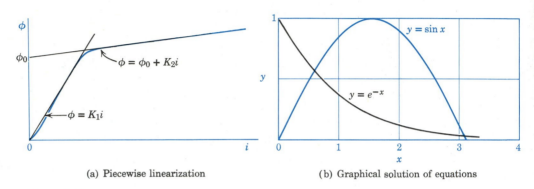

(a) Piecewise linearization   (b) Graphical solution of equations

Figure 2.26   Analysis based on graphical data.

Graphical Solution. In piecewise linearization we use graphical data to obtain analytical expressions that hold over limited regions. Sometimes it is desirable to plot given analytical functions to obtain a solution graphically. For example, the equation $e^{-x} - \sin x = 0$ is difficult to solve analytically. If the functions $y = e^{-x}$ and $y = \sin x$ are plotted as in Fig. 2.26b, the simultaneous solutions of these two equations are represented by points whose coordinates satisfy both equations. Therefore, the intersections of the two curves are the solutions to the original equation and can be read directly from the curves. This method is also useful when the characteristic of some portion of the circuit is available in graphical form or in a table of experimental data.

Networks with One Nonlinear Element

If there is a single nonlinear resistance in an otherwise linear resistive network, and if the v-i characteristic of the nonlinear element is known, a relatively simple method of solution is available. This situation occurs frequently in practice and the method of attack deserves emphasis.

The first step is to replace all except the nonlinear element with the Thévenin equivalent shown in Fig. 2.27a as V_T and R_T. The combination of V_T and R_T is equivalent "insofar as a load is concerned." For any value of load resistance, the terminal voltage is

$$V = V_T - R_T I \tag{2-41}$$

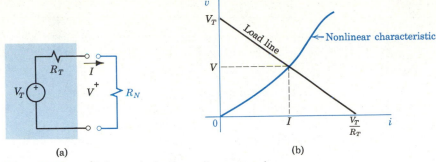

(a) (b)

Figure 2.27 Load-line analysis of a nonlinear network.

This equation is that of a straight line, the *load line*, with intercepts V_T and V_T/R_T and slope $-R_T$. (For the original network, the intercepts are V_{OC} and I_{SC}.) The graph of the nonlinear characteristic is v as a function of i for the load. The simultaneous satisfaction of these two relations, which yields the values of V and I at the terminals, is the intersection of the two curves (Example 10).

EXAMPLE 10

The nonlinear element whose characteristic is given in Fig. 2.27b is connected in the circuit of Fig. 2.28a. Determine the current I.

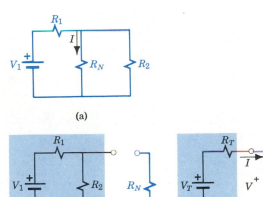

(a)

(b) (c)

Figure 2.28 Load-line analysis.

First, the nonlinear element is isolated (Fig. 2.28b) and the Thévenin equivalent determined. Considering R_1 and R_2 as a voltage divider,

$$V_T = V_{OC} = \frac{R_2}{R_1 + R_2} V_1$$

The short-circuit current is just $I_{SC} = V_1/R_1$. Therefore, the equivalent resistance is

$$R_T = \frac{V_{OC}}{I_{SC}} = \frac{R_2 V_1}{R_1 + R_2} \frac{R_1}{V_1} = \frac{R_1 R_2}{R_1 + R_2}$$

which is also the resistance seen looking in at the terminals with the voltage source removed.

Next, the load line is drawn with intercepts V_T and V_T/R_T as shown in Fig. 2.27b. The intersection of the load line with the nonlinear characteristic gives the desired current I.

For transistors, motors, and some control system components, the *v-i* characteristics are represented by a family of curves. The "load" may be the linear element represented by R_T with power supplied by a battery of voltage V_T. Once the load line is drawn, the effect of variations in operating conditions (signals) is clearly visible. Note that if V_T varies during operation, the load line moves parallel to itself.

SUMMARY

- The algebraic sum of the currents into a node is zero ($\Sigma i = 0$).
  The algebraic sum of the voltages around a loop is zero ($\Sigma v = 0$).

- The general procedure for formulating equations for circuits is:
  1. Arbitrarily assume a consistent set of currents and voltages.
  2. Write the element equations by applying the element definitions and write the connection equations by applying Kirchhoff's laws.
  3. Combine the element and connection equations to obtain the governing circuit equation in terms of the unknowns.

  The resulting simultaneous equations can be solved by:
  Successive substitution to eliminate all but one unknown, or
  Use of determinants and Cramer's rule.

- Skillful use of loop currents or node voltages may greatly reduce the number of unknowns and simplify the solution.
  Checking is essential because of the many opportunities for mistakes in sign and value.

- Two one-ports are equivalent if they present the same v-i characteristics.
  Two passive one-ports are equivalent if they have the same input resistance.
  For resistances in series, $R_{EQ} = R_1 + R_2 + \cdots + R_n$.
  For conductances in parallel, $G_{EQ} = G_1 + G_2 + \cdots + G_n$.

- For the voltage divider and current divider,

$$V_2 = \frac{R_2}{R_1 + R_2} V \quad \text{and} \quad I_2 = \frac{G_2}{G_1 + G_2} I = \frac{R_1}{R_1 + R_2} I$$

- Insofar as a load is concerned, any one-port network of resistance elements and energy sources can be replaced by a series combination (Thévenin) of an ideal voltage source V_T and a resistance R_T, or by a parallel combination (Norton) of an ideal current source I_N and a conductance G_N, where $V_T = V_{OC}$, $R_T = V_{OC}/I_{SC}$, $I_N = I_{SC}$, and $G_N = I_{SC}/V_{OC} = 1/R_T$.

- A series combination of voltage V and resistance R can be replaced by a parallel combination of current $I = V/R$ and conductance $G = 1/R$.

- If cause and effect are linearly related, the principle of superposition applies, and the total effect of several causes acting simultaneously is equal to the sum of the effects of the individual causes acting one at a time.

- Networks involving nonlinear elements can be solved in various ways, depending on the nature of the problem and the form of the data.
  If there is a single nonlinear element in an otherwise linear network, construction of a load line permits a simple graphical solution.

REVIEW QUESTIONS

1. Write out in words a physical interpretation of Eq. 2-18.
2. Outline a three-step procedure for applying the node-voltage method.
3. Outline a three-step procedure for applying the loop-current method.
4. If two passive one-ports are equivalent, is the same power dissipated in both? Prove your answer.
5. What is the difference between "passive" and "active" networks?
6. Under what circumstances is Thévenin's theorem useful?
7. Is the same power dissipated in an active circuit and the Norton equivalent?
8. How are the Norton and Thévenin components related?
9. Cite two nonelectrical examples where superposition applies.
10. Under what circumstances is the superposition principle useful?
11. How are voltage and current sources "removed"?
12. Cite nonelectrical examples of linear and nonlinear devices.
13. Give two engineering examples where nonlinearity is desirable.
14. The v-i characteristic of an active device is $v = 100 - 2i - i^2$. Outline two methods for determining the current that flows when a resistance R is connected to the device.
15. What is meant by "piecewise linearization"? A "load line"?

EXERCISES

1. In Fig. 2.29, $i_1 = 2$ A and $i_2 = -3$ A. Predict i_3.

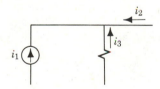

Figure 2.29

2. In Fig. 2.30, $v = 12$ V, $R_1 = 4$ Ω, and $R_2 = 2$ Ω. Calculate current i and voltages v_1 and v_2 across the resistances.

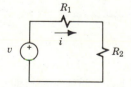

Figure 2.30

3. In Fig. 2.31, determine i_1 and i_2 in terms of the other quantities.

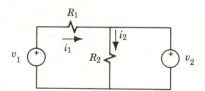

Figure 2.31

4. In Fig. 2.32, given $I_a = 3$ A, $I_b = 2$ A, and $I_c = -8$ A, determine i_1, i_2, and v.

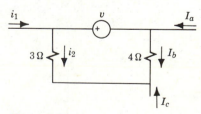

Figure 2.32

5. In Fig. 2.32, given $I_a = -4$ A, $I_c = 1$ A, and $v = 18$ V, determine i_1, i_2, and I_b.

6. In Fig. 2.33, $i_1 = 5$ A, $i_2 = 10 \cos t$ A, and $v_C = 5 \sin t$ V. Find:
   (a) i_L and v_L
   (b) v_{ac}, v_{ab}, and v_{cd}
   (c) The energy stored in L and C, all as functions of time.

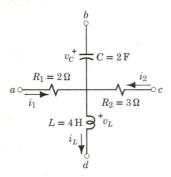

Figure 2.33

7. Solve Exercise 6 for $i_1 = 10\,e^{-2t}$ A, $i_2 = 4$ A, and $v_C = 3\,e^{-2t}$ V.

8. In Fig. 2.34, evaluate:
   (a) i_1, i_2, v_{ba}, and v_{de}
   (b) The power dissipated in the 4-Ω resistance.
   (c) The power supplied by the current source.

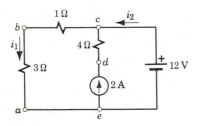

Figure 2.34

9. Device X requires 4 V at 1.5 mA, whereas device Y operates at 2 V and 1 mA. The two devices are to be operated from a 9-V battery as shown in Fig. 2.35. Design the circuit; i.e., specify the values of R_1 and R_2.

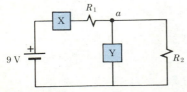

Figure 2.35

10. Given the circuit of Fig. 2.36, calculate the voltage across the 3-Ω resistance and the voltage across the current source using: (a) element currents, (b) loop currents, (c) node voltages. Include an adequate check.

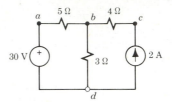

Figure 2.36

11. Given the circuit of Fig. 2.37, calculate the current I using: (a) element currents, (b) loop currents, (c) node voltages. Include an adequate check.

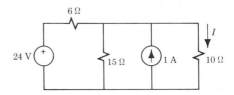

Figure 2.37

12. Given the circuit of Fig. 2.38, calculate the current in the 2-Ω resistance using: (a) element currents, (b) loop currents, (c) node voltages. Draw a conclusion about the methods.

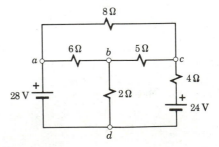

Figure 2.38

13. Given the circuit of Fig. 2.38, calculate the voltage across the 5-Ω resistance using: (a) element currents, (b) loop currents, (c) node voltages.

14. In Fig. 2.39, V_O is an initial voltage on C. Using only the specified variables, formulate the voltage equation for loop *abcda* with the switch closed.

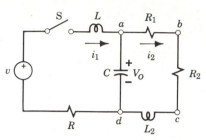

Figure 2.39

15. The heating element in a "hotplate" consists of two resistances that can be connected separately, in series, or in parallel to provide four "heat" settings.
    (a) Draw a wiring diagram showing R_1 as an inside loop and R_2 as an outside loop with terminals for series or parallel connection.
    (b) For 120-V operation and $R_1 = 30\ \Omega$, specify R_2 for a "low" setting of 160 W and calculate the other three power values.

16. Find current I in the *ladder* network of Fig. 2.40 by network reduction. Reduce series and parallel combinations to a single equivalent resistance, find the battery current, then use current division to find I.

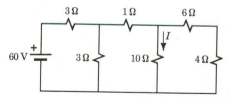

Figure 2.40

17. Find current I in the *ladder* network of Fig. 2.40 by the following ingenious method: Assume the current is 1 A and work backward to find the necessary source voltage; determine actual I by simple proportion.

18. Applying the current divider principle to Fig. 2.40, write down an expression for current I as a function of $V\ (= 60\ V)$ and solve for I.

19. Replace the circuit of Fig. 2.40 by a simpler equivalent that will enable you to find the current in the 1-Ω resistor by a single calculation. Calculate the current.

20. In Fig. 2.41, $R_2 = 2\ k\Omega$, $R_1 = 10\ k\Omega$, and $v = 24\ V$.
    (a) Predict v_2.
    (b) A practical voltmeter can be represented by a 10-kΩ resistance in series with an ideal (resistanceless) meter calibrated to indicate the voltage across the 10-kΩ meter resistance. Predict the reading of the voltmeter when connected across R_2.

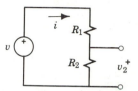

Figure 2.41   Voltage divider.

21. Repeat Exercise 20(b) if the effective voltmeter resistance is 100 kΩ, and draw a conclusion about the desirable internal resistance of a voltmeter.

22. The open-circuit voltage of an audio oscillator is 5 V. The terminal voltage drops to 4 V when a 2000-Ω resistor is connected across the terminals. Determine the internal resistance of the oscillator.

23. In Fig. 2.37, use Thévenin's theorem to find the current in the 15-Ω resistance.

24. In Fig. 2.40, replace the 10, 6, and 4-Ω resistances by a single equivalent resistance. Then replace the 60-V source and the two 3-Ω resistances by a Thévenin equivalent. Use the voltage divider principle to predict current I.

25. In Fig. 2.42, determine the current in the 3-Ω resistance without using simultaneous equations. (*Hint:* First replace all except the 3-Ω resistance and the 35-V battery by a Thévenin equivalent.)

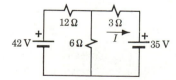

Figure 2.42

26. In Fig. 2.43, with switch S closed, the ammeter AM reads 60 mA. Predict the ammeter reading with switch S open. (*Hint:* Does the AM reading with S closed give you a clue to the Norton equivalent?)

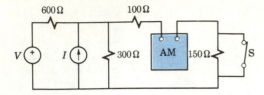

Figure 2.43

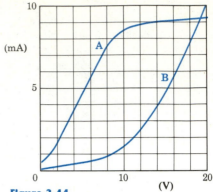

Figure 2.44

27. In Fig. 2.37, use Norton's theorem to find current I.

28. In Fig. 2.42, use Norton's theorem to find the current in the 6-Ω resistance.

29. Repeat Exercise 10 using superposition.

30. Repeat Exercise 12 using superposition.

31. Use superposition to find the current in the 15-Ω resistance in Fig. 2.37.

32. In Fig. 2.37, use superposition to find current I.

33. Find the current in the 12-Ω resistance in Fig. 2.42 by using superposition.

34. Measurements on a human nerve cell indicate an open-circuit voltage of 80 mV, and then a current of 5 nA through a 6-MΩ load. You are to predict the behavior of this cell with respect to an external load.
(a) What simplifying assumption must be made?
(b) Devise an appropriate model.
(c) Predict the current through a 10-MΩ load.

35. A device that can be represented by the mathematical model $v = 6i + 3i^2$ is connected in series with a source that can be represented by 36 V in series with 6 Ω.
(a) Predict the resulting device current mathematically.
(b) Sketch the v-i characteristic for $0 < i < 3$ and predict the device current graphically. Compare to the result for part (a).

36. Device A in Fig. 2.44 is connected in series with a 15-V battery and a 2-kΩ resistance. Predict the current that flows in two ways:
(a) Reproduce the i-v characteristic and represent device A by piecewise linearization.
(b) Use the load-line method. Compare results.

37. For device A in Fig. 2.44,
(a) Reproduce the v-i characteristic and represent it by piecewise linearization.
(b) Predict the approximate current in response to an applied voltage $v = 15 + 5 \cos \omega t$ V.

38. Assume that device B in Fig. 2.44 can be represented by the equation $i(\text{mA}) = a_1 v + a_2 v^2$ in the region $10 \le v \le 18$ V.
(a) Evaluate a_1 and a_2 by simultaneous solution of two equations for i at the two values of v.
(b) Use your mathematical model to determine i for $v = 14$ V and check graphically.

39. Device B in Fig. 2.44 is connected in parallel with a 5-mA current source and a 3-kΩ resistance; predict the voltage and the current in device B.

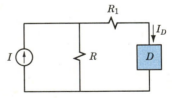

Figure 2.45

40. Device A in Fig. 2.44 is connected in the circuit of Fig. 2.45 where $R = 1000\ \Omega$ and $R_1 = 600\ \Omega$.
(a) If $I = 16$ mA, predict the device current I_D.
(b) Specify the value of I so that the power dissipated in D is 60 mW.

PROBLEMS

1. Given the circuit of Fig. 2.46, consider various methods to calculate the current in the 2-Ω resistance. Use the method that appears to require the least algebra.

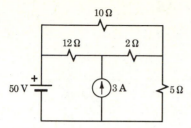

Figure 2.46

2. In Fig. 2.47, find the voltage across the 20-Ω resistor using loop currents. If possible, choose the loop currents so that only one equation is necessary.

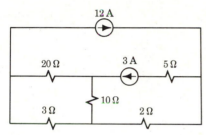

Figure 2.47

3. (a) Draw and label a circuit showing inductors L_1 and L_2 connected in series and derive an equation for the equivalent inductance of the combination.

(b) Repeat for inductances L_1 and L_2 connected in parallel.

4. (a) Draw and label a circuit showing capacitors C_1 and C_2 connected in series and derive an equation for the equivalent capacitance of the combination.

(b) Repeat for capacitances C_1 and C_2 connected in parallel.

5. A signal generator, consisting of a complicated network of nearly linear passive and active elements, supplies a variable load resistance R_L.

(a) Use Thévenin's theorem to replace the generator by a combination of V_G and R_G.

(b) Prove the *maximum power transfer* theorem, which states that: "For maximum power transfer from a source to a load, the resistance of the load should be made equal to the Thévenin equivalent resistance of the source."

6. In an amplifier a transistor can be represented by the two-port model in Fig. 2.48, where $R_1 = 10$ kΩ, $R_2 = 75$ kΩ, $R_3 = 150$ kΩ, and $\beta = 100$. (βI_1 is a "controlled" current source.)

(a) Simplify the model and voltage source and predict the voltage amplification V_L/V_S as a function of R_L.

(b) Calculate V_L/V_S for $R_L = 20$ kΩ.

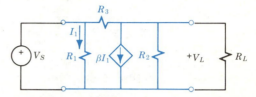

Figure 2.48

7. The behavior of a certain germanium diode is defined by $i = I_S(e^{bv} - 1)$ where $I_S = 10$ μA and $b = 40$ V^{-1}.

(a) Expand the exponential term in a power series and represent $i(v)$ using the first three terms of Eq. 2-39.

(b) Predict the current if a voltage $v = +0.05$ V is applied to the diode in series with a 500-Ω resistance.

(c) Check the accuracy of the method by calculating the current using the diode voltage from part (b) in the given exponential equation.

8. Devices A and B in Fig. 2.44 are connected in series and the combination in series with a 30-V battery and a 5-kΩ resistance.

(a) Plot the composite v-i characteristic for A and B in series.

(b) Predict i, v_A, and v_B by the load-line method.

(c) Predict the power dissipated in devices A and B.

9. The output characteristics of two generators are:

I(A)	0	25	50	75	100
V_1(V)	120	119	117	113	105
V_2(V)	120	118	113	105	90

The two generators are connected in parallel to supply power to a load resistance $R_L = 1.0$ Ω. Plot the characteristics and determine graphically the current and power supplied by each generator.

3

Signal Processing Circuits

Signal Waveforms

Periodic Waveforms

Ideal Amplifiers

Ideal Diodes

Waveshaping Circuits

A principal application of electronics is information processing. Typically, the information is available in the form of *signals*. An electrical signal may be a voltage or current varying with time in a manner that conveys information; the pattern of ones and zeros in a digital code and the varying currents in a telephone receiver are well-known examples. In processing such information, electronic *devices* are used to detect and amplify weak signals or to convert signals from one form to another—from a high-frequency signal that is easily radiated by an antenna to a low-frequency signal that is audible to humans, for example. The electronic *circuits* that carry signals between devices range in size from the intercontinental cable circuits that span oceans to the microscopic integrated circuits built into a tiny calculator chip of silicon.

In this chapter we look at some elementary signal processing circuits. First we study some common signal waveforms and learn how they can be handled. Next we examine the characteristics of ideal amplifiers and diodes and see how they perform their basic functions. Then we use the circuit principles of Chapters 1 and 2 in learning to analyze and design some electronic circuits that are widely used in communication and control.

SIGNAL WAVEFORMS

Certain patterns of time variation or *waveforms* are of special significance in electronics because they are encountered so frequently. Some of these are shown in Fig. 3.1.

A *direct* or *continuous* voltage is supplied by a storage battery or a direct-current (dc) generator. A *step* of current flows when a switch is thrown, suddenly applying a direct voltage to a resistance; the current in the battery of a pocket radio is an

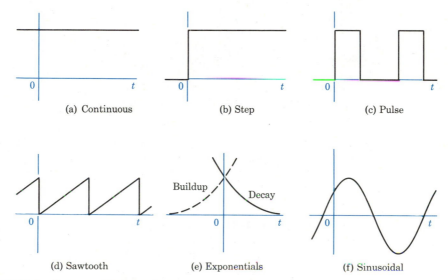

(a) Continuous   (b) Step   (c) Pulse

(d) Sawtooth   (e) Exponentials   (f) Sinusoidal

Figure 3.1 Common signal waveforms.

example. A *pulse* of current flows if the switch is turned **ON** and then **OFF**; the flow of information in a computer consists of very short pulses. A *sawtooth* wave increases linearly with time and then resets; this type of voltage variation causes the electron beam to move repeatedly across the screen of a television picture tube. A decaying *exponential* current flows if energy is stored in the electric field of a capacitor and allowed to leak off through a resistor; in an unstable system, a voltage may build up exponentially. A *sinusoidal* voltage is generated when a coil is rotated at a constant speed in a uniform magnetic field; also, oscillating circuits are frequently character-ized by sinusoidal voltages and currents.

Exponentials

Exponential and sinusoidal waveforms are easy to generate and easy to analyze because of their simple derivatives and integrals; we shall pay them special attention because they are encountered frequently and also because they are useful in the analysis of more complicated waves, such as the electrocardiograms of Fig. 3.2.

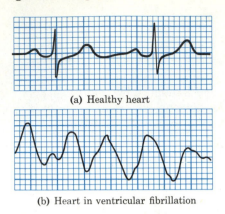

(a) Healthy heart

(b) Heart in ventricular fibrillation

Figure 3.2 Electrocardiograms.

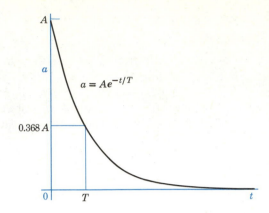

Figure 3.3 The decaying exponential.

The General Exponential. In general, an exponentially decaying quantity (Fig. 3.3) can be expressed as

$$a = A\,e^{-t/T} \tag{3-1}$$

where a = instantaneous value
A = amplitude or maximum value
e = base of natural logarithms = 2.718 . . .
T = time constant in seconds
t = time in seconds

The temperature of a hot body in cool surroundings, the angular velocity of a freely spinning bicycle wheel, and the current of a discharging capacitor can all be approximated by decaying exponential functions of time.

Time Constant. Since the exponential factor in Eq. 3-1 only *approaches* zero as t increases without limit, such functions theoretically last forever. In the same sense, all radioactive disintegrations last forever. To distinguish between different rates of disintegration, physicists use the term "half-life," where the half-life of radium, say, is the time required for any amount of radium to be reduced to one-half of its initial amount. In the case of an exponentially decaying current, it is mo onvenient to use the value of time that makes the exponent -1. When $t = T =$ the *time constant*, the value of the exponential factor is

$$e^{-t/T} = e^{-1} = \frac{1}{e} = \frac{1}{2.718} = 0.368 \tag{3-2}$$

In other words, after a time equal to the time constant, the exponential factor is reduced to approximately 37% of its initial value.

EXAMPLE 1

A capacitor is charged to a voltage of 6 V and then, at time $t = 0$, it is switched across a resistor and observations are made of current as a function of time. The experimental data are plotted in Fig. 3.4a. Draw the circuit, determine the time constant for this circuit, and derive an exponential equation for the current.

A smooth curve is drawn through the experimental points. The amplitude of the exponential is determined by noting that at $t = 0$, $i = I_0 = 6$ mA.

The time constant is determined by noting that when $i = I_0 e^{-1} = 6(0.368) \cong 2.2$ mA, $t = T \cong 2$ ms.

Then $1/T = 1/0.002 = 500$ and the current equation is

$$i = I_0 e^{-t/T} = 6 e^{-500t} \text{ mA} \qquad (3\text{-}3)$$

for time $t > 0$, i.e., *after* switching.

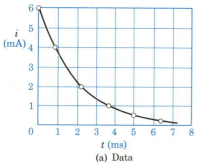

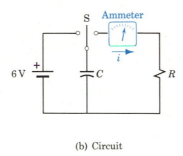

(a) Data

(b) Circuit

Figure 3.4   Observation of time constant.

Normalized Exponentials. All exponential equations can be reduced to a common form by expressing the variables in terms of dimensionless ratios or *normalizing* the variables. When this is done, Eq. 3-1 becomes

$$\frac{a}{A} = e^{-t/T} \qquad (3\text{-}4)$$

Note that after 2 time constants, $t = 2T$ and the function is down to $1/e^2$ times its initial value, or $a/A = (0.368)^2 = 0.135$. After 5 time constants, $t = 5T$ and the normalized function is down to $1/e^5 = 0.0067$. This value is less than the error expected in many electrical measurements, and it may be assumed that after 5 time constants an exponential current or voltage is practically negligible.

When plotted on semilog graph paper, an exponential function becomes a straight line with slope equal to the reciprocal of the time constant. As another interpretation of the time constant, it can be demonstrated (see Problem 1) that, if the exponential decay continued at its *initial rate,* the duration of the function would be just equal to T.

Exponential Voltages and Currents

Exponentials are important because many natural phenomena follow an exponential time variation. Also, the simplicity of the mathematical relations between exponentials and their derivatives and integrals makes them particularly valuable in circuit analysis, as in Example 2.

EXAMPLE 2

In Fig. 3.5a, the current $i = 5\,e^{-2t}$ A.
(a) Determine the voltage $v(t)$.

$$v_R = Ri = 4 \times 5\,e^{-2t} = 20\,e^{-2t} \text{ V}$$

$$v_L = L\frac{di}{dt} = 1(-2)5\,e^{-2t} = -10\,e^{-2t} \text{ V}$$

$$v = v_R + v_L = 20\,e^{-2t} - 10\,e^{-2t} = +10\,e^{-2t} \text{ V}$$

(b) If the current $i = 5\,e^{-2t}$ A flows in an initially uncharged 0.25-F capacitance ($v_C = 0$ at $t = 0$ in Fig. 3.5b), determine the voltage $v_C(t)$.

$$v_C(t) = \frac{1}{C}\int_0^t i\,dt + V_O = \frac{1}{0.25}\int_0^t 5\,e^{-2t}\,dt + 0$$

$$= \left.\frac{5}{0.25(-2)}e^{-2t}\right]_0^t = 10 - 10\,e^{-2t} \text{ V}$$

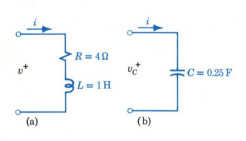

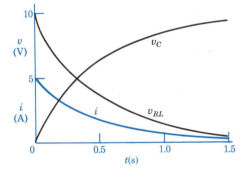

Figure 3.5 Response to exponentials.

Sinusoids

There are three principal reasons for emphasis on sinusoidal functions of time. First, many natural phenomena are sinusoidal in character; the vibration of a guitar string, the projection of a satellite on the rotating earth, and the current in an oscillating circuit are examples. Second, sinusoids are important in the generation and transmission of electric power and in the communication of intelligence; because the derivatives and integrals of sinusoids are themselves sinusoidal, a sinusoidal source always produces sinusoidal responses in any linear circuit. Third, other periodic waves can be represented by a series of sinusoidal components by means of Fourier analysis; this means that techniques for handling sinusoids are useful in predicting the behavior of circuits involving pulses and sawtooth waves as well. Clearly, an efficient technique for analyzing circuits involving sinusoidal voltages and currents is essential. Since the application of Kirchhoff's laws involves summations of currents and voltages, we need a convenient method for adding quantities that are continually varying functions of time.

The General Sinusoid. In general, a sinusoidally varying quantity (Fig. 3.6) can be expressed as

$$a = A \cos (\omega t + \alpha) \tag{3-5}$$

where a = instantaneous value
       A = amplitude or maximum value
       ω = frequency in radians per second (omega)
       t = time in seconds
       α = phase angle in radians (alpha)

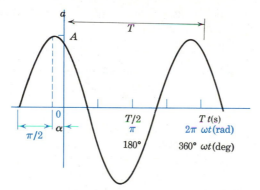

Figure 3.6   The general sinusoid.

Since a variation in angle of 2π rad corresponds to one complete cycle,

$$f = \frac{\omega}{2\pi} = \textit{frequency} \text{ in cycles per second or hertz}^\dagger \text{ (Hz)} \tag{3-6}$$

The time for one complete cycle is

$$T = \frac{1}{f} = \textit{period} \text{ in seconds} \tag{3-7}$$

The same sinusoid can be expressed by the sine-wave function

$$a = A \sin \left(\omega t + \alpha + \frac{\pi}{2} \right) \tag{3-8}$$

since $\sin (x + \pi/2) = \cos x$ for all values of x. In either case, the instantaneous value varies from positive maximum to zero to negative maximum to zero repeatedly; a current that varies sinusoidally is called an *alternating current*.

---

†Heinrich Hertz, a German physicist, first demonstrated the existence of radio waves.

EXAMPLE 3

A sinusoidal current with a frequency of 60 Hz reaches a positive maximum of 20 A at $t = 2$ ms (Fig. 3.7). Write the equation of current as a function of time.

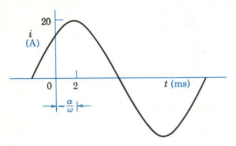

Figure 3.7   A sinusoidal current.

In the form of Eq. 3-5, $A = 20$ A and $\omega = 2\pi f = 2\pi(60) = 377$ rad/s. The positive maximum is reached when $(\omega t + \alpha) = 0$ or any multiple of 2π. Letting $(\omega t + \alpha) = 0$,

$$\alpha = -\omega t = -377 \times 2 \times 10^{-3} = -0.754 \text{ rad}$$

or

$$\alpha = -0.754 \times \frac{360°}{2\pi} = -43.2°$$

Therefore, the current equation is

$$i = 20 \cos (377t - 43.2°) \text{ A}$$

Note that the angle in parentheses is a convenient hybrid; to avoid confusion, the degree symbol is essential.

Sometimes it is convenient to associate the sinusoidal variation with the projection of a radial line of length A rotating with frequency ω from an initial or phase angle α. For radial line A of Fig. 3.8, the horizontal projection is $a = A \cos (\omega t + \alpha)$. The vertical projection of radial line A' is $a' = A' \sin (\omega t + \alpha + \pi/2)$, which is just equal to a; clearly, either a horizontal projection or a vertical projection can be used.

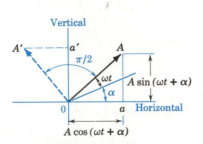

Figure 3.8   Projections of a rotating line.

PERIODIC WAVEFORMS

By definition, a *periodic* function of time is one in which

$$f(t + nT) = f(t) \tag{3-9}$$

where T is the period in seconds and n is any integer. The sinusoidal current shown in Fig. 3.9a is periodic with a period of $T = 1/f = 2\pi/\omega$ seconds since

$$i = 10 \cos \omega\left(t + \frac{2\pi}{\omega}\right) = 10 \cos(\omega t + 2\pi) = 10 \cos \omega t$$

Always looking for ways to simplify calculations, we ask: What single value of current can be used to represent this continuously varying current?

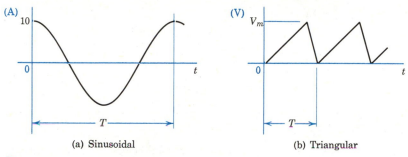

(a) Sinusoidal   (b) Triangular

Figure 3.9   Examples of periodic functions.

The value to be used to represent a varying current depends on the function to be performed by the current. If it is used to actuate a relay, the maximum or *peak* value of 10 A is critical. If it is rectified (so that current flow is always in the same direction) and used to deposit silver in an electrolytic plating operation, the *average* value of 6.37 A is the significant quantity. If it is used to develop power in a resistor, the *effective* value of 7.07 A is the significant quantity. The term "peak value" is self-explanatory, but the other terms deserve consideration.

Average Value

The average value of a varying current $i(t)$ over the period T is the steady value of current I_{av} that in period T would transfer the same charge Q. By definition,

$$I_{av}T = Q = \int_{t}^{t+T} i(t)\, dt = \int_{0}^{T} i\, dt$$

or

$$I_{av} = \frac{1}{T} \int_{0}^{T} i\, dt \qquad (3\text{-}10)$$

In a similar way, average voltage is defined as

$$V_{av} = \frac{1}{T} \int_{0}^{T} v\, dt \qquad (3\text{-}11)$$

In dealing with periodic waves, it is understood that the average is over one complete cycle (or an integral number of cycles) unless a different interval is specified. For example, the average value of any triangular wave (Fig. 3.9b) is equal to half the peak value since the area under the curve ($\int v\, dt$) for one cycle is $\frac{1}{2} V_m T$.

For a sinusoidal wave, the average value over a cycle is zero; the charge transferred during the negative half-cycle is just equal and opposite to that transferred during the positive half-cycle. In certain practical problems we are interested in the *half-cycle* average (Fig. 3.10) given by

$$I_{\text{half-cycle}} = \frac{1}{\frac{1}{2}T} \int_{-T/4}^{+T/4} I_m \cos \frac{2\pi t}{T} \, dt$$

$$= \frac{2I_m}{T} \left(\frac{T}{2\pi} \right) \sin \frac{2\pi t}{T} \Bigg]_{-T/4}^{+T/4}$$

$$= \frac{I_m}{\pi} \left[\sin \left(\frac{\pi}{2} \right) - \sin \left(-\frac{\pi}{2} \right) \right]$$

$$= \frac{2}{\pi} I_m = 0.637 I_m \qquad (3\text{-}12)$$

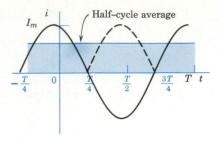

Figure 3.10 The half-cycle average.

Effective Value

In many problems we are interested in the energy-transfer capability of an electric current. By definition, the average value of a varying power $p(t)$ is the steady value of power P_{av} that in period T would transfer the same energy W. If

$$P_{\text{av}}T = W = \int_t^{t+T} p(t) \, dt = \int_0^T p \, dt \quad \text{then} \quad P_{\text{av}} = P = \frac{1}{T} \int_0^T p \, dt \qquad (3\text{-}13)$$

By convention, P always means average power and no subscript is necessary.

If electrical power is converted into heat in a resistance R,

$$P = \frac{1}{T} \int_0^T p \, dt = \frac{1}{T} \int_0^T i^2 R \, dt = I_{\text{eff}}^2 R \qquad (3\text{-}14)$$

where I_{eff} is defined as the steady value of current that is equally *effective* in converting power. Solving Eq. 3-14,

$$I_{\text{eff}} = \sqrt{\frac{1}{T} \int_0^T i^2 \, dt} = I_{\text{rms}} \qquad (3\text{-}15)$$

and I_{eff} is seen to be the "square root of the mean squared value" or the *root-mean-square* current I_{rms}.

The effective or rms value of a sinusoidal current can be found from Eq. 3-15. Where $i = I_m \cos (2\pi/T)t$,

$$I_{\text{rms}}^2 = \frac{1}{T} \int_0^T I_m^2 \cos^2 \frac{2\pi t}{T} \, dt = \frac{1}{T} \frac{I_m^2}{2} \int_0^T \left(1 + \cos \frac{4\pi t}{T} \right) dt = \frac{I_m^2}{2}$$

Therefore, the effective value of a sinusoidal current is

$$I_{\text{rms}} = \frac{I_m}{\sqrt{2}} = 0.707 I_m \qquad (3\text{-}16)$$

In Fig. 3.11, note that the mean value of $\cos^2 \omega t$ is just equal to 0.5 since the average of the $\cos 2\omega t$ term is zero over a full cycle. In words, Eq. 3-16 says: "For sinusoids, the effective value is $1/\sqrt{2}$ times the maximum value." For all other functions Eq. 3-15 must be used.

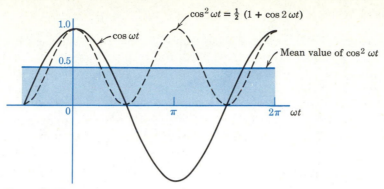

Figure 3.11 Calculating the root-mean-square value of a sinusoid.

In the United States, voltage of 120 V effective at a frequency of 60 Hz is standard for residential power. The instantaneous voltage is $v(t) = 120 \sqrt{2} \cos 2\pi 60t \cong 170 \cos 377t$ V.

Meter Readings

Basically, a measuring instrument converts a physical effect into an observable quantity. The cathode-ray oscilloscope (CRO) converts an applied voltage into a deflection of the spot of impact of an electron beam. The beam has little inertia and will follow rapid voltage variations, and therefore the CRO indicates *instantaneous* values (see Figs. 10.5 and 10.10).

In the common dc ammeter (Fig. 3.12a), a current in the meter coil (suspended in a magnetic field) produces a torque opposed by a spiral spring. Although the torque is directly proportional to the instantaneous current, the high inertia of the movement (coil, support, and needle) prevents rapid acceleration and the observed deflection is proportional to the *average* current.

In a common type of ac ammeter (Fig. 3.12b), the magnetic field is set up by the current itself and the torque produced is proportional to the square of the current. Because of the inertia of the movement, the observed deflection is proportional to the

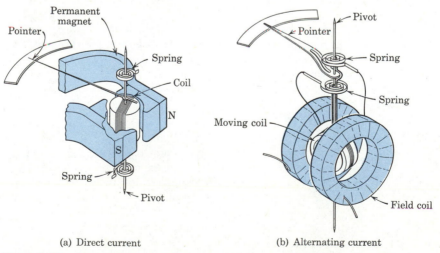

(a) Direct current   (b) Alternating current

Figure 3.12 Ammeter construction.

average squared current and the scale is calibrated to indicate *rms* values.

If a periodic voltage is applied to a rectifier in series with a capacitance, the capacitance will tend to charge up until the applied voltage reaches its maximum. When the applied voltage decreases, the rectifier prevents a current reversal and the capacitance voltage remains at the *peak* value, which can be read by an appropriate dc instrument. The design and use of such instruments are described in Chapter 22, and this brief mention just emphasizes the practical nature of the values defined in this section.

Root-Mean-Square Rating

The root-mean-square value has another significance in specifying the "rating" of an electric motor for a varying duty cycle. The interpretation of "rating" depends on the type of device being rated. A fish line rated at "10-lb test" can be expected to fail at slightly over 10 lb of static pull. If an automobile is rated at "200 hp," it means that a well-tuned sample operated under laboratory conditions at the optimum speed can develop a maximum of 200 hp for a limited time.[†] In contrast, a truck engine or an electric motor is usually rated at the power it can produce on a continuous basis over a long period of time without excessive deterioration.

Actually, most "50-hp" electric motors can develop two or three times this amount of power for short periods of time. If operated with an overload continuously, however, the excessive losses (proportional to i^2R) raise the operating temperature and the insulation deteriorates rapidly. If the rough approximation is made that current drawn is proportional to power output and heat generated is proportional to current-squared, the rms horsepower is the critical rating.

Note that Example 4 is based on the assumption that the thermal capacity of the motor permits an averaging of the heating rate over a short period (7 min). If the duty cycle were in hours instead of minutes, a 15-hp motor would be required.

[†]One horsepower = 746 W = 550 ft-lb/s. If a 3220-lb automobile actually had 200 hp available for acceleration, it would accelerate from 0 to 60 mph in 3.6 s.

EXAMPLE 4

An electric motor is to be specified for the duty cycle shown in Fig. 3.13. Full power of 15 hp is required for 2 min, the power decreases linearly for the next 3 min, the motor idles for 2 min, and then the cycle repeats.

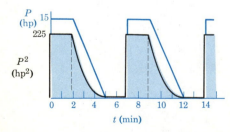

Figure 3.13   Rms horsepower calculation.

Assuming that the heat to be dissipated is proportional to the square of the power output, the problem is to find the rms horsepower. From Eq. 3-15,

$$\text{rms hp} = \sqrt{\frac{1}{T} \Sigma \, (\text{hp}^2 \times \text{time})} \qquad (3\text{-}17)$$

where the integral has been replaced by its equivalent, a summation. The form of Eq. 3-17 indicates a simple approach to this type of problem since $\Sigma \, (\text{hp}^2 \times \text{time})$ is the area under the curve of hp^2 versus time. As shown in Fig. 3.13, the rectangular area is 225 hp$^2 \times$ 2 min. Since the area under a parabola is $\frac{1}{3}$ that of the enclosing rectangle, the parabolic area is 225 hp$^2 \times$ 3 min/3. Then the rms hp is

$$\sqrt{\tfrac{1}{7}(225 \times 2 + \tfrac{1}{3} \times 225 \times 3)} = \sqrt{96.4} = 9.8 \text{ hp}$$

The next larger standard rating is 10 hp and this rating would be specified.

IDEAL AMPLIFIERS

Amplification is one of the principal functions of electronic circuits. When voltage or current signals are applied to the input terminals of an amplifier, larger voltage or current signals are available at the output terminals. In the circuit model of a real amplifier in Fig. 3.14a, input resistance R_i determines how much input current results from an input voltage v_i. The diamond symbol indicates that Av_i is a "controlled voltage source"; the voltage available in one part of the circuit is determined by a variable in another part of the circuit. Here the voltage appearing in the output circuit, consisting of output resistance R_o and load resistance R_L, is controlled by the input voltage v_i. With $R_L = \infty$, output current $i_o = 0$, and voltage Av_i appears at the output terminals. Factor A is the open-circuit voltage amplification or "gain" of the amplifier.

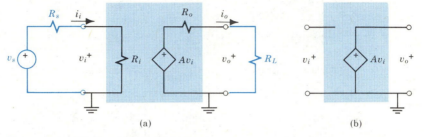

(a)   (b)

Figure 3.14   Real and ideal amplifiers.

In a practical case, the fraction of the source voltage v_s that appears at the input terminals is $R_i/(R_s + R_i)$. The fraction of the internally developed voltage appearing at the output terminals is $R_L/(R_o + R_L)$. The gain of the real amplifier is

$$A_r = \frac{v_o}{v_s} = \frac{Av_s[R_i/(R_s + R_i)][R_L/(R_L + R_o)]}{v_s} = A \cdot \frac{R_i}{R_s + R_i} \cdot \frac{R_L}{R_L + R_o} \quad (3\text{-}18)$$

For good performance, R_i should be large and R_o should be small. R_i should be large so that $v_i \simeq v_s$ and is unaffected by R_s; R_o should be small so that $v_o \simeq Av_i$ and is unaffected by R_L. In the *ideal amplifier* of Fig. 3.14b, $R_i = \infty$ and $R_o = 0$; therefore, $v_o = Av_i = Av_s$ or voltage gain = $v_o/v_s = v_o/v_i = A$. In this ideal amplifier A is assumed to be constant, independent of the magnitude or the frequency of the input signal.

EXAMPLE 5

A real amplifier is characterized by: $R_i = 1\ \text{M}\Omega$, $R_o = 10\ \Omega$, and $A = 100$. Determine the voltage gain in a circuit where $R_s = R_L = 1\ \text{k}\Omega$.

By Eq. 3-18, the gain of the real amplifier is

$$A_r = A\,\frac{R_i}{R_s + R_i} \cdot \frac{R_L}{R_L + R_o}$$

$$= 100\,\frac{10^6}{10^3 + 10^6} \cdot \frac{10^3}{10^3 + 10}$$

$$= 100\,\frac{1000 \times 10^3}{1001 \times 10^3} \cdot \frac{1000}{1010} \simeq 99$$

With little error, this amplifier can be assumed to be ideal.

Operational amplifiers (op amps), first used in analog computers to perform mathematical operations, are now available in high-performance, low-cost, integrated circuit form. In the *differential* amplifier of Fig. 3.15a, the output voltage is A times $v_p - v_n$, the difference in potential between positive and negative input terminals. In the conventional symbol of Fig. 3.15b, it is understood that all voltages are measured with respect to ground and the ground terminals at input and output are omitted for convenience.

Differential op amps in the form of Fig. 3.15 are available with very high R_i, very low R_o, and voltage gains of 10^5 or more for less than a dollar. They are widely used in amplification, instrumentation, and waveform generation. In many applications

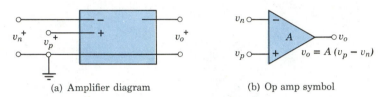

| (a) Amplifier diagram | (b) Op amp symbol |

Figure 3.15 Operational amplifier diagram and symbol.

inexpensive op amps can be assumed to be "ideal" circuit elements. The design of an amplifier, for example, is reduced to selecting the commercial unit that is nearly ideal in the parameters that are significant in a given situation.

Basic Inverting Circuit[†]

In the amplifier circuit of Fig. 3.16, the input signal is applied to the negative or inverting terminal and the positive or noninverting terminal is grounded. The input voltage v_1 is applied in series with resistance R_1, and the output voltage v_o is fed back through resistance R_F. Because the op amp gain A is very large, $v_i = -v_o/A \cong 0$.

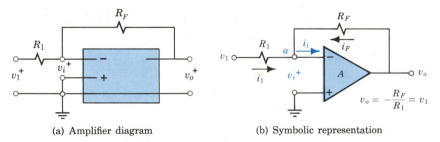

| (a) Amplifier diagram | (b) Symbolic representation |

Figure 3.16 Basic inverting amplifier circuit.

Because input resistance R_i is very large, $i_i = v_i/R_i \cong 0$. For an ideal amplifier, closely approximated by a practical op amp, $v_i = 0$ and $i_i = 0$; therefore, the sum of the currents into node a is

$$i_1 + i_F = \frac{v_1}{R_1} + \frac{v_o}{R_F} = 0 \qquad (3\text{-}19)$$

[†]Unfortunately, the term "operational amplifier" is applied to the basic high-gain amplifier as well as to the functional circuit in which it is employed. To lessen confusion, we shall start off by specifying "circuit" when the latter is intended. Later on, we shall use op amp for both, as is done in practice.

and

$$v_o = -\frac{R_F}{R_1} v_1 \quad \text{or} \quad \frac{v_o}{v_1} = A_F = -\frac{R_F}{R_1} \tag{3-20}$$

This relation describes an *inverting amplifier* circuit; the circuit gain with feedback is $A_F = -R_F/R_1$.

From this analysis, we can draw the following conclusions:

1. For an ideal amplifier, $v_i = 0$; input current $i_1 = v_1/R_1$ and is independent of R_F. Therefore, the input is isolated from the output. The input resistance of this feedback circuit is just $R_{iF} = v_1/i_1 = R_1$.
2. For an ideal amplifier with infinite gain or a practical op amp with very high gain, the gain of the amplifier circuit A_F is determined by the external resistors R_1 and R_F only.

The first conclusion leads to a simple circuit model for the inverting amplifier circuit; the output voltage $v_o = -R_F(v_1/R_1) = -i_1 R_F$ is represented by a controlled voltage source in Fig. 3.17. The second conclusion indicates an easy way to design an amplifier.

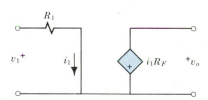

Figure 3.17 Circuit model for an inverting amplifier.

EXAMPLE 6

Design an amplifier with a voltage gain of -100, an input resistance of 1000 Ω, and a flat response over the audio-frequency range.

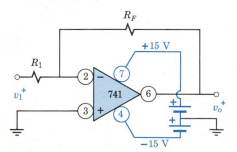

Figure 3.18 Amplifier design.

In this case, "design" means selecting an appropriate op amp, specifying the feedback network, and providing the necessary power supply.

The type 741 (see Appendix p. 731) is a high-performance op amp suitable for a variety of analog applications at audio frequencies. It has dual in-line pin connections as shown in Fig. 3.18 and operates from ±15-V power supplies or batteries.

The network design proceeds as follows:

$$R_1 = R_{iF} = 1000 \ \Omega$$

$$R_F = -R_1 \times A_F = R_1 \times 100 = 100 \ \text{k}\Omega$$

For noncritical applications, this is an adequate circuit.

The gain $A_F = -R_F/R_1$ is determined by external resistors and is independent of A as long as A is high.

Summing Circuit

If the circuit of Fig. 3.16 is modified to permit multiple inputs, the operational amplifier can perform addition. In Fig. 3.19, the input current is supplied by several voltages through separate resistances. In the practical case, $v_i = 0$ and $i_i = 0$ and the circuit model is as shown in Fig. 3.19b. The sum of the input currents is just equal and opposite to the feedback current or

$$i = i_1 + i_2 + \cdots + i_n = -i_F \tag{3-21}$$

Hence

$$\frac{v_1}{R_1} + \frac{v_2}{R_2} + \cdots + \frac{v_n}{R_n} = -\frac{v_o}{R_F}$$

or

$$v_o = -\left(\frac{R_F}{R_1} v_1 + \frac{R_F}{R_2} v_2 + \cdots + \frac{R_F}{R_n} v_n\right) \tag{3-22}$$

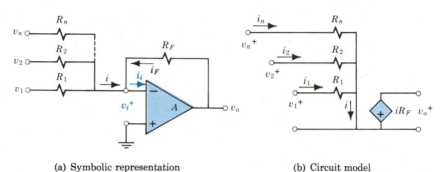

(a) Symbolic representation (b) Circuit model

Figure 3.19 Summing circuit; $v_o = -[v_1(R_F/R_1) + v_2(R_F/R_2) + \cdots + v_n(R_F/R_n)]$.

As would be expected from looking at the circuit model, the output of the *summing circuit* is the weighted sum of the inputs. Note that a signal can be subtracted by first passing it through an analog *inverter* consisting of an inverting amplifier with $R_F = R_1$.

Basic Noninverting Circuit

If the basic amplifier is reconnected as in Fig. 3.20, the behavior is modified. Here the signal is applied to the noninverting + terminal and a fraction of the output signal is fed back to the − terminal. Since $v_i = 0$ and $i_i = 0$, the governing equations are

$$v_1 = v_a + v_i = v_a \quad \text{and} \quad v_a = \frac{R_1}{R_1 + R_F} v_o \tag{3-23}$$

where R_1 and R_F constitute a voltage divider across the output voltage. The gain of the *noninverting amplifier* circuit is

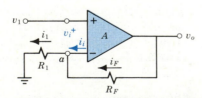

Figure 3.20 The noninverting amplifier circuit.

$$A_F = \frac{v_o}{v_1} = \frac{R_1 + R_F}{R_1} \tag{3-24}$$

Here again the gain is determined by the feedback network elements R_1 and R_F. For the noninverting circuit, however, the gain is positive and equal to or greater than unity.

EXAMPLE 7

An op amp is used in the circuit of Fig. 3.21, where input signal v_s is a function of time (a sinusoid or a combination of sinusoids). Derive an expression for $i_o(t)$.

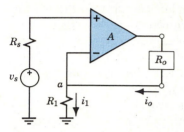

Figure 3.21 Voltage-current converter.

Because no current flows into an ideal op amp, there is no drop across R_s, and voltage $v_s(t)$ appears between the + terminal and ground.

Because there is no input voltage across an ideal op amp, $v_a = v_s = i_1 R_1$. But current i_1 must be supplied by the feedback current i_o. Hence

$$i_o(t) = i_1(t) = \frac{1}{R_1} v_s(t)$$

This noninverting circuit is serving as a voltage-to-current converter. The output current is proportional to the input voltage and independent of R_s and R_o.

IDEAL DIODES

A useful addition to our repertoire of electronic circuit elements is the *diode* or *rectifier*. A diode is a two-terminal device that acts as a switch; it permits current to flow readily in one direction but it tends to prevent the flow of current in the other direction. The direction of easy current flow is indicated by the arrowhead in the symbol of Fig. 3.22a. This is a nonlinear circuit element, and it is useful because it is *not* linear. However, the analysis of circuits containing diodes may be complicated because network theorems based on linearity cannot be used. Figure 3.22b shows the *i-v* characteristic of a semiconductor diode consisting of a junction of dissimilar materials that we will study later. Just as we use ideal models to represent real *R, L,* and *C* components, so we can approximate a real diode by the ideal characteristics shown in Fig. 3.23b. When the anode is positive with respect to the cathode, that is, when voltage *v* is positive, the "switch" is closed and unlimited current *i* can flow with no voltage drop. In contrast, when the cathode is positive with respect to the anode, the "switch" is open and no current flows even for large negative values of voltage *v*.

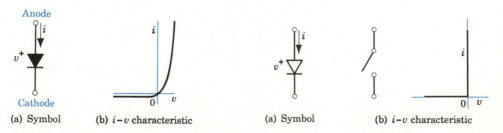

(a) Symbol   (b) *i–v* characteristic   (a) Symbol   (b) *i–v* characteristic

Figure 3.22 A real diode.   **Figure 3.23** An ideal diode.

Rectifiers

The nonlinear characteristic of a diode is used to convert alternating current into unidirectional, but pulsating, current in the process called *rectification*. The pulsations are removed in a frequency-selective circuit called a *filter*. Rectifier circuits employ one, two, or four diodes to provide various degrees of rectifying effectiveness. Filter circuits use the energy-storage capabilities of inductors and capacitors to smooth out the pulsations and to provide a steady output current. A combination of AC source, rectifier, and filter is called a *power supply*.

Half-Wave Rectifier. Ideally, a diode should conduct current freely in the forward direction and prevent current flow in the reverse direction. Practical diodes only approach the ideal. Semiconductor diodes, for example, present a small but appreciable voltage drop in the forward direction and permit a finite current to flow in the reverse direction. For most calculations, the reverse current flow is negligibly small and the forward voltage drop can be neglected with little error.

A practical circuit for *half-wave rectification* is shown in Fig. 3.24a. A *transformer* supplied from 120-V, 60-Hz house current provides the desired operating

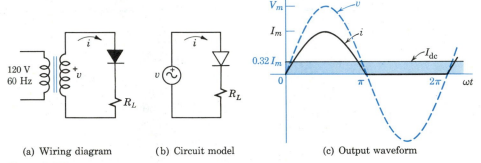

(a) Wiring diagram   (b) Circuit model   (c) Output waveform

Figure 3.24   A half-wave rectifier.

voltage, which is applied to a series combination of diode and load resistance R_L. (The transformer, consisting of two multiturn coils wound on a common iron core, provides a voltage "step-down" in direct proportion to the turn ratio.) For approximate analysis, the actual diode is represented by an ideal diode; the internal resistance of the transformer is neglected. For $V = V_m \sin \omega t$, the resulting current is

$$
\begin{cases}
i = \dfrac{v}{R_L} = \dfrac{V_m \sin \omega t}{R_L} & \text{for } 0 \leq \omega t \leq \pi \\[3mm]
i = 0 & \text{for } \pi \leq \omega t \leq 2\pi
\end{cases}
\tag{3-25}
$$

as shown in Fig. 3.24c.

The purpose of rectification is to obtain a unidirectional current. The dc component of the load current is the average value or

$$
I_{dc} = \frac{1}{2\pi} \int_0^{2\pi} i \, d(\omega t) = \frac{1}{2\pi} \int_0^{\pi} \frac{V_m \sin \omega t}{R_L} \, d(\omega t) + 0
$$

$$
= \frac{1}{2\pi} \frac{V_m}{R_L} \left[-\cos \omega t \right]_0^{\pi} = \frac{V_m}{\pi R_L} = \frac{I_m}{\pi}
\tag{3-26}
$$

The current through the load resistance consists of half-sinewaves, and the dc component is approximately 30% of the maximum value.

EXAMPLE 8

An ideal diode is connected in the circuit of Fig. 3.25. For $v = 170 \sin \omega t$ V, predict the current through $R = 5$ kΩ.

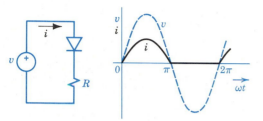

Figure 3.25 Application of diode.

For $0 < \omega t < \pi$, v is positive, the diode switch is closed, and

$$i(t) = \frac{v}{R} = \frac{170 \sin \omega t}{5000} = 34 \sin \omega t \text{ mA}$$

For $\pi < \omega t < 2\pi$, v is negative, the diode switch is open, and current $i = 0$. The resulting current is shown by the solid line. The sinusoidal voltage wave has been *rectified*.

The dc component of the rectified current is (by Eq. 3-26)

$$I_{dc} = \frac{I_m}{\pi} = \frac{34}{\pi} = 10.8 \text{ mA}$$

Full-Wave Rectifier. The bridge rectifier circuit of Fig. 3.26 provides a greater dc value from the same transformer voltage. When the transformer voltage $v = v_{ad}$ is positive, the current flow is along path *abcd* as shown and a half-sinewave of current results. When the applied voltage reverses, the voltage $v_{da} = -v_{ad}$ is positive and the

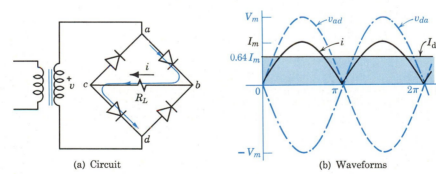

(a) Circuit   (b) Waveforms

Figure 3.26 A full-wave bridge rectifier.

current flow is along path *dbca*. The current through the load resistance is always in the same direction, and the dc component is twice as large as in the half-wave rectifier or

$$I_{dc} = \frac{2}{\pi} \frac{V_m}{R_L} = \frac{2 I_m}{\pi} \tag{3-27}$$

The bridge circuit is disadvantageous because four diodes are required, and two diodes and their power-dissipating voltage drops are always in series with the load. The full-wave rectifier circuit of Fig. 3.27 uses a more expensive transformer to

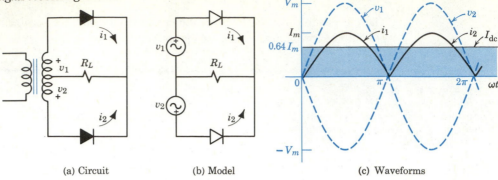

| (a) Circuit | (b) Model | (c) Waveforms |

Figure 3.27 A full-wave rectifier with phase inverter.

produce the same result with only two diodes and with higher operating efficiency. The second output winding on the transformer provides a voltage v_2 that is 180° out of phase with v_1; such a *center-tapped* winding serves as a *phase inverter*. While v_1 is positive, current i_1 is supplied through diode 1; while v_1 is negative, no current flows through diode 1 but v_2 is positive and, therefore, current i_2 is supplied through diode 2. The current through the load resistance is $i_1 + i_2$ and $I_{dc} = 2I_m/\pi$ as in the bridge circuit.

Filters

The desired result of rectification is direct current, but the output currents of the rectifier circuits described obviously contain large alternating components along with the dc component. Using full-wave instead of half-wave rectification reduces the ac component, but the remaining *ripple voltage* across R_L is still unsatisfactory for most electronic applications.

Capacitor Filter. The ripple voltage can be greatly reduced by a *filter* consisting of a capacitor shunted across the load resistor. The capacitor can be thought of as a "tank" that stores charge during the period when the diode is conducting and releases charge to the load during the nonconducting period.

If the diode is nearly ideal and if the steady state has been reached, the operation is as shown in Fig. 3.28. At time $t = 0$, the source voltage v is zero but the load

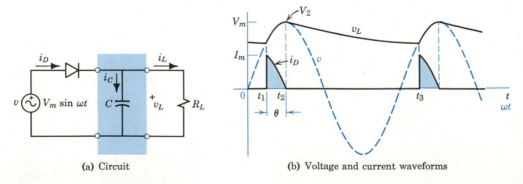

| (a) Circuit | (b) Voltage and current waveforms |

Figure 3.28 A capacitor filter.

voltage $v_L = v_C$ is appreciable because the previously charged capacitor is discharging through the load. At $t = t_1$, the increasing supply voltage v slightly exceeds v_L and the diode conducts. The diode current i_D rises abruptly to satisfy the relation $i_C = C\,dv/dt$ and then decreases to zero; the diode switches off when v drops below v_L. During the charging period, $t_1 < t < t_2$,

$$v_L = V_m \sin \omega t \tag{3-28}$$

During the discharging period, $t_2 < t < t_3$, the capacitor voltage decays exponentially so that

$$v_L = V_2\,e^{-(t-t_2)/R_L C} \tag{3-29}$$

with a time constant defined by the RC circuit. At time t_3, the supply voltage again exceeds the load voltage and the cycle repeats.

The load current i_L is directly proportional to load voltage v_L. Because i_L never goes to zero, the average value or dc component is relatively large as compared to the half-wave rectifier alone and the ac component is correspondingly lower. The ripple voltage is greatly reduced by the use of the capacitor.

Capacitor Filter—Approximate Analysis. If the time constant $R_L C$ is large compared to the period T of the supply voltage, the decay in voltage $v_C = v_L$ will be small, and the ripple voltage $V_r = \Delta v_C$ will be small. The magnitude of the ripple can be estimated by assuming that V_r is small, the charging interval $t_2 - t_1$ is small, and v_C is nearly constant. Under this assumption, all the load current is supplied by the capacitor, and the charge transferred to the load is

$$Q = I_{dc}T = C\,\Delta v_C = CV_r \tag{3-30}$$

Solving for the ripple voltage,

$$V_r = \frac{I_{dc}T}{C} = \frac{I_{dc}}{fC} = \frac{V_{dc}}{fR_L C} \tag{3-31}$$

This relation holds for a half-wave rectifier; a similar equation can be derived for a full-wave rectifier. By using this approximation, the performance of an existing filter can be predicted or a filter can be designed to meet specifications as in Example 9.

EXAMPLE 9

A load ($R_L = 3330\ \Omega$) is to be supplied with 50 V at 15 mA with a ripple voltage no more than 1% of the dc voltage. Design a rectifier–filter to meet these specifications.

Assuming a 120-V, 60-Hz supply and the half-wave rectifier with capacitor filter of Fig. 3.28a, Eq. 3-31 is applicable. Solving for C,

$$C = \frac{V_{dc}/V_r}{fR_L} = \frac{100}{60 \times 3330} = 500\ \mu F$$

For a peak value of 50 V, the rms value of transformer secondary voltage should be $50/\sqrt{2} = 35.4$ V. The transformer turn ratio should be

$$\frac{N_1}{N_2} = \frac{V_1}{V_2} = \frac{120}{35.4} = 3.39$$

WAVESHAPING CIRCUITS

A major virtue of electronic circuits is the ease, speed, and precision with which voltage and current waveforms can be controlled. Some of the basic waveshaping functions are illustrated by a radar pulse-train generator. The word *radar* stands for *ra*dio *d*etection *a*nd *r*anging. A very short burst of high-intensity radiation is transmitted in a given direction; a return echo indicates the presence, distance, direction, and speed of a reflecting object. The operation of a radar system requires a precisely formed series of timing pulses. Typically, these may be of 5 μs duration with a repetition rate of 500 pulses per second. Starting with a 500-Hz sinusoidal generator, the pulse train could be developed as shown in Fig. 3.29. Can you visualize some relatively simple electronic circuits that would perform the indicated functions?

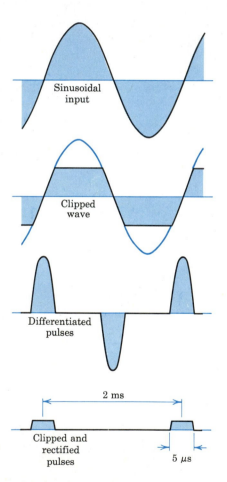

Figure 3.29   Generation of a train of timing pulses.

Clipping

A *clipping* circuit provides an output voltage v_2 equal to (or proportional to) the input voltage v_1 up to a certain value V; above V the wave is clipped off. If both positive and negative peaks are to be clipped, the desired *transfer characteristic* v_2

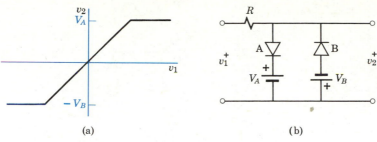

(a) (b)

Figure 3.30 Clipping characteristic and circuit.

versus v_1 is as shown in Fig. 3.30a. The switching action of diodes is used to provide clipping in Fig. 3.30b. The *bias* voltages are set so that diode A conducts whenever $v_1 > V_A$ and diode B conducts whenever $v_1 < -V_B$. When $-V_B < v_1 < V_A$, neither diode conducts and voltage v_1 appears across the output terminals. When either diode is conducting, the difference between v_1 and v_2 appears as a voltage drop across R. The effect of an asymmetric clipping circuit on a sinewave is illustrated in Example 10.

EXAMPLE 10

A sinewave $v_1 = 20 \sin \omega t$ V is applied to the circuit of Fig. 3.31a. Predict the output voltage v_2.

In this circuit the first diode and the 10-V battery provide clipping for voltages greater than $+10$ V. With $V_B = 0$, the second diode provides clipping of all negative voltages or rectification. The circuit characteristic and resulting output are as shown in Fig. 3.31b.

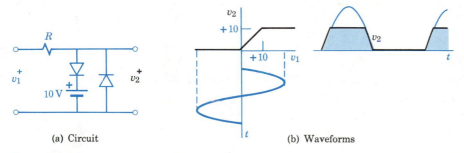

(a) Circuit (b) Waveforms

Figure 3.31 A clipping and rectifying circuit.

Clamping

To provide satisfactory pictures in television receivers, the peak values of certain variable signal voltages must be held or *clamped* at predetermined levels. In passing through ordinary amplifiers, the dc reference level is lost and a *clamper* or *dc restorer* is used.

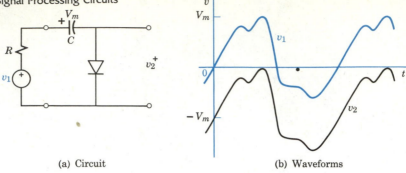

(a) Circuit   (b) Waveforms

Figure 3.32   A diode-clamping circuit.

In the circuit of Fig. 3.32a, if R is small the capacitor tends to charge up to the positive peak value of the input wave, just as in the half-wave rectifier with capacitor. When the polarity of v_1 reverses, the capacitor voltage remains at V_m because the diode prevents current flow in the opposite direction. Neglecting the small voltage across R, the output voltage across the diode is

$$v_2 = v_1 - V_m \tag{3-32}$$

The signal waveform is unaffected, but a dc value just equal to the peak value of the signal has been introduced. The positive peak is said to be clamped at zero.

If the amplitude of the input signal changes, the dc voltage across C also changes (after a few cycles) and the output voltage again just touches the axis. If the diode is reversed, the negative peaks are clamped at zero. If a battery is inserted in series with the diode, the reference level of the output may be maintained at voltage V_B.

Clamping and rectifying are related waveshaping functions performed by the same combination of diode and capacitor. In the rectifier the variable component is rejected and the dc value is transmitted; in the clamper the variable component is transmitted and the dc component is restored.

EXAMPLE 11

Design a circuit that will clamp the minimum point of any periodic signal to -5 V.

By Eq. 3-32, the output is to be

$$v_2 = v_1 + (V_{\min} - 5)$$

Therefore, the capacitor must charge up to the voltage $v_C = V_{\min} - 5$ with the polarity shown in Fig. 3.33.

When the input signal is negative with a magnitude greater than 5 V, current must flow through the diode, which must be connected as shown.

Figure 3.33   Designing a clamping circuit.

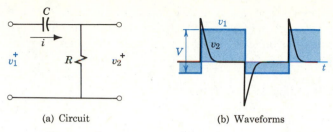

(a) Circuit   (b) Waveforms

Figure 3.34   Differentiating circuit.

Differentiating

Within limits the simple circuit of Fig. 3.34 provides an output that is the derivative of the input. For the special case of a rectangular input voltage wave, the output voltage is proportional to the capacitor charging current in response to a step input voltage. In this case a linear circuit transforms a rectangular wave into a series of short pulses if the time constant RC is small compared to the period of the input wave.

The general operation of this circuit is revealed if we make some simplifying assumptions. Applying Kirchhoff's voltage law to the left-hand loop,

$$v_1 = v_C + v_R \cong v_C \tag{3-33}$$

if v_R is small compared to v_C. Then

$$v_2 = v_R = Ri = RC\frac{dv_C}{dt} \cong RC\frac{dv_1}{dt} \tag{3-34}$$

The output is approximately proportional to the derivative of the input.

Integrating

From our previous experience with circuits we expect that if differentiating is possible, integrating is also. In Fig. 3.35, a square wave of voltage has been applied long enough for a cyclic operation to be established. The time constant RC is a little greater than the half-period of the square wave. The capacitor C charges and discharges on alternate half-cycles, and the output voltage is as shown.

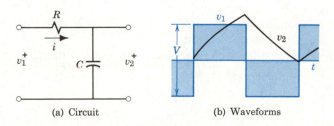

(a) Circuit   (b) Waveforms

Figure 3.35   Integrating circuit.

If the time constant RC is large compared to the period T of the square waves, only the straight portion of the exponential appears and the output is the *sawtooth* wave in which voltage is directly proportional to time. In general,

$$v_1 = v_R + v_C \cong v_R = iR \tag{3-35}$$

if v_C is small compared to v_R (i.e., $RC > T$). Then

$$v_2 = \frac{1}{C} \int i\, dt \cong \frac{1}{RC} \int v_1\, dt \tag{3-36}$$

and the output is approximately proportional to the integral of the input. If it is necessary, the magnitude of the signal can be restored by linear amplification.

Op Amp Integrator and Differentiator

In the previous discussion of op amps we assumed that the feedback network is purely resistive. In general, however, the network may contain capacitances, inductances, and resistances. Because of the high amplifier gain in combination with feedback, the mathematical operations performed are precise.

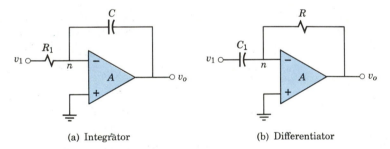

(a) Integrator (b) Differentiator

Figure 3.36 Op amp integration and differentiation circuits.

In Fig. 3.36a, the feedback element is a capacitor. For the op amp, $v_i \cong 0$, $i_i \cong 0$, and the sum of the currents into node n is

$$\frac{v_1}{R_1} + C\frac{dv_o}{dt} = 0 \tag{3-37}$$

Integrating each term with respect to time and solving,

$$v_o = -\frac{1}{R_1 C} \int v_1\, dt \tag{3-38}$$

and the device is an *integrator*. The analog integrator is very useful in computing, signal processing, and signal generating.

If the resistance and capacitance are interchanged as in Fig. 3.36b, the sum of the currents is

$$C_1 \frac{dv_1}{dt} + \frac{v_o}{R} = 0 \tag{3-39}$$

Solving,

$$v_o = -RC_1 \frac{dv_1}{dt} \tag{3-40}$$

and the output voltage is proportional to the derivative of the input. For practical reasons involving instability and susceptibility to noise, the differentiator is not so useful as the integrator.

EXAMPLE 12

Predict the output voltage of the circuit shown in Fig. 3.37 where the block represents an ideal amplifier with A very large.

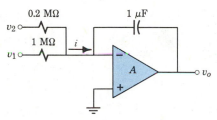

0.2 MΩ   1 μF

v_2

1 MΩ i

v_1

A

v_o

Figure 3.37   Integrator application.

For A very large, v_i is very small and representation as an integrator is accurate. Here $i = i_1 + i_2$ and

$$v_o = -\frac{1}{C} \int \left(\frac{v_1}{R_1} + \frac{v_2}{R_2} \right) dt$$

$$= -\frac{1}{CR_1} \int \left(v_1 + v_2 \frac{R_1}{R_2} \right) dt$$

$$= -\int (v_1 + 5v_2) \, dt$$

The output is the integral of a weighted sum.

SUMMARY

- Exponentials and sinusoids are important waveforms because: they occur frequently, they are easy to handle mathematically, and they are used to represent other waves.
  The general decaying exponential is $a = A \, e^{-t/T}$.
  The time constant T is a measure of the rate of decay.
  For $t = T$, $a/A = 1/e = 0.368$; for $t = 5T$, $a/A = 0.0067$ (negligible).
  The general sinusoid is $a = A \cos (\omega t + \alpha)$.
  Frequency $\omega = 2\pi f$ rad/s; $f = 1/T$ Hz, where period T is in seconds.

- In a periodic function of time, $f(t + T) = f(t)$.
  The average value of a periodic current is $I_{av} = (1/T) \int_0^T i \, dt$.
  For a sinusoid, the half-cycle average is $2I_m/\pi = 0.637I_m$.
  The effective or rms value of a periodic current is $I = \sqrt{(1/T) \int_0^T i^2 \, dt}$.
  For a sinusoid, the effective value is $I = I_m/\sqrt{2} = 0.707I_m$.

- An ideal amplifier is characterized by: infinite input resistance, zero output resistance, and constant gain; an ideal op amp has infinite gain.
  Feedback circuits are used with op amps to obtain: inverting and noninverting amplifiers, summing circuits, integrators, and differentiators.

- Essentially, a diode discriminates between forward and reverse voltages.
  The ideal diode presents zero resistance in the forward direction and infinite resistance in the reverse direction; it functions as a selective switch.

- A rectifier converts alternating current into unidirectional current.
  In halfwave rectification with a resistive load, $I_{dc} \cong V_m/\pi R_L = I_m/\pi$.
  A bridge circuit or phase inverter permits full-wave rectification; $I_{dc} = 2I_m/\pi$.

- A capacitor filter stores charge on voltage peaks and delivers charge during voltage valleys; the ripple voltage (half-wave) is $V_r \cong V_{dc}/fCR_L$.

- Waveforms can be shaped easily, rapidly, and precisely.
  A diode–resistor–battery circuit can perform clipping.
  A diode and peak-charging capacitor can clamp signals to desired levels.
  An op amp can perform differentiation and integration precisely.

REVIEW QUESTIONS

1. Write the general equation of an exponentially *increasing* function.
2. Why is the "time constant" useful?
3. Justify the statement: "In a linear circuit consisting of resistances, inductances, and capacitances, if any voltage or current is an exponential all voltages and currents will be exponentials with the same time constant."
4. Write a statement for sinusoids similar to the foregoing statement for exponentials and justify it.
5. Distinguish between average and effective values.
6. Why are ac ammeters calibrated to read rms values?
7. What is the reading of a dc ammeter carrying a current $i = 10 \cos 377t$ A?
8. What is "amplification"? "Gain"?
9. Why are *high* input resistance and *low* output resistance desirable in an amplifier?
10. What is a differential amplifier? What are its advantages?
11. What is an operational amplifier?

12. Why is the gain of an amplifier circuit independent of the gain of the op amp employed?
13. In circuit analysis of an ideal op amp with feedback, what values are assumed for v_i and i_i? Why?
14. How can an op amp be used to add two analog signals?
15. Why do we wish to replace actual devices with fictitious models?
16. Explain the operation of a full-wave bridge rectifier.
17. Explain the operation of a capacitor filter after a full-wave rectifier.
18. During capacitor discharge, why doesn't some of the current flow through the diode?
19. What is the effect on the ripple factor of the circuit in Fig. 3.28 of doubling R_L? Of halving C? Of doubling V_m?
20. Explain the operation of a diode clipper.
21. Sketch a clamping circuit and explain its operation.
22. Explain in words the operation of a differentiator and an integrator.

EXERCISES

1. Sketch the following signals and write mathematical expressions for $v(t)$:
   (a) A step of 2 V occurring at $t = 1$ s.
   (b) A sawtooth of amplitude 50 V that resets 20 times a second.
   (c) A train of pulses of amplitude 100 V, duration 10 μs, and frequency 40 kHz.
   (d) An exponential voltage with an initial amplitude of 5 V that decays to 1 V in 80 ms. Predict its value at $t = 20$ ms.

2. An exponential voltage has a value of 15.5 V at $t = 1$ s and a value of 2.1 V at $t = 5$ s.
   (a) Derive the equation for $v(t)$.
   (b) How many time constants elapse between $t = 5$ s and $t = 9$ s?
   (c) Predict the value at $t = 9$ s.

3. An exponential current has a value of 8.7 mA at $t = 2$ s and value of 3.2 mA at $t = 5$ s. Write the equation for $i(t)$ and predict the value at 11 s.

4. Throwing a switch in an experimental circuit develops a voltage $v = 6 e^{-500t}$ V across a parallel

combination of $R = 40\ \Omega$, and $C = 20\ \mu F$. Predict the current in each element and the total current.

5. Throwing a switch in an experimental circuit introduces a transient voltage $v = 120\ e^{-10t}$ V. If measurements in the circuit are accurate to 0.5%, approximately how long does the switching transient "last"?

6. A current $i = 2\ e^{-250t}$ mA flows through a series combination of $R = 3$ kΩ and $L = 4$ H. Determine the voltage across each element and the total voltage.

7. A 1-MΩ resistor is connected across a 500-μF capacitor initially charged to 100 V. Predict how long it will take for the voltage to drop to 14 V.

8. Sketch, approximately to scale, the following functions:
   (a) $i_1 = 2 \cos (377t + \pi/4)$ A
   (b) $v_1 = 10 \cos (300t - \pi/3)$ V
   (c) $i_2 = 1.5 \sin (400t - 60°)$ A

9. Express the following sinusoidal signals as specific functions of time:
   (a) A current with a frequency of 100 Hz and an amplitude of 10 mA and passing through zero with positive slope at $t = 1$ ms.
   (b) A voltage with a period of 20 μs and passing through a positive maximum of 5 V at $t = 5\ \mu s$.
   (c) A current reaching a positive maximum of 2 A at $t = 5$ ms and the next negative maximum at $t = 25$ ms.
   (d) A voltage having a positive maximum of 50 V at $t = 0$ and decreasing to a value of 25 V at $t = 2$ ms.

10. A sinusoidal signal has a value of -5 A at time $t = 0$ and reaches its first maximum (negative) of -10A at $t = 2$ ms. Write the equation for $i(t)$ and predict the value at $t = 5$ ms. Determine the frequency and period.

11. A current $i = 2 \cos (2000t - \pi/4)$ A flows through a series combination of $R = 20\ \Omega$ and $L = 10$ mH.
    (a) Determine the voltage across each element.
    (b) Sketch the two voltages, $v = f(\omega t)$, approximately to scale and estimate the amplitude and phase angle of $v_R + v_L$.

12. For each of the periodic voltage and current signals shown in Fig. 3.38:
    (a) Calculate average values of voltage, current, and power.
    (b) Calculate effective values of voltage and current.

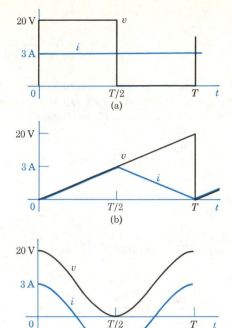

Figure 3.38

13. Calculate average and effective values of voltage and current for each of the periodic signals shown:
    (a) In Fig. 3.39a.
    (b) In Fig. 3.39b.

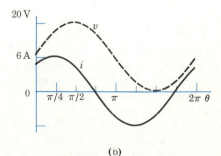

Figure 3.39

14. A signal consists of a series of positive rectangular pulses of magnitude V, duration T_d, and period T. Where $D = T_d/T$, derive expressions for average and effective values in terms of D and V.

15. The signal of Fig. 3.40a is applied to the circuit shown. The rectifier permits current flow only in the direction indicated by the triangle. Predict the readings on the two meters, if they are:
(a) DC ammeters
(b) AC ammeters of the types described in the text

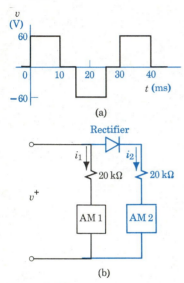

(a)

(b)

Figure 3.40

16. Repeat Exercise 15 if the voltage v is as shown in Fig. 3.41.

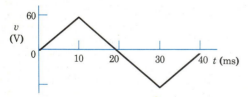

Figure 3.41

17. The duty cycle of an electric motor is: At $t = 0$ (motor idling and consuming negligible power), load increases abruptly to 60 hp and holds for 10 s, load decreases uniformly to 20 hp over 10 s and then is removed; motor idles for 40 s. Sketch the duty cycle (hp versus time) and specify the proper motor (available in 10-, 20-, 30-, 50-, and 100-hp ratings).

18. An op amp has a very high gain, a very high input resistance, and a very low output resistance. Design and draw an amplifier circuit (with $R_F = 3$ MΩ) providing:
(a) $A_F = -150$
(b) $A_F = -30$
(c) $A_F = +40$
(d) $A_F = +120$

19. In Fig. 3.19a, $R_1 = 1$ kΩ, $R_2 = 2$ kΩ, $R_n = 3$ kΩ, and $R_F = 90$ kΩ. If $v_1 = +1$ mV, $v_2 = -5$ mV, and $v_n = +2$ mV, predict v_o.

20. (a) Derive an expression for voltage v_M in terms of the other voltages in Fig. 3.42. Describe in words the function performed by this circuit.
(b) Design an op amp circuit to perform the same function.

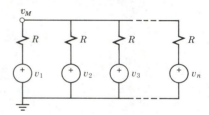

Figure 3.42

21. (a) Using an ideal amplifier, design a circuit (let $R_F = 1000$ Ω) to provide an output $v_o(t) = -v_1(t) - 10\,v_2(t)$.
(b) Redesign the circuit to provide an output $v_o(t) = +v_1(t) - 10\,v_2(t)$.

22. Given the op amp in Fig. 3.43:
(a) Define v_o in terms of the given inputs.
(b) Describe the operation performed assuming sinusoidal inputs.

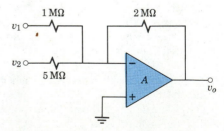

Figure 3.43

23. Given the amplifier circuit of Fig. 3.44:
(a) Define v_o in terms of v_1.
(b) Define the operation performed.

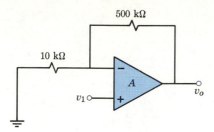

Figure 3.44

24. In Fig. 3.45, $v = 10 \sin \omega t$ V and $V_B = 6$ V. Under what circumstances will current i flow? Sketch v and i as functions of time on the same axes.

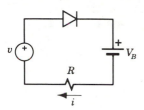

Figure 3.45

25. Explaining your reasoning and stating any simplifying assumptions, predict current I in Fig. 3.46.

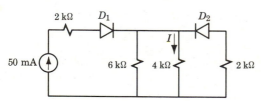

Figure 3.46

26. Four ideal diodes are connected in the fullwave rectifier of Fig. 3.26 where $R_L = 1000$ Ω. Specify the amplitude of v so that the *average* current through R_L is 20 mA.

27. An ideal diode is used in a half-wave rectifier with power supplied at 120 V (rms) and 60 Hz. For a load $R_L = 2000$ Ω, predict I_{dc}, V_{dc}, and the power delivered to the load.

28. Repeat Exercise 27 assuming a full-wave bridge rectifier circuit.

29. A diode is connected in series with a 30-V rms source to charge a 12-V battery with an internal

resistance of 0.1 Ω. Specify the series resistance necessary to limit the peak current to 2 A. Estimate the time required to recharge a 10-A·hr battery.

30. In Fig. 3.28, $C = 100$ μF and $R_L = 10$ kΩ. For $V_m = 20$ V at 60 Hz, predict:
(a) The dc load current through R_L.
(b) The percent ripple in v_L.
(c) The reading of a dc ammeter in series with R_L if C is disconnected.

31. A "load" requires 10 mA at 30 V dc with no more than 0.5 V ripple.
(a) Draw the circuit of a power supply consisting of a transformer with 120-V 60-Hz input, half-wave rectifier, capacitor filter, and effective R_L; specify C and the turn ratio of the transformer.
(b) Repeat for a full-wave bridge rectifier circuit. Draw a conclusion.

32. In Fig. 3.30b, $R = 1$ kΩ, $V_A = 2$ V, and $V_B = 5$ V. Sketch v_2 for $v_1 = 6 \sin \omega t$ V.

33. In Fig. 3.47, $v_1 = 10 \sin \omega t$ V.
(a) Sketch $v_1(t)$ and $v_2(t)$ on the same graph.
(b) Draw the transfer characteristic v_2 versus v_1.
(c) What function is performed by this circuit?

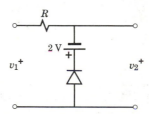

Figure 3.47

34. For $v_1 = 10 \sin \omega t$ V, $R = 1$kΩ, and $V = 4$ V in Fig. 3.48a, sketch $v_2(t)$.

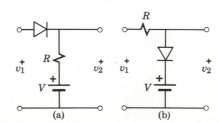

Figure 3.48

35. For $v_1 = 10 \sin \omega t$ V, $R = 1$ kΩ, and $V = 4$ V in Fig. 3.48b, sketch $v_2(t)$.

36. The periodic voltage v_1 of Fig. 3.49 is applied to the input of Fig. 3.47. Show the input signal v_i and the output signal v_o on the same graph.

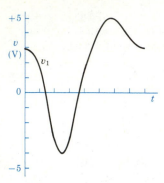

Figure 3.49

37. Repeat Exercise 36 for the circuit of Fig. 3.50.

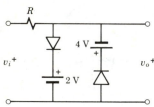

Figure 3.50

38. The periodic voltage v_1 of Fig. 3.49 is applied to the input of Fig. 3.51. Show the input signal v_i and the output signal v_o on the same graph.

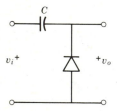

Figure 3.51

39. Repeat Exercise 38 for the circuit of Fig. 3.52 where $V_B = 2$ V.

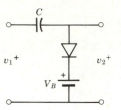

Figure 3.52

40. In Fig. 3.52, $v_1 = 10 \sin \omega t$ V and $V_B = 4$ V. Assuming v_1 has been applied a "long" time, plot v_1, v_C, and v_2 on the same time axis.

41. In Fig. 3.53, when voltage $v_1 = V_m(0.5 + \sin \omega t)$ V, the high-resistance dc voltmeter VM reads 70 V.
 (a) What functions are performed by each section of the circuits?
 (b) How is v_2 related to v_1? Show v_1, v_2, and v_3 on the same graph.
 (c) Determine V_m and define the function of this instrument.

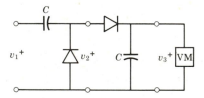

Figure 3.53

42. Devise a circuit using ideal amplifiers to provide an output $v_o = 10 \int v_1 \, dt - 5v_2$.

43. Devise a sweep circuit using an op amp to provide an output voltage proportional to time. Show a reset switch to ensure that $v_o = 0$ at $t = 0$.

PROBLEMS

1. Demonstrate analytically that the tangent to any exponential function at time t intersects the time axis at $t + T$ where T is the time constant.

2. The voltage across a 10-μF capacitor is observed on an oscilloscope. In 0.5 s after a 10-V source is removed, the voltage has decayed to 1.35 V.

Derive and label a circuit model for this capacitor.

3. A current $i = 2 \cos 2000t$ A flows through a series combination of $L = 30$ mH and $C = 5$ μF. (C is initially uncharged.)

   (a) Determine the total voltage across the combination as $f(t)$.

   (b) Repeat part (a) for L increased to 50 mH and interpret this result physically.

4. A voltage consists of a dc component of magnitude V_0 and a sinusoidal component of effective value V_1; show that the effective value of the combination is $(V_0^2 + V_1^2)^{1/2}$.

5. The circuit of Fig. 3.54 is a practical means for obtaining a high dc voltage without using a transformer. Stating any simplifying assumptions, predict voltages V_1 and V_L.

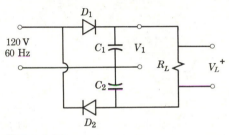

Figure 3.54

6. A diode and battery are used in the "voltage regulator" circuit of Fig. 3.55, where $R_S = 1000$ Ω and $R_L = 2000$ Ω. If V_1 increases from 16 to 24 V (a 50% increase), calculate the corresponding variation in load voltage V_L. Is the "regulator" doing its job?

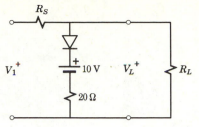

Figure 3.55

7. A voltage v_1 varies from -3 to $+6$ V. For a certain purpose, this voltage must be limited to a maximum value of $+4.0$ V. A diode, a 6-V battery, and assorted resistances are available. Design a suitable circuit.

8. The output of a flowmeter is $v = Kq$, where q is in cm^3/s and $K = 20$ mV·s/cm^3. The effective output resistance of the flowmeter is 2000 Ω. Design a circuit that will develop an output voltage $V_o = 10$ V (to trip a relay) after 200 cm^3 have passed the metering point.

9. In Fig. 3.56, input voltages A and B are restricted to either 0 or $+5$ V. Tabulate the four possible combinations of A and B and the corresponding values of output voltage V_o. Define in words the output in terms of the inputs. Why is this called an **OR** circuit? Where is it useful?

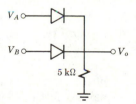

Figure 3.56

4

Natural Response

First-Order Systems

Second-Order Systems

Impedance Concepts

Poles and Zeros

In predicting the behavior of electrical circuits or mechanical systems, we must take into account two different sources of energy. The behavior determined by an *external* energy source we call a *forced* response; the behavior due to *internal* energy storage we call a *natural* response. For example, a stretched string may be forced to vibrate at any frequency by an externally applied alternating force; if the string is plucked and released, however, it will vibrate at its characteristic frequency, its *natural* frequency, because of the internal energy stored in the plucking. As another example, an automobile may go fast or slow, forward or backward, as energy from the engine is controlled by the operator; but if the moving automobile is taken out of gear, it will coast to rest in a predictable manner as the internal energy is dissipated.

In an electrical circuit, the external energy source or *forcing function* may be continuous or sinusoidal, or it may have any of the other waveforms shown in Fig. 3.1. Energy may be stored internally in the electric field of a capacitor or in the magnetic field of an inductor. In a chemical system, energy may be stored internally in the temperature of a liquid, in the potential energy of a compressed gas, or in the kinetic energy of a moving fluid. In the general problem, both external and internal energy sources are present and both forced and natural responses must be considered.

Forced responses can be maintained indefinitely by the continual input of energy. In contrast, because of unavoidable dissipation in electrical resistance or mechanical friction, natural responses tend to die out. After the natural behavior has become negligibly small, conditions are said to have reached the *steady state*. The period beginning with the initiation of a natural response and ending when the natural response becomes negligible is called the *transient* period.

Although the natural behavior of a system is not itself a response to any external function, a study of natural behavior will reveal characteristics of a system that will be useful in predicting the forced response as well. Our approach is first to investigate the natural behavior, then to study the forced behavior, and finally to consider the general case of forced and natural behavior. In this chapter, the emphasis is on the natural behavior of electrical circuits with some applications to mechanical and thermal systems. We start with relatively simple systems, derive the basic concepts, and then formulate a general procedure. The methods developed are powerful and they are applicable to problems in many branches of engineering.

EQUATIONS OF FIRST-ORDER SYSTEMS

In general, the behavior of an electrical circuit or any other physical system can be described by an integro-differential equation. In electrical circuits, the governing equation is obtained by applying the experimental laws for circuit elements and combinations of elements, taking into account external and internal energy sources.

The Governing Equation

In the circuit of Fig. 4.1, for example, a coil of inductance L and resistance R carrying a current I_O is suddenly shorted by throwing switch S at time $t = 0$. The equation governing current i as a function of time t after the switch is thrown can be

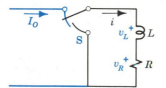

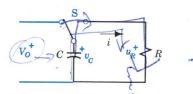

Figure 4.1 *LR* example.   **Figure 4.2** *RC* example.

obtained by applying Kirchhoff's voltage law to the shorted series circuit. For all time $t > 0$, the sum of the voltages around the closed path (clockwise) must be zero or

$$\Sigma v = 0 = -v_L - v_R = -L\frac{di}{dt} - Ri$$

or

$$L\frac{di}{dt} + Ri = 0 \tag{4-1}$$

As another example, in Fig. 4.2 capacitance C with initial voltage V_O (or initial charge $Q_O = CV_O$) is suddenly shorted across a resistance R at time $t = 0$. For all time $t > 0$, current $i(t)$ must satisfy Kirchhoff's voltage law or

$$\Sigma v = 0 = v_C - v_R = V_O - \frac{1}{C}\int_0^t i\, dt - Ri$$

Differentiating to eliminate the integral and the constant and rearranging the terms,

$$R\frac{di}{dt} + \frac{1}{C}i = 0 \tag{4-2}$$

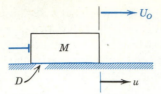

Figure 4.3 *MD* example.

In Fig. 4.3, the mass M is sliding on a surface for which the coefficient of friction is D. The driving force is removed at $t = 0$ when the velocity is U_O (or M is given an initial momentum MU_O). If the mass is allowed to coast, the only forces acting on M are those due to inertia and friction. Applying d'Alembert's principle (p. 38), for all time $t > 0$, velocity $u(t)$ is defined by

$$\Sigma f = 0 = -M\frac{du}{dt} - Du$$

or

$$M\frac{du}{dt} + Du = 0 \tag{4-3}$$

These examples result in analogous, first-order, linear, homogeneous, ordinary differential equations—*analogous* because they are of the same mathematical form, *first-order* because that is the order of the highest derivative present, *linear* because the variable and its derivatives appear only to the first power, *homogeneous* because there are no constant terms or forcing functions, and *ordinary* because there is a single independent variable t.

The homogeneous equation contains only terms relating to the circuit itself; the solution of this equation is independent of any forcing function and, therefore, it is the natural response of the circuit. This type of equation occurs so frequently that we must learn a convenient method for solving it.

Solving the Equation

The solution of a differential equation consists in finding a function that satisfies the equation. Consider the equation for the circuit of Fig. 4.2,

$$R\frac{di}{dt} + \frac{1}{C}i = 0 \tag{4-2}$$

This equation says that the combination of a function $i(t)$ and its derivative di/dt must equal zero; the function and its derivative must cancel somehow. This implies that the form of the function and that of its derivative must be the same and suggests an exponential function. If we let

$$i = A\,e^{st} \qquad \text{then} \qquad \frac{di}{dt} = sA\,e^{st} \tag{4-4}$$

where A is an amplitude and s is a frequency.[†] Substituting these expressions into the homogeneous equation,

$$RsA \ e^{st} + \frac{1}{C}A \ e^{st} = 0$$

Therefore, for all values of time t,

$$\left(Rs + \frac{1}{C}\right) A \ e^{st} = 0 \qquad (4\text{-}5)$$

In looking for solutions to this equation, $A = 0$ is one possibility, but this is the trivial case. The other possibility is to let $(Rs + 1/C) = 0$. This yields

$$s = -1/RC$$

and

$$i = A \ e^{(-1/RC)t} \qquad (4\text{-}6)$$

To evaluate the amplitude A, we return to the original equation

$$V_O - \frac{1}{C}\int_0^t i \ dt - Ri = 0 \qquad (4\text{-}7)$$

and consider a time $t = 0^+$, the first instant of time after $t = 0$. The integral of current over time represents the flow of charge and it takes a finite time for charge to be transferred. At $t = 0^+$, no time has elapsed and $\int_0^t i \ dt = 0$. Therefore the initial conditions are that

$$V_O - 0 - Ri(0^+) = 0 \qquad \text{or} \qquad i(0^+) = \frac{V_O}{R} \qquad (4\text{-}8)$$

By Eq. 4-6, at $t = 0^+$,

$$i(0^+) = I_O = A \ e^{s(0)} = A \ e^0 = A$$

Therefore

$$A = I_O = \frac{V_O}{R}$$

and, for all values of time after the switch is closed,

$$i = \frac{V_O}{R} e^{-t/RC} \qquad (4\text{-}9)$$

The physical explanation of this mathematical expression is based on the fact that resistor current is proportional to voltage, whereas capacitor current is proportional to rate of change of voltage. Upon closing the switch, the full voltage of the capacitor is applied across the resistor and initial current is high ($I_O = V_O/R$). But this current flow removes charge and reduces the voltage ($dv_C/dt = i/C$) across the capacitor and resistor and therefore the current is reduced. As the voltage decreases, the current is reduced, the voltage changes more slowly, and the current becomes ever smaller.

[†]Note that the units of s are "per second." The reason for calling s a "frequency" is explained on p. 102.

As another approach to evaluating i_O, we recall that the energy stored in a capacitance is $\frac{1}{2}Cv^2$, and this energy cannot be changed instantaneously since that would involve an infinite power. Therefore, the voltage on a capacitance cannot be changed instantaneously and we conclude that the voltage just after switching (at $t = 0^+$) must equal the voltage just before switching, V_O. When the switch is closed, this voltage appears across the resistance and

$$v_C = V_O = v_R = i(0^+)R$$

Hence,

$$i(0^+) = I_O = \frac{V_O}{R}$$

as before. This approach is used in Example 1.

EXAMPLE 1

Switch S (Fig. 4.4a) is arranged to disconnect the source and simultaneously, at time $t = 0$, short-circuit the coil of 2 H inductance and 10 Ω resistance. If the initial current in the coil I_O is 20 A, predict the current i after 0.2 s has elapsed.

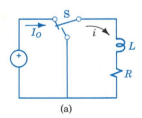

(a)

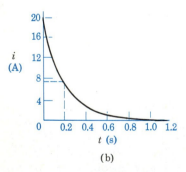

(b)

Figure 4.4 Exponential current decay in an *RL* circuit.

How long will it take for the natural behavior current to become zero?

1. By Kirchhoff's voltage law,

$$\Sigma v = 0 = -v_L - v_R = -L\frac{di}{dt} - Ri$$

2. The homogeneous equation is

$$L\frac{di}{dt} + Ri = 0$$

3. Assuming an exponential solution, we write

$$i = A\,e^{st}$$

where s and A are to be determined.

4. Substituting into the homogeneous equation,

$$LsA\,e^{st} + RA\,e^{st} = (sL + R)A\,e^{st} = 0$$

If $sL + R = 0$, $s = -\dfrac{R}{L}$ and $i = A\,e^{-(R/L)t}$

5. The energy stored in an inductance, $\frac{1}{2}Li^2$, cannot change instantaneously. Therefore, the current in the coil just after the switch is thrown must equal the current just before. At $t = 0^+$,

$$i = I_O = A\,e^0 = A \quad\text{or}\quad A = I_O = 20$$

Hence the solution is

$$i = I_O\,e^{-(R/L)t} = 20\,e^{-(10/2)t} = 20\,e^{-5t}\ \text{A} \qquad (4\text{-}10)$$

As shown in Fig. 4.4b, after 0.2 s the current is

$$i = 20\,e^{-5 \times 0.2} = 20 \times 0.368 = 7.36\ \text{A}$$

The current decreases continually, but never becomes zero.

General Procedure

Analysis of the steps followed in Example 1 suggests a general procedure for determining the natural behavior of an electrical circuit:

1. Write the governing equation using Kirchhoff's laws.
2. Reduce this to a homogeneous differential equation.
3. Assume an exponential solution with undetermined constants.
4. Determine the exponents from the homogeneous equation.
5. Evaluate the coefficients from the given conditions.

In a dynamic mechanical system, the first step usually involves the application of d'Alembert's principle and the equation for forces in equilibrium. Chemical engineering systems involving mass and energy transport may be analyzed in a similar way. Any such lumped linear system can be described by an ordinary differential equation that can be reduced to the homogeneous equation by eliminating constant terms and forcing functions. The solution of such an equation is always expressible in terms of exponential functions. The evaluation of the coefficients, which may be the most difficult step, is usually based on consideration of the energies stored in the system.

If current is the rate of transfer of charge and if the current in Example 1 flows "forever," will an infinite charge be transferred? To answer this question, express the total charge as the integral of current over an infinite time, or

$$q = \int_0^t i \, dt = \int_0^\infty 20 \, e^{-5t} \, dt$$

$$= -\frac{20}{5} e^{-5t} \Big]_0^\infty$$

$$= -\frac{20}{5} (0 - 1) = 4 \text{ C}$$

Although the current flows "forever," the rate of flow is decreasing in such a way that the total charge transferred approaches a finite limit (Fig. 4.5). The same is true of the energy dissipated in the resistance R (see Problem 4).

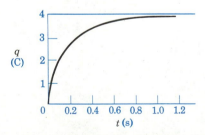

Figure 4.5 Total charge transfer.

Time Constant

The term "first-order system" refers to a system involving only a single energy-storage element for which the governing equation is a first-order differential equation.

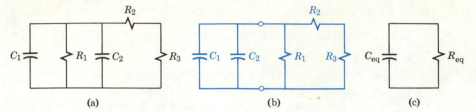

Figure 4.6 An *RC* circuit with a single time constant.

The natural behavior obtained by solving the homogeneous equation will always be similar in form to the decay of current or voltage in an *RL* or *RC* circuit where

$$i = I_O e^{-t/T} \qquad \text{or} \qquad v = V_O e^{-t/T} \qquad (4\text{-}11)$$

For the *RL* circuit of Example 1, the *time constant* is $T = L/R = 2/10 = 0.2$ s. For the *RC* circuit of Fig. 4.2, the current and the voltage reach $e^{-1} = 0.368$ times the initial values at $T = RC$.

The circuit of Fig. 4.6 is a special case in that the two capacitances can be combined and replaced by an equivalent capacitance $C_{\text{eq}} = C_1 + C_2$. The three resistances can be combined (R_1 in parallel with $R_2 + R_3$) and replaced by an equivalent resistance $R_{\text{eq}} = R_1(R_2 + R_3)/(R_1 + R_2 + R_3)$. The time constant for every voltage or current in this circuit is $T = R_{\text{eq}}C_{\text{eq}}$. (See Exercise 9a.)

NATURAL RESPONSE OF SECOND-ORDER SYSTEMS

If there is more than one energy-storage element, the behavior of the system is more complicated but the analysis follows the same procedure as for the simpler systems. In an *RLC* circuit, energy may be stored in the inductance or in the capacitance. In the series circuit of Fig. 4.7, assume that an initial voltage V_O exists on the capacitance C. With the switch S open there is no current in the inductance L and therefore no energy storage in its magnetic field. The resistance R is incapable of energy storage.

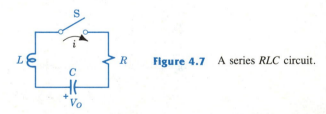

Figure 4.7 A series *RLC* circuit.

Following the general procedure for determining natural behavior, we first write the governing equation using Kirchhoff's voltage law. Traversing the loop (after the switch is closed) in a clockwise direction, we have

$$\Sigma v = 0 = -L\frac{di}{dt} - Ri + V_O - \frac{1}{C}\int_0^t i \, dt$$

By differentiating and rearranging terms, the homogeneous equation is

$$L\frac{d^2i}{dt^2} + R\frac{di}{dt} + \frac{1}{C}i = 0 \tag{4-12}$$

The presence of the second energy-storage element has increased the order of the highest derivative appearing in the homogeneous equation. This equation is typical of a *second-order* system.

The Characteristic Equation

Assuming an exponential solution, let $i = A\ e^{st}$. Substituting in the homogeneous equation yields

$$s^2 LA\ e^{st} + sRA\ e^{st} + \frac{1}{C}A\ e^{st} = 0$$

which is satisfied when

$$Ls^2 + Rs + \frac{1}{C} = 0 \tag{4-13}$$

This equation is noteworthy. For this series *RLC* circuit, it is the sum of three terms, each relating to a circuit element. It contains no voltages or currents, only terms characteristic of the circuit. Also, from this equation we expect to obtain information about the character of the natural behavior; it is called the *characteristic equation,* and it is very useful in circuit analysis.

The roots of the characteristic equation (Eq. 4-13) are

$$s_1 = -\frac{R}{2L} + \sqrt{\left(\frac{R}{2L}\right)^2 - \frac{1}{LC}}$$

and

$$s_2 = -\frac{R}{2L} - \sqrt{\left(\frac{R}{2L}\right)^2 - \frac{1}{LC}} \tag{4-14}$$

If either $i_1 = A_1\ e^{s_1 t}$ or $i_2 = A_2\ e^{s_2 t}$ satisfies a linear homogeneous equation, i.e., makes it zero, then the sum of the two terms also satisfies the equation. The most general solution then is

$$i = A_1\ e^{s_1 t} + A_2\ e^{s_2 t} \tag{4-15}$$

where A_1 and A_2 are determined by the initial conditions and s_1 and s_2 are determined by the circuit constants.

The values of R, L, and C are all real and positive, but the values of s_1 and s_2 may be real, complex, or purely imaginary.[†] The clue to the character of the natural response of a second-order system is found in the quantity under the radical sign—the *discriminant*. If the discriminant is positive, the roots of the characteristic equation will be real, negative, and distinct. If the discriminant is zero, the two roots will be

---

[†]Imaginary and complex numbers are explained in the Appendix. pp. 722 and 723.

real, negative, and identical. If the discriminant is negative, the roots will be complex numbers or, in the special case where $R = 0$, purely imaginary.

Roots Real and Distinct. If the discriminant in Eq. 4-14 is positive, s_1 and s_2 are real, negative, and distinct, and the natural behavior is the sum of two decaying exponential responses. Example 2 illustrates this case.

EXAMPLE 2

Given the circuit values shown in Fig. 4.8a, determine and plot the current response as a function of time after the switch is closed.

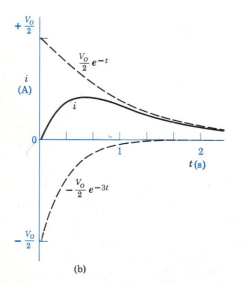

$L = 1\,\text{H}, \; C = \tfrac{1}{3}\,\text{F}, \; R = 4\,\Omega$
(a)

(b)

Figure 4.8 Natural response of an overdamped second-order system.

Roots of the characteristic equation (Eqs. 4-14) are

$$s_1, s_2 = -\frac{4}{2 \times 1} \pm \sqrt{\left(\frac{4}{2 \times 1}\right)^2 - \frac{1}{1 \times \frac{1}{3}}} = -2 \pm 1$$

The general solution for current (Eq. 4-15) is

$$i = A_1\, e^{-t} + A_2\, e^{-3t} \qquad (4\text{-}16)$$

To evaluate the constants, consider the initial conditions. At the instant the switch was closed, the voltage across the capacitance $v_C = -V_O$ and the current in the inductance $i_L = 0$. Because energy cannot be changed instantaneously, these values must hold just after closing the switch, at $t = 0^+$, say.

At $t = 0^+$, the current in each element in this series circuit must be zero or (Eq. 4-16)

$$i = i_O = 0 = A_1\, e^0 + A_2\, e^0 = A_1 + A_2$$

Therefore, $A_2 = -A_1$.

Also, the voltage iR across the resistance is zero. Therefore, $v_L = -v_C = V_O$ and

$$L\frac{di}{dt} = V_O \qquad \text{or} \qquad \frac{di}{dt} = \frac{V_O}{L}$$

From Eq. 4-16, at $t = 0^+$,

$$\frac{di}{dt} = \frac{V_O}{L} = -A_1\, e^0 - 3A_2\, e^0 = -A_1 - 3A_2$$

$$= -A_1 + 3A_1 = +2A_1$$

Solving,

$$A_1 = +\frac{V_O}{2L} = +\frac{V_O}{2} \qquad \text{and} \qquad A_2 = -A_1 = -\frac{V_O}{2}$$

The specific equation for the current response, plotted in Fig. 4.8b, becomes

$$i = \frac{V_O}{2}\, e^{-t} - \frac{V_O}{2}\, e^{-3t}$$

EXAMPLE 3

In the circuit of Example 2, let $L = 1$ H, $C = \frac{1}{17}$ F, and $R = 2\ \Omega$. Derive an expression for the natural response in Fig. 4.9.

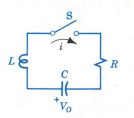

Figure 4.9   An *RLC* circuit with complex roots.

From Eq. 4-14, now

$$s = -\frac{R}{2L} \pm \sqrt{\left(\frac{R}{2L}\right)^2 - \frac{1}{LC}}$$

$$= -\frac{2}{2 \times 1} \pm \sqrt{\left(\frac{2}{2 \times 1}\right)^2 - \frac{1}{1 \times \frac{1}{17}}} = -1 \pm \sqrt{-16}$$

Therefore, where j is equivalent to $\sqrt{-1}$,[†]

$$s_1 = -1 + j4 \qquad \text{and} \qquad s_2 = -1 - j4$$

and the natural response is

$$i = A_1 e^{(-1+j4)t} + A_2 e^{(-1-j4)t} \qquad (4\text{-}17)$$

Roots Complex. If the discriminant in Eq. 4-14 is negative, s_1 and s_2 are complex conjugates and the natural behavior is the exponentially damped sinusoid defined by Eq. 4-17 in Example 3. However, the physical interpretation of Eq. 4-17 is not immediately clear. To gain insight into this type of response, let us consider the case of complex roots in a general way. The mathematics will be simplified by arbitrarily defining three new terms; for the circuit described by Eqs. 4-14,

$$\alpha = \frac{R}{2L} \qquad \omega_n^2 = \frac{1}{LC} \qquad \omega^2 = \frac{1}{LC} - \frac{R^2}{4L^2} = \omega_n^2 - \alpha^2 \qquad (4\text{-}18)$$

Now,[‡]

$$s_1 = -\alpha + j\omega \qquad \text{and} \qquad s_2 = -\alpha - j\omega$$

and the general form of Eq. 4-17 is

$$i = A_1 e^{(-\alpha+j\omega)t} + A_2 e^{(-\alpha-j\omega)t} \qquad (4\text{-}19)$$

By Euler's equation (p. 723), $e^{j\theta} = \cos\theta + j\sin\theta$; therefore,

$$e^{j\omega t} = \cos\omega t + j\sin\omega t \qquad \text{and} \qquad e^{-j\omega t} = \cos\omega t - j\sin\omega t$$

Factoring out the $e^{-\alpha t}$ and substituting for $e^{j\omega t}$ yields

$$i = e^{-\alpha t}[(A_1 + A_2)\cos\omega t + j(A_1 - A_2)\sin\omega t]$$

where A_1 and A_2 are constants that may be complex numbers. In fact, since these are physical currents, the coefficients of the $\cos\omega t$ and $\sin\omega t$ terms, $(A_1 + A_2)$ and $j(A_1 - A_2)$, must be real numbers that could be called B_1 and B_2 so that

$$i = e^{-\alpha t}[B_1\cos\omega t + B_2\sin\omega t] \qquad (4\text{-}20)$$

[†]Engineers use j instead of i for imaginary numbers to avoid confusion with the symbol for current. Charles P. Steinmetz, who first used combinations of real and imaginary numbers in circuit analysis, called them "general numbers." See *IEEE Spectrum*, April 1965.

[‡]Since for this case ω^2 is a negative number, the imaginary terms of the roots will be equal and opposite or the roots are always *complex conjugates*; in other words, complex roots always appear in conjugate pairs.

But the sum of a cosine function and a sine function must be another sine function (properly displaced in phase), so Eq. 4-20 can be rewritten[†] to give a natural response of

$$i = A\, e^{-\alpha t} \sin\,(\omega t + \theta) \tag{4-21}$$

This is evidently a sinusoidal function of time with an exponentially decaying amplitude, a so-called "damped sinusoid." Factor ω is seen to be the *natural frequency* of the oscillation in radians per second. Factor α is the *damping coefficient* (per second), and $A\, e^{-\alpha t}$ defines the envelope of the response. If α is large, the response dies out rapidly. Amplitude A and phase angle θ are constants to be determined from initial conditions. Useful relations include: i_L and v_C cannot suddenly change, $di_L/dt = v/L$, and $dv_C/dt = i/C$.

[†]To obtain Eq. 4-21 from Eq. 4-20, let $B_1 = A \sin \theta$ and $B_2 = A \cos \theta$.

EXAMPLE 4

(a) Rewrite Eq. 4-17 in Example 3 as a damped sinusoid and determine the natural response after closing the switch in Fig. 4.9.

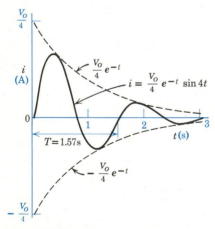

Figure 4.10   Response of an oscillatory second-order system.

(b) Determine the period of the oscillation and the damping coefficient.

For

$$\alpha = \frac{R}{2L} = 1 \quad \text{and} \quad \omega = \sqrt{\frac{1}{LC} - \frac{R^2}{4L^2}} = 4$$

the natural response (Eq. 4-21) is

$$i = A\, e^{-t} \sin\,(4t + \theta)$$

and

$$\frac{di}{dt} = A\, e^{-t}\, 4 \cos\,(4t + \theta) - A\, e^{-t} \sin\,(4t + \theta)$$

To evaluate the constants, we recall that just after the switch is closed the current must be zero (since $i_L = 0$) and the rate of change of current must be $V_O/L = V_O$ (since $-v_C = V_O = v_L = L\, di/dt$). At $t = 0^+$,

$$i = 0 = A\, e^0 \sin\,(0 + \theta) = A \sin \theta$$

A is finite; therefore, $\theta = 0$ and

$$\frac{di}{dt} = V_O = A\, e^0 4 \cos\,(0) - A\, e^0 \sin\,(0) = 4A$$

Therefore, $A = V_O/4$ and, as shown in Fig. 4.10,

$$i = \frac{V_O}{4}\, e^{-t} \sin 4t \tag{4-22}$$

The period is

$$T = \frac{1}{f} = \frac{2\pi}{\omega} = \frac{2\pi}{4} = 1.57 \text{ s}$$

The damping coefficient is $\alpha = 1 \text{ s}^{-1}$; in one second the envelope is down to $1/e$ times its initial value.

Roots Real and Equal. A system described by an equation with complex roots is said to be "oscillatory" or "underdamped"; in contrast, when the roots are real the system is said to be "overdamped." The limiting condition for oscillation, called the "critically damped" case, occurs when the roots are real and equal. For this case, the discriminant is equal to zero. Physically this is not an important case since it merely represents the borderline between the two regimes.

Mathematically it is an interesting case, requiring a special form of solution. If the roots are equal, $s_2 = s_1 = s$. Following the general procedure,

$$i = A_1\, e^{st} + A_2\, e^{st} = (A_1 + A_2)\, e^{st} = A\, e^{st}$$

but this is not sufficiently general for a second-order differential equation. For this case, a second term must be included[†] and the general solution is

$$i = A_1\, e^{st} + A_2 t\, e^{st} \tag{4-23}$$

For the *RLC* circuit of Example 4 with $L = 1$ H and $C = 0.25$ F, the discriminant is zero for

$$R^2 = \frac{4L^2}{LC} = \frac{4}{0.25} = 16 \qquad \text{or} \qquad R = R_{\text{critical}} = \sqrt{16} = 4\ \Omega$$

Plotting this response is left for the student. (See Exercise 18.)

Roots in the Complex Plane. We have observed that the character of the natural response of a second-order system is determined by the roots of the characteristic equation. Where the roots are real, negative, and distinct, the response is the sum of two decaying exponentials and the response is said to be overdamped. Where the roots are complex conjugates, the natural response is an exponentially decaying sinusoid and the response is said to be underdamped or oscillatory.

For a general view of this behavior that will give added insight into the response of higher-order systems, consider the effect on the location of the roots of a change in one system parameter. In general, roots are located in the complex plane, the location being defined by coordinates measured along the real or σ (sigma)[‡] axis and the imaginary or $j\omega$ axis. This is referred to as the *s* plane or, since *s* has the units of frequency, as the *complex frequency plane*. Specifically, let us consider the locus of the roots for a series *RLC* circuit as *R* varies from zero to infinity.

For this circuit the characteristic equation (Eq. 4-13) is

$$Ls^2 + Rs + \frac{1}{C} = 0$$

and the roots are

$$s = -\frac{R}{2L} \pm j\sqrt{\frac{1}{LC} - \frac{R^2}{4L^2}} = -\alpha \pm j\omega$$

[†]Boyce, W. E., and R. C. DiPrima, *Elementary Differential Equations*, 3rd ed., John Wiley & Sons, New York, 1977, or any other differential equations text.

[‡]The damping coefficient α was defined as $R/2L$, a positive number so that the decaying nature of $e^{-\alpha t}$ would be evident; positive values of α correspond to negative values of σ and lie in the left half of the complex plane.

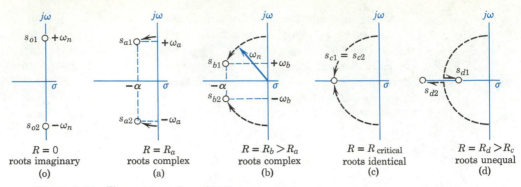

Figure 4.11 The root locus for an RLC circuit with R as the parameter.

For $R = 0$, $s = \pm j\sqrt{1/LC} = \pm j\omega_n$, defining values of the *undamped natural frequency* ω_n. In other words, for $R = 0$, $\alpha = 0$ and the response is oscillatory with no damping. The roots corresponding to the undamped case are located on the imaginary axis of Fig. 4.11o. For $R = R_a$, small but finite, a real part of s appears and the values of s are complex conjugates. As R increases to R_b, the roots move along a curved path to points s_b. The radial distance from the origin to either root is given by

$$\sqrt{\alpha^2 + \omega^2} = \sqrt{\left(-\frac{R}{2L}\right)^2 + \left(\frac{1}{LC} - \frac{R^2}{4L^2}\right)} = \sqrt{\frac{1}{LC}} = \omega_n \qquad (4\text{-}24)$$

a constant. Therefore, the locus is a circular arc of radius ω_n.

The critical value of resistance is defined by $\omega = 0$ or, in other words, where the discriminant equals zero. For this condition $\sigma = -R/2L = -\sqrt{1/LC}$. At this value of $R = R_c$, the roots coincide on the real axis. As R increases to R_d, s_{d_1} moves toward the origin and s_{d_2} moves out along the negative real axis. As R increases without limit, s_1 approaches the origin and s_2 increases without limit.

With a different circuit parameter the locus takes on a different shape (see Exercise 23). In any such plot, called a *root locus*, the location of the roots determines the character of the natural response, and the effect of changes in a parameter is clearly visible to the initiated. The first step in developing this powerful tool is mastery of a new concept—*impedance*.

IMPEDANCE CONCEPTS

If we limit the discussion to exponential waveforms, some interesting and valuable relations between voltage and current can be established. Actually this is not a severe restriction; already we have used exponentials to represent sinusoids, and exponentials with $s = 0$ can be used to represent direct currents. The range of exponential functions for positive and negative, large and small, real values of s is indicated in Fig. 4.12.

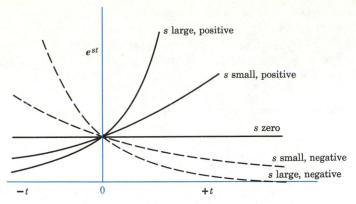

Figure 4.12 The range of exponential functions for real values of s.

The key property of an exponential function is that its time derivative is also an exponential. For example, if

$$i = I_O \, e^{st}, \qquad \frac{di}{dt} = sI_O \, e^{st} = si \qquad (4\text{-}25)$$

or if

$$v = V_O \, e^{st}, \qquad \frac{dv}{dt} = sV_O \, e^{st} = sv \qquad (4\text{-}26)$$

This property greatly facilitates calculating the response of circuits containing resistance, inductance, and capacitance because of the simple voltage-current relations that result.

Impedance to Exponentials

The ratio of voltage to current for exponential waveforms is defined as the *impedance* Z. For a resistance, $v = Ri$ and

$$Z_R = \frac{v}{i} = \frac{Ri}{i} = R \text{ in ohms} \qquad (4\text{-}27)$$

For an inductance, $v = L(di/dt) = sLi$ and

$$Z_L = \frac{v}{i} = \frac{sLi}{i} = sL \text{ in ohms} \qquad (4\text{-}28)$$

For a capacitance, $i = C(dv/dt) = sCv$ and

$$Z_C = \frac{v}{i} = \frac{v}{sCv} = \frac{1}{sC} \text{ in ohms} \qquad (4\text{-}29)$$

The impedances Z_R, Z_L, and Z_C are constants of proportionality between voltages and currents that are exponential functions of time t and frequency s. It can be demonstrated that in each case the dimensions of impedance are the same as those of resistance. (Can you do it?) The general relation

$$v = Zi \qquad (4\text{-}30)$$

corresponds to Ohm's law for purely resistive circuits, but it must be emphasized that impedance is defined *only for exponentials* and for waveforms that can be represented by exponentials.[†] With this restriction, impedances can be combined in series and parallel just as resistances are, and network theorems can be extended to include circuits containing L and C.

EXAMPLE 5

In the circuit of Fig. 4.13, $R_1 = 2\ \Omega$, $C = 0.25$ F, and $R = 4\ \Omega$. Using the impedance concept, find the currents i and i_C for a voltage $v = 6\ e^{-2t}$ V.

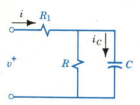

Figure 4.13   The impedance concept in circuit analysis.

Following the rules for resistive networks,

$$Z = Z_1 + \frac{Z_R Z_C}{Z_R + Z_C} = R_1 + \frac{R(1/sC)}{R + (1/sC)}$$

$$= 2 + \frac{4/(-2 \times 0.25)}{4 + 1/(-2 \times 0.25)} = 2 + \frac{-8}{2} = -2\ \Omega$$

Then

$$i = \frac{v}{Z} = \frac{6\ e^{-2t}}{-2} = -3\ e^{-2t}$$

Using the current-divider and impedance concepts,

$$i_C = \frac{Z_R \cdot i}{Z_R + Z_C} = \frac{4 \cdot i}{4 - 2} = -6\ e^{-2t}$$

The Impedance Function

For the circuit of Fig. 4.14, the governing equation is

$$L\frac{di}{dt} + Ri + \frac{1}{C}\int i\ dt = v \qquad (4\text{-}31)$$

For an exponential current $i = I_o\ e^{st}$, this becomes

$$Lsi + Ri + \frac{1}{sC}i = v$$

and

$$Z = \frac{v}{i} = sL + R + \frac{1}{sC} \qquad (4\text{-}32)$$

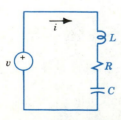

Figure 4.14   The impedance of an *RLC* circuit.

[†]Assume another waveform such as $i = I_o t$; then $v_L = L(di/dt) = LI_o$ and the ratio $v_L/i = LI_o/I_o t = L/t$ is a function of time and *not* a constant as it is for exponentials.

Note that when the characteristic equation (Eq. 4-13) is divided through by s, the right-hand side of Eq. 4-32 appears; but Eq. 4-32 was obtained by expressing the impedance as a ratio of an exponential voltage to an exponential current. The same relation could also have been obtained by considering that the resultant of impedances in series is the algebraic sum of the impedances or

$$Z = Z_L + Z_R + Z_C = Z(s)$$

where $Z(s)$ signifies the impedance to exponentials of the form $A\ e^{st}$.

Because the impedance function $Z(s)$ contains the same information as the characteristic equation, it is a useful concept in predicting the natural behavior of a system and it can be extended to include the prediction of steady-state response. As indicated in Example 6, the impedance function can be obtained easily in a circuit for which the governing equation may be quite complicated (i.e., it may consist of a set of simultaneous integro-differential equations).

EXAMPLE 6

Given the circuit of Example 5 with $v = V_O\ e^{st}$, determine and plot $Z(s)$ for real values of s.

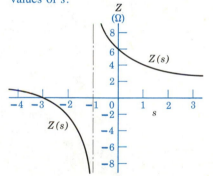

Figure 4.15 The impedance function for the circuit of Fig. 4.13.

A great advantage of the impedance concept is the ease with which impedances may be combined. Here

$$Z(s) = Z_1 + \frac{Z_R Z_C}{Z_R + Z_C} = R_1 + \frac{R(1/sC)}{R + 1/sC}$$

$$= R_1 + \frac{R}{RsC + 1} = \frac{sR_1RC + R_1 + R}{sRC + 1}$$

For $R_1 = 2\ \Omega$, $R = 4\ \Omega$, and $C = 0.25$ F,

$$Z(s) = \frac{2s + 6}{s + 1} = 2\frac{s + 3}{s + 1}$$

From the graph of Fig. 4.15 and from the expression for $Z(s)$ it is seen that $Z(s) = 0$ at $s = -3$ and that $Z(s)$ increases without limit as s approaches -1.

POLES AND ZEROS

In Example 6, $Z(s)$ is plotted for real values of s. Since each value of s corresponds to a particular exponential function, the impedance $Z(s_1)$ is the ratio of voltage to current for the particular exponential function of time $i = I_O\ e^{s_1 t}$. For instance, to determine the opposition to direct current flow, the impedance is calculated for $s = 0$ since $i = I_O\ e^{(0)t} = I_O$, a direct current. In Example 6, $Z(0) = 6\ \Omega$, the equivalent of the two resistances in series. This is physically correct (Fig. 4.13) because under steady conditions with no change in the voltage across the capacitor, no current flows in C and the circuit, in effect, consists only of the two resistors.

Every point in the complex frequency plane defines an exponential function and the complete plane represents all such functions. The magnitude of the impedance, $|Z(s)|$, can be plotted as a vertical distance above the s plane and, in general, the result will be a complicated surface.

EXAMPLE 7

Plot the magnitude of the impedance function of the circuit of Example 5 for real values of s and for imaginary values of s, and then sketch the impedance surface.

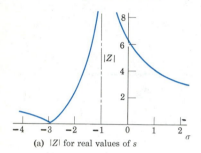

(a) $|Z|$ for real values of s

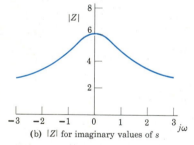

(b) $|Z|$ for imaginary values of s

The magnitude of $Z(s)$ for real values of s corresponds to the graph of Fig. 4.15 with negative quantities plotted above the axis. For imaginary values of $s(\pm j1, \pm j2, \text{etc.})$ the magnitude of $Z(s)$ is calculated and plotted in a similar way. For $s = -j$,

$$Z(s) = 2\frac{3 - j}{1 - j} \quad \text{and} \quad |Z| = \frac{2\sqrt{10}}{\sqrt{2}} \cong 4.5 \ \Omega$$

For the complex value $s = -1 - j2$,

$$Z(s) = 2\frac{2 - j2}{0 - j2} \quad \text{and} \quad |Z| = \frac{2\sqrt{8}}{\sqrt{4}} \cong 2.8 \ \Omega$$

Using the profiles of Fig. 4.16a and b and additional points in the complex plane, the impedance surface is sketched in Fig. 4.16c. The points of zero and infinite impedance are plotted in Fig. 4.16d.

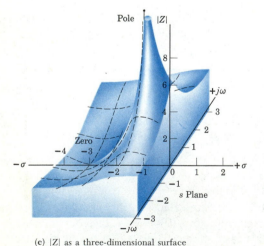

(c) $|Z|$ as a three-dimensional surface

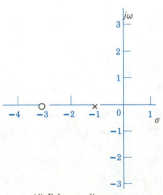

(d) Pole–zero diagram

Figure 4.16 Graphical representation of the impedance function.

Typically, the impedance function in three dimensions has the appearance of a tent pitched on the s plane. The height of the tent (magnitude of Z) becomes very great (approaches infinity) for particular values of s appropriately called *poles*. The tent touches the ground at particular values of s called *zeros*. The equation of the surface is usually quite complicated, but the practical use of the impedance function in network analysis or synthesis is relatively simple for two reasons. In the first place, we are usually interested in only a single profile of the surface; for example, the profile along the $j\omega$ axis indicates the response to sinusoidal functions of various frequencies. Second, just as two points define a straight line and three points define a circle, the poles and zeros uniquely define the impedance function except for a constant scale factor. As a result, the *pole-zero diagram* of Fig. 4.16d contains the essential information of the impedance function shown in Fig. 4.16c. The locations of the poles (X) and zeros (O), which are relatively easy to find, tell us a great deal about the natural response.

Physical Interpretation of Poles and Zeros

In the general relation between exponential voltage and current

$$v = Zi$$

what is the meaning of "zero impedance"? If the impedance is small, a given current can exist with a small voltage applied. Carrying this idea to the limit, under the conditions of zero impedance[†] a current can exist with *no applied voltage*; but a current with no applied voltage is, by definition, a natural current response. We conclude, therefore, that each zero, $s = s_1$, of the impedance function for any circuit designates a possible component, $I_1\, e^{s_1 t}$, of the natural response current of that circuit. This conclusion is supported by the fact that for the circuit of Fig. 4.14, the zeros of the impedance function (Eq. 4-32) are identical with the roots of the characteristic equation (Eq. 4-13). Knowing the locations of the impedance zeros, we can immediately identify the exponential components of the natural current. (See Example 8 on p. 108.)

If an impedance zero indicates the possibility of a current without a voltage, what is the significance of an impedance pole? Since $v = Zi$, if the impedance is very large, a given voltage can exist with only a small current flowing. Carrying this idea to the limit, under conditions of infinite impedance a voltage can exist with *no current flow*, in other words, a natural voltage. We conclude that each pole, $s = s_a$, of the impedance function of a circuit designates a possible component, $V_a\, e^{s_a t}$, of the natural voltage response of that circuit. (See Example 9.)

[†]To get the "feel" of zero impedance in a physical system, perform the following mental experiment. Grasp the bob of an imaginary pendulum between your thumb and forefinger. Applying a force, cause the bob to move along its arc with various motions. For motions against gravity or against inertia, appreciable force is required; the mechanical impedance is appreciable. To maintain a purely sinusoidal motion at the natural frequency of the pendulum only a little force (just that needed to make up friction losses) is required; the mechanical impedance to such a motion is small. To maintain a motion approximating an exponentially decaying sinusoid of just the right frequency, *no* force is required; for this motion, velocity is possible with no applied force. In other words, to this motion mechanical impedance is *zero*!

EXAMPLE 8

Given the circuit shown in Fig. 4.17, determine the poles and zeros of impedance. If energy is stored in the circuit in the form of an initial voltage V_O on the capacitor, predict the current i that will flow when the switch S is closed.

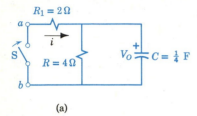

(a)

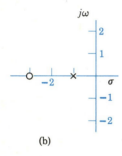

(b)

Figure 4.17 A pole-zero diagram of impedance.

As previously determined, the impedance function looking into the circuit at terminals ab is

$$Z(s) = 2\frac{s + 3}{s + 1}$$

Where $s = -1$, the denominator is zero and
$$Z(s) = \infty; \text{ therefore,}$$
$$s = -1 \text{ is a pole.}$$
Where $s = -3$, the numerator is zero and
$$Z(s) = 0; \text{ therefore,}$$
$$s = -3 \text{ is a zero.}$$

If the impedance is zero, a current can exist with no external forcing voltage. Therefore, the natural current behavior is defined by $s = s_1 = -3$ or

$$i = I_1 e^{-3t}$$

As before, I_1 is evaluated from initial data. At the instant the switch is closed, V_O appears across R_1 (tending to cause a current opposite to that assumed) and

$$i_o = I_1 e^0 = I_1 = -\frac{V_O}{R_1}$$

Hence

$$i = -\frac{V_O}{R_1} e^{-3t}$$

is the natural response current.

EXAMPLE 9

Returning to Example 8, assume that energy is stored in the form of an initial voltage V_O on the capacitor (perhaps by means of the voltage source shown in Fig. 4.18). Predict the voltage v that will appear across terminals ab when the switch S is opened.

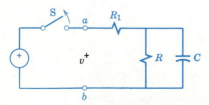

Figure 4.18 Natural voltage response.

With the external energy source removed, only a natural behavior voltage can appear. Such a voltage can exist with no current flow only if the impedance is infinite. For a pole at $s = s_a = -1$, the natural voltage behavior is

$$v = V_a e^{-t}$$

V_a is evaluated from initial data. At the instant the switch is opened, the current in R_1 goes to zero and the voltage across terminals ab is just $v_R = v_C = + V_O$. Hence,

$$v = V_O e^{-t}$$

is the natural response voltage. This result can be checked by letting $v = i_2 R$ where i_2 is the natural current that would flow in R if R were suddenly connected across the charged capacitor.

The General Impedance Function

Typical impedance functions include (for the circuit of Example 6)

$$Z(s) = R_1 + \frac{R(1/sC)}{R + (1/sC)} = \frac{sR_1RC + R_1 + R}{sRC + 1}$$

and (for the circuit of Fig. 4.14)

$$Z(s) = sL + R + \frac{1}{sC} = \frac{s^2LC + sRC + 1}{sC}$$

The impedance function for any network, no matter how complicated, consisting of resistances, inductances, and capacitances, can be reduced to the ratio of two polynomials in s. In general, we can write

$$Z(s) = K\frac{s^n + \cdots + k_2s^2 + k_1s + k_0}{s^m + \cdots + c_2s^2 + c_1s + c_0} \qquad (4\text{-}33)$$

Although it may not be easy, this can always be factored into

$$Z(s) = K\frac{(s - s_1)(s - s_2) \cdots (s - s_n)}{(s - s_a)(s - s_b) \cdots (s - s_m)} \qquad (4\text{-}34)$$

When $s = s_1, s_2, \ldots, s_n$, $Z(s) = 0$; therefore, these are zeros.

When $s = s_a, s_b, \ldots, s_m$, $Z(s) = \infty$; therefore, these are poles.

If the network is known, the impedance function can be written and factored and the pole-zero diagram constructed. Conversely, from the pole-zero diagram the impedance function can be formulated (except for scale factor K in Eq. 4-34). Theoretically, a network can then be designed or "synthesized," but practically this is not always easy. The complicated problems are so difficult that entire books have been written on the subject of network synthesis. In this book, however, we are concerned primarily with network analysis and the problems are more straightforward.

The General Admittance Function

Impedance is defined for exponentials as the ratio of voltage to current. The reciprocal of impedance is *admittance*, a useful property defined as the ratio of exponential current in amperes to voltage in volts so that

$$Y = \frac{i}{v} = \frac{1}{Z} \qquad (4\text{-}35)$$

measured in siemens. For an ideal resistance, the admittance is just the conductance or

$$Y_R = \frac{i_R}{v_R} = \frac{i}{Ri} = \frac{1}{R} = G$$

For exponential voltages and currents, the admittances of ideal inductive and capacitive circuit elements are

$$Y_L = \frac{i_L}{v_L} = \frac{i}{L(di/dt)} = \frac{i}{sLi} = \frac{1}{sL} \tag{4-36}$$

$$Y_C = \frac{i_C}{v_C} = \frac{C(dv/dt)}{v} = \frac{sCv}{v} = sC$$

Admittance is particularly useful in analyzing circuits that contain elements connected in parallel. Since current is directly proportional to admittance ($i = Yv$), admittances in parallel can be added directly just as conductances in parallel are added. For example, the total admittance of a parallel *GCL* circuit is

$$Y_L(s) = G + sC + 1/sL$$

The admittance function $Y(s)$ for any network consisting of lumped passive elements also can be reduced to the ratio of two polynomials in s. In the standard factored form this becomes

$$Y(s) = \frac{1}{Z(s)} = \frac{1}{K} \frac{(s - s_a)(s - s_b) \cdots (s - s_m)}{(s - s_1)(s - s_2) \cdots (s - s_n)} \tag{4-37}$$

Note that the admittance function has poles where the impedance function has zeros, and vice versa. The pole-zero diagram for the admittance function contains the same information as the pole-zero diagram for the impedance function, but the diagrams are labeled differently.

General Procedure for Using Poles and Zeros

The pole-zero concept is a powerful tool in determining the natural behavior (and, as we shall see in Chapter 5, the forced behavior) of any linear physical system. Modified to take advantage of this concept, the general procedure for determining the natural behavior of an electrical circuit is:

1. Write the impedance or admittance function for the terminals of interest.
2. Determine the poles and zeros, and plot the pole-zero diagram.
3. (a) For the terminals short-circuited, the natural behavior current is

$$i = I_1 e^{s_1 t} + I_2 e^{s_2 t} + \cdots + I_n e^{s_n t} \tag{4-38}$$

where $s_1, s_2, \ldots, s_n$ are zeros of the impedance function or poles of the admittance function.

(b) For the terminals open-circuited, the natural behavior voltage is

$$v = V_a e^{s_a t} + V_b e^{s_b t} + \cdots + V_m e^{s_m t} \tag{4-39}$$

where $s_a, s_b, \ldots, s_m$ are poles of the impedance function or zeros of the admittance function.

4. Evaluate the coefficients from the initial conditions (Example 10).

EXAMPLE 10

(a) In the circuit of Fig. 4.19a, the voltage source has been connected for a time long enough for steady-state conditions to be reached. At time $t = 0$, switch S is opened. Predict the open-circuit voltage across terminals ab.

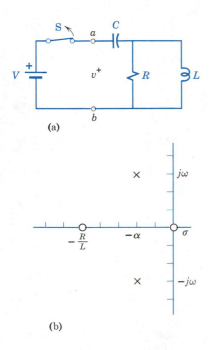

(a)

(b)

(b) Derive an expression for current i_R if the terminals ab are short-circuited by means of switch S_2 as shown in Figure 4.19c.

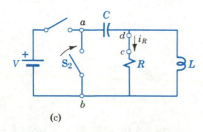

(c)

Figure 4.19 The use of poles and zeros.

Following the general procedure,

1. $Z_{ab}(s) = \dfrac{1}{sC} + \dfrac{RsL}{R + sL} = \dfrac{R + sL + s^2RLC}{sC(R + sL)}$

$$= R\dfrac{s^2 + (1/RC)s + 1/LC}{s(s + R/L)} = \dfrac{1}{Y(s)}$$

In the general form of Eq. 4.37,

$$Y_{ab}(s) = \dfrac{1}{R}\dfrac{(s - 0)(s - [-R/L])}{(s - s_1)(s - s_2)}$$

2. The admittance function has zeros at $s = 0$ and $s = -R/L$, and poles at $s = s_1$ and $s = s_2$ where

$$s_1, s_2 = -\dfrac{1}{2RC} \pm \sqrt{\dfrac{1}{4R^2C^2} - \dfrac{1}{LC}}$$

Assuming complex poles, the pole-zero diagram of admittance is as shown in Fig. 4.19b.

3. For terminals ab open-circuited, the natural behavior voltage is defined by the zeros of $Y(s)$ or

$$v = V_a e^0 + V_b e^{-(R/L)t}$$

4. At $t = 0^-$, the current in the inductance is constant and $v_L = L\,di/dt = 0 = v_R$, ∴ all the voltage V appears across the capacitance and the current in the inductance is zero. The second component of voltage (which reflects the *possibility* of energy storage in the inductance) is zero. The only energy storage is in the capacitor; at $t = 0^+$,

$$v = V_a e^0 = V$$

or the open-circuit voltage across terminals ab is just V. (There is no way for this ideal capacitor to discharge.)

We interpret the current of interest, i_R, as a short-circuit current, in this case at terminals cd. To this current the impedance is

$$Z_{cd}(s) = R + \dfrac{sL/sC}{sL + 1/sC} = \dfrac{s^2RLC + sL + R}{s^2LC + 1}$$

For $R = 1\ \Omega$, $L = 0.2$ H, and $C = 0.1$ F,

$$Z_{cd}(s) = \dfrac{0.02s^2 + 0.2s + 1}{0.02s^2 + 1} = \dfrac{s^2 + 10s + 50}{s^2 + 50}$$

The zeros are $s_1, s_2 = -5 \pm j5$. Therefore,

$$i_R = I_R e^{-5t} \sin(5t + \theta)$$

The zeros of $Z_{cd}(s)$ and $Z_{ab}(s)$ are the same because the circuits are the same for ab short-circuited. The poles differ, however, because the open circuits are quite different.

SUMMARY

- Forced behavior is the response to external energy sources.
  Natural behavior is the reponse to internal stored energy.

- Many physical systems with one energy-storage element can be described adequately by first-order integro-differential equations. The general procedure for determining the natural behavior of a linear system is:

  1. Write the governing integro-differential equation.
  2. Reduce this to a homogeneous differential equation.
  3. Assume an exponential solution with undetermined constants.
  4. Determine the exponents from the homogeneous equation.
  5. Evaluate the coefficients from the given conditions.

- In a second-order system with two energy-storage elements, the character of the natural response is determined by the discriminant.
  If the discriminant is *positive*, the response is *overdamped* and is represented by the sum of two decaying exponentials: $a = A_1 e^{s_1 t} + A_2 e^{s_2 t}$.
  If the discriminant is negative, the response is *oscillatory* and is represented by the damped sinusoid: $a = A e^{-\alpha t} \sin(\omega t + \theta)$.
  If the discriminant is *zero*, the response is *critically damped* and is represented by the sum of two different terms: $a = A_1 e^{st} + A_2 t e^{st}$.

- Impedance Z (ohms) and admittance Y (siemens) are defined for exponentials.

  $$\text{Where } v = Zi, \ Z_R = R, \ Z_L = sL, \text{ and } Z_C = 1/sC.$$

  $$\text{Where } i = Yv, \ Y_R = G, \ Y_L = 1/sL, \text{ and } Y_C = sC.$$

- Impedances and admittances in complicated networks are combined in the same way as resistances and conductances, respectively.
  The impedance function $Z(s)$ and the admittance function $Y(s)$ contain the same information as the characteristic equation.

- The pole-zero diagram contains the essential information of the impedance function or the admittance function.
  A zero of the impedance function indicates the possibility of a current without an applied voltage; therefore, a natural current.
  A pole of the impedance function indicates the possibility of a voltage without an applied current; therefore, a natural voltage.

- Using the pole-zero concept, the general procedure is:

  1. Write the impedance or admittance function for the terminals of interest.
  2. Determine the poles and zeros of impedance or admittance.
  3. Use the poles and zeros to identify possible components of natural voltage or current.
  4. Evaluate the coefficients from the given conditions.

REVIEW QUESTIONS

1. Cite an example of natural behavior in each of the following branches of engineering: aeronautical, chemical, civil, industrial, and mechanical.
2. To what extent is the natural behavior of a system influenced by the waveform of the forcing function that stores energy in the system?
3. Discuss the possibility of a positive exponent appearing in the natural response of a passive network. In any physical system.
4. Outline the procedure for determining the natural behavior of a mechanical system in translation.
5. In contrast to the exponential behavior of its idealized model, an actual coasting automobile comes to a complete stop long before an infinite time. Why?
6. Measurements on a certain very fast transient are difficult and the results include large random errors. In determining the time constant for this system, would a linear or a semilog plot of experimental values be preferable? Why?
7. What is the physical difference between a "first-order" system and a "second-order" system?
8. In what sense is the "characteristic equation" characteristic of the circuit?
9. In what sense does the "discriminant" discriminate?
10. What important initial information is available from an inductor in a circuit? From a capacitor? Explain.

11. In the case of a practical resistor, can the current be changed instantaneously? Can the voltage? Explain.
12. How do the waveforms represented by $i_1 = I_O e^{j4t}$ and $i_2 = I_O(e^{j4t} + e^{-j4t})$ differ? Could both be plotted on the same graph? Explain.
13. A damped sinusoid has a period, but is it a "periodic wave"? Explain.
14. Why is the natural response determined by the homogeneous equation instead of by the governing integro-differential equation?
15. What is the meaning of a "complex" frequency?
16. Give three reasons why so much attention is devoted to exponential functions in network analysis?
17. If, for exponentials, si corresponds to di/dt, what is the significance of $(1/s)i$?
18. What is the value of $|Z|$ (height of the "tent") in Fig. 4.16 for large real values of s? For large imaginary values? What is the limit of $|Z|$ as s approaches infinity?
19. In regard to *natural* response, is there any information available in the impedance function (Fig. 4.16c) that is not in the pole-zero diagram (Fig. 4.16d)? Explain.
20. Devise and sketch a physical system in which there is force without velocity (corresponding to an infinite mechanical impedance) and in which the force decays exponentially with time.

EXERCISES

1. In Fig. 4.20, switch S has been closed a long time.
   (a) If conditions have reached the steady state, what is the voltage across L? The current through R? The current through L?
   (b) If switch S is opened at time $t = 0$, write the equation governing current i as a function of time.
   (c) Solve for $i(t)$ and sketch $i(t)$ on a graph.
   (d) Determine $v(t)$ and sketch it on the graph of part (c).

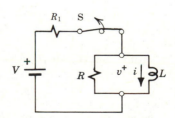

Figure 4.20

2. The energy stored in a 100-μF capacitance is 80 mJ. At time $t = 0$, a 5-kΩ resistance is connected across the capacitance.
   (a) What is the initial voltage on the charged capacitance?
   (b) Draw the circuit showing a switch that is closed at $t = 0$, and predict the voltage across C and the current through R at $t = 0^+$.
   (c) Express the current *out* of C and the current *into* R in terms of v_C and write the governing equation in terms of v_C.
   (d) Assume an exponential solution for $v_C(t)$, show that it satisfies the governing equation and the initial condition, and predict v_C at $t = 1$ s.
   (e) From your expression for $v_C(t)$, derive an expression for $i_R(t)$, evaluate the time constant T, and predict i_R at $t = 1.5$ s.
   (f) Calculate the total energy dissipated in resistance R. Explain the significance of this value.

3. In Fig. 4.21, I is a constant-current source and the switch has been in position 1 a long time.
   (a) What is the voltage across L? The current through R? The current through L?
   (b) Just *after* moving S to position 2, what is the current through L? The current through R? The voltage across R_1 and L? The rate of change of current through L?
   (c) Write the equation governing voltage v as a function of time after moving S to position 2.
   (d) Solve for $v(t)$ and sketch it on a graph.
   (e) Repeat parts (c) and (d) for $i_L(t)$.

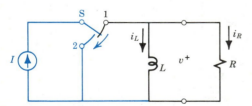

Figure 4.21

4. In Fig. 4.22, C_2 is uncharged and switch S has been in position 1 for a long time
   (a) Under steady-state conditions, what is the current in C_1? The current in R_1? The voltage across C_1?
   (b) If switch S is thrown to position 2 at time $t = 0$, write the equation governing current i as a function of time.

   (c) Assuming $C_1 = C_2$, solve for $i(t)$ and sketch $i(t)$ on a graph.
   (d) Determine $v_{C_2}(t)$ and sketch it on the graph of part (c).

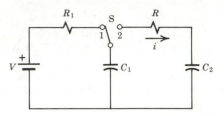

Figure 4.22

5. Two bicycle wheels appear to be similar in dimension and construction, but one has a time constant twice that of the other. Which will be easier to pedal? Why?

6. In Fig. 4.23, switch S is in position 1 for a long time and moved to position 2 at $t = 0$. For the resulting parallel circuit, determine $v(t)$ following the five-step procedure on p. 95.

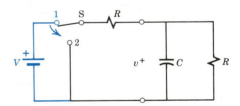

Figure 4.23

7. In Fig. 4.24, $V = 10$ V, $R_1 = 10\ \Omega$, $R_2 = 5\ \Omega$, and $L = 5$ H. Switch S is closed a long time and then opened at $t = 0$.
   (a) Determine the values of i_1 and i at $t = 0^+$.
   (b) Sketch approximately to scale i_1 and i as functions of time before and after $t = 0$.

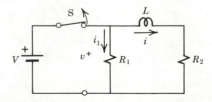

Figure 4.24

8. After switch S in Fig. 4.22 is thrown to position 2, the source-free circuit consists of two capacitances and a resistance.

(a) What single capacitance C_{eq} is equivalent to C_1 and C_2 in series?

(b) What is the time constant of the circuit in which i flows?

(c) Referring to Eq. 4-11, derive an expression for $i(t)$ in terms of the circuit parameters.

9. In the circuit of Fig. 4.6a, $C_1 = 2 \, \mu\text{F}$, $C_2 = 3 \, \mu\text{F}$, $R_1 = 5 \, \text{k}\Omega$, $R_2 = 2 \, \text{k}\Omega$, and $R_3 = 3 \, \text{k}\Omega$.

(a) Predict the time constant for voltages and currents in this circuit.

(b) Voltage $V = 21$ V is applied across R_3 for a long time and removed at $t = 0$. Define polarities and derive specific equations for $v_{C1}(t)$ and $i_{R2}(t)$.

10. In Fig. 4.25, energy is stored in the circuit in the form of initial currents I_1 and I_2 in the inductances.

(a) Write the governing equations as a pair of simultaneous equations in i_1 and i_2 and their derivatives.

(b) Outline a procedure for determining the natural response currents $i_1(t)$ and $i_2(t)$ from the governing equations.

(c) Would you expect this to be an easy or difficult procedure? (An easier method is suggested in Exercise 37.)

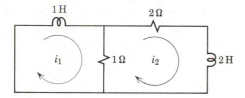

Figure 4.25

11. In Fig. 4.26, switch S_1 has been in position 1 and switch S_2 has been closed for a long time before $t = 0$. The 2-μF capacitor is initially uncharged.

(a) For $t < 0$, find i_L, v_L, i_{C1}, and v_{C1}. (Define voltage polarities consistent with current directions.)

(b) At $t = 0$, switch S_1 is thrown to position 2 and switch S_2 is opened. For $t = 0^+$, find i_L, i_R, i_{C1}, v_{C1}, v_R, v_{C2}, and v_L.

(c) For $t = 0^+$, find di_R/dt, di_L/dt, dv_{C1}/dt, and dv_{C2}/dt.

12. In a series RLC circuit, $R = 1250 \, \Omega$, $L = 62.5$ H, and $C = 250 \, \mu\text{F}$. Energy is stored in the form of an initial voltage $V_O = 25$ V on the capacitor.

(a) Draw the circuit, define polarities, and calculate the numerical values of i, v_R, and v_L at $t = 0^+$.

(b) Write the "governing" equation, the "homogeneous" equation, and the "characteristic" equation for this circuit.

(c) Calculate di/dt and d^2i/dt^2 at $t = 0^+$.

(d) Will the natural response be under, over, or critically damped?

(e) Solve for $i(t)$, evaluating the constants from initial conditions.

13. The characteristic equation of a circuit is $10s^2 + Rs + 10 = 0$. What values of R will allow an oscillatory behavior in the circuit?

14. In the circuit of Exercise 12, at $t = 0$ the capacitance voltage is $V_O = 20$ V and initial current $I_O = 40$ mA is flowing in the inductance (tending to discharge the capacitance). Draw the circuit and calculate the numerical values of i, v_C, v_L, v_R, and di/dt at $t = 0^+$.

15. In a series RLC circuit, $R = 666 \, \Omega$, $L = 8.33$ H, and $C = 100 \, \mu\text{F}$. Energy is stored in the form of a 2-A current in the inductor. At $t = 0$, the exciting current is removed and the inductor shorted across R and C (initially uncharged) in series.

(a) Draw the circuit and calculate the numerical values of i, v_C, v_R, and v_L at $t = 0^+$.

(b) Calculate di/dt and d^2i/dt^2 at $t = 0^+$.

(c) For this system, write the "governing" equation, the "homogeneous" equation, and the "characteristic" equation.

(d) Will the natural response be under, over, or critically damped?

(e) Solve for $i(t)$, evaluating all constants from initial conditions.

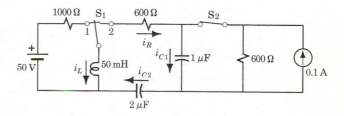

Figure 4.26

16. Explain the natural response of an oscillatory *RLC* circuit on a physical basis in terms of the energy interchanges that occur. (You may wish to reason by analogy with an oscillating pendulum.)

17. In Fig. 4.27, switch S has been in position 1 for a "long" time. At time $t = 0$, the switch is moved to position 2.
    (a) For time $t = 0^+$, evaluate v_C, i, and v_L.
    (b) Derive an expression for current i as a function of time $t > 0$ in terms of the *given* quantities.

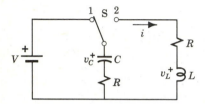

Figure 4.27

18. An *RLC* series circuit has an initial current I_O in the inductor and no voltage on the capacitor.
    (a) For the case of critical damping, derive a general expression for the natural current response.
    (b) For $I_O = 2$ A, $L = 1$ H, $C = 0.25$ F, and $R = 4$ Ω, determine and plot to scale the response current as a function of time.

19. In the circuit of Fig. 4.28, $V = 10$ V, $L = 1$ H, $R = 5$ Ω, and $C = 0.1$ F. The switch is thrown to position 2 at $t = 0$.
    (a) Assuming the switch was in position 1 a "long" time, evaluate i_L, i_R, and $v_C = v$ just

before the switch is thrown and just *after* the switch is thrown. Where is energy stored in the system?
    (b) By summing the currents out of node a, write the equation governing the behavior of this circuit for $t > 0$ in terms of $v = v_{ab}$. Use this equation and information from part (a) to evaluate dv/dt at $t = 0^+$. (*Hint:* What is the value of $\int v \, dt$ at $t = 0^+$?)
    (c) From the "governing" equation, write the "homogeneous" equation and the "characteristic" equation.
    (d) Use the roots of the characteristic equation to write the voltage response $v(t)$ with undetermined constants, but in a form that has physical significance.
    (e) Evaluate the coefficients by writing simultaneous equations using information about v and dv/dt at $t = 0^+$.
    (f) Sketch the envelope of the equation, note the intercepts on the time axis, and sketch in $v(t)$ approximately to scale.

20. In the circuit of Fig. 4.29, $L = 0.5$ H, $R = 400$ Ω, and $C = 0.5$ μF. The capacitor has an initial voltage $V_O = 30$ V, and the switch is closed at $t = 0$.
    (a) Determine the numerical values of v, i_R, i_L, and i_C just after the switch is closed, i.e., at $t = 0^+$.
    (b) Calculate dv/dt and d^2v/dt^2 at $t = 0^+$.
    (c) For this system, write the "governing" equation in terms of v, the "homogeneous" equation, and the "characteristic" equation.
    (d) Will the response be under, over, or critically damped?
    (e) Solve for $v(t)$ and evaluate all constants from initial conditions.
    (f) Plot enough points to sketch, approximately to scale, the voltage as a function of time. (You may prefer to do this in terms of the components.)

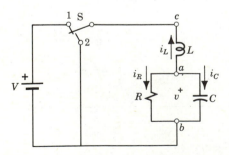

Figure 4.28

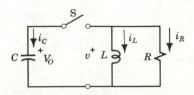

Figure 4.29

21. Repeat Exercise 20 with $L = 64$ mH, $R = 400\ \Omega$, and $C = 0.5\ \mu$F. The switch is closed at $t = 0$ with $V_O = 30$ V.

22. Draw a "map of s-land" similar to but larger than Fig. 4.30. In each square sketch the response expected for a root of the characteristic equation located at that point in the complex plane. Follow the example drawn for a root with positive values of σ and ω, showing an exponentially increasing sinusoidal function of time, $e^{(\sigma + j\omega)t}$. Clearly indicate differences in the expected response corresponding to differences in σ and ω.

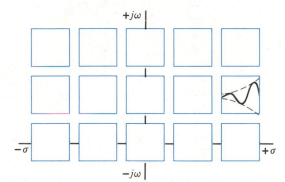

Figure 4.30

23. Determine and sketch the root locus for a series RLC circuit in which:
   (a) C is the changing parameter.
   (b) L is the changing parameter.

24. In the circuit of Fig. 4.31, $R_1 = 2\ \Omega$, $R = 4\ \Omega$, and $L = 2$ H.
   (a) Determine enough points to plot $Z(s)$.
   (b) What is the impedance of this circuit to direct current?
   (c) If a voltage $v = V_O\,e^{st}$ is acting where $V_O = 1$ V, what current i flows for $s = -4$? For $s = -3$? For $s = -2$?
   (d) Assume current $i = 2\,e^{-4t}$ A is flowing and calculate the necessary voltage $v(t)$.

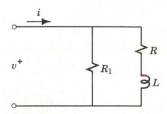

Figure 4.31

25. In Fig. 4.24, $R_1 = R_2 = R$. Switch S is closed for a "long" time and then opened at $t = 0$.
   (a) Express impedance $Z(s)$ seen by current i.
   (b) Obtain the characteristic equation from $Z(s)$.
   (c) Derive an expression for $i(t)$.

26. In Fig. 4.22, $C_1 = C_2 = C$ and C_2 is initially uncharged.
   (a) Express the impedance $Z(s)$ seen by current i after the switch is moved to position 2.
   (b) Obtain the characteristic equation from $Z(s)$.
   (c) Derive an expression for $i(t)$.

27. Given the circuit of Fig. 4.32:
   (a) Define in words the impedance at terminals ab.
   (b) Determine the impedance $Z_{ab}(s)$.
   (c) For $i = 5\,e^{-3t}$ A, evaluate $Z_{ab}(s)$ and determine v_C the voltage across the capacitance by using the impedance concept.
   (d) For $i = 5$ A dc, repeat part (c) and check your answer by reasoning from basic principles of L and C.

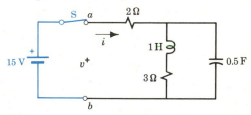

Figure 4.32

28. For the circuit of Fig. 4.32:
   (a) Write the impedance function and sketch the pole-zero diagram.
   (b) Write the admittance function and sketch the pole-zero diagram.

29. For the circuits in Fig. 4.33a and b:
   (a) Write the impedance function and sketch the pole-zero diagram.
   (b) Write the admittance function and sketch the pole-zero diagram.

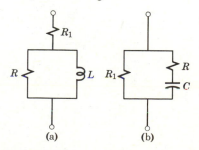

Figure 4.33

30. For a certain network, the pole-zero diagram of impedance is as shown in Fig. 4.34. Energy is stored in the network and then at time $t = 0$ a switch is thrown to disconnect all external energy sources.

(a) Write expressions (with undetermined constants) for the short-circuit natural response current and the open-circuit natural response voltage.

(b) Write an expression for $Z(s)$ for this network.

(c) Synthesize the network, using one element each of R, L, and C. (*Hint:* Let $C = 0.01$ F and determine the necessary values of R and L.)

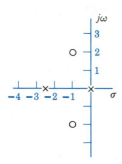

Figure 4.34

31. For a certain network, the pole-zero diagram of admittance is as shown in Fig. 4.35. Energy is stored in the network and then at time $t = 0$ a switch is thrown to disconnect all external en-

ergy sources. Write expressions (with undetermined constants) for:

(a) The short-circuit natural response current,

(b) The open-circuit natural response voltage.

(c) What combination of R, L, and C could be the network?

32. For the circuit of Exercise 19 (Fig. 4.28) with the switch in position 2:

(a) Evaluate the admittance of the circuit as measured at terminals ab in the form of Eq. 4-37. (It may help to redraw the circuit.)

(b) Determine the poles and zeros of admittance and plot a pole-zero diagram.

(c) Write the natural voltage response across these terminals with undetermined constants.

(d) Compare the result of part (c) with that for Exercise 19(d).

(e) For the initial conditions given, determine $v(t)$.

33. In Fig. 4.32, switch S is closed for a long time and then opened at $t = 0$. Follow the general procedure using the pole-zero concept to determine $v(t)$.

34. In Fig. 4.32, switch S is closed for a long time and then opened at $t = 0$. Identify terminals cd at which the capacitor current i_C would be a short-circuit current. Redraw the circuit showing these terminals and determine $Z_{cd}(s)$. Derive an expression for $i_C(t)$ following the general procedure.

35. In the circuit of Fig. 4.36, $V_O = 10$ V, $C = 0.5$ F, $R_1 = R_2 = 2\ \Omega$, and $L = 1$ H.

(a) Outline a step-by-step procedure using the immittance concept for obtaining the natural response current i upon closing switch S.

(b) Derive an expression for $i(t)$ with two undetermined constants.

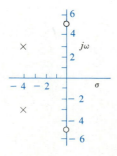

Figure 4.35

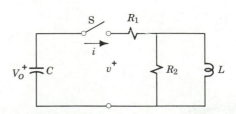

Figure 4.36

36. In the circuit of Fig. 4.37, switch S is closed for a long time and opened at $t = 0$.
  (a) Determine $Y_{ab}(s)$.
  (b) For $L = 1$ H, $C = 0.1$ F, and $R = 2$ Ω, plot the pole-zero diagram of admittance.
  (c) For $I = 10$ A, predict $v(t)$.

37. The necessity for solving simultaneous equations in Exercise 10 is avoided if the impedance concept is employed. Open up the circuit to the left of the 1-H inductance and designate a pair of terminals at which i_1 would be a short-circuit current.
  (a) Write the impedance function $Z(s)$, determine the poles and zeros, and plot the pole-zero diagram.
  (b) For initial currents $I_1 = I_2 = 3$ A, determine $i_1(t)$ and sketch it approximately to scale.

38. Use the pole-zero concept to find $v(t)$ in the circuit of Exercise 21.
  (a) Write the admittance function for the terminals across which voltage v appears. (*Hint:* Redraw the circuit.)
  (b) Determine the poles and zeros of $Y(s)$ and plot.
  (c) Write an expression for $v(t)$ with two undetermined constants.
  (d) What is the natural frequency of this circuit in Hz?

39. In the circuit of Fig. 4.38, $I = 2$ A, $L = 2$ H, $R_1 = 4$ Ω, $R_2 = 2$ Ω, and $C = 0.25$ F. Switch S is in position 1 for a long time and then thrown to position 2 at $t = 0$.
  (a) Determine v_C and dv_C/dt at $t = 0^+$.
  (b) Draw the pole-zero diagram of admittance $Y_{ab}(s)$.
  (c) Determine $v_C(t)$.

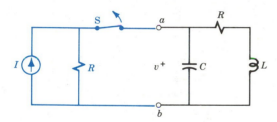

Figure 4.37

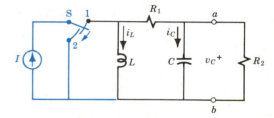

Figure 4.38

PROBLEMS

1. In Fig. 4.39, switch S is opened at $t = 0$ after being closed a long time. Predict current i as a function of time after $t = 0$. Sketch $i(t)$ just before $t = 0$ and for a significant time after.

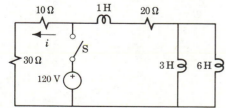

Figure 4.39

2. A vacuum-insulated bottle contains W kg of a liquid of specific heat C_p (J/kg · K) at a temperature τ_0 (K above the outside temperature). The effective thermal conductance of the container is U (J/s · K temperature difference).
  (a) Stating all assumptions, express the heat stored in the liquid and the rate of heat conduction through the container.
  (b) Predict the temperature of the liquid as a function of time.

3. An automobile weighing 3220 lb is moving at a velocity of 50 ft/s. At time $t = 0$, the car is shifted into neutral and allowed to coast. In

2.5 s, the velocity drops to 45 ft/s. Predict the distance the auto will coast on a level road.

4. In an *RC* circuit undergoing natural behavior, where is energy stored? Where is energy dissipated? Using the expression for current, show that the total energy dissipated during natural behavior is just equal to the energy stored initially.

5. Energy is to be stored in an inductor and dissipated in a 20-Ω resistor. The power in the resistor must be cut in half every second.
   (a) Design the inductor, i.e., specify *L* of an ideal inductance.
   (b) If a practical inductor introduces 1 Ω of resistance for every 10 H of inductance, specify *L*.

6. A certain application requires an electrical circuit with the pole-zero diagram of admittance shown in Fig. 4.40.
   (a) Create a possible circuit with these general characteristics.
   (b) Design the circuit by specifying the component values.

7. The circuit of Fig. 4.41 is useful in the design of control systems. In this case, $C = 1\ \mu F$, $R = 1\ M\Omega$, and the initial voltage (on the right-hand capacitor only) is $V_O = 20$ V. Determine the current $i(t)$ after switch S is closed.

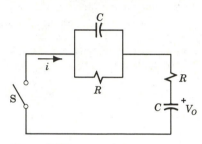

Figure 4.41

8. In Fig. 4.25, energy is stored in the circuit in the form of initial currents $I_1 = 10$ A and $I_2 = 2$ A in the inductances. Determine the current in the 1-Ω resistance as a function of time.

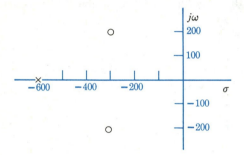

Figure 4.40

5

Forced Response

Response to Forcing Functions

AC Circuit Analysis

Analogs and Duals

Internal energy storage gives rise to natural behavior; regardless of how the energy is stored or where, the natural response of a circuit is determined by the characteristics of the circuit itself. In contrast, the form of the forced response due to an external energy source is dependent on the form of the forcing function; for example, a sinusoidal voltage always produces sinusoidal currents. Having mastered the technique of predicting natural behavior, we are now ready to tackle the problem of determining the response to some important forcing functions or waveforms. Then in Chapter 6 we shall use our knowledge of natural and forced response to determine the complete response of an electrical circuit, employing an approach that is applicable to nonelectrical problems as well.

In limiting consideration to just the forced response, we are assuming that natural behavior is absent. Either there was no natural response, or sufficient time has elapsed since the initiation of any transient effect for the natural response to become negligibly small. This condition exists in many practical problems.

The material in this chapter represents another step in a cumulative process. We use the accepted definitions, the fundamental element and connection laws, the methods of representing signal waveforms, and the impedance concept to solve a new set of problems. Although impedance was defined for exponentials, the general interpretation permits its application to continuous and sinusoidal waves as well. To simplify the analysis of circuits involving sinusoidal voltages and currents, we derive a new tool, the *phasor*.

RESPONSE TO FORCING FUNCTIONS

By definition, impedance is the ratio of voltage to current for exponentials of the form $i = I_0 e^{st}$. In general,

$$v = Z(s)i \qquad (5\text{-}1)$$

a special form of Ohm's law where impedance $Z(s)$ is in ohms.

For resistance, $\qquad v_R = Ri \quad$ and $\quad Z_R(s) = v/i = R.$

For inductance, $\quad v_L = L\,di/dt = sLi \quad$ and $\quad Z_L(s) = v/i = sL.$

For capacitance, $\quad i_C = C\,dv/dt = sCv \quad$ and $\quad Z_C(s) = v/i = 1/sC.$

Impedances to exponentials can be combined in series and parallel just as resistances are.

Response to Exponentials

In the circuit of Fig. 5.1a, an exponential forcing function $i = I_0 e^{-\alpha t}$ flows in the series combination of R and L. There is no information on how the current was initiated; we do know that such a current could not be established by closing a switch at $t = 0$ because it takes time to establish a current in an inductance. Assuming such a current does exist, what is the response of the circuit in terms of voltages across the elements?

Since $Z_R(s) = R$,

$$v_R = Z_R(s)i = RI_0 e^{-\alpha t}$$

Since $Z_L(s) = sL = -\alpha L$,

$$v_L = Z_L(s)i = -\alpha L I_0 e^{-\alpha t}$$

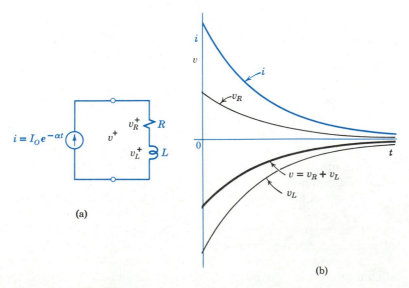

(a)

(b)

Figure 5.1 Forced response of an RL circuit to an exponential current.

The total voltage is the sum of the two in series or

$$v = v_R + v_L = (R - \alpha L)I_O e^{-\alpha t}$$

As expected, the response to an exponential forcing function is itself exponential of the same "frequency" α or the same time constant $T = 1/\alpha$. The negative sign of v_L is due to the tendency of an inductance to oppose a change in current. Assuming that $\alpha L > R$, the total voltage is negative as shown in Fig. 5.1b.

The impedance $Z(s) = R + sL$ and the response $v(t)$ vary widely with changes in parameter s. For a given magnitude I_O, the inductive voltage v_L is directly proportional to s and changes sign with s. For the special case of $s = -R/L$, $Z(s) = 0$ or the circuit presents no opposition to this particular current function. Our interpretation is that, since current is possible with no forcing voltage, this is a natural response current. Two other interesting cases are for $s = 0$ and $s = j\omega$.

Response to Direct Currents

A unidirectional current of constant magnitude is called a continuous or direct current, abbreviated dc. In common usage a voltage of the same form is called a *dc voltage*, and we sometimes talk of *dc currents* in contrast to alternating or *ac currents*. Direct-current energy supplied by chemical batteries, thermocouples, solar cells, or rotating dc generators is used for a wide variety of purposes including ship propulsion, electrolytic refining, electron beam acceleration, and computer operation. (In circuit diagrams dc voltage sources appear so often that the special symbol of Fig. 5.2 is used; the longer line indicates the positive terminal.)

For the special case of $s = 0$ (see Fig. 4.12), the general exponential becomes

$$i = I_O e^{0t} = I_{dc} = I^\dagger \tag{5-2}$$

The dc values of impedance $Z(s)$, for the case of $s = 0$, we designate $Z(0)$. For resistance, inductance, and capacitance these impedances are:

$$Z_R(0) = R \qquad Z_L(0) = 0 \qquad Z_C(0) = \infty \tag{5-3}$$

The same results are obtained by considering the fundamental element equations. In a resistance, $v = Ri$ for all waveforms; for a direct current I, a direct voltage $V = RI$ appears across a resistance. The impedance $v/i = V/I = R$ as expected.

When a direct current $i = I$ flows in an inductance, $di/dt = 0$ and, therefore, $v = L(di/dt) = 0$. The impedance $v/i = 0$ or an inductance has the characteristic of a short circuit. We say:

An inductance looks like a short circuit to a dc current.

When a direct voltage is applied to an inductance carrying no initial current,

$$i = \frac{1}{L} \int_0^t v \, dt = \frac{V}{L} t \tag{5-4}$$

and the current increases linearly with time.

---

†Capital I and V are used to designate constant values, such as effective values or dc magnitudes, and lower case i and v are used to designate variable quantities, particularly functions of time.

When a direct voltage $v = V$ appears across a capacitance, $dv/dt = 0$ and, therefore, $i = C(dv/dt) = 0$. The impedance $v/i = \infty$ or we say:

A capacitance looks like an open circuit to a dc voltage.

When a direct current $i = I$ flows in an initially uncharged capacitance,

$$v = \frac{1}{C} \int_0^t i \, dt = \frac{I}{C} t \tag{5-5}$$

and the voltage increases linearly with time.

EXAMPLE 1

Switch S in Fig. 5.2a has been closed for a long time. Determine the current i and the voltages across R_1, R_2, and R_3.

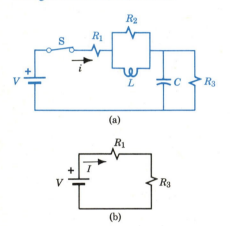

(a)

(b)

Figure 5.2 Forced response to a dc voltage.

We assume that after the "long time" specified all natural response has died away and only a forced response exists. For a direct forcing voltage the response is a direct current. To a direct current, inductance L looks like a short circuit effectively removing R_2 from the circuit; therefore, the voltage across R_2 is zero.

To a direct voltage, capacitance C looks like an open circuit and it can be removed from the circuit; therefore, all the current i flows through R_3. For the dc case, the circuit looks like Fig. 5.2b.

The total series resistance is $R_1 + R_3$ and

$$i = I = \frac{V}{R_1 + R_3}$$

By the voltage divider rule,

$$V_1 = R_1 I = \frac{R_1}{R_1 + R_3} V$$

and

$$V_3 = R_3 I = \frac{R_3}{R_1 + R_3} V$$

Representation of Sinusoids

If a sinusoidal source supplies a linear circuit consisting of resistances, inductances, and capacitances, all voltages and currents will be sinusoids of the same frequency. To perform the addition and subtraction of voltages and currents required in the application of Kirchhoff's laws, we need a technique for combining quantities that are continuously changing. The convenient and general method that we shall use is based on complex algebra and an ingenious transformation of functions of time into constant quantities.

The combination of a real number and an imaginary number defines a point in the complex plane (Fig. 5.3) and also defines a *complex number*. The complex number may be considered to be the point or the directed line segment to the point; both interpretations are useful. (Imaginary numbers and complex algebra are explained in the Appendix.)

Given the complex number $\mathbf{W}$ of magnitude M and direction θ, in *rectangular* form,

$$\mathbf{W} = a + jb$$

or

$$\mathbf{W} = M(\cos\theta + j\sin\theta)$$

by inspection of Fig. 5.3. By Euler's theorem (see Appendix p. 723),

$$\cos\theta + j\sin\theta = e^{j\theta} \qquad (5\text{-}6)$$

Therefore,

$$\mathbf{W} = M e^{j\theta} \qquad (5\text{-}7)$$

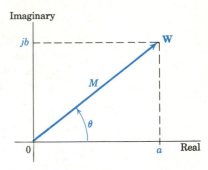

Figure 5.3   A complex number

This is called the *exponential* or *polar* form and can be written symbolically as

$$\mathbf{W} = M\underline{/\theta} \qquad (5\text{-}8)$$

which is read "magnitude M at angle θ." (In this book, complex numbers are always in boldface type.)

To convert complex numbers to *polar* form from rectangular,

$$M = \sqrt{a^2 + b^2} \qquad \theta = \arctan\frac{b}{a} \qquad (5\text{-}9)$$

To convert complex numbers to *rectangular* form from polar,

$$a = M\cos\theta \qquad b = M\sin\theta \qquad (5\text{-}10)$$

The complex constant $\mathbf{W} = M e^{j\theta}$ is represented by a fixed radial line. If the line rotates at an angular velocity ω, as shown in Fig. 5.4, $\mathbf{W}$ is a complex function of time and

$$\mathbf{W}(t) = M e^{j(\omega t + \theta)} \qquad (5\text{-}11)$$

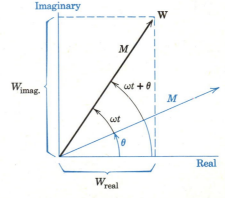

The projection of this line on the real axis is

$$W_{\text{real}} = f_1(t) = M\cos(\omega t + \theta)$$

and the projection on the imaginary axis is

$$W_{\text{imag}} = f_2(t) = M\sin(\omega t + \theta)$$

Figure 5.4   A complex function of time.

Either component could be used to represent a sinusoidal quantity; in this book, we shall always work with the real components.

Phasor Representation

If an instantaneous voltage is described by a sinusoidal function of time such as

$$v(t) = V_m\cos(\omega t + \theta) = \sqrt{2}\, V\cos(\omega t + \theta) \qquad (5\text{-}12)$$

where V_m is the amplitude and V is the effective value, then $v(t)$ can be interpreted as the "real part of" a complex function or

$$v(t) = \text{Re}\,\{V_m\,e^{j(\omega t+\theta)}\} = \text{Re}\,\{(V\,e^{j\theta})(\sqrt{2}\,e^{j\omega t})\} \qquad (5\text{-}13)$$

In the second form of Eq. 5-13, the complex function in braces is separated into two parts; the first is a complex constant, and the second is a function of time that implies rotation in the complex plane and includes the conversion factor relating effective to maximum values (Eq. 3-16). The first part we define as the *phasor* **V** where

$$\mathbf{V} = V\,e^{j\theta} = V\underline{/\theta} \qquad (5\text{-}14)$$

The phasor $\mathbf{V} = V\,e^{j\theta}$ is called a *transform* of the voltage $v(t)$. It is obtained by transforming a function of time into a complex constant that retains the essential information: effective value and phase angle. The term $e^{j\omega t}$ indicates rotation at angular velocity ω, but this is the same for all voltages and currents associated with a given source and therefore this term may be put to one side until needed.

EXAMPLE 2

(a) Write the equation of the current shown in Fig. 5.5a as a function of time and represent the current by a phasor.

The current reaches a positive maximum of 10 A at $\pi/6$ rad or $30°$ before $\omega t = 0$; therefore, the phase angle θ is $+\pi/6$ rad and the equation is

$$i = 10\,\cos\left(\omega t + \frac{\pi}{6}\right)\text{A}$$

This could also be expressed as

$$i(t) = \text{Re}\,\{10\,e^{j(\omega t+\pi/6)}\} = \text{Re}\,\{7.07\,e^{j(\pi/6)}\sqrt{2}\,e^{j\omega t}\}\text{ A}$$

By comparison with the defining Eqs. 5-13 and 5-14, the current phasor 5.5b is

$$\mathbf{I} = 7.07\,e^{j(\pi/6)}\text{ A}$$

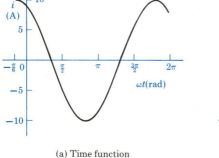

(a) Time function

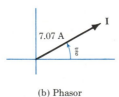

(b) Phasor

Figure 5.5 Phasor representation.

(b) Represent, by a phasor, the current $i = 20\sqrt{2}\,\sin(\omega t + \pi/6)$ A.

Our definition of phasor is based on the cosine function as the real part of the complex function of time. Since $\cos(x - \pi/2) = \sin x$, we first write

$$i = 20\sqrt{2}\,\cos\left(\omega t + \frac{\pi}{6} - \frac{\pi}{2}\right) = 20\sqrt{2}\,\cos\left(\omega t - \frac{\pi}{3}\right)\text{A}$$

Then the current phasor is $\mathbf{I} = 20\,e^{j(-\pi/3)}$ A.

Phasor Diagrams

Electrical engineering deals with practical problems involving measurable currents and voltages that can exist in and across real circuit components such as those in your TV set. In solving circuit problems, however, we choose to deal with abstractions rather than physical currents and voltages. Learning to read and write and interpret such abstractions is an essential part of an engineering education. Working with the abstract concept that we call a phasor will provide some good experience in this important activity.

The phasor has much in common with the vector that is so valuable in solving problems in mechanics. As you may recall, vector representation of forces permits the use of graphical methods for the composition and resolution of forces. Also, and this may be more important, the use of vector representation—in free-body diagrams, for example—provides a picture of the physical relationships that is not provided by the algebraic equations alone. In the same way, *phasor diagrams* can be used for graphical solution, as a quick check on the algebraic solution, and to gain new insight into voltage and current relations.

EXAMPLE 3

Given $v_1 = 150\sqrt{2}\cos(377t - \pi/6)$ V and $\mathbf{V}_2 = 200\underline{/+60°}$ V, find $v = v_1 + v_2$.

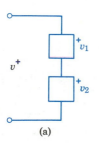

(a)

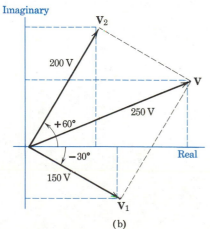

(b)

Figure 5.6   Use of phasor diagrams.

The expression for v_1 is a mathematical abstraction carrying a great deal of information: a voltage varies sinusoidally with time, completing $377/2\pi = 60$ Hz and reaching a positive maximum value of $150\sqrt{2}$ V when $t = (\pi/6)/377 = 0.00139$ s. To work with sinusoid v_1, we first *transform* the time function to phasor $\mathbf{V}_1 = 150\underline{/-30°}$ V.

The expression for $\mathbf{V}_2$ is "an abstraction of an abstraction"; part of the information is specified and part is only implied. The sinusoidal character is implied by the boldface phasor notation. The effective value is clearly 200 V and the phase angle is 60°. Since v_1 and v_2 are associated, we assume that they have the same angular frequency or $\omega_2 = 377$ rad/s. Reading "between the lines," we conclude that $v_2 = 200\sqrt{2}\cos(377t + \pi/3)$ V.

The phasor diagram of Fig. 5.6b shows the complex quantities $\mathbf{V}_1$ and $\mathbf{V}_2$ plotted to scale. Using the parallelogram law for adding vectors, the construction is as shown. Scaling off the resultant, it appears that $\mathbf{V}$ has a magnitude of about 250 V at an angle of approximately 23°.

For algebraic addition, the rectangular form is convenient, and we write (Eq. 5-10)

$$\mathbf{V}_1 = 150\cos(-30°) + j150\sin(-30°) = 130 - j75 \text{ V}$$

$$\mathbf{V}_2 = 200\cos 60° + j200\sin 60° = 100 + j173 \text{ V}$$

By the rules for equality and addition (see pp. 723 and 724),

$$\mathbf{V} = \mathbf{V}_1 + \mathbf{V}_2 = 230 + j98 = 250\underline{/23.1°}\,\text{V}$$

After the inverse transformation on $\mathbf{V}$,

$$v = v_1 + v_2 = 250\sqrt{2}\cos(377t + 23.1°) \text{ V}$$

Looking at Fig. 5.6, we see that in Example 3 the algebraic addition involves the same components as the graphical addition, and we can conclude that phasor diagrams can be used for graphical calculations with the phasors treated just like vectors. A second virtue of the phasor diagram is that a quick sketch only approximately to scale provides a good check on algebraic calculations. Mistakes in sign, decimal point, or angle are easily spotted.

AC CIRCUIT ANALYSIS

Although an "alternating" current can be nonsinusoidal, unless otherwise stated we assume that an alternating current or an "ac voltage" is sinusoidal. Most electrical energy is generated, transmitted, and consumed in the form of alternating current. Also, much of the dc energy used in power and communication applications is first generated as ac energy and then converted to dc form. The number of applications of ac circuits is so great that we can justify taking time to learn some short cuts in solving such circuits.

The representation of sinusoids by exponentials leads to the phasor concept whereby functions of time are transformed into constant quantities for easy manipulation and display. Also, we know that in dealing with exponentials the impedance concept provides a convenient method for finding forced response. Therefore, we expect exponential representation of sinusoids to lead to a convenient method for analyzing circuits with sinusoidal sources and for displaying the results.

To confirm our expectation, let us consider the response of circuit elements to sinusoidal signals in terms of the defining element equations.

Element Impedance

In general, the ratio of an exponential voltage to the corresponding current is a measure of the opposition to the flow of current called the *impedance*.

When a current $i = \sqrt{2}I \cos \omega t$ represented by phasor $I \underline{/0°}$ flows through a resistance R, the voltage is given by

$$v_R = Ri = \sqrt{2}RI \cos \omega t = \sqrt{2}V_R \cos \omega t \qquad (5\text{-}15)$$

represented by phasor $V_R \underline{/0°}$. In this case the opposition to current flow is

$$\frac{V_R \underline{/0°}}{I \underline{/0°}} = \frac{RI \underline{/0°}}{I \underline{/0°}} = R \underline{/0°} \qquad (5\text{-}16)$$

When the current $i = \sqrt{2}I \cos \omega t$ flows through an inductance L, the voltage is given by

$$v_L = L\frac{di}{dt} = \sqrt{2}\,\omega LI \,(-\sin \omega t) = \sqrt{2}\omega LI \cos (\omega t + 90°) \qquad (5\text{-}17)$$

represented by phasor $V_L \underline{/90°}$. The opposition to current flow is

$$\frac{V_L \underline{/90°}}{I \underline{/0°}} = \frac{\omega LI \underline{/90°}}{I \underline{/0°}} = \omega L \underline{/90°} = j\,\omega L \qquad (5\text{-}18)$$

When the voltage $v = \sqrt{2}V \cos \omega t$ appears across a capacitance C, the current through the capacitance is given by

$$i_C = C\frac{dv}{dt} = \sqrt{2}\omega CV(-\sin \omega t) = \sqrt{2}\omega CV \cos (\omega t + 90°) \qquad (5\text{-}19)$$

represented by phasor $I_C\underline{/90°}$. The opposition to current flow is

$$\frac{V\underline{/0°}}{I_C\underline{/90°}} = \frac{V\underline{/0°}}{\omega CV\underline{/90°}} = \frac{1}{\omega C}\underline{/-90°} = -j\frac{1}{\omega C} \qquad (5\text{-}20)$$

We now see that opposition to the flow of sinusoidal current can also be represented by an impedance. Just as each passive circuit element has associated with it an impedance to exponentials $Z(s)$ and an impedance to direct currents $Z(0)$, it also has an impedance to sinusoids $Z(j\omega) = \mathbf{Z}$.

For resistance, $\qquad Z_R(j\omega) = \mathbf{Z}_R = R = R\underline{/0°}$

For inductance, $\qquad Z_L(j\omega) = \mathbf{Z}_L = j\omega L = \omega L\underline{/90°} \qquad (5\text{-}21)$

For capacitance, $\quad Z_C(j\omega) = \mathbf{Z}_C = \dfrac{1}{j\omega C} = j\left(-\dfrac{1}{\omega C}\right) = \left(\dfrac{1}{\omega C}\right)\underline{/-90°}$

The quantity R is called the *ac resistance*[†] and is measured in ohms. The corresponding term for an inductance is called the *inductive reactance* $X_L = \omega L$ in ohms. For a capacitance, the corresponding term is called *capacitive reactance* $X_C = -1/\omega C$ in ohms. (The significance of the minus sign is indicated in Fig. 5.8 by the 180° phase difference between $\mathbf{V}_C$ and $\mathbf{V}_L$.)

Circuit Laws in Terms of Phasors

The element equations define the voltage-current characteristics of the passive elements shown in Fig. 5.7. The sign convention indicates that for the assumed reference direction of positive current, the corresponding reference polarity for voltage is as shown. Such a convention is necessary because the actual current and voltage are both changing sign periodically, but not simultaneously.

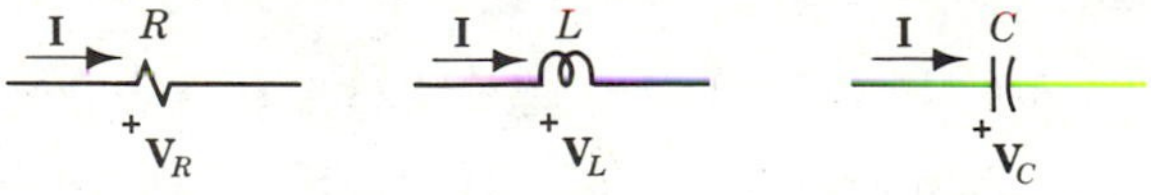

$$\mathbf{V}_R = R\mathbf{I} \qquad\qquad \mathbf{V}_L = j\omega L\mathbf{I} \qquad\qquad \mathbf{V}_C = \frac{1}{j\omega C}\mathbf{I}$$

$$\mathbf{I}_R = \frac{1}{R}\mathbf{V}_R \qquad\quad \mathbf{I}_L = \frac{1}{j\omega L}\mathbf{V}_L \qquad\quad \mathbf{I}_C = j\omega C\,\mathbf{V}_C$$

Figure 5.7 Element equations in terms of phasors.

The voltage-current relations are illustrated in Fig. 5.8 where the same current $i(t) = \sqrt{2}I \cos (\omega t + \theta)$ flows through three elements in series. For the resistance,

$$\mathbf{V}_R = R\mathbf{I} = R\underline{/0°} \cdot I\underline{/\theta} = RI\underline{/\theta}$$

[†]The ac resistance of a physical resistor may be different from the dc resistance.

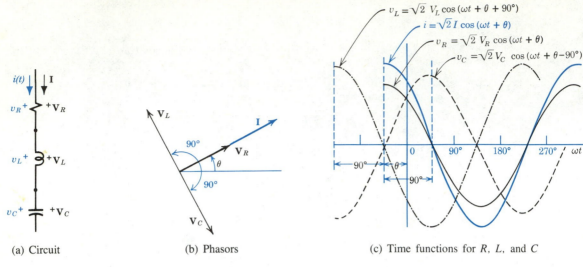

(a) Circuit   (b) Phasors   (c) Time functions for R, L, and C

Figure 5.8 Voltage-current relations in ac circuits.

Therefore, the voltage and current phasors for R are drawn at the same angle. The voltage and current are shown as functions of time in Fig. 5.8c; the resistance voltage and current waves reach maximum and zero values simultaneously. We say:

The voltage across a resistance is in phase with the current through it.

For an inductance,

$$\mathbf{V}_L = j\omega L \mathbf{I} = \omega L \underline{/90^\circ} \cdot I \underline{/\theta} = \omega L I \underline{/\theta + 90^\circ}$$

Therefore, the voltage phasor for L is 90° *ahead* of the current phasor. When the inductance voltage is shown as a function of time (Fig. 5.8c), it is clear that the voltage wave reaches its maximum and zero values at an *earlier* time than the current wave. We say:

The voltage across an inductive reactance leads the current through it by 90°.

For a capacitance,

$$\mathbf{V}_C = (1/j\omega C)\,\mathbf{I} = (1/\omega C)\underline{/-90^\circ} \cdot I \underline{/\theta} = (1/\omega C)I \underline{/\theta - 90^\circ}$$

Therefore, the voltage phasor is 90° *behind* the current phasor. When the capacitance voltage is shown as a function of time (Fig. 5.8c), it is clear that the voltage wave reaches its maximum and zero values at a *later* time than the current wave. We say:

The voltage across a capacitive reactance lags the current through it by 90°.

(These phrases emphasize that the *voltage across an element* is specified with respect to the *current through that element* and not with respect to any other current.)

The connection equations for sinusoidal forced response are based on Kirchhoff's laws. Assuming that all sources in a linear circuit are sinusoids of the same frequency, all steady-state voltages and currents are sinusoids of that same frequency. Kirchhoff's voltage law $\Sigma v = 0 = v_1 + v_2 + \cdots + v_k$ becomes, in terms of phasors,

$$\Sigma \mathbf{V} = 0 = \mathbf{V}_1 + \mathbf{V}_2 + \cdots + \mathbf{V}_k \tag{5-22}$$

EXAMPLE 4

Voltage $v = 120\sqrt{2}\cos(1000t + 90°)$ V is applied to the circuit of Fig. 5.9a where $R = 15\,\Omega$, $C = 83.3\ \mu$F, and $L = 30$ mH. Find $i(t)$.

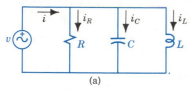

(a)

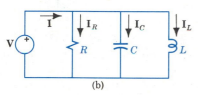

(b)

Figure 5.9   Response to a sinusoid.

The first step is to transform $v(t)$ into phasor $\mathbf{V} = 120\underline{/90°}$ V. Then

$$\mathbf{I}_R = \frac{1}{R}\mathbf{V} = \frac{1}{15}120\underline{/90°} = 8\underline{/90°} = 0 + j8\ \text{A}$$

$$\mathbf{I}_C = j\omega C\mathbf{V} = (0.0833\underline{/90°})(120\underline{/90°})$$
$$= 10\underline{/180°} = -10 + j0\ \text{A}$$

$$\mathbf{I}_L = \frac{1}{j\omega L}\mathbf{V} = \frac{120\underline{/90°}}{30\underline{/90°}} = 4\underline{/0°} = 4 + j0\ \text{A}$$

By Kirchhoff's current law, $\Sigma\mathbf{I} = 0$ or

$$\mathbf{I} = \mathbf{I}_R + \mathbf{I}_C + \mathbf{I}_L = (0 - 10 + 4) + j(8 + 0 + 0)$$
$$= -6 + j8 = 10\underline{/127°}\ \text{A}$$

The final step is to transform phasor $\mathbf{I}$ into

$$i(t) = 10\sqrt{2}\cos(1000t + 127°)\ \text{A}$$

In other words, for steady-state sinusoidal functions, Kirchhoff's voltage law holds for voltage phasors and $\Sigma\mathbf{V} = 0$ around any closed loop. In Fig. 5.8b total phasor voltage $\mathbf{V}$ could be found as the phasor sum $\mathbf{V}_R + \mathbf{V}_L + \mathbf{V}_C$ (see Exercise 15). Following a similar line of reasoning, Kirchhoff's current law holds for current phasors and $\Sigma\mathbf{I} = 0$ into any node (see Example 4).

AC Impedance

In general, phasor voltage and phasor current are related by a *complex impedance* $\mathbf{Z} = Z(j\omega)$ or

$$\mathbf{V} = \mathbf{Z}\mathbf{I} \tag{5-23}$$

which is sometimes referred to as the Ohm's law of sinusoidal circuits. For sinusoidal signals, any two-terminal network is completely defined by the impedance $\mathbf{Z} = \mathbf{V}/\mathbf{I}$ at the terminals (Fig. 5.10). In general, $\mathbf{Z}$ is complex and can be written as

$$\mathbf{Z} = \frac{\mathbf{V}}{\mathbf{I}} = Z\underline{/\phi_Z} = R + jX \tag{5-24}$$

where

$$Z = \sqrt{R^2 + X^2} \quad \text{and} \quad \phi_Z = \arctan\frac{X}{R}$$

The real part of $\mathbf{Z}$ is called *resistance* in ohms and is designated R; in the series *RLC* circuit of Fig. 5.10c, the real part of $\mathbf{Z}$ is the resistance of a particular element, but in general this is *not* true. The imaginary part of $\mathbf{Z}$ is called *reactance* in ohms and is designated X; for the series *RLC* circuit,

$$X = X_L + X_C = \omega L - 1/\omega C$$

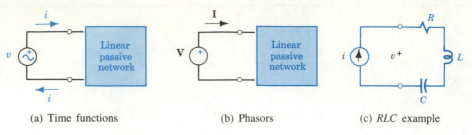

(a) Time functions   (b) Phasors   (c) *RLC* example

Figure 5.10 Representation of a two-terminal network.

Impedance $\mathbf{Z}$ is measured in ohms when phasors $\mathbf{V}$ and $\mathbf{I}$ are in volts and amperes, respectively. Note that $\mathbf{Z}$ is a complex quantity, like a phasor, but $\mathbf{Z}$ is *not* a phasor. The term phasor is reserved for quantities representing sinusoidal varying functions of time. The chief virtue of complex impedance is that although it expresses the relation between two time-varying quantities, $\mathbf{Z}$ itself is *not* a function of time.

Impedance $\mathbf{Z}$ is defined for sinusoidal waveforms. With this restriction, impedances can be combined in series and parallel just as resistances are. When we take advantage of this concept, differential equations are replaced by algebraic equations; steady-state ac circuit problems become only slightly more difficult to solve than corresponding dc circuit problems.

EXAMPLE 5

A voltage $v = 12\sqrt{2}\cos 5000t$ V is applied to the circuit of Fig. 5.11. Find the individual and combined impedances and the current $i(t)$.

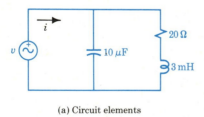

(a) Circuit elements

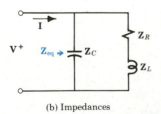

(b) Impedances

Figure 5.11 Calculation of current in an ac circuit.

$\mathbf{Z}_R = R = 20 = 20 + j0 = 20\underline{/0°}\ \Omega$

$\mathbf{Z}_L = j\omega L = j5000 \times 0.003 = j15\ \Omega = 15\underline{/90°}\ \Omega$

$\mathbf{Z}_C = -j\dfrac{1}{\omega C} = \dfrac{-j}{5000 \times 10^{-5}} = -j20 = 20\underline{/-90°}\ \Omega$

$\mathbf{Z}_{RL} = \mathbf{Z}_R + \mathbf{Z}_L = 20 + j15 = 25\underline{/37°}\ \Omega$

$\mathbf{Z}_{eq} = \dfrac{\mathbf{Z}_{RL} \cdot \mathbf{Z}_C}{\mathbf{Z}_{RL} + \mathbf{Z}_C} = \dfrac{25\underline{/37°} \cdot 20\underline{/-90°}}{20 + j15 - j20} = \dfrac{500\underline{/-53°}}{20 - j5}$

$= \dfrac{500\underline{/-53°}}{20.6\underline{/-14°}} = 24.3\underline{/-39°} = 18.9 - j15.3\ \Omega$

(Note that R_{eq} is *not* equal to R.)

$\mathbf{I} = \dfrac{\mathbf{V}}{\mathbf{Z}} = \dfrac{12\underline{/0°}}{24.3\underline{/-39°}} = 0.49\underline{/39°}$ A

$i(t) = 0.49\sqrt{2}\cos(5000t + 39°)$ A

Admittance

The steady-state sinusoidal response of any two-terminal network is completely defined by the impedance $\mathbf{Z} = \mathbf{V}/\mathbf{I}$ at the terminals. Another way of characterizing a network is in terms of the admittance $\mathbf{Y}$, defined as the ratio of phasor current to voltage or

$$\mathbf{Y} = \frac{\mathbf{I}}{\mathbf{V}} = \frac{1}{\mathbf{Z}}$$

where $\mathbf{Y}$ is the complex admittance in siemens. Specifically, for R, L, and C elements,

$$\mathbf{Y}_R = \frac{1}{R\,\underline{/0°}} = G\,\underline{/0°} \qquad \mathbf{Y}_L = \frac{1}{j\omega L} = \frac{1}{\omega L}\,\underline{/-90°} \qquad \mathbf{Y}_C = j\omega C = \omega C\,\underline{/90°}$$

In general, $\mathbf{Y}$ is complex and can be written as

$$\mathbf{Y} = Y\,\underline{/\phi_Y} = G + jB \qquad\qquad (5\text{-}25)$$

where

$$Y = \sqrt{G^2 + B^2} \qquad \text{and} \qquad \phi_Y = \arctan\frac{B}{G}$$

The real part of $\mathbf{Y}$ is called *conductance* in siemens and is designated G; in general, G is *not* the conductance of a particular element. The imaginary part of $\mathbf{Y}$ is called *susceptance* in siemens and is designated B. A positive B is associated with a predominantly capacitive circuit.

EXAMPLE 6

In Example 5 (Fig. 5.11), voltage $\mathbf{V} = 12\,\underline{/0°}$ V is applied to the parallel combination of $\mathbf{Z}_C = 20\,\underline{/-90°}\,\Omega$ and $\mathbf{Z}_{RL} = 25\,\underline{/37°}\,\Omega$. Use admittances to calculate branch and total currents and show them on a phasor diagram.

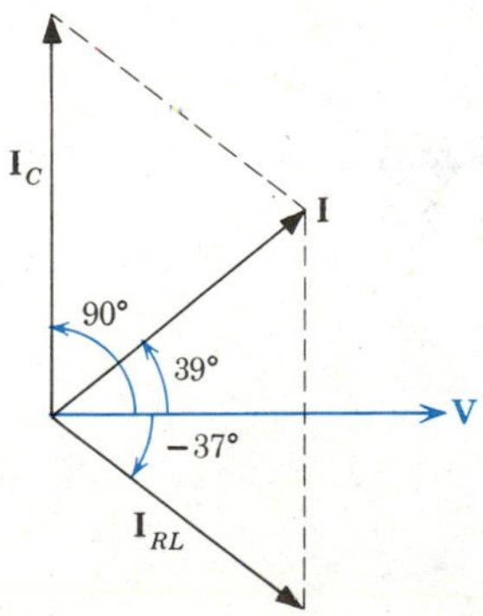

Figure 5.12 Phasor diagram of currents calculated from admittances.

The admittances are

$$\mathbf{Y}_C = \frac{1}{\mathbf{Z}_C} = \frac{1}{20\,\underline{/-90°}} = 0.05\,\underline{/+90°} = 0 + j0.05 \text{ S}$$

$$\mathbf{Y}_{RL} = \frac{1}{\mathbf{Z}_{RL}} = \frac{1}{25\,\underline{/37°}} = 0.04\,\underline{/-37°} = 0.032 - j0.024 \text{ S}$$

$$\mathbf{Y}_{eq} = \mathbf{Y}_C + \mathbf{Y}_{RL} = 0 + j0.05 + 0.032 - j0.024$$
$$= 0.032 + j0.026 = 0.041\,\underline{/39°} \text{ S}$$

The currents are

$$\mathbf{I}_C = \mathbf{Y}_C\mathbf{V} = 0.05\,\underline{/+90°} \times 12\,\underline{/0°} = 0.6\,\underline{/+90°} \text{ A}$$

$$\mathbf{I}_{RL} = \mathbf{Y}_{RL}\mathbf{V} = 0.04\,\underline{/-37°} \times 12\,\underline{/0°} = 0.48\,\underline{/-37°} \text{ A}$$

$$\mathbf{I} = \mathbf{I}_C + \mathbf{I}_{YL} = 0 + j0.6 + 0.384 -- j0.288$$

$$= 0.384 + j0.312 = 0.49\,\underline{/39°} \text{ A}$$

Alternatively, the total current is

$$\mathbf{I} = \mathbf{Y}_{eq}\mathbf{V} = 0.041\,\underline{/39°} \times 12\,\underline{/0°} = 0.49\,\underline{/39°} \text{ A}$$

The phasor diagram in Fig. 5.12 provides a visual check on the magnitudes and phases of the currents.

Since **Z** and **Y** are complex quantities, they obey the rules of complex algebra and can be represented by plane vectors. In contrast to phasors, **Z** and **Y** can lie in only the first and fourth quadrants of the complex plane because R and G are always positive for passive networks.

General Circuit Analysis

In talking about circuit analysis in general, we need a term that includes both impedance and admittance. For this purpose we shall use the word *immittance*, derived from *im*pedance and ad*mittance*.

We are now ready to formulate a general procedure applicable to a variety of problems in steady-state sinusoidal circuits. While the exact sequence depends on the particular problem (see Example 7), the following steps are usually necessary:

1. Transform time functions to phasors and convert element values to complex immittances.
2. Combine immittances in series or parallel to simplify the circuit.
3. Determine the desired response in phasor form, using connection equations.
4. Draw a phasor diagram to check computations and to display results.
5. Transform phasor results to time functions if required.

ANALOGS AND DUALS

One justification for this intensive study of electrical circuit analysis is that the techniques developed are equally applicable to analogous situations in other fields. By definition:

Analogous systems are described by the same integro-differential equation or set of equations.

The solution of the equation describing one physical system automatically yields the solution for any analogous system. The solution of an electrical circuit provides results that are directly applicable to analogous mechanical, hydraulic, or thermal circuits.

Analogs

As an illustration, consider the mass M sliding at velocity u with a friction coefficient D on a surface as shown in Fig. 5.13a (same as Fig. 4.3). By d'Alembert's principle, the governing equation is

$$Du + M\frac{du}{dt} = f \tag{5-26}$$

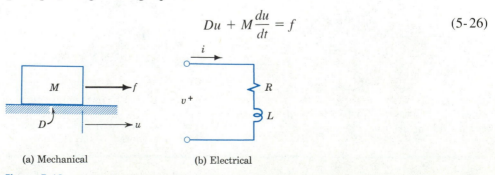

(a) Mechanical   (b) Electrical

Figure 5.13   Examples of analogous circuits.

EXAMPLE 7

The circuit of Fig. 5.14a represents a "load" consisting of C, R, and L supplied by a generator over a transmission line approximated by L_T. For a desired load voltage $v = 28.3 \cos(5000t + 45°)$ V, determine the current $i(t)$ and the branch currents.

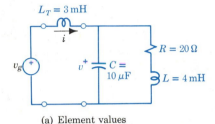

$L_T = 3\,\text{mH}$

(a) Element values

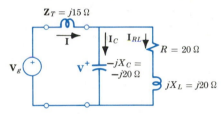

$\mathbf{Z}_T = j15\,\Omega$

(b) Complex impedances

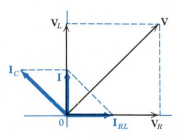

(c) Current phasors

Figure 5.14 General circuit analysis.

In phasor form, $v(t)$ becomes $\mathbf{V} = 20\underline{/45°}$ V.

$$\mathbf{Z}_T = j\omega L_T = j5000 \times 3 \times 10^{-3} = j15\,\Omega$$

$$\mathbf{Y}_C = j\omega C = j5000 \times 10^{-5} = j0.05\ \text{S}$$

$$\mathbf{Z}_C = 1/\mathbf{Y}_C = -j20\,\Omega$$

$$\mathbf{Z}_L = j\omega L = j5000 \times 4 \times 10^{-3} = j20\,\Omega$$

The diagram of Fig. 5.14b displays these values.

Since $\mathbf{I} = \mathbf{YV}$, we need the admittance of the parallel circuit constituting the load.

$$\mathbf{Y}_{RL} = \frac{1}{\mathbf{Z}_{RL}} = \frac{1}{20 + j20} = \frac{1}{20\sqrt{2}\underline{/45°}}$$

$$= 0.025\sqrt{2}\underline{/-45°} = 0.025 - j0.025\ \text{S}$$

$$\mathbf{Y} = \mathbf{Y}_C + \mathbf{Y}_{RL} = j0.05 + (0.025 - j0.025)$$

$$= 0.025 + j0.025 = 0.025\sqrt{2}\underline{/+45°}\ \text{S}$$

$$\mathbf{I} = \mathbf{YV} = (0.025\sqrt{2}\underline{/45°})(20\underline{/45°}) = 0.7\underline{/90°}\ \text{A}$$

$$i(t) = 0.7\sqrt{2}\cos(5000t + 90°)\ \text{A}$$

The phasor diagram should be drawn as the calculations proceed, to provide a continuous check as well as a final display of results. In addition to the given V and the calculated I, we know

$$\mathbf{I}_C = \mathbf{Y}_C\mathbf{V} = (0.05\underline{/90°})(20\underline{/45°}) = 1\underline{/135°}\ \text{A}$$

$$\mathbf{I}_{RL} = \mathbf{Y}_{RL}\mathbf{V} = (0.035\underline{/-45°})(20\underline{/45°}) = 0.7\underline{/0°}\ \text{A}$$

$$\mathbf{V}_R = \mathbf{R}\mathbf{I}_{RL} = (20\underline{/0°})(0.7\underline{/0°}) = 14\underline{/0°}\ \text{V}$$

$$\mathbf{V}_L = \mathbf{Z}_L\mathbf{I}_{RL} = (20\underline{/90°})(0.7\underline{/0°}) = 14\underline{/90°}\ \text{V}$$

These are shown on the phasor diagram of Fig. 5.14c. The construction lines verify the connection equations, $\mathbf{I} = \mathbf{I}_C + \mathbf{I}_{RL}$ and $\mathbf{V} = \mathbf{V}_R + \mathbf{V}_L$. Other checks are also available: $\mathbf{I}_C$ leads $\mathbf{V}$ by 90°, $\mathbf{V}_R$ is in phase with $\mathbf{I}_{RL}$, and $\mathbf{V}_L$ leads $\mathbf{I}_{RL}$ by 90°.

This looks familiar; in form it is similar to the equation resulting from the application of Kirchhoff's voltage law to the circuit of Fig. 4.1 (repeated here as Fig. 5.13b)

$$Ri + L\frac{di}{dt} = v \tag{5-27}$$

Mathematically, these two integro-differential equations are the same and, therefore, the circuits are analogous; the solutions of the two equations are identical if one set of symbols is substituted for the other.

For exponential functions we expect a *motional impedance* $Z(s) = D + sM$. The natural behavior is defined by $Z(s) = 0$ or $s = -D/M$ and velocity $u = U_0 e^{-(D/M)t}$ decays exponentially with a time constant M/D. A steady pull $f = F$ on the mass corresponds to a "direct" force; then $Z(0) = D$, and steady velocity $U = F/D$ is the forced response to a direct force.

For a sinusoidal pull $f = \sqrt{2}F \cos(\omega t + \theta)$, a phasor approach is indicated where $\mathbf{F} = F\underline{/\theta}$. The motional impedance is

$$Z(j\omega) = \mathbf{Z} = D + j\omega M = Z\underline{/\phi} \tag{5-28}$$

where

$$Z = \sqrt{D^2 + (\omega M)^2} \qquad \text{and} \qquad \phi = \arctan \frac{\omega M}{D} \tag{5-29}$$

Then

$$\mathbf{U} = \frac{\mathbf{F}}{\mathbf{Z}} = \frac{F\underline{/\theta}}{Z\underline{/\phi}} = U\underline{/\theta - \phi} \tag{5-30}$$

and

$$u(t) = \sqrt{2}U \cos(\omega t + \theta - \phi) \tag{5-31}$$

In this case we have solved the mechanical circuit by using the techniques developed for electrical circuits. Recognizing that the two circuits are analogous, we could have gone directly to the solution by making appropriate substitutions in Eq. 5-24. The analogous terms are identified by a comparison of Eqs. 5-26 and 5-27; the results are tabulated in the first two rows of Table 5-1.

Table 5-1   Analogous Quantities

Mechanical	M	D	K	f	u	x
Electrical	L	R	C	v	i	q
Rotational	J	D_r	K_r	τ	ω	θ
Thermal		R_T	C_p	τ	q_T	w_T
Electrical II	C	G	L	i	v	$\lambda^\dagger$

†The term λ (lambda) is called flux linkage and is equal to $\int v\, dt$. (See Eq. 21-1.)

The mechanical analog for capacitance C can be obtained by comparing the governing equations for the circuits in Fig. 5.15a and b. Here friction coefficient D

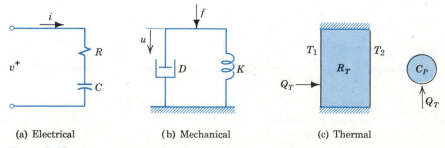

(a) Electrical   (b) Mechanical   (c) Thermal

Figure 5.15   Other examples of analogous circuits.

(analogous to R) is introduced by the *dash pot* and compliance K is introduced by the spring. From the equations

$$v = Ri + \frac{1}{C} \int i\, dt \quad \text{and} \quad f = Du + \frac{1}{K} \int u\, dt \qquad (5\text{-}32)$$

it is seen that compliance K is the analog of capacitance C. Recalling that displacement $x = \int u\, dt$ and charge $q = \int i\, dt$, we see that x and q are analogous.

Two important equations in heat transfer by conduction are

$$q_T = \frac{\tau}{R_T} \quad \text{and} \quad q_T = C_p \frac{d\tau}{dt} = \frac{dw_T}{dt} \qquad (5\text{-}33)$$

where q_T = rate of heat flow in joules per second (Fig. 5.15c),
  R_T = thermal resistance in seconds · kelvins per joule,
  τ = temperature difference = $T_1 - T_2$ in kelvins,
  C_p = thermal capacitance in joules per kelvin,
  w_T = thermal energy in joules.

The first equation expresses the fact that the rate of heat transfer by conduction is directly proportional to the temperature difference. The second provides for the heat absorbed by a body under conditions of changing temperature. Comparison of these equations with those describing electric circuit components leads to the thermal analogs shown in Table 5-1. Note that in this analog there is no term analogous to mass or inductance; heat flow does not exhibit any momentum or inertia effect.

Equations 5-33 have little practical value because they imply lumped thermal effects, whereas heat conduction is always a distributed phenomenon. A further complication is that R_T and C_p vary widely with temperature. In spite of these limitations, writing the equations and identifying the analogous terms are valuable because of the insights gained when the techniques of circuit theory are applied (see Problem 8).

Electrical analogs have been used to study heat flow in power transistors, the production and absorption of neutrons in a nuclear reactor, and the behavior of diverse acoustical and hydraulic systems. In recognition of the ease and effectiveness of the electronic analog computer, the principles and operation of this versatile engineering tool are described in Chapter 16.

Duals

For the series combination of R and L (Fig. 5.16a on p. 138),

$$\mathbf{Z} = R + j\omega L = Z\underline{/\phi} = \sqrt{R^2 + (\omega L)^2}\ \underline{/\arctan \omega L/R} \qquad (5\text{-}34)$$

For a current $\mathbf{I} = I\underline{/\theta}$,

$$\mathbf{V} = \mathbf{ZI} = Z\underline{/\phi} \cdot I\underline{/\theta} = V\underline{/\theta + \phi} \qquad (5\text{-}35)$$

The total voltage across R and L in series leads the current by phase angle ϕ, something less than 90°.

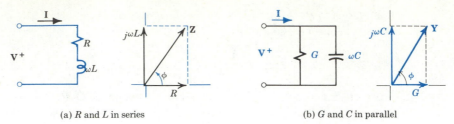

(a) R and L in series   (b) G and C in parallel

Figure 5.16   Series and parallel circuits and their immittance diagrams.

For the parallel combination of G and C (Fig. 5.16b),

$$\mathbf{Y} = G + j\omega C = Y\underline{/\phi} = \sqrt{G^2 + (\omega C)^2}\;\underline{/\arctan \omega C/G} \tag{5-36}$$

For a voltage $\mathbf{V} = V\underline{/\theta}$,

$$\mathbf{I} = \mathbf{Y}\mathbf{V} = Y\underline{/\phi} \cdot V\underline{/\theta} = I\underline{/\theta + \phi} \tag{5-37}$$

The total current through G and C in parallel leads the voltage by phase angle ϕ, something less than 90°.

At this time an alert student should be aware of a certain amount of repetition in the discussion of electrical circuits. The equations and diagrams for the series RL circuit of Fig. 5.16a are similar to those for the parallel GC circuit of Fig. 5.16b. With some changes in symbols, the equations and diagrams for a parallel GCL circuit (Fig. 5.9) would be just like those for the series RLC circuit of Fig. 5.8. Apparently, when we solve one electrical circuit, we automatically obtain the solution to another. How can we take advantage of this interesting fact?

In drawing up the table of mechanical-electrical analogs, we used the series RLC circuit (Fig. 5.8a). If, instead, we write the integro-differential equation for a parallel GCL circuit similar to Fig. 5.9, we get the different set of analogs shown in the fifth row of Table 5-1. It is always possible to draw two electrical circuits analogous to a given physical system, and the corresponding terms in the two circuits are related in accordance with the principle of *duality*.

> *When the set of transforms that converts one system into another also converts the second into the first, the systems are said to be duals.*

The series RLC circuit and the parallel GCL circuit are said to be *duals* because, by using the set of transforms indicated in Table 5-2, either circuit can be converted into the other. In general, the loop equations of a planar network have the same form as the node equations of its dual.[†] Some of the dual relations that exist in electrical networks are listed in Table 5-3.

A knowledge of duality enables us to double the benefit from any circuit analysis we perform. Sometimes it is advantageous to construct the dual of a given circuit instead of working with the given circuit itself. The first step (see Fig. 5.17) is to place a node in each loop of the given circuit and one more node outside; these are the nodes

---

[†]In the language of network topology, a *planar* network is one which can be drawn on a sphere; every planar network has a dual.

Table 5-2
Dual Quantities

L	R	C	v	i	Z	X
C	G	L	i	v	Y	B

Table 5-3 Dual Relations

Loop current	Node voltage
Kirchhoff's voltage law	Kirchhoff's current law
Series connection	Parallel connection
Current source	Voltage source
Short circuit	Open circuit

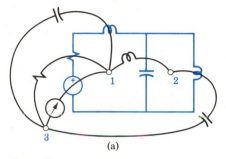
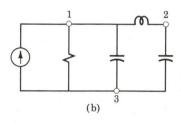

Figure 5.17 Construction of a dual circuit.

of the dual circuit. Then, through each element of the given circuit, draw a line terminating on the new nodes. Finally, on each line place the dual of the element through which the line is drawn; these lines are the branches of the dual circuit. The procedure is illustrated in Example 8.

EXAMPLE 8

The voltage divider of Fig. 5.18 is a useful device. Write an expression for V_2 in terms of V. Then draw the dual circuit and state (without derivation) the dual relation.

For the *voltage divider*,

$$\frac{V_2}{V} = \frac{Z_2}{Z_1 + Z_2} \tag{5-38}$$

The *voltage* across Z_2 is to the total *voltage* as the *impedance* Z_2 is to the sum of the *impedances*.

For the current divider, the dual relation is

$$\frac{I_2}{I} = \frac{Y_2}{Y_1 + Y_2} \tag{5-39}$$

The current through Y_2 is to the total current as the admittance Y_2 is to the sum of the admittances.

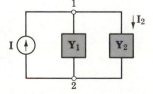

(a) Voltage divider

Figure 5.18 Deriving the dual of a circuit.

(b) Current divider

Generality of the Impedance Approach

The impedance approach is a good illustration of the power of a general method. Although the impedance concept was developed for use in determining natural response, it turns out to be applicable to forced response as well. Although it was defined in terms of exponential functions, with a proper interpretation impedance can also be used with direct and sinusoidal currents. Furthermore, when an electrical circuit has been solved, we automatically have the solution to its dual; through the analysis of an electrical circuit, we gain insight into the behavior of other analogous systems. Mastery of this important concept and the associated techniques is well worthwhile.

SUMMARY

- Impedance is defined for exponentials of the form $i = I_0\, e^{st}$.
  Impedances are combined in series and parallel, just as resistances are.
  In general, the forced response is governed by $v = Z(s)i$ where

$$Z_R(s) = R \qquad Z_L(s) = sL \qquad Z_C(s) = 1/sC$$

- For $s = 0$, $i = I_O = I$, a direct current. In this case, $Z_R(0) = R$.
  $Z_L(0) = 0$. $\therefore$ an inductance looks like a short circuit to a direct current.
  $Z_C(0) = \infty$. $\therefore$ a capacitance looks like an open circuit to a direct voltage.

- A sinusoidal function of time $a = A_m \cos(\omega t + \theta)$ can be interpreted as the real part of the complex function $A_m\, e^{j(\omega t + \theta)} = A\, e^{j\theta} \cdot \sqrt{2}\, e^{j\omega t}$.
  The complex constant $A\, e^{j\theta}$ is defined as phasor **A**, the transform of $a(t)$.
  Phasor calculations follow the rules of complex algebra.

- Phasor voltage and current are related by the complex impedance $\mathbf{Z} = Z(j\omega)$ or the complex admittance $\mathbf{Y} = Y(j\omega)$.

$$\mathbf{V} = \mathbf{ZI} \quad \text{where} \quad Z_R(j\omega) = R \quad Z_L(j\omega) = j\omega L \quad Z_C(j\omega) = 1/j\omega C$$

$$\mathbf{I} = \mathbf{YV} \quad \text{where} \quad Y_R(j\omega) = G \quad Y_L(j\omega) = 1/j\omega L \quad Y_C(j\omega) = j\omega C$$

For phasors, Kirchhoff's laws are written: $\Sigma \mathbf{V} = 0 \qquad$ and $\qquad \Sigma \mathbf{I} = 0$.

- The sinusoidal response of a two-terminal network is completely defined by

$$\mathbf{Z} = Z\underline{/\phi_Z} = R + jX \qquad \text{or} \qquad \mathbf{Y} = Y\underline{/\phi_Y} = G + jB$$

where $\phi_Z = \tan^{-1}(X/R)$ $\phi_Y = \tan^{-1}(B/G)$
  R = ac resistance in ohms G = ac conductance in siemens
  X = reactance in ohms B = susceptance in siemens
  $\quad = \omega L$ or $-1/\omega C$ $\quad = \omega C$ or $-1/\omega L$

- To determine the forced response to sinusoids:
  1. Transform time functions to phasors and evaluate complex immittances.
  2. Combine immittances in series or parallel to simplify the circuit.
  3. Determine the desired response in phasor form.
  4. Draw a phasor diagram to check values and display results.
  5. Transform phasors to time functions if required.

- Analogous systems are described by similar integro-differential equations. Corresponding terms in the equations are analog quantities. The solution of a problem is applicable to all analogous problems.

- Two networks are duals if the set of transforms that converts the first into the second also converts the second into the first. The loop equations of a planar network have the same form as the node equations of its dual.

REVIEW QUESTIONS

1. Define impedance.

2. Derive the dc element impedances for R, L, and C from the values of $Z(j\omega)$, letting ω approach zero as a limit.

3. In terms of steady-state response to a direct current or voltage, what is the effect of an inductance? Of a capacitance?

4. What is meant by a "long time" after closing a switch?

5. What two major advantages result from representing sinusoids by exponentials?

6. Write out in words a definition of a phasor.

7. How can a phasor, a constant quantity, represent a variable function of time?

8. Given $i_1 = 10 \cos (1000t + \pi/2)$ and $i_2 = 5 \cos 2000t$, is the phasor method applicable to finding $i_1 + i_2$? Explain.

9. Describe the quantity designated by the code symbols $\mathbf{I} = 20\underline{/\pi/4}$, $\omega = 60$.

10. Sketch three phasors on the complex plane and show the graphical construction to obtain $\mathbf{V}_1 + \mathbf{V}_2 - \mathbf{V}_3$.

11. Write an equation for $i(t)$ in terms of $\mathbf{I}$.

12. How can we neglect the imaginary part of the exponential function used in phasor representation?

13. What is the distinction between a phasor and a complex quantity?

14. Outline in words the reasoning that leads to the relation $\Sigma \mathbf{I} = 0$ into any node.

15. Define admittance, reactance, susceptance, and immittance.

16. A two-terminal circuit consists of a capacitance C in parallel with a series combination of resistance R and inductance L. In the admittance $\mathbf{Y} = G + jB$, is $G = 1/R$? What is the value of G?

17. Starting with element immittances on the complex plane, show the graphical determination of $\mathbf{Y}$ and $\mathbf{Z}$ for the circuit in the preceding question.

18. Sketch a sinusoidal voltage wave $v(t)$ and a current $i(t)$ that *leads* the voltage by 45°. Draw the corresponding phasors. If v and i are the inputs to a two-terminal circuit, is it predominantly capacitive or inductive?

19. Given $\mathbf{I} = 10\underline{/150°}$ and $\mathbf{V} = 200\underline{/-150°}$, is the associated $\mathbf{Z}$ inductive or capacitive? What is the phase angle ϕ?

20. Draw up a table summarizing the voltage-current phase angle relations for inductive and capacitive reactances and susceptances.

21. Define analog and dual.

22. Describe a hydraulic system and draw its electrical analog.

23. Draw a mechanical dual of the D-M combination in Fig. 5.14. It may help to write the dual equation first.

24. Draw a phasor diagram showing $\mathbf{U}$ and $\mathbf{F}$ of Eq. 5-30.

25. Why is practical heat conduction "always a distributed phenomenon"?

26. Without reference to the text, list 6 electrical quantities or phrases and their duals.

EXERCISES

1. A voltage $v = 12 e^{-200t}$ V appears across a parallel combination of $R = 3$ kΩ, $C = 50\mu$F, and $L = 2$ H. Determine the current in each element and the total current.

2. In Fig. 5.19, $L = 2$ H, $R = 3\Omega$, and $C = 0.2$ F. If $v_R = 6 e^{-2t}$ V, predict i_R, v_L, and i.

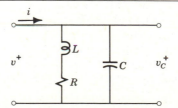

Figure 5.19

3. In Fig. 5.19, $L = 5$ H, $R = 20\,\Omega$, and $C = 0.2$ F.
   (a) Calculate sufficient points to plot $Y(s)$ for real values of s.
   (b) Determine response current i to voltage $v = 10\,e^{st}$ where $s = 0$, $s = -2$, $s = -4$, and $s = -5$.
   (c) What is the impedance of the circuit to direct current?

4. In Fig. 5.20, determine current I a long time after switch S is closed by:
   (a) Calculating $Z(s)$ and letting s approach 0 as a limit;
   (b) Modifying the circuit in accordance with the rules in color on pp. 123 and 124.

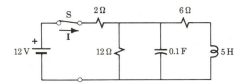

Figure 5.20

5. In Exercise 4, what is the voltage on the capacitance after a "long" time?

6. In the circuit of Fig. 5.21, I_s is a dc current.
   (a) If S has been open for a long time, what are i_R and i_C?
   (b) If S is closed at $t = 0$, find the initial values of i_L, i_C, di_L/dt, and di_R/dt.
   (c) A long time after S is closed, what are i_R, i_C, and i_L?

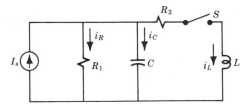

Figure 5.21

7. Given the circuit of Fig. 5.19 where $R = 3$ kΩ, $L = 1$ H, and $C = 0.25$ μF, determine the impedance presented to:
   (a) A direct current $i = I = 20$ mA.
   (b) An exponential current $i = 20\,e^{-4000t}$ mA.
   (c) A sinusoidal current
$$i = 20\sqrt{2}\cos(4000t + 60°) \text{ mA.}$$

8. (a) Sketch, approximately to scale, the following functions:
$$v_1 = 20\cos(628t + \pi/3) \text{ V}$$
$$v_2 = 30\sqrt{2}\sin(628t + 60°) \text{ V}$$
$$i = 7.07\cos(377t - \pi/4) \text{ A}$$
   (b) Represent the time functions by phasors in polar and rectangular form.

9. Find:
   (a) The fourth roots of (-1),
   (b) The cube roots of $(-1 - j\sqrt{3})$.

10. Given $\mathbf{A} = 3 + j4$, $\mathbf{B} = 2\,e^{j(\pi/6)}$, $\mathbf{C} = 4\underline{/-90°}$, sketch the three quantities on the complex plane and find:
   (a) $\mathbf{AB}$, (b) $\mathbf{B} - \mathbf{C}$, (c) $\mathbf{C/A}$, (d) $\mathbf{CA}^*$,
   (e) $\mathbf{B} + \mathbf{AC}$, (f) $(\mathbf{B} + \mathbf{C})/\mathbf{A}$, (g) $\mathbf{B}^2$,
   (h) $\sqrt{\mathbf{AC}}$, (i) $(\mathbf{C/B})^{1/3}$.

11. Write the time functions represented by:
   (a) $\mathbf{I} = 3 + j4$
   (b) $\mathbf{V} = 2\,e^{-j(\pi/6)}$
   (c) $\mathbf{I} = 4\underline{/-90°}$

12. In Fig. 5.22, $v_1 = 8\cos(\omega t - 30°)$ V and $v_2 = 10\cos(\omega t + 45°)$ V. Estimate $\mathbf{V}$ graphically from a phasor diagram drawn to scale and then calculate $\mathbf{V}$ from the component phasors. Determine $v(t)$.

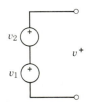

Figure 5.22

13. In Fig. 5.23, $i_1 = 4\cos(\omega t + 30°)$ A and $i_2 = 6\cos(\omega t + 60°)$ A.
   (a) Determine $\mathbf{I}$ graphically from a phasor diagram drawn to scale and check by an analytical phasor solution, and determine $i(t)$.
   (b) Expand the trigonometric expressions and determine $i(t)$ using trigonometric identities. Compare the difficulty of the two methods.

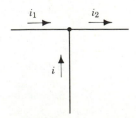

Figure 5.23

14. A current $i = 2\sqrt{2}\cos(2000t - \pi/4)$ A flows through a series combination of $R = 20\,\Omega$ and $L = 10$ mH.
(a) Determine the voltage across each element.
(b) Sketch the two voltages, $v = f(\omega t)$, approximately to scale and estimate the amplitude and phase angle of $v_R + v_L$.
(c) Perform the addition of voltages in phasor form and compare the sum to the estimate in part (b).

15. A current $i(t) = 2\sqrt{2}\cos(5000t + 30°)$ mA flows in a series combination of $R = 2309\,\Omega$ and $L = 0.8$ H. Determine total voltage $v(t)$:
(a) By finding the individual phasor voltages and then adding them before transforming **V** to $v(t)$.
(b) By finding the total impedance and using **V = ZI**.

16. A voltage $v(t) = 20\sqrt{2}\cos(1000t - 50°)$ V is applied across a parallel combination of $R = 2$ kΩ and $C = 0.5\ \mu$F. Determine total current $i(t)$:
(a) By finding the individual phasor currents and then adding them before transforming **I** to $i(t)$.
(b) By finding the total admittance and using **I = YV**.

17. In Fig. 5.24, $R = 2\,\Omega$, $L = 0.2$ H, and $i_R = 10\sqrt{2}\cos(10t + 45°)$ A.
(a) On a phasor diagram, show $\mathbf{I}_R$ and $\mathbf{V}_R$ and draw a dashed arrow to show the direction of $\mathbf{V}_C$.
(b) If the phasor $\mathbf{V} = V\underline{/0°}$, show **V** and $\mathbf{V}_C$ on the diagram.
(c) Show $\mathbf{V}_L$ and determine $v(t)$.

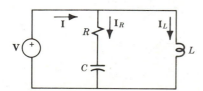

Figure 5.24

18. When a voltage $v = 20\sqrt{2}\cos 120t$ V is applied across an impedance, the current is $i = 4\sqrt{2}\cos(120t + 37°)$ A; find the real and imaginary components of the complex impedance **Z**.

19. Given the loop equations

$$\begin{cases} \mathbf{V}_1 = \mathbf{I}_1(R_1 + jX_C) + j\mathbf{I}_2X_C \\ 0 = j\mathbf{I}_1X_C + \mathbf{I}_2(R_2 + jX_L + jX_C) \end{cases}$$

draw the network and show the loop currents I_1 and I_2 with proper polarities.

20. Assuming the applied voltage v is represented by the phasor $V\underline{/0°}$, construct a labeled phasor diagram showing all voltages and currents in: (a) Fig. 5.25, (b) Fig. 5.21, (c) Fig. 5.19. (*Note:* Assume relative values of R, X_L, and X_C.)

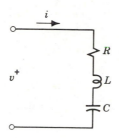

Figure 5.25

21. In Fig. 5.26, $R = 2$ kΩ and $C = 2.5\ \mu$F.
(a) Determine the two components of a *series* circuit having the same terminal admittance at $\omega = 1000$ rad/s.
(b) Repeat part (a) for $\omega = 500$ rad/s.

Figure 5.26

22. In Fig. 5.19, $L = 0.2$ H, $R = 3$ kΩ, and $C = 10$ nF.
(a) Determine the two components of a series circuit having the same terminal impedance at $\omega = 20$ krad/s.
(b) Repeat for $\omega = 10$ krad/s.

23. An inductor of 8-Ω reactance and 6-Ω resistance is connected in parallel with a capacitor of 8 Ω reactance across a 120-V, 60-Hz line. Calculate the total admittance, total current, and branch currents in phasor form. Show $\mathbf{I}_C$, $\mathbf{I}_L$, and **I** on a phasor diagram using $V\underline{/0°}$ as a reference.

24. In the circuit of Fig. 5.27, $v = 5\sqrt{2}\cos 5t$ V, $R = 2\,\Omega$, $L = 1$ H, and $C = 0.05$ F.
   (a) Find the input admittance $Y(j\omega)$.
   (b) Determine $\mathbf{I}_R$, $\mathbf{I}$, and $\mathbf{I}_L$ and show them with $\mathbf{V}$ and $\mathbf{V}_C$ on a phasor diagram.

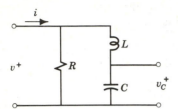

Figure 5.27

25. A voltage $v(t) = 10\sqrt{2}\cos(2t + 60°)$ V is applied to the circuit of Exercise 2. Predict $i(t)$.

26. A current $i_C = 6\sqrt{2}\cos 1000t$ A flows in the $250\text{-}\mu$F capacitance of Fig. 5.28.
   (a) Determine $\mathbf{V}_C$.
   (b) If total current $\mathbf{I} = 10\underline{/-37°}$ A, is reactance X inductive or capacitive? Explain your reasoning.
   (c) Show $\mathbf{V}_C$ and $\mathbf{I}$ on a phasor diagram. Add $\mathbf{I}_R$ and $\mathbf{I}_X$, leaving magnitudes undefined.
   (d) Specify R (in ohms) and the circuit element (L or C) represented by X.

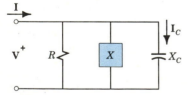

Figure 5.28

27. In the circuit of Fig. 5.27, $R = 5\,\Omega$, $C = 20\,\mu$F, and $L = 8.06$ mH. The steady-state voltage across L is $v_L = 145\sqrt{2}\cos(2000t + 90°)$ V.
   (a) Calculate phasors $\mathbf{I}_L$ and $\mathbf{V}_C$.
   (b) Predict phasors $\mathbf{I}_R$ and $\mathbf{I}$.
   (c) Show all sinusoidal voltages and currents on a clearly labeled phasor diagram.
   (d) Specify $i(t)$.

28. In Fig. 5.28, the total current is $\mathbf{I} = 10\underline{/0°}$ A and the current through $X_C = -20\,\Omega$ is $\mathbf{I}_C = 4\underline{/127°}$ A. Show these on a phasor diagram. Drawing the phasor diagram as you go along, determine R and reactance X. Is X inductive or capacitive?

29. In Fig. 5.29, $L_1 = L = 2$ mH, $C = 500\,\mu$F, and $R = 2\,\Omega$.
   (a) For $v = 10\sqrt{2}\cos(1000t + 90°)$ V, determine the input impedance.
   (b) Calculate currents $\mathbf{I}$, $\mathbf{I}_R$, and $\mathbf{I}_L$ and show them on a phasor diagram along with $\mathbf{V}$, $\mathbf{V}_{L_1}$, and $\mathbf{V}_C$.

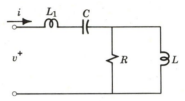

Figure 5.29

30. The load in Fig. 5.30 consists of $G = 0.08$ S in parallel with $B_C = 0.06$ S. It is fed by a current source $\mathbf{I} = 45\underline{/0°}$ A and a voltage source $\mathbf{V} = 190\underline{/0°}$ V. If $R = 10\,\Omega$ and $X_L = 4\,\Omega$, predict $\mathbf{I}_1$ and $\mathbf{I}_2$, using the node-voltage method.

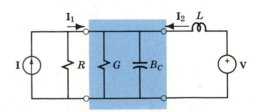

Figure 5.30

31. Write the governing equations for the electrical circuit, the analogous equations for a mechanical (translation) circuit, and draw the mechanical analog for: (a) Fig. 5.26, (b) Fig. 5.25, (c) Fig. 5.19, (d) Fig. 5.27.

32. Write the governing equations for the mechanical circuit, the analogous equations for an electrical circuit, and draw the electrical analog for (a) Fig. 5.31a, (b) Fig. 5.31b, (c) Fig. 5.31c. (*Note:* $u = dx/dt$ is measured from the same reference as x.)

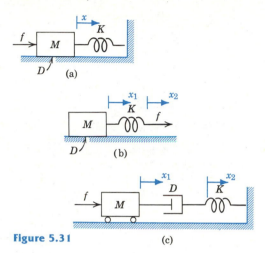

Figure 5.31

33. In Fig. 5.31a, $M = 2$ kg, $k = 0.1$ m/N, and $f = 20 \cos 5t$ N.
(a) If $D = 0$, predict $u(t)$ and $x(t)$.
(b) If $D = 6$ N $\cdot$ m/s, predict $u(t)$ and $x(t)$.

PROBLEMS

1. In the circuit of Fig. 5.33, the sources have been connected for a long time. What does this imply regarding i_C and v_L? Calculate the current in the 10-kΩ resistance, the charge on the capacitance, and the energy stored in the inductance.

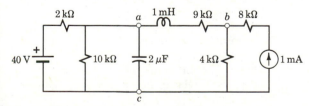

Figure 5.33

2. In Fig. 5.25, $R = 2\,\Omega$, $L = 3$ H, and $C = 0.5$ F. Determine the forced response $i(t)$ for $v(t) = 10t^2$ V where t is in seconds.

34. Write the dual of the following statement: "In a series *RLC* circuit with a constant voltage applied, the current is a maximum at the frequency for which the inductive reactance is equal to the capacitive reactance."

35. For the circuit of Fig. 5.32:
(a) Write the governing equations in terms of summations of currents (Σi) into nodes a and b.

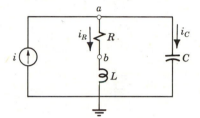

Figure 5.32

(b) Replace each quantity in the two equations by its dual.
(c) Construct the circuit for which the equations of part (b) are the governing equations.
(d) Following the procedure outlined on page 138, construct the dual of the circuit in part (c) and compare with the original.

36. Construct the dual of Fig. 5.20.

3. Write the first six terms of the series expansion for e^x. Evaluate each term for $x = j1$. On the complex plane, plot each term to scale, adding onto the preceding term. Measure the sum of the six terms and express it in polar form. Is the result what you expected?

4. When the rate of decrease of a quantity (dq/dt) is directly proportional to the quantity itself (q), the quantity varies as an exponential function of time.
(a) Solve the differential equation $dq/dt = -kq$ by separating the variables, integrating both sides, and solving for $q(t)$.
(b) If q is electrical charge, draw an electrical circuit for which this differential equation is a mathematical model.
(c) From your solution to part (a), derive an expression for $i(t)$ in terms of the initial charge Q_O and the circuit parameters.

5. In Fig. 5.34, $\mathbf{V}_1 = 30 + j10$ V, $\mathbf{V}_2 = 30 + j0$ V, $L = 1$ H, $\omega = 1000$ rad/s, $R_1 = R_2 = 1$ kΩ, and $C_1 = C_2 = 1$ μF.
   (a) Use loop currents to find $\mathbf{I}_1$ and $\mathbf{I}_2$ through $\mathbf{V}_1$ and $\mathbf{V}_2$.
   (b) Use node voltages to find $\mathbf{V}_b$.

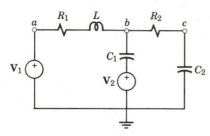

Figure 5.34

6. A 120-V capacitor motor used in a refrigerator can be modeled by the circuit shown in Fig. 5.35. The main winding is $R + jX = 1 + j8$ Ω; the auxiliary winding is $R_a + jX_a = 30 + j50$ Ω. The mechanical output is represented electrically by $R_o = 10$ Ω.
   (a) For good torque production, the current in the auxiliary winding should lead the current in the main winding by 90°. Design the capacitor; i.e., specify the appropriate value of C.
   (b) Assuming the power in R_a and R represents losses, estimate the efficiency of this motor.

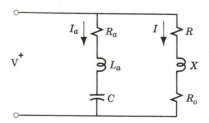

Figure 5.35

7. A "load" of $R = 1000$ Ω is to be supplied by an amplifier designed to provide maximum power to a load of 500 Ω. An experienced engineer suggests the "coupling circuit" of Fig. 5.36. Design the coupling circuit for proper operation at $f = 2000$ Hz.

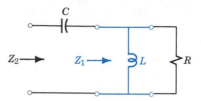

Figure 5.36

8. An unheated house has an effective overall thermal resistance $R_T = 0.001°$ F · hr/Btu. The effective overall thermal capacitance of the contents of the house is $C_p = 6000$ Btu/°F. The outside temperature varies from 100°F at 2 p.m. to 50°F at 2 a.m. Making and stating any necessary assumptions, predict the minimum temperature of the interior and the time at which it will occur.

9. The circuit of Fig. 5.34 has the parameters given in Problem 5. Carefully outlining your reasoning, determine the configuration and the parameters of the dual of Fig. 5.34.

10. At small amplitudes, the period of a pendulum is 0.5 s. After 10 oscillations the amplitude of the motion is roughly $\frac{1}{3}$ of the initial amplitude.
   (a) Write the governing equation of the system, assuming a restoring force proportional to displacement x and a damping force proportional to the velocity u.
   (b) Draw an electrical circuit analogous to the pendulum and identify the analogous quantities.
   (c) Compute the approximate locations of the poles and zeros and plot a pole-zero diagram.
   (d) Describe in words the significance of the poles in relation to the motion of the pendulum.

CHAPTER 6

Complete Response

General Procedure

First-Order Circuits

Second-Order Circuits

Impulse Response

The natural response of a circuit is due to energy stored in inductances or capacitances. Working from the differential equation, we developed a method of using the poles and zeros of the impedance (or admittance) function to indicate the character of the natural response. The forced response is produced by external energy sources such as batteries or generators. Expanding the impedance concept, we then used exponential representation to transform integro-differential equations into algebraic equations, and finding the forced response was reduced to a routine procedure.

In our previous work only one of these responses was considered at a time. In determining natural response (Fig. 6.1a), we assumed energy had been stored by an external source and then that source was removed. In determining forced response (Fig. 6.1b), we assumed that a sufficient time had elapsed so that all natural response components had died away, or at least had become negligibly small. In general, however, there is a transient period during which behavior is not so simple; to determine the current that flows after switch S in Fig. 6.1c is closed we must determine the *complete response*.

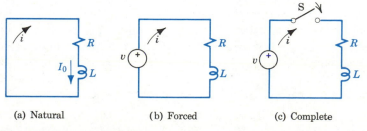

(a) Natural   (b) Forced   (c) Complete

Figure 6.1 Components of circuit response.

Components of the Complete Response

Considering the problem from an energy viewpoint, we note that the natural response tends to decay exponentially as the stored energy is dissipated. The forced response continues indefinitely because energy is supplied to make up any losses. Since we are interested in the transient period in which both effects are present, we conclude that the complete response is some combination of natural and forced responses.

To provide a firm basis for analysis we write the differential equation, using the element equations and the connection equation. Assuming a linear circuit and a sinusoidal forcing function at all times after closing the switch in Fig. 6.2,

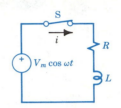

$$L\frac{di}{dt} + Ri = V_m \cos \omega t \qquad (6\text{-}1)$$

Figure 6.2 Complete response.

Any expression that satisfies this equation is a solution of the equation; one such solution is the forced response current

$$i_f = I_m \cos (\omega t + \phi) \qquad (6\text{-}2)$$

For the related homogeneous equation

$$L\frac{di}{dt} + Ri = 0 \qquad (6\text{-}3)$$

a solution is the natural response current

$$i_n = I_n\, e^{-(R/L)t} \qquad (6\text{-}4)$$

But if the value of i_n from Eq. 6-4 makes the left-hand side of Eq. 6-1 just equal to zero, then the governing equation is satisfied by $i_f + i_n$, or

$$
\begin{aligned}
i &= i_f + i_n \\
&= I_m \cos (\omega t + \phi) + I_n\, e^{-(R/L)t} \qquad (6\text{-}5)
\end{aligned}
$$

must also be a solution of Eq. 6-1. We conclude that the complete response is the *sum* of the forced and natural responses.[†]

Having reached this conclusion, we are in the fortunate position of being able to solve a new group of problems without developing any new theory. The techniques of finding forced and natural response can be applied directly to finding the complete response of a variety of circuits with a variety of driving functions or *excitations*.

In this group are some very practical and important engineering problems. In the design of some systems, the transient behavior is the critical factor; for example, whether the control system of a supersonic aircraft is satisfactory may depend upon its ability to respond smoothly to a sudden disturbance. In other systems, only the transient behavior has any significance; as an illustration, a radar system must generate, transmit, receive, and interpret sharp pulses of energy that never reach a steady state. In this chapter first we develop a general procedure, then we apply it to some

---

[†]Mathematicians call these the *particular* solution and the *complementary* solution, respectively. These two components can be added only if the circuit is linear; see Chapter 2, p. 46.

first-order circuits with dc and sinusoidal driving functions, and finally we consider the complete response of circuits with more than one energy-storage element.

GENERAL PROCEDURE

Our general procedure consists of finding the forced and natural response components and then combining them properly. What are the essential characteristics of each component and how are they determined?

Characteristics of the Components

Each component has a distinctive *form* to be discovered and an *amplitude* to be determined. For any excitation that can be described as an exponential, the *form* of the forced response is the same as the form of the forcing function; a direct voltage causes a direct current and a sinusoidal current produces a sinusoidal voltage. The *amplitude* of the forced response is determined by the magnitude of the forcing function and the impedance of the circuit; for direct currents the impedance of interest is $Z(0)$ and for sinusoids it is $Z(j\omega)$.[†]

The *form* of the natural response is governed by the circuit itself. Basically, the form is obtained from the homogeneous differential equation, but our approach is to use the poles and zeros of an immittance function to identify the natural components of voltage and current. The *amplitude* of the natural response is determined from energy considerations. From this viewpoint, the amplitude is just that required to provide for the difference between the actual initial energy storage (in L and C) and that indicated by the forcing function. This idea will become more clear when applied to specific examples.

Procedure

The complete response of any linear two-terminal circuit to exponentials is obtained as follows:

1. *Write the appropriate impedance or admittance function.* The function $Z(s)$ or $Y(s)$ carries all the information of the integro-differential equation based on element and connection equations. Choose the terminals across which the desired voltage appears (as an open-circuit voltage) or into which the desired current flows (as a short-circuit current).
2. *Determine the forced response from the forcing function and the proper immittance.* The form of the forcing function indicates the value of s in $Z(s)$ or $Y(s)$ and the approach used in determining the forced response. For direct currents, apply Ohm's law and $V = RI$. For sinusoids, use phasor methods and $\mathbf{V} = \mathbf{ZI}$.
3. *Identify the natural components from poles or zeros of $Z(s)$ or $Y(s)$.* Working with impedance, for example, use the poles to obtain possible components of natural response voltage and the zeros to obtain possible current components. Write these components with undetermined amplitudes.

[†]Exponential forcing functions are not common in practical problems and they are not discussed in detail here.

4. *Add the forced and natural responses and evaluate the undetermined constants.*
Obtain the necessary information from initial conditions. From energy considerations we know that inductance currents and capacitance voltages cannot
change instantaneously; the energy distribution just after a switch is closed, for
example, must be the same as that just before the switch was closed.

FIRST-ORDER CIRCUITS

The general procedure is outlined in electrical terms and the emphasis in the
following examples is primarily on electric circuits. It should be apparent, however,
that the method is applicable to nonelectrical problems as well. The step function that
is produced by closing an electrical switch is analogous to suddenly opening a valve,
applying a force, or igniting a flame; the complete responses of hydraulic, mechanical,
and thermal circuits are similar to those of electric circuits.

Step Response of an RL Circuit

Closing switch S in Fig. 6.3 at time $t = 0$ results in the application of a step
voltage of magnitude V to the series combination of resistance R and inductance L. To
obtain the complete response current i during the transient period we follow the
procedure outlined.

1. *Write the impedance function.*
For this series circuit, $Z(s) = R + sL$.
2. *Determine the forced response.*
For a direct voltage source, $s = 0$ and $Z(0) = R$.
Then $i_f = V/R$ is the forced response.
3. *Identify the natural components.*
Setting $Z(s) = 0$, $s = -R/L$.
Therefore, $i_n = A\,e^{-(R/L)t}$ is a possible natural response.
4. *Evaluate the undetermined constants.*
In general, $i = i_f + i_n = V/R + A\,e^{-(R/L)t}$. Just *before* closing the switch the
current in the inductance was zero; therefore, just *after* closing the switch the
current must still be zero. (We call this time $t = 0^+$.) At $t = 0^+$,

$$i = I_0 = 0 = \frac{V}{R} + A\,e^{-(R/L)(0)} = \frac{V}{R} + A\,e^0 = \frac{V}{R} + A$$

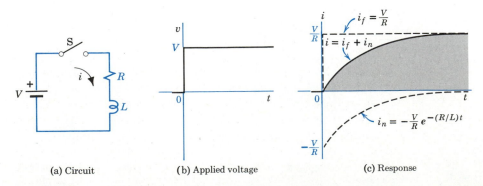

(a) Circuit   (b) Applied voltage   (c) Response

Figure 6.3   Response of an *RL* circuit with a step voltage applied.

Therefore, $A = 0 - V/R = -V/R$, and the complete response is

$$i = i_f + i_n = \frac{V}{R} - \frac{V}{R} e^{-(R/L)t} \qquad (6\text{-}6)$$

Forced, natural, and complete responses are plotted in Fig. 6.3c. The forced response, a current step, has the same form as the forcing function, a voltage step. The natural response is a decaying exponential; the sign is negative because the actual initial energy storage is less than that called for by the forcing function. The complete response current is a continuous function as are all inductance currents.

EXAMPLE 1

In Fig. 6.4a, switch S_1 is open and S_2 is closed. At $t = 0$, switch S_1 is closed and a short time later at $t = t'$ switch S_2 is opened. Determine and plot $i(t)$ for the transient period.

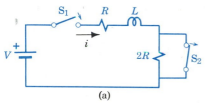

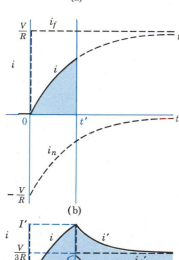

Figure 6.4 Complete response.

Since natural response is initiated at two different times, the problem will be solved in two parts following the four-step procedure.

For $0 < t < t'$, with S_2 closed,

1. The impedance function is $Z(s) = R + sL$.
2. The forced response is $i_f = V/R$.
3. The natural response is $i_n = A\,e^{-(R/L)t}$.
4. The complete response is $i = i_f + i_n$

$$= V/R + A\,e^{-(R/L)t}$$

At $t = 0^+$, $i = 0 = V/R + A$. $\therefore A = -V/R$ and

$$i = V/R - (V/R)\,e^{-(R/L)t} \qquad (6\text{-}7)$$

as shown in Fig. 6.4b.

For $t' < t < \infty$, with S_2 open,

1. The impedance function is $Z'(s) = R + 2R + sL$.
2. The forced response is $i_f' = V/3R$.
3. The natural response is $i_n' = A'\,e^{-(3R/L)(t-t')}$.
4. The complete response is

$$i' = i_f' + i_n' = V/3R + A'\,e^{-(3R/L)(t-t')} \qquad (6\text{-}8)$$

At $t = t'$ just *before* opening switch S_2, by Eq. 6-7

$$i = (V/R)(1 - e^{-(R/L)t'}) = I'$$

At $t = t'^+$ just *after* opening S_2, the current must be the same; by Eq. 6-8,

$$i' = I' = V/3R + A'\,e^0. \ \therefore A' = I' - V/3R$$

When this value is substituted in Eq. 6-8,

$$i' = V/3R + (I' - V/3R)\,e^{-(3R/L)(t-t')} \qquad (6\text{-}9)$$

This response is plotted in Fig. 6.4c, assuming $t' \cong L/R$, one time constant.

Note that the amplitude of the natural response in each case is determined by the difference between the actual current and the current called for by the forcing function.

In an *RL* circuit the initial slope di/dt can be obtained from the differential equation

$$L\frac{di}{dt} + Ri = V$$

by noting that at $t = 0^+$, $i = 0$, $Ri = 0$, and

$$\frac{di}{dt} = \frac{V - Ri}{L} = \frac{V}{L}$$

As the current increases, Ri increases and the slope decreases. After several time constants, Ri approaches V and di/dt approaches zero, or the current approaches the steady-state value V/R.

AC Switching Transients

A long time after an ac source is connected to a circuit, the current reaches the steady state with a predictable amplitude and phase relation. In a practical situation, the transient behavior just after the switch is closed cannot be predicted because the natural response depends on the particular point in the cycle at which the switch is closed. Since the "switching transient" may in some cases exceed the rating of the generator and associated transmission equipment, it is essential that the power engineer be able to predict the response under the worst possible conditions.

The method for analyzing switching transients is illustrated in Example 2. Note that the initial current is twice the peak current reached under steady-state conditions. Is this the worst possible condition? Since the initial current is proportional to the initial voltage ($i = v/R$), the largest current will flow if the switch is closed at a voltage maximum as in the Example. Is there a point during the cycle at which the switch could be closed with no natural response component? If so, can you find it? (See Exercise 10.)

Series-Parallel Circuits

The step response of the parallel circuit of Fig. 6.6 can be obtained by applying the principle of duality to the result for the circuit of Fig. 6.3. By inspection of Eq. 6-6 and reference to Tables 5-2 and 5-3, we write

$$v = IR - IR\,e^{-t/RC} \qquad\qquad (6\text{-}10)$$

and we have received another dividend from our investment of time and effort in a general principle.

The circuit of Fig. 6.7 (see p. 154) contains another element, and the differential equation approach would be more complicated than for a simple series or parallel circuit. Our impedance approach, however, handles such problems easily, as shown by Example 3. Note the importance of identifying the proper terminals for determining natural response. These are marked in Fig. 6.7a for investigating v.

EXAMPLE 2

A load consisting of a series combination of $R = 5\ \Omega$ and $C = 306\ \mu F$ (Fig. 6.5) is connected to a voltage $v = 6000 \cos 2\pi 60t$ V at time $t = 0$. Find current i as a function of time after closing the switch.

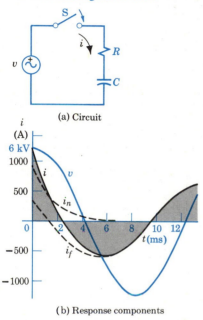

(a) Circuit

(b) Response components

Figure 6.5   An ac switching transient.

1. The impedance function is $Z(s) = R + 1/sC$.
2. The forced response is determined for $Z(j\omega)$.
   Here $\mathbf{Z} = R - j1/\omega C$
   $$= 5 - j1/377 \times 306 \times 10^{-6}$$
   $$= 5 - j8.66 = 10\underline{/-60°}\ \Omega$$
   In phasor notation, the forced response is

   $$\mathbf{I}_f = \mathbf{V}/\mathbf{Z} = \frac{6000}{\sqrt{2}}\underline{/0°}\Big/10\underline{/-60°} = \frac{600}{\sqrt{2}}\underline{/+60°}$$

   and
   $$i_f = 600 \cos (377t + 60°)\ \text{A}$$

3. The natural response is $i_n = I_0\ e^{-t/RC}$.
4. The complete response is $i = i_f + i_n$ as before. However, there is no requirement of continuity of current in a capacitance. Instead, we make use of the fact that voltage on a capacitance cannot suddenly change. Assuming that the capacitance is initially uncharged (the practical case), at $t = 0^+$ the voltage $v_C = 0$ and the full applied voltage appears across R. At $t = 0^+$,

   $$i = I_0 = V/R = 6000 \cos 0°/5 = 1200$$

   $$= i_f + i_n = 600 \cos 60° + I_0\ e^0$$

Therefore, $I_0 = 1200 - 300 = 900$ A
The complete response is

$$i = 600 \cos (377t + 60°) + 900\ e^{-654t}\ \text{A}$$

as plotted in Fig. 6.5b.

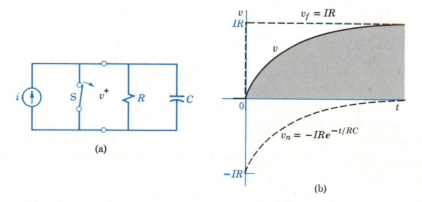

(a)

(b)

Figure 6.6   The step response of a parallel RC circuit.

EXAMPLE 3

The step current in Fig. 6.7 is obtained from a dc source of amplitude I initially shorted by switch S. At $t = 0$, the switch is opened. Predict the voltage v as a function of time.

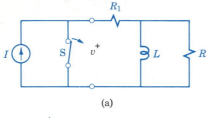

(a)

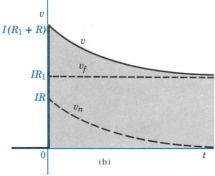

(b)

Figure 6.7 The step response of a series-parallel circuit.

1. The impedance function at the terminals across which v appears is

$$Z(s) = R_1 + \frac{R(sL)}{R + sL} = \frac{R_1 R + (R_1 + R)sL}{R + sL}$$

2. The forced response is

$$v_f = IZ(0) = I\frac{R_1 R}{R} = IR_1$$

3. The natural response voltage is indicated by the pole of the impedance function. $Z(s) = \infty$ for $R + sL = 0$; therefore, $s = -R/L$ and

$$v_n = A\,e^{-(R/L)t}$$

4. The complete response is

$$v = v_f + v_n = IR_1 + A\,e^{-(R/L)t}$$

If switch S has been closed a long time, $i_L = 0$. Just after opening the switch, $i_L = 0$ and all the current flows through R_1 and R in series. At $t = 0^+$,

$$v = I(R_1 + R) = v_f + v_n = IR_1 + A\,e^0$$

Solving, $A = IR$ and

$$v = IR_1 + IR\,e^{-(R/L)t}$$

as shown in Fig. 6.7b.

If i_L were the current of interest, that branch would be opened and the impedance measured "looking in" there with all energy sources "removed." Current sources are removed by open-circuiting and voltage sources by short-circuiting. For determining i_{Ln}, $Z(s) = sL + R$ since open-circuiting I (with switch S open) removes R_1 from consideration.

As an alternative approach, Thévenin's theorem could be used to replace all except L. The resulting equivalent circuit would be a simple series combination of V_T, R_T, and L. The forced and natural components could be identified by inspection or by comparison to Fig. 6.3.

SECOND-ORDER CIRCUITS

For a circuit with two different energy-storage elements, the governing differential equation is of the second order. The characteristic equation has two roots or, in other words, the impedance function has two zeros and two poles (one may be at infinity). Working from either interpretation, we conclude that the natural response contains two components and that there are two arbitrary coefficients to be determined. The general procedure for finding the complete response is applicable, although the determination of the constants from initial conditions may be difficult.

Determination of the constants is difficult because each problem is different and no step-by-step procedure can be set down in advance. Where there are two unknown constants, two equations must be written and these are usually for i and di/dt (or v and dv/dt). Useful clues are that i_L and v_C cannot suddenly change and that $v_L = L\, di/dt$ and $i_C = C\, dv/dt$. When these clues are imaginatively combined with Kirchhoff's laws to fit the given circuit, the route to the solution becomes clear.

Step Response of an RLC Series Circuit

As an illustration of the approach to solving second-order circuits, consider the case of a voltage step applied to a series combination of RLC (Fig. 6.8). In the typical case, C is initially uncharged and L carries no initial current. The step is obtained from a voltage source V suddenly connected by means of switch S at time $t = 0$. The problem is to find the complete response current i as a function of time. The impedance function is $Z(s) = R + sL + 1/sC$ (step 1) and the forced response is $i_f = V/Z(0) = V/\infty = 0$ (step 2). The natural components (step 3) are obtained by setting $Z(s) = 0$. For

$$s^2 L + sR + \frac{1}{C} = 0 \qquad (6\text{-}11)$$

the roots of the characteristic equation are

$$s = -\frac{R}{2L} \pm \sqrt{\frac{R^2}{4L^2} - \frac{1}{LC}} = -\alpha \pm j\omega \qquad (6\text{-}12)$$

using the nomenclature of Chapter 4. For roots real and distinct (Eq. 4-12),

$$i_n = A_1 e^{s_1 t} + A_2 e^{s_2 t} \qquad (6\text{-}13)$$

For roots real and equal (Eq. 4-18),

$$i_n = A_1 e^{st} + A_2 t\, e^{st} \qquad (6\text{-}14)$$

For roots complex (Eqs. 4-15, 4-16),

$$i_n = e^{-\alpha t}(B_1 \cos \omega t + B_2 \sin \omega t) \qquad (6\text{-}15)$$

or

$$i_n = A\, e^{-\alpha t} \sin (\omega t + \theta) \qquad (6\text{-}16)$$

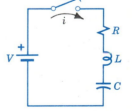

Figure 6.8 An RLC series circuit with a voltage step.

For any case, there are two constants to be determined from initial conditions (step 4). Assuming i_L is initially zero, at $t = 0^+$,

$$i = i_f + i_n = 0 \qquad (6\text{-}17)$$

Assuming v_C is initially zero, at $t = 0^+$, $v_C = 0$ and $v_R = Ri = 0$, and therefore the full applied voltage appears across L or

$$L\frac{di}{dt} = V \qquad \text{and} \qquad \frac{di}{dt} = \frac{V}{L} \qquad (6\text{-}18)$$

Equations 6-17 and 6-18 provide two conditions for evaluating the constants.[†]
For illustration, consider the case of roots real and distinct. From Eq. 6-13, in general

$$\frac{di}{dt} = s_1 A_1 \, e^{s_1 t} + s_2 A_2 \, e^{s_2 t}$$

At $t = 0^+$,

$$\frac{di}{dt} = s_1 A_1 + s_2 A_2 = \frac{V}{L} \tag{6-19}$$

and

$$i = i_n = A_1 + A_2 = 0 \tag{6-20}$$

Constants A_1 and A_2 are obtained by solving Eqs. 6-19 and 6-20 simultaneously. The slightly different approach when roots are complex and there is initial energy storage is illustrated in Example 4.

[†]Equation 6-18 can be obtained from the governing equation $V = Ri + L \, di/dt + \int i \, dt/C$ by noting that at $t = 0^+$, $i = 0$. ∴ $Ri = 0$; also $\int dt = 0$. ∴ $\int i \, dt/C = 0$. Hence, $V = 0 + L \, di/dt + 0$.

EXAMPLE 4

In the circuit of Fig. 6.9, $V = 20$ V, $R = 4\ \Omega$, $L = 2$ H, and $C = 0.01$ F. After being closed for a long time, switch S is opened at $t = 0$. Find and plot the complete response current.

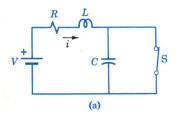

(a)

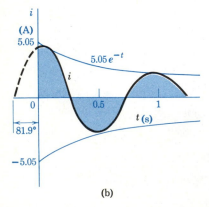

(b)

Figure 6.9 The step response of an RLC circuit with energy storage.

Following the general procedure (the steps should be familiar by now), after S is opened

$$Z(s) = R + sL + 1/sC = 4 + 2s + 100/s$$

$$Z(0) = \infty; \text{ therefore } i_f = 0$$

Setting $Z(s) = 0$,

$$s = -\frac{4}{4} \pm \sqrt{1^2 - \frac{100}{2}} = -1 \pm \sqrt{-49}$$

Therefore, $s = -1 \pm j7$ are the two roots. For roots complex (Eq. 6-16),

$$i_n = A \, e^{-t} \sin (7t + \theta) = i \tag{6-21}$$

At $t = 0^+$, $i = i_L(0^-) = V/R = 20/4 = 5$ and

$$5 = i(0) = A \, e^0 \sin (0 + \theta) \therefore A \sin \theta = 5$$

As another initial condition, note that $v_C = 0$, and $v_R = iR = V$; therefore,

$$v_L = V - v_R - v_C = V - V - 0 = 0 = L \, di/dt$$

Hence, at $t = 0^+$, $di/dt = 0$. From Eq. 6-21,

$$di/dt = -A \, e^{-t} \sin (7t + \theta) + 7 A \, e^{-t} \cos (7t + \theta)$$

$$0 = -A \sin \theta + 7 A \cos \theta$$

∴ $\sin \theta / \cos \theta = 7 = \tan \theta$ or $\theta = \arctan 7 = 81.9°$ and $A = 5/\sin 81.9° = 5.05$.

Therefore, as shown in Fig. 6.9b,

$$i = 5.05 \, e^{-t} \sin (7t + 81.9°) \text{ A} \tag{6-22}$$

The damping coefficient α indicates a time constant for the envelope of 1 s and the frequency is 7 rad/s or 1.1 Hz.

AC Switching Transients

The sudden application of a sinusoidal forcing function to a series-parallel circuit containing two different energy-storage elements represents the most general case of complete response to be treated in this book. Although no new principles are involved, Example 5 is instructive as an illustration of the techniques employed. Note how $Y(s)$ and $Y(j\omega)$ are used to determine the forced response and the form of the natural response. The amplitude and phase angle of the natural response are determined from energy considerations; at $t = 0$, there is no energy stored in the capacitor ($v_C = 0$) or in the inductor ($i_L = 0$).

EXAMPLE 5

In the circuit of Fig. 6.10a, a current $i = 8.5 \cos 4t$ A is applied to the parallel combination by opening switch S at time $t = 0$. Given that $C = \frac{1}{17}$ F, $R = 2 \ \Omega$, and $L = 1$ H, find and plot voltage v as a function of time.

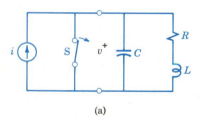

(a)

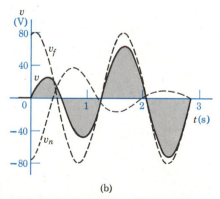

(b)

Figure 6.10   Buildup of a sinusoidal voltage in an *RLC* circuit.

Following the same general procedure,

$$Y(s) = sC + \frac{1}{R + sL} = \frac{s^2LC + sRC + 1}{R + sL}$$

$$Y(j\omega) = \frac{1 - \omega^2 LC + j\omega RC}{R + j\omega L} = \frac{1 - \frac{16}{17} + j\frac{8}{17}}{2 + j4}$$

$$= 0.106 \underline{/19.5°} \text{ S}$$

$$\mathbf{V}_f = \frac{\mathbf{I}}{\mathbf{Y}} = \frac{(8.5/\sqrt{2})\underline{/0°}}{0.106\underline{/19.5°}} = (80/\sqrt{2})\underline{/-19.5°} \text{ V}$$

For $Y(s) = 0$, $LCs^2 + RCs + 1 = 0$ and

$$s = -R/2L \pm \sqrt{(R/2L)^2 - 1/LC}$$

$$= -1 \pm \sqrt{1 - 17} = -1 \pm j4$$

Therefore, $v_n = A\,e^{-t}\cos(4t + \theta)$, and

$$v_f + v_n = 80 \cos(4t - 19.5°) + A\,e^{-t}\cos(4t + \theta)$$

At $t = 0^+$, $v_C = v = 0$, or

$$0 = 80 \cos(-19.5°) + A \cos \theta \therefore A \cos \theta = -75.5$$

Also $i_L = 0$. $\therefore$ all the current flows in C and

$$\frac{8.5}{C} = \frac{dv}{dt} = -320 \sin(-19.5°) - A \cos \theta - 4A \sin \theta$$

Substituting the value for $A \cos \theta$ and solving,

$$A \sin \theta = 9.4 \text{ and } A \sin \theta / A \cos \theta = -9.4/75.5$$

$$\therefore \theta = \arctan(-9.4/75.5) = -7.1° \text{ and}$$

$$A = A \cos \theta / \cos \theta = -75.5/\cos(-7.1°) = -76$$

Therefore, the complete response is (Fig. 6.10b)

$$v = 80 \cos(4t - 19.5°) - 76\,e^{-t}\cos(4t - 7.1°) \text{ V}$$

Example 5 is a special case in that the frequency of the source was selected to be equal to the frequency of the natural response. Under these conditions the voltage oscillations build up from zero and reach the steady-state amplitude in a few cycles. The more general case is illustrated by Exercise 22.

PULSE AND IMPULSE RESPONSE

It is appropriate at this point to introduce an abstract concept that is of minor importance in a first course but of great importance in advanced courses. The mechanical analog of this electrical effect is the familiar hammer blow on a mass, giving the mass momentum. Because the analog is well understood, because all the necessary electrical foundation has been laid, and because this discussion rounds out the treatment of complete response, it is included here. However, mastery of this concept is not essential to study of the material in this book and this section may be omitted.

Pulse Response

An interesting conclusion from Example 4 is that a step voltage can produce a damped sinusoidal response in an *RLC* circuit. In an analogous way the sudden lateral displacement of a guitar string can produce vibrations that die away exponentially. More commonly the string is rapidly displaced and released or "plucked." The corresponding electrical forcing function is the *pulse*.

As shown in Fig. 6.11, a pulse can be generated by the combination of a positive step and, a short time later, a negative step. This representation is useful because we are familiar with step responses. This is a *rectangular pulse* with an amplitude V and a duration t'.

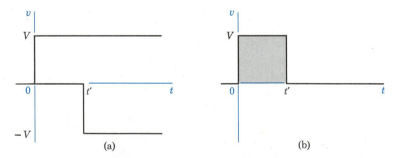

(a)   (b)

Figure 6.11   A rectangular pulse formed from two steps.

The response of the *RC* circuit (Fig. 6.12) to such a forcing function is obtained as the response to two voltage steps. There is no forced response because $Z_C(0) = \infty$, and the complete response to the first step is

$$i = i_n = \frac{V}{R} e^{-t/RC} \qquad 0 < t < t' \qquad (6\text{-}23)$$

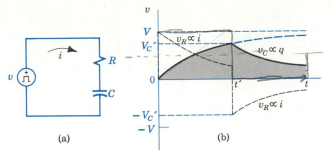

Figure 6.12 The pulse response of an *RC* circuit.

Assuming the capacitor is initially uncharged, the capacitance voltage at time t is

$$v_C = \frac{1}{C}\int_0^t i\, dt = \frac{1}{C}\int_0^t \frac{V}{R} e^{-t/RC}\, dt = \frac{1}{C}\frac{V}{R}(-RC)\, e^{-t/RC}\Big]_0^t$$

$$= -V\, e^{-t/RC}\Big]_0^t = V(1 - e^{-t/T}) \tag{6-24}$$

where $T = RC$ is the time constant.

If a negative step is applied at $t = t'$, the current response has the *form* of Eq. 6-23. Since the capacitance voltage cannot suddenly change, at $t = t'^+, v_C = V_C'$ as evaluated from Eq. 6-24 and shown in Fig. 6.12b. The current reverses as the capacitance discharges and

$$i' = i_n' = -\frac{V_C'}{R} e^{-(t - t')/RC} \qquad t' < t < \infty \tag{6-25}$$

Since i is proportional to v_R at all times, a separate current curve is not shown. The charge q on the capacitor is equal to Cv_C and varies as v_C.

Practical Pulses

Pulses are widely used in data processing and in instrumentation. The signals received, processed, and stored in digital computers are usually in pulse form. Rectangular pulses are formed and manipulated by electronic switches operating at very high speeds to provide millions of arithmetic operations per second.

A great deal can be learned about a device by observing its response to short and long pulses. For this purpose, laboratory *pulsers* are available that will generate pulses with duration times of a few nanoseconds to several seconds and repetition rates up to many million pulses per second. To determine the low-frequency characteristics of a device, it might be represented by the model in Fig. 6.13a on p. 160. When a long pulse is applied at the input, the output ($v_2 = v_R$) will "droop" as in Fig. 6.12b; the amount of droop indicates the time constant of the device. (See Exercise 25.) At high frequencies, the same device might be represented by the model in Fig. 6.13b. When a fast-rising pulse is applied at the input, the output ($v_2 = v_{C'}$) will rise more slowly; to the experienced engineer, the "rise time" indicates the high-frequency characteristics of the device (see Problem 6).

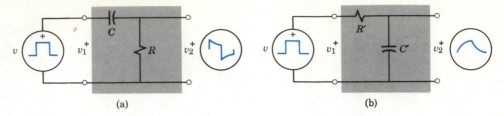

Figure 6.13   Device models for use in pulse analysis and typical response.

Impulse Function

The effect of the pulse in Fig. 6.12 is to charge the capacitor, and the pulse response is an exponential discharge current. The amount of charge stored is dependent upon the amplitude of the pulse V and the duration T_P. The product VT_P is the area of the voltage-time curve and is a significant characteristic of the pulse; the same charge can be stored by a shorter pulse with a greater amplitude.

The effect of varying the pulse dimensions while keeping the area constant is shown in Fig. 6.14. If a steady current $I_O = V_O/R$ flowed for a time $T_P = 0.2T = 0.2RC$, the charge stored would be

$$Q_O = I_O T_P = \frac{V_O}{R} T_P \Rightarrow \frac{V_O(0.2RC)}{R} = 0.200 C V_O \tag{6-26}$$

Actually, by Eq. 6-24, for the pulse of Fig. 6.14a,

$$Q_a = C V_C = C V_O (1 - e^{-0.2}) = 0.181 C V_O = 0.905 Q_O$$

For a duration $T_P/2$ and amplitude $2V_O$ (same area),

$$Q_b = C V_C = C 2V_O (1 - e^{-0.1}) = 0.19 C V_O = 0.95 Q_O$$

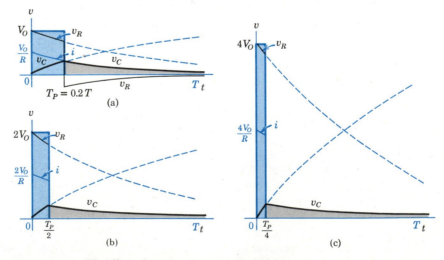

Figure 6.14   The effect of varying the pulse dimensions.

For a duration $T_P/4$ and amplitude $4V_O$,

$$Q_C = CV_C = C4V_O(1 - e^{-0.05}) = 0.195CV_O = 0.975Q_O$$

From these calculations we conclude that as the pulse becomes shorter the actual charge stored approaches the charge stored by a rectangular current pulse I_OT_P. The limiting case of a pulse of area I_OT_P is the *impulse*, a pulse of infinite amplitude for an infinitesimal time but with a finite magnitude. The magnitude of a current impulse is $M_P = \int i\ dt$ and the form of $i(t)$ is not important; in fact, in many problems the shape of $i(t)$ is not known, but the response to such an impulse can be determined precisely.

Impulse Response

Three analogous expressions for impulses are

$$\int f\ dt = MU \qquad \int i\ dt = CV \qquad \int v\ dt = LI \qquad (6\text{-}27)$$

In each case it is assumed that initially the element is at "rest" with no initial velocity, charge, or current. The effect of a force impulse on a mass is to impart momentum. What is the effect of a current impulse on a capacitance? Since $\int i\ dt = CV$, the effect is to store charge or to establish a voltage on the capacitance. The effect of a voltage impulse on an inductance is to impart *electrokinetic momentum* (LI) or to establish a current. Since there is no forced response to an impulse—because $i(t)$ is zero for all time after $t = 0$—the complete response is the natural response of the circuit with stored energy. (See Example 6 on p. 162.)

Practical Impulses

The impulse concept can be used to provide initial energy storage without the use of switches; in general network analysis this is advantageous.[†] Also, in the analysis of ideal circuits impulses occur as part of the complete response; suddenly connecting an ideal current source to an ideal inductor produces an impulse of voltage. Note that the derivative of a step function is an impulse.

Of more practical significance is the fact that pulses approximating impulses occur in many actual circuits. Sometimes such pulses are unavoidable, as in the case of suddenly applying a voltage generator to a capacitive circuit (see Example 2 in this chapter). Another example is the charge built up on a power transmission line during an electric storm which, when released by a lightning stroke, produces an intense voltage pulse that may destroy the insulation of transformers on the line. In other apparatus, pulses serve valuable functions; a short timing pulse is used to trigger the sweep of the electron beam across the screen of a television receiver. If the duration of any pulse is sufficiently short, it can be treated as an impulse by using the approach outlined here.

[†]Another use of impulse functions is in the so-called "operational method" of circuit analysis which is closely related to the Laplace transform method used in more advanced books. See footnote p. 237.

EXAMPLE 6

The series RL circuit of Fig. 6.15a is "hit" by a voltage impulse of M_P volt-seconds at time $t = 0$. What fraction of the voltage impulse $v(t)$ appears across L? Determine the current response $i(t)$.

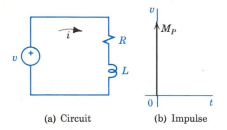

(a) Circuit   (b) Impulse

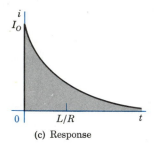

(c) Response

Figure 6.15 Impulse response.

The convention for designating an impulse is a vertical arrow drawn at the appropriate instant, as in Fig. 6.15b; the length of the arrow is not significant since the amplitude of all impulses is infinite. (Note the distinction between the *amplitude* of the function and the *magnitude* of the impulse.)

The effect of a voltage impulse is to establish a finite current in L. Since i is finite, $Ri = v_R$ is finite, whereas $v(t)$ is infinite; therefore practically all the voltage impulse appears across L.

By Eq. 6-27, the current established is

$$I = \frac{1}{L} \int v \, dt = \frac{M_P}{L} = I_O$$

The impulse response is then the natural response consisting of the exponential decay of the initial current I_O, or

$$i = i_n = I_O \, e^{-(R/L)t} = \frac{M_P}{L} e^{-(R/L)t} \qquad (6\text{-}28)$$

as shown in Fig. 6.15c. There is no forced response because $v(t)$ is zero for all time after $t = 0$.

If, for example, M_P is approximated by a pulse of 1 MV for 1 ms, $M_P = 10^6 \times 10^{-3} = 10^3$ V $\cdot$ s. In a 20-H inductor with 20-Ω resistance, $T = L/R = 1$ s and

$$I_O = \frac{M_P}{L} = \frac{10^3}{20} = 50 \text{ A}$$

SUMMARY

- The complete response is the sum of the forced response and the natural response:

$$i = i_f + i_n \qquad \text{or} \qquad v = v_f + v_n$$

- The *form* of the forced response is the same as the form of the forcing function; the *amplitude* is determined by the magnitude of the forcing function and the impedance or admittance.

- The *form* of the natural response is determined by the poles or zeros of the impedance or admittance function; the *amplitude* is determined by the difference between the initial energy storage and that indicated by the forcing function.

- The general procedure for finding complete response is:
  1. Write the appropriate impedance or admittance function.
  2. Determine the forced response from the forcing function and the proper immittance.
  3. Identify the natural components from poles and zeros of $Z(s)$ or $Y(s)$.
  4. Add the forced and natural responses and evaluate the undetermined constants from initial conditions.

- In determining constants from initial conditions, make use of continuity of current in an inductance and voltage across a capacitance. In second-order circuits, also make use of the expressions:

$$\frac{di}{dt} = \frac{v}{L} \quad \text{and} \quad \frac{dv}{dt} = \frac{i}{C}$$

- A rectangular pulse can be generated by the combination of a positive step and, a short time later, a negative step.

- An impulse is the limiting case of a pulse of area M_P, amplitude M_P/T_P, and duration T_P as T_P approaches zero.
  For a current impulse, the magnitude $M_P = \int i\, dt = CV$.
  For a voltage impulse, the magnitude $M_P = \int v\, dt = LI$.

- The effect of an impulse is to store energy in an energy-storage element; the impulse response of a circuit is the natural response of the circuit with that initial stored energy.

REVIEW QUESTIONS

1. What is the justification for saying that the complete response is equal to the sum of the forced and natural responses?

2. What determines the form of the forced response? The amplitude?

3. What determines the form of the natural response? The amplitude?

4. How are natural response components related to pole-zero diagrams?

5. Is it possible to determine the natural response components completely before determining the forced response? Explain.

6. Draw a circuit consisting of R and L in parallel, indicate some initial energy storage, and outline the procedure you would follow in determining the complete response to a forcing function.

7. Cite two specific examples of step functions occurring in nonelectrical situations.

8. Write the differential equation for obtaining v in Fig. 6.7 and outline the procedure for finding the natural response. Compare the difficulty of this method with that of solving $Z(s) = 0$.

9. List four commonly employed relations for evaluating undetermined constants from initial conditions.

10. Cite two specific examples of short pulses occurring in nonelectrical situations.

11. What is the distinction between a pulse and an impulse?

12. How can frequency response be deduced from pulse response?

13. In the Summary to Chapter 5, statements are made regarding how L and C "look" to direct currents; formulate similar statements for impulses.

14. Why is there never a "forced response" to an impulse?

15. Why is the actual function $i(t)$ unimportant in determining the response to the impulse $\int i\, dt$?

EXERCISES

1. Demonstrate by substitution that Eq. 6-5 is a solution of Eq. 6-1.
2. (a) Write the governing equation for the circuit of Fig. 6.3a.
   (b) Demonstrate that Eq. 6-6 is a solution of the governing equation.
3. The complete response of a system is

$$a(t) = A_1 \cos \omega t + A_2 \, e^{-bt} \cos (2\omega t + \theta).$$

   In terms of the discussion on p. 148:
   (a) Identify the components of the response.
   (b) What determines A_1? A_2?
   (c) Write a general expression for the forcing function in this system.
   (d) What is the "order" of the system?
4. In Fig. 6.16, a step input is applied to an RC combination at $t = 0$.
   (a) Derive equations for $i(t)$ and $v_C(t)$ if C is initially uncharged.
   (b) For $V = 20$ V, $R = 2$ kΩ, and $C = 5$ μF, plot $i(t)$ and $v_C(t)$.
   (c) Repeat parts (a) and (b) for an initial voltage $v_C = -5$ V.

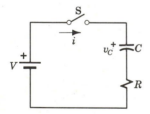

Figure 6.16

5. A battery can be modeled as an ideal source of 20 V in series with a resistance $R_S = 100$ Ω. At time $t = 0$, the battery is connected to a series combination of $R = 400$ Ω and $C = 100$ μF. Predict and sketch the current $i(t)$.
6. In Fig. 6.17, a step input of current is applied to an RL combination by opening switch S at $t = 0$.

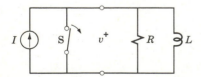

Figure 6.17

(a) Derive equations for $v(t)$ and $i_L(t)$.
(b) For $I = 2$ A, $R = 2$ Ω, and $L = 6$ H, plot $v(t)$ and $i_L(t)$.

7. In Fig. 6.18, switch S is moved to position 1 at $t = 0$. Following the general procedure:
   (a) Predict and sketch $i(t)$.
   (b) Predict and sketch $v_C(t)$ on the graph of part (a). Check by Kirchhoff's voltage law.

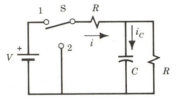

Figure 6.18

8. In Fig. 6.18, switch S is moved to position 1 at $t = 0$.
   (a) Predict and sketch $i_C(t)$.
   (b) A long time later, switch S is moved to position 2. Predict and sketch $i_C(t)$.
9. By means of the circuit in Fig. 6.19 the voltage v_2 is to be brought to 13 V in 3 ms after closing S_1 and held there by closing S_2. If $R = 10$ kΩ and $C = 1$ μF, "design" the circuit by specifying voltage V and resistance R_2. Sketch $v_2(t)$.

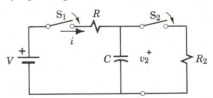

Figure 6.19

10. Determine the time t at which the switch in Example 2 (Fig. 6.5a) could be closed with no natural response component.
11. In Fig. 6.2, $v = 20 \cos 1000t$ V, $R = 5$ Ω, and $L = 8.66$ mH. Switch S is closed at $t = 0$.
    (a) Determine $i(t)$ and sketch the components versus time.
    (b) Determine the time t at which the switch could be closed without any natural component of current.

12. In Fig. 6.18, $R = 500\ \Omega$, $C = 2\ \mu F$, and the dc source is replaced by an ac source $v = 28.28 \cos 1000t$ V. The switch is moved to position 1 at $t = 0$. Following the general procedure, predict and sketch $i(t)$.

13. (a) At time $t = 0$, switch S in Fig. 6.20 is closed in position 1. Take advantage of Thévenin's theorem to evaluate critical quantities and sketch, approximately to scale on a clearly labeled graph, capacitance voltage $v_C(t)$ for $t > 0$.

(b) At time $t_2 = 3\ RC$, switch S is suddenly thrown to position 2. Sketch $v_C(t)$ for $t > 3\ RC$ on the same graph.

(c) Without detailed derivation, write the equation for $v_C(t)$ after $t = 3\ RC$.

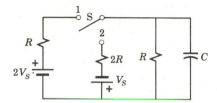

Figure 6.20

14. The switch in Fig. 6.21 is opened at $t = 0$. Determine:

(a) The admittance function appropriate for investigating $v(t)$.

(b) The forced response component of v.

(c) The natural response component of v.

(d) The complete response $v(t)$.

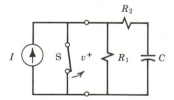

Figure 6.21

15. Repeat Exercise 14 for $v_C(t)$.

16. A step voltage is applied to the series RLC circuit of Fig. 6.22 by closing the switch. For $V = 10$ V, $R = 150\ \Omega$, $L = 0.5$ H, and $C = 100\ \mu F$, predict and sketch the complete current response. Describe the response in words.

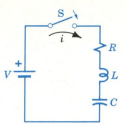

Figure 6.22

17. In the circuit of Fig. 6.22, $V = 70$ V, $R = 4\ \Omega$, $L = 2$ H, and $C = 0.01$F; there is no initial energy storage. Predict and plot the complete response current $i(t)$.

18. (a) Demonstrate by substitution that Eq. 6-14 is a solution of Eq. 4-12.

(b) Repeat Exercise 17 for $V = 60$ V, $R = 40\ \Omega$, $L = 4$ H, and $C = 0.01$ F.

19. In Fig. 6.23, switch S is closed at $t = 0$. Determine the initial and final values of:

(a) Current i and voltage v.

(b) Rates of change di/dt and dv/dt.

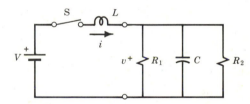

Figure 6.23

20. In Fig. 6.23, $V = 12$ V, $L = 0.8$ H, $R_1 = 3\ \Omega$, $R_2 = 6\ \Omega$, and $C = 0.25$ F. Switch S is closed at $t = 0$. Following a logical procedure, predict $i(t)$.

21. In Fig. 6.24, $R = 5\ \Omega$, $L = 1.6$ H, and $C = 0.025$ F. Switch S is closed at $t = 0$.

(a) Predict *qualitatively* the voltage $v(t)$ and sketch it on a graph.

(b) Predict *quantitatively* $v(t)$ for $V = 3$ V.

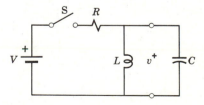

Figure 6.24

22. A voltage $v = 20 \cos 5t$ V is applied at $t = 0$ to a series RLC circuit where $R = 12 \ \Omega$, $L = 2$ H, and $C = 0.1$ F. Predict and sketch the complete current response.

23. In Fig. 6.25, switches S_1 and S_2 have been closed for a long time.
   (a) Where in this circuit is energy stored? How much?
   (b) If switch S_2 is opened at $t = 0$, predict i_L, v_C, i, and i_C at $t = 0^+$.
   (c) Derive an expression for $i(t)$ with two undetermined constants.
   (d) Obtain a specific expression for $i(t)$.

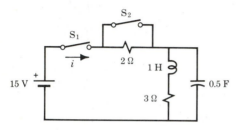

Figure 6.25

24. In Fig 6.25, switches S_1 and S_2 have been open for a long time. Switch S_1 is closed at $t = 0$.
   (a) Working logically, derive an expression for $v_C(t)$ with two undetermined constants.
   (b) Obtain a specific expression for $v_C(t)$.

25. A pulse of 1 ms duration is applied to a device represented by the model of Fig. 6.13a. As observed on an oscilloscope, the output signal "droops" 30%; i.e., the voltage at the end of the pulse is only 70% of the initial magnitude (see Fig. 6.12). Sketch the input and output waveforms and determine the time constant of the model.

26. (a) How does an impulse differ from a pulse?
   (b) What is the effect of a voltage impulse on an inductance?
   (c) If a series combination of $L = 10$ H and $R = 2 \ \Omega$ is "hit" by an impulse of 20 V · s, determine the current response $i(t)$.

27. A parallel combination of $R = 4000 \ \Omega$ and $C = 500 \ \mu$F is "hit" by an impulse of 10 A · s. Before and after the impulse, the current through the terminals is zero or the terminals may be considered open. Determine:
   (a) The charge carried by the impulse.
   (b) The voltage distribution at $t = 0^+$.
   (c) The current response in the parallel circuit.

28. In Fig. 6.26, $L = 1$ H and $C = 1 \ \mu$F; there is no initial energy storage. At $t = 0$, a rectangular pulse of amplitude 500 V and duration 10 μs is applied. Stating any simplifying assumptions, determine the current just after $t = 10 \ \mu$s and the current $i(t)$.

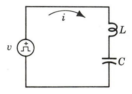

Figure 6.26

PROBLEMS

1. You need a light that flashes automatically 2 times per second. Available is a gaseous discharge lamp that is nonconducting until the applied voltage reaches 30 V and then is highly conducting. Also available are a 60-V battery and a 10-μF capacitor. Devise a simple RC circuit that will provide automatic flashing and specify the appropriate value of R.

2. An ac generator can be represented by a series combination of $v = 2000 \cos 500t$ V and $R = 5\ \Omega$. At time $t = 0$, it is connected across $R = 40\ \Omega$ and $C = 50\ \mu$F in *parallel*.
   (a) Draw and label the circuit diagram.
   (b) Determine and graph generator current $i(t)$, showing components.
   (c) Compare the peak current during the transient period with the maximum reached during steady-state operation.

3. In Fig. 6.27, switch S has been in position 1 a long time, applying generator voltage $v = 5 \cos 1000t$ V to the series circuit consisting of $R = 30\ \Omega$ and $L = 40$ mH. To achieve a certain desired effect, at an instant when the generator voltage is a maximum (at $t = 0$, for example), the switch is thrown to position 2 and $V = 5$ V dc is applied. Predict and sketch $i(t)$ before and after the switch is thrown.

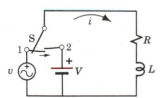

Figure 6.27

4. In Fig. 6.28, switch S has been open for a long time; at time $t = 0$, switch S is closed. You are to predict the current $i(t)$.
   (a) Draw the pole-zero diagram of impedance for the right-hand portion of the circuit.
   (b) Derive an expression for $i(t)$, following a logical, step-by-step procedure.
   (c) Sketch $i(t)$ during the transient period. Indicate your time scale.

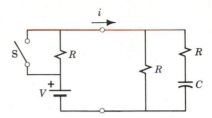

Figure 6.28

5. An engineering student is designing an electric automobile for a cross-country race in which reliability is very important. Whenever the switch in the motor field circuit is opened (Fig. 6.29), he observes an arc across the contacts. He reasons that the arc can be eliminated by placing a capacitor across the switch.
   (a) Evaluate his reasoning qualitatively.
   (b) Considering the effect of C on the circuit and explaining your reasoning, recommend an optimum value for C.

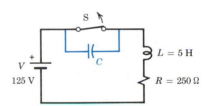

Figure 6.29

6. The high-frequency response of a device (represented by the model of Fig. 6.13b) is to be inferred from the pulse response. When a 10-μs pulse is applied to the input, the rise time of the output is observed on an oscilloscope to be 2 μs. Sketch the input and output waveforms and determine the time constant of the model. Predict the frequency at which the amplitude of the output voltage V_2 will drop to $1/\sqrt{2}$ times that of the input voltage V_1.

7. The door chime of Fig. 22.21a is hit by a plunger. Draw a simplified mechanical model of the chime (using lumped elements) and the analogous electrical circuit. Sketch the vertical velocity of the center of the chime as a function of time.

7

Steady-State AC Circuits

Power Calculations

Frequency Response

Resonance

Three-Phase Circuits

The forced response to a periodic function is a steady-state behavior even though voltages and currents are continuously varying. Most electrical power is generated, transmitted, and utilized in the form of steady-state alternating currents. Also, much of the communication of information by wire or by radio is in the form of ac signals ranging in frequency from a few cycles per second to billions of cycles per second. Some of the most common problems encountered in electrical engineering involve the analysis of ac circuits, and many books have been written on this important subject.

These are practical problems arising in the design and application of electrical apparatus. They are of concern to the astronautical engineer planning a satellite transmitter installation, the chemical engineer selecting remote process controls, the construction engineer planning a temporary power line, the mechanical engineer building a vibration amplifier, and the industrial engineer improving plant efficiency.

Previous chapters in this book provide the necessary foundation of basic laws, phasor representation, complex immittance, and circuit analysis. Building on that foundation, we now develop techniques and approaches that transform difficult problems into routine exercises. First we study the variation of power in circuit elements and learn to make practical power calculations. Then we consider circuits in which the frequency of the forcing function is varied and study the interesting phenomenon of resonance. Finally we acknowledge the fact that most of the billions of kilowatt-hours of electrical energy used each year is in polyphase form and take a quick look at three-phase circuits.

POWER CALCULATIONS

Since voltage is the energy per unit charge and current is the charge per unit time, the basic expression for electrical energy per unit time or *power* is

$$p = vi = i^2R = \frac{v^2}{R} \qquad (7\text{-}1)$$

in a resistance. For sinusoidal variation of voltage and current, the instantaneous value of power is a periodic function. Because average power P is usually the important quantity, we have defined the effective or rms current so that power in a resistance is given by

$$P = \frac{1}{T}\int_t^{t+T} i^2R \, dt = I^2R = \frac{V^2}{R} \qquad (7\text{-}2)$$

where $I = I_{rms}$ and $V = V_{rms}$. This is the most commonly used value of current (or voltage) and is written without a subscript. Other values are defined in terms of the effective value; for example, a "110-volt 60-cycle" house circuit that actually *measures* 120 V rms on an ac voltmeter supplies a voltage $v = \sqrt{2} \, 120 \cos 2\pi 60t$ V.

Power is measured with a *wattmeter*. In a common type of wattmeter, the voltage v applied to the "voltage coil" (the moving coil of Fig. 3.12b) establishes a magnetic field strength directly proportional to v. The current i flowing in the "current coil" (the field coil) reacts with the magnetic field to produce a torque proportional to the instantaneous vi product. The deflection of the needle is proportional to the developed torque, and the inertia of the meter movement performs the desired averaging operation. The resulting deflection is proportional to the average vi product and the scale is calibrated to read average power in watts. For sinusoidal voltages and currents applied to the proper coils, the wattmeter reading is equal to $VI \cos \theta$ (Eq. 7-7).

Reactive Power

Since power in a resistance is proportional to the square of current or voltage, the power is always positive and energy is dissipated throughout the entire cycle. In contrast, inductance and capacitance store but do not dissipate energy. When current through an inductance is increasing, energy ($w_L = \frac{1}{2}Li^2$) is transferred from the circuit to the magnetic field; but when the current decreases, this energy is returned. Similarly, when the voltage across a capacitance is increasing, energy ($w_C = \frac{1}{2}Cv^2$) is transferred from the circuit to the electric field and power is *positive*. When the voltage decreases, this energy is returned and power is *negative*.

From the basic element relations, for an inductance L carrying a current

$$i = I_m \cos(\omega t - \pi/2) = \sqrt{2}I \sin \omega t$$

the voltage is

$$v_L = L \, di/dt = \sqrt{2} \, \omega LI \cos \omega t$$

and the instantaneous power (Fig. 7.1) is

$$p_L = iv_L = 2I^2\omega L \sin \omega t \cos \omega t = I^2X_L \sin 2\omega t \qquad (7\text{-}3)$$

where $X_L = \omega L$ is the inductive reactance. The amplitude of the power variation is I^2X_L; the $\sin 2\omega t$ factor indicates the periodic return of energy to the circuit.

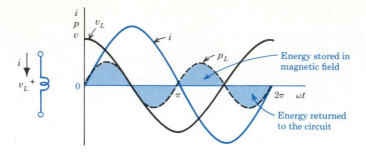

Figure 7.1 The variation of power in an inductance.

For a capacitance C carrying the same current, the voltage is

$$v_C = (1/C) \int i \, dt + V_0 = -\sqrt{2}\,(1/\omega C)I \cos \omega t$$

and the instantaneous power (Fig. 7.2) is

$$p_C = iv_C = 2I^2(-1/\omega C) \sin \omega t (\cos \omega t) = I^2 X_C \sin 2\omega t \qquad (7\text{-}4)$$

where $X_C = -1/\omega C$ is the capacitive reactance. We conclude that in a reactive element there is no net energy transfer or the average power is zero. However, there is a periodic storage and return of energy and the amplitude of the power variation is

$$P_X = I^2 X \qquad (7\text{-}5)$$

This useful quantity is called *reactive power* because of its similarity to *active power* $P = I^2 R$. Reactive power can also be expressed as V^2/X.

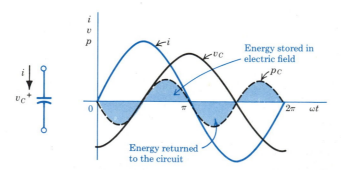

Figure 7.2 The variation of power in a capacitance.

Power Factor

In the general case of alternating current supplied to a complex impedance (Fig. 7.3), voltage and current differ in phase by an angle θ. In phasor notation, $\mathbf{V} = \mathbf{ZI} = (Z\underline{/\theta})\mathbf{I}$; in terms of effective values, $V = ZI$. The instantaneous power is

$$p = vi = \sqrt{2}\,V \cos (\omega t + \theta) \cdot \sqrt{2}I \cos \omega t = 2VI \cos (\omega t + \theta) \cos \omega t$$

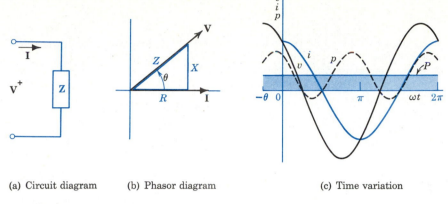

(a) Circuit diagram   (b) Phasor diagram   (c) Time variation

Figure 7.3   Power in a general ac circuit.

Recalling that $2 \cos A \cos B = \cos (A - B) + \cos (A + B)$, we write

$$p = VI \cos \theta + VI \cos (2\omega t + \theta) \qquad (7\text{-}6)$$

Integrating over a cycle, or noting that the term $VI \cos (2\omega t + \theta)$ equals zero over a cycle, we find that the average power is

$$P = VI \cos \theta \qquad (7\text{-}7)$$

In a circuit containing resistance only, voltage and current are in phase, $\theta = 0$, and $P = VI$ in watts[†] where V and I are effective values in volts and amperes, respectively. In general, the average power P is VI multiplied by a factor that can never exceed unity. By definition, the *power factor* is

$$\text{pf} = \cos \theta = \frac{P}{VI} \qquad (7\text{-}8)$$

In a typical distribution system, components are connected in parallel across a common voltage. An inductive component in which the current lags the voltage (as in Fig. 7.3) is said to have a *lagging power factor;* a capacitive component is said to have a *leading power factor* because the current leads the voltage.

Since $X = Z \sin \theta$ (reactance X is the imaginary component of the impedance) and $V = ZI$, Eq. 7-5 becomes

$$P_X = I^2X = I^2Z \sin \theta = VI \sin \theta \qquad (7\text{-}9)$$

The quantity $\sin \theta$ is sometimes called the *reactive factor* because of its resemblance to the power factor. The units of reactive power P_X are *volt-amperes reactive* (abbreviated VAR) to emphasize the difference between this quantity and the power in watts.

[†]To quote an early power engineer: "In 1880 we didn't know what watts were; volt-amperes were a measure of power. If anything happened so that the power did not get into the volt-amperes—well, that was the manufacturer's fault. Nobody raised any questions, and nobody answered any." P. N. Nunn, *General Electric Review,* September 1956, p. 43.

EXAMPLE 1

A coil is to be represented by a linear model consisting of inductance L in series with resistance R. In the laboratory, when a 60-Hz current of 2 A (rms) is supplied to the coil (Fig. 7.4), the voltmeter across the coil reads 26 V (rms). A wattmeter indicates 20 W delivered to the coil. Determine L and R.

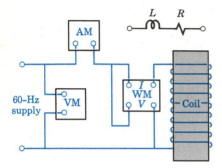

Figure 7.4   Power measurement.

From the basic expression for power (Eq. 7-2),

$$R = \frac{P}{I^2} = \frac{20}{2^2} = 5\,\Omega$$

From the definition of power factor (Eq. 7-8),

$$\theta = \cos^{-1}\frac{P}{VI} = \cos^{-1}\frac{20}{2 \times 26} = 67.4°$$

From the definition of reactive factor (Eq. 7-9),

$$P_X = VI \sin \theta = 26 \times 2 \sin 67.4° = 48 \text{ VAR}$$

From the expression for reactive power (Eq. 7-5),

$$X_L = \frac{P_X}{I^2} = \frac{48}{2^2} = 12\,\Omega$$

$$L = \frac{X_L}{\omega} = \frac{X_L}{2\pi f} = \frac{12}{2\pi 60} = 0.0318 \text{ H}$$

Check:

$$V = ZI = \sqrt{R^2 + (\omega L)^2}\,I$$
$$= \sqrt{5^2 + (377 \times 0.0318)^2} \cdot 2 = 13 \times 2 = 26 \text{ V}$$

Complex Power

The definition of ac power $VI \cos \theta$ and reactive power $VI \sin \theta$ provides the basis for a new abstract concept that is useful in solving industrial problems in power generation and consumption. Consider an induction motor carrying a current I and represented by the linear model of Fig. 7.5a. The phasor diagram, using effective values, shows current $\mathbf{I}$ lagging voltage $\mathbf{V}$ by an angle θ in this inductive load. The phasor diagram is repeated in Fig. 7.6b, which also shows the real and imaginary components of the supply voltage. If we *divide* the magnitude of each voltage phasor by I, we obtain a similar triangle, the *impedance triangle* (Fig. 7.6a). If we *multiply*

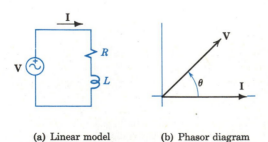

(a) Linear model   (b) Phasor diagram

Figure 7.5   Representation of an industrial load.

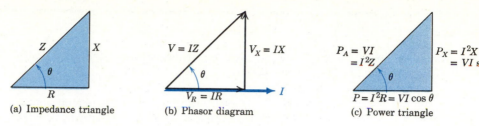

(a) Impedance triangle   (b) Phasor diagram   (c) Power triangle

Figure 7.6   Derivation of the complex power triangle.

each voltage magnitude by I, we obtain another similar triangle, the *power triangle* (Fig. 7.6c).

The hypotenuse of the power triangle is a new quantity *apparent power* defined by

$$\mathbf{P}_A = VI \cos \theta + jVI \sin \theta = VI \underline{/\theta} \qquad (7\text{-}10)$$

or by

$$\mathbf{P}_A = P + jP_X = P_A \underline{/\theta} \qquad (7\text{-}11)$$

Apparent power is measured in *volt-amperes* or *kilovolt-amperes* (abbreviated VA or kVA). In Fig. 7.6c we can visualize apparent power in VA as a complex quantity with a real component equal to power in watts and an imaginary component equal to reactive power[†] in VAR, pronounced "vars."

In deriving Eq. 7-8 for power factor and Eq. 7-9 for reactive power, we assumed that "voltage and current differ in phase by an angle θ." To eliminate an ambiguity, we now specify that:

Phase angle θ is the angle associated with the equivalent impedance.

When the current lags the voltage, θ is positive and therefore reactive power $VI \sin \theta$ is positive. (The power $VI \cos \theta$ is always positive if $|\theta| \leq 90°$.) With this interpretation, lagging pf reactive power is positive and leading pf reactive power is negative. In Example 1, the meter readings do not indicate whether θ is positive or negative; our convention indicates that θ is positive and P_X is identified as a positive 48 VAR.

The apparent power is a practical measure of the capacity of ac equipment. In a transformer used to "step down" the voltage from a distribution potential of 4000 V to a safe handling potential of 120 V, the allowable output is limited by transformer heating due to losses. The losses (discussed in detail in Chapter 21) are determined by voltage and current and are unaffected by power factor. The size of the transformer, for example, required to supply a given industrial load is determined by the apparent power in volt-amperes instead of by the power in watts.

The concept of complex power provides another approach to solving steady-state ac problems. Calculations follow the rules of complex algebra, and vector techniques and graphical methods are applicable. Powers in different portions of a network can be added directly, and reactive powers can be added with proper attention to sign. Drawing a power triangle is a valuable aid in visualization. The approach is illustrated in Example 2 on p. 174.

[†]Instead of P_X, the conventional symbol for reactive power is Q; but already we use this symbol for charge and in the next section we shall use Q for another important quantity. To reduce (but not eliminate) confusion, we shall be unconventional here.

EXAMPLE 2

An industrial load consists of the following: 30 kW of heating and 150 kVA (input) of induction motors operating at 0.6 lagging pf. Power is supplied to the plant at 4000 V. Determine the total current and plant power factor.

Load	P_A (kVA)	θ (°)	pf	P (kW)	P_X (kVAR)
Heating	30	0	1.0	30	0
Motors	150	53	0.6	90	120
Plant	170	45	0.71	120	120

(a) Table of power quantities

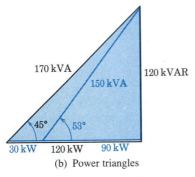

170 kVA 150 kVA 120 kVAR
45° 53°
30 kW 120 kW 90 kW

(b) Power triangles

Figure 7.7 Power factor calculation.

In general, apparent power $\mathbf{P}_A$ is given by

$$P + jP_X = P_A\underline{/\theta}$$

Note that if any two of the four quantities are known, the other two can be calculated. To keep track of the quantities involved, set up the table of Fig. 7.7a where the given quantities are in color.

Assuming that the pf of the heating load is unity, $\theta_H = 0$ and

$$P_A = P_H/\cos\theta_H = 30/1.0 = 30 \text{ kVA}$$

$$P_{XH} = P_A \sin\theta_H = 30 \times 0 = 0$$

The power "triangle" for this load is the horizontal line segment labeled 30 kW in Fig. 7.7b.

For the motor, $\theta = \cos^{-1} 0.6 = 53°$ and the power components are

$$P_M = P_A \cos\theta_M = 150 \cos 53° = 90 \text{ kW}$$

$$P_{XM} = P_A \sin\theta_M = 150 \sin 53° = 120 \text{ kVAR}$$

From the power triangle, for the plant

$$P = P_H + P_M = 30 + 90 = 120 \text{ kW}$$

$$P_X = P_{XH} + P_{XM} = 0 + 120 = 120 \text{ kVAR}$$

Then

$$\mathbf{P}_A = 120 + j120 = 170\underline{/45°} \text{ kVA}$$

The plant power factor is $\cos\theta = 0.71$. The plant current is

$$\frac{P_A}{V} = \frac{170 \text{ kVA}}{4 \text{ kV}} = 42.5 \text{ A}$$

Power Factor Correction

To supply the plant in Example 2, the local power company must install a transformer capable of supplying 170 kVA (neglecting provision for overloads) and wire large enough to carry 42.5 A. In addition to the 120 kW of power supplied to the plant, the power company must generate power to supply the I^2R losses in all the generation, transmission, and distribution equipment.

To supply the same power to another plant operating at unity pf would require a transformer rated at only 120 kVA and wire rated at $120/4 = 30$ A, and the I^2R losses in all equipment would be smaller. It is reasonable to expect that the charge for supplying power to this second plant would be lower, reflecting the lower cost to the power company.

One plan for doing this is to include a power factor penalty clause in the rate schedule. The customer with low power factor pays a higher rate per kilowatt-hour (kWh). The usual cause of low power factor is lagging pf equipment such as induction motors, furnaces, and welders. If a power factor penalty is imposed, it may be economically justifiable to "correct" the situation by adding capacitors that draw a leading pf reactive power. Such capacitors operate at nearly zero power factor and are rated in kVAR at a specified voltage rather than in terms of the capacitance provided. (See Example 3.)

EXAMPLE 3

The power factor of the plant in Example 2 is to be corrected to 0.9 lagging. Specify the necessary auxiliary equipment.

Load	P_A (kVA)	θ (°)	pf	P (kW)	P_X (kVAR)
Previous	170	45	0.71	120	120
Desired	133.3	25.8	0.9	120	58
Correction	62	90	0.0	0	−62

(a) Table of power quantities

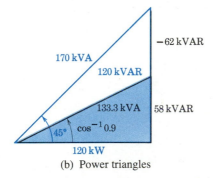

(b) Power triangles

Figure 7.8 Power factor correction.

The power of 120 kW is unaffected by adding reactive power equipment. A table of given values (in color) and calculated values is shown in Fig. 7.8. The magnitude of the apparent power desired after correction is

$$P_A = \frac{P}{\cos\theta} = \frac{120}{0.9} = 133.3 \text{ kVA}$$

and the new angle is $\theta = \cos^{-1} 0.9 = 25.8°$.
    Therefore,

$$P_X = P_A \sin\theta = 133.3 \sin 25.8° = 58 \text{ kVAR}$$

The necessary correction is $58 - 120 = -62$ kVAR.

The new power triangle is shown in Fig. 7.8b. (Note that *apparent powers* cannot be combined algebraically.) The nearest commercial unit is 60 kVAR rated at 4000 V and would be so specified. Since

$$P_X = V^2/X_C = V^2(-\omega C)$$

the capacitance for 60-Hz operation would be

$$C = -P_X/\omega V^2$$

$$= 62 \times 10^3/377 \times 16 \times 10^6 = 10.3 \ \mu\text{F}$$

FREQUENCY RESPONSE

The different responses of inductance and capacitance to sinusoidal signals can be exploited in communication circuits. Since $X_L = \omega L$ is directly proportional to frequency whereas $X_C = -1/\omega C$ is of opposite sign and inversely proportional to frequency, inductive and capacitive elements can be combined in networks that are *frequency selective*. For example, in a radio receiver a parallel combination of L and C is used to select a single signal from the myriad radio waves that are intercepted by the antenna. The extreme degree of frequency selectivity achievable is one of the key properties of electric circuits.

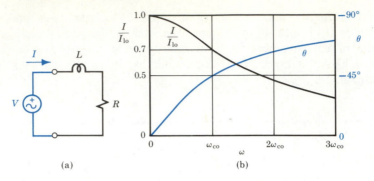

Figure 7.9 A frequency selective circuit.

Simple Filters

The simple RL circuit of Fig. 7.9a is often employed to discriminate between input signals of different frequencies. For an input voltage of constant amplitude V and varying frequency $\omega = 2\pi f$, the impedance is $\mathbf{Z} = R + j\omega L$. At low frequencies, the current is limited by the resistance alone and $I_{lo} \cong V/R$. At high frequencies, the ωL term predominates and $I_{hi} \cong V/\omega L$ or current is inversely proportional to frequency. In general,

$$\mathbf{I}(\omega) = \frac{\mathbf{V}}{\mathbf{Z}} = \frac{\mathbf{V}}{R + j\omega L} = \frac{\mathbf{V}/R}{1 + j\omega L/R} \qquad (7\text{-}12)$$

or, since $\mathbf{I}_{lo} = \mathbf{V}/R$,

$$\frac{\mathbf{I}}{\mathbf{I}_{lo}} = \frac{1}{1 + j\omega L/R} = \frac{1}{\sqrt{1 + (\omega L/R)^2}} \underline{/\tan^{-1}(-\omega L/R)} = \frac{I}{I_{lo}}\underline{/\theta} \qquad (7\text{-}13)$$

as shown in Fig. 7.9b. At the intermediate frequency

$$\omega_{co} = \frac{R}{L} \qquad (7\text{-}14)$$

resistive and reactive components of impedance are equal and

$$\frac{\mathbf{I}}{\mathbf{I}_{lo}} = \frac{1}{1 + j1} = \frac{1}{\sqrt{2}}\underline{/-45°} = 0.707\underline{/-45°} \qquad (7\text{-}15)$$

At this "70% current point," the power delivered to the circuit ($I^2 R$) drops to one-half the maximum value. This *half-power frequency* is an easily calculated measure of the frequency response. Because signals of frequencies below ω_{co} are passed on to the resistance, this RL circuit is said to be a "low-pass filter" with a "cutoff frequency" of ω_{co} (see Exercise 13).

The simple RC circuit of Fig. 7.10a can serve as a high-pass filter because at high frequencies the current is limited by the resistance alone and $I_{hi} = V/R$. In general,

$$\mathbf{I} = \frac{\mathbf{V}}{\mathbf{Z}} = \frac{\mathbf{V}}{R - j1/\omega C} = \frac{\mathbf{V}/R}{1 - j1/\omega CR} \qquad (7\text{-}16)$$

or, in dimensionless form,

$$\frac{\mathbf{I}}{\mathbf{I}_{hi}} = \frac{1}{1 - j1/\omega CR} = \frac{1}{\sqrt{1 + (1/\omega CR)^2}} \underline{/\tan^{-1}(1/\omega CR)} = \frac{I}{I_{hi}}\underline{/\theta} \qquad (7\text{-}17)$$

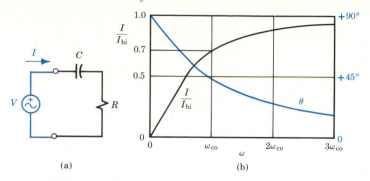

(a)

(b)

Figure 7.10 A simple high-pass filter.

as shown in Fig. 7.10b. Here the cutoff or half-power frequency is defined by $(1/\omega CR) = 1$ or

$$\omega_{co} = \frac{1}{RC} \qquad (7\text{-}18)$$

(Note that having analyzed the frequency response of series RL and RC circuits, we have also analyzed their duals. See Exercise 14.)

L-Section Filters

Since the RL circuit of Fig. 7.9a is a low-pass filter, it could be used in a power supply (see p. 76) to reduce the ripple while passing the dc components of the rectified wave without opposition. However, inductors in the sizes required for low ripple are relatively expensive; for a given investment, the combination of inductance and capacitance shown in Fig. 7.11b gives better results. Such a two-port is called an L-*section* because of its geometry.

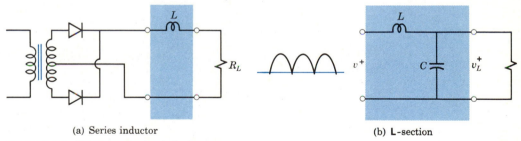

(a) Series inductor

(b) L-section

Figure 7.11 Filters employing inductors.

If the inductance L is greater than a certain critical value[†] L_c where

$$L_c \cong \frac{R_L}{6\pi f} \cong \frac{R_L}{1000} \qquad (7\text{-}19)$$

for 60-Hz operation, the flywheel effect of the inductance keeps the current flowing continuously (instead of in short pulses) and the output voltage changes only slightly with changes in load resistance R_L. Predicting the filtering action under this condition is a straightforward application of ac circuit analysis.

[†]See Millman and Halkias, *Electronic Devices and Circuits*, McGraw-Hill Book Co., New York, 1967.

As represented in Fig. 7.11b, the output of a full-wave rectifier consists of a series of half-sinewaves. The Fourier series representation of such a periodic wave is

$$v = \frac{2}{\pi} V_m (1 - \tfrac{2}{3} \cos 2\omega t - \tfrac{2}{15} \cos 4\omega t - \tfrac{2}{35} \cos 6\omega t - \cdots) \qquad (7\text{-}20)$$

The first term, a constant, represents the dc component of relative magnitude 1. The second term is a second harmonic of relative magnitude $\tfrac{2}{3}$ and frequency 2ω. The succeeding terms in this infinite series are smaller and, furthermore, they are easier to filter out. Our analysis is concerned with the second harmonic term because, if that is filtered out effectively, all the higher harmonics are removed even more effectively (see Exercise 17).

A common problem is to determine the ripple factor in the voltage across R_L for given values of L and C. These values are not critical and it is convenient to make simplifying assumptions that reduce the work involved. In the circuit of Fig. 7.12 let us assume that:

1. The input voltage to the filter is $v = (2/\pi)V_m - (4/3\pi)V_m \cos 2\omega t$, representing the dc and second harmonic terms of the rectified voltage.
2. The resistance of inductor L is negligibly small and, therefore, the full dc component of voltage appears across R_L.
3. The reactance $1/2\omega C$ is small compared to R_L (as it must be for good filtering) and, for the parallel combination, $Z_{RC} = Z_{23} \cong -1/2\omega C$.
4. The reactance of the inductor $2\omega L$ is large compared to Z_{RC} (as it must be for good filtering), and $Z_{13} \cong (2\omega L - 1/2\omega C) \cong 2\omega L$.

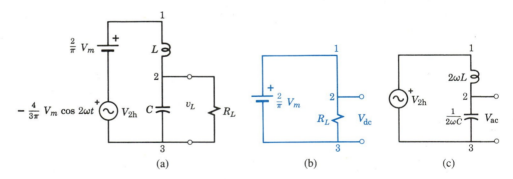

Figure 7.12 Analysis of an L-section filter by superposition.

Since the filter is linear, the principle of superposition is applicable and the circuit of Fig. 7.12a can be replaced by an equivalent combination of two simpler circuits. For the effective values shown on the voltage divider of Fig. 7.12c (based on assumptions 3 and 4),

$$\frac{V_{ac}}{V_{2h}} = \frac{Z_{23}}{Z_{13}} \cong \frac{1/2\omega C}{2\omega L} \cong \frac{1}{4\omega^2 LC} \qquad (7\text{-}21)$$

or the output is

$$V_{ac} = \frac{1}{4\omega^2 LC} V_{2h} = \frac{1}{4\omega^2 LC} \frac{4V_m}{3\pi\sqrt{2}} \qquad (7\text{-}22)$$

The desired result of rectification and filtering is direct current. As a measure of the effectiveness of the process, we define the *ripple factor r* where

$$r = \frac{I_{ac}}{I_{dc}} = \frac{V_{ac}}{V_{dc}} = \frac{\text{rms value of ac components}}{\text{dc component}} \qquad (7\text{-}23)$$

For the L-section filter, the second harmonic is the only significant ripple component, so the ripple factor is

$$r = \frac{V_{ac}}{V_{dc}} = \frac{4V_m}{4\omega^2 LC(3\pi\sqrt{2})} \cdot \frac{\pi}{2V_m} = \frac{\sqrt{2}/3}{4\omega^2 LC} = \frac{0.47}{4\omega^2 LC} \qquad (7\text{-}24)$$

The ω^2 factor in the denominator of Eq. 7-22 justifies our assumption that higher harmonics can be neglected. Example 4 illustrates ripple calculation.

EXAMPLE 4

A load requiring 50 mA at 20 V ($R_L = 400\,\Omega$) is to be supplied from a full-wave rectifier with an L-section filter consisting of $L = 20$ H and $C = 40\ \mu$F. Check the assumptions made in deriving Eq. 7-24 and determine the ripple factor.

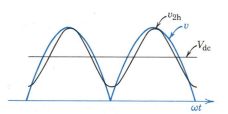

Figure 7.13   Calculating ripple factor from the Fourier components of the filter input.

By Eq. 7-19, $L_c = 400/1000 = 0.4$ H. Therefore, $L > L_c$ and the ac analysis is applicable.

Assuming 60-Hz operation, for the second harmonic,

$$X_L = 2\omega L = 2(2\pi\,60)20 = 15{,}080\,\Omega$$

$$X_C = -\frac{1}{2\omega C} = -\frac{1}{2(2\pi 60)40 \times 10^{-6}} = -33\,\Omega$$

and assumptions 3 and 4 appear to be justified.

The rms value of the second harmonic input component in Fig. 7.13 is, by Eq. 7-20,

$$V_{2h} = \frac{2}{3\sqrt{2}}V_{dc} \cong 9.4 \text{ V}$$

The filter reduces this to (Eq. 7-21)

$$V_{ac} \cong \frac{1/2\omega C}{2\omega L}V_{2h} = \frac{33}{15{,}080}9.4 = 0.021 \text{ V}$$

The ripple factor is

$$r = \frac{V_{ac}}{V_{dc}} = \frac{0.021}{20} \cong 0.001 \cong 0.1\%$$

The same result is obtained directly from Eq. 7-24.

To reduce the ripple factor further, a second L-section can be added before the load resistor. For two identical sections, the denominator of Eq. 7-24 becomes $(4\omega^2 LC)^2$.

Bridged-T Filters

A more complicated frequency response is provided by the *bridged-T filter* of Fig. 7.14. At low frequencies, the capacitances are effectively open circuits and the input voltage $\mathbf{V}_i$ is passed through via R_4. At high frequencies, C_1 and C_2 are effectively shorted and $\mathbf{V}_i$ is passed through directly. At an intermediate frequency ω_0 determined

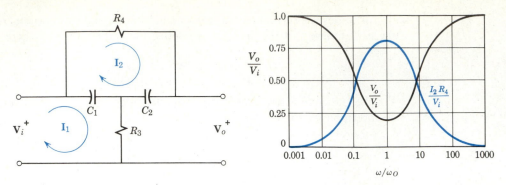

Figure 7.14 The frequency response of a bridged-T filter.

by the parameters of the circuit, the signal following one path is partially cancelled by the signal following the other and the output voltage V_o is reduced.

In analyzing the response, it is convenient to find loop current I_2 and express V_o as $V_i - I_2 R_4$. Assuming a high-impedance load so that the output current is negligible, we write

$$\Sigma V = 0 = V_i - I_1(Z_1 + Z_3) + I_2 Z_1 \tag{7-25}$$

$$\Sigma V = 0 = I_1 Z_1 - I_2(Z_1 + Z_2 + Z_4) \tag{7-26}$$

From Eq. 7-26, $I_1 = I_2(Z_1 + Z_2 + Z_4)/Z_1$; substituting in Eq. 7-25 yields

$$I_2 = \cfrac{V_i}{\cfrac{Z_1 + Z_2 + Z_4}{Z_1}(Z_1 + Z_3) - Z_1} = \cfrac{V_i}{Z_2 + Z_4 + Z_3 + \cfrac{Z_2 Z_3}{Z_1} + \cfrac{Z_3 Z_4}{Z_1}} \tag{7-27}$$

Therefore,

$$I_2 Z_4 = \cfrac{V_i}{1 + \cfrac{Z_3}{Z_4} + \cfrac{Z_2 Z_3}{Z_1 Z_4} + \cfrac{Z_2}{Z_4} + \cfrac{Z_3}{Z_1}} = \cfrac{V_i}{1 + \cfrac{Z_3}{Z_4}\left(1 + \cfrac{Z_2}{Z_1}\right) + \cfrac{Z_2}{Z_4}\left(1 + \cfrac{Z_3 Z_4}{Z_1 Z_2}\right)}$$

Expressing the impedances in terms of circuit parameters leads to

$$I_2 R_4 = \cfrac{V_i}{1 + \cfrac{R_3}{R_4}\left(1 + \cfrac{\omega C_1}{\omega C_2}\right) - j\cfrac{1}{\omega C_2 R_4}(1 - \omega^2 C_1 C_2 R_3 R_4)} \tag{7-28}$$

This expression becomes purely real for $\omega^2 C_1 C_2 R_3 R_4 = 1$ defining the critical frequency

$$\omega_O = \frac{1}{\sqrt{C_1 C_2 R_3 R_4}} \tag{7-29}$$

The filter response is described by

$$\frac{V_o}{V_i} = 1 - \frac{I_2 R_4}{V_i} = 1 - \cfrac{1}{1 + \cfrac{R_3}{R_4}\left[1 + \cfrac{C_1}{C_2}\right] + j\cfrac{1}{\omega C_2 R_4}\left[\left(\cfrac{\omega}{\omega_O}\right)^2 - 1\right]} \tag{7-30}$$

By proper choice of the circuit parameters, a specified band of frequencies can be filtered out. (See Exercise 19.) The advantages of the bridged-T filter are its simplicity and low cost. If we are willing to use more complicated circuits and more expensive elements (op amps, as in Chapter 16), we can obtain more complete filtering and sharper cutoff.

RESONANCE

The slender suspension bridge across the Tacoma Narrows in Washington showed tendencies to oscillate up and down during construction and was nicknamed "Galloping Gertie." On November 7, 1940, only a few months after construction was completed, the oscillation began to build up under a moderate wind and then abruptly changed to a writhing motion. Within an hour the violent twisting had torn the multimillion dollar bridge to pieces. Ten years later, after much study, experiment, and wind-tunnel research, a new bridge designed to be stable in winds up to 120 mph was constructed on the same site.

This is an example of *resonance,* a phenomenon characteristic of low-loss second-order structures and systems. At a particular *resonant frequency,* the impedance is small and the forced response may be very great even for moderate applied forces or voltages. There are many engineering applications of resonance. In an internal combustion engine with a "tuned" intake manifold, the resonance effect is used to increase the amount of fuel–air mixture delivered to the cylinders at high speed. In a refrigerator, the reciprocating compressor is mounted on a support designed to minimize the vibrations transmitted to the cabinet. In an electrical circuit with inductance and capacitance in series, there is always a frequency at which the two reactances just cancel, resulting in the low-impedance characteristic of series resonance.

Series Resonance

For a series *RLC* circuit the impedance is

$$\mathbf{Z} = R + j\omega L + \frac{1}{j\omega C} = R + j\left(\omega L - \frac{1}{\omega C}\right) \qquad (7\text{-}31)$$

$$\mathbf{Z} = \sqrt{R^2 + \left(\omega L - \frac{1}{\omega C}\right)^2} \bigg/ \tan^{-1}\frac{\omega L - 1/\omega C}{R} \qquad (7\text{-}32)$$

The frequency at which the reactances just cancel and the impedance is a pure resistance is the *resonant frequency* ω_O.

In Fig. 7.15, where $\theta_Z = 0$ and $(\omega_O L - 1/\omega_O C) = 0$, the resonant frequency is

$$\omega_O = \frac{1}{\sqrt{LC}} \frac{\text{rad}}{\text{s}} \qquad \text{or} \qquad f_O = \frac{1}{2\pi\sqrt{LC}} \text{ Hz} \qquad (7\text{-}33)$$

By definition, at series resonance the impedance $Z_O = R$ and $\theta_Z = 0$. At frequencies lower than ω_O, the capacitive reactance term predominates; the impedance increases rapidly as frequency decreases, and θ_Z approaches $-90°$. At frequencies higher than ω_O, the inductive reactance term predominates; the impedance increases rapidly with frequency, and θ_Z approaches $+90°$. (Can you visualize the curve of θ_Z as a function of ω?)

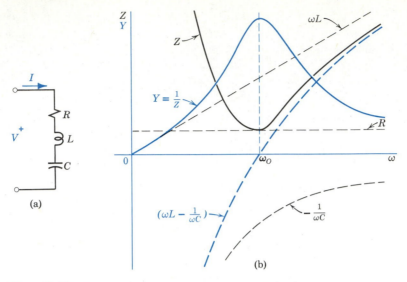

Figure 7.15 The frequency response of a series *RLC* circuit.

Also shown in Fig. 7.15 is the variation with frequency of admittance $Y = 1/Z$. The admittance is high at resonance, where $Y_O = 1/R$, and it decreases rapidly as frequency is changed from ω_O. Since $I = YV$, the variation of current with a constant amplitude voltage has the same shape as the admittance curve. If voltages of various frequencies are applied to a series *RLC* circuit, a voltage of frequency ω_O will be favored and a relatively large current will flow in response to this particular voltage. The circuit is *frequency selective*.

EXAMPLE 5

The variable capacitor behind a radio dial provides a capacitance of 1 nF when the dial setting is "80." Calculate the inductance of the associated coil.

A dial setting of "80" means a frequency of 800 kHz. The coil-capacitor combination provides resonance at this frequency; therefore, $\omega_O L = 1/\omega_O C$ and

$$L = \frac{1}{\omega_O^2 C} = \frac{1}{(2\pi f_O)^2 C}$$

$$= \frac{1}{(2\pi \times 8 \times 10^5)^2 \times 1 \times 10^{-9}} = 39.6 \,\mu\text{H}$$

Phasor Diagram Interpretation

Additional information about the phenomenon of resonance is revealed by phasor diagrams drawn for various frequencies. Since current is common to each element in this series circuit, phasor **I** (rms value) is drawn as a horizontal reference in Fig. 7.16. The voltage $\mathbf{V}_R$ across the resistance is in phase with the current, and voltages $\mathbf{V}_L$ and $\mathbf{V}_C$ across the inductance and capacitance, respectively, lead and lag the current. For $\omega = \omega_O$, $\mathbf{V}_L$ and $\mathbf{V}_C$ just cancel (Why?) and the voltage across the resistance is exactly

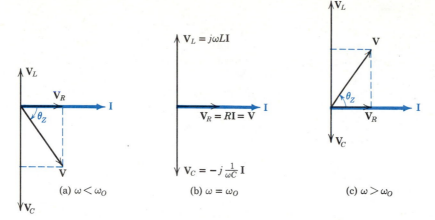

Figure 7.16 Phasor diagrams for a series *RLC* circuit.

equal to the applied voltage **V**. For $\omega < \omega_O$, $\mathbf{V}_C$ is greater than $\mathbf{V}_L$ and the sum of the three voltages is **V**. In comparison to the case for $\omega = \omega_O$, a larger **V** is required for the same current or the admittance is lower; θ_Z is negative indicating a leading power factor. For $\omega > \omega_O$, $\mathbf{V}_L$ is greater than $\mathbf{V}_C$ and the current lags **V** (the sum of the three component voltages) by angle θ_Z.

For a given current **I**, the impedance **Z** is directly proportional to the voltage **V**. Therefore, the voltage diagrams of Fig. 7.16 can be transformed into impedance diagrams. For $\omega < \omega_O$, the impedance is predominantly capacitive. For $\omega = \omega_O$, the impedance is a pure resistance and has a minimum value. For $\omega > \omega_O$, the impedance is predominantly inductive and increases with frequency.

EXAMPLE 6

A generator supplies a variable frequency voltage of constant amplitude $= 10$ V (rms) to the circuit of Fig. 7.17 where $R = 5\ \Omega$, $L = 4$ mH, and $C = 0.1\ \mu$F. The frequency is to be varied until a maximum rms current flows. Predict the maximum current, the frequency at which it occurs, and the resulting voltages across the inductance and capacitance.

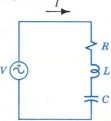

Figure 7.17 A series resonant circuit.

For a series *RLC* circuit, maximum current corresponds to maximum admittance, which occurs at the resonant frequency. At $\omega = \omega_O$, $\omega_O L = 1/\omega_O C$, and $Y_O = 1/Z_O = 1/R$. Therefore,

$$I = VY = \frac{V}{R} = \frac{10}{5} = 2 \text{ A}$$

and

$$\omega_O = \frac{1}{\sqrt{LC}} = \frac{1}{\sqrt{4 \times 10^{-3} \times 10^{-7}}}$$

$$= 5 \times 10^4 \text{ rad/s}$$

Then

$$V_L = \omega L I = 5 \times 10^4 \times 4 \times 10^{-3} \times 2 = 400 \text{ V}$$

and

$$V_C = \frac{I}{\omega C} = \frac{2}{5 \times 10^4 \times 10^{-7}} = 400 \text{ V}$$

As expected, V_L is just equal to V_C and the phasor diagram is similar to that in Fig. 7.16b.

Quality Factor Q

How can the voltage across one series element be greater than the voltage across all three, as in Example 4? The answer is related to the fact that L and C are energy-storage elements and high *instantaneous* voltages are possible. As indicated in Fig. 7.16b, phasors $\mathbf{V}_L$ and $\mathbf{V}_C$ are 180° out of phase; instantaneous voltages are also 180° out of phase, and a high positive voltage across L is cancelled by a high negative voltage across C (see Exercise 22). The voltages do exist, however, and they can be measured with ordinary voltmeters. The ability to develop high element voltages is a useful property of resonant circuits, and we need a convenient measure of this property.

At resonance, $\omega_0 L = 1/\omega_0 C$ and $I = I_0 = V/R$. Therefore, the reactance voltages are

$$V_L = \omega_0 L I_0 = \omega_0 L \frac{V}{R} = \frac{\omega_0 L}{R} V \tag{7-34}$$

and

$$V_C = \frac{I_0}{\omega_0 C} = \frac{V}{\omega_0 CR} = \frac{\omega_0 L}{R} V = V_L \tag{7-35}$$

By definition, in a series circuit

$$\frac{\omega_0 L}{R} = \frac{1}{\omega_0 CR} = Q \tag{7-36}$$

where Q is the *quality factor*, a dimensionless ratio.

In a practical circuit, R is essentially the resistance of the coil since practical capacitors have very low losses in comparison to practical inductors. Hence, Q is a measure of the energy-storage property (L) in relation to the energy dissipation property (R) of a coil or a circuit. A well-designed coil may have a Q of several hundred, and an RLC circuit employing such a coil will have essentially the same Q since the capacitor contributes very little to R. In practical coils, the losses (and therefore R in the circuit model) increase with frequency and $Q = \omega L/R$ is fairly constant over a limited range of frequencies.

Equation 7-34 indicates that for a given value of current there is a "resonant rise in voltage" across the reactive elements equal to Q times the applied voltage. In addition to indicating the magnitude of this resonant rise, Q is a measure of the frequency selectivity of the circuit. A circuit with high Q discriminates sharply; a low-Q circuit is relatively unselective. The solid admittance curve of Fig. 7.18a is redrawn from Fig. 7.15; for the same L and C, the admittance curve near resonance depends on R. Increasing the resistance by 50% reduces Y_0 to $\frac{2}{3}$ of its original value and the selectivity is reduced. Decreasing the resistance to $\frac{2}{3}$ of its original value (and thereby increasing Q by a factor of $\frac{3}{2}$) increases Y_0 and increases selectivity.

The effect of changing one reactive element while keeping the other and the resistance constant is indicated by Fig. 7.18b. Although ω_0 is shifted, the general shape of the response curve is still the same. A *general resonance equation* that describes series RLC circuits with various parameters and various resonant frequencies would be very valuable. Is it possible that such a general equation also describes parallel resonant circuits? Let us investigate that possibility.

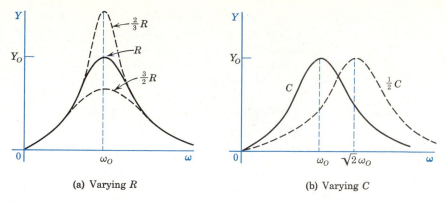

(a) Varying R (b) Varying C

Figure 7.18 The effect of circuit parameters on frequency response.

Parallel Resonance

Fig. 7.19 shows a parallel *GCL* circuit. For this circuit the admittance is

$$\mathbf{Y} = G + j\omega C + \frac{1}{j\omega L} = G + j\left(\omega C - \frac{1}{\omega L}\right) \tag{7-37}$$

$$\mathbf{Y} = \sqrt{G^2 + \left(\omega C - \frac{1}{\omega L}\right)^2} \bigg/ \tan^{-1}\frac{\omega C - 1/\omega L}{G} \tag{7-38}$$

The frequency at which the susceptances just cancel and the admittance is a pure conductance is the resonant frequency ω_O. Where $\theta_Y = 0$ and $(\omega_O C - 1/\omega_O L) = 0$, the resonant frequency is

$$\omega_O = \frac{1}{\sqrt{CL}} \frac{\text{rad}}{\text{s}} \qquad \text{or} \qquad f_O = \frac{1}{2\pi\sqrt{CL}} \text{ Hz} \tag{7-39}$$

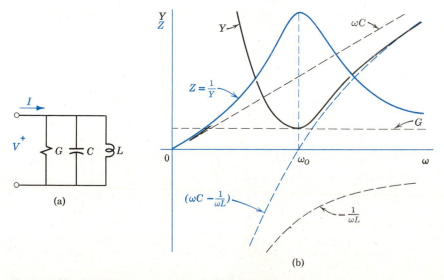

(a)

(b)

Figure 7.19 The frequency response of a parallel *GCL* circuit.

By definition, at parallel resonance the admittance $Y_O = G$ and $\theta_Y = 0$. At frequencies lower than ω_O, the inductive susceptance term predominates; the admittance increases rapidly as frequency decreases and θ_Y approaches $-90°$. At frequencies higher than ω_O, the capacitive susceptance term predominates; the admittance increases rapidly with frequency and θ_Y approaches $+90°$.

These equations and descriptive statements look familiar; as a matter of fact, they were written by applying the duality transforms (Tables 5-2 and 5-3) to the corresponding equations (7-32 and 7-33) and statements for series resonance.[†] Since the circuits are duals, we expect duality in the frequency response curves; Fig. 7.19b is a relabeled version of Fig. 7.15b. (You may wish to translate for yourself the paragraph starting: "Also shown in Fig. 7.15 . . ." on p. 182).

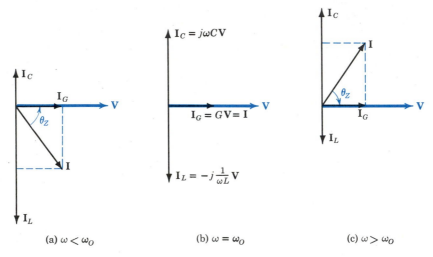

(a) $\omega < \omega_O$ (b) $\omega = \omega_O$ (c) $\omega > \omega_O$

Figure 7.20 Phasor diagrams for a parallel *GCL* circuit.

The phasor diagrams for a parallel *GCL* circuit are shown in Fig. 7.20. In this case there is a "resonant rise in current" in the reactive elements. At resonance, $V = V_O = I/G$. Therefore, the reactance currents are

$$I_C = \omega_O C V_O = \omega_O C \frac{I}{G} = \frac{\omega_O C}{G} I = Q_P I \qquad (7\text{-}40)$$

and

$$I_L = \frac{V_O}{\omega_O L} = \frac{I}{\omega_O L G} = \frac{\omega_O C}{G} I = Q_P I \qquad (7\text{-}41)$$

Basic Definition of Q

The expressions for Q in a parallel circuit are quite different from those in a series circuit; in fact, if the same circuit elements are reconnected, $Q_P = 1/Q_S$. But basically

---

[†]See Skilling, *Electrical Engineering Circuits*, 2nd ed., John Wiley & Sons, New York, 1965, pp. 232 ff.

the phenomena described by Q are the same in both circuits; apparently we need a more basic definition of Q. In essence, quality factor is a measure of the energy-storage property of a circuit in relation to its energy dissipation property. A definition based on this concept is

$$Q = 2\pi \frac{\text{maximum energy stored}}{\text{energy dissipated per cycle}} \qquad (7\text{-}42)$$

In a pendulum driven at its resonant frequency, the only energy input goes to supply friction losses; the energy stored is transferred back and forth from potential to kinetic. In the analogous electric circuit, the input impedance at resonance is a pure resistance and the stored energy is transferred back and forth between the magnetic field of the inductance and the electric field of the capacitance. It can be shown (see Problem 4) that at the resonant frequency the total stored energy is a constant. At an instant when the voltage across the capacitance is zero, the current in the inductance is maximum and all the stored energy is in the inductance. At an instant when v_C is maximum, i_L is zero and all the stored energy is in C. (Is this confirmed by the phasor diagrams of Figs. 7.16 and 7.20?)

Knowing the energy distribution, we can derive specific values of Q from the basic definition. For a series RLC circuit, the common current is $i = \sqrt{2}I \cos \omega t$, and all the stored energy is in L when $i = \sqrt{2}I$ or

$$W_{\text{stored}} = W_L = \tfrac{1}{2}Li^2 = LI^2$$

The energy dissipated per cycle is the energy per second (average power) divided by the frequency in hertz, or

$$W_{\text{diss/cycle}} = \frac{P}{f_O} = \frac{I^2 R}{f_O}$$

By Eq. 7-42,

$$Q_S = 2\pi \frac{LI^2}{I^2 R/f_O} = 2\pi f_O \frac{L}{R} = \frac{\omega_O L}{R} \qquad (7\text{-}43)$$

Following a similar line of reasoning, the quality factor for a simple parallel circuit can be obtained (see Exercise 26). The great advantage of this basic definition of Q is that it is also applicable to more complicated lumped circuits, to distributed circuits such as transmission lines, and to nonelectrical systems. Note that for a reactive *component* such as a coil, $Q = \omega L/R$, a function of frequency; for a resonant *system*, Q is evaluated at ω_O and is a constant. (See Example 7 on p. 188.)

Normalized Response

The frequency selectivity demonstrated in Example 7 is characteristic of all resonant circuits, and the response curves of all such circuits have the same general shape. Now we wish to derive a simple but general equation that describes series and parallel circuits, resonant at high and low frequencies, with large and small energy dissipation. To eliminate the effect of specific parameter values, we use dimensionless

EXAMPLE 7

A typical radio receiver employs parallel resonant circuits in selecting the desired station. Using the values from Example 5 ($C = 1$ nF and $L = 39.6$ μH), the resonant frequency is again 800 kHz and the dial setting is "80." If the *GCL* linear model has a Q of 100, what is the value of G?

If two signals of the same amplitude (50 μA rms) but different frequencies (f_1 at "80" and f_2 at "85") are introduced from the antenna, what are the respective voltages developed across the tuned circuit?

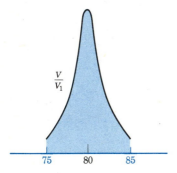

Figure 7.21 Frequency selectivity.

For a Q of 100, Eq. 7-40 yields

$$G = \frac{\omega_o C}{Q_P} = \frac{2\pi \times 8 \times 10^5 \times 10^{-9}}{100} = 50.3 \ \mu S$$

and the equivalent resistance is

$$R = \frac{1}{G} \cong 20,000 \ \Omega$$

At the resonant frequency f_1,

$$V_1 = \frac{I_1}{Y_o} = \frac{I_1}{G} = \frac{50 \times 10^{-6}}{50.3 \times 10^{-6}} \cong 1 \ V$$

At frequency $f_2 = 850$ kHz, $\omega = 2\pi \times 8.5 \times 10^5$ rad/s,

$$Y_2 = \sqrt{G^2 + (\omega C - 1/\omega L)^2}$$
$$= \sqrt{(50.3)^2 + (5341 - 4728)^2} = 615 \ \mu S$$

and

$$V_2 = \frac{I_2}{Y_2} = \frac{50 \times 10^{-6}}{615 \times 10^{-6}} \cong 0.08 \ V$$

For the same impressed current, the voltage response for the desired signal is over 12 times as great as that for the undesired signal (Fig. 7.21), and the power developed in G is over 150 times as great since $P = V^2 G$.

---

ratios. For illustration we choose to work with the admittance of a series *RLC* circuit, but the result has a more general interpretation. For the series circuit of Fig. 7.22a,

$$\mathbf{Z} = R + j\left(\omega L - \frac{1}{\omega C}\right) = \frac{1}{\mathbf{Y}}$$

and, at resonance,

$$\mathbf{Z}_0 = R = \frac{1}{\mathbf{Y}_o}$$

The dimensionless ratio of the admittance at any frequency ω to that at the resonant frequency ω_o is

$$\frac{\mathbf{Y}}{\mathbf{Y}_o} = \frac{\mathbf{Z}_0}{\mathbf{Z}} = \frac{R}{R + j\left(\omega L - \frac{1}{\omega C}\right)} = \frac{1}{1 + j\left(\frac{\omega L}{R} - \frac{1}{\omega CR}\right)} \qquad (7\text{-}44)$$

Introducing the factor ω_o/ω_o and letting the dimensionless ratio $\omega_o L/R = Q = 1/\omega_o CR$, the admittance ratio for a series *RLC* circuit is

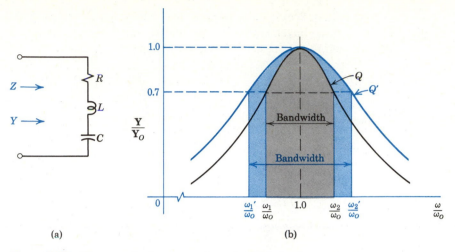

(a)   (b)

Figure 7.22   The normalized response of a series RLC circuit.

$$\frac{\mathbf{Y}}{\mathbf{Y}_O} = \frac{1}{1 + j\left(\dfrac{\omega L}{R} \cdot \dfrac{\omega_0}{\omega_0} - \dfrac{1}{\omega CR} \cdot \dfrac{\omega_0}{\omega_0}\right)} = \frac{1}{1 + jQ\left(\dfrac{\omega}{\omega_0} - \dfrac{\omega_0}{\omega}\right)} \qquad (7\text{-}45)$$

This simple equation also describes the impedance ratio $\mathbf{Z}/\mathbf{Z}_O$ for a parallel GCL circuit; the even simpler reciprocal equation describes the impedance ratio for a series circuit and the admittance ratio for a parallel circuit.

Bandwidth

The curves in Fig. 7.22b have been normalized; all variables are dimensionless ratios. The difference in the two curves shown is due to a difference in selectivity. A convenient quantitative measure of selectivity is defined by letting the imaginary term in the denominator of Eq. 7-45 be equal to $\pm j1$. Where

$$\left(\frac{\omega}{\omega_0} - \frac{\omega_0}{\omega}\right) = \pm\frac{1}{Q}$$

then

$$\frac{\mathbf{Y}}{\mathbf{Y}_O} = \frac{1}{1 \pm j1} = \frac{1}{\sqrt{2}}\ \underline{/\mp 45°} \qquad (7\text{-}46)$$

Imposing this condition defines two frequencies, ω_1 and ω_2, such that

$$\frac{\omega_1}{\omega_0} - \frac{\omega_0}{\omega_1} = -\frac{1}{Q} \qquad \text{and} \qquad \frac{\omega_2}{\omega_0} - \frac{\omega_0}{\omega_2} = +\frac{1}{Q} \qquad (7\text{-}47)$$

Frequencies ω_1 and ω_2 are called the lower and upper "70% points," because at these frequencies the magnitude Y/Y_0 is $1/\sqrt{2} = 0.707$. For a given applied voltage, current is proportional to admittance and these are also called "70% current points." More generally, these are *half-power frequencies* because power is proportional to the square of the current. (Do ω_1 and ω_2 also designate half-power frequencies in parallel circuits?)

The frequency range between the half-power points, $\omega_2 - \omega_1$, is called the *bandwidth*, a convenient measure of the selectivity of the circuit. Solving Eqs. 7-47 and selecting the consistent roots, we obtain

$$\omega_1 = \omega_O\sqrt{1 + \left(\frac{1}{2Q}\right)^2} - \frac{\omega_O}{2Q} \quad \text{and} \quad \omega_2 = \omega_O\sqrt{1 + \left(\frac{1}{2Q}\right)^2} + \frac{\omega_O}{2Q} \quad \text{(7-48)}$$

Subtracting the first equation from the second, the bandwidth is

$$\omega_2 - \omega_1 = \frac{\omega_O}{Q} \quad \text{or} \quad f_2 - f_1 = \frac{f_O}{Q} \quad \text{(7-49)}$$

Note that the bandwidth is inversely proportional to the quality factor Q, a very convenient result. (See Example 8.)

EXAMPLE 8

A series circuit is to be resonant at 800 kHz. Specify the value of C required for a given inductor ($L = 40 \ \mu\text{H}$ and $R = 4.02 \ \Omega$) and predict the bandwidth. Assume that the capacitor is ideal, i.e., introduces no resistance.

For the specified resonant frequency,

$$C = \frac{1}{\omega_O^2 L} = \frac{1}{(2\pi 8 \times 10^5)^2 \times 4 \times 10^{-5}} = 0.99 \text{ nF}$$

and the Q of the circuit or the inductor is

$$Q = \frac{\omega_O L}{R} = \frac{2\pi 8 \times 10^5 \times 4 \times 10^{-5}}{4.02} = 50$$

Then the bandwidth is, by Eq. 7-49,

$$f_2 - f_1 = \frac{f_O}{Q} = \frac{800 \times 10^3}{50} = 16 \text{ kHz}$$

In the typical case where $Q \geq 10$, the factor $\sqrt{1 + (1/2Q)^2} \cong 1$ and

$$\omega_1 = \omega_O - \frac{\omega_O}{2Q} \quad \text{and} \quad \omega_2 = \omega_O + \frac{\omega_O}{2Q} \quad \text{(7-50)}$$

In other words, the resonance curve is symmetric and the half-power points are equidistant from the resonant frequency, or

$$\omega_2 - \omega_O = \omega_O - \omega_1 = \frac{\omega_O}{2Q} \quad \text{(7-51)}$$

This simple relation makes it easy to sketch the response curve of a series or parallel resonant circuit by plotting three points.

Practical Resonant Circuits

In focusing attention on principles, we neglected some practical aspects of resonant circuits. The practical parallel circuit consists of an inductor in parallel with a capacitor, and the only resistance is usually that due to losses in the inductor. The

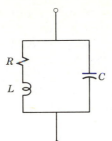

Figure 7.23 Model of a practical parallel circuit.

linear model of Fig. 7.23 is the appropriate representation of the actual circuit; if Q is 20 or more, the general equations (Eqs. 7-36, 39, 40, 41, and 51) hold. Example 9 illustrates practical circuit design.

EXAMPLE 9

Design a parallel circuit to be resonant at 800 kHz with a bandwidth of 32 kHz. Use the inductor from Example 8 with $L = 40$ μH and $R = 4.02$ Ω and predict the impedance at resonance.

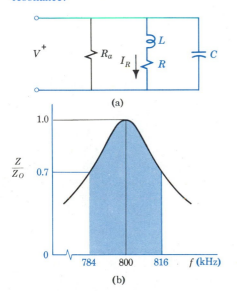

Figure 7.24 Design of a parallel resonant circuit.

The desired quality factor is

$$Q = \frac{f_o}{f_2 - f_1} = \frac{800 \times 10^3}{32 \times 10^3} = 25$$

Since $Q > 20$, Eq. 7-39 holds for this practical circuit and $C = 1/\omega_o{}^2 L = 0.99$ nF as before. From Example 8, the Q of the inductor is 50. Therefore, to double the bandwidth we must halve the Q and double the losses. One possibility is to add a 4.02-Ω resistance in series with R.

Another possibility is shown in Fig. 7.24a. To double the losses, add R_a so that

$$P_a = \frac{V^2}{R_a} = I_R{}^2 R \cong \left(\frac{V}{\omega_o L}\right)^2 R$$

or

$$R_a = \frac{(\omega_o L)^2}{R} \cdot \frac{R}{R} = Q^2 R = 50^2 \times 4.02 = 10{,}050 \text{ Ω}$$

The parallel combination of inductor, capacitance, and resistance can be replaced by the model of Fig. 7.19a where (Eqs. 7-37 and 7-41),

$$G = Y_o = \frac{1}{Z_o} = \frac{1}{Q\omega_o L}$$

Therefore,

$$Z_O = Q\omega_o L = 25 \times 2\pi 8 \times 10^5 \times 4 \times 10^{-5} = 5025 \text{ Ω}$$

The emphasis in this discussion is on a variable frequency ω. In many cases the circuit is "tuned" by varying a capacitor or, less frequently, an inductor. This corresponds to varying ω_O in Eq. 7-45, and the same general resonance behavior is observed.

The specific values of the circuit parameters are usually determined by economic considerations. At audiofrequencies (20 to 20,000 Hz), inductors with iron cores are used to obtain inductances in the order of henrys. At radio broadcast frequencies (above 500,000 Hz), air-core inductors with inductances in the order of microhenrys are resonated with variable air-dielectric capacitors with capacitances in the order of nanofarads. Reasonable values of resistance and quality factor are obtained in properly designed inductors using copper wire. At higher frequencies, undesired resonance may occur because of the inductance of a short lead (a small fraction of a microhenry) and the capacitance between two adjacent conductors (a few picofarads).

THREE-PHASE CIRCUITS[†]

The generation and transmission of electrical power is more efficient in polyphase systems employing combinations of two, three, or more sinusoidal voltages. In addition, polyphase circuits and machines possess some unique advantages; for example, power in a three-phase circuit is constant rather than pulsating as it is in a single-phase circuit. Also three-phase motors start and run much better than single-phase motors. The most common form of polyphase system employs three *balanced* voltages, equal in magnitude and differing in phase by $360°/3 = 120°$. The discussion here is restricted to balanced three-phase circuits. First we see how such voltages are generated and connected, and then we learn to analyze the resulting circuits.

Three-Phase Voltage Generation

The elementary ac generator of Fig. 7.25 consists of a rotating magnet and a stationary winding. The turns of the winding are spread along the periphery of the machine. The voltage generated in each turn of the four-turn winding is slightly out of phase with the voltage generated in its neighbor because it is cut by maximum magnetic flux density an instant earlier or later. (Machine construction and voltage generation are discussed in Chapter 22.) The voltages in the four turns are in series and, therefore, they add to produce voltage $v_{a'a}$.

If the winding were continued around the machine, the voltage generated in the last turn would be 180° out of phase with that in the first and they would cancel, producing no useful effect. For this reason, a winding is commonly spread over no

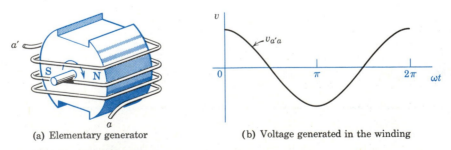

(a) Elementary generator   (b) Voltage generated in the winding

Figure 7.25   Generation of an alternating voltage.

[†]This section can be postponed and considered in connection with polyphase machines in Chapter 24.

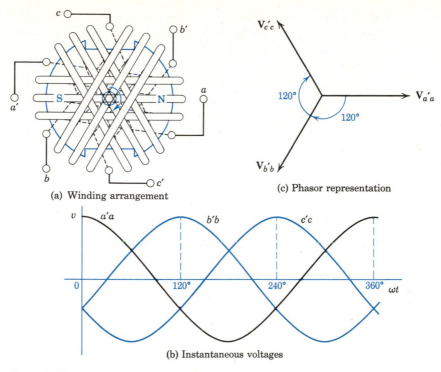

(a) Winding arrangement

(c) Phasor representation

(b) Instantaneous voltages

Figure 7.26   Balanced three-phase voltages.

more than one-third of the periphery; the other two-thirds can be used to generate two other similar voltages. The three sinusoids (sinusoids are obtained with a proper winding distribution and magnet shape) generated by the three similar windings are shown in Fig. 7.26. Defining $v_{a'a}$ as the potential of terminal a' with respect to terminal a, we describe the voltages as

$$v_{a'a} = \sqrt{2}\,V \cos \omega t \qquad\qquad \mathbf{V}_{a'a} = V\,\underline{/0°}$$

$$v_{b'b} = \sqrt{2}\,V \cos (\omega t - 120°) \qquad \mathbf{V}_{b'b} = V\,\underline{/-120°} \qquad (7\text{-}52)$$

$$v_{c'c} = \sqrt{2}\,V \cos (\omega t - 240°) \qquad \mathbf{V}_{c'c} = V\,\underline{/-240°}$$

The three similar portions of a three-phase system are called "phases," a slightly different use of the word. Because the voltage in phase $a'a$ reaches its maximum first, followed by that in phase $b'b$, and then by that in phase $c'c$, we say the *phase rotation* is *abc*. This is an arbitrary convention; for any given machine the phase rotation may be reversed by reversing the direction of rotation of the magnet or by interchanging the labels on two of the three-phase windings. We shall assume the phase rotation is *abc* unless otherwise stated.

Delta Connection

Three separate *single-phase* loads can be connected to the three windings of a generator and supplied with power independently. The three-phase system running along the rear property lines in residential areas could be operated in this way. The first customer could be connected across phase a, the second across phase b, and the third across phase c. However, savings in wire and other benefits are gained by interconnecting the three phases.

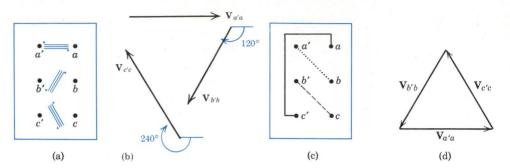

Figure 7.27 Three-phase generator and Δ connection.

Consider the terminal board of a three-phase generator as shown in Fig. 7.27a; the relative orientation of the windings is indicated in color. The corresponding voltage phasors are shown in Fig. 7.27b. If terminal c' is connected to terminal a (solid line in Fig. 7.27c), voltage $a'c$, equal to voltage $a'a$ plus voltage $c'c$, appears across terminals $a'c$. In phasor notation,

$$\mathbf{V}_{a'c} = \mathbf{V}_{a'a} + \mathbf{V}_{c'c}$$

If then terminal b' is connected to terminal c (as shown by the dashed line), the voltage appearing across terminals $a'b$ is

or

$$v_{a'b} = v_{a'a} + v_{c'c} + v_{b'b}$$

$$\mathbf{V}_{a'b} = \mathbf{V}_{a'a} + \mathbf{V}_{c'c} + \mathbf{V}_{b'b} \tag{7-53}$$

The graph of Fig. 7.26b shows that at any instant the sum of the three voltages is zero or $v_{a'b} = 0$, and this is verified by the sum of the balanced phasors in Fig. 7.26c. (The sum of a set of symmetric complex quantities is always zero. See Exercise 36.)

Since the voltage between terminals a' and b is zero, these terminals can be connected (dotted line) and the result is a system of voltages connected in *delta*. The same three windings can be reconnected as in Fig. 7.28 to yield another delta connection. In either Δ (delta), a voltage appears across each pair of terminals; three wires

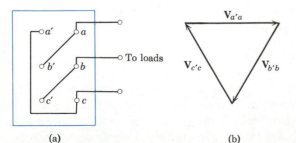

(a)   (b)   Figure 7.28 Another Δ connection.

connected to terminals a, b, and c can supply three single-phase loads. This is the arrangement commonly employed to serve residential customers. Can three such wires carry as much power as six wires of the same size supplying separate single-phase loads? (See Problem 7.)

Wye Connection

The same machine can be reconnected to give a quite different result. If terminals a', b', and c' are connected together (Fig. 7.29a) a, b, and c become the output terminals of a *wye* or *star* connection. Connecting a, b, and c together would yield

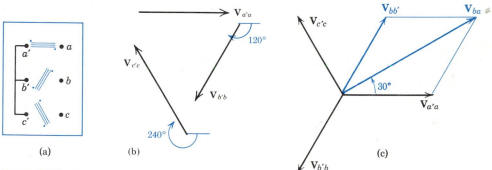

Figure 7.29　Three-phase generator and Y connection.

another Y (wye). In the Δ connection, the voltage across the output terminals is just equal to the voltage of a winding. In the Y connection, however, the voltage across terminals ba is (from Fig. 7.29c)

$$\mathbf{V}_{ba} = \mathbf{V}_{a'a} - \mathbf{V}_{b'b} = \mathbf{V}_{a'a} + \mathbf{V}_{bb'} = 2\mathbf{V}_{a'a}\cos 30°\underline{/30°} = \sqrt{3}\,\mathbf{V}_{a'a}\underline{/30°}$$

In general,

$$\mathbf{V}_{\text{line}} = \sqrt{3}\,\mathbf{V}_{\text{phase}}\underline{/30°} \tag{7-54}$$

and we say that:

In a Y connection, the line-to-line voltage is $\sqrt{3}$ times the phase voltage and is displaced 30° in phase; the line current is just equal to the phase current.

Delta-Circuit Calculations

Given a three-phase source and load, a common problem is to calculate the various currents and voltages and the power supplied to the load. In Fig. 7.30, the load consists of three equal impedances connected in Δ. The internal connection of the source is unimportant; it could be Δ or Y. The only requirement is that there be available at terminals a, b, and c three balanced voltages.

Although three-phase circuits are no more difficult to solve than single-phase circuits, there are many opportunities for confusion in dealing with three of everything. A clearly labeled schematic wiring diagram should be drawn and a phasor diagram should be used as a guide and a check. Calculations are simplified and mistakes reduced if, at first, the wiring diagram and the phasor diagram have the same orientation. In Fig. 7.31, if $\mathbf{V}_{ab}$ is taken as the horizontal reference, phasor $\mathbf{V}_{ab}$ is in

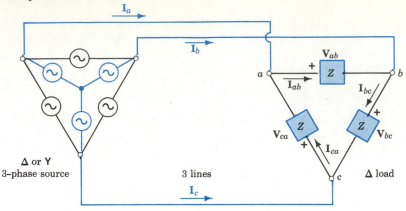

Figure 7.30 A delta-connected load with Δ or Y source.

the same direction as a line going from a to b.[†] Phasor $\mathbf{V}_{bc}$ is in the same direction as a line from b to c and is 120° "behind" $\mathbf{V}_{ab}$ (Fig. 7.31b). Similarly, $\mathbf{V}_{ca}$ is 240° behind $\mathbf{V}_{ab}$.

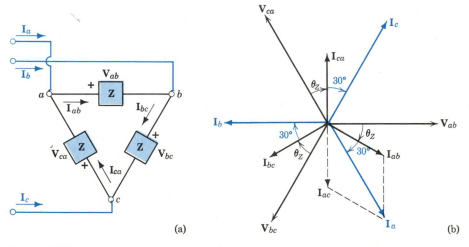

Figure 7.31 Phasor diagram for a Δ load.

In a delta load the voltages across the three impedances (the phase voltages) are identical with the voltages between the "lines" from source to load (the line voltages). The currents in each phase (labeled consistently with the phase voltages in Fig. 7.31a) are obtained in the customary way. In phasor representation, the *phase* currents are

$$\mathbf{I}_{ab} = \frac{\mathbf{V}_{ab}}{\mathbf{Z}_{ab}} = \frac{V\underline{/0°}}{Z\underline{/\theta_Z}} = \frac{V}{Z}\underline{/0° - \theta_Z}$$

$$\mathbf{I}_{bc} = \frac{\mathbf{V}_{bc}}{\mathbf{Z}_{bc}} = \frac{V\underline{/-120°}}{Z\underline{/\theta_Z}} = \frac{V}{Z}\underline{/-120° - \theta_Z} \qquad (7\text{-}55)$$

$$\mathbf{I}_{ca} = \frac{\mathbf{V}_{ca}}{\mathbf{Z}_{ca}} = \frac{V\underline{/-240°}}{Z\underline{/\theta_Z}} = \frac{V}{Z}\underline{/-240° - \theta_Z}$$

[†]In general, v_{ab} is the voltage of a with respect to b. If Z is a pure resistance, v_{ab} and i_{ab} are directly proportional or "in phase"; phasors $\mathbf{V}_{ab}$ and $\mathbf{I}_{ab}$ coincide.

The *line* currents are not equal to the phase currents. For example,

$$\mathbf{I}_a = \mathbf{I}_{ab} + \mathbf{I}_{ac} = \mathbf{I}_{ab} - \mathbf{I}_{ca}$$

This addition is indicated in Fig. 7.31b. Since $\mathbf{I}_{ac} = -\mathbf{I}_{ca}$,

$$\mathbf{I}_a = 2I_{ab}\cos 30°\underline{/-\theta_Z - 30°} = \sqrt{3}\,\mathbf{I}_{ab}\underline{/-30°}$$

$\mathbf{I}_b$ and $\mathbf{I}_c$ are obtained similarly and the result is a set of three balanced line currents. In general,

$$\mathbf{I}_{\text{line}} = \sqrt{3}\,\mathbf{I}_{\text{phase}}\underline{/-30°} \tag{7-56}$$

and we say that:

In a Δ connection, the line current is $\sqrt{3}$ times the phase current and is displaced $-30°$ in phase; the line-to-line voltage is just equal to the phase voltage.

EXAMPLE 10

A load consisting of three identical impedances $\mathbf{Z} = 10\underline{/-45°}\ \Omega$ in Δ is connected to a three-phase, 220-V source. Determine phase and line currents and draw a labeled phasor diagram.

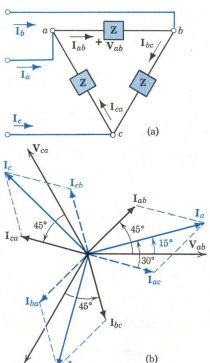

(a)

(b)

Figure 7.32 Delta load calculations.

The source is assumed to supply balanced three-phase voltages with a line-to-line value of 220 V rms and phase rotation *abc*. If $\mathbf{V}_{ab}$ is taken as the horizontal reference, a suitable wiring diagram is shown in Fig. 7.32a. The phase currents are labeled consistently in a clockwise direction and the line currents are all assumed *into* the load to preserve the balance.

Obviously, the phase voltages are just equal to the line voltages. The phase currents are

$$\mathbf{I}_{ab} = \frac{\mathbf{V}_{ab}}{\mathbf{Z}} = \frac{220\underline{/0°}}{10\underline{/-45°}} = 22\underline{/+45°}\ \text{A}$$

$$\mathbf{I}_{bc} = \frac{\mathbf{V}_{bc}}{\mathbf{Z}} = \frac{220\underline{/-120°}}{10\underline{/-45°}} = 22\underline{/-75°}\ \text{A}$$

$$\mathbf{I}_{ca} = \frac{\mathbf{V}_{ca}}{\mathbf{Z}} = \frac{220\underline{/-240°}}{10\underline{/-45°}} = 22\underline{/-195°}\ \text{A}$$

The current in line *a* is

$$\mathbf{I}_a = \mathbf{I}_{ab} + \mathbf{I}_{ac} = 22\underline{/+45°} + 22\underline{/-15°}$$
$$= 22\sqrt{3}\underline{/45° - 30°} = 38\underline{/+15°}\ \text{A}$$

By symmetry,

$$\mathbf{I}_b = 38\underline{/-105°}\ \text{A} \quad\text{and}\quad \mathbf{I}_c = 38\underline{/-225°}\ \text{A}$$

The voltages, phase currents, and line currents are shown on the phasor diagram in Fig. 7.32b.

Power Calculations

The total power in a balanced three-phase load is the sum of three equal phase powers or

$$P_{\text{total}} = 3P_p = 3V_p I_p \cos \theta \tag{7-57}$$

where $\cos \theta$ is the power factor of the load or θ is the angle between phase voltage V_p and phase current I_p. As illustrated by Example 10, θ is *not* the angle between line voltage and line current. It is easier to measure line quantities, however, so an expression for total power in terms of V_l and I_l is useful.

In a Δ load, $V_l = V_p$ and $I_l = \sqrt{3}\,I_p$; therefore,

$$P_\Delta = 3V_p I_p \cos \theta = 3V_l \frac{I_l}{\sqrt{3}} \cos \theta = \sqrt{3}\,V_l I_l \cos \theta \tag{7-58}$$

In a Y load, $V_l = \sqrt{3}\,V_p$ and $I_l = I_p$; therefore,

$$P_Y = 3V_p I_p \cos \theta = 3\frac{V_l}{\sqrt{3}} I_l \cos \theta = \sqrt{3}\,V_l I_l \cos \theta \tag{7-59}$$

The same expression holds for power in a Δ or a Y connected load (see Example 11).

The measurement of power in a practical situation is not straightforward because phase currents in a Δ-connected load and phase voltages in a Y-connected load are usually not accessible. However, some clever engineer observed that two wattmeters, properly connected, can measure the total power in any three-phase system, balanced or unbalanced. Referring to Fig. 7.33, it is clear that three wattmeters with their current coils connected in series with the lines and their voltage coils connected from line to neutral would measure the three phase powers and their sum would be the total power. It can be shown that, since we are interested only in potential differences, the location of the common terminal or "neutral" of the three voltage coils is unimportant; the sum of the three wattmeter readings is always the total power. If now the "neutral" of the wattmeter coils is connected to line c, one wattmeter will sense no voltage; therefore it will read zero and can be dispensed with.

If the *two-wattmeter method* is applied in Example 11, one wattmeter W_1 reads the product of $\mathbf{I}_a$ and $\mathbf{V}_{ac}$ *(not $\mathbf{V}_{ca}$)* and the cosine of the angle between them (Fig. 7.34) or

$$W_1 = I_a V_{ac} \cos \theta_1$$
$$= 12.7 \times 220 \times \cos 15°$$
$$= 2698 \text{ W}$$

and the other wattmeter reads

$$W_2 = I_b V_{bc} \cos \theta_2$$
$$= 12.7 \times 220 \times \cos 75°$$
$$= 723 \text{ W}$$

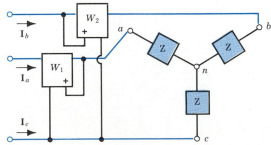

Figure 7.33 The two-wattmeter method.

and the total is $W_1 + W_2 \cong 3420$ W as calculated. (The derivation of the general relation is left for the curious student; see Problem 9.)

EXAMPLE 11

Three equal impedances of $\mathbf{Z} = 10\underline{/+45°}\ \Omega$ are connected in Y across a 220-V supply. Determine phase voltages, phase and line currents, and phase and total power. Draw a labeled phasor diagram.

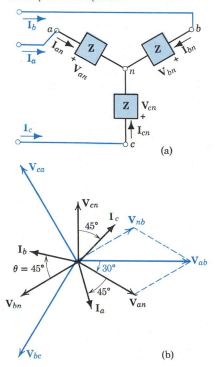

Figure 7.34 Power calculations.

A properly oriented wiring diagram is shown in Fig. 7.34a. By inspection, line current I_a = phase current I_{an}. (Point n is the "neutral" and is important in unbalanced systems.) From Eq. 7-54,

$$\mathbf{V}_p = \frac{\mathbf{V}_l}{\sqrt{3}}\underline{/-30°} \qquad \text{or} \qquad \mathbf{V}_{an} = \frac{\mathbf{V}_{ab}}{\sqrt{3}}\underline{/-30°}$$

In this problem,

$$\mathbf{V}_{an} = (220/\sqrt{3})\underline{/0° - 30°} = 127\underline{/-30°}\ \text{V}$$

This is consistent with the orientation of the wiring diagram; the other phase voltages are determined by inspection and drawn on the phasor diagram as

$$\mathbf{V}_{bn} = 127\underline{/-150°}\ \text{V} \qquad \text{and} \qquad \mathbf{V}_{cn} = 127\underline{/-270°}\ \text{V}$$

Then

$$\mathbf{I}_a = \mathbf{I}_{an} = \frac{\mathbf{V}_{an}}{\mathbf{Z}_{an}} = \frac{127\underline{/-30°}}{10\underline{/+45°}} = 12.7\underline{/-75°}\ \text{A}$$

By symmetry,

and

$$\mathbf{I}_b = \mathbf{I}_{bn} = 12.7\underline{/-195°}\ \text{A}$$
$$\mathbf{I}_c = \mathbf{I}_{cn} = 12.7\underline{/-315°}\ \text{A}$$

as shown on the phasor diagram. The phase power is

$$P_p = V_p I_p \cos \theta = 127 \times 12.7(0.707) = 1140\ \text{W}$$

The total power is, by Eq. 7-59,

$$P_t = \sqrt{3}\ V_l I_l \cos \theta = \sqrt{3} \times 220 \times 12.7(0.707) \cong 3420\ \text{W}$$

Check: $\qquad P_t = 3P_p = 3 \times 1140 \cong 3420\ \text{W}$

As illustrated in the examples, polyphase circuits are complicated, and clearly labeled wiring and phasor diagrams are necessary to minimize confusion. In a given phase, calculations are handled just as in single-phase circuits. The symmetry of balanced three-phase systems reduces the work required.

SUMMARY

- For sinusoids: Power (average) $P = VI \cos \theta = I_R^2 R = V_R^2/R$ in W
  Reactive power $P_X = VI \sin \theta = I_X^2 X = V_X^2/X$ in VAR

  Complex power $\mathbf{P}_A = P + jP_X$
  Power factor $= \cos \theta = P/VI$
  Reactive factor $= \sin \theta = P_X/VI$

- To determine total apparent power, add power and reactive power components separately and combine vectorially; display on power triangles.

- In L-section filters, the inductor opposes current changes and the capacitor absorbs voltage peaks; ac circuit analysis shows that the ripple is

$$r = \frac{V_{ac}}{V_{dc}} \cong \frac{0.47}{4\omega^2 LC} \qquad \text{(with full-wave rectification)}$$

- A circuit containing inductance and capacitance is in resonance if the terminal voltage and current are in phase. At the resonant frequency the power factor is unity, and the impedance and admittance are purely real.
  For series RLC or parallel GCL circuits, $\omega_O = 1/\sqrt{LC}$.

- The frequency selectivity of a resonant circuit is determined by bandwidth.

$$BW = \omega_2 - \omega_1 = \frac{\omega_O}{Q} \qquad \text{where} \qquad Q = 2\pi \frac{\text{maximum energy stored}}{\text{energy dissipated per cycle}}$$

$$\text{For } Q \geq 10, \quad \omega_1 = \omega_O - \frac{\omega_O}{2Q} \qquad \text{and} \qquad \omega_2 = \omega_O + \frac{\omega_O}{2Q}$$

- For resonant circuits:

$$\text{Series:} \quad Q = \frac{\omega_O L}{R} = \frac{1}{\omega_O C R} \qquad \text{and} \qquad V_L = V_C = QV \text{ at } \omega_O$$

$$\text{Parallel:} \quad Q = \frac{\omega_O C}{G} = \frac{1}{\omega_O L G} \qquad \text{and} \qquad I_L = I_C = QI \text{ at } \omega_O$$

$$\text{Series } \frac{\mathbf{Y}}{\mathbf{Y}_O} = \text{parallel } \frac{\mathbf{Z}}{\mathbf{Z}_O} = \frac{1}{1 + jQ\left(\dfrac{\omega}{\omega_O} - \dfrac{\omega_O}{\omega}\right)}$$

- A balanced three-phase system consists of three equal single-phase sources connected in Δ or Y supplying three equal loads connected in Δ or Y. For balanced three-phase systems in

$$\Delta: \quad \mathbf{V}_{line} = \mathbf{V}_{phase} \qquad \text{and} \qquad \mathbf{I}_{line} = \sqrt{3}\,\mathbf{I}_{phase}\underline{/-30°}$$

$$Y: \quad \mathbf{I}_{line} = \mathbf{I}_{phase} \qquad \text{and} \qquad \mathbf{V}_{line} = \sqrt{3}\,\mathbf{V}_{phase}\underline{/+30°}$$

- In analyzing balanced three-phase circuits:
  Draw a carefully oriented and clearly labeled wiring diagram.
  Sketch the phasor diagram as a guide and as a check.
  Analyze one phase and use symmetry for the other phases.
  Total power $= 3V_{phase}I_{phase} \cos\theta = \sqrt{3}\,V_{line}I_{line} \cos\theta$.

REVIEW QUESTIONS

1. What is the frequency of the power pulsations in a 60-Hz circuit?

2. What is the interpretation of "negative power p" delivered "to" a passive circuit?

3. What is the interpretation of "negative reactive power P_X"?

4. Sketch a sinusoidal voltage, a sinusoidal current leading by about 45°, and the corresponding instantaneous power curve.

5. What is measured in VA? In VAR?
6. Is power a vector or a scalar quantity? What is "complex power"?
7. Outline the steps in adding two apparent powers.
8. Why is the local public utility interested in the pf of a plant?
9. Cite an example of resonance from acoustics, aeronautics, and hydraulics.
10. In terms of energy storage, what is the condition necessary for resonance?
11. What actually happens when you turn the tuning dial of a radio?
12. How is the resonant frequency ω_o related to the undamped natural frequency ω_n?
13. Above ω_o, is a series RLC circuit capacitive or inductive? Below ω_o?
14. How can the voltage across one series element in an RLC circuit be greater than the total voltage across all three?
15. Why is Q called the "quality factor"?
16. Using the duality transformations, translate the next full paragraph after Eq. 7-33.
17. How could you determine the Q of a pendulum experimentally?
18. Name two important virtues of resonant circuits.
19. Do ω_1 and ω_2 defined by Eq. 7-47 designate half-power frequencies in parallel circuits?
20. Why is Y/Y_o called "relative response"?
21. If $\mathbf{I}_1 = I\underline{/30°}$ is one current in a balanced three-phase system, what are the other two?
22. Given $\mathbf{V}_{12} = V\underline{/90°}$, define the other two voltages of a balanced three-phase system and show them connected in Δ and Y.
23. Explain by phasor diagrams the $\sqrt{3}$ in the V and I relations in Y and Δ circuits.
24. In analyzing a Δ load, why is it unimportant whether the source is Δ or Y?
25. In a "440-V, three-phase" system, what voltage is 440 V? Is this rms?
26. What is the advantage of using the same orientation for wiring and phasor diagrams?

EXERCISES

1. When a current $i = 7.07 \cos 500t$ A flows in a certain device, the voltage developed is $v = 283 \cos (500t + 45°)$ V.
   (a) Sketch current and voltage waveforms approximately to scale.
   (b) Sketch the instantaneous power on the graph of part (a).
   (c) Compute the average power and the reactive power and show them on a graph.
   (d) Determine the impedance of the device and specify the series components of a linear model.
2. An unknown impedance Z is connected across a "110-V, 60-Hz" line with appropriate instruments. The voltmeter reads 120 V, the ammeter reads 25 A, and the wattmeter reads 1800 W.
   (a) Draw and label the circuit.
   (b) Determine the power, apparent power, reactive power, and power factor.

3. Predict the reading of a wattmeter in which:
   (a) A current $i = 10\sqrt{2} \cos (\omega t + 45°)$ A flows in the current coil and a voltage $v = 220\sqrt{2} \cos (\omega t - 15°)$ V is applied to the voltage coil.
   (b) A current $i = 5 \cos (\omega t - 120°)$ A flows in the current coil and a voltage $v = 20 \cos (\omega t - 75°)$ V is applied to the voltage coil.
4. In the circuit of Fig. 7.35, ammeter AM reads 10 A and voltmeter VM reads 250 V.
   (a) Where in this circuit is average power dissipated? Predict the reading on wattmeter WM.
   (b) Predict total reactive power and overall power factor. Leading or lagging?
5. A load connected across a 200-V, 60-Hz line draws 10 kW at a leading pf of 0.5.
   (a) Determine the current and the reactive power.

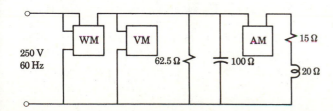

250 V
60 Hz

WM   VM   62.5 Ω   100 Ω   AM   15 Ω

20 Ω

Figure 7.35

(b) What series combination of circuit elements is equivalent to this load?

(c) What parallel combination of circuit elements is equivalent to this load?

6. Repeat Exercise 5 for a load drawing 20 kW at 0.5 pf lagging.

7. A 1-hp motor is rated at 120 V, 10 A, and 0.8 pf.

(a) Calculate the rated efficiency.

(b) Calculate the reactive power.

(c) Calculate the power lost in the motor.

8. Meters are connected to read the current and power into a load impedance $\mathbf{Z}$ from a 200-V, 60-Hz source.

(a) Draw the wiring diagram.

(b) If the ammeter reads 5 A and the wattmeter reads 600 W, determine the power, reactive power, apparent power, and power factor.

(c) When a small pure capacitive reactance X_C is connected in series with $\mathbf{Z}$, the ammeter reads 4 A. Is $\mathbf{Z}$ capacitive or inductive? Predict the new wattmeter reading.

9. An industrial load consists of 100 kVA of induction motors operating at 0.6 pf lagging and 20 kW of resistance heating. Sketch the total plant power diagram. Calculate the total kVA and the plant power factor.

10. Specify the reactive kVA of capacitors required to correct the power factor of the plant in Exercise 9 to 0.85 lagging. If this plant is supplied by a transformer at voltage V, compare the magnitudes of the I^2R losses in the transformer in the original and corrected cases.

11. A typical industrial rate schedule includes a penalty for low power factor. List two reasons why the utility is concerned about your power factor.

12. An industrial load consists of the following: 10 kW of lighting at 100 V, a resistance heating unit that draws 150 A at 200 V, and an induction furnace that requires 360 kW and 400 kVAR at 2000 V. Power is supplied to the plant at 40 kV and 60 Hz.

(a) Express the total complex power in rectangular and polar form.

(b) Show each load and the total on a power diagram.

(c) Calculate the total line current.

(d) Specify the unit to be connected in parallel with the three loads to "correct" the power factor to 0.95.

13. For a load resistance $R = 10$ kΩ, design a sim-

ple RL low-pass filter with a cutoff frequency at $f = 1000$ Hz.

14. (a) Draw the dual of Fig. 7.9a and sketch the frequency-response curves; describe the frequency-response characteristics of the new circuit in words.

(b) Repeat for the dual of Fig. 7.10a.

15. Given a load resistance R, an inductance L, and a capacitance C:

(a) Arrange L, C, and R in two different ways to make a high-pass filter.

(b) Arrange L, C, and R in two different ways to make a low-pass filter.

16. The circuit of Fig. 7.36 is one type of filter. The input consists of four sinusoidal components with $\omega_1 = 10$, $\omega_2 = 100$, $\omega_3 = 1000$, and $\omega_4 = 10,000$ rad/s.

(a) Derive a general expression for the transfer ratio V_o/V_i.

(b) For $C = 10\ \mu F$, $L = 1$ H, and $R = 10$ kΩ, evaluate the ratio V_o/V_i for each component and draw a conclusion.

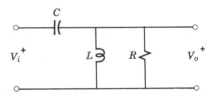

Figure 7.36

17. The output of a 60-Hz full-wave rectifier is applied to the filter in Fig. 7.11. At 60 Hz (NOTE!), the filter reactances are $X_L = 2000$ Ω and $X_C = 200$ Ω. The dc voltage is 300 V across $R_L = 2$ kΩ.

(a) Predict the ac ripple voltage (rms) in the output.

(b) Predict the rms value of the next higher harmonic in the output.

18. Design a 60-Hz full-wave rectifier circuit to supply 200 V at 50 mA with less than 1% ripple factor. Use an L-section filter with $C = 10\ \mu F$. Assuming that the rectifier output voltage is represented by the first two terms of the Fourier series:

(a) Calculate the required transformer secondary voltage (rms), stating any necessary assumptions.

(b) Specify the required value of inductance L. (Is your approach valid?)

(c) Determine the factor by which the filter has reduced the magnitude of the ac voltage component.

(d) Predict, without detailed calculation, the ac voltage across R_L if two L-sections in cascade are used.

19. In the bridged-T filter of Fig. 7.14, $C_2 = 5\ \mu F = 5C_1$ and $R_4 = 5\ k\Omega = 5R_3$. Analyze the frequency response; i.e., determine ω_O and calculate just enough points to sketch V_o/V_i vs ω/ω_O.

20. Examine the dial of a radio and determine the frequency range of the AM broadcast band. Calculate the range of capacitance needed with a $100\text{-}\mu H$ inductance to tune over the band.

21. Write the duals of: "impedance," "reactance," "resistance," "inductance," and "capacitance." Write out a statement (50 words or less) describing the behavior of a series resonant circuit including each of the above words, underlined. Obtain a statement of the behavior of a parallel resonant circuit by substituting the duals.

22. A 10-nF capacitor is connected in series with a coil of 2 mH inductance and 5 Ω resistance.

(a) Determine the resonant frequency.

(b) If a voltage of 2 V is to appear across the capacitor at resonance, what total voltage must be applied to the series combination?

(c) How can the voltage across one series component be larger than the total applied voltage? Sketch v_C and v_L as functions of ωt.

23. A resonant circuit consists of a capacitance C in series with an inductor of inductance L and resistance R. Voltage V is applied.

(a) Define the resonant frequency ω_O and the quality factor Q.

(b) Derive an expression for V_C at ω_O in terms of Q and V.

(c) Derive an expression for the voltage across the inductor at ω_O in terms of Q and V.

(d) Draw a conclusion regarding the relative magnitudes of V_C and V_{LR} for $Q > 10$.

(e) Derive an expression for the phase angle between inductor voltage and current in terms of Q.

24. A signal source supplies $i = 20\sqrt{2}\ \cos 2000t$ mA to a series RLC circuit and meters arranged to read V_{total} and V_C. C is adjusted until the total

voltage across R, L, and C is a minimum, 2 V; the voltage across C alone is observed to be 40 V.

(a) Draw the circuit of this "Q meter."

(b) Calculate the Q of the circuit.

(c) Calculate C, L, and R.

25. In Fig. 7.37, $I = 2$ mA at $\omega = 5 \times 10^6$ rad/s, $G = 5\ \mu S$, and $L = 2$ mH.

(a) For what value of C is the voltage V a maximum?

(b) Predict the magnitude of this maximum voltage.

(c) For $I = 2$ mA at $\omega = 2 \times 10^6$ rad/s (C unchanged), predict V.

(d) Under the condition of part (a), predict the magnitude of I required to establish a current of 1 A in L.

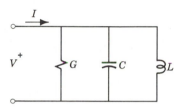

Figure 7.37

26. (a) Using the basic definition, derive an expression for Q for the circuit of Fig. 7.37.

(b) Show that, in any resonant circuit, at the half-power frequencies the resistance is just equal to the magnitude of the reactance.

27. Given the circuit of Fig. 7.36:

(a) Derive an expression for input impedance $Z(j\omega)$.

(b) Define the resonant frequency and derive an expression for ω_O.

(c) For a certain value of R, the Q of this circuit is 50; predict the new value of Q if R is doubled.

28. A coil that can be modeled as $L = 1$ H in series with $R = 1\ \Omega$ is connected in parallel with a capacitor that can be modeled as $C = 100\ \mu F$ in series with $R = 1\ \Omega$.

(a) Draw the circuit, define the resonant frequency for this parallel combination, and calculate ω_O.

(b) Calculate the quality factors for the coil, the capacitor, and the combination at ω_O.

(c) For an applied voltage $V = 20$ V rms at the resonant frequency, calculate Z, I_{total}, I_L, and I_C.

29. A series circuit has a resonant frequency of 1000 Hz; at 1050 Hz, the power factor is 0.707. Calculate the Q.

30. Design a series *RLC* circuit for a resonant frequency of 200 krad/s and a bandwidth of 5 krad/s. The available coil has an inductance of 2.5 mH and a Q of 50.

31. In Fig. 7.38, the vertical scale is current in milliamperes.
(a) Design a circuit with the specified current response to an input voltage of constant amplitude $V = 62.8$ V.
(b) Draw and label your circuit.

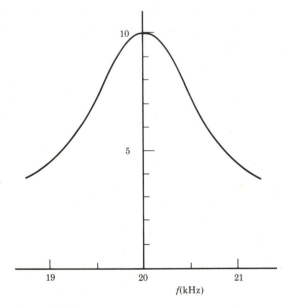

10

5

19   20   21

f(kHz)

Figure 7.38

32. For the parallel circuit of Fig. 7.23:
(a) Derive an expression for admittance as a function of ω.

(b) For $R = 25$ Ω and $L = 0.05$ mH, specify C for resonance at 10^7 rad/s.
(c) For your circuit, estimate the Q, estimate ω_O following the suggestion on p. 191, compare to the specified ω_O, and draw a conclusion.

33. Design a parallel resonant circuit to provide an impedance at resonance of 50,000 Ω, a resonant frequency of 2 MHz, and a Q of 80.

34. In Fig. 7.38, the vertical scale is voltage in volts.
(a) Design a circuit to give the voltage response shown for an input current $I = 1$ mA.
(b) Draw the circuit as you would construct it in the lab, labeling the components.

35. The terminals of a three-phase generator (Fig. 7.39) are numbered from 1 to 6. Measurements indicate the voltages are as follows:

$$v_{12} = 100\sqrt{2} \cos \omega t \text{ V}$$
$$v_{34} = 100\sqrt{2} \cos (\omega t + 60°) \text{ V}$$
$$v_{56} = 100\sqrt{2} \cos (\omega t + 120°) \text{ V}$$

(a) Draw a delta voltage system using these voltages, clearly indicating what connections are to be made on the panel and which are the output terminals.
(b) Repeat for a wye voltage system.

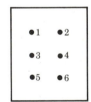

• 1   • 2

• 3   • 4

• 5   • 6

Figure 7.39

36. Given the circuit of Fig. 7.40 with:

$\mathbf{V_1} = 240\underline{/90°}$ V $\mathbf{V_3} = 240\underline{/-150°}$ V

$\mathbf{V_2} = 240\underline{/-30°}$ V $\mathbf{Z} = 40\underline{/-30°}$ Ω

(a) Calculate phasors $\mathbf{I_1}$, $\mathbf{I_2}$, and $\mathbf{I_3}$.

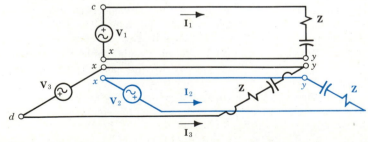

Figure 7.40

(b) If terminals x are connected together and terminals y are connected together, what will be current $\mathbf{I}_{yx}$?

(c) Under the condition of part (b), what will be the voltage $\mathbf{V}_{cd}$?

(d) Draw the phasors of $\mathbf{V}_1$, $\mathbf{V}_2$, $\mathbf{V}_3$, and $\mathbf{I}_1$, $\mathbf{I}_2$, $\mathbf{I}_3$ on a single phasor diagram with all phasors starting from a common point.

37. Three resistors of 20 Ω each are connected in Δ across a 3-ϕ, 440-V line. Calculate phase and line currents.

38. Repeat Exercise 37 with the resistors connected in Y.

39. (a) Three wires enter a building. The voltages with respect to ground are measured to be: 130, 0, and 130 V. The voltage between the two outside wires is 260 V. Describe this electrical distribution system in terms your roommate would understand.

(b) Repeat part (a) if the three voltages with respect to ground are 130, 225, and 130 V, and the voltage between any pair of wires is 260 V.

40. In urban areas it is desirable to supply "110-V" appliances and "220-V" appliances from the same bank of transformers. A convenient compromise is a Y-connected source with the neutral grounded. If a line-to-line voltage of 208 V is acceptably close to "220," what voltage is available from line to ground?

41. Three equal impedances are connected in Δ across a 240-V, 3-ϕ line. One of the three line-to-line voltages is $240\,/60°$ V and one of the phase currents is $6\,/200°$ A.

(a) Identify the other voltages and phase currents and show all on a phasor diagram (all phasors starting from the same point).

(b) Determine the phase impedance in complex form.

(c) Determine the three line currents in phasor form.

42. Three equal impedances of $20\,/-30°$ Ω are connected in Δ across a 240-V, 3-ϕ line.

(a) Draw a clearly labeled wiring diagram properly oriented.

(b) Calculate phase and line currents.

(c) Draw a labeled phasor diagram of voltages and currents.

(d) Calculate the total power.

43. Repeat Exercise 42 with the impedances connected in Y.

44. Repeat Exercise 42 for three equal impedances $\mathbf{Z} = 6 + j8$ Ω connected in Δ.

45. Repeat Exercise 44 with the impedances connected in Y.

46. Design a three-phase, 15-kW heater for operation on a 240-V line by specifying the resistance and power rating of the three identical resistors connected in Y.

47. Repeat Exercise 46 assuming Δ-connected resistors.

48. A balanced Δ capacitive load is connected as shown in Fig. 7.31a. A voltmeter across line ab reads 300 V. An ammeter in line a reads 20 A. A wattmeter with voltage coil across ab and current coil in line a reads 4240 W. Sketch the phasor diagram and determine the complex phase impedance $\mathbf{Z}$.

49. For the load in Example 10:

(a) Redraw Fig. 7.32a to show the WM connections in a power measurement by the two-wattmeter method.

(b) Draw the phasors of the voltages and currents sensed by the meters.

(c) Predict the WM readings and compare to the result of Eq. 7-58.

PROBLEMS

1. The monthly rate schedule for a given plant is: first 5000 kWh @ 5¢/kWh, next 20,000 kWh @ 4¢, next 75,000 kWh @ 3¢, and excess at 2¢. The total charge for energy (kWh) is decreased or increased by 0.5% for each 1% that the average pf is greater or less than 85%.

For a certain manufacturing plant operating 16 hours per day, the monthly consumption is 180,000 kWh and 160,000 kVARh (lagging).

(a) Compute the average monthly "power" bill.

(b) Analyze a proposal by the electrical engineer to improve plant pf to 85% by adding capacitors ($20 per kVAR with 25% annual fixed charges covering depreciation, maintenance, etc.).

2. An industrial load is represented in Fig. 7.41 by $R = 6$ Ω and $X_L = 8$ Ω. The load voltage $\mathbf{V}$ is $250\,/0°$ V.

(a) Calculate the load current, power, reactive power, and power factor.

(b) Calculate the generator voltage $\mathbf{V}_G$ required at the input end of the transmission line (represented by the series impedance $\mathbf{Z}_T = 1 + j3\ \Omega$) and the power lost in transmission P_T.

(c) If capacitor $X_C = 12.5\ \Omega$ is connected in parallel by closing switch S, calculate $\mathbf{I}_C$, the new load current $\mathbf{I}$, and the new power factor. Show $\mathbf{V}$, $\mathbf{I}_L$, $\mathbf{I}_C$, and $\mathbf{I}$ on a phasor diagram.

(d) Calculate the new generator voltage and the new transmission power loss.

(e) What two advantages do you see for "improving the power factor" by adding a parallel capacitor?

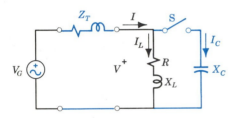

Figure 7.41

3. For the circuit Fig. 7.42, derive an expression for $Z(s)$ as the ratio of two polynomials in s (Eq. 4-33). Demonstrate that for $L/R = RC$ or $R = \sqrt{L/C}$, $Z(j\omega) = R$. Describe the frequency response of this circuit. Derive an expression for the complete response of the circuit to a step voltage of magnitude V.

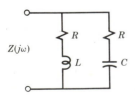

Figure 7.42

4. Prove that the energy stored in a series RLC circuit excited at its resonant frequency is a constant and not a function of time.

5. In a parallel GCL circuit, $L = 2$ mH and $Q = 50$ at $f_O = 400$ kHz.

(a) Define resonance for this circuit and determine the value of C.

(b) Determine the half-power frequencies and the bandwidth of the resonant circuit.

(c) If a total current $I = 50\ \mu$A flows into the parallel circuit, predict the voltage across the circuit.

(d) The practical parallel resonant circuit consists of a capacitor C in parallel with an inductor with inductance $L = 2$ mH and a resistance R. Estimate the value of R at $f = 400$ kHz if the Q of the practical circuit is 50.

6. For the practical tuning circuit of Fig. 7.23:

(a) Define the resonant frequency ω_O.

(b) Derive a general expression for ω_O in terms of $Q = \omega L/R$ and ω_S, the resonant frequency for the same elements in series.

7. For the same line-to-line voltage and the same current per conductor, compare the power transmission capacity of six wires supplying three single-phase resistive loads with the capacity of three wires supplying one three-phase resistive load. Draw a conclusion regarding the economics of three-phase systems.

8. Three identical impedances $Z_Y \underline{/\theta_Y}$ are connected in Y across a balanced three-phase line. Determine the magnitude Z_Δ and phase angle θ_Δ of the impedances which, when connected in Δ across the same line, would draw the same power and reactive power.

9. Demonstrate the validity of the two-wattmeter method of power measurement. For Fig. 7.34, draw phasors $\mathbf{V}_{ac}$, $\mathbf{V}_{bc}$, $\mathbf{V}_{an}$, $\mathbf{V}_{bn}$, $\mathbf{I}_a$, and $\mathbf{I}_b$ for any power factor angle θ. Express the wattmeter readings in general terms and show that their sum agrees with Eqs. 7-58 and 59.

10. When power is measured in a balanced three-phase load, the two wattmeter readings are 3.8 kW and 1.9 kW. What is the power factor of the load?

8

General Network Analysis

One-Port Networks

Two-Port Networks

In analyzing specific circuits, we developed some techniques that can be applied to circuits in general. Important principles that are broadly applicable to circuits of considerable complexity we called *network theorems*. In Chapter 2 we used some of these principles to simplify one-port resistive networks and to draw general conclusions about their behavior. In this chapter we extend those theorems to the case of ac circuits and then introduce two new theorems and show how they are applied. As before, we demonstrate the validity of the concepts and leave general proofs to more advanced courses.

Another important class of circuits includes *two-port* or *three-terminal* networks. Transformers and amplifiers are two-ports; energy or information is introduced at the *input* and abstracted at the *output*. The last half of this chapter is devoted to the analysis and application of some common two-port networks.

ONE-PORT NETWORKS

The theorems applied in Chapter 2 to two-terminal resistive networks are easily extended to include general ac one-ports.

Equivalence

Two one-ports are equivalent if they present the same V-I characteristics.

Two passive one-ports are equivalent if they have the same input impedance or the same input admittance; for sinusoidal excitation and response, for example, this means the same $\mathbf{Z}$ or $\mathbf{Y}$. Ordinarily, two real networks are equivalent at one frequency only.

Network Reduction

Replacing a complicated network with a relatively simple equivalent is advantageous in network analysis. Already we have used the fact that in passive circuits impedances and admittances are readily combined. For impedances in series,

$$\mathbf{Z}_{EQ} = \mathbf{Z}_1 + \mathbf{Z}_2 + \cdots + \mathbf{Z}_n \tag{8-1}$$

and for admittances in parallel,

$$\mathbf{Y}_{EQ} = \mathbf{Y}_1 + \mathbf{Y}_2 + \cdots + \mathbf{Y}_n \tag{8-2}$$

Dividers

The useful voltage divider and its dual the current divider become, for ac circuits,

$$\mathbf{V}_2 = \frac{\mathbf{Z}_2}{\mathbf{Z}_1 + \mathbf{Z}_2}\mathbf{V} = \frac{\mathbf{Z}_2}{\mathbf{Z}_{EQ}}\mathbf{V} \quad \text{and} \quad \mathbf{I}_2 = \frac{\mathbf{Y}_2}{\mathbf{Y}_1 + \mathbf{Y}_2}\mathbf{I} = \frac{\mathbf{Y}_2}{\mathbf{Y}_{EQ}}\mathbf{I} \tag{8-3}$$

When a particular voltage or current is desired, using these relations is preferable to writing and solving loop or node equations.

Linearity and Superposition

The concept of linearity and the theorem of superposition are inseparable; the first is essential to the second, and the second defines the first. We say that function f is linear in x if

$$f(kx) = kf(x) \quad \text{and} \quad f(x_1 + x_2) = f(x_1) + f(x_2) \tag{8-4}$$

These defining equations paraphrase the superposition theorem (p. 45), which is particularly useful in predicting the response of a network to a complex signal.

EXAMPLE 1

The signal shown in Fig. 8.1a is applied to a filter circuit. Predict the output.

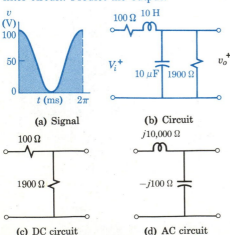

(a) Signal

(b) Circuit

(c) DC circuit

(d) AC circuit

Figure 8.1 Superposition application.

The signal can be described as

$$v_i = 50 + 50 \cos 1000t = V_i + V_{mi} \cos 1000t \text{ V}$$

For the dc component, the voltage divider output is

$$V_o = \frac{R_2}{R_1 + R_2}V_i = \frac{1900}{100 + 1900}50 = 47.5 \text{ V}$$

For the ac component the resistances are negligible in comparison to the reactances; therefore, the output amplitude is

$$V_{mo} \cong \frac{Z_C}{Z_L + Z_C}V_{mi} = \frac{-j100}{j10{,}000 - j100}50 = -0.5 \text{ V}$$

Using superposition, the total output is

$$v_o = 47.5 - 0.5 \cos 1000t \text{ V}$$

Active Networks

Active networks containing ac energy sources can be represented by simpler equivalents using Thévenin's theorem or Norton's theorem.

Insofar as a load is concerned, any one-port network of linear elements and energy sources can be replaced by a series combination of an ideal voltage source and a linear impedance or a parallel combination of an ideal current source and a linear admittance.

Here the Thévenin equivalent is a phasor voltage $\mathbf{V}_T$ in series with a complex impedance $\mathbf{Z}_T$; the Norton equivalent is a phasor current $\mathbf{I}_N$ in parallel with a complex admittance $\mathbf{Y}_N$. The parameters are: $\mathbf{V}_T = \mathbf{V}_{OC}$, $\mathbf{Z}_T = \mathbf{V}_{OC}/\mathbf{I}_{SC}$, $\mathbf{I}_N = \mathbf{I}_{SC}$, and $\mathbf{Y}_N = \mathbf{I}_{SC}/\mathbf{V}_{OC} = 1/\mathbf{Z}_T$. At the convenience of the circuit analyst, the equivalents are interchangeable.

EXAMPLE 2

Predict the current $\mathbf{I}$ in Fig. 8.2a in response to a voltage $\mathbf{V} = 10\underline{/0°}$ V. The impedance magnitudes are given in ohms.

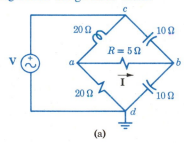

(a)

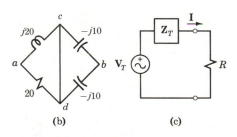

(b)   (c)

Figure 8.2 Thévenin's theorem application in an ac circuit.

This bridge network cannot be simplified by series-parallel combination of impedances.

One approach is to replace all except the 5-Ω resistance by a Thévenin equivalent. With R removed, $\mathbf{V}_a$ and $\mathbf{V}_b$ can be determined by voltage division and then $\mathbf{V}_T$ by subtraction.

$$\mathbf{V}_a = \frac{20}{20 + j20} \cdot 10\underline{/0°} = \frac{20 \times 10\underline{/0°}}{20\sqrt{2}\underline{/45°}} = 5 - j5$$

$$\mathbf{V}_b = \frac{-j10}{-j10 - j10} 10\underline{/0°} = \frac{10 \times 10\underline{/0°}}{20} = 5 + j0$$

$$\mathbf{V}_T = \mathbf{V}_{OC} = \mathbf{V}_a - \mathbf{V}_b = -j5 = 5\underline{/-90°}\ V$$

With the voltage source removed ($\mathbf{V} = 0$) by shorting node c to ground,

$$\mathbf{Z}_T = \mathbf{Z}_{ab} = \mathbf{Z}_{ac}\|\mathbf{Z}_{ad} + \mathbf{Z}_{cb}\|\mathbf{Z}_{db} = \frac{20(j20)}{20 + j20} + \frac{-j10}{2}$$

$$= 10 + j10 - j5 = 10 + j5\ \Omega$$

Then

$$\mathbf{I} = \frac{\mathbf{V}_T}{R + \mathbf{Z}_T} = \frac{5\underline{/-90°}}{5 + 10 + j5} = 0.32\underline{/-108.4°}\ A$$

Maximum Power Transfer

One of the distinguishing characteristics of the engineer is a concern with optimization. The engineer solves real problems and his or her solutions are always compromises. Efficiency costs money; safety increases complexity; performance adds weight; improvement takes time. In striving for the optimum design, the engineer provides the highest efficiency per dollar, the most powerful performance per pound, or the best results within the deadline. In addition, he or she must cope with the fact that the cheapest energy is in short supply and the most efficient device may be environmentally unacceptable.

Frequently the engineer has the problem of "matching" one component of a system to another to obtain optimum results. For example, an automobile transmission must be matched to the engine and, in another sense, the passing gear provides a match between the engine and the load represented by acceleration up a steep hill. The bicycle-gear ratio represents a simple solution to the problem of matching a load to a source. Why does a racing bicycle have 10 gear ratios?

The optimum load for an electric source depends on the desired result. A dc generator (Fig. 8.3a) has certain *fixed losses* that are present whether or not a load is

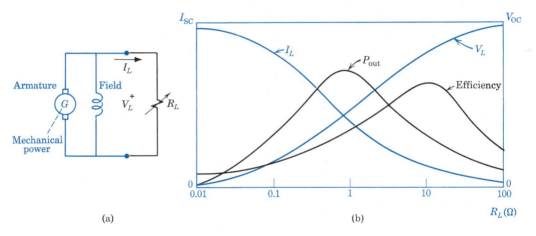

(a)   (b)

Figure 8.3 A dc generator with variable load resistance.

connected and *variable losses* that are directly related to the output current. As shown in the graph, for R_L small, the output current I_L is high, but the output voltage V_L is low, and therefore power P is low; for R_L large, the output voltage is high, but the current is low, and therefore power is low. At an intermediate value of R_L the power output is a maximum. The efficiency, output/input, increases as the load resistance increases from zero. Because of the fixed losses, however, a maximum efficiency occurs at a particular value of R_L. If the available power is large and efficiency is the important criterion, a relatively large value of load resistance is optimum; this is true in most applications of electrical power equipment. If efficiency is less important and the power available is limited, then a lower value of load resistance is optimum; this is true in many communication applications.

With the aid of Thévenin's theorem we can draw a general conclusion about the conditions for maximum power transfer. We know that any network of linear elements and energy sources (and, approximately, any real generator and its associated circuitry) can be represented by a series combination of an ideal voltage **V** and an impedance **Z**. In the simplest case, these are the open-circuit generator voltage $\mathbf{V}_G$ and

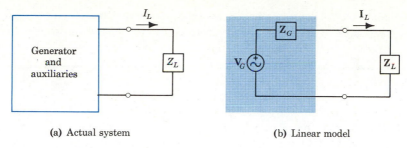

(a) Actual system

(b) Linear model

Figure 8.4   Power transfer analysis.

the internal impedance $\mathbf{Z}_G$ (Fig. 8.4b). The effective value of current delivered to the load is

$$I_L = \frac{V_G}{|\mathbf{Z}_G + \mathbf{Z}_L|} = \frac{V_G}{\sqrt{(R_G + R_L)^2 + (X_G + X_L)^2}}$$

The power transferred to the load is

$$P_L = I_L^2 R_L = \frac{V_G^2 R_L}{(R_G + R_L)^2 + (X_G + X_L)^2} \tag{8-5}$$

For a given generator, V_G, R_G, and X_G are fixed; R_L and X_L can be adjusted for optimum results. For maximum power transfer, the value of P_L in Eq. 8-5 is to be maximized. The general procedure is to differentiate with respect to the variable and set the derivative equal to zero, but Eq. 8-5 indicates that for any value of R_L, P_L is maximum for $X_L = -X_G$. Hence, the first requirement for maximum power transfer is that the reactance of the load be made equal and opposite to the equivalent reactance of the source.

Under this condition,

$$P_L = \frac{V_G^2 R_L}{(R_G + R_L)^2} = V_G^2 R_L (R_G + R_L)^{-2}$$

and for maximum P_L,

$$\frac{dP_L}{dR_L} = V_G^2[(R_G + R_L)^{-2} - 2R_L(R_G + R_L)^{-3}] = 0$$

Solving,

$$2R_L = R_G + R_L \qquad \text{or} \qquad R_L = R_G$$

Hence, the second requirement is that the resistance of the load be made equal to the equivalent resistance of the source. In general, for maximum power transfer the impedance of the load should be adjusted so that

$$\mathbf{Z}_L = R_L + jX_L = R_G - jX_G = \mathbf{Z}_G^* \tag{8-6}$$

This principle can be expressed in words as follows:

For maximum power transfer the impedance of the load should be made equal to the complex conjugate of the Thévenin equivalent impedance of the source.

Note that the source impedance is *not* made equal to the load impedance; increasing the source resistance can only decrease the power transfer to a given load.

EXAMPLE 3

A generator represented by a voltage $V_G = 100\underline{/0°}$ V and an internal impedance $\mathbf{Z}_G = 10 + j20$ Ω is connected to a load $\mathbf{Z}_L = 20 + jX_L$ by means of a transmission network $\mathbf{Z}_T = R_T - j15$ Ω. Specify the values of R_T and X_L for maximum power transfer to the load.

From the viewpoint of the load, the "source" consists of the generator and the transmission line. In this general case (Fig. 8.5b), the "source" has one variable element R_T, and the load has one variable element X_L.

If P_L is to be maximized, then I_L must be maximized and R_T must be *zero* because the addition of any resistance can only decrease I_L.

Insofar as the load is concerned, the equivalent source impedance is then

$$\mathbf{Z}_{\text{EQ}} = \mathbf{Z}_G + \mathbf{Z}_T = 10 + j20 + R_T - j15$$

$$= 10 + 0 + j(20 - 15) = 10 + j5 \ \Omega$$

By Eq. 8.6, for maximum I_L and maximum P_L,

$$X_L = -X_{\text{EQ}} = -5 \ \Omega$$

If a variation in R_L were permitted, then, for maximum power transfer from this source, R_L would be made equal to $R_{\text{EQ}} = 10$ Ω.

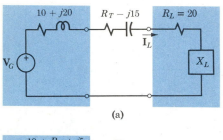

(a)

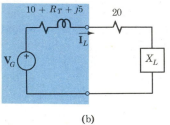

(b)

Figure 8.5 Maximizing power.

Impedance matching is used when the available power is limited and efficiency is not important, for example, when connecting a loudspeaker to a given amplifier. Under the conditions for maximum power transfer, the efficiency is just 50% since $I^2R_L = I^2R_G$ and half the power is dissipated internally in the equivalent source. No electrical appliance is designed to draw maximum power from a wall outlet, which typically has an equivalent internal impedance of a small fraction of an ohm. Instead, an electrical load is usually designed to draw a definite amount of power and, for high efficiency and low voltage drop, the impedance of the source is made as low as is economically feasible.

Reciprocity Theorem

Another interesting principle (which we need for a development later in the chapter) can be demonstrated by the circuit of Fig. 8.6a. For convenience in analysis,

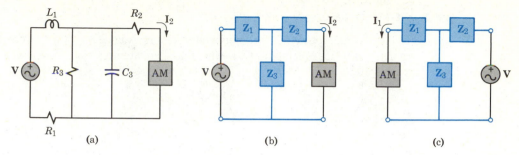

Figure 8.6 Reciprocity in linear networks.

the circuit elements are replaced (Fig. 8.6b) by their equivalent impedances evaluated at the frequency of the voltage source **V**. Using the current-divider concept, the current read by the ammeter in branch 2 is

$$\mathbf{I}_2 = \frac{\mathbf{Z}_3}{\mathbf{Z}_2 + \mathbf{Z}_3} \cdot \frac{\mathbf{V}}{\mathbf{Z}_1 + \dfrac{\mathbf{Z}_2 \mathbf{Z}_3}{\mathbf{Z}_2 + \mathbf{Z}_3}} = \frac{\mathbf{Z}_3}{\mathbf{Z}_1 \mathbf{Z}_2 + \mathbf{Z}_1 \mathbf{Z}_3 + \mathbf{Z}_2 \mathbf{Z}_3} \mathbf{V} \tag{8-7}$$

Suppose the voltage source and the ammeter are interchanged as in Fig. 8.6c. How will the current indicated by the ammeter compare to that in the original circuit? Using the current-divider concept again,

$$\mathbf{I}_1 = \frac{\mathbf{Z}_3}{\mathbf{Z}_1 + \mathbf{Z}_3} \cdot \frac{\mathbf{V}}{\mathbf{Z}_2 + \dfrac{\mathbf{Z}_1 \mathbf{Z}_3}{\mathbf{Z}_1 + \mathbf{Z}_3}} = \frac{\mathbf{Z}_3}{\mathbf{Z}_1 \mathbf{Z}_2 + \mathbf{Z}_2 \mathbf{Z}_3 + \mathbf{Z}_1 \mathbf{Z}_3} \mathbf{V} \tag{8-8}$$

The current is exactly the same! This calculation illustrates a principle, called the *reciprocity theorem*, which can be proved in general form. One statement of the reciprocity theorem is as follows:

> *In any passive, linear network, if a voltage* **V** *applied in branch 1 causes a current* **I** *to flow in branch 2, then voltage* **V** *applied in branch 2 will cause current* **I** *to flow in branch 1.*

The ratio of a voltage $\mathbf{V}_1$ in one part of a network to a current $\mathbf{I}_2$ in another part is called the *transfer impedance* $\mathbf{Z}_{12} = \mathbf{V}_1 / \mathbf{I}_2$. This network parameter has all the properties of an impedance; for example, the pole-zero diagram of the transfer impedance $\mathbf{Z}_{12}$ determines the character of the response i_2 to an excitation v_1. (In a similar way, the frequency response of an amplifier is determined by a *transfer function* relating output to input variables.) An important conclusion of the reciprocity theorem is:

> *In passive, linear networks, the transfer impedance* $\mathbf{Z}_{12}$ *is just equal to the reciprocal transfer impedance* $\mathbf{Z}_{21}$.

In Fig. 8.6, the transfer impedance is

$$\mathbf{Z}_{12} = \frac{\mathbf{V}_1}{\mathbf{I}_2} = \mathbf{Z}_{21} = \frac{\mathbf{V}_2}{\mathbf{I}_1} = \frac{\mathbf{Z}_1 \mathbf{Z}_2 + \mathbf{Z}_2 \mathbf{Z}_3 + \mathbf{Z}_3 \mathbf{Z}_1}{\mathbf{Z}_3} \tag{8-9}$$

TWO-PORT NETWORKS

If we attempt to find the equivalent series resistance R_{ab} of the circuit in Fig. 8.7a by network reduction methods, we find that there are no resistances in series or parallel

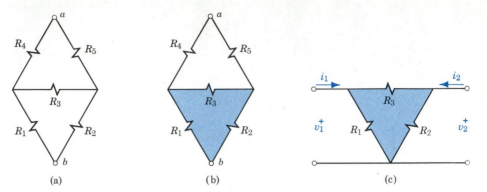

Figure 8.7   Analysis of a bridge circuit.

to be combined. There is something different about this circuit. (What is it?) The network reduction formulas previously derived are applicable to two-terminal networks only; in this circuit there is a *three-terminal network* that cannot be replaced by a two-terminal equivalent. However, if the three-terminal network shaded in Fig. 8.7b is transformed into a properly selected three-terminal equivalent network, the difficulty is removed and the equivalent series resistance can be found.

Two-Ports

The three-terminal network is a special case of the general *two-port* suggested by Fig. 8.7c. By convention, $+i_1$ and $+i_2$ are assumed *into* the network. Note that the current at the third terminal is defined by i_1 and i_2, and the third voltage is defined by v_1 and v_2. In general, of the four variables only two are independent and we may write

$$\begin{cases} v_1 = f_1(i_1, i_2) \\ v_2 = f_2(i_1, i_2) \end{cases} \quad \text{or} \quad \begin{cases} i_1 = \phi_1(v_1, v_2) \\ i_2 = \phi_2(v_1, v_2) \end{cases} \tag{8-10}$$

Equations similar to Eqs. 8-10 are useful in characterizing real devices such as transistors and motors. For the special case of ideal elements, the functions are linear and Eqs. 8-10 become

$$\begin{cases} v_1 = k_1 i_1 + k_2 i_2 \\ v_2 = k_3 i_1 + k_4 i_2 \end{cases} \quad \text{or} \quad \begin{cases} i_1 = k_5 v_1 + k_6 v_2 \\ i_2 = k_7 v_1 + k_8 v_2 \end{cases} \tag{8-11}$$

For sinusoidal excitation, the voltages and currents are represented by phasors, the constants are complex impedances or admittances, and Eqs. 8-11 become

$$\begin{cases} \mathbf{V}_1 = \mathbf{Z}_{11}\mathbf{I}_1 + \mathbf{Z}_{12}\mathbf{I}_2 \\ \mathbf{V}_2 = \mathbf{Z}_{21}\mathbf{I}_1 + \mathbf{Z}_{22}\mathbf{I}_2 \end{cases} \quad \text{or} \quad \begin{cases} \mathbf{I}_1 = \mathbf{Y}_{11}\mathbf{V}_1 + \mathbf{Y}_{12}\mathbf{V}_2 \\ \mathbf{I}_2 = \mathbf{Y}_{21}\mathbf{V}_1 + \mathbf{Y}_{22}\mathbf{V}_2 \end{cases} \tag{8-12}$$

For linear elements, $\mathbf{Z}_{12} = \mathbf{Z}_{21}$ and $\mathbf{Y}_{12} = \mathbf{Y}_{21}$ by the theorem of reciprocity, and we conclude that any linear two-port can be characterized by a set of three parameters.

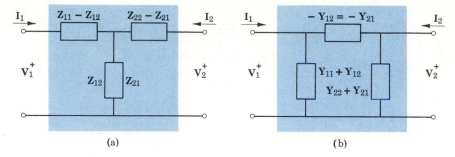

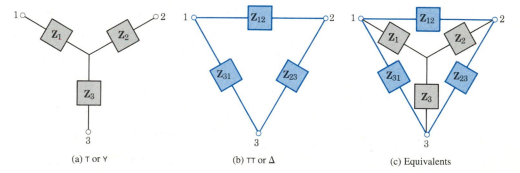

Figure 8.8   Equivalent T and TT networks.

Comparing Eq. 8-12a with the circuit of Fig. 8.8a, we see that one possible arrangement of the impedance parameters is in a T configuration as shown. Writing loop equations for Fig. 8.8a,

$$V_1 = (Z_{11} - Z_{12} + Z_{12})I_1 + Z_{12}I_2 = Z_{11}I_1 + Z_{12}I_2$$

$$V_2 = Z_{21}I_1 + (Z_{22} - Z_{21} + Z_{21})I_2 = Z_{21}I_1 + Z_{22}I_2$$

as required. Comparing Eq. 8-12b with the circuit of Fig. 8.8b, we see that one possible arrangement of the admittance parameters is in a TT configuration as shown. This can be confirmed by writing the appropriate node equations. Other circuit arrangements are possible and will be developed when needed.

T-TT Transformation

For our present purposes, let us consider the T and TT configurations as three terminal networks. As redrawn in Fig. 8.9, these can also be called Y and Δ networks.

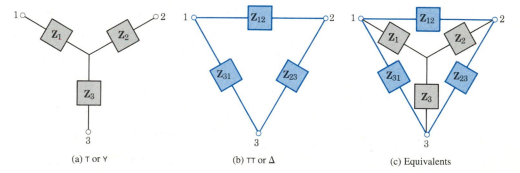

(a) T or Y   (b) TT or Δ   (c) Equivalents

Figure 8.9   Equivalent three-terminal networks.

Here all elements are designated by their impedances (rather than admittances) and lettered in a consistent scheme.

We specified that two passive two-terminal networks are equivalent if they have the same terminal impedance; if this is true, one can be substituted for the other.

Extending this concept of equivalence to three-terminal networks, we say that the T and π networks are equivalent if the impedance at any pair of terminals of one network is just equal to the impedance at the corresponding pair of terminals of the other. On the basis of this equality we can determine the conditions under which one network can be substituted for the other. As noted previously, in practical networks the equivalence may hold for only a single frequency.

The procedure is straightforward, but the algebra is laborious. Equating the impedances of the two networks between terminals 1 and 2 (terminal 3 open),

$$\mathbf{Z}_1 + \mathbf{Z}_2 = \frac{\mathbf{Z}_{12}(\mathbf{Z}_{23} + \mathbf{Z}_{31})}{\mathbf{Z}_{12} + \mathbf{Z}_{23} + \mathbf{Z}_{31}} \qquad (8\text{-}13)$$

Between terminals 2 and 3,

$$\mathbf{Z}_2 + \mathbf{Z}_3 = \frac{\mathbf{Z}_{23}(\mathbf{Z}_{31} + \mathbf{Z}_{12})}{\mathbf{Z}_{12} + \mathbf{Z}_{23} + \mathbf{Z}_{31}} \qquad (8\text{-}14)$$

Between terminals 3 and 1,

$$\mathbf{Z}_3 + \mathbf{Z}_1 = \frac{\mathbf{Z}_{31}(\mathbf{Z}_{12} + \mathbf{Z}_{23})}{\mathbf{Z}_{12} + \mathbf{Z}_{23} + \mathbf{Z}_{31}} \qquad (8\text{-}15)$$

Note the symmetry of these three equations; it is due to the inherent symmetry of the networks. The second equation can be written directly from the first by a cyclical change in subscripts.

To solve for the T (or Y) elements in terms of the π (or Δ) elements, we first eliminate $\mathbf{Z}_3$ by subtracting Eq. 8-15 from Eq. 8-14. Then this result is subtracted from Eq. 8-13 to eliminate $\mathbf{Z}_2$. The result is

$$\mathbf{Z}_1 = \frac{\mathbf{Z}_{12}\mathbf{Z}_{31}}{\mathbf{Z}_{12} + \mathbf{Z}_{23} + \mathbf{Z}_{31}} \qquad (8\text{-}16)$$

By symmetry (Fig. 8.9c) we conclude that:

The impedance of the equivalent Y element is the product of the adjacent Δ elements divided by the sum of the Δ elements.

This is called the Δ-Y (delta-wye) or π-T (pi-tee) transformation.

If instead we solve for the equivalent Δ (or π) elements in terms of the Y (or T) elements, the result is

$$\mathbf{Z}_{12} = \frac{\mathbf{Z}_1\mathbf{Z}_2 + \mathbf{Z}_2\mathbf{Z}_3 + \mathbf{Z}_3\mathbf{Z}_1}{\mathbf{Z}_3} \qquad (8\text{-}17)$$

Again we invoke symmetry and conclude that:

The impedance of the equivalent Δ element is the sum of the products of the Y elements divided by the opposite Y element.

Another formulation of Eq. 8-17 is obtained if we recognize and take advantage of the duality relation between the Y and Δ networks. From Eq. 8-16 we can write directly

$$\mathbf{Y}_{12} = \frac{\mathbf{Y}_1\mathbf{Y}_2}{\mathbf{Y}_1 + \mathbf{Y}_2 + \mathbf{Y}_3} \qquad (8\text{-}18)$$

EXAMPLE 4

Three equal impedances $\mathbf{Z} = 10\underline{/60°}\ \Omega$ are connected in Y to form a three-phase load (Fig. 8.10). Determine the impedances of the equivalent Δ.

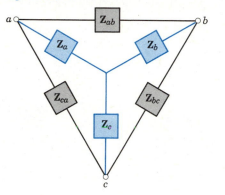

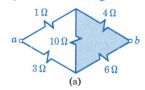

Figure 8.10 Y to Δ conversion.

For this balanced network, Eq. 8-17 yields

$$\mathbf{Z}_\Delta = \mathbf{Z}_{ab} = \frac{\mathbf{Z}_a\mathbf{Z}_b + \mathbf{Z}_b\mathbf{Z}_c + \mathbf{Z}_c\mathbf{Z}_a}{\mathbf{Z}_c}$$

$$= \frac{3(\mathbf{Z}_Y)^2}{\mathbf{Z}_Y} = 3\mathbf{Z}_Y = 30\underline{/60°}\ \Omega$$

The impedances of the equivalent Δ load are just equal to three times the corresponding impedances of the balanced Y load.

The phase angle of the impedances and the power factor of the load are unchanged.

These are not equations to be memorized; rather they are formulas to be applied in specific problems. A common application is in the simplification of networks where two-terminal network reduction methods do not work.

EXAMPLE 5

Determine the equivalent series resistance of the bridge network of Fig. 8.11a.

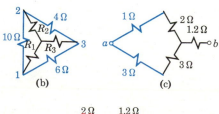

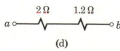

Figure 8.11 Δ to Y transformation.

The 10-Ω resistance is part of two three-terminal circuits that cannot be simplified by series-parallel combination.

First, the shaded Δ is replaced by an equivalent Y (Fig. 8.11b), using Eq. 8-16.

$$Z_1 = \frac{Z_{12}Z_{31}}{Z_{12} + Z_{23} + Z_{31}} = \frac{10 \times 6}{10 + 6 + 4} = \frac{60}{20} = 3\ \Omega$$

$$Z_2 = \frac{Z_{23}Z_{12}}{Z_{12} + Z_{23} + Z_{31}} = \frac{4 \times 10}{10 + 6 + 4} = \frac{40}{20} = 2\ \Omega$$

$$Z_3 = \frac{Z_{31}Z_{23}}{Z_{12} + Z_{23} + Z_{31}} = \frac{6 \times 4}{10 + 6 + 4} = \frac{24}{20} = 1.2\ \Omega$$

When this substitution is made, resistances in series and parallel can be combined to give

$$Z_{ab} = \frac{(3 + 3)(1 + 2)}{(3 + 3) + (1 + 2)} + 1.2$$

$$= \frac{6 \times 3}{6 + 3} + 1.2 = 2 + 1.2 = 3.2\ \Omega$$

Coupling Circuits

The two-port is frequently used as a *coupling circuit* to tie one component of a system to another in an optimum way. The π circuit of Fig. 8.12a is used to couple two stages of amplification; the input voltage v_1 contains ac signals and a dc component inherent in the operation of the transistors that provide amplification. The output voltage contains only ac signals because the coupling capacitor C_C blocks the dc component.

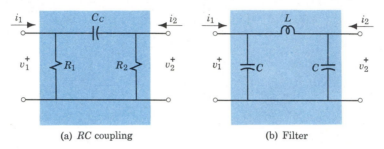

(a) *RC* coupling   (b) Filter

Figure 8.12   Two-ports as coupling circuits.

The inverse function is performed by the *low-pass filter* (Fig. 8.12b). This is another form of the special circuit coupling a rectifier and its load. AC components of the rectified wave are blocked by the high inductive reactance and shunted to ground through the low capacitive reactance; dc power flows unimpeded to the device connected at the output terminals.

The two-ports examined so far provide *conductive* coupling, so-called because there is a direct electrical connection between the output and the input. Another possibility is to employ a magnetic field to couple energy from one component to another; in this important category are transformers and induction motors. In the transformer, the output is electrically isolated from the input. In the induction motor, the rotating winding is mechanically and electrically isolated from the stationary winding. From a circuit viewpoint, these important practical features are described in terms of the circuit parameter called "mutual inductance."

Mutual Inductance

As you may recall from physics, the voltage v in volts induced in a coil of n turns encircling a changing magnetic flux ϕ in webers is

$$v = n\frac{d\phi}{dt} \tag{8-19}$$

If the flux is due to current flowing in the coil itself (Fig. 8.13a), the flux is proportional to the current and the number of turns or $\phi = kni$ and

$$v = n\frac{d(kni)}{dt} = kn^2\frac{di}{dt} = L\frac{di}{dt} \tag{8-20}$$

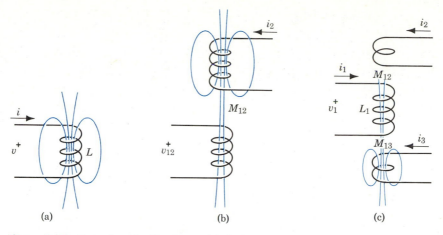

Figure 8.13 Examples of self and mutual inductance.

where L is the *self-inductance* in henrys. Up to this point, only this form of inductance has been discussed, so we have referred to L simply as "inductance."

If the changing flux is due to a current i_2 in a second coil (Fig. 8.13b), the voltage v_{12} induced in coil 1 is

$$v_{12} = M_{12}\frac{di_2}{dt} \qquad (8\text{-}21)$$

where M_{12} is the *mutual inductance* in henrys. In the same way, a voltage v_{13} can be induced in coil 1 due to a changing current i_3 in a third coil. For a linear system, superposition applies and the voltage appearing across coil 1 in Fig. 8.13c is

$$v_1 = v_{11} + v_{12} + v_{13} = L_1\frac{di_1}{dt} + M_{12}\frac{di_2}{dt} + M_{13}\frac{di_3}{dt} \qquad (8\text{-}22)$$

The magnitude and sign of mutual inductance M_{12} depend on the proximity and the orientation of the two coils. If the orientation is such that the flux contributed by current i_2 subtracts from that due to current i_1, the corresponding voltage term v_{12} is negative. If it is desired that mutual inductance M_{12} always be a positive quantity, then a negative sign must be used with the $M_{12}(di_2/dt)$ term in Eq. 8-22. There are conventional methods of designating coil terminals so that this sign is determined automatically.

Inductive Coupling

A practical form of magnetic or inductive coupling is shown in Fig. 8.14a. To increase the coupling, that is, to increase the mutual inductance M, the two real coils represented by this linear model may be wound on a common core with favorable magnetic properties, but the two coils are electrically isolated. This inductively coupled two-port appears in a wide variety of practical forms. In a radio receiver, a few turns of fine wire on an insulating form couple the antenna to the first amplifier stage. In a power system, many turns of heavy wire wound on an iron core several meters tall couple the generator to the high-voltage transmission line.

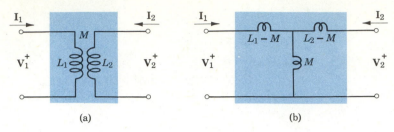

Figure 8.14 An inductively coupled circuit and the equivalent T circuit.

For purposes of analysis it may be desirable to replace the inductively coupled two-port by the equivalent T circuit of Fig. 8.14b. For sinusoids, the voltage developed in the input of Fig. 8.14a by the input current is $V_{11} = j\omega L_1 I_1$ where $j\omega L_1$ is the self-impedance Z_{11}. The voltage V_{12} coupled into the input by a current I_2 is $V_{12} = j\omega M_{12} I_2$ where $j\omega M_{12}$ is a mutual impedance Z_{12}. By the reciprocity theorem, for linear networks $Z_{12} = Z_{21}$ or $M_{12} = M_{21} = M$ as shown. Then

$$V_1 = j\omega L_1 I_1 + j\omega M I_2 = Z_{11} I_1 + Z_{12} I_2$$
$$V_2 = j\omega M I_1 + j\omega L_2 I_2 = Z_{21} I_1 + Z_{22} I_2$$

$$(8\text{-}23)$$

Comparing Eqs. 8-23 with the circuit of Fig. 8.14b, we see that the T circuit shown is equivalent. In the equivalent conductively coupled circuit, the common element is a self-inductance of magnitude M.

In Example 6, the effect of the secondary circuit on the input impedance is as if an impedance $(\omega M)^2/Z_{22}$ were connected in series with the primary. At resonance, Z_{22}

EXAMPLE 6

A series RLC circuit is inductively coupled to an input circuit consisting of R_1 and L_1 (Fig. 8.15). Determine the total input impedance.

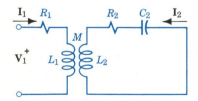

Figure 8.15 Coupled impedance.

The "primary" voltage is the sum of the $I_1 Z_{11}$ voltage drop and the voltage induced by the "secondary" current or

$$V_1 = I_1 Z_{11} + j\omega M I_2 = I_1(R_1 + j\omega L_1) + j\omega M I_2$$

For the secondary loop, self-impedance Z_{22} consists of L_2, R_2, and C_2. Therefore,

$$V_2 = j\omega M I_1 + Z_{22} I_2 = 0$$

$$= j\omega M I_1 + \left(R_2 + j\omega L_2 - j\frac{1}{\omega C_2}\right) I_2 \qquad (8\text{-}24)$$

Solving Eq. 8-24 for I_2 and substituting

$$V_1 = I_1 Z_{11} + j\omega M\left(-\frac{j\omega M I_1}{Z_{22}}\right)$$

Then the input impedance is

$$Z_{in} = \frac{V_1}{I_1} = Z_{11} + \frac{(\omega M)^2}{Z_{22}} \qquad (8\text{-}25)$$

is a relatively small pure resistance, but the impedance coupled into the primary may be a relatively large resistance.

Transformer Coupling

If the two coils of a transformer[†] are wound on the same highly permeable iron core, both coils are linked by essentially the same magnetic flux ϕ. Because $v = N\, d\phi/dt$, a changing flux in the core induces voltages in the windings propor-

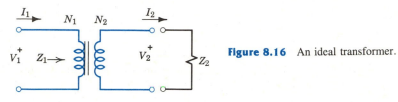

Figure 8.16   An ideal transformer.

tional to the number of turns. In the ideal transformer of Fig. 8.16 (closely approximated by a real transformer),

$$\frac{V_2}{V_1} = \frac{N_2}{N_1} \tag{8-26}$$

The current ratio can be predicted by noting that in an ideal transformer there is no power loss, and the power in must equal the power out. Therefore, a *step-down* in current must accompany a *step-up* in voltage or

$$\frac{I_2}{I_1} = \frac{V_1}{V_2} = \frac{N_1}{N_2} \tag{8-27}$$

If an impedance Z_2 is connected to the *secondary*, the impedance seen at the *primary* is, by Eqs. 8-26 and 8-27,

$$Z_1 = \frac{V_1}{I_1} = \frac{(N_1/N_2)V_2}{(N_2/N_1)I_2} = \left(\frac{N_1}{N_2}\right)^2 \frac{V_2}{I_2} = \left(\frac{N_1}{N_2}\right)^2 Z_2 \tag{8-28}$$

On the basis of this brief discussion, we can see three important advantages of the transformer as a coupling device:

1. AC power is easily and efficiently transformed from the optimum generating voltage of, say, 25 kV up to an efficient long-distance transmission voltage of 500 kV, down to an optimum urban distribution voltage of 50 kV, and then down to a safe utilization voltage of 120 or 240 V.
2. Only a changing flux induces voltage; direct currents are not "transformed." In an amplifier, for example, a transformer passes on the ac signal but filters out dc currents that might adversely affect some amplifier components.
3. The transformer transforms impedance and facilitates impedance matching. By proper choice of the turns ratio, the impedance of a given load can be made to appear to be optimum for a given source. For example, the low resistance of a loudspeaker can be matched to the much higher output resistance of an amplifier.

[†]For a detailed discussion see pages 594–598.

SUMMARY

- Two passive one-ports are equivalent if they have the same input impedance $\mathbf{Z}$ or the same input admittance $\mathbf{Y}$.

  In general, two real networks are equivalent at one frequency only.

  For impedances in series, $\mathbf{Z}_{EQ} = \mathbf{Z}_1 + \mathbf{Z}_2 + \cdots + \mathbf{Z}_n$

  For admittances in parallel, $\mathbf{Y}_{EQ} = \mathbf{Y}_1 + \mathbf{Y}_2 + \cdots + \mathbf{Y}_n$

  Voltage division: $\mathbf{V}_2 = \dfrac{\mathbf{Z}_2}{\mathbf{Z}_{EQ}} \mathbf{V}$ Current division: $\mathbf{I}_2 = \dfrac{\mathbf{Y}_2}{\mathbf{Y}_{EQ}} \mathbf{I}$

- If cause and effect are linearly related, the principle of superposition applies and the total effect of several signals acting simultaneously is equal to the sum of the effects of the individual signals acting one at a time.

- Insofar as a load is concerned, any one-port network of linear elements and energy sources can be replaced by a series combination of an ideal voltage source $\mathbf{V}_T$ and a linear impedance $\mathbf{Z}_T$ (Thévenin's equivalent) or by a parallel combination of an ideal current source $\mathbf{I}_N$ and a linear admittance $\mathbf{Y}_N$ (Norton's equivalent).

- For maximum power transfer from a given source, the impedance of the load should be made equal to the complex conjugate of the Thévenin equivalent impedance of the source.

- The reciprocity theorem says: In any passive, linear network, if a voltage $\mathbf{V}$ applied in branch 1 causes a current $\mathbf{I}$ to flow in branch 2, then voltage $\mathbf{V}$ applied in branch 2 will cause current $\mathbf{I}$ to flow in branch 1.

  In such a network, transfer impedance $\mathbf{Z}_{12} = \mathbf{V}_1/\mathbf{I}_2 = \mathbf{V}_2/\mathbf{I}_1 = \mathbf{Z}_{21}$.

- Any linear two-port network can be characterized by a set of three parameters and represented by an equivalent T (Y) or π (Δ) network.

  For sinusoidal excitation, the voltages and currents are represented by phasors, the constants are complex impedances or admittances, and Eqs. 8-13 become

$$\begin{cases} \mathbf{V}_1 = \mathbf{Z}_{11}\mathbf{I}_1 + \mathbf{Z}_{12}\mathbf{I}_2 \\ \mathbf{V}_2 = \mathbf{Z}_{21}\mathbf{I}_1 + \mathbf{Z}_{22}\mathbf{I}_2 \end{cases} \quad \text{or} \quad \begin{cases} \mathbf{I}_1 = \mathbf{Y}_{11}\mathbf{V}_1 + \mathbf{Y}_{12}\mathbf{V}_2 \\ \mathbf{I}_2 = \mathbf{Y}_{21}\mathbf{V}_1 + \mathbf{Y}_{22}\mathbf{V}_2 \end{cases}$$

  For linear elements, $\mathbf{Z}_{12} = \mathbf{Z}_{21}$ and $\mathbf{Y}_{12} = \mathbf{Y}_{21}$ by the theorem of reciprocity.

- Coupling circuits are two-ports used to connect one component of a system to another in an optimum way.

  If two ideal coils are inductively coupled, mutual inductance M exists and the voltage across coil 1 is

$$v_1 = L_1 \frac{di_1}{dt} + M \frac{di_2}{dt}$$

- Transformer coupling optimizes power transfer, filters out dc currents, and permits impedance matching. For an ideal transformer,

$$\frac{V_2}{V_1} = \frac{N_2}{N_1} \qquad \frac{I_2}{I_1} = \frac{N_1}{N_2} \qquad \frac{Z_1}{Z_2} = \left(\frac{N_1}{N_2}\right)^2$$

REVIEW QUESTIONS

1. What is the difference between a "dual" and an "equivalent"?
2. Explain in words the two equations used to define linearity.
3. Under what circumstances is Thévenin's theorem useful?
4. Is the same power dissipated in an active circuit and the Norton equivalent?
5. How are voltage and current sources "removed"?
6. Energy is to be added to a ball by throwing. For maximum energy would you choose a marble, a golf ball, a baseball, or a medicine ball? Why?
7. Why is the gear ratio different on men's and women's bicycles?
8. Why is maximum power transfer sometimes un-desirable? Cite a practical example.
9. Explain how impedances are matched by using a transformer.
10. How is the fact that $\mathbf{Z}_{12} = \mathbf{Z}_{21}$ related to the reciprocity theorem?
11. Draw a T network of resistances and add a resistance between terminals 1 and 2. Outline the procedure for reducing this to a π network.
12. Is the same power dissipated in a Δ three-phase load and the equivalent Y load?
13. Sketch, qualitatively, the frequency response curve of V_2/V_1 versus ω for the coupling circuit of Fig. 8.12a.
14. Why is inductive coupling advantageous in a motor?

EXERCISES

1. A one-port consists of $\mathbf{Z}_1 = 20\,\underline{/53.1°}$ Ω in series with $\mathbf{Z}_2 = 9\,\underline{/-90°}$ Ω.
   (a) Find R and X of an equivalent series combination.
   (b) Find G and B of an equivalent parallel combination.
2. For voltage $\mathbf{V} = 120\,\underline{/0°}$ V applied across the $\mathbf{Z}_1$ and $\mathbf{Z}_2$ of Exercise 1, predict $\mathbf{V}_1$ and $\mathbf{V}_2$.
3. A one-port consists $\mathbf{Z}_1 = 10\,\underline{/+30°}$ Ω and $\mathbf{Z}_2 = 20\,\underline{/-45°}$ Ω connected in parallel.
   (a) Find R and X of an equivalent series combination.
   (b) Find G and B of an equivalent parallel combination.
4. For $\mathbf{I} = 10\,\underline{/0°}$ A into the one-port of Exercise 3, find $\mathbf{I}_1$ and $\mathbf{I}_2$.
5. In Fig. 8.17, the current in the 12-Ω inductive reactance is $12\,\underline{/90°}$ A. Find the current in L.
6. In Fig. 8.17, the voltage across the capacitive reactance is $20\,\underline{/-56.75°}$ V. Find the voltage across L.

7. In Fig. 8.17, the reactances are given in ohms for $\omega = 1000$ rad/s. Find the steady-state current in L if:
   (a) $\mathbf{V}$ is replaced by $v = 120 + 120\sqrt{2}\,\cos 1000t$ V.
   (b) $\mathbf{V}$ is replaced by $v = 120\sqrt{2}\,\cos 1000t + 120\,\cos 4000t$ V.
8. In Fig. 8.18, $\mathbf{V} = 120\,\underline{/0°}$ V, $\mathbf{I} = 8\,\underline{/0°}$ A, and $\mathbf{Z} = 10\,\underline{/0°}$ Ω. Use superposition to find the current through $\mathbf{Z}$.

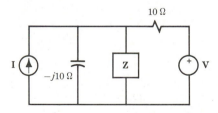

Figure 8.18

9. In Fig. 8.17, $\mathbf{V} = 20\,\underline{/-90°}$ V and the load consists of an inductive reactance $X_L = 60$ Ω.
   (a) Using Norton's theorem, replace the remainder of the network by a simple combination of current source and admittance and determine the current in X_L.
   (b) Check your answer using loop equations.

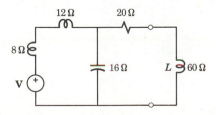

Figure 8.17

10. In Fig. 8.18, $\mathbf{I} = 8\underline{/0°}$ A, $\mathbf{V} = 60\underline{/90°}$ V, and $\mathbf{Z} = 10\underline{/90°}$ Ω. Replace $\mathbf{V}$ and the 10-Ω resistance by a Norton equivalent and predict $\mathbf{I}_Z$.

11. In the circuit of Exercise 10, replace all except $\mathbf{Z}$ by a Thévenin equivalent and predict $\mathbf{I}_Z$.

12. In Fig. 8.19, a variable load R_L is supplied through a transmission line $R_T = 15$ Ω from a voltage generator $V = 600$ V and a current generator $I = 10$ A.
    (a) Replace the two generators by a single equivalent current generator.
    (b) Replace the generators and transmission line by a single equivalent voltage generator.
    (c) Determine the power delivered to a load $R_L = 25$ Ω.
    (d) Determine the power loss in R_T.

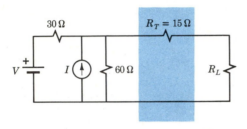

Figure 8.19

13. Repeat part (c) of Exercise 12 using superposition.

14. Two measurements on a human nerve cell indicate an open-circuit voltage of 80 mV and a current of 5 nA through a 6-MΩ load. You are to predict the behavior of this cell with respect to an external load.
    (a) What simplifying assumption must be made?
    (b) What is the maximum power available from the cell?

15. A load $R_L = 20$ Ω is supplied through a transmission line R_T from a generator $V_G = 240$ V, $R_G = 30$ Ω as shown in Fig. 8.20. Specify the value of R_T for:
    (a) Maximum power delivered *to* the load.
    (b) Maximum power supplied *by* the generator.
    (c) Maximum power dissipated in R_T.

16. In the circuit of Fig. 8.18, $\mathbf{I} = 8\underline{/0°}$ A and $\mathbf{V} = 120\underline{/0°}$ V. Specify the value of $\mathbf{Z}$ to absorb maximum power and calculate $P_{\max}$.

17. A current $\mathbf{I} = 2\underline{/0°}$ A flows between terminals 1 and 2 in a complicated linear network when a voltage $\mathbf{V} = 10\underline{/90°}$ V is applied between ter-

minals 3 and 4. If a voltage $\mathbf{V} = 25\underline{/+45°}$ V is applied between terminals 1 and 2, what current would be observed to flow between terminals 3 and 4? State the theorems used.

18. In Fig. 8.17, an ammeter placed in series with L reads I_L.
    (a) Calculate the transfer impedance $\mathbf{Z}_{12} = \mathbf{V}/\mathbf{I}_L$.
    (b) Redraw the circuit to show $\mathbf{V}$ in series with L and the ammeter to read the current in the 8-Ω inductive reactance.
    (c) Calculate the new value of $\mathbf{V}/\mathbf{I}$ and draw a conclusion.

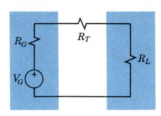

Figure 8.20

19. For the two-port of Fig. 8.21, evaluate the parameters k_5, k_6, k_7, k_8 (Eq. 8-11b). Determine the parameters of an equivalent π network following the scheme of Fig. 8.8b.

20. From Eqs. 8-13, 14, and 15, derive Eq. 8-16 for the equivalent Y element $\mathbf{Z}_1$.

21. Derive Eq. 8-17 for the equivalent Δ element $\mathbf{Z}_{12}$.

22. In Fig. 8.21, the right-hand 2-Ω resistance is removed (shorted out). For the new T network, derive the equivalent π.

23. In the circuit of Fig. 8.22, $\mathbf{V} = 50\underline{/0°}$ V.
    (a) Find the equivalent impedance and calculate $\mathbf{I}$.
    (b) Calculate $\mathbf{I}_R$ and explain your result.

24. The input to the coupling circuit of Figure 8.12a is $v_1 = 10 + 10 \cos 10t + 10 \cos 1000t$. If $R_1 = R_2 = 10$ kΩ and $C_C = 1$ μF, predict v_2. In what sense is this a *filter*?

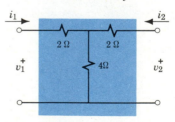

Figure 8.21

25. The input to the coupling circuit of Figure 8.12b is $v_1 = 10 + 10 \cos 500t + 10 \cos 2000t$. If $C = 10 \, \mu F$ and $L = 10$ H, predict v_2. In what sense is this a *filter*?

26. When two coils are connected in series, the total inductance is measured to be 18 mH. When the connections to one coil are reversed, the total series inductance is 30 mH.
(a) What is the mutual inductance?
(b) If the coils are similar, except that coil 2 has twice as many turns as coil 1, what are L_1 and L_2?

27. In Fig. 8.14a, $L_1 = 1$ H, $L_2 = 2$ H, and $M = 0.5$ H. A capacitance $C = 8 \, \mu F$ is connected across the output and a current $\mathbf{I}_1 = 15 \underline{/0°}$ A at

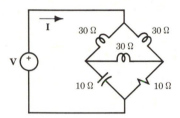

Figure 8.22

$\omega = 500$ rad/s is applied. Determine the voltage $\mathbf{V}_2$ and the input impedance $\mathbf{Z}_1$.

28. For the circuit of Fig. 8.15, sketch the magnitude of the input impedance as a function of frequency near the series resonant frequency of the secondary.

29. An ideal transformer (with 5 turns in the primary winding for every turn in the secondary) can deliver continuously a 10-A current at 250 V to a load. If the load is a pure resistance, determine:
(a) The primary voltage and current.
(b) The power input to the primary.
(c) The impedance seen at the primary terminals.

30. A loudspeaker with an input resistance $R_L = 8 \, \Omega$ is to be matched for maximum power transfer to an amplifier with an output resistance $R_o = 1800 \, \Omega$.
(a) Specify the turns ratio for an appropriate transformer.
(b) For open-circuit amplifier voltage $V_o = 36$ V, calculate the power transferred to the load with the transformer.
(c) Calculate the power to the loudspeaker if it were connected directly across the amplifier output and compare to the result of part (b).

PROBLEMS

1. A telephone amplifier with an output resistance $R_o = 600 \, \Omega$ is to feed a transmission line that can be represented by an infinite ladder of resistances as in Fig. 8.23. Design the transmission line; i.e., specify R for maximum power transfer.

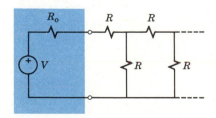

Figure 8.23

2. In Fig. 8.24, $i_1 = 15 + 10 \cos 2000t$ A, $R_1 = 4 \, \Omega$, $C = 125 \, \mu F$, $L = 1$ mH, $R = 2 \, \Omega$, and $v = 20 \cos 2000t$ V. Clearly indicating the methods used, take advantage of network theorems to determine the steady-state current i.

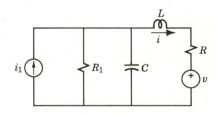

Figure 8.24

3. Telephone transmission lines can be represented by the ladder network shown in Fig. 8.25 where L and C are the inductance and capacitance per unit length. If the line is long, the impedance becomes equal to the *characteristic impedance* Z_O, the impedance looking into an *infinite* line.

(a) If the line is infinitely long, what is the impedance of the section to the right of terminals aa?

(b) Use this fact to evaluate the characteristic impedance of this line.

(c) Let the unit length over which L and C are measured approach zero and evaluate Z_O.

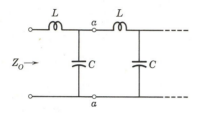

Figure 8.25

4. A load Z is supplied from two energy sources $I = 10\,\underline{/90°}$ A and $V = 200\,\underline{/0°}$ V as shown in Fig. 8.26. Specify the value of Z to absorb maximum power, and predict the power absorbed by the specified value of Z.

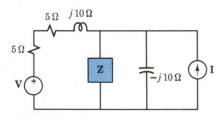

Figure 8.26

5. In Fig. 8.27, load resistance R_L is equal to $R_G/2$.

(a) Determine the power transferred to R_L if connected directly to the generator.

(b) To increase power transfer, the coupling circuit shown is suggested as an "impedance-matching" device. Define the relation between X_1, X_2 and X_3 that must be satisfied if power transfer is to be maximum.

(c) Outline a procedure for designing the coupling circuit.

(d) By what factor would the power transferred be increased by proper impedance matching?

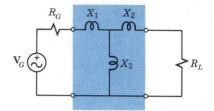

Figure 8.27

6. Simplify the circuit of Fig. 8.2a by T-π transformation and determine the current in the 20-Ω resistance.

7. A signal generator has an internal resistance $R_G = 500\ \Omega$. Design a coupling circuit (specify values of L and C in Fig. 8.28) to provide maximum power transfer at $\omega = 10{,}000$ rad/s to a load $R_L = 5000\ \Omega$.

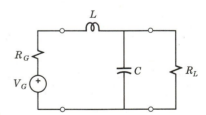

Figure 8.28

8. A coil with an inductance of 50 μH and a Q of 50 is inductively coupled ($M = 50\ \mu$H) to a second coil with $L_2 = 100\ \mu$H and $Q = 50$ (see Fig. 8.15). The secondary is tuned to series resonance at $f_o = 1.59$ MHz by means of C_2. Predict the input impedance Z_1 at the resonant frequency.

9

Introduction to Systems

Systems Engineering

Block Diagrams

Feedback Circuits

Transfer Functions

A system is a combination of diverse but interacting elements integrated to achieve an overall objective. The elements may be electrical devices, assembly lines, or human beings concerned with processing materials, information, or energy. The objective may be to guide a space vehicle, to control a chemical process, or to provide a valuable service.

In the health care system we call a hospital, for example, the human elements include doctors, nurses, and technicians using a great variety of instruments and apparatus and employing information in the form of observations and records. Patients are admitted, observed, prepared, treated, charged, and discharged. Actions by the doctor produce changes in the patient that are reported back to the doctor who compares patient condition to a reference and makes a decision that is communicated to the dispensary that transmits medicine to the nurse for administering to the patient.

SYSTEMS ENGINEERING

The increasing complexity of man-made systems and the growing body of principles and techniques for predicting system behavior have resulted in the activity called *systems engineering*. The systems engineer takes an overall view; he or she is interested in specifying the performance of the system, creating a combination of components to meet the specifications, describing the characteristics of components and their interconnections, and evaluating system performances. Systems engineers leave the design of components and connecting links to others; their focus is on external characteristics—performance, reliability, and cost.

Guidance System

From the standpoint of the systems engineer, a device or a subsystem can be represented by a simple *block* (or model) labeled to indicate the function performed (without indicating *how* it is performed). In the space vehicle guidance system in Fig. 9.1, transducer T converts the desired heading θ_R into reference signal V_R, and sensor

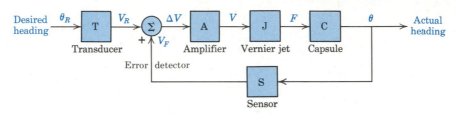

Figure 9.1   A space vehicle guidance system.

S converts the actual heading θ into signal V_F, which is fed back to the input. Error detector (or summer) Σ compares the actual heading with the desired heading and amplifier A magnifies the difference. An actuator (not shown) opens vernier jet J, which accelerates capsule C into the new heading. The elements and their outputs are diverse in nature, but no one element can be considered alone; the interconnections, particularly the feedback loop, determine the overall behavior. For harmonious operation and optimum performance the elements must be matched in such characteristics as impedance and frequency response.

A *feedback* loop for returning part of the output to the input in order to improve performance is an important component in many systems. In Chapter 3 we used feedback to provide stable operation of op amps. The concept of feedback is not new; it is inherent in many biological systems, and it is found in well-established social systems. The formulation of the principles of feedback is credited to H.S. Black in his studies in 1934; exploitation of the insight he provided has been widespread in engineering and electronics.

In electrical systems there are two general categories of feedback applications. One is concerned with the use of feedback for purposes of comparison and subsequent control, as in the vehicle guidance system (Fig. 9.1) and the automatic voltage regulator (Fig. 25.15). The analysis of the behavior of such control systems is the subject of Chapter 25. A second application of feedback is to improve the performance or characteristics of a device or system. Through the intentional use of feedback, the gain of amplifiers can be made much greater, or more stable, or less sensitive to change, or less dependent on frequency (Chapter 19). On the other hand, amplifier designers must take steps to prevent unintentional feedback from degrading the performance of their products. In this chapter we study the basic concept of feedback, learn how to reduce complicated systems to the basic form, investigate the general benefits of *positive* and *negative* feedback, and then round out our study of circuits by developing the concept of a transfer function.

Amplifier with Feedback

An electronic amplifier is represented by the circuit model of Fig. 9.2a. The transistors and circuit elements have been replaced by equivalent components where

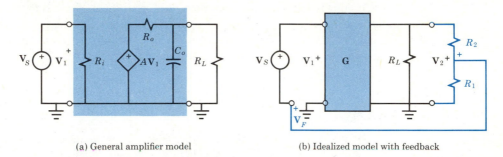

(a) General amplifier model   (b) Idealized model with feedback

Figure 9.2   An electronic amplifier with voltage feedback.

R_i and R_o are the input and output resistances, C_o is an output capacitance, and A is an open-circuit voltage amplification factor. Such an amplifier with feedback can be represented by the model of Fig. 9.2b. The voltage gain of the amplifier alone, called the *forward gain*, is defined by

$$\mathbf{G} = \frac{\mathbf{V}_2}{\mathbf{V}_1} \tag{9-1}$$

where $\mathbf{V}_1$ and $\mathbf{V}_2$ are phasors and $\mathbf{G}$ is a complex function of frequency.

The voltage divider consisting of R_1 and R_2 provides a means of tapping off a portion $\mathbf{V}_F$ of the output voltage $\mathbf{V}_2$ and feeding it back into the input, "closing the loop." The *feedback factor* is

$$\mathbf{H} = \frac{\mathbf{V}_F}{\mathbf{V}_2} \tag{9-2}$$

In general, $\mathbf{H}$ is a complex quantity although here $\mathbf{H} = R_1/(R_1 + R_2)$ is a real number.

With a fraction of the output fed back in series with input voltage $\mathbf{V}_S$, the amplifier input $\mathbf{V}_1$ is the sum of the signal $\mathbf{V}_S$ and the feedback voltage $\mathbf{V}_F$ or

$$\mathbf{V}_1 = \mathbf{V}_S + \mathbf{V}_F \tag{9-3}$$

The output voltage is now

$$\mathbf{V}_2 = \mathbf{G}\mathbf{V}_1 = \mathbf{G}(\mathbf{V}_S + \mathbf{V}_F) = \mathbf{G}(\mathbf{V}_S + \mathbf{H}\mathbf{V}_2) = \mathbf{G}\mathbf{V}_S + \mathbf{G}\mathbf{H}\mathbf{V}_2$$

Solving,

$$\mathbf{G}_F = \frac{\mathbf{V}_2}{\mathbf{V}_S} = \frac{\mathbf{G}}{1 - \mathbf{G}\mathbf{H}} \tag{9-4}$$

where $\mathbf{G}_F$ is the gain with feedback. Here "gain" is a voltage ratio; in general, gain relates two signals, which may be different in character, i.e., an output displacement in terms of an input current.

Equation 9-4 is the basic equation for system gain with feedback, and by studying this equation some of the possibilities of such a system can be foreseen. If the complex quantity **GH** is positive, as it approaches unity the gain increases without limit. If **GH** is just unity, the gain is infinite and there is the possibility of output with no input; the amplifier has become a self-excited *oscillator*. If **GH** is negative, the gain is reduced, but accompanying benefits may offset the loss in gain. For example, if **GH** is very large with respect to unity, $\mathbf{G}_F \cong -\mathbf{G}/\mathbf{GH} = -1/\mathbf{H}$, or the gain is independent of the gain of the amplifier itself. This is the condition we created to obtain precise amplification with any op amp whose forward gain is "high."

BLOCK DIAGRAMS

Before investigating the specific behavior of feedback systems, we need a simplified notation that will permit us to disregard the details of circuits and devices and focus our attention on system aspects. Also, our approach should emphasize the paths whereby signals are transmitted.

Block Diagram Representation

Let us represent linear two-port devices by blocks labeled to indicate the functions performed and use single-line inputs and outputs labeled to indicate the signals at those points. The amplifier whose small-signal circuit model is within the colored area in Fig. 9.2a is represented symbolically by a rectangle labeled **G** (Table 9-1). Such a symbol can represent any two-port device for which the output signal is a known function of the input signal. In general, the ratio of output signal to input signal is called the *transfer function*.

Table 9-1   Block Diagram Symbols and Defining Equations

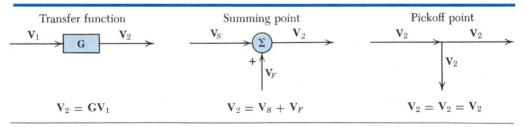

Transfer function	Summing point	Pickoff point
$\mathbf{V}_2 = \mathbf{GV}_1$	$\mathbf{V}_2 = \mathbf{V}_S + \mathbf{V}_F$	$\mathbf{V}_2 = \mathbf{V}_2 = \mathbf{V}_2$

To be represented by such a block, a device must be linear. Also, it is assumed that each device has infinite input impedance and zero output impedance so that devices may be interconnected without loading effects. In the system of Fig. 9.2b, $R_1 + R_2$ must be large with respect to R_L so that the amplifier gain is unaffected by the feedback connection. If this amplifier is to provide the input to another device that draws an appreciable current, a transistor amplifier, for example, transfer function **G** must be recalculated, taking into account the finite input impedance of the next device. When these conditions are met, the blocks are said to be *unilateral*; linear unilateral blocks may be interconnected freely.

Two different types of connection are possible (Table 9-1). A *summing point* has two or more inputs and a single output. The output is the sum or difference of the inputs, depending on the signs; a plus sign is assumed if no sign is indicated. In contrast, from a *pickoff point* the signal can be transmitted undiminished in several directions. Physically this is illustrated by a voltage that can serve as input to several devices with high input impedances or by a tachometer that senses the speed transmitted by a shaft without diminishing the speed.

Using block diagram notation, the amplifier of Fig. 9.2b is represented as in Fig. 9.3. In this case, all variables are voltages but, in general, different variables may

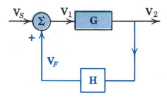

Figure 9.3 Block diagram of the basic feedback system.

appear on the diagram. (All signals entering a summing point must have the same units.) The amplifier block represents Eq. 9-1 and the feedback block represents Eq. 9-2. In this case, **G** is a voltage gain, but in general it is a transfer function. The summing point fulfills Eq. 9-3, and Eq. 9-4 gives the gain of the system. This basic *closed-loop* system appears so frequently that Eq. 9-4 should be memorized.

BLOCK DIAGRAM ALGEBRA

Analysis of the behavior of complicated systems can be simplified by following a few rules for manipulating block diagram elements. These rules are based on the definitions of block diagram symbols and could be derived each time, but it is more convenient to work from the rules.

1. *Any closed-loop system can be replaced by an equivalent open-loop system.*
2. *The gain of cascaded blocks is the product of the individual gains.*
3. *The order of summing does not affect the sum.*
4. *Shifting a summing point beyond a block of gain G_A requires the insertion of G_A in the variable added.*
5. *Shifting a pickoff point beyond a block of gain G_A requires the insertion of $1/G_A$ in the variable picked off.*

These rules and the corollaries of rules 4 and 5 are illustrated in Example 1.

EXAMPLE 1

Determine the overall gain of the system shown in Fig. 9.4a.

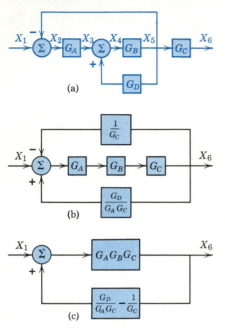

(a)

(b)

(c)

Figure 9.4   Block diagram algebra.

It would be possible to determine the gain by writing the equation for each element in turn and solving simultaneously. A preferable procedure is to simplify the system by block diagram algebra.

Applying rule 5, the pickoff point is moved beyond G_C and $1/G_C$ is inserted in the upper and lower loops.

Applying the corollary of rule 4 permits shifting the summing point for the lower loop before G_A and $1/G_A$ is inserted in that loop (Fig. 9.4b).

Applying rule 3, the two feedback loops are combined. Applying rule 2, the three cascaded blocks are combined. The result is in the basic form (Fig. 9.4c).

Applying rule 1, the overall gain can be written by inspection as

$$G_{61} = \frac{X_6}{X_1} = \frac{G_A G_B G_C}{1 - (G_A G_B)(G_D/G_A - 1)}$$

FEEDBACK CIRCUITS

By using the rules of block diagram algebra, all linear feedback networks can be reduced to the basic system of Fig. 9.3 for which the gain is $\mathbf{G}_F = \mathbf{G}/(1 - \mathbf{GH})$. The behavior of the system can be predicted by examining the quantity $1 - \mathbf{GH}$. In a complicated network, the denominator will be a polynomial, as in Example 1, and its poles and zeros will correspond to critical points in the behavior pattern.

Positive Feedback

For $\mathbf{GH}$ positive but less than unity, the denominator is less than unity and the system gain is greater than the forward gain $\mathbf{G}$. This condition of *positive feedback* was used in *regenerative receivers* to provide high gain in the days when electronic amplifiers were poor and signals were weak. Since operation with positive feedback becomes less stable as GH approaches unity, regenerative receivers required constant adjustment.

If the value of **GH** approaches $1 + j0$, the denominator approaches zero and the overall gain increases without limit. An amplifier of infinite gain can provide an output with no input; such a device is called an *oscillator* or signal generator and is an important part of many electronic systems.

Oscillation. In the amplifier of Fig. 9.2a, the output tends to decrease as the frequency increases because of the filtering effect of R_oC_o. This is analogous to the low-pass filter of Fig. 7.9. As we shall see in Chapter 19, the high-frequency gain can be expressed as

$$\mathbf{A} = A(j\omega) = \frac{A_O}{1 + j\omega C_o R_o} = \frac{A_O}{1 + j\omega/\omega_2}$$

where A_O is a constant representing the gain at moderate frequencies for which $\omega/\omega_2 \ll 1$, and ω_2 is the *upper cutoff frequency* at which **A** drops to 70% of A_O. At a somewhat higher frequency where $\omega = \sqrt{3}\,\omega_2$,

$$\mathbf{A}_H = \frac{A_O}{1 + j\sqrt{3}} = \frac{A_O}{2}\underline{/-\tan^{-1}\sqrt{3}} = \frac{A_O}{2}\underline{/-60°}$$

For a three-stage amplifier consisting of three similar sections (Fig. 9.5), at this

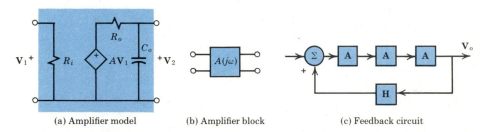

(a) Amplifier model   (b) Amplifier block   (c) Feedback circuit

Figure 9.5   A three-stage amplifier with feedback.

frequency the overall gain is

$$\mathbf{G} = \mathbf{A}_H^3 = \left(\frac{A_O}{2}\right)^3 \underline{/3(-60°)} = \left(\frac{A_O}{2}\right)^3 \underline{/-180°} = -\left(\frac{A_O}{2}\right)^3 + j0$$

Assuming feedback is present and $\mathbf{H} = 0.008$,

$$\mathbf{GH} = -0.008\left(\frac{A_O}{2}\right)^3 + j0 = -\left(\frac{A_O}{10}\right)^3 + j0$$

Where $\mathbf{G}_F = \mathbf{G}/(1 - \mathbf{GH})$, for oscillation $\mathbf{G}_F$ must be infinite or the denominator must be zero or

$$\mathbf{GH} = -\left(\frac{A_O}{10}\right)^3 + j0 = +1 + j0 \qquad \text{and} \qquad A_O = -10$$

These results are interpreted as follows: In a three-stage amplifier with a per-stage gain of -10, if 0.8% of the output voltage is fed back to the input, oscillation is possible at a frequency for which the per-stage phase shift is $60°$. In practice, there is a range of frequencies that will be amplified, but one of these will be maximized and the level of oscillation will build up until A_O is just equal to -10. In this case $\mathbf{G}$ is the frequency-dependent element; in other cases $\mathbf{H}$ may determine the frequency at which $\mathbf{GH} = 1 + j0$. (See Chapter 19.)

Another way of looking at the phenomenon of oscillation is to recognize that in any impulsive input, such as turning on a switch, components of energy corresponding to all frequencies are present. Also small voltages are generated by the random motion of electrons in conductors. Any such action can initiate the small signal, which builds up and up through amplification and feedback to the equilibrium level. For small signals, A_O is usually quite large (much greater than -10); as the signal level increases, the gain decreases due to nonlinearity and the amplitude of oscillation stabilizes at the equilibrium level.

The preceding illustration indicates that a circuit designed for amplification may produce oscillation if some small part of the output is coupled or fed back to the input with the proper phase relation. Opportunities for such coupling exist in the form of stray electric and magnetic fields or through mutual impedances of common power supplies. Amplifier designers use shielding and filtering to minimize coupling and reduce the possibility of oscillation.

Negative Feedback

Negative feedback, where the amplitude and phase of the signal fed back are such that the overall gain of the amplifier G_F is less than the forward gain G, may be used to advantage in amplifier design. Since additional voltage or current gain can be obtained easily with electronic devices, the loss in gain is not important if significant advantages are obtained. Among the possible improvements in amplifier performance are:

1. Gain can be made practically independent of device parameters; gain can be made insensitive to temperature changes, aging of components, and random variations in parameters.
2. Gain can be made practically independent of reactive elements and, therefore, insensitive to frequency.
3. Gain can be made selective to discriminate against noise, distortion, or system disturbances.
4. The input and output impedances of an amplifier can be greatly improved.

These applications of feedback are investigated in detail in Chapter 19. In connection with this introduction to systems, we look at just two examples.

Stabilizing Gain.   For precise performance or long-term operation, system behavior must be predictable and stable. However, amplifier components such as transistors and op amps vary widely in their critical parameters. By employing negative feedback and sacrificing gain (which can be made up readily), stability can be improved.

EXAMPLE 2

An amplifier with an initial forward gain $G_1 = 1000$ undergoes a 10% reduction in gain when a noisy transistor is replaced with another of the same type. Predict the stabilizing effect of a feedback loop where $H = -0.009$.

Using the basic equation for gain with feedback, the initial gain is

$$G_{F1} = \frac{G_1}{1 - G_1 H} = \frac{1000}{1 - 1000(-0.009)} = 100$$

If forward gain is reduced by 10% to $G_2 = 900$,

$$G_{F2} = \frac{G_2}{1 - G_2 H} = \frac{900}{1 - 900(-0.009)} = 99$$

With feedback the change in gain is 1% or only $\frac{1}{10}$ as great as it is in the amplifier alone.

It can be shown (see Problem 2) that, for small changes in G, the fractional change in overall gain with feedback is equal to the change in G divided by $1 - GH$. By definition, the *sensitivity* of gain to change is

$$\frac{dG_F/G_F}{dG/G} = \frac{1}{1 - GH} \tag{9-5}$$

In words, the sensitivity of the gain to change is reduced by the factor $1 - GH$.

Reducing Time Constants. The speed of response of control elements is usually a limiting factor in the performance of control systems. An actuating motor with inertia takes a finite time to come up to a given speed, and it takes a finite time for the level in a chemical tank to drop a given amount. By using feedback principles, the speed of response can be greatly increased or, in other words, the time constant of the system can be reduced. The control system applications of this concept are treated in Chapter 25. Here we are interested in the application to electrical circuits.

To illustrate the powerful effect of feedback on the time constant of an electrical circuit, consider the RC two-port of Fig. 9.6a. For exponential functions, the transfer function can be derived as

$$\frac{v_2}{v_1} = \frac{(1/sC)i}{(R + 1/sC)i} = \frac{1}{1 + sRC} \tag{9-6}$$

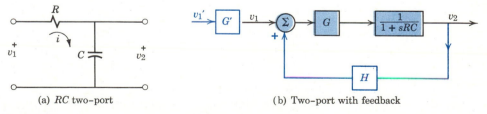

(a) RC two-port   (b) Two-port with feedback

Figure 9.6   Reduction of a time constant by means of feedback.

The time constant RC is a measure of the time for v_2 to approach v_1 if v_1 is a suddenly applied dc voltage, for example. (Or it is a measure of the time required for a motor to reach operating speed or a liquid level to drop to a new value.)

If amplification (G) and feedback (H) are provided, the system is described by the block diagram of Fig. 9.6b. For this system,

$$\frac{v_2}{v_1} = \frac{\dfrac{G}{1 + sRC}}{1 - \dfrac{GH}{1 + sRC}} = \frac{G}{1 + sRC - GH} = \frac{G}{1 - GH + sRC} \tag{9-7}$$

If an amplifier is added ahead of v_1 so that

$$G' = \frac{v_1}{v_1'} = \frac{1 - GH}{G}$$

the new transfer function is

$$\frac{v_2}{v_1'} = \frac{v_2}{v_1} \cdot \frac{v_1}{v_1'} = \frac{G}{(1 - GH) + sRC} \cdot \frac{1 - GH}{G} = \frac{1}{1 + s\dfrac{RC}{1 - GH}} \tag{9-8}$$

and the time constant has been reduced by the factor $(1 - GH)$. As in the preceding illustration, a desirable result has been achieved at the expense of amplification.

TRANSFER FUNCTIONS

In the previous discussion we defined the transfer function of a two-port as the ratio of output signal to input signal. For exponential signals of the form Ve^{st}, we know that the complete response current in a one-port is determined by the admittance $Y(s)$. Let us now extend that concept and show that the response of a two-port is determined by the transfer function $T(s)$.

The Transfer Function Concept

Applying the voltage-divider principle to the circuit of Fig. 9.7, the ratio of the output voltage to the input voltage is

$$T(s) = \frac{1/sC}{R + 1/sC} = \frac{1/RC}{s + 1/RC} \tag{9-9}$$

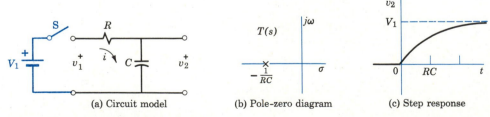

(a) Circuit model   (b) Pole–zero diagram   (c) Step response

Figure 9.7 Step response of an RC circuit.

where $T(s)$ is the voltage transfer function. To determine the output voltage for a step input voltage using our customary procedure (p. 150), we note that for a dc step, $s = 0$ and $T(s) = T(0) = 1$. Therefore, the forced response would be just $v_{2f} = V_1$. Since a pole of the transfer function appears at $s = -1/RC$, we expect the natural response to contain a term $Ae^{-t/RC}$. The complete response would be

$$v_2 = v_{2f} + v_{2n} = V_1 + Ae^{-t/RC} \tag{9-10}$$

Assuming the capacitor is initially uncharged at $t = 0^+$, $v_C = 0$; therefore,

$$v_2 = 0 = V_1 + Ae^0 \quad \text{and} \quad A = -V_1$$

As shown in Fig. 9.7c, the complete response would be

$$v_2 = V_1 - V_1 e^{-t/RC} \tag{9-11}$$

To check this result, we note that the complete response current for the one-port is $i = (V_1/R)\, e^{-t/RC}$ since $Y(0) = 0$ and $i_f = 0$. In this circuit, $v_2 = v_C = v_1 - iR$; hence,

$$v_2 = V_1 - V_1\, e^{-t/RC}$$

which is identical with the expression obtained by using the transfer function concept.

From this simple illustration we conclude that:

The transfer function is derived and interpreted just as an admittance or impedance function.

Note that an immittance is the ratio of two quantities measured at the same port, whereas a transfer function is the ratio of two quantities measured at different ports. The two quantities may be entirely different in physical nature.

For the special case of sinusoidal excitation, the transfer function (Eq. 9-9) becomes

$$\frac{\mathbf{V_2}}{\mathbf{V_1}} = \mathbf{T}(j\omega) = \frac{1}{1 + j\omega CR} = \frac{1}{\sqrt{1 + (\omega CR)^2}} \,\underline{/-\tan^{-1} \omega CR} \tag{9-12}$$

where $\mathbf{V_2}$ and $\mathbf{V_1}$ are the output and input phasors. It was pointed out in Chapter 5 that the phasor $\mathbf{V_1} = V_1\, e^{j\theta_1}$ is a *transform* of the sinusoidal function of time $v_1 = \sqrt{2}V_1 \cos(\omega t + \theta_1)$; this transform represents the sinusoidal function. In this chapter let us employ transforms[†] of exponential functions as well so that $V_1(s)$ might represent a function $v_1(t) = A_1\, e^{s_1 t} + A_2\, e^{s_2 t} + \cdots$ defined for $t > 0$. Using this notation, we rewrite Eq. 9-9 as

$$\frac{V_2(s)}{V_1(s)} = T(s) = \frac{1/RC}{s + 1/RC} \tag{9-13}$$

The value of the transfer function concept and the transform notation in control system analysis is illustrated in Example 3 on p. 238.

[†]The Laplace transformation is a mathematical operation that can be used to transform functions of time into functions of the complex variable s. Here the transfer function $T(s)$ is the ratio of the Laplace transforms of voltages $v_1(t)$ and $v_2(t)$ where v_1 and v_2 are defined for $t > 0$ and are equal to zero for $t \leq 0$. Powerful methods are available for determining the complete response of a system directly from the transfer function using Laplace transformation. See Gardner and Barnes, *Transients in Linear Systems*, John Wiley & Sons, New York 1942.

EXAMPLE 3

The torque developed in a certain electric motor is directly proportional to magnetic field current i. Part of the developed torque τ_d is used in overcoming friction and the remainder is available for acceleration. Determine the response of the motor to a step input of field voltage V.

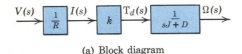

(a) Block diagram

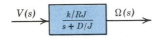

(b) Transfer function

Figure 9.8 Transfer function of a motor.

The governing equations (see Table 5-1, p. 136) and the corresponding transfer functions are

$$i = \frac{v}{R} \qquad\qquad \frac{I(s)}{V(s)} = \frac{1}{R}$$

$$\tau_d = ki \qquad\qquad \frac{T_d(s)}{I(s)} = k$$

$$\tau_d = J\frac{d\omega}{dt} + D\omega \qquad\qquad \frac{\Omega(s)}{T_d(s)} = \frac{1}{sJ + D}$$

The block diagram is as shown in Fig. 9.8a.
The overall transfer function is

$$\frac{\Omega(s)}{V(s)} = \frac{I(s)}{V(s)} \times \frac{T_d(s)}{I(s)} \times \frac{\Omega(s)}{T_d(s)} = \frac{k/R}{sJ + D} = \frac{k/RJ}{s + D/J}$$

For $s = 0$, $\Omega(0) = (k/RD)V$ and, by analogy with Eq. 9-11,

$$\omega(t) = \frac{kV}{RD} - \frac{kV}{RD}\, e^{-(D/J)t}$$

The motor speed ω approaches as a limit the speed governed by rotational friction D alone.

Advantages in Using Transfer Functions

In systems analysis, one possible approach is to write a set of differential equations and solve them simultaneously. A more convenient approach is based on the transfer function concept. The following observations are based on Figs. 9.1 and 9.8 and the accompanying discussion:

Systems usually consist of combinations of cascaded elements, and block diagram representation is convenient.

If each element is characterized by its transfer function, an overall transfer function can be determined by using the rules of block diagram algebra.

The individual transfer functions consist of relatively simple algebraic factors and the overall function is just a combination of these factors.

Analogous transfer functions are used in characterizing electrical, mechanical, hydraulic, and pneumatic elements and the interpretation of one is applicable to all.

The transfer function approach permits the determination of the transient response or the steady-state sinusoidal response.

Looking at this list, we can understand why the transfer function approach is used universally.

Dynamic Response .

To be useful, a control system must be able to cope with changing conditions such as a sudden change in load or a disturbance at some point in the system. In evaluating a system, one approach is to determine the dynamic behavior described by the complete response to a step function. An alternative approach is to consider the frequency response for steady-state sinusoidal inputs. In either approach we rely on the transfer function.

Since transfer functions are derived from linear ordinary differential equations, devices characterized by transfer functions must be represented by models made up of lumped, linear elements. The circuit models we have used to represent electrical and mechanical devices satisfy this requirement, so transfer functions can be deter-mined directly as in Example 3. Once the transfer function is obtained, the complete step response can be determined following the procedure outlined in Chapter 6.

First-Order Systems. The order of a system is defined by the highest power of s in the denominator of the transfer function. For the RC circuit of Fig. 9.9, the transfer

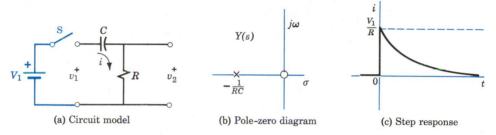

| (a) Circuit model | (b) Pole–zero diagram | (c) Step response |

Figure 9.9   Dynamic response of a first-order system.

function is (using the voltage-divider principle)

$$T(s) = \frac{V_2(s)}{V_1(s)} = \frac{R}{R + 1/sC} = \frac{s}{s + 1/RC} \tag{9-14}$$

Since s appears in the denominator to the first power only, this is a first-order system. For $s = 0$, $T(s) = T(0) = 0$ and the forced response is zero. A pole of the transfer function appears at $s = -1/RC$, so the natural response contains a term $A\,e^{-t/RC}$.

Assuming the capacitor is initially uncharged, at $t = 0^+$, $v_C = 0$, and

$$v_2 = V_1 = A_1\,e^0$$

As shown in Fig. 9.9c, the voltage response is iR or

$$v_2 = V_1\,e^{-t/RC} \tag{9-15}$$

First-order transfer functions of the form of Eq. 9-14 and Eq. 9-9 occur frequently in control systems, and it is worthwhile to generalize from the results obtained in these two specific cases. For a first-order system having a transfer function of the form

$$T(s) = \frac{ks}{s + \alpha} \tag{9-16}$$

the step response $y(t)$ is always of the form

$$y(t) = Ye^{-\alpha t} \tag{9-17}$$

For a system having a transfer function of the form

$$T(s) = \frac{k}{s + \alpha} \tag{9-18}$$

the step response $y(t)$ is always of the form

$$y(t) = Y(1 - e^{-\alpha t}) \tag{9-19}$$

The presence of a zero at $s = 0$ in the first transfer function (Eq. 9-16) completely changes the step response.

Second-Order Systems. In Example 3 it was assumed that field current i was directly proportional to v and limited only by resistance R. Actually, the winding used to produce the magnetic field necessary for motor operation has a high inductance. How does the motor respond if this inductance is taken into account? The answer to this question is found in the form of the new transfer function, and its derivation illustrates the convenience of our approach.

The new relation for the field circuit is $v = (R + sL)i$ and the transfer function relating field current and voltage (actually an admittance) is

$$T_1(s) = \frac{I(s)}{V(s)} = \frac{1}{sL + R} \tag{9-20}$$

For the motor as a transducer, the transfer function relating developed torque and field current is

$$T_2(s) = \frac{\mathrm{T}_d(s)}{I(s)} = k \tag{9-21}$$

For the mechanical side of the motor, the transfer function relating shaft velocity and developed torque is

$$T_3(s) = \frac{\Omega(s)}{\mathrm{T}_d(s)} = \frac{1}{sJ + D} \tag{9-22}$$

These three transfer functions are shown in Fig. 9.10a.

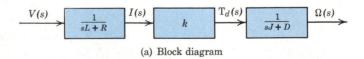

V(s)
1/(sL + R)
I(s)
k
Td(s)
1/(sJ + D)
Ω(s)
(a) Block diagram
V(s)
k/((sL + R)(sJ + D))
Ω(s)
(b) Overall transfer function
jω
σ
−D/J
−R/L
(c) Pole-zero diagram

Figure 9.10 A field-excited shunt motor, a second-order system.

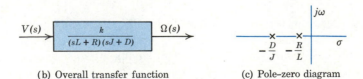

Applying the rules of block diagram algebra, we obtain the overall function

$$T(s) = \frac{I(s)}{V(s)} \times \frac{T_d(s)}{I(s)} \times \frac{\Omega(s)}{T_d(s)} = \frac{\Omega(s)}{V(s)} = \frac{k}{(sL + R)(sJ + D)} \tag{9-23}$$

What is the character of the step response of this system? There are two poles of the transfer function corresponding to two roots of the denominator (Fig. 9.10c). In other words, since s appears in the denominator to the second power, this is a second-order system. Combining the forced response with the natural response, we anticipate that the total step response is of the form

$$\omega(t) = \frac{k}{RD}V + A_1 e^{-Rt/L} + A_2 e^{-Dt/J} \tag{9-24}$$

Even without evaluating A_1 and A_2, we can see that in this case there are two time constants influencing the response. The fact that field current builds up slowly tends to make the response sluggish. Also, the inertia of the motor and load limits the rate of change of shaft velocity. Since a rapid response is usually desired in a control system, this is not a desirable mode of operation.

In this case, the two poles are real and negative. From our experience with electric circuits, we expect that in another case there might be complex roots and an entirely different form of response. The other cases are considered in detail in Chapter 25.

Frequency Response

Another useful interpretation of the transfer function is in terms of the steady-state response, i.e., amplitude and phase angle as functions of frequency. In some cases, a laboratory measurement of frequency response is the best method of determining the system characteristics. Given a transfer function, the frequency response is obtained directly by substituting $s = j\omega$. For the system described by $T(s) = 1/(s\tau + 1)$ where τ (tau) is the *time constant*, the frequency response is

$$\frac{V_2}{V_1} = \mathbf{T} = \frac{1}{1 + j\omega\tau} = \frac{1}{\sqrt{1 + (\omega\tau)^2}} \underline{/\tan^{-1} - \omega\tau} \tag{9-25}$$

The calculation of overall gain in complex systems or multistage amplifiers is simplified by using a logarithmic unit. By definition, the power gain in *bels* (named in honor of Alexander Graham Bell) is the logarithm to the base 10 of the power ratio. A more convenient unit is the *decibel* where

$$\text{Gain in decibels} = \text{dB} = 10 \log \frac{P_2}{P_1} \tag{9-26}$$

Since for a given resistance, power is proportional to the square of the voltage,

$$\text{Gain in decibels} = 10 \log \frac{V_2^2}{V_1^2} = 20 \log \frac{V_2}{V_1} \tag{9-27}$$

In general, the input and output resistances of an amplifier are not equal. However, the decibel is such a convenient measure that it is used as a unit of voltage (or current) gain,[†] regardless of the associated resistances.

EXAMPLE 4

Express in decibels (dB) the relative response of a resonant circuit at the lower half-power frequency (Eq. 7-47).

Since this is a half-power frequency, the relative response in dB is

$$10 \log \frac{P_1}{P_0} = 10 \log 0.5 \cong -3 \text{ dB}$$

Alternatively, at f_1 the response is 0.707 of the resonant frequency value. Therefore, the relative gain in dB is

$$20 \log \frac{V_1}{V_0} = 20 \log 0.707 \cong -3 \text{ dB}$$

The response is "down 3 dB" at the cutoff frequencies.

If we follow the procedure of H.W. Bode and plot gain in decibels using the gain at zero frequency as a base, the *Bode chart* of Eq. 9-25 is as shown in Fig. 9.11c. For $\omega\tau \gg 1$, the gain is inversely proportional to $\omega\tau$ and the gain curve on this log-log plot is a straight line with a slope of -20 dB/decade. The gain curve can be approximated by asymptotes that intersect at the *breakpoint* or *corner frequency*. At the breakpoints the phase angle is $-45°$ and the gain is actually down 3 dB. One important virtue of the Bode chart is that the response curve for the *product* of two transfer functions is just the *sum* of the two response curves when gains are plotted in dB.

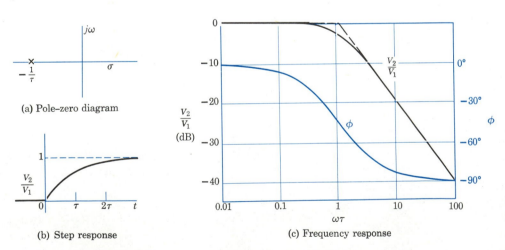

(a) Pole–zero diagram

(b) Step response

(c) Frequency response

Figure 9.11   Interpretations of the transfer function $T(s) = 1/(s\tau + 1)$.

[†]The decibel may be used to indicate power *level* P_2 by specifying a reference level P_1, commonly taken as 1 mW into a 600-Ω resistance. A power level of 10 W would be described as $10 \log (10/0.001) = 40$ dBm or "40 dB above 1 mW."

EXAMPLE 5

Laboratory data for the response curve of an amplifier are plotted as small circles in Fig. 9.12a. Determine the transfer function of the amplifier.

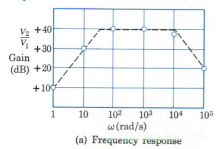

(a) Frequency response

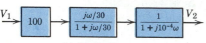

(b) Block diagram

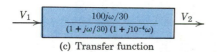

(c) Transfer function

Figure 9.12 Amplifier performance in terms of the transfer function.

It appears that at high and low frequencies the response drops off at about 20 dB/decade. On the basis of this evidence (and some experience!), the dashed straight-line approximation is drawn. The high-frequency response with a breakpoint ($\omega\tau_2 = 1$) at 10^4 rad/s corresponds to a transfer function of the form

$$T_2 = \frac{1}{1 + j\omega\tau_2} = \frac{1}{1 + j(\omega/\omega_2)} = \frac{1}{1 + j10^{-4}\omega}$$

The low-frequency response indicates that, for $\omega\tau_1 \ll 1$, the gain is directly proportional to frequency. With a breakpoint ($\omega\tau_1 = 1$) at 30 rad/s, this corresponds to a transfer function of the form

$$T_1 = \frac{j\omega\tau_1}{1 + j\omega\tau_1} = \frac{j\omega/30}{1 + j\omega/30}$$

For the middle frequencies around $\omega = 10^3$, $T_2 \cong 1$ and $T_1 \cong 1$. By Eq. 9-27 the mid-frequency gain of the amplifier is

$$A = \frac{V_2}{V_1} = \text{antilog}\ \frac{40}{20} = 10^2 = 100$$

The amplifier can be represented by the three ideal elements shown in Fig. 9.12b or by the single transfer function

$$T = \frac{100j\omega/30}{(1 + j\omega/30)(1 + j10^{-4}\omega)} \tag{9-28}$$

SUMMARY

■ A system is a combination of diverse but interacting elements integrated to achieve an overall objective.
  In a closed-loop system, a portion of the output is fed back and compared to or combined with the input. The gain of the basic closed-loop system is

$$G_F = \frac{G}{1 - GH}$$

■ Complicated systems that can be analyzed into linear, unilateral two-port elements are conveniently represented by simple diagrams consisting of blocks, summing points, and pickoff points.
  Block diagrams can be manipulated according to a set of simple rules.

■ The quantity $1 - GH$ is the key to the behavior of closed-loop systems.
  For $|1 - GH| < 1$, positive feedback occurs and the system gain is greater than the forward gain.

For $|1 - GH| = 0$, the system gain increases without limit and oscillation is possible.

For $|1 - GH| > 1$, negative feedback occurs and the system gain is less than the forward gain.

■ The benefits of negative feedback outweigh the loss in gain:
Amplifier gain can be made insensitive to device parameters or frequency.
Input and output impedances of amplifiers can be improved.
Noise and distortion can be discriminated against.
Time constants of components can be modified.

■ The transfer function is the ratio of two exponential functions of time; it is derived and interpreted just as an immittance function.
The transform $X(s)$ represents the exponential function $x(t)$ for $t > 0$.
Transfer functions can be determined directly from circuit models.
Transfer functions follow the rules of block diagram algebra.

■ Frequency response is obtained directly from the transfer function.
Step response is the sum of forced response (for $s = 0$) and natural response determined by the poles of the denominator of $T(s)$.

■ The system transfer function may be written in terms of pole and zero locations or in terms of steady-state characteristics ($s = j\omega$).
On a Bode chart (dB versus log frequency), the response curve for the product of two transfer functions is the sum of the individual response curves.

REVIEW QUESTIONS

1. Cite an example of feedback in a biological system and describe its operation.
2. Cite an example of feedback in a social system and describe its operation.
3. Cite an example of feedback in each of the following branches of engineering: aeronautical, chemical, civil, industrial, and mechanical.
4. If the speaker of a public address system is moved farther away from the microphone, what is the effect on the possibility of a "squeal," the frequency of the squeal if it does develop, and the amplitude of the squeal?
5. Differentiate between the two general categories of feedback applications in electrical systems described in the introduction to this chapter.
6. From memory, draw and label a basic feedback system and write the expression for overall gain.
7. What is the physical meaning for $GH = 1$? What is the ac *power* gain around the closed loop for this situation?
8. List three conditions that must be met if a system is to be represented by a block diagram.

9. Write out the corollaries to block diagram algebra rules 4 and 5.
10. Draw a complicated system with two pickoff points and two summing points. Using the rules of block diagram algebra, reduce the system to an open-loop equivalent.
11. Define positive feedback and negative feedback. Give an example of each.
12. Explain how a multistage audio amplifier may become an oscillator at a certain frequency.
13. List five specific benefits that may result from negative feedback and indicate where they would be important.
14. List three reasons why the gain of a transistor amplifier without feedback might change in normal operation.
15. Explain on a physical basis the operation of feedback in reducing the time constant in Fig. 9.6. Consider voltages at various times at various points.
16. What information on system performance is provided by step response?

17. Define "transform," "transfer function," and "step response."

18. Outline the procedure for determining step response from the transfer function.

19. What information is obtained from the poles of $T(s)$? The zeros?

20. What form of step response is indicated by complex poles of $T(s)$?

EXERCISES

1. In Fig. 9.2b, the forward gain of the amplifier is $1000\underline{/180°}$, $R_1 = 10$ kΩ, and $R_2 = 190$ kΩ.
   (a) For $\mathbf{V}_1 = 0.01\underline{/0°}$ V, calculate the output voltage, the feedback factor, and the voltage fed back.
   (b) For a source voltage of $0.51\underline{/0°}$ V, calculate $\mathbf{V}_1$ and the output voltage and the ratio of output to source voltage.
   (c) Calculate the gain with feedback from the forward gain and the feedback factor, and compare it to the result of part (b).

2. A driver on a freeway wishes to keep his left fender two feet from the white line. Draw a labeled block diagram for the closed-loop system he uses.

3. Reduce the block diagram of Fig. 9.13 to the basic feedback system of Fig. 9.3 and indicate the values of G and H. Determine the gain of the system.

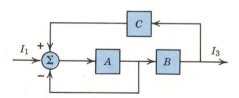

Figure 9.13

4. Repeat Exercise 3 for the diagram of Fig. 9.14.

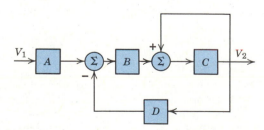

Figure 9.14

5. Given a system described by the equations

$$V_5 = BV_3 \qquad V_3 = AV_2$$
$$V_2 = V_1 - V_4 \qquad V_4 = CV_3$$

where V_1 is the input and V_5 is the output:
   (a) Draw a labeled block diagram representing this system.
   (b) Determine the gain V_5/V_1.

6. Given a system described by the equations

$$V_2 = V_1 + V_4 - AV_3$$
$$V_3 = BV_2 \qquad V_4 = CV_3$$

where V_1 is the input and V_4 is the output:
   (a) Draw a labeled block diagram representing this system.
   (b) Determine the gain V_4/V_1.

7. Given a system described by the equations

$$V_2 = V_1 - I_1R_1 \qquad I_1 = I_2 + I_3$$
$$V_2 = I_2R_2 = I_3R_3$$

where V_1 is the input, I_3 is the output, and each R is a block:
   (a) Draw a block diagram representing this system. (How many summing points?)
   (b) Draw the corresponding electrical circuit. (What do the first two equations represent in an electrical circuit?)
   (c) Calculate I_3/V_1 from the block diagram (by first reducing it to the basic system of Fig. 9.3) and then from the circuit.

8. A feedback system is represented by Fig. 9.3 with $\mathbf{H} = 0.005\underline{/0°}$. Define the conditions necessary for oscillation.

9. A feedback system is represented by Fig. 9.13 with $\mathbf{A} = 50\underline{/90°}$ and $\mathbf{B} = 20\underline{/90°}$. Determine the value of $\mathbf{C}$ necessary for oscillation.

10. For frequencies below $\omega = 2000$ rad/s, the gain of an amplifier is described by $\mathbf{G} = 0.1\omega\underline{/180°}$. For oscillation at $\omega = 1500$ rad/s, what value of $\mathbf{H}$ is required?

11. An electronic amplifier with other desirable properties provides a gain G of only 5.
   (a) Design a network to provide positive feedback and an overall gain of 100.
   (b) If G increases by 2%, say, what is the new G_F? Comment on the stability of this network.

12. Draw a noninverting op amp circuit and represent it in block diagram form as a basic feedback circuit. Define H and derive an expression for $G_F = V_2/V_1$ by using the block diagram. For op amp gain A very large, derive the standard expression for gain with feedback for an ideal op amp circuit.

13. Draw the block diagram representation of a noninverting amplifier (see Exercise 12).
   (a) Assuming $A = 10^5$ and $R_1 = 10 \text{ k}\Omega$, design a negative feedback amplifier to provide an overall gain of 100.
   (b) If A increases by 2%, what is the new G_F? Compare this result to the result of Exercise 11(b).

14. A cheap amplifier ordered from a catalog is supposed to have a voltage gain of 100, but when delivered, it proves to have a gain of 70. In fact, when first turned on it has a gain of only 50, which rises to 70 over the course of a minute or two. A second amplifier of the same type shows an initial gain of 90, rising to 110 during warmup.

   Design and diagram an arrangement to obtain a fairly stable gain of 100 by cascading the two amplifiers. Specify the feedback factor. Predict the percentage change in gain of your amplifier system during warmup.

15. Given the circuit in Fig. 9.15:
   (a) Define and derive the transfer function.
   (b) Sketch the pole-zero diagram of $T(s)$.
   (c) Sketch the step response.

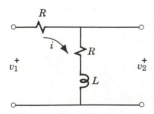

Figure 9.15

16. Repeat Exercise 15 for the system of Fig. 9.16.

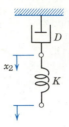

Figure 9.16

17. Repeat Exercise 15 for the system of Fig. 9.17.

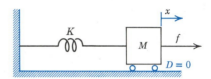

Figure 9.17

18. Given $T(s) = (s + 6)/(s + 2)$:
   (a) Sketch the pole-zero diagram.
   (b) Determine the step response.

19. Determine the response to a step of amplitude 10 (no initial energy storage) given:
   (a) $T(s) = 3s/(s + 3)$
   (b) $T(s) = 5/(s + 2)$

20. Determine the transfer function given the unit step response:
   (a) $x(t) = 5 - 5 e^{-2t}$
   (b) $i(t) = 6 e^{-t/50}$

21. (a) Explain why the natural response of a system is defined by the poles of the transfer function rather than the zeros.
   (b) Given $T(s) = (s + a)/(s^2 + sb)$, write an expression for the natural response.

22. The input to one unit of a communication system is 5 W. Calculate the dB gain if the output is:
   (a) 100 W
   (b) 20 W
   (c) 0.25 W
   (d) 0.05 W

23. In a communication system, a microphone provides an input of 20 mV. The preamplifier provides 26 dB of voltage gain, a cable introduces 12 dB of loss, the coupling circuit introduces

6 dB of loss, and the final amplifier provides 40 dB of gain.

(a) Calculate the output voltage of each element.

(b) Calculate the overall gain in dB and the output voltage.

24. The power loss in 1 km of coaxial telephone cable is 1.5 dB.

(a) What is the power loss in 2 km of telephone cable? Three?

(b) What is the total power loss from New York to Los Angeles?

(c) What signal power input is required at New York to deliver 1 μW at Los Angeles? Would amplifiers be desirable?

25. Plot on a Bode diagram (gain in decibels referred to $\omega = 0$) for $0.1 < \omega < 1000$ rad/s.

(a) $G_a = 1/(1 + j\omega)$

(b) $G_b = 1/(20 + j2\omega)$

(c) $G_c = 1/(1 + j\omega)(20 + j2\omega)$

26. A real amplifier can be represented by a combination of a potentiometer, an ideal amplifier (with a constant voltage gain $\mathbf{A} = \mathbf{V}_2/\mathbf{V}_1 = 1000\underline{/0°}$), and an RC coupling circuit as shown in Fig. 9.18.

(a) Derive the transfer function for the coupling circuit in terms of radian frequency and time constant.

(b) Sketch and label the frequency response of the coupling circuit on a Bode chart.

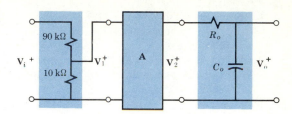

Figure 9.18

(c) Sketch and label the frequency responses of the potentiometer and of the ideal amplifier.

(d) Sketch and label the frequency response of the real amplifier.

(e) For output capacitance $C_o = 0.001$ μF, design the coupling circuit (i.e., specify R_o) for an upper breakpoint of 20 kHz.

27. An amplifier has a constant voltage gain of 1000 at frequencies up to $\omega = 100$ rad/s and an upper breakpoint at $\omega_2 = 1000$ rad/s.

(a) Draw a clearly labeled Bode plot of the frequency response. Write the transfer function $G(j\omega)$.

(b) To obtain a better high-frequency response, it is suggested that feedback be used with $H = -0.004$. Predict the new gain at $\omega = 100$ rad/s.

(c) Predict the new breakpoint ω_2 with $H = -0.004$.

(d) Draw the new response on a Bode chart.

PROBLEMS

1. An amplifier with a gain $G_1 = 200$ is subject to a 20% change in gain due to part replacement. Design an amplifier (in block diagram form) that provides the same overall gain but in which similar part replacement will produce a change in gain of only 0.1%. Use amplifier G_1 as part of your design.

2. Consider the basic feedback system of Fig. 9.3. Assuming small changes in G, derive a general expression for the fractional change in overall gain with feedback (dG_F/G_F) in terms of the change in forward gain (dG/G) and the values of G and H.

3. In order to reduce the time constant of the circuit in Fig. 9.19, a fraction k of the voltage across R is to be fed back. The feedback voltage is combined with a new input voltage V_0 and introduced

into an amplifier of gain A which supplies V_1.

(a) Draw a labeled wiring diagram and a block diagram showing the system with feedback.

(b) Determine the step response of the system and compare the time constants with and without feedback.

(c) Design the system (specify V_0, k, and A) so that the step response with feedback has the same amplitude but is 10 times as fast.

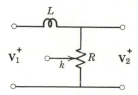

Figure 9.19

4. For the mechanical system in Fig. 9.20:
   (a) Derive the transfer function relating position x_2 to force f.
   (b) For $M = 1$ N, $D = 1$ N $\cdot$ s/m, and $K = 0.5$ m/N, sketch the pole-zero diagram and the step response.

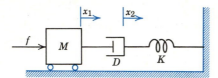

Figure 9.20

5. The low-frequency response of an amplifier is defined by $\mathbf{A} = A_O/(1 - j\omega_1/\omega)$ where $\omega_1 = 200$ rad/s and $A_O = 200$.
   (a) Plot a couple of points and sketch a graph of A/A_O versus ω with ω on a logarithmic scale.
   (b) Assume that a feedback loop with $H = -0.04$ is provided and repeat part (a). Label this graph "With feedback."

6. In the amplifying system of Fig. 9.21, $A = -1000$ and $H = 0.03$, both independent of frequency, $R = 10$ kΩ, and $C = 1.59$ nF.
   (a) Determine the upper cutoff frequency (corner frequency) of the system with switch S open.
   (b) Repeat part (a) with S closed.
   (c) For two identical systems with S closed in cascade, predict the overall gain and the new upper cutoff frequency. Comment on the effect of feedback.

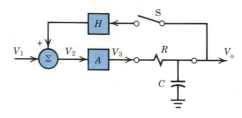

Figure 9.21

P A R T **II**

Electronics

10

Cathode-Ray Tubes

Electron Motion

Cathode-Ray Tubes

Oscilloscopes

The history of electronics starts with the discovery of cathode rays and continues into tomorrow. It is a fascinating story of individual contributions by mathematicians, physicists, engineers, and inventors and of the painstaking building of a solid technology.

While Hittorf and Crookes were studying cathode rays (1869), Maxwell was developing his mathematical theory of electromagnetic radiation. Soon after Edison observed electronic conduction in a vacuum (1883), Hertz demonstrated (1888) the existence of the radio waves predicted by Maxwell. At the time of J. J. Thomson's measurement of e/m (1897), Marconi was becoming interested in wireless and succeeded in spanning the Atlantic (1901). While Einstein was generalizing from the photoelectric effect, Fleming was inventing the first electron tube (1904), a sensitive diode detector utilizing the Edison effect. DeForest's invention of the triode (1906) made it possible to amplify signals electronically and led to Armstrong's sensitive regenerative detector (1912) and the important oscillator.

Zworykin's invention of the picture tube (1924) and the photomultiplier (1939) opened new areas of electronics. Watson-Watts' idea for radio detection and ranging, or *radar,* developed rapidly under the pressure of World War II. Postwar demands and increased knowledge of semiconductor physics led to the invention of the transistor by Shockley, Bardeen, and Brattain (1947) and the silicon solar cell by Pearson (1954). Kilby's integrated circuit (1958) permitted building active and passive circuit elements simultaneously and Noyce (1958) provided the technique for fabricating a complete network containing many semiconductor devices on a single monolithic chip. Einstein had pointed out (1917) the possibility of stimulated emission of radiation, and Townes used this concept (1953) to achieve *maser* operation. Townes and Schawlow (1958) described a device that could extend the maser principle into the optical region, and Maiman succeeded in doing so (1960) with his pulsed ruby *laser*.

Eckert and Mauchly started the computer revolution (1946) with ENIAC, their 18,000-vacuum tube electronic digital computer; Hoff (1969) matched the computational ability of ENIAC with his microprocessor on a single silicon chip. The availability of tiny, cheap, powerful computers in the 1970s revolutionized computation, communication, information processing, and control.

The discovery of cathode rays marked the beginning of the electronic era, and the invention of the multielectrode vacuum tube brought electronics into our daily living. As we learned to build efficient devices for generating and controlling streams of electrons, these electron tubes were applied in communication, entertainment, industrial control, and instrumentation. In many of these applications, the tube has been replaced by the semiconductor devices emphasized in the remainder of Part II. However, sophisticated versions of the basic cathode-ray tube are widely used in oscilloscopes, data display devices, television cameras and receivers, and radar scanners.

In this chapter we study some of the physical principles that underlie the operation of electronic devices. First we examine the behavior of electrons in a vacuum and derive the equations for motion in electric and magnetic fields. Then we consider electron emission, acceleration, and deflection in a cathode-ray tube. Finally, we see how cathode-ray oscilloscopes can be used for precise observations over a wide range of conditions.

ELECTRON MOTION

The model of the electron as a negatively charged particle of finite mass but negligible size is satisfactory for many purposes. Based on many careful measurements, the accepted values for charge and mass of the electron are

$$e = 1.602 \times 10^{-19} \text{ C} \cong 1.6 \times 10^{-19} \text{ C}$$

$$m = 9.109 \times 10^{-31} \text{ kg} \cong 9.1 \times 10^{-31} \text{ kg}$$

In contrast, the hydrogen ion, which carries a positive charge of the same magnitude, has a mass approximately 1836 times as great. If the mass, charge, and initial velocity are known, the motion of individual electrons and ions in electric and magnetic fields can be predicted using Newton's laws of mechanics.

Motion in a Uniform Electric Field

A uniform electric field of strength $\mathcal{E}$ is established between the parallel conducting plates of Fig. 10.1a by applying a potential difference or voltage. By definition (Eq. 1-5)

$$\mathcal{E} = -\frac{dv}{dl} = -\frac{V_b - V_a}{L} \text{ volts/meter} \tag{10-1}$$

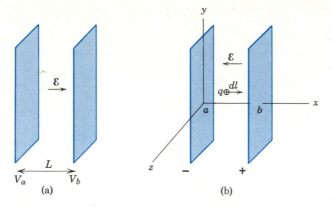

Figure 10.1 An electric charge in a uniform electric field.

if the spacing is small compared to the dimensions of the plates. Note that if $V_b > V_a$, $V_b - V_a$ is a positive quantity and the electric field is negative (directed in the $-x$ direction in Fig. 10.1b).

By definition (Eq. 1-4), the electric field strength is the force per unit positive charge. Therefore, the force in newtons on a charge q in coulombs is

$$\mathbf{f} = q\,\mathcal{E} \tag{10-2}$$

Considering the energy dw gained by a charge q moving a distance dl against the force of the electric field (Eq. 1-3), the voltage of point b with respect to point a is

$$V_{ba} = \frac{1}{q}\int_a^b dw = \frac{1}{q}\int_a^b f\,dl = \frac{1}{q}\int_a^b q(-\mathcal{E})\,dl = -\int_a^b \mathcal{E}\,dl \tag{10-3}$$

In words, the voltage between any two points in an electric field is the line integral of the electric field strength. Equation 10-3 is the corollary of Eq. 10-1.

In general, an electron of charge $-e$ in an electric field $\mathcal{E}$ experiences a force

$$f_x = (-e)(\mathcal{E}) = -e\,\mathcal{E} = ma_x$$

and an acceleration

$$a_x = \frac{f_x}{m} = -\frac{e\,\mathcal{E}}{m} \tag{10-4}$$

We see that an electron in a uniform electric field moves with a constant acceleration. We expect the resulting motion to be similar to that of a freely falling mass in the earth's gravitational field. Where u is velocity and x is displacement, the equations of motion are

$$u_x = \int_0^t a_x\,dt = a_x t + U_O = -\frac{e\,\mathcal{E}}{m}t + U_O \tag{10-5}$$

$$x = \int_0^t u_x\,dt = \frac{a_x t^2}{2} + U_O t + X_O = -\frac{e\,\mathcal{E}}{2m}t^2 + U_O t + X_O \tag{10-6}$$

Application of these equations is illustrated in Example 1.

EXAMPLE 1

A voltage V_D is applied to an electron deflector consisting of two horizontal plates of length L separated a distance d, as in Fig. 10.2. An electron with initial velocity U_O in the positive x direction is introduced at the origin. Determine the path of the electron and the vertical displacement at the time it leaves the region between the plates.

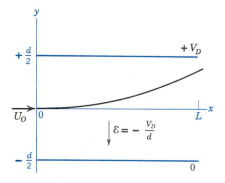

Figure 10.2 Calculation of electron deflection in a uniform electric field.

Assuming no electric field in the x direction and a uniform electric field $\mathcal{E}_y = -V_D/d$, the accelerations are

$$a_x = 0 \quad \text{and} \quad a_y = -\frac{e\mathcal{E}_y}{m} = \frac{eV_D}{md}$$

There is no acceleration in the x direction and the electron moves with constant velocity to the right. There is a constant upward acceleration and the electron gains a vertical component of velocity. The path is determined by

$$x = U_O t \quad \text{and} \quad y = -\frac{e\mathcal{E}}{2m}t^2 = \frac{eV_D}{2md}t^2$$

Eliminating t,

$$y = \frac{eV_D}{2mdU_O^2}x^2 \tag{10-7}$$

or the electron follows a parabolic path.

At the edge of the field, $x = L$ and the vertical displacement is

$$y_L = \frac{eV_D}{2mdU_O^2}L^2$$

If V_D exceeds a certain value, displacement y exceeds $d/2$ and the electron strikes the upper plate.

Energy Gained by an Accelerated Electron

When an electron is accelerated by an electric field it gains kinetic energy at the expense of potential energy, just as does a freely falling mass. Since voltage is energy per unit charge, the potential energy "lost" by an electron in "falling" from point a to point b is, in joules,

$$\text{PE} = W = q(V_a - V_b) = -e(V_a - V_b) = eV_{ba} \tag{10-8}$$

where V_{ba} is the potential of b with respect to a.

The kinetic energy gained, evidenced by an increase in velocity, is just equal to the potential energy lost, or

$$\text{KE} = \tfrac{1}{2}mu_b^2 - \tfrac{1}{2}mu_a^2 = \text{PE} = eV_{ba} \tag{10-9}$$

This important equation indicates that the kinetic energy gained by an electron in an electric field is determined only by the voltage difference between the initial and final points; it is independent of the path followed and the electric field configuration. (We assume that the field does not change with *time*.)

Frequently we are interested in the behavior resulting from a change in the energy of a single electron. Expressed in joules, these energies are very small; a more convenient unit is suggested by Eq. 10-8. An *electron volt* is the potential energy lost by 1 electron falling through a potential difference of 1 volt. By Eq. 10-8,

$$1 \text{ eV} = (1.6 \times 10^{-19} \text{ C})(1 \text{ V}) = 1.6 \times 10^{-19} \text{ J} \tag{10-10}$$

For example, the energy required to remove an electron from a hydrogen atom is about 13.6 eV. The energy imparted to an electron in a linear accelerator may be as high as 24 BeV (24 billion electron volts).

For the special case of an electron starting from rest $(u_a = 0)$ and accelerated through a voltage V, Eq. 10-9 can be solved for $u_b = u$ to yield

$$u = \sqrt{2(e/m)V} = 5.93 \times 10^5 \sqrt{V} \text{ m/s} \tag{10-11}$$

In deriving Eq. 10-11 we assume that mass m is a constant; this is true only if the velocity is small compared to the velocity of light, $c \cong 3 \times 10^8$ m/s.

EXAMPLE 2

Find the velocity reached by an electron accelerated through a voltage of 3600 V.

Assuming that the resulting velocity u is small compared to the velocity of light c, Eq. 10-11 applies and

$$u = 5.93 \times 10^5 \sqrt{3600} = 3.56 \times 10^7 \text{ m/s}$$

In this case,

$$\frac{u}{c} = \frac{3.56 \times 10^7}{3 \times 10^8} \cong 0.12$$

At the velocity of Example 2 the increase in mass is appreciable, and the actual velocity reached is about 0.5% lower than that predicted. For voltages above 4 or 5 kV, a more precise expression should be used (see Problem 1).

Motion in a Uniform Magnetic Field

One way of defining the strength of a magnetic field (Eq. 1-6) is in terms of the force exerted on a unit charge moving with unit velocity normal to the field. In general, the force in newtons is

$$\mathbf{f} = q\mathbf{u} \times \mathbf{B} \tag{10-12}$$

where q is charge in coulombs and $\mathbf{B}$ is magnetic flux density in teslas (or webers/meter^2). The vector cross product is defined by the right-hand screw rule illustrated in Fig. 10.3. Rotation from the direction of $\mathbf{u}$ to the direction of $\mathbf{B}$ advances the screw in the direction of $\mathbf{f}$. The magnitude of the force is $quB \sin \theta$ and the direction is always normal to the plane of $\mathbf{u}$ and $\mathbf{B}$.

Equation 10-12 is consistent with three observable facts:

1. A charged particle at rest in a magnetic field experiences no force $(u = 0)$.
2. A charged particle moving parallel with the magnetic flux experiences no force $(\theta = 0)$.
3. A charged particle moving with a component of velocity normal to the magnetic flux experiences a force that is normal to u and therefore the magnitude of velocity (or speed) is unchanged.

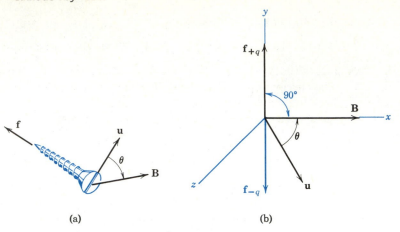

Figure 10.3 The right-hand screw rule defining the vector cross product.

From the third statement we conclude that no work is done by a magnetic field on a charged particle and its kinetic energy is unchanged.

Figure 10.4 shows an electron entering a finite region of uniform flux density. For an electron ($q = -e$) moving in the plane of the paper ($\theta = 90°$), the force is in the direction shown with a magnitude

$$f = eU_oB \tag{10-13}$$

Applying the right-hand rule, we see that the initial force is downward and, therefore, the acceleration is downward and the path is deflected as shown. A particle moving with constant speed and constant normal acceleration follows a circular path. The centrifugal force due to circular motion must be just equal to the centripetal force due to the magnetic field, or

$$m\frac{U_o^2}{r} = eU_oB$$

and the radius of the circular path is

$$r = \frac{mU_o}{eB} \tag{10-14}$$

The dependence of radius on the mass of the charged particle is the principle underlying the *mass spectrograph* for studying isotopes.

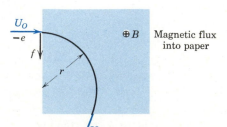

Magnetic flux
into paper

Figure 10.4 Electron motion in a uniform magnetic field.

Motion in Combined $\mathcal{E}$ and B Fields

In the general case, both electric and magnetic fields are present and exert forces on a moving charge. The total force is

$$\mathbf{f} = q(\mathcal{E} + \mathbf{u} \times \mathbf{B}) \tag{10-15}$$

The special cases previously described can be derived from this general relation. If both $\mathcal{E}$ and B are present, the resulting motion depends on the relative orientation of $\mathcal{E}$ and B, and also on the initial velocities. An interesting case is that in which an electron starts from rest in a region where $\mathcal{E}$ and B are mutually perpendicular. The electron is accelerated in the direction of $-\mathcal{E}$, but as soon as it is in motion there is a reaction with B and the path starts to curve. Curving around, the electron is soon traveling in the $+\mathcal{E}$ direction and experiences a decelerating force that brings it to rest. Once the electron is at rest, the cycle starts over; the resulting path is called a *cycloid* and resembles the path of a point on a wheel as it rolls along a line.

If the fields are not uniform, mathematical analysis is usually quite difficult. A case of practical importance is the "electron lens" used in electron microscopes and for electric field focusing of the electron beam in the cathode-ray tube.

CATHODE-RAY TUBES

The television picture tube and the precision electron display tube used in oscilloscopes are modern versions of the evacuated tubes used by Crookes and Thomson to study cathode rays. The electrons constituting a "cathode ray" have little mass or inertia, and therefore they can follow rapid variations; their ratio of charge to mass is high, so they are easily deflected and controlled. The energy of high-velocity electrons is readily converted into visible light; therefore, their motion is easily observed. For these reasons, the "C-R tube" or CRT is a unique information processing device; from our standpoint, it is also an ingenious application of the principles of electron motion and electron emission.

Electron Emission

The Bohr model of the atom is satisfactory for describing how free electrons can be obtained in space. As you may recall, starting with the hydrogen atom consisting of a single proton and a single orbital electron, models of more complex atoms are built up by adding protons and neutrons to the nucleus and electrons in orbital groups or shells. In a systematic way, shells are filled and new shells started. The chemical properties of an element are determined by the *valence* electrons in the outer shell. Good electrical conductors like copper and silver have one highly mobile electron in the outer shell.

Only certain orbits are allowed, and atoms are stable only when the orbital electrons have certain discrete energy levels. Transfer of an electron from an orbit corresponding to energy W_1 to an orbit corresponding to a lower energy W_2 results in the radiation of a *quantum* of electromagnetic energy of frequency f given by

$$W_1 - W_2 = hf \tag{10-16}$$

where h is Planck's constant $= 6.626 \times 10^{-34}$ J $\cdot$ s.

The energy possessed by an orbital electron consists of the kinetic energy of motion in the orbit and the potential energy of position with respect to the positive ion representing all the rest of the neutral atom. If other atoms are close (as in a solid), the energy of an electron is affected by the charge distribution of the neighboring atoms. In a crystalline solid, there is an orderly arrangement of atoms and the permissible electron energies are grouped into *energy bands*. Between the permissible bands there may be ranges of energy called *forbidden bands*.

For an electron to exist in space it must possess the energy corresponding to motion from its normal orbit out to an infinite distance; the energy required to move an electron against the attractive force of the net positive charge left behind is the *surface barrier energy* W_B. Within a metal at absolute zero termperature, electrons possess energies varying from zero to a maximum value W_M. The minimum amount of work that must be done on an electron before it is able to escape from the surface of a metal is the *work function* W_W where

$$W_W = W_B - W_M \qquad (10\text{-}17)$$

For copper, $W_W = 4.1$ eV, while for cesium $W_W = 1.8$ eV.

The energy required for electron emission may be obtained in various ways. The "beta rays" given off spontaneously by *radioactive* materials (along with alpha and gamma rays) are emitted electrons. In *photoelectric* emission, the energy of a quantum of electromagnetic energy is absorbed by an electron. In *high-field* emission, the potential energy of an intense electric field causes emission. In *secondary* emission, a fast-moving electron transfers its kinetic energy to one or more electrons in a solid surface. All these processses have possible applications, but the most important process is *thermionic* emission in which thermal energy is added by heating a solid conductor.

The temperature of an object is a measure of the kinetic energy stored in the motion of the constituent molecules, atoms, and electrons. The energies of the individual constituents vary widely, but an average energy corresponding to temperature T can be expressed as kT, where $k = 8.62 \times 10^{-5}$ eV/K is the Boltzmann constant. At a temperature above absolute zero, the distribution of electron energy in a metal is modified and some electrons possess energies appreciably above W_M. Statistical analysis shows that the probability of an electron receiving sufficient energy to be emitted is proportional to $e^{-W_W/kT}$. At high temperatures, many electrons possess energies

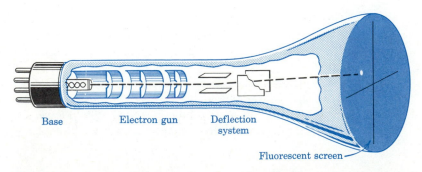

Base   Electron gun   Deflection system   Fluorescent screen

Figure 10.5 The essential components of a cathode-ray tube.

greater than W_B and emission current densities of the order of 1 A/cm^2 are practical. Commercial cathodes make use of special materials that combine low work function with high melting point.

CRT Components

The essential components of a CRT are shown in Fig. 10.5; an *electron gun* produces a focused beam of electrons, a *deflection system* determines the direction of the beam, and a *fluorescent screen* converts the energy of the beam into visible light.

Electron Gun. Electrons are emitted from the hot cathode (Fig. 10.6a) and pass through a small hole in the cylindrical control electrode; a negative voltage (with respect to the cathode) on this electrode tends to repel the electrons and, therefore, the voltage applied controls the intensity of the beam. Electrons passing the control electrode experience an accelerating force due to the electric field established by the positive voltages V_F and V_A on the focusing and accelerating anodes. The space between these anodes constitutes an electron lens (Fig. 10.6b); the electric flux lines

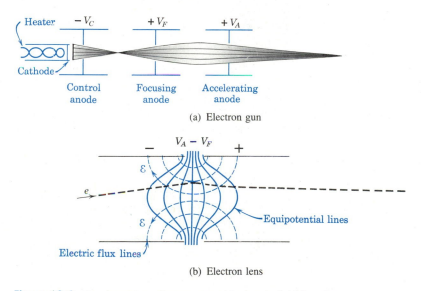

(a) Electron gun

(b) Electron lens

Figure 10.6 An elementary electron gun with electric field focusing.

and equipotential lines resulting from the voltage difference $V_A - V_F$ provide a precise focusing effect. A diverging electron is accelerated forward by the field, and at the same time it receives an inward component of velocity that brings it back to the axis of the beam at the screen.

Deflection System. In a television picture tube, the beam is moved across the screen 15,750 times per second, creating a picture consisting of 525 horizontal lines of varying intensity. The deflection of the beam may be achieved by a magnetic field or an electric field. Figure 10.7 shows the beam produced in the electron gun entering the vertical deflection plates of an electric field system.

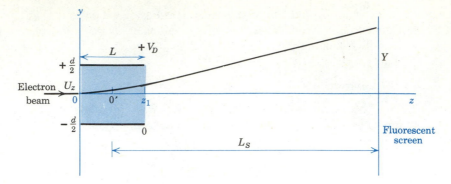

Figure 10.7   An electric field vertical deflection system.

The beam deflection at the screen can be calculated using our knowledge of electron motion. For the coordinate system of Fig. 10.7, Eq. 10-11 gives an axial velocity

$$U_z = \sqrt{2eV_A/m} \tag{10-18}$$

where V_A is the total accelerating potential. Within the deflecting field, the parabolic path (Eq. 10-7) is defined by

$$y = \frac{eV_D}{2mdU_z^2} z^2 = kz^2 \tag{10-19}$$

The slope of the beam emerging from the deflecting field at $z = z_1 = L$ is

$$\frac{dy}{dz} = 2kz = 2kL \tag{10-20}$$

The equation of the straight-line path followed by the beam to the screen is

$$y - y_1 = \frac{dy}{dz}(z - z_1) = 2kL(z - z_1)$$

where $z_1 = L$ and $y_1 = kz_1^2 = kL^2$. Substituting these values and solving,

$$y = 2kL(z - L) + kL^2 = 2kL\left(z - \frac{L}{2}\right) \tag{10-21}$$

Since for $y = 0$, $z = L/2$, Eq. 10-21 leads to the conclusion that the electron beam appears to follow a straight-line path from a virtual source at $0'$. At the screen, $z = L_S + L/2$ and (by Eqs. 10-19 and 10-18) the deflection is

$$Y = 2kLL_S = \frac{eV_D LL_S}{mdU_z^2} = \frac{eV_D LL_S}{md} \cdot \frac{m}{2eV_A} = \frac{LL_S}{2dV_A} V_D \tag{10-22}$$

or the vertical deflection at the screen is directly proportional to V_D, the voltage applied to the vertical deflecting plates. A second set of plates provides horizontal deflection.

Fluorescent Screen.   Part of the kinetic energy of the electron beam is converted into luminous energy at the screen. Absorption of kinetic energy results in an immediate *fluorescence* and a subsequent *phosphorescence*. The choice of screen material

EXAMPLE 3

Determine the *deflection sensitivity* in centimeters of deflection per volt of signal for a CRT in which $L = 2$ cm, $L_S = 30$ cm, $d = 0.5$ cm, and the total accelerating voltage is 2 kV.

From Eq. 10-22, the deflection sensitivity is

$$\frac{Y}{V_D} = \frac{LL_S}{2dV_A} = \frac{0.02 \times 0.3}{2 \times 0.005 \times 2000}$$

$$= 0.0003 \text{ mV} = 0.03 \text{ cm/V}$$

To obtain a reasonable deflection, say 3 cm, a voltage of 100 V would be necessary. In a practical CRO, amplifiers are provided to obtain reasonable deflections with input signals of less than 0.1 V.

depends on the application. For laboratory oscilloscopes, a medium persistence phosphor with output concentrated in the green region is desirable; the eye is sensitive to green and the persistence provides a steady image of a repeated pattern. For color television tubes, short persistence phosphors that emit radiation at various wavelengths are available. For radar screens a very long persistence is desirable.

OSCILLOSCOPES

The CRT provides a controlled spot of light whose x and y deflections are directly proportional to the voltages on the horizontal and vertical deflecting plates. The cathode-ray oscilloscope (CRO), consisting of the tube and appropriate auxiliary apparatus, is a precise and flexible laboratory instrument. The input characteristics are ideal for measurement purposes and the frequency response may extend to many megahertz.

The block diagram of Fig. 10.8 indicates the essential components of a CRO. A signal applied to the "Vert" terminal causes a proportional vertical deflection of the

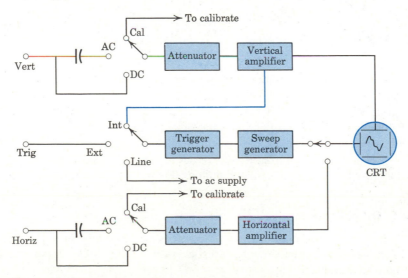

Figure 10.8 Basic components of a cathode-ray oscilloscope.

spot; the calibration of the *vertical amplifier* can be checked against an internal calibrating signal. The *attenuator* precisely divides large input voltages. An *intensity control* varies the accelerating potentials in the electron gun, and a *focus control* determines the potentials on the focusing electrodes.

The *sweep generator* causes a horizontal deflection of the spot proportional to time; it is *triggered* to start at the left of the screen at a particular instant on the internal vertical signal ("Int"), an external signal ("Ext"), or the ac supply ("Line"). Instead of the sweep generator, the *horizontal amplifier* can be used to cause a deflection proportional to a signal at the "Horiz" terminal.

The particular point on the triggering waveform that initiates the sawtooth sweep is determined by setting the *slope* and *level* controls. In Fig. 10.9, the input to the

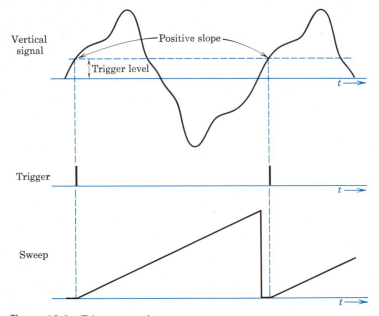

Figure 10.9 Trigger operation.

vertical amplifier provides the triggering waveform. The level control determines the instantaneous voltage level at which a trigger pulse is produced by the trigger generator. With the slope switch in the "+" position, triggering occurs only on a positive slope portion of the triggering waveform. By proper adjustment of these two controls, it is possible to initiate the sweep consistently at almost any point in the triggering waveform. For example, if slope is set to "−" and level is set to "0," the sweep will begin when the triggering signal goes down through zero.

CRO Applications

In its practical form, with controls conveniently arranged for precise measurements over a wide range of test conditions, the CRO is the most versatile laboratory instrument.

Voltage Measurement. An illuminated scale dividing the screen into 1-cm divisions permits use of the CRO as a voltmeter. With the vertical amplifier sensitivity set at 0.1 V/cm, say, a displacement of 2.5 cm indicates a voltage of 0.25 V. Currents can be determined by measuring the voltage across a known resistance.

Time Measurement. A calibrated sweep generator permits time measurement. With the sweep generator set at 5 ms/cm, two events separated on the screen by 2 cm are separated in time by 10 ms or 0.01 s. By measuring the period of a wave, the frequency can be determined by calculation.

Waveform Display. A special property of the CRO is its ability to display high-frequency or short-duration waveforms. If voltages varying with time are applied to vertical (y) and horizontal (x) input terminals, a pattern is traced out on the screen; if the voltages are periodic and one period is an exact multiple of the other, a stationary pattern can be obtained. A sawtooth wave from the sweep generator (Fig. 10.10a)

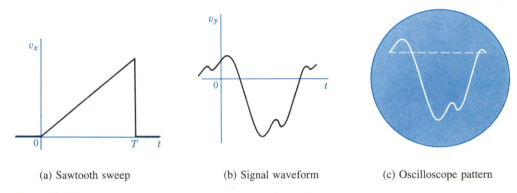

(a) Sawtooth sweep (b) Signal waveform (c) Oscilloscope pattern

Figure 10.10 Display of a repeated waveform on an oscilloscope.

applied to the x-deflection plates provides an x-axis deflection directly proportional to time. If a signal voltage wave is applied to the vertical-deflection plates, the projection of the beam on the y-axis is directly proportional to the amplitude of this signal. If both voltages are applied simultaneously, the pattern displayed on the screen is the signal as a function of time (Fig. 10.10c). A *blanking circuit* turns off the electron beam at the end of the sweep so the return trace is not visible. Nonrepetitive voltages are made more visible by using a long persistence fluorescent material, or a high-speed camera can be used for a permanent record.

X-Y Plotting. The relation between two periodic variables can be displayed by applying a voltage proportional to x to the horizontal amplifier and one proportional to y to the vertical amplifier. The characteristics of diodes or transistors are quickly displayed in this way. The hysteresis loop of a magnetic material can be displayed by connecting the induced emf (proportional to B) to the vertical amplifier and an iR drop (proportional to H) to the horizontal amplifier. Since the two amplifiers usually have a common internal ground, some care is necessary in arranging the circuits.

Phase-Difference Measurement. If two sinusoids of the same frequency are connected to the X and Y terminals, the phase difference is revealed by the resulting

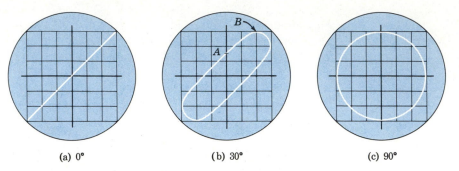

(a) 0°   (b) 30°   (c) 90°

Figure 10.11   Phase difference as revealed by CRO patterns.

pattern (Fig. 10.11). For applied voltages $v_x = V_x \cos \omega t$ and $v_y = V_y \cos (\omega t + \theta)$, it can be shown that the phase difference is

$$\theta = \sin^{-1}\frac{A}{B} \tag{10-23}$$

where A is the y deflection when the x deflection is zero and B is the maximum y deflection. The parameters of the circuit are useful in determining whether the angle is leading or lagging.

Frequency Comparison. When the frequency of the sinusoid applied to one input is an exact multiple of the frequency of the other input, a stationary pattern is obtained. For a 1:1 ratio, the so-called Lissajous patterns are similar to those in Fig. 10.11. For 1:2 and 2:3 ratios, the Lissajous figures might be as shown in Fig. 10.12.

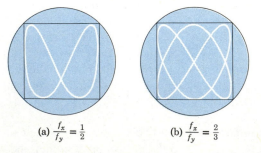

(a) $\dfrac{f_x}{f_y} = \dfrac{1}{2}$ (b) $\dfrac{f_x}{f_y} = \dfrac{2}{3}$

Figure 10.12   Lissajous figures for comparing the frequencies of two sinusoids.

For a stationary pattern, the ratio of the frequencies is exactly equal to the ratio of the number of tangencies to the enclosing rectangle. Patterns can be predicted by plotting x and y deflections from the two signals at corresponding instants of time.

Square-Wave Testing. The unique convenience of a CRO is illustrated in the measurement technique called *square-wave testing*. Any periodic wave can be represented by a Fourier series of sinusoids. As indicated in Fig. 10.13b, the sum of the first three odd harmonics (with appropriate amplitude and phase) begins to approximate a square wave; additional higher harmonics would increase the slope of the

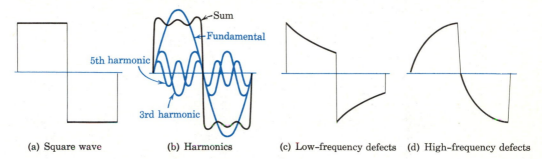

(a) Square wave   (b) Harmonics   (c) Low-frequency defects   (d) High-frequency defects

Figure 10.13 Square-wave testing with a CRO.

leading edge and smooth off the top. If a square-wave input to a device under test results in an output resembling Fig. 10.13c, it can be shown that low-frequency components have been attenuated and shifted forward in phase. If the output resembles Fig. 10.13d, it can be shown that high-frequency components have been attenuated and shifted backward in phase. The frequency response of an amplifier, for example, can be quickly determined by varying the frequency of the square-wave input until these distortions appear; the useful range of the amplifier is bounded by the frequencies at which low-frequency and high-frequency defects appear. As another example, two devices giving similar responses to a square-wave input can be expected to respond similarly to other waveforms.

CRO Features

Every year sees new advances in CRO design as the instrument manufacturers strive to meet new needs of the laboratory and the field by taking advantage of newly developed devices and techniques. Among the features offered by modern CROs are:

Differential Inputs. Each amplifier channel has two terminals beside the ground terminal. By amplifying only the difference between two signals, any common signal such as hum is rejected. (See p. 439.)

Optional Probes. To improve the input characteristics of a CRO or to expand its functions, the input stage can be built into a small unit (connected to the CRO by a shielded cable) that can be placed at the point of measurement. *Voltage probes* insert impedances in series with the CRO input to increase the effective input impedance. *Current probes* use transformer action or the Hall effect to convert a current into a proportional voltage for measurement or display.

Dual Channels. The waveforms of two different signals can be displayed simultaneously by connecting the signals alternately to the vertical deflection system. The more expensive *dual beam* feature requires separate electron guns and deflection systems but permits greater display flexibility.

Delayed and Expanded Sweep. To display a magnified version of a selected small portion of a waveform, an auxiliary delayed sweep with a faster sweep speed can be used.

Storage. If the input signal is a single, nonrepetitive event, the image can be *stored* by using a long persistence phosphor. By placing a storage mesh directly behind the phosphor and controlling the rate at which the charge pattern leaks off the mesh, a *variable persistence* is obtained.

Sampling. To display signals at frequencies beyond the limits of the CRO components, very short *sample* readings are taken on successive recurrences of the waveform. Each amplitude sample is taken at a slightly later instant on the waveform and the resulting dots appear as a continuous display. With this sampling technique, 18-GHz signals can be displayed.

Digital Readout. The addition of a built-in *microprocessor* provides direct digital readout of time interval, frequency, or voltage in addition to the conventional CRO display. The operator sets two markers to indicate a horizontal or vertical displacement on the waveform; the microprocessor, a computer-on-a-chip, interrogates the function switches and the scale switches, calculates the desired variable, and converts it to digital form for display by light-emitting diodes.

Oscilloscopes of the Future

As Barney Oliver points out,[†] the three revolutions in electronics—from vacuum tubes to transistors in the 1950s, from transistors to integrated circuits in the 1960s, and from ICs to large-scale integration in the 1970s—have been reflected in instrumentation. The availability of hundreds of active elements in a tiny chip at low cost permitted fresh design approaches that resulted in greatly expanded precision, sophistication, and reliability of instruments. A possibility for the future is an open-ended instrument that accepts either preprogrammed or user-programmable instructions and into which the user enters his or her choice of parameters, functions, frequency ranges, data reduction operations, and display characteristics by means of a single keyboard as we now enter numbers in a hand-held calculator.

[†]IEEE *Spectrum,* special issue on instrumentation, p. 44, November 1974.

SUMMARY

- Individual charged particles in electric and magnetic fields obey Newton's laws. For electrons (of primary interest here),

$$\mathbf{f} = -e(\boldsymbol{\mathcal{E}} + \mathbf{u} \times \mathbf{B})$$

- In any electric field, KE gained = PE lost = $W_{ab} = eV_{ab}$.
  For an electron starting from rest (and for $V < 4$ kV),

$$u = \sqrt{2(e/m)V} = 5.93 \times 10^5 \sqrt{V} \text{ m/s}$$

  In a uniform electric field where $\mathcal{E}_x = \mathcal{E}$, $f_x = -e\mathcal{E}$, and

$$a_x = -\frac{e\mathcal{E}}{m} \qquad u_x = -\frac{e\mathcal{E}}{m}t + U_O \qquad x = -\frac{e\mathcal{E}}{2m}t^2 + U_O t + X_O$$

- In any magnetic field, $\mathbf{f}$ is normal to $\mathbf{u}$ and no work is done.
  In a uniform magnetic field, the path is circular with $r = mU_O/eB$.

- Electron emission from a solid requires the addition of energy equal to the work function; this energy can be obtained in various ways.
  The probability of an electron possessing energy eV_T varies as $e^{-eV_T/kT}$.

- A cathode-ray tube consists of an electron gun producing a focused beam, a magnetic or electric deflection system, and a fluorescent screen for visual display.
  For deflecting voltage V_D, the deflection is

$$Y = \frac{LL_s}{2dV_A} V_D$$

- A cathode-ray oscilloscope includes display tube, intensity and focus controls, amplifiers and attenuators, sweep generator, and triggering circuit.
  A cathode-ray oscilloscope measures voltage and time, displays waveforms, and compares phase and frequency.

REVIEW QUESTIONS

1. What quantities are analogous in the equations of motion of a mass in a gravitational field and motion of an electron in an electric field?
2. Define the following terms: electron-volt, electron gun, vector cross product, work function, space-charge limited, and virtual cathode.
3. Justify the statement that "no work is done on a charged particle by a steady magnetic field."
4. Describe the motion of an electron starting from rest in a region of parallel electric and magnetic fields.
5. Sketch the pattern expected on a CRT screen when the deflection voltages are $v_x = V \sin \omega t$ and $v_y = V \cos 2\omega t$.
6. What is the effect on a TV picture of varying the voltage on the control anode (Fig. 10.6)? On the accelerating anode?
7. Sketch a magnetic deflection system for a CRT.
8. Explain qualitatively the process of thermionic emission.
9. How can a CRO be used to measure voltage? Frequency? Phase?
10. Explain the operation of CRO sweep, trigger, and blanking circuits.

EXERCISES

1. An electron is accelerated from rest by a potential of 200 V applied across a 5-cm distance under vacuum. Calculate the final velocity and the time required for transit. Repeat for a hydrogen ion. Express the energy gained by each particle in electron-volts.

2. In Fig. 10.14, an electron is introduced at point P in an evacuated space. It is accelerated from rest toward anode 1 at voltage V_1 and passes through a small hole at point b. In terms of the given quantities and the properties of an electron:

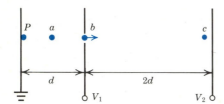

Figure 10.14

(a) What is the acceleration at point a?
(b) What is the velocity at point b?
(c) What voltage V_2 would just bring the electron to rest at point c?

3. In a region where $\mathcal{E} = 200$ V/m, an electron is accelerated from rest. Predict the velocity, kinetic energy gained, displacement, and potential energy lost after 2 ns. Compare the PE lost ($e\,\Delta V$) and the KE gained ($\frac{1}{2}mu^2$).

4. An electron is to be accelerated from rest to a velocity $u = 12 \times 10^6$ m/s after traveling a distance of 5 cm.
(a) Specify the electric field required and estimate the time required.
(b) Repeat part (a) for a hydrogen ion.
(c) Express the energy gained by each particle in joules.

5. For the electron of Exercise 3, derive expressions for velocity as a function of time and as a function of distance.

6. An electron with an energy of 200 eV is projected at an angle of 45° into the region between two parallel plates carrying a voltage V and separated a distance $d = 5$ cm (see Fig. 10.15).

(a) If $V = -200$ V, determine where the electron will strike.
(b) Determine the voltage V at which the electron will just graze the upper plate.

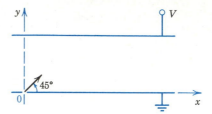

Figure 10.15

7. If the polarity of the upper plate in Fig. 10.15 is reversed so that $V = +200$ V, determine where the electron will strike.

8. Two electrons, e_1 traveling at velocity u and e_2 traveling at velocity $2u$, enter an intense magnetic field directed out of the paper (Fig. 10.16). Sketch the paths of the electrons.

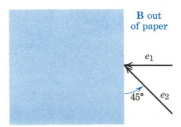

Figure 10.16

9. An electron with a velocity of 30 km/s is injected at right angles to a uniform magnetic field where $B = 0.02$ T into the paper.
(a) Sketch the path of the electron.
(b) Predict the radius of the path and the time required to traverse a semicircle.

10. In a *mass spectrograph* isotopes of charge e and unknown mass are accelerated through voltage V and injected into a transverse magnetic field B. Derive an expression for the mass m in terms of the radius r of the circle described by the particle.

11. In a cyclotron (see any physics book), hydrogen ions are accelerated by an electric field, curved around by a magnetic field, and accelerated

again. If B is limited to 2 T, what cyclotron diameter is required for a velocity half that of light?

12. An electron moves in an electric field $\mathcal{E} = 100$ V/m with a velocity of 10^6 m/s normal to the earth's magnetic field ($B = 5 \times 10^{-5}$ T). Compare the electric and magnetic forces with that due to the earth's gravitational field. Is it justifiable to neglect gravitational forces in practical problems?

13. The work function of copper is 4.1 eV and the melting point is 1356 K. Thorium has a work function of 3.5 eV and melts at 2120 K. Compare the factor of $e^{-W_W/kT}$ for these two metals at 90% of their respective melting points and decide which is more likely to be used as a cathode.

14. The Richardson-Dushman equation indicates that cathode emission current is proportional to $T^2 e^{-W_W/kT}$. For a special cathode coating with a work function of 1 eV, the emission current density at 1100 K is 200 mA/cm². Estimate the emission current densities at 800 and 1200 K.

15. In the CRT deflection system of Fig. 10.7, the accelerating voltage is 1000 V, the deflecting plates are 2.5 cm long and 1 cm apart, and the distance to the screen is 40 cm.
    (a) Determine the velocity of the electrons striking the screen.
    (b) Determine the deflecting voltage required for a deflection of 4 cm.

16. The vertical deflecting plates of a CRT are 2 cm long and 0.5 cm apart; length L_S in Fig. 10.7 is 50 cm. The acceleration potential is 5 kV.
    (a) Determine the deflection sensitivity.
    (b) What is the maximum allowable deflection voltage?

(c) For a brighter picture, the acceleration potential is increased to 10 kV. What is the new deflection sensitivity?

17. A CRO with a 10×10 cm display has vertical amplifier settings of 0.1, 0.2, and 0.5 V/cm and horizontal sweep settings of 1, 2, and 5 ms/cm. Assuming the sweep starts at $t = 0$, select appropriate settings and sketch the pattern observed when a voltage $v = 2 \sin 60\pi t$ V is applied to the vertical input.

18. Repeat Exercise 17 for an applied voltage $v = 0.5 \cos 1000t$ V.

19. On the CRO of Exercise 17, determine the frequency of a square-wave signal that occupies:
    (a) 2.5 cm per cycle at a sweep setting of 5 ms/cm.
    (b) 4.0 cm for 10 cycles at a sweep setting of 1 ms/cm.

20. On a CRO vertical and horizontal amplifiers are set at 1 V/cm. Sketch the pattern observed when the voltages applied to the vertical and horizontal inputs are:
    (a) $v_r = 5 \cos 400t$ and $v_h = 5 \cos 100t$ V.
    (b) $v_r = 5 \cos 200t$ and $v_h = 5 \cos (700t - \pi/2)$ V.

21. Design a circuit for displaying the v-i characteristic of a diode on a CRO. Assume that the horizontal and vertical amplifiers have a common internal ground.

22. The CRO of Exercise 21 has an 8×8-cm screen. Anticipating an I-V characteristic similar to that shown in Fig. 11.8, specify the significant CRO settings.

PROBLEMS

1. As Einstein pointed out, the actual mass m of a moving particle is $m = m_O/\sqrt{1 - u^2/c^2}$ where m_O is the rest mass, u is velocity, and c is the velocity of light. Since energy and mass are equivalent, the potential energy lost in falling through a voltage V must correspond to an increase in mass $m - m_O$.
    (a) Equate the potential energy lost to the equivalent energy gained and calculate the voltage required to accelerate an electron to 99% of the speed of light.

(b) Calculate the electron velocity for $V = 24$ kV, a typical value in a television picture tube, and determine the percentage of error in Eq. 10-11.

2. The magnetic deflecting *yoke* of a TV picture tube provides a field $B = 0.002$ T over an axial length $l = 2$ cm. The *gun* provides electrons with a velocity $U_z = 10^8$ m/s. Determine:
    (a) The radius of curvature of the electron path in the magnetic field.

(b) The approximate angle at which electrons leave the deflecting field.

(c) The distance from yoke to screen for a 3-cm *positive* deflection.

(d) A general expression for deflection Y in terms of acceleration potential V_a and compare with Eq. 10-22. (Assume small angular deflections where $Y/L = \tan \alpha \cong \alpha$.)

3. How long is an electron in the deflecting region of the CRT of Exercise 16? If the deflecting voltage should not change more than 10% while deflection is taking place, approximately what frequency limit is placed on this deflection system?

4. The useful range of an amplifier is bounded by frequencies f_1 and f_2 at which low-frequency and high-frequency defects occur.

(a) In a certain amplifier, signals at $3f_1$ are amplified linearly, but at f_1 sinusoidal components are reduced to 70% of their relative value and shifted forward 45° in phase. For a square-wave input of frequency f_1, draw the fundamental and third harmonic components in the output and compare their sum to Fig. 10.13c.

(b) In the same amplifier, signals at $f_2/3$ are amplified linearly, but at f_2 sinusoidal components are reduced to 70% of their relative value and shifted backward 45° in phase. For a square-wave input at frequency $f_2/3$, repeat part (a) and compare to Fig. 10.13d.

11

Semiconductor Diodes

Conduction in Solids

Doped Semiconductors

Junction Diodes

Special Purpose Diodes

The cathode-ray tube is a good example of the virtues and disadvantages of vacuum tubes. By using carefully constructed electric and magnetic fields, the flow of mobile electrons can be precisely controlled, and the result is a versatile information processing device. However, the high power necessary for thermionic emission and the large dimensions required for practical operation are serious disadvantages. In contrast, semiconductor electronic devices require no cathode power and their dimensions are extremely small. The emphasis in this book will be on semiconductor diodes and transistors in discrete and integrated forms.

First we consider conduction in solids where electron motion is influenced by the fixed ions of a conductor or the doping atoms added to a semiconductor. Using our knowledge of conduction mechanisms, we then examine the operation of semiconductor diodes and develop a quantitative understanding of diode junction behavior. Finally we derive circuit models for real diodes and look at some applications where the ideal diode representation of Chapter 3 is inadequate. We shall use the knowledge of conduction and junction phenomena gained here in analyzing the operation of transistors in Chapter 12 and the operation of more sophisticated devices in Chapters 13 and 14.

CONDUCTION IN SOLIDS

Conduction occurs in a vacuum if free electrons are available to carry charge under the action of an applied field. In an ionized gas, positively charged ions as well as electrons contribute to the conduction process. In a liquid, the charge carriers are positive and negative ions moving under the influence of an applied field. Solids vary widely in the type and number of charge carriers available and in the ease with which the carriers move under the action of applied fields. In electronic engineering we are interested in *insulators*, which have practically no available charge carriers; *conductors*, which have large numbers of mobile charge carriers; and *semiconductors*, which have conductivities intermediate between those of insulators and conductors.

Metallic Conductors

In a typical metal such as copper or silver, the atoms are arranged in a systematic array to form a *crystal*. The atoms are in such close proximity that the outer, loosely bound electrons are attracted to numerous neighboring nuclei and, therefore, are not closely associated with any one nucleus (Fig. 11.1). These *conduction electrons* are visualized as being free to wander through the crystal structure or *lattice*. At absolute zero temperature, the conduction electrons encounter no opposition to motion and the resistance is zero.

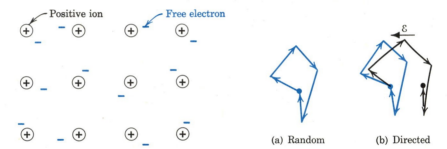

(a) Random   (b) Directed

Figure 11.1 A simplified two-dimensional representation of a metallic crystal.

Figure 11.2 The motion of an electron in a metallic crystal.

At ordinary temperatures, the electron-deficient atoms called *ions* possess kinetic energy in the form of vibration about neutral positions in the lattice; this vibrational energy is measured by temperature. There is a continual interchange of energy between the vibrating ions and the free electrons in the form of elastic and inelastic collisions. The resulting electron motion is random, there is no net motion, and the net current is zero (Fig. 11.2a).

If a uniform electric field of intensity $\mathcal{E}(V/m)$ is applied, the electrons are accelerated; superimposed on the rapid random motion, there is a small component of velocity in the direction of $-\mathcal{E}$. At each inelastic collision with an ion, the electron loses its kinetic energy; it then accelerates again, gains a component of velocity in the $-\mathcal{E}$ direction, and loses its energy at the next inelastic collision (Fig. 11.2b). The time between collisions is determined by the random velocity and the length of the mean free path. On the average, the electrons gain a directed *drift velocity u* that is directly proportional to $\mathcal{E}$ (since the acceleration is constant for an average increment of time), and

$$u = \mu(-\mathcal{E}) \qquad (11-1)$$

where μ (mu) is the *mobility* in meters per second/volts per meter or $m^2/V \cdot s$.

The resulting flow of electrons carrying charge $-e$ at drift velocity u constitutes a current. If there are n free electrons per cubic meter, the current density $J\,(A/m^2)$ is

$$J = n(-e)u = ne\mu\mathcal{E} = \sigma\mathcal{E} \tag{11-2}$$

where

$$\sigma = ne\mu$$

is the *conductivity* of the material in siemens per meter. The reciprocal of σ (sigma) is the *resistivity* ρ (rho) in ohm-meters. Equation 11-2 is another formulation of Ohm's law; it says that the current density is directly proportional to the voltage gradient. For a conductor of area $A\,(m^2)$ and length l (m),

$$I = JA = \sigma A \mathcal{E} = \sigma A \frac{V}{l} = \sigma \frac{A}{l} V \tag{11-3}$$

where $\sigma A/l$ is the conductance G in siemens. Alternatively,

$$R = \frac{V}{I} = \frac{1}{G} = \frac{l}{\sigma A} = \rho \frac{l}{A} \tag{11-4}$$

where R is the resistance in ohms.

EXAMPLE 1

The resistivity of copper is 1.73×10^{-8} $\Omega \cdot m$ at 20°C. Find the average drift velocity in a copper conductor with a cross-sectional area of 10^{-6} m^2 carrying a current of 4 A.

For copper with an atomic weight of 63.6 and a density of 8.9 g/cm^3, by Avogadro's law the number of atoms per $meter^3$ is

$$n_A = \frac{6.022 \times 10^{23} \text{ atoms/g-atom} \times 8.9 \times 10^6 \text{ g/m}^3}{63.6 \text{ g/g-atom}}$$

$$= 8.43 \times 10^{28}/m^3$$

Assuming that there is one free electron per atom, $n = n_A$. By Eq. 11-2, the velocity is

$$u = \frac{J}{ne} = \frac{I/A}{ne} = \frac{4 \times 10^6}{8.43 \times 10^{28} \times 1.6 \times 10^{-19}}$$

$$\cong 3 \times 10^{-4} \text{ m/s}$$

The average drift velocity in a good conductor is very slow compared to the random thermal electron velocities of the order of 10^5 m/s at room temperatures. As temperature increases, random thermal motion increases, the time between energy-robbing collisions decreases, mobility decreases, and therefore conductivity decreases. It is characteristic of metallic conductors that resistance increases with temperature.

Semiconductors

The two semiconductors of greatest importance in electronics are silicon and germanium. These elements are located in the fourth column of the periodic table and have four valence electrons. The crystal structure of silicon or germanium follows a tetrahedral pattern with each atom sharing one valence electron with each of four neighboring atoms. The *covalent bonds* are shown in the two-dimensional representation of Fig. 11.3. At temperatures close to absolute zero, the electrons in the outer shell are tightly bound, there are no free carriers, and silicon is an insulator.

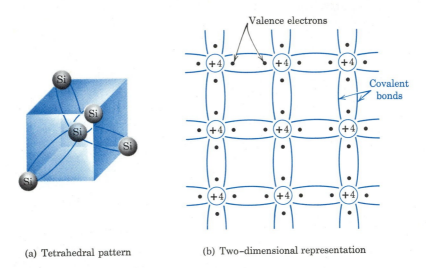

(a) Tetrahedral pattern   (b) Two–dimensional representation

Figure 11.3   Arrangement of atoms in a silicon crystal.

The energy required to break a covalent bond is about 1.1 eV for silicon and about 0.7 eV for germanium. At room temperature (300 K), a few electrons have this amount of thermal energy and are excited into the conduction band (see p. 258), becoming free electrons. When a covalent bond is broken (Fig. 11.4), a vacancy or

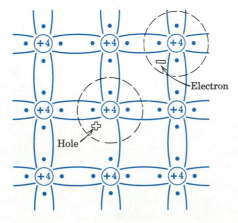

Figure 11.4   Silicon crystal with one covalent bond broken.

hole is left. The region in which this vacancy exists has a net positive charge; the region in which the freed electron exists has a net negative charge. In such semiconductors *both electrons and holes contribute to electrical conduction.* If a valence electron from another covalent bond fills the hole (without ever gaining sufficient energy to become "free"), the vacancy appears in a new place and the effect is as if a positive charge (of magnitude *e*) has moved to the new location.[†]

Consideration of hole behavior as the motion of a positively charged particle of definite mass and mobility is consistent with the quantum mechanical properties of valence electrons. Attempting to describe this behavior in terms of classical physics leads to results that are contrary to the observed facts. For our purposes we shall consider that:

Conduction in semiconductors is due to two separate and independent particles carrying opposite charges and drifting in opposite directions under the influence of an applied electric field.

With two charge-carrying particles, the expression for current density is

$$J = (n\mu_n + p\mu_p)e\mathcal{E} = \sigma\mathcal{E} \tag{11-5}$$

where *n* and *p* are the concentrations of electrons and holes (number/m³) and μ_n and μ_p are the corresponding mobilities. Therefore, the conductivity is

$$\sigma = (n\mu_n + p\mu_p)e \tag{11-6}$$

Some of the properties of silicon and germanium are shown in Table 11-1.

Since holes and electrons are created simultaneously, in a pure semiconductor the number of holes is just equal to the number of conduction electrons, or

$$n = p = n_i \tag{11-7}$$

where n_i is the *intrinsic* concentration. New electron-hole pairs are generated continuously and, under equilibrium conditions, they are removed by recombination at the same rate. At ordinary operating temperatures, only a very small fraction of the valence electrons are in the conduction state at any instant.

Table 11-1 Properties of Silicon and Germanium at 300 K[‡]

	Silicon	Germanium
Energy gap (eV)	1.1	0.67
Electron mobility μ_n (m²/V·s)	0.135	0.39
Hole mobility μ_p (m²/V·s)	0.048	0.19
Intrinsic carrier density n_i (/m³)	1.5×10^{16}	2.4×10^{19}
Intrinsic resistivity ρ_i (Ω·m)	2300	0.46
Density (g/m³)	2.33×10^6	5.32×10^6

[‡]Quoted by Esther M. Conwell in "Properties of Silicon and Germanium II," *Proc. IRE*, June 1958, p. 1281.

[†]There is an analogy in our economic system. In payment for a sweater, you may give a retailer $30 in cash or accept a bill from him for $30. You feel the same with $30 *subtracted* from your wallet or with $30 *added* to your debts. If you send the bill to your father for payment, the "vacancy appears in a new place" and the effect is as if your father had sent you $30. The sum of cash to you and bills to your father constitutes the net financial "current" from your father to you.

EXAMPLE 2

Estimate the relative concentration of silicon atoms and electron-hole pairs at room temperature, and predict the intrinsic resistivity.

By Avogadro's law, the concentration of atoms is

$$n_A = \frac{6.022 \times 10^{23} \text{ atoms/g-atom} \times 2.33 \times 10^6 \text{ g/m}^3}{28.09 \text{ g/g-atom}}$$

$$\cong 5 \times 10^{28} \text{ atoms/m}^3$$

From Table 11-1, the intrinsic concentration at 300 K is 1.5×10^{16} electron-hole pairs per cubic meter; therefore, there are

$$\frac{n_A}{n_i} \cong \frac{5 \times 10^{28}}{1.5 \times 10^{16}}$$

$$\cong 3.3 \times 10^{12} \text{ silicon atoms per electron-hole pair}$$

Since in the pure semiconductor $n = p = n_i$, by Eq. 11-6 the intrinsic conductivity is

$$\sigma_i = (n\mu_n + p\mu_p)e = (\mu_n + \mu_p)n_i e$$

$$= (0.135 + 0.048) \times 1.5 \times 10^{16} \times 1.6 \times 10^{-19}$$

$$= 4.4 \times 10^{-4} \text{ S/m}$$

and the intrinsic resistivity is

$$\rho_i = \frac{1}{\sigma_i} = \frac{1}{4.4 \times 10^{-4}} \cong 2280 \ \Omega \cdot \text{m}$$

The result of Example 2 confirms the fact that intrinsic silicon is a poor conductor as compared to copper ($\sigma = 5.8 \times 10^7$ S/m); semiconductors are valuable, not for their conductivity, but for two unusual properties. First, the concentration of free carriers, and consequently the conductivity, increases exponentially with temperature (approximately 5% per degree Celsius at ordinary temperatures). The *thermistor*, a temperature-sensitive device that utilizes this property, is used in instrumentation and control, and it also may be used to compensate for the *decrease* in conductivity with temperature in the metallic parts of a circuit. Second, the conductivity of a semiconductor can be increased greatly, and to a precisely controlled extent, by adding small amounts of impurities in the process called *doping*. Since there are two types of mobile charge carriers, of opposite sign, extraordinary distributions of charge carriers can be created. The semiconductor diode and the transistor utilize this property.

DOPED SEMICONDUCTORS

The startling effect of doping can be illustrated by adding to silicon or germanium an impurity element from the adjacent third or fifth column of the periodic table (see Table 11-2). Assume that a small amount of a *pentavalent* element (antimony, phosphorus, or arsenic) is added to otherwise pure silicon. Such a doping element has five valence electrons and the effective ionic charge is $+5e$. When a pentavalent atom

Table 11-2 Semiconductor Elements in the Periodic Table
(with atomic number and atomic weight)

III $(+3)$	IV $(+4)$	V $(+5)$
5 **B** BORON 10.82	6 **C** CARBON 12.01	7 **N** NITROGEN 14.008
13 **Al** ALUMINUM 26.97	14 **Si** SILICON 28.09	15 **P** PHOSPHORUS 31.02
31 **Ga** GALLIUM 69.72	32 **Ge** GERMANIUM 72.60	33 **As** ARSENIC 74.91
49 **In** INDIUM 114.8	50 **Sn** TIN 118.7	51 **Sb** ANTIMONY 121.8

replaces a silicon atom in the crystal lattice (Fig. 11.5), only four of the valence electrons are used to complete the covalent bonds; the remaining electron, with the addition of a small amount of thermal energy, becomes a free electron available for conduction. The resulting material is called an *n-type* semiconductor because of the presence of negative charge carriers in an electrically neutral crystal. The net positive charges left behind are immobile and cannot contribute to current.

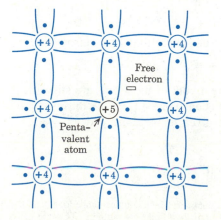

Figure 11.5 Effect of *n*-type doping.

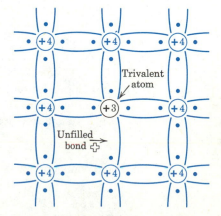

Figure 11.6 Effect of *p*-type doping.

If a small amount of a *trivalent* element (aluminum, boron, gallium, or indium) is added to otherwise pure silicon, a *p-type* semiconductor is obtained. When a trivalent atom replaces a silicon atom in the crystal lattice (Fig. 11.6), only three valence electrons are available to complete covalent bonds. If the remaining unfilled covalent bond is filled by a valence electron from a neighboring atom, a mobile hole is created and there is the possibility of current conduction by the motion of positive charges. Used in this way, a trivalent atom is called an *acceptor* atom because it accepts an electron. (The net negative charge created is immobile.) In the same sense, a pentavalent atom is called a *donor* atom.

Conduction

By adding donor or acceptor atoms in small amounts, the conductivity of a semiconductor can be increased enormously. To predict the magnitude of this effect we must understand quantitatively the process of generation and recombination. The *thermal generation rate g* (electron-hole pairs/s·m³) depends on the properties of the material and is a function of temperature. The energy required for a valence electron to become a free electron, or the energy needed in the "generation" of a conduction electron, can be expressed in electron volts as eV_g. The average energy corresponding to temperature T can be expressed as kT. Statistical analysis shows that the probability of a valence electron receiving sufficient energy to become free is proportional to $e^{-eV_g/kT}$; the rate of thermal generation (and also the rate of thermionic emission; see p. 258) involves this factor and, therefore, is highly temperature dependent.

In a semiconductor the mobile electrons and holes tend to recombine and disappear. If there are few electron-hole pairs in existence, the *rate of recombination* is low; if there are many, the rate is high. If, as in *n*-type material, there are few holes but many electrons, the rate of recombination is high due to the large number of electrons. In general,

$$R = rnp \tag{11-8}$$

where R = recombination rate (electron-hole pairs/s·m³) and
    r = a proportionality constant for the material.

This version of the *mass-action law* says that the rate of recombination is dependent on the numbers of reacting elements present.[†]

Under equilibrium conditions, the rate of generation is just equal to the rate of recombination or

$$g = R = rnp \tag{11-9}$$

In the pure crystal the intrinsic concentrations of electrons and holes are equal, or $n_i = p_i$ and

$$g = R = rn_ip_i = rn_i^2 \tag{11-10}$$

Even in a doped crystal, the great bulk of the atoms are still silicon (or germanium) and the thermal generation rate is unchanged from the intrinsic value. Therefore, Eq. 11-10 is a general relation at a given temperature and

$$np = n_i^2 \tag{11-11}$$

in doped semiconductors as well.

[†]As a crude analogy, consider a dance in which couples dance together for a while, then separate and wander among the other couples until they find new partners. The rate of recombination depends on the number of unattached partners available.

A second fundamental relation is based on the fact that the total crystal must be electrically neutral. There are two types of charge to consider: *immobile* ions and *mobile* carriers. Practically all donor or acceptor atoms are ionized at ordinary temperatures leaving immobile positive and negative ions in concentrations N_d and N_a. In addition there are mobile positive and negative carriers in concentrations p and n. For electrical neutrality in any doped crystal,

$$p + N_d = n + N_a \qquad (11\text{-}12)$$

In n-type material created by adding donor impurity to intrinsic semiconductor, $N_a = 0$ and

$$n_n = N_d + p_n \cong N_d \qquad (11\text{-}13\text{a})$$

in the practical case where $N_d \gg p_n$. (See Example 3.) The concentration of holes in n-type material is, by Eq. 11-11,

$$p_n = \frac{n_i^2}{n_n} \cong \frac{n_i^2}{N_d} \qquad (11\text{-}13\text{b})$$

The corresponding equations for p-type material are

$$p_p \cong N_a \qquad \text{and} \qquad n_p \cong \frac{n_i^2}{N_a} \qquad (11\text{-}14)$$

These relations are used in predicting the effect of doping in Example 3.

EXAMPLE 3

Predict the effect on the electrical properties of a silicon crystal at room temperature if every 10-millionth silicon atom is replaced by an atom of indium.

Since there are 5×10^{28} silicon atoms/m^3, the necessary concentration of acceptor atoms is

$$N_a = 5 \times 10^{28} \times 10^{-7} = 5 \times 10^{21} \cong p_p$$

The intrinsic concentration is $p_i = n_i = 1.5 \times 10^{16}/\text{m}^3$. Therefore, there are $(5 \times 10^{21}) \div (1.5 \times 10^{16}) = 3.3 \times 10^5$ as many holes. The new electron concentration is

$$n_p = \frac{n_i^2}{p_p} \cong \frac{n_i^2}{N_a} = \frac{2.25 \times 10^{32}}{5 \times 10^{21}} = 4.5 \times 10^{10}/\text{m}^3$$

Therefore, the electron concentration has been reduced by the factor $(1.5 \times 10^{16}) \div (4.5 \times 10^{10}) = 3.3 \times 10^5$ and Eq. 11-14a is justified.

The new conductivity is

$$\sigma_p = (n_p\mu_n + p_p\mu_p)e \cong N_a(\mu_p)e$$

$$= 5 \times 10^{21} \times 0.048 \times 1.6 \times 10^{-19} = 38.4 \text{ S/m}$$

as compared to 0.00044 S/m in intrinsic silicon. In other words, the resistivity of the doped silicon is

$$\rho_p = 1/\sigma_p = 1/38.4 = 0.026 \ \Omega \cdot \text{m}$$

as compared to the intrinsic value $\rho_i = 2300 \ \Omega \cdot \text{m}$ in Table 11-1.

We conclude from Example 3 that a small concentration of impurity greatly modifies the electrical properties of a semiconductor. Even more startling effects occur in regions where the impurity concentration is highly nonuniform, as it is at a junction of n-type and p-type materials.

Diffusion

If the doping concentration is nonuniform, the concentration of charged particles is also nonuniform, and it is possible to have charge motion by the mechanism called *diffusion*. In an analogous situation, if a bottle of perfume is opened at one end of an apparently still room, in a short time some of the perfume will have reached the other end. For mass transfer by diffusion, two factors must be present. There must be random motion of the particles and there must be a concentration gradient. The thermal energy of apparently still air provides the first, and opening a bottle of concentrated perfume provides the second. Note that no force is required for this type of transport; the net transfer is due to a statistical redistribution.

If a nonuniform concentration of randomly moving electrons or holes exists, diffusion will occur. If, for example, the concentration of electrons on one side of an imaginary plane is greater than that on the other, and if the electrons are in random motion, after a finite time there will be a net motion of electrons from the more dense side. This net motion of charge that constitutes a *diffusion current* is proportional to the concentration gradient dn/dx. The diffusion current density due to electrons is given by

$$J_n = eD_n \frac{dn}{dx} \tag{11-15}$$

where D_n = the diffusion constant for electrons (m^2/s). If dn/dx is positive, the resulting motion of negatively charged electrons in the $-x$ direction constitutes a positive current in the $+x$ direction. The corresponding relation for holes is

$$J_p = -eD_p \frac{dp}{dx} \tag{11-16}$$

Note that in each case the charged particles move away from the region of the greatest concentration, but the motion is not due to any force of repulsion. Like mobility, diffusion is a statistical phenomenon and it is not surprising to find that they are related by the Einstein equation

$$\frac{\mu_n}{D_n} = \frac{\mu_p}{D_p} = \frac{e}{kT} \tag{11-17}$$

or that the factor e/kT, which appeared in thermionic emission and thermal generation, shows up again here.

JUNCTION DIODES

Semiconductors, pure or doped, p-type, or n-type, are bilateral; current flows in either direction with equal facility. If, however, a p-type region exists in close proximity to an n-type region, there is a carrier density gradient that is unilateral;

current flows easily in one direction only. The resulting device, a *semiconductor diode*, exhibits a very useful control property.

pn Junction Formation

The fabrication of semiconductor devices is a developing technological art. For satisfactory results, the entire device must have a common crystalline structure. The pure crystals necessary may be "grown" by touching a "seed" crystal to the surface of molten silicon and then slowly raising it (Fig. 11.7a). The silicon crystallizes onto the seed and "grows" in the same crystalline orientation as the seed. If the melt contains a donor element such as arsenic, an *n*-type semiconductor is obtained. If at a certain point in the growing process an excess of acceptor element is added to the melt, the subsequent crystal will be *p*-type. The metallurgical boundary between the two materials is called a *grown junction*.

If a dot of trivalent indium is placed on an *n*-type silicon wafer (sliced from a grown crystal) and heated to the proper temperature, a small drop of molten mixture will form. Upon cooling, the alloy solidifies into *p*-type silicon that follows the crystalline pattern of the host crystal (Fig. 11.7b). The resulting device is an *alloy-junction* diode. By careful control of the alloying procedure it is possible to obtain diodes in which the transition from *p*-type to *n*-type material occurs in a distance corresponding to a few atomic layers.

Precisely controlled geometries and improved device properties are possible with the *planar* technology illustrated in Fig. 11.7c. First a thin layer of an inert oxide (such as glassy silicon dioxide) is formed on the surface of the *n*-type semiconductor wafer by heating in the presence of oxygen. Then the oxide layer is removed from a selected area by etching through a mask. When the heated semiconductor is exposed to boron gas, for example, acceptor atoms penetrate the selected area of the crystalline structure by solid-state diffusion and a precisely defined *pn* junction is formed. (The unetched oxide layer is impervious.) In another technique, the junction is formed by deposition of the doped material from a vapor. The atoms formed by this *epitaxial growth* (grown "upon, in the same form") follow the orientation of the underlying *substrate* crystal, but they may possess entirely different electrical characteristics.

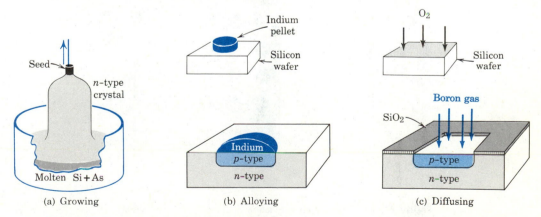

(a) Growing (b) Alloying (c) Diffusing

Figure 11.7 Methods of forming *pn* junctions.

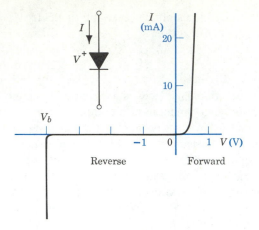

Figure 11.8 Junction diode symbol and *I-V* characteristic for germanium.

pn Junction Behavior

The external behavior of a typical *pn* junction diode is shown in Fig. 11.8. To use a diode effectively, we must understand (at least qualitatively) the internal mechanism that results in these remarkable properties. An understanding of diode behavior is directly transferable to the analysis of more complicated devices, such as the transistor, so a detailed study at this point is justified. We assume that the diode consists of a single crystal with an abrupt change from *p*- to *n*-type at the junction in the *y-z* plane; only variations with respect to *x* are significant.

The distribution of charges near the junction in such a diode is represented in Fig. 11.9. The circles represent immobile ions with their net charges; the separate + and − signs represent mobile holes and electrons. We imagine that a junction is created instantaneously; because of the density gradient at the junction, holes diffuse from the

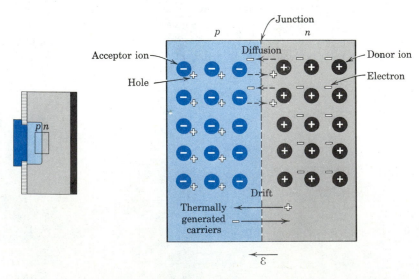

Figure 11.9 Schematic representation of charges and conduction mechanisms in a section of a *pn* diode.

p-type material to the right across the junction and recombine with some of the many free electrons in the n-type material. Similarly, electrons diffuse from the n-type material to the left across the junction and recombine. The diffusion of holes leaves *uncovered* bound negative charges on the left of the junction, and the diffusion of electrons leaves *uncovered* bound positive charges on the right.

To avoid repeating every statement, we call holes in the p-region and electrons in the n-region *majority carriers*. Thus the result of the diffusion of majority carriers is the uncovering of bound charge in the *depletion* or *transition region* and the creation of an internal electric field $\mathcal{E}$. Electrons in the p-region and holes in the n-region are called *minority carriers*. Under the action of the electric field, minority carriers drift across the junction. With two types of carriers (reduced from four by the use of the terms "majority" and "minority") and two different conduction mechanisms (drift and diffusion), we expect the explanation of the behavior displayed in Fig. 11.8 to be complicated.

Open Circuit. The simpler representation of the diode in Fig. 11.10 shows only the uncovered charge in the transition region; the remainder of the open-circuited diode is electrically neutral. In a material of permittivity ϵ with the distribution of charge density shown, there is an electrical field (Gauss' law)

$$\mathcal{E} = \frac{1}{\epsilon} \int \rho \, dx \tag{11-18}$$

Since $\mathcal{E} = -dv/dx$ (Eq. 10-1), the potential distribution across the junction is obtained by integrating the field. The potential barrier or "potential hill" of height V_O acts to oppose diffusion of majority carriers (holes to the right and electrons to the left) and to encourage drift of minority carriers across the junction.

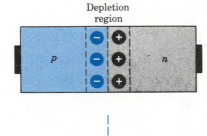

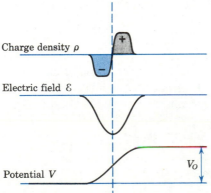

Figure 11.10 An open-circuited pn diode.

Under open-circuit conditions the net current is zero; the tendency of majority carriers to diffuse as a result of the density gradient is just balanced by the tendency of minority carriers to drift across the junction as a result of the electric field. The height of the potential barrier and the magnitude of V_O are determined by the necessity for equilibrium; for no net current, the diffusion component must be just equal to the drift component. The equilibrium value V_O is also called the *contact potential* for the junction. This potential, typically a few tenths of a volt, cannot be used to cause external current flow. Connecting an external conductor across the terminals creates two new contact potentials that just cancel that at the junction.[†]

Forward Bias. The diffusion component of current is very sensitive to the barrier height; only those majority carriers having kinetic energies in excess of eV_O diffuse across the junction. (The probability of a carrier possessing sufficient energy can be expressed in terms of $e^{-eV_O/kT}$.) If equilibrium is upset by a decrease in the junction potential to V_O-V (Fig. 11.11), the probability of majority carriers possessing sufficient energy to pass the potential barrier is greatly increased. The result is a net flow of carriers. Holes from p-type material cross the junction and are injected into n-type material where they represent minority carriers; electrons move in the reverse direction.

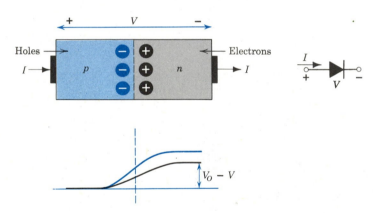

Figure 11.11 A forward-biased *pn* diode.

In other words, a reduction in the potential barrier encourages carriers to flow from regions where they are in the majority to regions where they are in the minority. If the junction potential is reduced by the application of an external *biasing* voltage V, as shown in Fig. 11.11, a net current I flows. (The small voltage drop across the neutral diode body at moderate currents is neglected in this discussion.) Under forward bias, the current is sustained by the continuous generation of majority carriers and recombination of the injected minority carriers.

[†]An analogy may be helpful. In a vertical column of gas, molecules drift downward under the action of gravity, but they diffuse upward because of their random thermal energy. A pressure difference exists between top and bottom, but there is no net flow. The pressure difference cannot be used to cause any external flow.

Reverse Bias. If the junction potential is raised to $V_O + V$ by the application of a reverse bias (Fig. 11.12), the probability of majority carriers possessing sufficient energy to pass the potential barrier is greatly decreased. The net diffusion of majority carriers across the junction, and therefore the minority carrier injection current, is reduced to practically zero by a reverse bias of a tenth of a volt or so.

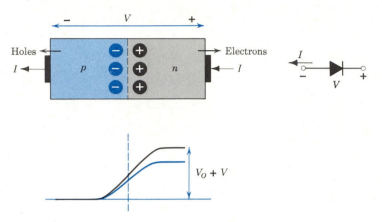

Figure 11.12   A reverse-biased *pn* diode.

A very small reverse current does flow. The minority carriers thermally generated in the material adjacent to the transition region drift in the direction of the electric field. Since all minority carriers appearing at the edge of the transition region are forced across by the field (holes "roll down" a potential hill; electrons "roll up"), this reverse current depends only on the rate of thermal generation and is independent of the barrier height. (This current reaches its maximum value I_s or "saturates" at a low value of reverse bias.)

At ordinary temperatures the reverse current is very small, measured in μA (germanium) or nA (silicon). Ideally the reverse current in a diode should be zero; a low value of I_s is one of the principal advantages of silicon over germanium. Practically I_s is an important parameter that is sensitive to temperature. It is directly proportional to the thermal generation rate $g = rn_i^2$ which varies exponentially with temperature. As a rough approximation near room temperature, for silicon or germanium diodes, I_s doubles with each increase of $10°$ C.

General Behavior

Calculation of the current flow in a *pn* junction diode is usually based on a detailed analysis of the minority-carrier density gradients at the edges of the transition region.[†] For our purposes here, we can assume that the diode current consists of an injection component I_i sensitive to junction potential and a reverse saturation com-

[†]See pp. 26–38 of Charles A. Holt, *Electronic Circuits*, John Wiley & Sons, New York, 1978.

ponent I_s independent of junction potential. As we expect from our previous experience with thermionic emission and thermal generation, the injection process is related to the statistical probability of electrons and holes having sufficient energy. Under a forward bias voltage V, this probability increases; the critical factor becomes

$$e^{-e(V_O-V)/kT} = e^{-eV_O/kT} \cdot e^{eV/kT} = A_1 e^{eV/kT} \qquad (11\text{-}19)$$

At a given temperature, the injection current I_i is of the form $A_1 A_2 e^{eV/kT} = A e^{eV/kT}$, and the reverse current is negative and nearly constant at value I_s. The total current is

$$I = I_i - I_s = A e^{eV/kT} - I_s \qquad (11\text{-}20)$$

Under open-circuit conditions where $V = 0$,

$$I = 0 = A e^0 - I_s \quad \therefore A = I_s$$

In general, then, the junction diode current is

$$I = I_s(e^{eV/kT} - 1) \qquad (11\text{-}21)$$

At room temperature (say, $T = 20°$ C $= 293$ K), $e/kT \cong 40$ V^{-1} and a convenient form of the equation is

$$I = I_s(e^{40V} - 1) \qquad (11\text{-}22)$$

EXAMPLE 4

The current of a germanium diode at room temperature is 100 μA at a voltage of -1 V. Predict the magnitude of the current for voltages of -0.2 V and $+0.2$ V at room temperature.

The current at -1 V can be assumed to be the reverse saturation current or $I_s = 100$ μA.
   At $V = -0.2$ V, $e^{40V} = e^{-8} \cong 0$.
Therefore, $I \cong -I_s = -100$ μA.
   At $V = +0.2$ V, $e^{40V} = e^8 \cong 3000$.
Therefore, $I = 100(3000 - 1) \times 10^{-6} \cong +300$ mA.

   Repeat the prediction for operation at 20° C above room temperature.

   At 20° C above room temperature, I_s will double twice and be approximately four times as great.
The new exponent will be $8 \times 293/(293 + 20) = 7.49$.
The two values of current will be $-I_s = -400$ μA
and $I = I_s e^{7.49} = +715$ mA, respectively.

We can draw some general conclusions from Example 4. Since $e^{40(-0.1)} = e^{-4} = 0.02$, for a reverse bias of more than 0.1 V or so, $I \cong -I_s$. Since $e^{40(0.1)} = e^{+4} = 55$, for a forward bias of more than 0.1 V or so, $I \cong I_s e^{40V}$.

Equation 11-21 holds quite well for germanium diodes, but it holds only approximately for silicon diodes. At normal currents, silicon diodes follow the relation

$$I = I_s(e^{eV/\eta kT} - 1) \qquad (11\text{-}23)$$

where factor η (eta) is approximately 2. For silicon diodes at high currents (and for germanium diodes in general), $\eta \cong 1$.

EXAMPLE 5

The voltage at which significant diode current flows is called the *threshold voltage*. Assuming that the criterion is 1% of rated current, estimate the threshold voltage at room temperature for two diodes rated at 1 A: a germanium diode with $I_s = 3$ μA and a silicon diode with $I_s = 60$ nA.

By Eq. 11-22, for the germanium unit,

$$V \cong \frac{1}{40}\ln\frac{I}{I_s} = \frac{1}{40}\ln\frac{10^{-2}}{3 \times 10^{-6}} \cong 0.2 \text{ V}$$

By Eq. 11-23, for the silicon unit with $\eta = 2$,

$$V \cong \frac{2}{40}\ln\frac{I}{I_s} = \frac{1}{20}\ln\frac{10^{-2}}{6 \times 10^{-8}} \cong 0.6 \text{ V}$$

These are typical values for practical diodes.

Breakdown Diodes

Along with the change in junction potential due to forward and reverse biasing, there is a significant change in the width of the transition or depletion region. For a given junction potential, a definite number of bound charges must be uncovered; therefore, the width of the depletion region depends on the density of doping. Under forward bias, the junction potential is reduced and the depletion region narrows; under reverse bias, the depletion region widens. (This behavior is utilized in the field-effect transistor described in the next chapter.)

If the reverse bias is increased sufficiently, a sudden increase in reverse current is observed (see Fig. 11.8).[†] This behavior is due to the *Zener* effect or the *avalanche* effect. In Zener breakdown, the electric field in the junction becomes high enough to pull electrons directly out of covalent bonds. The electron-hole pairs thus created then contribute a greatly increased reverse current. The avalanche effect occurs at voltages higher than Zener breakdown voltages. At these high voltages, carriers gain sufficient energy between collisions to knock electrons from covalent bonds; these too can reach ionizing energies in a cumulative process.

In either type of breakdown, the reverse current is large and nearly independent of voltage. If the power dissipated is within the capability of the diode, there is no damage; reduction of the reverse bias voltage below V_b reduces the current to I_s. By controlling doping densities, it is possible to manufacture Zener diodes[‡] with breakdown voltages from a few volts to several hundred. These are valuable because of their ability to maintain a nearly constant voltage under conditions of widely varying current.

[†] The alloyed silicon diode was invented by Gerald Pearson in 1952. On his way out to lunch one day, he suggested to his assistant at Bell Telephone Laboratories that he try to alloy electrodes to pure silicon crystals using a variety of metals. When Pearson returned, he discovered that the aluminum specimen exhibited the startling behavior of Fig. 11.8. He first ascribed the reverse current to the Zener effect, but it turned out to be what we now call *avalanche breakdown*.

[‡] In commercial practice, the term "Zener diodes" includes devices employing both effects.

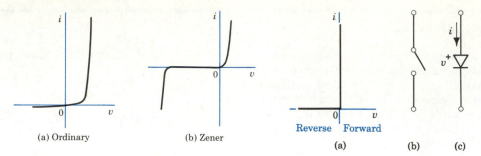

Figure 11.13 (a) Ordinary   (b) Zener

Figure 11.13   Real diode characteristics.

Reverse   Forward

(a)   (b)   (c)

Figure 11.14   Ideal diode models.

Circuit Models

The electrical model of a device is frequently called a *circuit model* because it is composed of ideal circuit elements. The models are derived from the characteristic curves of the real devices (Fig. 11.13) and, when properly devised, they can be used to predict accurately the behavior of real devices in practical applications.

The important diode characteristic is the discrimination between forward and reverse voltages. The ideal diode presents no resistance to current flow in the forward direction and an infinite resistance to current flow in the reverse direction. A selective switch, which is closed for forward voltages and open for reverse voltages, is equivalent to the ideal diode since it has the same current-voltage characteristic. Such a switch is represented in circuit diagrams by the symbol shown in Fig. 11.14c; the triangle points in the direction of forward current flow.

In a semiconductor diode, only a few tenths of a volt (0.3 V for germanium, 0.7 V for silicon) of forward bias is required for appreciable current flow. The combination of an ideal diode and a voltage source shown in Fig. 11.15b is usually satisfactory for predicting diode performance. If the reverse saturation current in a semiconductor diode is appreciable, this fact must be incorporated into the model. One possibility is to use a current source with a magnitude I_s directed as shown in Fig. 11.15c.[†]

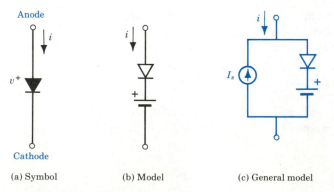

(a) Symbol   (b) Model   (c) General model

Figure 11.15   A semiconductor diode.

[†]For example, the manufacturers' ratings for the IN4004 subminiature silicon rectifier indicate conservative values as follows: 1 V drop at 1 A in the forward direction and only 10 μA reverse current under rated conditions.

EXAMPLE 6

A common application of silicon diodes is shown in Fig. 11.16. Predict I_2 as a function of V_1.

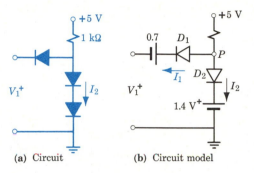

(a) Circuit   **(b)** Circuit model

Figure 11.16   Diode switching circuit.

As a first step in the analysis, replace the diodes by appropriate circuit models. The two diodes in series are represented by one ideal diode and a source of $2 \times 0.7 = 1.4$ V.

V_P cannot exceed 1.4 V, because at that voltage D_2 is forward biased and current I_2 flows readily.

Also, V_P cannot exceed $V_1 + 0.7$, because at that voltage D_1 is forward biased and current I_1 flows readily.

The critical value is $V_P = 1.4 = V_1 + 0.7$ or

$$V_1 = 1.4 - 0.7 = 0.7 \text{ V}$$

For $0 < V_1 < 0.7$ V, D_1 is forward biased and $I_2 = 0$.
For $V_1 > 0.7$ V, D_2 is forward biased and

$$I_2 = V/R = (5 - 1.4)/1000 = 3.6 \text{ mA}$$

Conclusion: A very small change in V_1 can switch I_2 from 0 to 3.6 mA.

The characteristic curve for a Zener diode can be linearized as in Fig. 11.17. When forward biased (v positive), current flows freely; the forward resistance is small and can be neglected. For reverse bias voltages greater than the breakdown potential, the resistance is estimated to be

$$R_Z = \frac{\Delta v}{\Delta i} = \frac{12 - 10}{0.01 - 0} = 200 \ \Omega$$

The circuit model includes a voltage source to indicate that reverse current does not flow until the negative voltage across the diode exceeds 10 V.

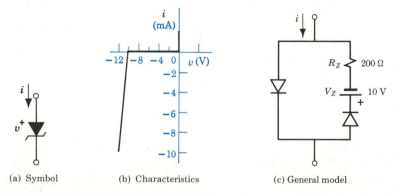

(a) Symbol   **(b)** Characteristics   **(c)** General model

Figure 11.17   Zener diode representation.

Zener-Diode Voltage Regulator

Filters are designed to minimize the rapid variations in load voltage due to cyclical variations in rectifier output voltage (Fig. 3.28). There are other reasons for load voltage variation. If the amplitude of the supply voltage (V_m) fluctuates, as it does in practice, the dc voltage will fluctuate. If the load current changes due to a change in R_L, the dc voltage will change due to the IR drop in the transformer, rectifier, and inductor (if any). A filter cannot prevent these types of variation and, if the load voltage is critical, a *voltage regulator* must be employed. Zener diodes are suitable for voltage regulation.

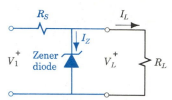

Figure 11.18 A Zener diode voltage regulator circuit.

A simple Zener-diode voltage regulator (Fig. 11.18) consists of a series voltage dropping resistance R_S and a Zener diode in parallel with the load resistance R_L. Voltage V_1 is the dc output of a rectifier–filter. The function of the regulator is to keep V_L nearly constant with changes in V_1 or I_L. The operation is based on the fact that, in the Zener breakdown region, small changes in diode voltage are accompanied by large changes in diode current (see Fig. 11.17). The large currents flowing through R_S produce voltages that compensate for changes in V_1 or I_L.

EXAMPLE 7

A load draws a current varying from 10 to 100 mA at a nominal voltage of 100 V. A regulator consists of $R_S = 200\ \Omega$ and a Zener diode represented by $V_Z = 100$ V and $R_Z = 20\ \Omega$ (Fig. 11.19).

For $I_L = 50$ mA, determine the variation in dc input voltage V_1 corresponding to a 1% variation in output voltage V_L.

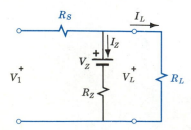

Figure 11.19 Performance of a voltage regulator.

To place the diode well into the Zener breakdown region, it is good design practice to use a minimum current $I_Z(\text{min}) = 0.1\,I_L(\text{max}) = 0.1 \times 100 = 10$ mA here. Then

$$V_L = V_Z + I_Z R_Z = 100 + 0.01(20) = 100.2 \text{ V}$$

and

$$V_1 = V_L + (I_L + I_Z)R_S$$
$$= 100.2 + (0.05 + 0.01)200 = 112.2 \text{ V}$$

If V_L increases by 1% or 1 V, $V_L' = 101.2$ V,

$$I_Z' = \frac{V_L' - V_Z}{R_Z} = \frac{101.2 - 100}{20} = 0.06 \text{ A}$$

and

$$V_1' = V_L' + (I_L + I_Z')R_S$$
$$= 101.2 + (0.05 + 0.06)200 = 123.2 \text{ V}$$

Conclusion: A change in V_1 of 11 V produces a change in V_L of only 1 V.

SPECIAL PURPOSE DIODES

The *pn* junction diode described by Eq. 11-21 is widely used in rectifying and waveshaping functions as discussed in Chapter 3. Other diodes,[†] similar in basic structure but presenting quite different external characteristics, can be explained in terms of our knowledge of junctions.

Tunnel Diodes

If the doping concentrations are increased, thereby increasing the concentration of uncovered charges, the width of the depletion region and the associated potential barrier are decreased. If the doping is greatly increased so that the depletion region is less than 10 nm wide, a new conduction mechanism is possible and the device characteristics are radically changed.

As explained by Leo Esaki in his 1958 announcement, for very thin potential barriers quantum mechanics theory indicates that there is a finite probability that an electron may *tunnel* through the barrier without ever possessing enough energy to climb over it. The *I-V* characteristics of an *Esaki diode* are shown in Fig. 11.20. The solid colored line indicates the tunneling effect. At voltages well below the threshold for normal forward current, electrons can tunnel from the *n* region to the *p* region if there are available holes of appropriate energy level into which the electron can move without ever possessing sufficient energy to become free.[‡]

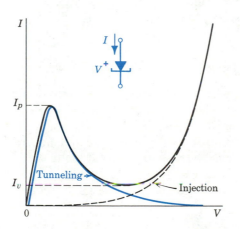

Figure 11.20 Tunnel diode characteristics.

The tunneling current increases with voltage until the effect of forward bias begins to alter the energy relationship and reduce the available holes. After the peak current I_p is reached, the tunneling current *decreases* with increased voltage and the normal injection current begins to dominate.

[†]Millman and Halkias, *Integrated Electronics*, McGraw-Hill Book Co., New York, 1972, has an excellent discussion in Chapter 3.
[‡]Millman and Halkias, op. cit.

The peak current I_p and the valley current I_v are stable operating points. Since tunneling is a wave phenomenon, electron transfer occurs at the speed of light and switching between I_p and I_v is fast enough for computer applications. Furthermore, between I_p and I_v there is a region in which the resistance $r = dV/dI$ is *negative*. By offsetting losses in L and C components, such a negative resistance permits oscillation and the tunnel diode is used as a very high-frequency oscillator.

Metal-Semiconductor Diodes

The leads to a semiconductor diode should make *ohmic contact* to provide for the easy flow of current without creation of additional potential barriers. Aluminum in contact with silicon acts as a p-type impurity, and the flow of hole current is easily accomplished by recombination with electrons supplied by the external circuit. Aluminum in contact with n-type material, however, would create a *rectifying contact* instead of the desired ohmic contact. In practice, therefore, contacts to n-type material are made of special alloys or a layer of heavily doped n^+ material is used to provide a low-potential transition between the semiconductor and the metal.

The rectifying contact created between metal and semiconductor differs from a pn junction in two important respects. First, in such a *Schottky diode* the forward voltage drop is only about half that across a pn junction at the same current. Second, since in an Al-n diode there are only majority carriers, switching is very fast because there is no wait for the recombination of injected minority carriers. These characteristics make Schottky diodes useful in integrated circuits.

Photodiodes

In a *photodiode* radiant energy is used to create electron-hole pairs in the region near the junction. Under reverse-biased conditions, the light-injected minority carriers cross the junction and augment the normal reverse saturation current due to thermally generated carriers. The resulting current (a fraction of a milliampere) is approximately proportional to the total illumination, and the ratio of "illuminated" current to "dark" current is very high. These characteristics are desirable in a phototransducer and photodiodes are used in the conversion of solar energy to electricity (Chapter 20).

Light-Emitting Diodes

The creation of electron-hole pairs is a reversible process; energy is released when an electron recombines with a hole. In silicon and germanium, recombination usually occurs at defects in the crystal that can *trap* a moving electron or hole, absorb its energy, and hold it until a recombination partner comes along. Only occasionally in silicon or germanium but frequently in a *III-V* compound such as gallium arsenide, an electron drops directly into a hole and a photon of energy is generated. Gallium-arsenide junctions providing optimum conditions for the generation of radiation in the visible range are called *light-emitting diodes* (LED). Under special conditions, the light emitted is coherent and the device is a *junction laser*.

SUMMARY

- In a metal, electrons of concentration n, charge e, and mobility μ constitute a current density $J = ne\mu\, \mathcal{E} = \sigma\mathcal{E}$, defining conductivity σ.
  In a semiconductor, thermally generated electrons and holes constitute a drift current density $J = (n\mu_n + p\mu_p)e\mathcal{E} = \sigma\mathcal{E}$.

- Low-density doping with a pentavalent (trivalent) element produces an n-type (p-type) semiconductor of greatly increased conductivity that is highly temperature dependent.
  For equilibrium, generation rate equals recombination rate and $np = n_i^2$.
  In n-type, $n_n \cong N_d$ and $p_n \cong n_i^2/N_d$; in p-type, $p_p \cong N_a$ and $n_p \cong n_i^2/N_a$.

- The diffusion current density due to nonuniform concentrations of randomly moving electrons and holes is
  $$J = J_n + J_p = eD_n\frac{dn}{dx} - eD_p\frac{dp}{dx}$$

- A pn junction diode is a single crystal with an abrupt change from p- to n-type material at the junction.
  Diffusion of majority carriers across the junction uncovers bound charge, creates a potential hill, and encourages an opposing drift of minority carriers.
  Forward biasing reduces the potential hill and encourages diffusion.
  Reverse biasing increases the potential hill and discourages diffusion.
  In general,
  $$I = I_s(e^{eV/kT} - 1)$$
  Normal reverse current is very small and nearly constant.
  At breakdown, reverse current is large and nearly independent of voltage.

- Essentially, a diode discriminates between forward and reverse voltages.
  Actual diodes differ significantly from ideal diodes.
  Circuit models of the required precision can be derived from the characteristic curves.

- A voltage regulator minimizes changes in dc load voltage.
  In Zener diodes small voltage changes cause large current changes that produce voltage drops to compensate for variations in V_{in} or I_{out}.

- Heavily doped Esaki tunnel diodes provide unique characteristics.
  Schottky metal-semiconductor diodes have special applications.
  Photodiodes use radiant energy to create electron-hole pairs.
  Light-emitting diodes use the energy released in direct recombination.

REVIEW QUESTIONS

1. How does resistance vary with temperature for a conductor? For a semiconductor? Why?
2. How does the speed of transmission of a "dot" along a telegraph wire compare with the drift velocity?
3. Define the following terms: mobility, covalent bond, electron-hole generation, recombination, doping, and intrinsic.
4. Describe the formation of a hole and its role in conduction.

5. How do you expect the term n_i^2 to vary with temperature?

6. In a semiconductor, what is the effect of doping with donor atoms on electron density? On hole density?

7. Is a semiconductor negatively charged when doped with donor atoms?

8. Is the approximation of Eq. 11-12 justified if $N_d = 100n_i$?

9. Why does the factor eV/kT appear in thermionic emission, electron-hole generation, and pn junction current flow?

10. Cite two nonelectrical examples of diffusion.

11. What are the two charge carriers and the two conduction mechanisms present in semiconductors?

12. Define the following terms: junction, uncovered charge, majority carriers, transition region, injection, and potential hill.

13. Describe the forward, reverse, and breakdown characteristics of a junction diode.

14. Describe with sketches the operation of an np diode.

15. How is contact potential utilized in a thermocouple?

16. Sketch curves near the origin to distinguish between the v-i characteristics of ideal and semiconductor diodes.

17. Why does reverse saturation current vary with temperature?

18. Describe the behavior of a Zener diode under forward and reverse bias.

19. Why do we wish to replace actual devices with fictitious models? Why *linear* models?

20. Cite an example from aeronautical, chemical, civil, industrial, and mechanical engineering of a process or a device that is customarily analyzed in terms of a mathematical or physical model.

21. Explain what is meant by a "linearized characteristic curve."

22. Given a *real* semiconductor diode in a half-wave rectifier circuit, sketch the voltage waveform across a load resistor. Repeat for an *ideal* diode.

23. Explain the operation of a voltage regulator using the diode of Fig. 11.8.

EXERCISES

1. If electrons drift at a speed of 0.5 mm/s in ordinary residential wiring (No. 12 copper wire, diameter = 0.081 in.), what current does this represent?

2. Given the resistance $R = \rho l/A$, derive an expression for resistivity ρ in terms of carrier density, mobility, and charge.

3. For $V = 1$ V, sketch a graph of $e^{-eV/kT}$ for $200 < T < 400$ K and write a brief statement describing the relation between thermal generation rate and temperature.

4. Estimate the relative concentration of germanium atoms and electron-hole pairs at room temperature and predict the intrinsic resistivity.

5. A silicon crystal is doped with 10^{20} aluminum atoms/m³.
   (a) Justify *quantitatively* the statement that: "The metallurgical properties of the doped crystal are not significantly altered but the electrical properties are greatly changed."
   (b) Determine the factor by which the conductivity has been changed by doping.

6. It is desired to increase the conductivity of silicon by a factor of 1000 by doping with arsenic.

Specify the doping density and identify the important charge carriers.

7. A silicon sample contains boron at a concentration of 2×10^{20} atoms/m³.
   (a) Estimate the hole and electron concentrations at room temperature. Is this p-type or n-type material?
   (b) Repeat part (a) for 300° C where $n_i = 3 \times 10^{21}$/m³.

8. Consider three specimens: a pure metal, a pure semiconductor, and a moderately doped semiconductor, at room temperature. Predict the effect on the conductivity of each specimen by a 10% increase in absolute temperature (K). Write an appropriate expression for conductivity for each; predict the new conductivity by selecting the most appropriate phrase from the following:

   (a) About the same.
   (b) Increased by about 10%.
   (c) Increased by more than 10%.
   (d) Decreased by about 10%.
   (e) Decreased by more than 10%.

   Explain, briefly, your reasoning.

9. A chip of intrinsic silicon is 1 mm × 2 mm in area and 0.1 mm thick.

   (a) What voltage is required for a current of 2 mA between opposite faces?
   (b) Repeat part (a) for a current of 2 mA between opposite 0.1 × 1 mm ends.

10. The block of Exercise 9 is doped by adding 1 atom of phosphorus per 10^7 atoms of silicon.

    (a) How many atoms per hole-electron pair are there in intrinsic silicon?
    (b) Is the doped material n-type or p-type? Why?
    (c) In the doped material, what is the density of majority and minority carriers? Can one be neglected?
    (d) What voltage is required for a current of 2 mA between faces?
    (e) By what factor has doping increased the conductivity?

11. Repeat Exercise 10, assuming that 1 in every 10^8 atoms of silicon is replaced by an atom of boron.

12. A certain block of pure silicon has a resistance of 200 Ω. It is desired to reduce this resistance to 0.2 Ω by doping with arsenic.

    (a) What are the charge carriers of importance in the doped semiconductor?
    (b) What fraction of the silicon atoms must be replaced by arsenic atoms?

13. The resistivity of a p-type silicon specimen is specified to be 0.12 Ω·m.

    (a) Estimate electron and hole concentrations.
    (b) Repeat for an n-type specimen.

14. A 5000-Ω resistor is to be fabricated (in an "integrated circuit") as a narrow strip of p-type silicon 4 μm thick. If the strip is 25 μm wide and 600 μm long, what concentration of acceptor atoms is required?

15. (a) Calculate the diffusion constants D_n and D_p for silicon at room temperature.
    (b) At a junction created by heating an n-type silicon crystal in the presence of boron gas (see Fig. 11.7c), the density of electrons drops from 5×10^{20} to $4 \times 10^{20}/m^3$ in a distance 0.1 μm. Estimate the diffusion current density across the junction.

16. In a germanium pn junction diode, the density of holes drops from 10^{21} to $0.9 \times 10^{21}/m^3$ in 2 μm. Estimate the diffusion current density due to holes across the junction at room temperature.

17. A diode consists of a left-hand portion doped with 10^{20} antimony atoms/m^3 and a right-hand portion doped with 4×10^{20} boron atoms/m^3.

    (a) Sketch the diode, indicating the junction and identifying the minority and majority carriers in each portion.
    (b) For electrical neutrality, how do the widths of the depletion layers in n and p regions compare?
    (c) Sketch a graph of charge density across the unbiased diode.

18. A semiconductor diode consists of a left-hand portion doped with arsenic and a right-hand portion doped with indium.

    (a) Sketch the diode, indicating the junction. Below the sketch, draw graphs of charge density, electric field, and potential distribution across the unbiased diode.
    (b) Draw labeled arrows indicating directions of drift and diffusion of electrons, holes, and net currents.
    (c) For forward bias, indicate the necessary polarity.

19. In a reverse-biased junction, when all the available charge carriers are being drawn across the junction, increasing the reverse junction voltage does not increase the reverse current.

    (a) At what voltage will the reverse current in a germanium pn junction at room temperature reach 98% of its saturation value?
    (b) If the reverse saturation current is 20 μA, calculate the current for a forward-bias voltage of the magnitude determined in part (a).

20. For a germanium diode carrying 10 mA, the required forward bias is about 0.2 V. Estimate the reverse saturation current and the bias voltages required for currents of 1 and 100 mA, respectively. Comment on the range of voltages required for a 100 to 1 change in current.

21. At a reverse voltage of 0.7 V, the reverse current in a germanium diode is 0.01 μA. Predict the current at voltages of -1.4, 0, and $+0.35$ V. Estimate the voltage at a forward current of 18 mA.

22. Draw a "rectifier" circuit and describe, with sketches, how an np junction diode can be used to rectify alternating currents. Emphasize the role of the majority and minority carriers and the junction.

23. A real semiconductor diode in the circuit of Fig. 3.25 has a reverse saturation current of 20 μA.

Predict the current i at room temperature under the following conditions:

(a) $v = +0.15$ V, $R = 0$.

(b) $v = -5.0$ V, $R = 1000$ Ω.

(c) $v = +5.0$ V, $R = 1000$ Ω.

24. (a) Represent the real diode of Fig. 11.21 by a circuit model consisting of an ideal diode and a voltage source.

(b) Use your model to estimate the current if the diode is placed in series with a 6-V battery and a 3-Ω resistance.

(c) Obtain a graphical solution and compare with the result of part (b).

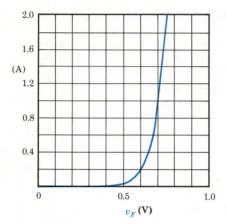

Figure 11.21 Silicon diode characteristics.

25. The diode of Fig. 11.21 is to be used in a rectifier circuit consisting of a voltage $v = 300 \cos \omega t$ V and a resistance $R_L = 200$ Ω. Devise an appropriately simple circuit model and predict I_{dc}.

26. Derive approximate circuit models for the diodes characterized in Fig. 11.22.

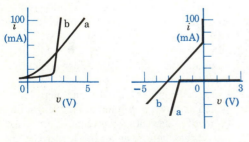

Figure 11.22   **Figure 11.23**

27. Derive circuit models for the devices characterized in Fig. 11.23.

28. The circuit of Fig. 11.24 contains three silicon diodes. Predict V_B for the following values of V_A: +2, +1, 0, and −1 V.

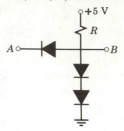

Figure 11.24

29. The circuit of Fig. 11.25 contains two silicon diodes.

(a) For switch S closed, estimate I_A and I_B for $V_A = -1, -0.1, +1$, and +2 V.

(b) For switch S open, estimate I_A and I_B for $V_A = +1$ V.

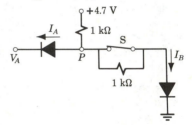

Figure 11.25

30. The germanium diode whose characteristics are shown in Fig. 11.8 is connected in the three different circuits of Fig. 11.26. In each case, select the appropriate model and predict the current i.

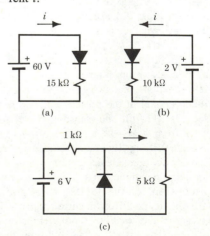

Figure 11.26

31. Devise a circuit model to represent, approximately, the Zener diode of Fig. 11.27.

32. The Zener diode whose reverse characteristics are given in Fig. 11.27 is connected in the circuit of Fig. 11.28 where $R_L = 10\ \Omega$.
(a) Replace the diode by a model and redraw the circuit.
(b) For $V_L = 14$ V and $V_1 = 26$ V, specify R_S.
(c) Under what circumstances could superposition concepts be used in solving this nonlinear circuit?
(d) If V_1 drops to 24 V, what is the new V_L?

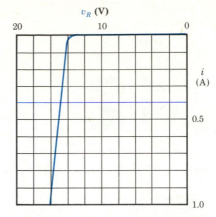

Figure 11.27 Zener region characteristics.

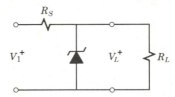

Figure 11.28

PROBLEMS

1. Stating any necessary assumptions, show that the spacing between atoms in copper and germanium crystals is on the order of 1 Å (0.1 nm).

2. Sketch in perspective a rectangular block of semiconductor of unknown type in a magnetic field B out of the paper. When a current I is introduced from right to left, it is observed that the upper surface exhibits a positive potential with respect to the lower (the Hall effect).
(a) Is the specimen a p-type or n-type semiconductor?
(b) How does this experiment justify the concept of the hole as a mobile positive charge?

3. If donor impurity atoms are added to an intrinsic semiconductor, how is the hole density affected? If 10^{16} atoms/m^3 of pentavalent antimony are added to pure silicon, predict the new hole density.

4. An experimental melt contains 100 g of pure germanium, 5×10^{-5} g of indium, and 3×10^{-5} g of antimony. Determine the conductivity of a sample of this germanium alloy.

5. For a nonlinear device, the incremental resistance is defined as $r = dv/di$. For an ideal semiconductor diode, derive a general expression for r in terms of I_s and an approximate expression in terms of I the forward current. Evaluate r for a reverse current of 5 μA and for a forward current at 5 mA.

6. The Zener diode of Fig. 11.16b is used in the voltage regulator circuit of Fig. 11.28, where $R_S = 10,000\ \Omega$ and $R_L = 20,000\ \Omega$. Derive an expression for V_L in terms of V_1. If V_1 increases from 16 to 24 V (a 50% increase), calculate the corresponding variation in load voltage V_L. Is the "regulator" doing its job?

7. Using your understanding of Zener and avalanche breakdown mechanisms and your knowledge of the effect of temperature on the energies of electrons and lattice atoms, predict (qualitatively) the effect of an increase in temperature on the breakdown voltage V_b for diodes governed by each of the two mechanisms.

12

Transistors and Integrated Circuits

Field-Effect Transistors

Bipolar Junction Transistors

Integrated Circuits

Thyristors

The ability of DeForest's multielectrode vacuum tube to amplify signals electronically caused a revolution in the field of communications. Amplifiers made it possible to generate high-frequency signals, send them around the world, and restore them to usable levels at the destination. Also, amplifiers provided the flexibility and sensitivity needed in electronic control and instrumentation. The long search for a semiconductor device capable of performing the same functions, but without the disadvantages of the thermionic tube, led to the discovery of the *transistor*.

The transistor is available in several basic forms for a wide variety of applications ranging from implantable heart pacers to giant digital computers. In its most common configuration, it may be a tiny portion of a complex integrated circuit or it may serve as a discrete switch controlling many amperes.

In this chapter we develop the principles necessary for the analysis and design of circuits containing transistors. We start with the simplest form of "unipolar" transistor and derive the *i-v* characteristics that determine its behavior. Then we study the more complicated "bipolar" transistor and derive its characteristics. Next we see how combinations of transistors, diodes, resistors, and capacitors can be fabricated into "integrated circuits." Finally we learn how multijunction thyristors can be used to control large amounts of power efficiently.

FIELD-EFFECT TRANSISTORS

The control of current in a vacuum tube by varying an electric field has its semiconductor counterpart in the *field-effect transistor* (FET). Shockley proposed the device in 1952, but a decade passed before the fabrication techniques necessary for dependable operation became available. Now such transistors are widely used in a variety of forms.

JFET

In the *junction-gate* form of FET or JFET (Fig. 12.1b), a thin conducting *channel* of finite conductance is established between two *pn* junctions. (The arrow in the

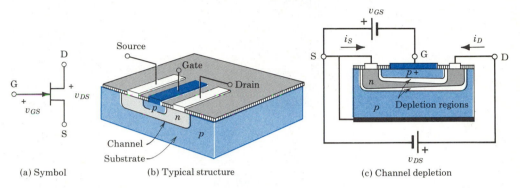

(a) Symbol   (b) Typical structure   (c) Channel depletion

Figure 12.1   An *n*-channel, depletion-mode JFET.

symbol always points toward *n*-type material.) The current from the *source* to the *drain* for a given voltage v_{DS} depends on the dimensions of the channel. If the *pn* junctions are reverse biased by applying a voltage to the *gate*, a depletion region containing no mobile carriers is formed and the width of the conducting channel is reduced. As shown by the low-voltage curves in Fig. 12.2a, the slope of the *i-v* characteristics, and therefore the conductance $\Delta i_D / \Delta v_{DS}$, is dramatically affected by the gate-source voltage v_{GS}.

The depletion region is asymmetric (Fig. 12.1c) because the reverse bias is highest at the drain end of the channel where the difference in potential between gate

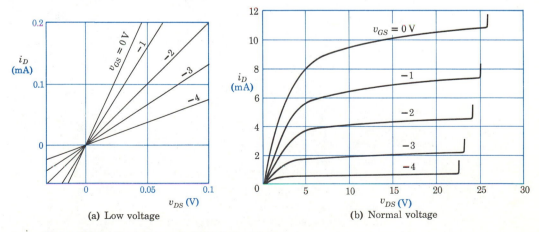

(a) Low voltage   (b) Normal voltage

Figure 12.2   Depletion-mode FET characteristics.

and channel is highest. For larger values of v_{DS}, the reverse bias increases until the two depletion regions almost meet, tending to "pinch off" the conducting channel. In Fig. 12.2b the *pinch-off voltage* V_p for $v_{GS} = 0$ is about 5 V. Above pinch-off, an increase in v_{DS} decreases the channel width, offsetting an increase in current density, and the i_D curve flattens out.

If a negative voltage is applied to the gate, pinch-off occurs at a lower drain-source voltage and the drain current is limited to a lower value. Above pinch-off, the current curves are relatively flat until the gate-drain voltage reaches the point where avalanche *breakdown* occurs. The normal operating range extends from pinch-off to breakdown.

MOSFET

In the metal-oxide-semiconductor FET or MOSFET, a thin layer of SiO_2 insulates the gate contact from the channel. The *n-channel enhancement-mode* transistor in Fig. 12.3 offers the highest performance. No channel is built into this device; here

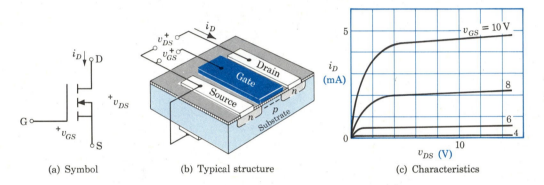

(a) Symbol (b) Typical structure (c) Characteristics

Figure 12.3 An *n*-channel enhancement-mode MOSFET.

a conducting channel is *induced* by an electric field established between the gate and the *p*-type substrate. With no gate voltage, little current flows through the two back-to-back *pn* junctions. With a small positive gate voltage, holes in the adjacent *p* material are repelled and a depletion layer formed. At a slightly more positive voltage, an *inversion layer* of mobile electrons is formed at the surface of the *p*-type region, which becomes *n*-type. (See Eq. 11-11.) The conductivity of the region has been "enhanced," the transistor has been "turned on," and current can flow between source and drain.

The drain current is not proportional to v_{DS}, however. As the potential at the drain end of the channel becomes more positive, the effective gate-to-channel voltage and the accompanying electric field are reduced. The carrier density in the inversion layer is reduced and the current levels off.

Similar *p*-channel devices are widely used. However, the greater mobility of electrons permits smaller *n*-type channels and therefore lower capacitances for the same resistance. *N*-channel transistors offer faster switching in digital systems and higher frequency response in amplifiers.

In another form of MOSFET, a narrow lightly doped conducting layer is diffused into the channel region (Fig. 12.4b). At $v_{GS} = 0$, an appreciable drain current flows. Application of the appropriate gate voltage can cause depletion *or* enhancement of the

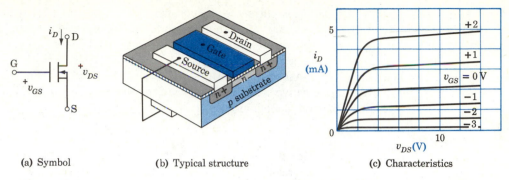

(a) Symbol   (b) Typical structure   (c) Characteristics

Figure 12.4 An n-channel depletion- or enhancement-mode (DE) MOSFET.

conducting channel. The characteristic curves are similar to those for the JFET with the added flexibility of permitting positive or negative control voltages. Such DE MOSFETs are available with either n- or p-type channels.[†]

Transfer Characteristics

The i-v characteristics of field-effect transistors show that output current is controlled by input voltage, and the FET can be used as a voltage-controlled *switch*. If the output current is sent through a resistance, the voltage developed may be much larger than the input voltage and the FET can be used as an *amplifier*. Since the characteristics of individual devices are never precisely known, approximate methods of analysis are acceptable. Within the normal operating region, that is, between pinch-off or turn-on and breakdown, drain current i_D is nearly independent of drain-source voltage v_{DS}, and the *transfer characteristics* are as shown in Fig. 12.5.

From theoretical analysis and practical measurement it can be shown that the transfer characteristics are approximately parabolic in all three cases. For the JFET, the drain current in the constant-current region is

$$i_{DS} = I_{DSS}(1 - v_{GS}/V_p)^2 \qquad (12\text{-}1)$$

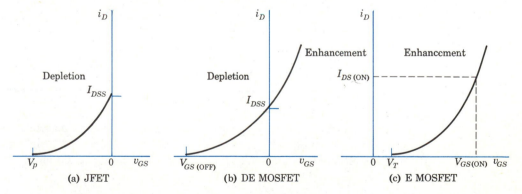

(a) JFET   (b) DE MOSFET   (c) E MOSFET

Figure 12.5 Transfer characteristics for three types of FET.

[†]The FET symbols are related to their distinctive characteristics. The gate terminal on a JFET represents a junction; the gap on MOSFETs identifies an insulated gate. The channel bar is continuous (normally conducting) for JFET and DE MOSFET, broken (normally "open") for the E MOSFET.

where i_{DS} = the drain current in the constant-current region,
I_{DSS} = the value of i_{DS} with gate shorted to source, and
V_p = the pinch-off voltage.

The depletion-*or*-enhancement MOSFET is also described by Eq. 12-1, and both positive and negative values of v_{GS} are permitted. For the enhancement-only MOSFET, the corresponding relation is

$$i_{DS} = K(v_{GS} - V_T)^2 \qquad (12\text{-}2)$$

where K is a device parameter and V_T is the turn-on or *threshold voltage*.

These simple relations are useful in predicting the dc behavior of FETs. In practice, the manufacturer usually specifies typical values of I_{DSS} and the *gate-source cutoff voltage* $V_{GS(OFF)}$, which is approximately equal to V_p since the same pinch-off effect is created between gate and channel. For the enhancement-only MOSFET, the manufacturer specifies V_T and a particular value of $I_{DS(ON)}$ corresponding to a specified value of $V_{GS(ON)}$.

EXAMPLE 1

For the *n*-channel enhancement-only MOSFET of Fig. 12.6, the manufacturer specifies $V_T = 4$ V and $I_{DS} = 7.2$ mA at $V_{GS} = 10$ V. For $V_{DD} = 24$ V and $R_G = 100$ MΩ, specify R_D for operation at $V_{DS} = 8$ V.

The arrowhead pointing toward the channel identifies an *n*-channel device.

Substituting the given data in Eq. 12-2,

$$I_{DS} = 0.0072 = K(V_{GS} - V_T)^2 = K(10 - 4)^2$$

$$\therefore K = 0.0072/(6)^2 = 0.0002 \text{ A/V}^2$$

Since $i_G = 0$, there is no drop across R_G; therefore, $V_{GS} = V_{DS} = 8$ V, and

$$I_D = 0.0002(8 - 4)^2 = 3.2 \text{ mA}$$

Around the drain-source loop,

$$\Sigma V = 0 = V_{DD} - I_D R_D - V_{DS}$$

or

$$R_D = \frac{V_{DD} - V_{DS}}{I_D} = \frac{24 - 8}{0.0032} = 5 \text{ k}\Omega$$

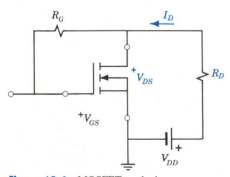

Figure 12.6 MOSFET analysis.

Because the gate-source junction of the JFET is reverse biased, the input signal current is very small; in other words, the input resistance is very high and little input power is required. In the MOSFET or *insulated-gate* FET, the input resistance may be as high as 10^{15} Ω.[†]

[†]The nearly infinite gate resistance creates a practical handling problem. If a small static charge (10^{-10} C) builds up on the gate-substrate capacitance (10^{-12} F), the voltage produced ($V = Q/C = 100$ V) may rupture the gate. External gate leads from an *IC* are usually protected by a Zener diode.

FETs are particularly useful in digital systems where thousands of units, some acting as resistors or capacitors, can be fabricated on a single silicon chip at very low cost. The primary advantages of the FET are its high element density and low power requirement in comparison to the more common "bipolar" transistor.

EXAMPLE 2

The MOSFET of Fig. 12.4 is to be used as an amplifier. Assuming $I_{DSS} = 2$ mA and $V_{GS(OFF)} \cong -4$ V, predict the voltage gain with a load resistance $R_L = 4$ kΩ.

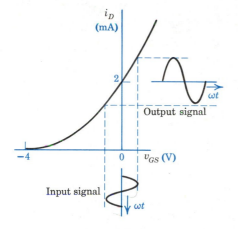

Figure 12.7 A MOSFET as an amplifier.

The transfer characteristic (Eq. 12-1) becomes

$$i_D = 0.002(1 + v_{GS}/4)^2$$

For a signal $v_{GS}(t) = V_m \sin \omega t$ (Fig. 12.7), the output voltage across $R_L = 4$ kΩ will be

$$v_L = i_D R_L = 8(1 + 0.25 V_m \sin \omega t)^2$$

$$= 8 + 4V_m \sin \omega t + 0.5V_m^2 \sin^2 \omega t$$

The output consists of an 8-V dc component, a $4V_m$ signal component at the input frequency, and a $0.5V_m^2$ "distortion" component resulting from the nonlinear (parabolic) transfer characteristic.

For small values of V_m, this is an amplifier with a voltage gain of 4.

For large values of V_m, this is a "square-law device" with special virtues described in Chapter 18.

BIPOLAR JUNCTION TRANSISTORS

The most widely employed electronic device is the *bipolar junction transistor* or BJT.[†] In contrast to the "unipolar" FET, in the "bipolar" device both majority and minority carriers play significant roles. In comparison to the FET, the BJT permits much greater gain and provides better high-frequency performance. Essentially, this transistor consists of two *pn* junctions in close proximity. Like the diode, it is formed of a single crystal, with the doping impurities distributed so as to create two abrupt changes in carrier density.

[†]Semiconductor experiments at Bell Telephone Laboratories led to new theoretical concepts, and William Shockley proposed an idea for a semiconductor amplifier that would critically test the theory. The actual device had far less amplification than predicted, and John Bardeen suggested a revision of the theory. In December 1947, Bardeen and Walter Brattain discovered a new phenomenon and created a novel device—the *point-contact transistor*. The next year Shockley invented the *junction transistor*.

Fabrication

The structure of an *npn* alloy-diffused transistor is shown in Fig. 12.8a. An *n*-type semiconductor chip less than 1 mm square serves as the *collector* and provides mechanical strength for mounting. A *p*-type *base* region is created by diffusion and a connection is provided by a metallic contact. An *n*-type *emitter* region is then alloyed to the base region. The result is a pair of *pn* junctions separated by a base region that is thinner than this paper.

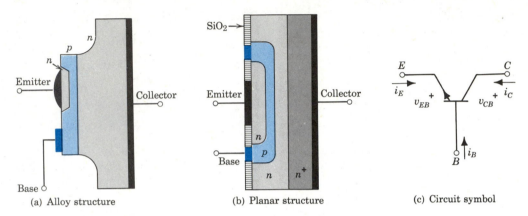

Figure 12.8 Types of *npn* bipolar junction transistors.

In the planar structure (Fig. 12.8b), a lightly doped film (*n*) is grown epitaxially upon a heavily doped (n^+) substrate. After oxidation of the surface, a window is opened by etching and an impurity (*p*) is allowed to diffuse into the crystal to form a junction. After reoxidation, a smaller window is opened to permit diffusion of the emitter region (*n*). Finally, contacts are deposited and leads attached.

In the conventional symbol (Fig. 12.8c), the emitter lead is identified by the arrow that points in the direction of positive charge flow in normal operation. (A *pnp* transistor is equally effective. The arrow in the symbol always points toward *n*-type material.) Although an *npn* transistor will work with either *n* region serving as the emitter, it should be connected as labeled because doping densities and geometries are intentionally asymmetric.

Operation

Figure 12.9a represents a thin horizontal slice through an *npn* transistor. The operation can be explained qualitatively in terms of the potential distributions across the junctions (Fig. 12.9b). The emitter junction is forward biased; the effect of bias voltage V_{EB} is to reduce the potential barrier at the emitter junction and to facilitate the injection of electrons (in this example) into the base where they are minority carriers. The collector junction is reverse biased; the effect of bias voltage V_{CB} is to increase the potential barrier at the collector junction. The base is so thin that almost all the electrons injected into the base from the emitter diffuse across the base and are swept

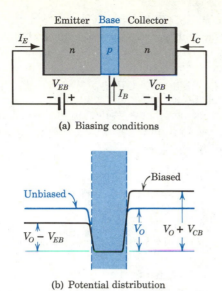

(a) Biasing conditions

(b) Potential distribution

Figure 12.9 Operation of an *npn* transistor.

up the potential hill into the collector where they recombine with holes "supplied" by the external battery. (Actually, electrons are "removed" by the external battery, leaving a supply of holes available for recombination.)

The net result is the transfer from the emitter circuit to the collector circuit of a current that is nearly independent of the collector-base voltage. As we shall see, this transfer permits the insertion of a large load resistance in the collector circuit to obtain voltage amplification. Alternatively, variation of the base current (the small difference between emitter and collector currents) can be used to control the relatively larger collector current to achieve current amplification or to perform the switching operation used in digital signal processing.

DC Behavior

The dc behavior of a BJT can be predicted on the basis of the motion of charge carriers across the junctions and into the base. With the emitter junction forward biased and the collector junction reverse biased, in so-called *normal* operation, the carrier motion of an *npn* transistor is as shown in the symbolic diagram of Fig. 12.10. The emitter current i_E (negative in an *npn* transistor) consists of electrons injected across the *np* junction and holes injected from the base. To optimize transistor performance, the base is doped relatively lightly, emitter efficiency γ (gamma) is almost

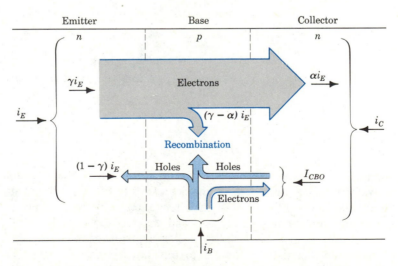

Figure 12.10 Carrier motion in the normal operation of an *npn* transistor.

unity, and most of the current consists of electrons injected from the emitter. Some of the injected electrons recombine with holes in the p-type base, but the base is made very narrow so that most of the electrons (minority carriers in this p region) diffuse across the base and are swept across the collector junction (up the potential hill). The factor α (alpha) varies from 0.90 to 0.999; a typical value is 0.98.

The electron current αi_E constitutes the major part of the collector current. In addition, there is a reverse current across the collector junction due to thermally generated minority carriers, just as in a diode (see Eq. 11-21). In a transistor, this *collector cutoff current* is given by

$$-I_{CBO}(e^{\,eV/kT} - 1) \cong I_{CBO} \qquad (12\text{-}3)$$

when the reverse biasing exceeds a few tenths of a volt. The total collector current is then

$$i_C = -\alpha i_E + I_{CBO} \qquad (12\text{-}4)$$

where α, the *forward current-transfer ratio,* increases slightly with increased collector biasing voltage v_{CB}.

The base current (Fig. 12.10) consists of holes diffusing to the emitter, positive charges to supply the recombination that occurs in the base, and the reverse saturation current to the collector. By Kirchhoff's law,

$$i_B = -i_E - i_C \qquad (12\text{-}5)$$

and i_B is the small difference between two nearly equal currents.

Common-Base Characteristics

The transistor connection of Fig. 12.11 is called the *common-base* (*CB*) configuration because the base is common to input and output ports. The *i-v* characteristics of a BJT in this configuration can be derived from our knowledge of diode

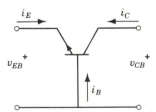

Figure 12.11 A transistor in the common-base configuration.

characteristics and transistor operation. Since the emitter-base section is essentially a forward-biased diode, the input characteristics in Fig. 12.12b are similar to those of the first quadrant of Fig. 12.12a; the effect of collector-base voltage v_{CB} is small. With the emitter open-circuited, $i_E = 0$ and the base-collector section is essentially a reverse-biased junction. For $i_E = 0$, $i_C \cong I_{CBO}$ (exaggerated in Fig. 12.12c) and the collector characteristic is similar to the third quadrant of Fig. 12.12a. For $i_E = -5$ mA, the collector current is increased by an amount $-\alpha i_E \cong +5$ mA (see Eq.

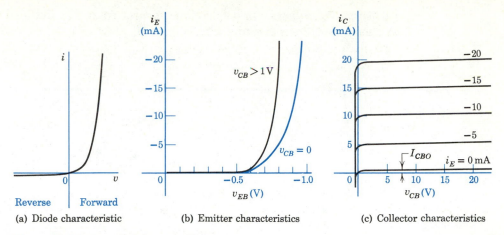

Figure 12.12 Common-base characteristics of an *npn* transistor.

12-4) and the curve is as shown. The slope of the curves in Fig. 12.12c is due to an effective increase in α as v_{CB} increases. Because the factor α is always less than 1, the common-base configuration is not good for practical current amplification.

Common-Emitter Characteristics

    If the *npn* transistor is reconnected in the *common-emitter (CE)* configuration of Fig. 12.13, current amplification is possible. The input current is now base current i_B,

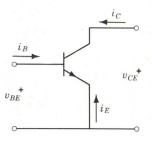

Figure 12.13 A transistor in the common-emitter configuration.

and emitter current $i_E = -(i_C + i_B)$; therefore, collector current is

$$i_C = -\alpha i_E + I_{CBO} = +\alpha(i_C + i_B) + I_{CBO}$$

Solving,

$$i_C = \frac{\alpha}{1 - \alpha} i_B + \frac{I_{CBO}}{1 - \alpha} \tag{12-6}$$

To simplify Eq. 12-6, we define the *current transfer ratio (CE)* as

$$\beta = \frac{\alpha}{1 - \alpha} \tag{12-7}$$

and note that the *collector cutoff current (CE)* is

$$\frac{I_{CBO}}{1 - \alpha} = (1 + \beta)I_{CBO} = I_{CEO} \tag{12-8}$$

The simplified equation for output (collector) current in terms of input (base) current and the current transfer ratio (beta) is

$$i_C = \beta i_B + I_{CEO} \qquad (12\text{-}9)$$

EXAMPLE 3

A silicon *npn* transistor with $\alpha = 0.99$ and $I_{CBO} = 10^{-11}$ A is connected as shown in Fig. 12.14. Predict i_C, i_E, and v_{CE}. (*Note:* It is convenient and customary in drawing electronic circuits to omit the battery, which is assumed to be connected between the +10-V terminal and ground.)

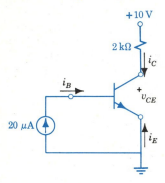

Figure 12.14 Transistor operation.

For this transistor,

$$\beta = \frac{\alpha}{1 - \alpha} = \frac{0.99}{1 - 0.99} = \frac{0.99}{0.01} = 99$$

and the collector cutoff current is

$$I_{CEO} = (1 + \beta)I_{CBO} = (1 + 99)10^{-11} = 10^{-9} \text{ A}$$

The collector current is

$$i_C = \beta i_B + I_{CEO} = 99 \times 2 \times 10^{-5} + 10^{-9} \cong 1.98 \text{ mA}$$

As expected for a silicon transistor, I_{CEO} is a very small part of i_C.

The emitter current is

$$i_E = -(i_B + i_C) = -(0.02 + 1.98)10^{-3} = -2 \text{ mA}$$

The collector-emitter voltage is

$$v_{CE} = 10 - i_C R_C \cong 10 - 2(\text{mA}) \times 2(\text{k}\Omega) = 6 \text{ V}$$

Since $v_{CB} = v_{CE} - v_{BE} \cong 6 - 0.7 = +5.3$ V, the *np* collector-base junction is reverse biased as required.

Typical common-emitter characteristics are shown in Fig. 12.15. The input current i_B is small and, for a collector-emitter voltage of more than a volt or so, depends only on the emitter-base junction voltage. For a silicon BJT, about 0.7 V of forward bias provides adequate base current.

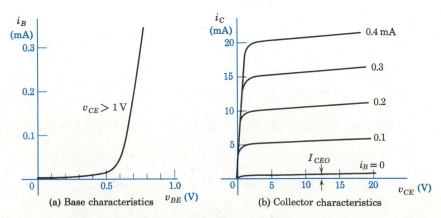

Figure 12.15 Common-emitter characteristics of an *npn* transistor.

The collector characteristics are in accordance with Eq. 12-9; for $i_B = 0$, the collector current is small and nearly constant at a value I_{CEO} (exaggerated in the drawing). For each increment of current i_B, the collector current is increased an amount βi_B. For $\alpha = 0.98$, $\beta = \alpha/(1 - \alpha) = 0.98/(1 - 0.98) = 49$, and a small increase in i_B corresponds to a large increase in i_C. A small increase in α produces a much greater change in β, and the effect of v_{CE} on i_C is more pronounced here than in the *CB* configuration (Fig. 12.12c).

Current Amplification

The collector characteristics of a bipolar junction transistor indicate the possibility of current amplification. Graphical analysis of an *npn* BJT amplifier is shown in Fig. 12.16. The input signal (assumed sinusoidal) is applied in parallel with the

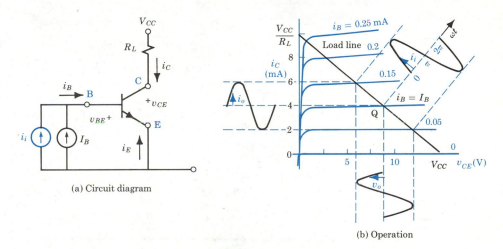

(a) Circuit diagram

(b) Operation

Figure 12.16 An elementary *npn* transistor current amplifier.

base-biasing current. The reverse bias on the collector is maintained by V_{CC}; note that most of this voltage appears across the collector junction since the voltage across the forward-biased emitter junction is quite small. (See Fig. 12.15a.) We anticipate that a signal current i_i will cause a variation in base current i_B that will produce a variation in collector current i_C; the varying component of i_C constitutes an amplified output current.

By Kirchhoff's voltage law around the output "loop,"

$$v_{CE} = V_{CC} - i_C R_L \tag{12-10}$$

This is the equation of the load line shown in Fig. 12.16b. For no signal input, $i_i = 0$ and $i_B = I_B$, the *quiescent* value. The quiescent point Q lies at the intersection of the load line and the characteristic curve for $i_B = I_B$. The intersections of the load line and the characteristic curves of the nonlinear transistor represent graphical solutions of Eq. 12-10 and define the instantaneous values of i_C and v_{CE} corresponding to values of the input signal current i_i. The signal output i_o is the sinusoidal component of i_C, the current in R_L. The current amplification, or current gain, is

$$A_I = \frac{i_o}{i_i} = \frac{I_{om}}{I_{im}} = \frac{I_o}{I_i} \tag{12-11}$$

depending on whether the current ratio is expressed in terms of instantaneous, maximum, or rms values.

EXAMPLE 4

For the elementary amplifier of Fig. 12.16, the collector-battery voltage $V_{CC} = 15$ V. The quiescent point Q is at $i_B = I_B = 0.1$ mA and $v_{CE} = V_{CE} = 9$ V. Specify R_L and predict the current gain and the output voltage for an input current $i_i = 0.05 \sin \omega t$ mA.

By Eq. 12-10, the load resistance is

$$R_L = \frac{V_{CC} - V_{CE}}{I_C} = \frac{15 - 9}{0.004} = 1500 \ \Omega$$

The load line intersects the collector current axis at

$$i_C = V_{CC}/R_L = 15/1500 = 0.01 = 10 \text{ mA}$$

For $\omega t = \pi/2$, the base current is $i_B = 0.15$ mA, the signal current is $i_B - I_B = 0.15 - 0.1 = 0.05$ mA, and the corresponding collector current is $i_C = 6$ mA. A signal current swing of 0.05 mA causes a collector current swing of $6 - 4 = 2$ mA. The current gain is

$$A_I = \frac{I_{om}}{I_{im}} = \frac{2 \times 10^{-3}}{5 \times 10^{-5}} = 40$$

The sinusoidal component of voltage across R_L is

$$v_o = R_L i_o = 1500 \times 0.002 \sin \omega t = 3 \sin \omega t \text{ V}$$

Switching

In contrast to the continuous signals amplified in the circuit of Fig. 12.16, computer information is generated, processed, and stored as discrete signals; each electronic element is either **ON** or **OFF**. Because a small base current can control a much larger collector current, the transistor is attractive as a possible control device. The fact that base current is an exponential function of base-emitter voltage permits a BJT to operate as a sensitive switching element in Example 5. In Chapter 13 we explore the application of transistors and diodes to logic gates and memory elements.

INTEGRATED CIRCUITS

In 1958 J. S. Kilby developed an "integrated circuit," a single monolithic chip of semiconductor in which active and passive circuit elements were fabricated by successive diffusions and depositions. Shortly thereafter Robert Noyce fabricated a complete circuit including the interconnections on a single chip. The ability to control

EXAMPLE 5

The transistor of Fig. 12.15 ($\beta \cong 50$ and $I_{CEO} \cong 1$ nA) is used in the *switching circuit* of Fig. 12.17. Predict the output voltage for input voltages of 0.2, 0.7, and 0.8 V.

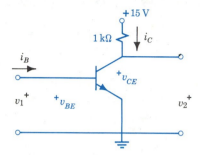

Figure 12.17 A transistor switch.

For $v_1 = v_{BE} = 0.2$ V, base current $i_B \cong 0$ and $i_C = I_{CEO} \cong 1$ nA. The voltage drop across the 1-kΩ resistor is only 1 μV and voltage $v_2 \cong 15$ V.

For $v_1 = v_{BE} = 0.7$ V, $i_B \cong 0.12$ mA and $i_C = \beta i_B = 50 \times 0.12 = 6$ mA. Therefore, voltage $v_2 = v_{CE} = 15 - 1000(0.006) = 9$ V.

For $v_1 = v_{BE} = 0.8$ V, $i_B \cong 0.4$ mA. Assuming $\beta = 50$, $i_C = \beta i_B = 50 \times 0.4 = 20$ mA. But this is impossible since only 15 V is available to drive collector current through the 1-kΩ resistor.

Figure 12.15b indicates the β concept is not valid for small values of v_{CE}; as i_B increases, v_{CE} approaches as a limit a value of less than 1 V.

Conclusion: A change in input from 0.2 to 0.8 V switches the output from 15 to 1 V.

precisely the dimensions and doping concentration of diffused or deposited regions combined with advances in photolithographic techniques have made possible the mass production of sophisticated devices of very small size but exceedingly high reliability.

With the exception of inductance, all the ordinary electronic circuit elements can be fabricated in semiconductor form. A resistor (Fig. 12.18a) is obtained in the form of a thin filament of specified conductivity, terminated in metallic contacts and isolated by a reverse-biased junction. Resistances from a few ohms to 20,000 Ω are possible. A low-loss capacitor is formed by depositing a thin insulating layer of silicon dioxide on a conducting region and then providing a metalized layer to form the second plate of the capacitor (Fig. 12.18b). Alternatively, a reverse-biased diode may be used to provide capacitance; capacitances up to 50 pF are possible. Typically, capacitors require the largest areas, resistors next, and active devices such as diodes, transistors, and FETs the smallest. Interconnections are provided by a layer of aluminum that is deposited, masked, and etched to leave the desired pattern.

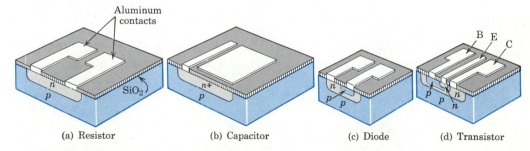

(a) Resistor (b) Capacitor (c) Diode (d) Transistor

Figure 12.18 Basic components of microcircuits.

Design Implications

The availability of sophisticated devices of high reliability in small packages at low cost has changed the character of electronic design. Instead of developing optimum circuits, buying components, and fabricating such units as amplifiers and digital registers, the design engineer selects mass-produced, high-performance functional units and incorporates them into specific products. Originally conceived as a means of reducing the size of complex equipment, integrated circuits now are equally noted for their low cost and high reliability.[†]

Since processing cost is proportional to area, circuits are designed to use more *active devices* (transistors and diodes) and fewer resistors and capacitors. (This is in sharp contrast to vacuum tube technology, where the active devices are the most expensive items by far.) The small circuit dimensions decrease the likelihood of pickup of unwanted noise; therefore signal levels can be lower. This in turn means lower voltages and lower power requirements. On the other hand, component values are hard to control precisely, and they may be voltage or temperature dependent. Also, unintentional capacitances may provide undesired coupling between circuit components.

Commercial integrated circuits (ICs) are usually available in three standard packages. The *multipin circular* type is the same size as a discrete transistor unit, but the chip it holds may contain hundreds of transistors and diodes. For use where size is important, the hermetically sealed *flat pack* occupies only 30 mm³. For convenience in use with printed-circuit boards, the *dual-in-line* package (DIP) provides easy plug-in operation. The cost of ICs ranges from less than a dollar to several dollars. In some cases, the cost of the package is greater than that of the chip.

Fabrication Processes

The fabrication of monolithic integrated circuits follows the planar technology used in transistor manufacture (Fig. 12.8b). A thin slice or *wafer* of p-type silicon 2 to 5 in. (50 to 125 mm) in diameter and 10 mils (0.01 in. or 0.25 mm) thick provides the *substrate*. The active and passive components are built within a thin n-type *epitaxial layer* on top (Fig. 12.19a). First the surface is oxidized by heating it in the presence of oxygen. Then the cooled wafer is coated with a photosensitive material called *photoresist* and covered with a mask of the desired geometry. Upon exposure, the light-sensitive material hardens and the unexposed material is dissolved away. An etching solution, in which the photoresist is insoluble, is then used to remove the SiO_2, leaving a pattern of windows through which the first diffusion takes place.

One method of insulating components from each other is to provide *diode isolation* by surrounding each element with a reverse-biased pn junction. The first diffusion provides p-type regions that extend through to the substrate (Fig. 12.19c). The individual circuit elements are then built within these isolated islands by successive steps involving oxidation, masking, etching, and diffusion of p- and n-type impurities. In the final step, aluminum is evaporated onto the wafer and selectively removed to provide interconnections and relatively large *bonding pads* to which leads are welded.

[†]See References at the end of this chapter.

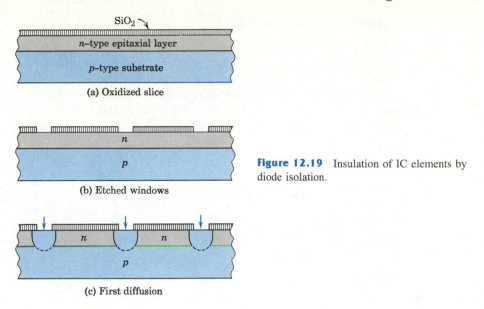

Figure 12.19 Insulation of IC elements by diode isolation.

(a) Oxidized slice

(b) Etched windows

(c) First diffusion

Integrated circuits are cheap because thousands of complex units can be fabricated simultaneously. After the circuit is designed functionally and tested experimentally, it is designed dimensionally and masks are drawn about 500 times full scale. Then each mask is reduced photographically to actual size and precisely replicated to provide a master mask containing hundreds of identical patterns. Each processed slice (Fig. 12.20) is *diced* to yield several hundred individual chips that are tested and encapsulated. If each of 500 chips contains 1000 components, and if 20 wafers are processed in a batch, a total of $500 \times 1000 \times 20 = 1$ million components are fabricated at once. Even if the yield of perfect chips is only 10%, the cost per component is very small.

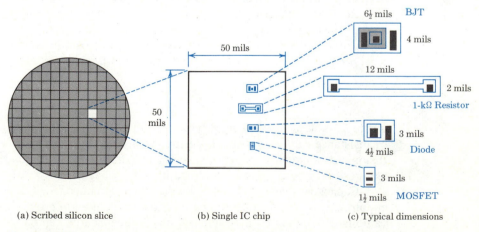

(a) Scribed silicon slice

(b) Single IC chip

(c) Typical dimensions

Figure 12.20 Typical integrated circuit details (after R.G. Hibberd).

Component Formation

Bipolar transistors in IC form are obtained by successive diffusions of boron (p-type) and phosphorus (n-type). A high concentration of phosphorus is necessary to overcome the previous boron diffusion and to create a high conductivity n^+-type emitter (Fig. 12.21a). Another n^+ area is created to provide a low-resistance contact with the collector. (To minimize the series resistance of the thin collector region, another n^+ layer is formed in the substrate before the epitaxial n layer is grown.)

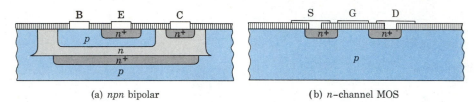

(a) npn bipolar   (b) n–channel MOS

Figure 12.21   Structure of IC transistors.

The MOS transistor in IC form (Fig. 12.21b) is identical with its discrete counterpart. MOS transistors can be smaller—and therefore cheaper—than bipolar transistors because they are self-isolating. The source, drain, and channel incorporate insulating pn junctions, and the gate is isolated by oxide. The fabrication process is simple in concept, but MOS device properties are sensitive to surface conditions, which must be very carefully controlled.

Complementary devices, one n-channel and one p-channel, can be combined in a single unit. These CMOS units require only a single power supply, use very little power, and offer special advantages in digital circuits.

Diodes may be formed in various ways. For example, the emitter-base junction of a transistor (with collector tied to the base) provides a low-voltage diode with fast response. If the emitter is omitted, the remainder of a transistor provides a collector-base diode with higher reverse-voltage rating but slower response.

Resistors are usually formed during transistor base diffusion and the conductivity is dependent upon the transistor requirements. The desired value of resistance is obtained by specifying the length l and width w of the conducting path. Since $R = \rho l/A = \rho l/tw$, for a given resistivity ρ and thickness t,

$$R = \frac{\rho}{t} \cdot \frac{l}{w} = R_S \frac{l}{w} \tag{12-12}$$

where R_S is the *sheet resistance*, typically 100 to 200 Ω. (See Example 6.) Low values of resistance can be obtained by using the higher conductivity n^+-type emitter diffusion. Also, the channel between source and drain of an MOS transistor can be used as a resistor whose value is governed by gate voltage.

Capacitors may be fabricated by the MOS technique (Fig. 12.18b), or the inherent property of a reverse-biased pn junction can be used to provide a capacitor whose value is determined by bias voltage. Junction capacitors can be formed at the same time as the collector junctions of transistors, but the capacitance per unit area is quite

EXAMPLE 6

Design a 4000-Ω resistor using stripes 0.5 mil wide with 0.5-mil spacing.

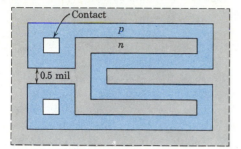

Figure 12.22 Layout of IC resistor.

Assuming $R_S = 100\ \Omega$, by Eq. 12-12,

$$l/w = R/R_S = 4000/100 = 40$$

Therefore the conducting path should be 40 squares long.

For optimum space utilization, the path is folded back on itself. The stripe shown in Fig. 12.22 is approximately 40 squares long (end effects are neglected) and occupies an area of about 32 mils². (This is greater than the area required for a standard transistor.)

Note: IC surface dimensions are usually given in mils although metric units are used elsewhere.

low (~0.2 pF/mil²). For MOS capacitors of area A, the capacitance is given by

$$C = \epsilon \frac{A}{t} \qquad (12\text{-}13)$$

where ϵ = permittivity of silicon dioxide $\cong 3.5 \times 10^{-11}$ F/m. For a minimum practical thickness of $t = 0.002$ mil, the capacitance is about 0.4 pF/mil².

Applications

The design of integrated circuits consists in planning an optimum geometrical layout of standard components and connections (Fig. 12.23). Since the fabrication of *microelectronic* devices requires sophisticated techniques and precision equipment, large volume production is necessary. Large volume means standardization of a relatively few products.

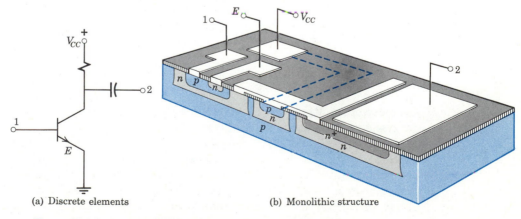

(a) Discrete elements

(b) Monolithic structure

Figure 12.23 An elementary amplifier in IC form (simplified).

Integrated circuits are used in great quantities in digital computers because of their small size, low power consumption, and high reliability. The complexity of digital ICs ranges from simple logic gates and memory units (Chapter 13) to large arrays capable of complete data processing. The small size of MOS elements has led to *large-scale integration* (LSI), in which thousands of elements are created on a single chip. Typical is the Intel 8086 microprocessor that is as fast as many mini-computers. This *n*-channel MOS device incorporates 29,000 transistors on a single chip 225 mils square. On the horizon is VLSI with a million active devices in a single package.

In space vehicles and in hearing aids, the light weight of ICs is important. Microelectronics also finds application in device arrays where it is advantageous to have many matched elements on a single chip. For example, 100 of the amplifiers in Fig. 12.23 could be put on a single chip and, since they were fabricated simultaneously, they could be expected to "track" well with changes in temperature. More complex circuits, with high performance made possible by the liberal use of active devices, are used in the versatile op amps that are widely employed in place of individually designed units (Fig. A9).

THYRISTORS

The diode is an automatic, unsophisticated switch; it is **ON** whenever the applied voltage provides forward bias. The transistor also can be used as a switch; collector current flows whenever there is adequate base current to turn the switch **ON**. Transistors are widely used as switches in digital circuits (Chapter 13) and in other cases where moderate amounts of power are to be controlled. A major disadvantage, however, is that transistors require high and continuous base current in the **ON** state. In contrast, the multilayered semiconductor devices called *thyristors* have the ability to control large amounts of power with only a minimum of control energy. As a result, they are commonly used in sophisticated applications to rectification (ac to dc), inversion (dc to ac), relaying, timing, ignition, and speed control at power levels ranging from a few milliwatts to hundreds of kilowatts.

Operation of Four-Layer Devices

The silicon *pnpn* or *four-layer diode* consists of three *pn* junctions in series as shown in Fig. 12.24. The symbol contains a 4 with the apex pointing in the direction of positive current flow. In the three-diode representation, for negative anode-to-cathode voltage V_{AK}, center junction J_2 is forward biased, but the outside junctions are reverse biased. As a result, this is a "reverse blocking" diode and the characteristic for negative V_{AK}, the **OFF** state, resembles that of a reverse-biased diode. No reverse current flows unless the avalanche breakdown voltage is exceeded.

For positive V_{AK}, the current at low voltages is limited by the reverse-bias behavior of junction J_2. As the applied voltage increases, the current increases slowly until the *breakover voltage* V_{BO} is reached. At this point the current rises abruptly and the voltage drops sharply; the diode has switched to the **ON** state.

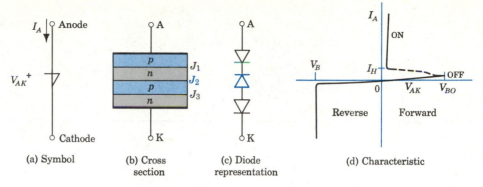

Figure 12.24 The four-layer diode.

The switching operation of the four-layer diode can be understood by viewing the device as two interconnected transistors. In Fig. 12.25, back-to-back *pnp* and *npn* transistors are shown physically displaced but electrically connected. By Eq. 12-4, the collector currents will be

$$I_{C1} = -\alpha_1 I_{E1} + I_{CBO1} = -\alpha_1 I_A + I_{CBO1}$$

$$I_{C2} = -\alpha_2 I_{E2} + I_{CBO2} = +\alpha_2 I_A + I_{CBO2} \qquad (12\text{-}14)$$

Summing the currents into transistor T1,

$$I_A + I_{C1} - I_{C2} = 0 = I_A - \alpha_1 I_A + I_{CBO1} - \alpha_2 I_A - I_{CBO2}$$

Solving, and recognizing that the cutoff currents of *pnp* and *npn* transistors are of opposite sign,

$$I_A = \frac{-I_{CBO1} + I_{CBO2}}{1 - (\alpha_1 + \alpha_2)} = \frac{I_{CO}}{1 - (\alpha_1 + \alpha_2)} \qquad (12\text{-}15)$$

For silicon at very low emitter currents, α (see Fig. 12.10) is small because of recombination of holes and electrons in the transition region. As I_A increases with an increase in V_{AK}, α_1 and α_2 increase. As the quantity $(\alpha_1 + \alpha_2)$ approaches unity,

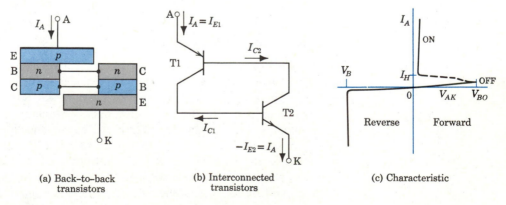

Figure 12.25 Analysis of four-layer diode operation.

the current tends to increase without limit and breakover occurs. After switching **ON**, the diode voltage is very small (~1 V) and the current is limited by the resistance of the external circuit. If the applied voltage is reduced, the switch remains **ON** until the current has reached a minimum *holding current* I_H. To switch the four-layer diode **OFF**, therefore, the current is reduced below I_H by reducing the applied voltage or increasing the series resistance.

EXAMPLE 7

Design a sawtooth wave oscillator for operation at 500 Hz. Available is a four-layer diode rated at $V_{BO} = 100$ V, $I_A = 20$ A, and $I_H = 20$ mA.

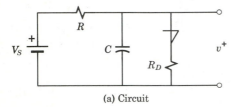

(a) Circuit

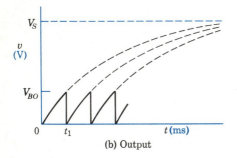

(b) Output

Figure 12.26   Sawtooth oscillator.

The breakover characteristic of a *pnpn* diode makes it suitable for timing functions. In the circuit of Fig. 12.26a, the capacitor charges up to V_{BO} and discharges through the diode and resistor R_D.

For proper operation $V_S > V_{BO}$, therefore, specify

$$V_S = 3V_{BO} = 3 \times 100 = 300 \text{ V}$$

To ensure turnoff, $V_S/R < I_H$ or

$$R > V_S/I_H = 300/0.02 = 15 \text{ k}\Omega$$

Therefore, specify $R = 25$ kΩ.

For $v/V_S = 1 - e^{-t_1/RC} = 100/300$,

$$t_1/RC \cong 0.4 = 1/RCf$$

Therefore,

$$C = 1/0.4Rf = 1/0.4 \times 25 \times 10^3 \times 500 = 0.2 \ \mu\text{F}$$

For $I_A = 20$ A, $R_D = V_{BO}/I_A = 100/20 = 5$ Ω. The output (Fig. 12.26b) approximates a sawtooth.

Operation of Silicon Controlled Rectifiers

In a *pnpn triode* a *gate* connection to the inner *p* layer permits the introduction of a *gate current* that switches this thyristor to the **ON** state at voltages lower than V_{BO} for the corresponding diode. Such a *silicon controlled rectifier* (SCR) provides flexible switching and, since the gate current is required only momentarily, very efficient control of large amounts of power. In Fig. 12.27b, the gate connection is to the base of the *npn* transistor section. (In the *silicon controlled switch*, SCS, gate signals may be applied to the base of the *pnp* transistor as well.) Injection of base current increases α and reduces the necessary breakover voltage. Therefore the firing voltage is a function of gate current I_G; for large values of I_G, the behavior approximates that of a simple *pn* diode.

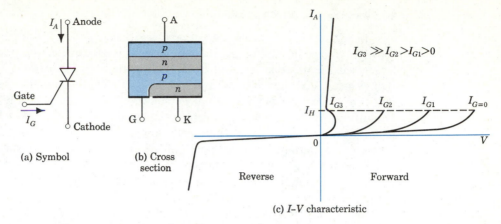

(a) Symbol

(b) Cross section

$I_{G3} \gg I_{G2} > I_{G1} > 0$

(c) *I–V* characteristic

Figure 12.27   The silicon controlled rectifier (SCR).

In the typical application, firing is determined by a current pulse applied at the desired instant (Fig. 12.28). The *conduction angle* θ can be varied from zero to 180° by controlling the *firing angle* ϕ. Assuming that V_{AK} is negligibly small in the **ON** state and that motor current i_M is directly proportional to applied voltage v, the current waveform will be as shown. If the average or dc current is the important controlled variable, this can be determined from

$$I_{dc} = \frac{1}{2\pi} \int_{\phi}^{\pi} I_m \sin \omega t \; d(\omega t) = \frac{I_m}{2\pi}(1 + \cos \phi) = \frac{I_m}{2\pi}(1 - \cos \theta) \quad (12\text{-}16)$$

Note that once the SCR has turned **ON**, gate current is no longer required. The SCR is turned **OFF** when the supply voltage goes through zero.

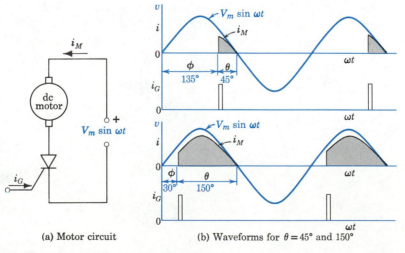

(a) Motor circuit

(b) Waveforms for $\theta = 45°$ and 150°

Figure 12.28   SCR speed control of a dc motor.

SCR Ratings and Applications

In comparison to power transistors, thyristors are designed with thick base layers for high voltage capability and the necessary low α characteristic. Currents may be larger because the total junction areas are available for conduction. Maximum SCR ratings up to 2000 V and 600 A (average) are available. In common sizes, gate current pulses of a few milliamperes (50 mA) requiring gate voltages of a few volts (1 to 2 V) will control anode currents several thousand times as large (100 A). For a typical *turn-on* time of 1 μs, the average gate power is very, very small. In general, the *turn-off* time (during which the anode voltage must be kept below the holding voltage) is several times as long as the turn-on time. In comparison to other power rectifiers, thyristors are compact and extremely reliable; they have no inherent failure mechanism.

The high power gain, efficiency, and reliability of thyristors have led to a host of applications. In the power control circuit of Fig. 12.29, an SCR provides

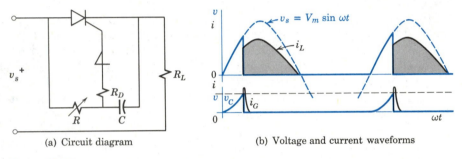

| (a) Circuit diagram | (b) Voltage and current waveforms |

Figure 12.29 SCR rectification with *pnpn* diode trigger.

rectification and a *pnpn* diode serves as a trigger timed by variable resistance R. With the SCR **OFF**, supply voltage v_s appears across RC, the capacitor begins to charge, and v_C increases at a rate determined by R. (Note the difference in response from that in the sawtooth oscillator of Fig. 12.26.) When v_C exceeds the breakover voltage of the *pnpn* diode, the capacitor discharges through current-limiting resistance R_D, providing a gate current pulse that fires the SCR. If full-wave rectification is desired, a bridge rectifier can be placed in front of the control circuit of Fig. 12.29a. For light dimming or ac motor speed control, two SCRs can be connected in inverse parallel to provide current on both half-cycles. The *triac* (*tri*ode *ac* switch), a popular five-layer thyristor, is essentially an integrated double-ended SCR in a single unit.

SUMMARY

- In a field-effect transistor, the gate voltage controls the channel conductivity.
  The mobile carrier density may be depleted or enhanced.
  In the constant-current region, transfer characteristics are parabolic.
  FETs are characterized by high input resistance.

- A bipolar junction transistor consists of two *pn* junctions in close proximity; normally, the emitter junction is forward biased, the collector reverse biased.
  In common-base operation,

$$i_C = -\alpha i_E + I_{CBO} \qquad \text{where } \alpha \cong 1$$

In common-emitter operation, a small base current controls the relatively larger collector current to achieve current amplification.

$$i_C = \beta i_B + I_{CEO} \qquad \text{where } \beta = \frac{\alpha}{1 - \alpha}$$

- Graphical analysis of nonlinear transistors is based on load-line construction.
  Variations in input quantities about a properly selected quiescent point result in larger output variations that represent amplification.

- A small change in input voltage can switch the output voltage from near zero to a high value as required in digital operations.

- Integrated circuits have changed the character of electronic design.
  Transistors, diodes, resistors, and capacitors are fabricated simultaneously by successive oxidation, masking, etching, and diffusion steps.
  For diffused resistors (up to 20 kΩ), $R = R_s l/w$.
  For MOS capacitors (up to 100 pF), $C = \epsilon A/t$.

- Thyristors control large amounts of power with minimum control energy.
  The *pnpn* diode turns **ON** when the breakover voltage is exceeded.
  The SCR triode thyristor can be turned **ON** by timed gate current pulses.
  Because they are compact, efficient, and reliable, SCRs are widely used in rectification, inversion, timing, relaying, and power control.

REFERENCES

1. Robert Hibberd, *Integrated Circuits,* McGraw-Hill Book Co., New York, 1969.
  An excellent introduction to IC design, technology, and applications; for engineers and technicians.

2. Charles Holt, *Electronic Circuits,* John Wiley and Sons, New York, 1978.

A complete treatment of technology and applications at the undergraduate level.

3. Jacob Millman, *Microelectronics,* McGraw-Hill Book Co., New York, 1979.
  Detailed treatment of many digital and analog devices and circuits.

REVIEW QUESTIONS

1. Explain the function of the source, gate, and drain in the JFET.
2. Sketch the cross section of a JFET at pinch off and explain the effect on i_D.
3. Explain the operation of an enhancement-type MOSFET.
4. Sketch, from memory, the transfer characteristics of a DE MOSFET.
5. Explain with sketches the operation of a *pnp* transistor.
6. Why must the base be narrow for BJT action?
7. Describe how amplification and switching are achieved by a transistor.
8. Explain common-base transistor characteristics in terms of diode characteristics.
9. Explain the role of the load line in graphical analysis of amplifiers.
10. List two ways in which the availability of ICs has changed electronic design.

11. Explain how system reliability is increased by using ICs.
12. How many diffusion steps are required in typical IC fabrication?
13. How does diode isolation provide element insulation?
14. Without reference to the text, outline the process of making an IC, including resistors, capacitors, and transistors.
15. What are the advantages of thyristors over *pn* diodes? What are the advantages of thyristors over transistors?
16. Explain the behavior of a *pnpn* diode as breakover is approached and reached.
17. Why doesn't removal of gate current turn **OFF** an SCR?

EXERCISES

1. The JFET of Fig. 12.2 is connected as shown in Fig. 12.30. Complete the diagram by indicating the proper polarities for V_{GG} and V_{DD}. For $V_{GG} = 3$ V and $V_{DD} = 20$ V, predict the drain current.

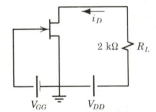

Figure 12.30

2. The JFET of Fig. 12.2 is connected as shown in Fig. 12.30. Indicate the proper polarities for V_{GG} and V_{DD}. If V_{DD} is 20 V and R_L is changed to 2.5 kΩ, specify V_{GG} for a drain current of 5 mA.
3. The MOSFET of Fig. 12.3 is to be used as a switch in the circuit of Fig. 12.31 where $V_{DD} = 10$ V and $R_L = 2.3$ kΩ. Draw the load

line defining the relation between the nonlinear device and the linear supply. For V_i switched from 2 to 10 V, calculate the corresponding values of I_D and V_o.

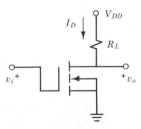

Figure 12.31

4. A JFET constructed like the one in Fig. 12.1b has the following typical properties: $I_{DSS} = 4$ mA and $V_{GS(OFF)} = -5$ V.
   (a) Sketch a set of i_D vs v_{DS} curves.
   (b) Draw transfer characteristic i_D vs V_{GS}.

(c) For $V_{GS} = -2$ V in Fig. 12.32, specify R_S for operation in the normal region.

(d) For the value of R_S in part (c) and $R_D = 3$ kΩ, select a reasonable value of v_{DS} and specify V_{DD}.

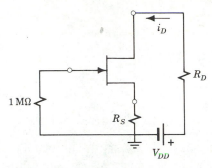

Figure 12.32

5. The FET of Fig. 12.2 is to be used as a voltage amplifier. An average gate voltage $V_{GS} = -2$ V is maintained and a load resistance $R_L = 5$ kΩ is inserted in the drain circuit of Fig. 12.1c. The input voltage is inserted in series with battery V_{GS} and the output voltage is taken across R_L. Estimate the output (drain) current variation for an input (gate-source) signal of 1 sin ωt V and the voltage amplification possibilities of this amplifier.

6. Assume that your roommate has a good knowledge of physics, including semiconductors, and mathematics, but has never studied electronics. Using labeled sketches and graphs, explain to him or her the operation of a p-channel enhancement-mode MOSFET.

7. The MOSFET of Fig. 12.3 is to be used as a passive load resistor by connecting the gate terminal to the drain.

(a) Draw a graph of i_D versus v_{DS} for this connection.

(b) Estimate the resistance offered to a dc current of 2 mA.

(c) Use piecewise linearization to represent this device by a combination of voltage source and resistance.

8. For an n-channel enhancement-mode MOSFET, threshold voltage $V_T = 1$ V and $I_{DS} = 8$ mA for $V_{GS} = 5$ V.

(a) Draw the transfer characteristic for the normal operating region.

(b) For $V_{GS} = +3$ V, $V_{DD} = 16$ V, and $R_D = 3$ kΩ, draw the circuit and predict the drain current and the drain-source voltage.

9. The MOSFET of Fig. 12.3 is to be used as an elementary amplifier.

(a) Draw an appropriate circuit showing bias voltages V_{GG} and V_{DD} and load resistor R_L.

(b) For $V_{DD} = 12$ V, $V_{GG} = 8$ V, and $R_L = 2$ kΩ, determine quiescent drain current I_D.

(c) Estimate the input signal voltage for an output $v_o = 1 \sin \omega t$ V.

10. The n-channel MOSFET shown in Fig. 12.33 can operate in the depletion or enhancement mode. In the normal operation (constant-current) region with gate shorted to source, the drain current is 8 mA, and the pinch-off voltage is -4 V. The MOSFET is to operate at $I_D = 2$ mA and $V_{DS} = 7$ V (the quiescent point).

(a) Sketch the $i_D - v_{GS}$ transfer characteristic and specify the quiescent value of v_{GS}.

(b) Complete the circuit design by specifying R_1 and R_2. (*Note:* Neglect any voltage drop across resistance R_G, which carries no current.)

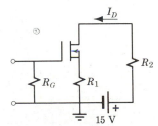

Figure 12.33

11. For a DE MOSFET, $I_{DSS} = 2$ mA and $V_P = -4$ V.

(a) Calculate V_{GS} for operation at $I_{DS} = 1$ mA in the constant-current region.

(b) For the circuit shown in Fig. 12.32, derive an expression for i_D as a $f(v_{DS})$.

(c) For operation at $I_D = 1$ mA and $V_{DS} = 8$ V with $R_D = 5$ kΩ, specify R_S and V_{DD}.

12. For a pnp BJT:

(a) Sketch the charge distribution when unbiased.

(b) Sketch the potential distribution when unbiased.

(c) Sketch the potential distribution when properly biased.

13. For high-gain amplification, high values of α and β are desirable. Considering the effect of doping on factor γ (Fig. 12.10), explain why it is desirable to dope the base relatively lightly in comparison to the emitter and collector regions.

14. Assume that your roommate has a good knowledge of physics, including semiconductors, and mathematics, but has never studied electronics. Using labeled sketches and graphs, explain to him or her the "normal" operation of a *pnp* junction transistor.

15. A BJT is connected as in Fig. 12.34.
    (a) Label the polarities of v_{EB} and v_{CB} to provide for *normal* operation.
    (b) For switch S *open*, sketch a graph of I_C versus $\pm V_{CB}$.
    (c) For switch S *closed* and $I_E = -2$ mA, repeat part (b).

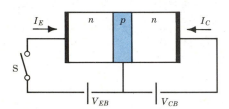

Figure 12.34

16. A BJT has the following parameters: $\alpha = 0.97$ and $I_{CBO} = 10\ \mu$A. For $0 < i_C < 10$ mA:
    (a) Sketch the common-base collector characteristics.
    (b) Sketch the common-emitter collector characteristics.

17. For a silicon BJT, the base current is 0.3 mA at a base-emitter voltage of 0.7 V. Assuming that the base-emitter junction follows the behavior of a theoretical diode (Eq. 11-22), predict the range of base currents for base-emitter voltages from 0.65 to 0.75 V. Comment on the assumption that $V_{BE} \cong 0.7$ V in many situations.

18. A silicon BJT is connected as shown in Fig. 12.35, where $Rc = 3.5$ kΩ. Stating any assumptions, predict I_C and specify R_B to establish V_{CE} at 5 V.

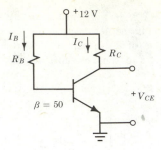

Figure 12.35

19. A silicon *pnp* BJT is connected in the circuit shown in Fig. 12.36.
    (a) Complete the drawing by putting the arrowhead on the emitter symbol and indicating the proper battery polarities for a forward-biased emitter-base junction and a reverse-biased collector-base junction.
    (b) Assuming $|V_{BE}| = 0.7$ V, calculate I_B for $V_{BB} = 4$ V and $R_B = 33$ kΩ.
    (c) If $V_{CC} = 15$ V, $R_L = 1.5$ kΩ, and $\beta = 60$, calculate I_C, I_E, and V_{CE}.

20. A silicon *npn* transistor is connected in the circuit shown in Fig. 12.36.
    (a) Complete the drawing by putting the arrowhead on the emitter symbol and indicating the proper battery polarities for a forward-biased emitter-base junction and a reverse-biased collector-base junction.

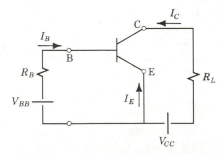

Figure 12.36

    (b) Assuming $|V_{BE}| = 0.7$ V, specify R_B for $V_{BB} = 4$ V and $I_B = 0.2$ mA.
    (c) If $V_{CE} = 8.5$ V, $R_L = 500$ Ω, and $\beta = 75$, calculate I_C, I_E, and V_{CC}.

21. A subcircuit commonly used in ICs is shown in Fig. 12.37. (Assume $\beta = 30$ for the transistor.)
    (a) Sketch the i-v characteristic of a silicon diode and estimate the voltage at point x with respect to point y if the diode is conducting.
    (b) If the base-emitter junction of the transistor is forward biased, estimate the voltage of point y and the voltage at point x with respect to ground.
    (c) Estimate the base and collector currents of the transistor and the output voltage V_o under these conditions.

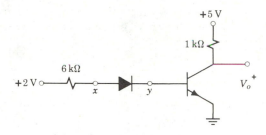

Figure 12.37

22. The silicon transistor of Fig. 12.15 is used in the circuit of Fig. 12.16 with $V_{BB} = 6.7$ V, $R_B = 60$ kΩ, $R_L = 2$ kΩ, and $V_{CC} = 20$ V.
    (a) Estimate β from the graph, calculate I_B, and predict I_C at the quiescent point.
    (b) Reproduce the collector characteristics and draw the "load line." Determine I_C graphically.
    (c) Find i_C (graphically) for $i_i = 0$, $+50$, and -50 μA.
    (d) If i_i is a sinusoid varying between $+50$ and -50 μA, write an expression for $i_C(t)$.
    (e) Estimate the current gain of this amplifier.

23. A certain IC device would cost $120,000 for the complete design and $2 each to manufacture. Predict:
    (a) The unit cost for a custom run of 100.
    (b) The unit cost for a standard run of 100,000.
    (c) The number of units at which design cost is 10% of the manufacturing cost.

24. It costs $60 to process a 100-mm wafer, of which only the central 80 mm is usable. If the final yield is 20%, estimate the fabrication cost of a single 1.5 mm $\times$ 1.5 mm IC chip.

25. Sketch the masks required to fabricate an IC transistor.

26. Assuming $R_S = 200$ Ω, lay out and determine the dimensions of a 10-kΩ diffused p-type resistor for an IC. Use 0.5-mil stripes with 0.5-mil spacing in an approximately square area.

27. Lay out and determine the dimensions of a 10-pF IC capacitor, assuming an oxide thickness of 0.003 mil.

28. A parallel RC circuit with a time constant of 2 μs is to be fabricated in IC form. Following the standards of Example 7, design and lay out the circuit with the resistance and capacitance occupying approximately equal areas.

29. In Fig. 12.26a, $V_S = 200$ V, $R = 10$ kΩ, and $C = 0.5$ μF. If V_{BO} is 50 V, predict the oscillator frequency.

30. Sketch waveforms of $v_S = V_S \sin \omega t$, $v_G = V_G \sin (\omega t - 45°)$, i_L, and v_{AK} for an SCR in series with v_S and R_L.

31. Redraw Fig. 12.25 to represent an SCR by interconnected transistors. Express I_{C1} and I_{C2} in terms of β_1 and β_2, and derive an expression for I_A. What is the condition for turn-on? Is this consistent with the conclusion reached for the $pnpn$ diode?

32. In Fig. 12.29, $V_m = 200$ V and $R_L = 10$ Ω.
    (a) For a conduction angle of 120°, determine average load current and power.
    (b) Repeat for a conduction angle of 75°.
    (c) If the **ON** voltage across the SCR is 1 V, calculate the power lost in the diode in part (a).

33. In Fig. 3.26a, $V_m = 300$ V and $R_L = 20$ Ω. It is desired to control the power delivered to the load without changing v or R_L.
    (a) Redraw the circuit showing how an SCR could be used for this purpose.
    (b) Sketch voltage and current waveforms for an SCR conduction angle of 45°.
    (c) Predict the power delivered to R_L for conduction angles of 45° and 135°.

PROBLEMS

1. Two MOSFETs (Fig. 12.3) are to be used as a switch in a digital computer. In the circuit of Fig. 13.2c, determine the i_D vs V_{DS} characteristic for the upper (load) transistor. For $V_{DD} = 10$ V, plot $V_{DD} - V_{DS(load)}$ on a reproduction of Fig. 12.3c. Estimate the output voltages for inputs of 2 and 10 V. Describe the switching operation.

2. In a conventionally arranged amplifier, the transistor terminals are labeled X, Y, Z. With no signal applied, it is observed that terminal X is 20 V positive with respect to terminal Y and 0.5 V positive with respect to terminal Z. Explain your reasoning:
   (a) Which terminal is the emitter?
   (b) Which terminal is the collector?
   (c) Which terminal is the base?
   (d) Whether X is p- or n-type material?

3. The variation in α with v_{CB} in a transistor (Fig. 12.12c) can be explained in terms of the effect of v_{CB} on the *effective width* of the base.
   (a) As the reverse bias increases, what is the effect on the width of the transition (or depletion) region adjacent to the base-collector junction?
   (b) How does a change in transition width change the effective base width?
   (c) How does base width affect α?
   (d) Why does an increase in v_{CB} produce the effect shown in Fig. 12.12b?

4. A transistor amplifier using a 2N3114 is to supply a 10-V (rms) output signal across a 10-kΩ load resistance R_L. "Design" an elementary amplifier by drawing an appropriate circuit and specifying the components. Estimate the input signal current required.

5. A nearly constant current source $I \cong 2$ mA is needed in an IC, and the circuit of Fig. 12.38 is proposed with $R = 1$ kΩ. Choose R_y so that the voltage across R_y is about 4 times V_{BE} so that small changes in V_{BE} do not affect the operation. Stating any assumptions, specify R_x. Comment on the "constancy" of this circuit if β changes from 50 to 200, say.

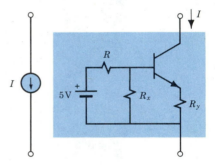

Figure 12.38

13

Logic Elements

Switching Logic

Electronic Switches

Transistor Logic Gates

Memory Elements

Digital Integrated Circuits

The audible portion of a radio program transmitted on a modulated carrier is in the form of *continuous* or *analog* signals; the signal amplitude or frequency can take any value within a wide range of values. In contrast, the information processed by a hand calculator is in the form of *discrete* or *digital* signals; each electronic component is either on or off. A major virtue of electronic circuits is the ease and speed with which digital signals can be processed, and the use of such signals in control, computation, and instrumentation is the most rapidly developing aspect of electronic engineering.

The processing of information in digital form requires special circuits, and the efficient design of digital circuits requires a special numbering system and even a special algebra. The circuits must provide for storing instructions and data, receiving new data, performing calculations, making decisions, and communicating the results. For example, an automatic airline reservation system must receive and store information from the airline regarding the number of seats available on flights all over the world, respond to inquiries from travel agents across the country, subtract the number of seats requested from the number available or add the number of seats canceled, handle 50 or so requests per minute, and keep no one waiting more than a minute.

Information processing is an important component of all branches of engineering and science. Aeronautical and chemical engineers may be designing automatic control systems. Civil and industrial engineers may be concerned with data on traffic flow or product flow. Mechanical engineers may be designing "smart" tools or products. Chemists and medical doctors may be interested in automated laboratory analysis.

Biologists and physicists may need remote control of precisely timed events. Everyone engaged in experimental work or in management can benefit from the new data processing techniques.

In this chapter we begin our study of digital systems by learning the basic functions performed by decision-making elements and seeing how semiconductor devices can be used in practical circuits. Then we look at basic memory elements and see how they can be constructed from bipolar and MOS transistors. Finally we examine more sophisticated memory elements with desirable operating features. In the next chapter we shall learn to design the data processing circuits and devices that are used in digital computers.

SWITCHING LOGIC

Digital computers, automatic process controls, and instrumentation systems have the ability to take action in response to input stimuli and in accordance with instructions. In performing such functions, an information processing system follows a certain *logic*; the elementary logic operations are described as **AND**, **OR**, and **NOT**. Tiny electronic circuits consuming very little power can perform such operations dependably, rapidly, and efficiently.

Gates

A *gate* is a device that controls the flow of information, usually in the form of pulses. First we consider gates employing magnetically operated switches called *relays*. If the switches are normally open, they close when input signals in the form of currents are applied to the relay coils.

In Fig. 13.1a, the lamp is turned on if switch A *and* switch B are closed; it is therefore called an **AND** circuit or **AND** gate. In Fig. 13.1b, the lamp is turned on if switch A *or* switch B is closed *or* if both are closed; it is called an **OR** circuit. In Fig. 13.1c, the switch is normally closed. For an input at C, there is **NOT** an output; the input has been *inverted* by the **NOT** gate.

In general, there may be several inputs and the output may be fed to several other logic elements. In a typical digital computer there are millions of such logic elements. Although the first practical digital computer (the Harvard Mark I, 1944) employed relays, the success of the modern computer is due to our taking advantage of the switching capability of semiconductor devices.

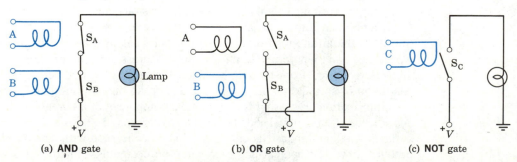

(a) **AND** gate (b) **OR** gate (c) **NOT** gate

Figure 13.1 Gate circuits using relays.

Electronic Switching

In the gates of Fig. 13.1, "data" are represented by **ON** and **OFF** switch positions or by the corresponding presence or absence of voltages. How can diodes and transistors be used as switches? If the input to the circuit in Fig. 13.2a is zero (i.e., if the input terminals are shorted), the diode is forward biased, current flows easily, and the output is a few tenths of a volt or approximately zero. If the input is 5 V or more, no diode current flows, and the output is $+5$ V. The **OFF–ON** positions of this diode switch correspond to output voltage levels near zero and $+5$ V.

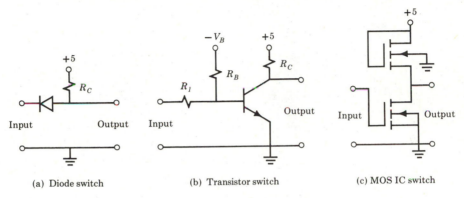

(a) Diode switch   (b) Transistor switch   (c) MOS IC switch

Figure 13.2 Elementary semiconductor switching circuits.

In the elementary transistor switch (Fig. 13.2b) with a zero input, the emitter junction is reverse biased by V_B, the collector current is very small, and the output is approximately $+5$ V. An input of $+5$ V forward biases the emitter junction, a large collector current flows, and the output is approximately zero. Again the output voltage levels are near zero and $+5$, but the signal has been inverted in the transistor switch.

The MOS transistor can operate as a binary switch controlled by changes in gate voltage. Since an MOS transistor can also function as a resistor, the arrangement of Fig. 13.2c is convenient in integrated circuits. With the gate of the upper transistor connected to the drain terminal, enhancement is high and the resistance of the "load" is determined by the channel dimensions. With no input signal, the lower transistor is **OFF** and the output is approximately $+5$ V. A positive input signal turns the "driver" transistor **ON** and the output falls to nearly zero.

Basic Logic Operations

To facilitate our work with logic circuits, we use some of the conventional symbols and nomenclature developed by circuit designers and computer architects. Each basic logic operation is indicated by a *symbol*, and its function is defined by a *truth table* that shows all possible input combinations and the corresponding outputs. Logic operations and variables are in boldface type. Electronic circuits are drawn as single-line diagrams with the ground terminal omitted; all voltages are with respect to ground so that "no input" means a shorted terminal. The two distinct voltage levels provided electronically represent the binary numbers **1** and **0** corresponding to the **TRUE** and **FALSE** of logic.

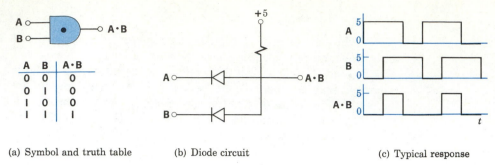

(a) Symbol and truth table   (b) Diode circuit   (c) Typical response

Figure 13.3 A two-input **AND** gate.

AND Gate. The symbol for an **AND** gate is shown in Fig. 13.3a, where **A·B** is read "**A AND B**." As indicated in the truth table, an output appears only when there are inputs at **A AND B**. If the inputs in Fig. 13.3b are in the form of positive voltage pulses (with respect to ground), inputs at A and B reverse bias both diodes, no current flows through the resistance, and there is a positive output (**1**). In general, there may be several input terminals. If any one of the inputs is zero (**0**), current flows through that forward-biased diode and the output is nearly zero (**0**). For two inputs varying with time, a typical response is shown in Fig. 13.3c.

OR Gate. The symbol for an **OR** gate is shown in Fig. 13.4a where **A + B** is read "**A OR B**." As indicated in the truth table, the output is **1** if input **A** OR input **B** is **1**. For no input (zero voltage) in Fig. 13.4b, no current flows, and the output is zero (**0**). An input of +5 V (**1**) at either terminal A or B or both (or at *any* terminal in the general case) forward biases the corresponding diode, current flows through the resistance, and the output voltage rises to nearly 5 V (**1**). For two inputs varying with time, the response is as shown.

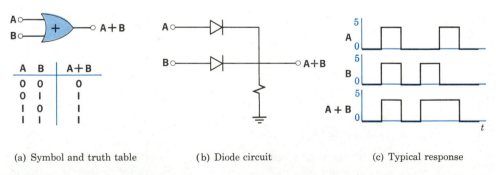

(a) Symbol and truth table   (b) Diode circuit   (c) Typical response

Figure 13.4 A two-input **OR** gate.

NOT Gate. The inversion inherent in a transistor circuit corresponds to a logic **NOT** represented by the symbol in Fig. 13.5a where $\overline{A}$ is read "**NOT A**." As indicated in the truth table, the **NOT** element is an *inverter*; the output is the *complement* of the single input. With no input (**0**), the transistor switch is held open by the negative base voltage and the output is +5 V (**1**). A positive input voltage (**1**) forward biases the

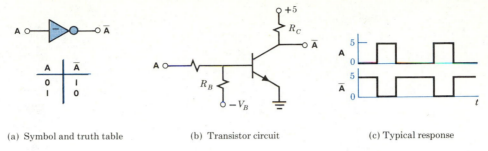

| (a) Symbol and truth table | (b) Transistor circuit | (c) Typical response |

Figure 13.5 A transistor **NOT** gate.

emitter junction, collector current flows, and the output voltage drops to a few tenths of a volt (**0**). For a changing input, the output response is as shown.

NOR Gate. In the diode **OR** gate of Fig. 13.4b, an input of $+5$ V at **A OR B** produces a voltage across R and a positive output voltage. But this output is less than the input (by the diode voltage drop), and after a few cascaded operations the signal would decrease below a dependable level. A transistor can be used to restore the level as in Fig. 13.6b; however, the inherent inversion results in a **NOT OR** or **NOR** operation. The small circle on the **NOR** element symbol and the bar in the $\overline{A + B}$ output indicate the inversion process.

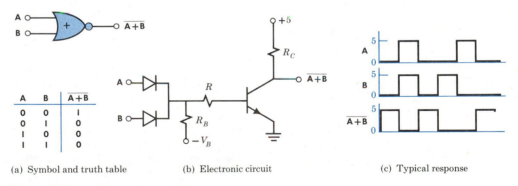

| (a) Symbol and truth table | (b) Electronic circuit | (c) Typical response |

Figure 13.6 A diode-transistor **NOR** gate.

In the **NOR** circuit with no input, the transistor switch is held **OFF** by the negative base voltage and the output is $+5$ V (**1**). A positive voltage at terminal A or B raises the base potential, forward biases the emitter junction, turns the transistor switch **ON**, and drops the output to nearly zero (**0**). In addition to restoring the signal level, the transistor provides a relatively low output resistance so that this **NOR** element can supply inputs to many other gates. Another advantage is that all the basic logic operations can be achieved by using only **NOR** gates.

NAND Gate. Diodes and a transistor can be combined to perform an inverted **AND** function. Such a **NAND** gate has all the advantages of a **NOR** gate and is very easy to fabricate, particularly in IC form. In a complex logic system it is convenient to use just one type of gate, even when simpler types would be satisfactory, so that gate characteristics are the same throughout the system.

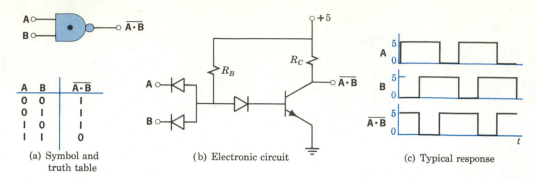

A	B	$\overline{A \cdot B}$
0	0	1
0	1	1
1	0	1
1	1	0

(a) Symbol and truth table

(b) Electronic circuit

(c) Typical response

Figure 13.7 A diode-transistor **NAND** gate.

The **NAND** gate function is defined by the truth table in Fig. 13.7. The small circle on the **NAND** element symbol and the bar on the $\overline{A \cdot B}$ output indicate the inversion process. With positive inputs at A and B (**1 AND 1**), the diodes are reverse biased, and no diode current flows; the positive base current causes heavy collector current and the output is approximately zero (**0**). If either A or B has a zero (**0**) input, at least one diode conducts to ground, the emitter-junction voltage drops below the critical value, no base current and hence no collector current flows, and the output is +5 V (**1**). For more positive cutoff, a second diode may be placed in series with the base of the transistor.

EXAMPLE 1

Use **NAND** gates to form a two-input **OR** gate.

The desired function is defined by the truth table of Fig. 13.8a. Comparing this with Fig. 13.7a, we see that if each input were inverted (replaced by its complement), the **NAND** gate would produce the desired result as indicated in the truth table.

To provide simple inversion, we tie both terminals of a **NAND** gate together (the first and last rows of Fig. 13.7a). The desired logic circuit is shown in Fig. 13.8b.

A	B	f	$\overline{A}$	$\overline{B}$	$\overline{\overline{A} \cdot \overline{B}}$
0	0	0	1	1	0
0	1	1	1	0	1
1	0	1	0	1	1
1	1	1	0	0	1

(a) Desired function

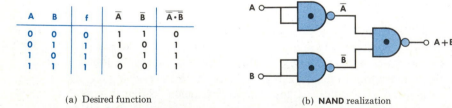

(b) **NAND** realization

Figure 13.8 Using **NAND** gates to form an **OR** gate.

In Example 1, we see that three **NAND** gates could be used to replace an **OR** gate. The combination of **NAND** gates is equivalent to an **OR** gate in that it performs the same logic operation. In digital nomenclature, the function **f** is defined by

$$f = A + B = \overline{\overline{A} \cdot \overline{B}}$$

We arrived at this relation by considering the desired and available truth tables. A "digital algebra" for the direct manipulation of such expressions is presented in Chapter 14. First, however, let us see how practical semiconductor logic elements function.

ELECTRONIC SWITCHES

Binary digital elements must respond to two-valued input signals and produce two-valued output signals. In practice the "values" are actually ranges of values separated by a *forbidden region*. The two discrete *states* may be provided electrically by switches, diodes, or transistors. They may be provided also by relays, or they may be provided optoelectronically by means of phototubes or semiconductor photo diodes. Magnetic cores in which the two states are represented by opposite directions of magnetization were used extensively in early computers. Purely fluid switches are sometimes used in hydraulic control applications. In applications requiring high speed and flexibility, however, electronic switches predominate.

Classification of Electronic Logic

Electronic logic circuits are classified in terms of the components employed. Basic operations can be performed by diode logic (DL), resistor-transistor logic (RTL), or diode-transistor logic (DTL). Currently popular are transistor-transistor logic (TTL), metal-oxide-semiconductor (MOS) and complementary MOS (CMOS), and emitter-coupled logic (ECL). In specifying a logic system, the designer selects the type of logic whose characteristics match the requirements.

Logic types vary in *signal degradation*, *fan-in*, *fan-out*, and *speed*. A major disadvantage of diode logic (Figs. 13.3 and 13.4) is that the forward voltage drop is appreciable, and the output signal is "degraded" in that the forbidden region is narrowed. The use of transistors minimizes degradation. The number of inputs that can be accepted is called the fan-in and is low (3 or 4) in DL and high (8 or 10) in TTL. The number of inputs that can be supplied by a logic element is called the fan-out. Fan-out depends on the output current capability (and the input current requirement) and varies from 4 in DL to 10 or more in TTL.

The speed of a logic operation depends on the time required to change the voltage levels, which is determined by the effective time constant of the element. In high-speed diodes, the charge storage is so low that response is limited primarily by wiring and load capacitances. In transistors in the **ON** state, base current is high and the charge stored in the base region is high; before the collector bias can reverse, this charge must be removed. Typically, 5 to 10 ns are required to process a signal. (In ECL, the charge stored is minimized and ECL gates can operate at rates up to 200 MHz.)

In practical design, the important factors are cost, speed, immunity to noise, power consumption, and reliability. In IC manufacture, a gate containing many highly uniform active components may cost no more than a discrete transistor and is no less reliable. As a result, sophisticated logic elements of greatly improved characteristics are now available at reasonable cost.

Diode Switches

The diode is an automatic "gate," **CLOSED** to reverse voltages and **OPEN** to forward voltages. Semiconductor diode and idealized diode characteristics are shown in Fig. 13.9a and b. The diode model of Fig. 13.9c with $V_F = 0.7$ V is suitable for the analysis of logic circuits using silicon diodes.

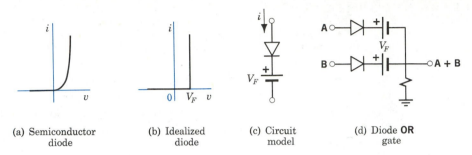

(a) Semiconductor diode (b) Idealized diode (c) Circuit model (d) Diode **OR** gate

Figure 13.9   The diode as a switch.

The highly nonlinear characteristic of the diode isolates the several inputs of a logic gate. The output of an **OR** gate (Fig. 13.9d) always follows the most positive of the input signals; diodes with less positive inputs are reverse biased. After several successive **OR** gates, the output signal will be significantly degraded. Eventually it is necessary to restore the signal, and a transistor inverter can be used for this purpose. In alternating **AND** and **OR** gates, the diode voltage drops tend to cancel and the logic level is maintained. This idea was used in designing complicated diode logic circuits.

Transistor Switches

The performance of logic circuits can be improved by taking advantage of the switching capability of the transistor. As you may recall, a transistor amplifier is operated in the *linear* or *normal* region with the emitter-base junction forward biased and the collector-base junction reverse biased. If both junctions are reverse biased, however, practically no collector current flows and the transistor is said to be operating in the *cutoff region*. In the basic switching circuit of Fig. 13.10a, if the input voltage is zero, there is no base current and operation is at point 1 in Fig. 13.10b. The collector current is practically zero ($I_C \cong I_{CEO}$) and the *switch* whose contacts are the collector and emitter terminals is **OPEN**. The cutoff current is exaggerated in Fig. 13.10b; a typical value of collector current of less than 1 μA with an applied voltage of 5 V corresponds to a dc *cutoff resistance* of more than 5 MΩ.

A positive voltage pulse applied to the input terminal forward biases the emitter-base junction, causes an appreciable base current, and moves operation to point 2. An increase in base current above 60 μA produces no further effect on collector current and the transistor is said to be operating in the *saturation region*. (This limit on collector current is represented in the dc transistor model of Fig. 13.12b by $I_C \leq \beta I_B$.) The voltage drop across the "switch" is called *collector saturation* voltage $V_{CE(sat)}$ and is typically a few tenths of a volt. Note that if V_{CE} is less than V_{BE}, the collector-base

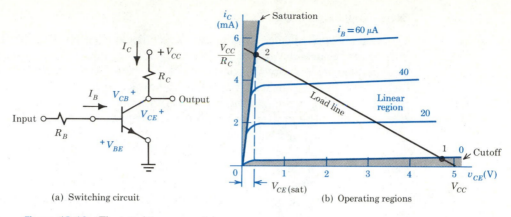

(a) Switching circuit

(b) Operating regions

Figure 13.10 The transistor as a switch.

junction is also forward biased. As indicated in Fig. 13.10b, a collector current of around 6 mA at a saturation voltage of 0.3 V corresponds to a *saturation resistance* of around 50 Ω; more typical values are 10–20 Ω. When the transistor switch is **CLOSED**, the collector current is determined primarily by the load resistance if $R_C \gg R_{sat}$, and $I_C \cong V_{CC}/R_C$.

TRANSISTOR LOGIC GATES

The transistor switch of Fig. 13.10 is basically an RTL inverter or **NOT** gate. When the input is near zero (logic **0**), the output is near +5 (logic **1**); a positive input (**1**) causes current flow and the output drops to $V_{CE(sat)}$(**0**). The *transfer characteristic*, that is, the relationship between input and output voltages, is shown in Fig. 13.11. The *threshold voltage* V_T is approximately 0.7 V.[†] Such switches can be connected in series or parallel to provide **NAND** or **NOR** gates.

[†]In practice, the circuit designer may hold V_{BE} below 0.4 V, say, to ensure cutoff and provide $V_{BE} = 0.8$ V to ensure saturation.

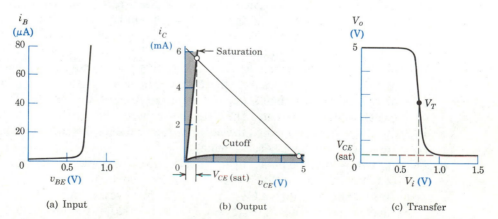

(a) Input

(b) Output

(c) Transfer

Figure 13.11 Characteristics of an RTL switch.

Diode-Transistor Logic

A more sophisticated, and faster, device is the DTL **NAND** gate of Fig. 13.12. It is faster because the charging currents accompanying a change in transistor state flow through the low forward resistances of diodes instead of through the higher series resistances necessary in RTL operation. A fast-acting Schottky diode may be added in parallel with the collector-base junction to limit the forward bias and reduce the turn-off delay. The greater sophistication is inexpensively realized in IC form.

The dc model of a silicon **NAND** gate is shown in Fig. 13.12b. Diodes A and B are represented by ideal diodes in series with 0.7-V sources. Series diodes D_S are replaced by one ideal diode and a 1.4-V source. The transistor input characteristics (Fig. 13.11a) resemble those of a junction diode, and the input circuit is modeled just as the other diodes. In general, the collector circuit is modeled as a controlled current

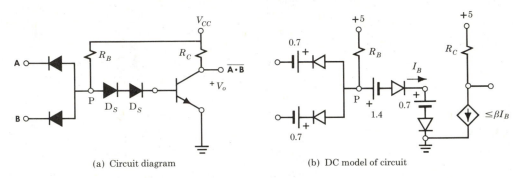

(a) Circuit diagram   (b) DC model of circuit

Figure 13.12 Operation of a DTL **NAND** gate.

source where $I_C \leq \beta I_B$ if I_{CEO} is neglected. Here the inequality sign is introduced to reflect the limit on collector current at saturation. Implicit in the transistor model is the fact that V_{CE} is always equal to or greater than $V_{CE(\text{sat})}$. The performance of the gate can be predicted on the basis of this model. (See Example 2.)

If this gate follows other DTL switches in the **OPEN** state, inputs A and B are approximately 5 V (**1**), and diodes A and B are reverse biased. Series diodes D_S are forward biased by V_{CC} and the high base current holds the transistor in saturation, so the output is about 0.3 V (**0**). The potential at point P is about $2 \times 0.7 + 0.7 = 2.1$ V. If now input A drops, at $V_P = V_T = 2.1 - 0.7 = 1.4$ V, diode A is forward biased and begins to conduct. The potential at point P drops, the transistor is cut off, and V_o rises to $+5$ V (**1**). Note that a change in input from 5 to 1.4 or 3.6 V is required to switch from the high state, and a change from 0.3 to 1.4 or 1.1 V is required to switch from the low state. The difference between the operating input voltage and the threshold voltage is called the *noise margin* since unwanted voltages lower than this value will not cause false switching.

Typically, the output of the **NAND** gate of Example 2 is connected to the inputs of other DTL gates. When V_o is high, the next input diodes are reverse biased and no "load" current is drawn. When V_o is low, however, each input diode draws current equal to $(V_{CC} - V_P)/R_B = (5 - 0.7 - 0.3)/5000 = 0.8$ mA in this example. The

EXAMPLE 2

For the DTL **NAND** gate of Fig. 13.12a, $V_{CC} = 5$ V and $R_B = R_C = 5$ kΩ. For the transistor, $\beta = 30$ and $V_{CE(sat)} = 0.3$ V. If input A is 5 V and input B is 0.3 V, determine the no-load output voltage V_o. *Note:* In dealing with nonlinear "threshold" devices, circuit equations involve inequalities. The practical approach is to assume a state (**ON, OFF**) and check to see if that state is consistent with the data.

For input $A = 5$ V and $B = 2$ V, determine the base current, the collector current, and the no-load output voltage.

The lowest input governs the state of a **NAND** gate. For $B = 0.3$ V, V_P cannot exceed $0.3 + 0.7 = 1.0$ V.

If current did flow through the series diodes,

$$V_{BE} = V_P - 2(0.7) = 1.0 - 1.4 = -0.4 \text{ V}$$

But a forward bias of $V_{BE} \geq +0.7$ is required,
∴ no base current flows, the transistor is cut off, and

$$V_o = V_{CC} - I_C R_C = 5 - 0 = 5 \text{ V}$$

For $B = 2$ V, V_p cannot exceed $2 + 0.7 = 2.7$ V. The critical value for turn on is $V_P = 3(0.7) = 2.1$ V. The lowest value governs ∴ base current flows and

$$I_B \cong (V_{CC} - 2.1)/R_B = (5 - 2.1)/5k = 0.58 \text{ mA}$$

But the largest possible no-load collector current is

$$I_C \cong V_{CC}/R_C = 5/5k = 1 \text{ mA}$$

which requires a base current of only

$$I_B = I_C/\beta = 1/30 \cong 0.033 \text{ mA}$$

∴ the transistor is in saturation and

$$V_o = V_{CE(sat)} = 0.3 \text{ V}$$

base current I_B must put the transistor into saturation and also supply additional collector current ($I_C = \beta I_B$) to "drive" the load represented by following gates. The maximum fan-out for DTL gates is about 8. (See Problem 2.)

Transistor-Transistor Logic

The emitter-base section of a transistor is essentially a *pn* junction diode. (See Fig. 12.12b.) Therefore, the input diodes of the DTL gate of Fig. 13.12a can be replaced by a multiemitter transistor. As shown in Fig. 13.13b, the collector-base junction of T_1 provides an offset voltage equal to that of one series diode. The resulting *transistor-transistor logic* gate requires less silicon area and is faster in operation than a similar DTL gate.

TTL and DTL gates are similar in operation. In Fig. 13.13b, if we assume that the output is **LOW**, then T_2 is **ON** and the emitter-base junction of T_2 and diode D_S must be forward biased. Therefore, voltage V_{P2} must be approximately $0.7 + 0.7 = 1.4$ V, and the base-collector junction of T_1 must be forward biased to supply adequate base current to T_2. This can be true only if V_p is approximately $1.4 + 0.7 = 2.1$ V, and this condition requires that both inputs A and B be **HIGH**, that is, greater than $2.1 - 0.7 = 1.4$ V.

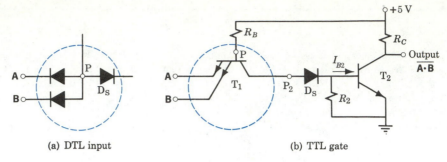

(a) DTL input   (b) TTL gate

Figure 13.13 Derivation of a basic TTL gate from a DTL gate.

If we assume that either A or B is **LOW** (0.3 V corresponding to $V_{CE(sat)}$ of a preceding stage), then a base-emitter junction of T_1 is forward biased and V_P cannot exceed $0.3 + 0.7 = 1.0$ V. Under this condition, D_S cannot be forward biased, I_{B2} cannot be adequate, and T_2 must be cut off, allowing the output to go **HIGH**. The logic relation is that the output is **LOW** for both inputs **HIGH** and **HIGH** for either input **LOW**. In other words, the output is **NOT** (**A AND B**) or $\overline{\text{A·B}}$ and this is a **NAND** gate.

A more sophisticated, and very popular, version of the TTL **NAND** gate is shown in Fig. 13.14. Here the output stage consisting of R_4, T_4, D, and T_3 is called a *totem pole*. The operation can be explained as follows. If we assume that the output is **LOW**, then T_3 is **ON**, the emitter-base junctions of T_3 and T_2 must be forward biased, and the collector-base junction of T_1 must be forward biased to supply base current to T_2 and, in turn, to T_3. This can be true only if V_P is approximately $0.7 + 0.7 + 0.7 = 2.1$ V, and this requires that inputs **A**, **B**, and **C** be **HIGH**.

Note that if T_2 is solidly **ON** (saturated) to supply adequate base current to T_3, then $V_{C2} = V_{B4}$ cannot exceed $V_{CE(sat)} + V_{BE3} = 0.3 + 0.7 = 1$ V. But for T_4 to be **ON**, the voltage at the base of T_4 would have to be $V_{C3} + V_D + V_{BE4} = 0.3 + 0.7 + 0.7 =$

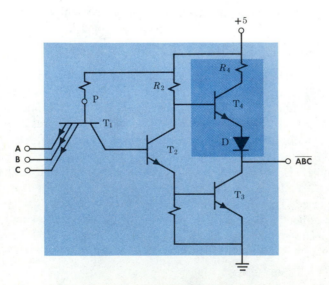

Figure 13.14 An integrated TTL **NAND** gate.

1.7 V. We conclude that with T_2 **ON**, T_4 must be **OFF**. This means that little current flows through R_4, and power loss is small when the output goes **LOW**.

If one or more inputs goes **LOW** (0.3 V, say), one base-emitter junction of T_1 is forward biased, and the voltage at point P cannot exceed $0.3 + 0.7 = 1.0$ V. Since $V_P = 2.1$ V is required to turn T_2 and T_3 **ON**, we know that T_2 and T_3 are **OFF**. With T_2 and T_3 essentially open circuits, V_{B4} rises and causes T_4 to conduct. Neglecting the small voltage drop due to I_{B4}, we find that the output voltage is equal to $V_{CC} - I_{B4} R_2 - V_{BE4} - V_D \cong 5 - 0 - 0.7 - 0.7 = 3.6$ V. The logic relation is that the output is **HIGH** if any input is **LOW**—the characteristic of a **NAND** gate.

For fast switching and high fan-out, the output resistance of a logic gate should be low. Since all capacitances associated with the inputs to the next logic level must be charged or discharged as the voltage level changes, the output circuit must be capable of *supplying* large currents in an upward transition or *sinking* large currents in a downward transition. The output resistance of a saturated transistor (T_3) is inherently low (about 10 Ω typically), providing the desired low output resistance when the output goes **LOW**. When the output is **HIGH**, T_4 is active and provides the low output resistance characteristic. In IC form, TTL gates are small, reliable, and cheap; because of their excellent characteristics they are used in a great variety of IC devices.

EXAMPLE 3

For the device shown in Fig. 13.15, $V_{CE(sat)} = 0.2$ V and the threshold voltage $V_{BE} = 0.8$ V. Identify the logic element, complete the truth table of voltages, and predict whether T_5 is **ON** or **OFF**.

V_A	V_B	V_C	V_x	V_y	V_z	T_5
2.0	2.4	3.4	2.4	1.0	0.2	**ON**
3.4	3.4	1.0	1.8	5	3.4	**OFF**

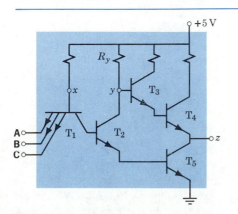

Figure 13.15 TTL gate analysis.

This is another form of three-input TTL **NAND** gate. Multi-emitter transistor T_1 acts as an **AND** gate; V_x is **HIGH** only if $A \cdot B \cdot C = 1$. In other words, the lowest input governs.

Our approach is to assume T_5 is **ON** and determine what voltages are required. For T_5 **ON**, T_2 must be **ON**; T_5, T_2, and the base-collector junction of T_1 must be forward biased, and V_x must be at least $0.8 + 0.8 + 0.8 = 2.4$ V. Therefore, all input voltages must exceed

$$V_x - V_{BE1} = 2.4 - 0.8 = 1.6 \text{ V}$$

In the first case, the lowest input is $V_A = 2.0$ V, and we conclude that T_5 is **ON** and

$$V_y = V_{BE5} + V_{CE2(sat)} = 0.8 + 0.2 = 1.0 \text{ V}$$

$$V_z = V_{CE5(sat)} = 0.2 \text{ V}$$

In the second case, $V_C = 1.0 < 1.6$ V; therefore, T_5 is **OFF** and

$$V_x = V_{ABC(min)} + V_{BE1} = 1.0 + 0.8 = 1.8 \text{ V}$$

$$V_y = V_{CC} - I_{B3}R_y \cong V_{CC} = 5 \text{ V}$$

$$V_z = V_y - V_{BE3} - V_{BE4} = 5 - 0.8 - 0.8 = 3.4 \text{ V}$$

With $V_z = V_{out}$ **HIGH**, T_4 supplies the small leakage currents of the connected gates.

Three-State Logic

In a complex digital control or computation system, logic devices "talk" to each other over a "party line" called a "system bus." For example, one of several input devices may communicate with one of several storage devices in a data processing unit over a data bus. This requires that devices *not* selected be disconnected from the bus. In *three-stage logic*, in addition to **0** and **1**, there is a "high-impedance state" in which the output terminal is essentially disconnected from the internal circuitry.

In the typical TTL gate of Fig. 13.14, assume the output is connected to a bus and consider **C** as a control input. If **C** is held **LOW**, inputs **A** and **B** have no effect; the inputs have been "disconnected" and T_3 turned **OFF**, providing a high impedance path from the output to ground. If, simultaneously, the base of T_4 is grounded, T_4 is **OFF**, and there is a high impedance path (very low current) from the output to +5 V. The output terminal has been isolated and, in effect, the unit has been disconnected from the bus. In a three-state gate, the circuit is arranged so that holding the **CONTROL** terminal **LOW** effectively grounds (through a transistor switch) an input emitter and the base of T_4.

Buffers

The performance of an electronic component may be adversely affected if it is heavily loaded, that is, if it is required to supply considerable power to another component. For good performance, an oscillator generating a precise frequency must be isolated from the device that it controls, or a register with limited fan-out must be separated from a bus supplying data to many devices. A *buffer* is an intermediate unit placed between the source of information and the load requiring power or current. In digital systems a buffer may be used to receive data at low current levels and make it available at higher current levels.

The symbol for a three-state inverting buffer is shown in Fig. 13.16. With several sources connected to a common bus, the **OUTPUT ENABLE** of all but one would be held **LOW**, effectively disconnecting them. With **OUTPUT ENABLE** held **HIGH** on one unit, that **INPUT** is complemented, and the **OUTPUT** is capable of driving the bus supplying data to one or more loads. Typical TTL buffers have a fan-out of 30.

INPUT ———————— OUTPUT

OUTPUT
ENABLE

Figure 13.16   Three-state inverting buffer.

Open-Collector TTL Gates

Another approach to the problem of interconnecting several devices is to use *open-collector* gates. If two standard TTL gates (Fig. 13.14) have their outputs electrically connected and one output is **LOW** while the other is **HIGH**, there is a low impedance path from V_{CC} through T_4 of the **HIGH** gate through T_3 of the **LOW** gate; the resulting high current will usually destroy T_3. However, in the **NAND** gate of Fig. 13.17 the collector is left "open." (In practice, it is tied to V_{CC} by *pull-up resistor R*

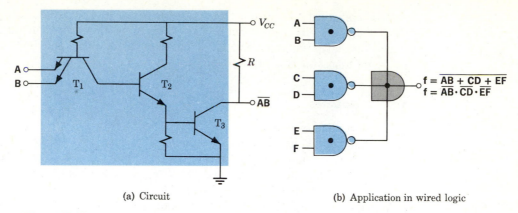

(a) Circuit   (b) Application in wired logic

Figure 13.17   Open-collector TTL gate.

to bring the output to nearly +5 V when T_3 cuts off.) Now the outputs of several **NAND** gates can be wired together, and the output of the combination will be **LOW** if any one of the individual outputs is **LOW**. In the "wired logic" shown symbolically in Fig. 13.17, the output is described as $f = \overline{AB} + \overline{CD} + \overline{EF}$ (wire-**NOR**) or as $f = \overline{AB} \cdot \overline{CD} \cdot \overline{EF}$ (wire-**AND**); the electrical connection has performed a logic function at no cost.

Emitter-Coupled Logic

Where very high speed is required, the emitter-coupled logic (ECL) gate of Fig. 13.18 can be used. The operation of the ECL unit is as follows. Assuming T_1 is

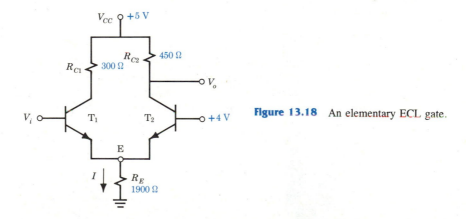

Figure 13.18   An elementary ECL gate.

conducting and $V_i = 4.4$ V, the base-emitter voltage for T_1 is about 0.7 V; therefore, $V_E \cong 4.4 - 0.7 = 3.7$ V with respect to ground and $I \cong 2$ mA. Since the base bias voltage on T_2 is inadequate for conduction, T_2 is **OFF**, and $V_o \cong 5$ V. If V_i drops to 3.6 V, say, T_2 begins to conduct and holds V_E at approximately $4 - 0.7 = 3.3$ V turning T_1 **OFF** and dropping V_o to $V_{CC} - I_C R_{C2} \cong V_{CC} - I R_{C2} = 5 - 0.00174$ $(450) = 4.2$ V (since $I = V_E/R_E = 3.3/1900 = 1.74$ mA).

In a practical ECL gate, a "level shifter" circuit restores the logic swing to 4.4 V (**HIGH**) and 3.6 V (**LOW**). It can be shown that in this "current-switching" circuit, the transistors never saturate, and the delay associated with removing charge from a saturated transistor is avoided.

MOS Logic Gates

The n-channel enhancement-mode MOS transistor has several virtues that make it useful in digital circuits; it requires only a small area, it is normally **OFF**, and it has a high input resistance. Furthermore, the basic gates can be fabricated from interconnected MOS structures only and, therefore, they are easy to produce in IC form. MOS logic is attractive in applications where low power consumption is important and extremely high switching speed is not required.

The characteristics in Fig. 13.19b indicate the possibility of using a MOSFET as a switch or inverter. An input voltage $V_i = v_{GS} \cong 0.5$ V (**LOW**) is insufficient for

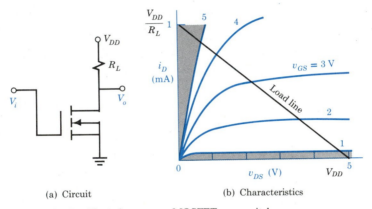

(a) Circuit   (b) Characteristics

Figure 13.19 The enhancement MOSFET as a switch.

turn-on and the output is $V_o = v_{DS} \cong 5$ V (**HIGH**). A **HIGH** input of $V_i = 5$ V drives the MOSFET into saturation, and $V_o \cong 0.5$ V (**LOW**). The output of one inverter can drive the input of another in logical operations.

In the inverter of Fig. 13.20, an enhancement-mode MOSFET is used as a *driver*

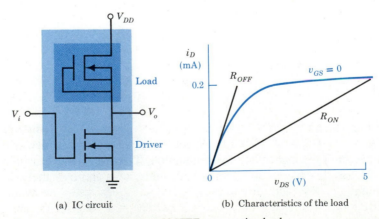

(a) IC circuit   (b) Characteristics of the load

Figure 13.20 The depletion MOSFET as a passive load.

and a depletion-mode device is used as a passive *load*. With the gate tied to the source, $v_{GS} = 0$ and the i_L-v_L characteristic is as shown. When the driver is **OFF**, $V_o \cong V_{DD}$ and for the load $v_{DS} = V_{DD} - V_o$ is low; the load can be modeled by the resistance R_{OFF}. When the driver is **ON** (V_i = **HIGH**), the output $V_o \cong 0$ and v_{DS} for the load $\cong V_{DD}$. The effective load resistance has increased, but the essential **NOT** function has been realized. These simple NMOS elements are used in complex LSI circuits.

CMOS Gates

A popular family of IC logic gates employs complementary symmetry, combining p- and n-channel enhancement-mode MOSFETs in the same chip. The basic building block of the CMOS gates is the inverter of Fig. 13.21a. When input voltage

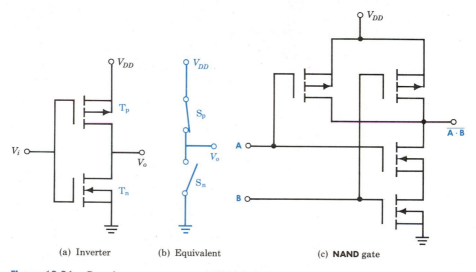

(a) Inverter   (b) Equivalent   (c) **NAND** gate

Figure 13.21   Complementary-symmetry (CMOS) logic gates.

$V_i = v_{GS}$ is **LOW**, the n-channel device T_n is **OFF**. For the p-channel device T_p, $v_{GS} = V_i - V_{DD} \cong -V_{DD}$ and, therefore, T_p is **ON**. This state corresponds to Fig. 13.21b with switch S_p closed and switch S_n open. Hence the output is $V_o = V_{DD}$ or **HIGH**.

As the input voltage increases, T_p turns **OFF** and T_n turns **ON**; the output goes **LOW**. By design, the threshold voltage is usually about $V_{DD}/2$. The high resistance presented by T_p **OFF** limits the current drain on the power supply to a very small value.

A two-input CMOS logic gate is shown in Fig. 13.21c. If either **A** or **B** is **LOW**, one of the series T_n devices is **OFF** and one of the parallel T_p devices is **ON**; the output is **HIGH**. If both **A** and **B** are **HIGH**, both T_n devices are **ON** and both T_p devices are **OFF**; the output is **LOW**. These are the logic relations of a **NAND** gate. More inputs and other functions can be created by using the same elements.

CMOS circuits consume significant power only during transitions from one state to another; the very low quiescent power consumption is the key factor in many applications. Most of the functions available in TTL form are available in CMOS as well. CMOS units can drive low-power TTL gates, but buffers are required to drive other TTL types.

MEMORY ELEMENTS

The outputs of the basic logic gates are determined by the present inputs; in response to the various inputs, these *combinational* circuits make "decisions." Along with these *decision* components we need *memory* components to store instructions and results; the outputs of such *sequential* circuits are affected by past inputs as well as present.

What characteristics are essential in a memory unit? A binary storage device must have two distinct states, and it must remain in one state until instructed to change. It must change rapidly from one state to the other, and the state value (**0** or **1**) must be clearly evident. The *bistable multivibrator* or *flip-flop*, a simple device that meets these requirements inexpensively and reliably, is used in all types of digital data processing systems.

A Logic Gate Memory Unit

First let us analyze a basic memory unit consisting of familiar logic gates. In the **NOR** gate flip-flop of Fig. 13.22, the output of each **NOR** gate is fed back into the input

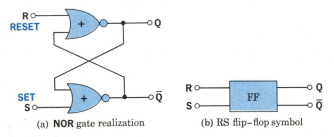

(a) **NOR** gate realization (b) RS flip–flop symbol

Figure 13.22 The flip-flop, a basic memory unit.

of the other gate. The operation is summarized in Table 13-1 where we assume to start that Q_0, the *present* state of the output **Q**, is **0** and inputs to the *set* terminal S and the *reset* (or *clear*) terminal R are both **0**.

To **SET** the flip-flop, a **1** is applied to S only. For $Q_0 = 0$, $\overline{Q} = \overline{Q_0 + S} = 0$, and $Q = \overline{R + \overline{Q}} = 1$, the output is inconsistent with the input, the system is *unstable*, and **Q** must *flip*. After **Q** changes, Q_0 (the present state **Q**) changes to **1**, and $\overline{Q}$ becomes $\overline{1 + 1} = 0$; hence $Q = \overline{0 + 0} = 1$, a *stable* state. (In this analysis, keep in mind that *if either input to a* **NOR** *gate is* **1**, *the output is* **0**.) Removing the input from S causes no change. We conclude that $Q = 1$, and $\overline{Q} = 0$ is the stable state after being **SET**. Applying another input to S produces no change.

To **RESET** the flip-flop, a **1** is applied to R only. This results in an unstable system and **Q** must *flop* to **0**. (You should perform the detailed analysis.) A change in Q_0 to **0** produces a stable output $Q = 0$. Removing the input to R or applying another input to R produces no change. We conclude that $Q = 0$ and $\overline{Q} = 1$ is the stable state after being **RESET**.

Table 13-1 Analysis of a Memory Unit

Action	Q_0	S	R	$\overline{Q}$	Q	Conclusion
Assume	0	0	0	1	0	This is a stable state.
Apply **1** to S	0	1	0	0	1	Unstable state; **Q** changes.
(Q_0 becomes **1**)	1	1	0	0	1	Stable.
Remove **1** from S	1	0	0	0	1	The stable state after **SET**.
Apply **1** to S again	1	1	0	0	1	No change in **Q**.
Remove **1** from S	1	0	0	0	1	The stable state after **SET**.
Apply **1** to R	1	0	1	0	0	Unstable state; **Q** changes.
(Q_0 becomes **0**)	0	0	1	1	0	Stable.
Remove **1** from R	0	0	0	1	0	The stable state after **RESET**.
Apply **1** to S and R	0	1	1	0	0	Unacceptable; $Q \neq \overline{Q}$.

Note that only a momentary input is required to produce a complete transition; this means that very short pulses can be used for triggering. Attempting to **SET** and **RESET** simultaneously creates an undesirable state with both **Q** and $\overline{Q} = 0$. This state ambiguity is unacceptable in a bistable unit and actual circuits are designed to avoid this condition.

A Transistor Flip-Flop

The operation of an electronic flip-flop is based on the switching and amplifying properties of a transistor. In the switch in Fig. 13.23a, the voltage divider R_A-R_B reverse biases the base-emitter junction and the transistor is in cutoff or the switch is **OPEN**; because of inversion, a positive voltage appears at the output terminal. If a positive signal is applied to the input, the base-emitter junction is forward biased, the switch is **CLOSED**, and the output voltage drops to zero (nearly).

An input signal making the + terminal of R_A more positive could also forward bias the base-emitter junction and drop the output voltage to zero. This switch is a form of **NOR** gate; a flip-flop can be created from two such switches.

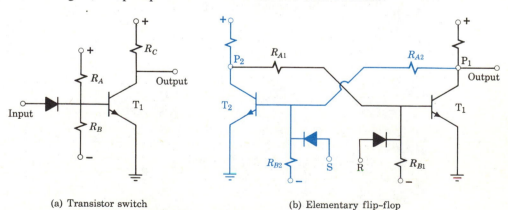

(a) Transistor switch

(b) Elementary flip–flop

Figure 13.23 A transistor RS flip-flop.

To follow the operation of the flip-flop (Fig. 13.23b), assume that T_1 is conducting (**CLOSED**) and T_2 is cut off (**OPEN**). With T_1 conducting, the potential of point P_1 is zero and, in combination with the negative voltage applied to R_{B2}, this ensures that T_2 is cut off. With T_2 cut off, the potential of point P_2 is large and positive, and this supplies the bias that ensures that T_1 is conducting and V_o is low. In logic terms, $\mathbf{Q = 0}$; this is a stable state that we may designate as the **0** state of this binary memory element.

A positive pulse applied to **RESET** terminal R has no effect since T_1 is already conducting. However, a positive pulse at **SET** terminal S causes T_2 to begin conducting, the potential of P_2 drops, the forward bias on T_1 is reduced, the potential of P_1 rises, the forward bias on T_2 increases, the potential of P_2 drops further, T_2 goes into saturation, and T_1 is cut off. The output voltage is high, $\mathbf{Q = 1}$, and this indicates another stable state that we may designate as the **1** state. If a flip-flop in the **1** state receives a positive pulse at R, transition proceeds in the opposite direction (since the device is symmetric) and the device is **RESET** to the **0** state. In a well-designed flip-flop, these changes in state take place in a few nanoseconds, and we see that this simple device satisfies all the requirements of a binary storage element.

Timing Waveforms

In the RS flip-flop of Fig. 13.22, a **1** input at the S input will **SET** the output **Q** to **1**. To **RESET** the flip-flop, a **1** is applied to input R. The duration of the input signal (as long as it exceeds a certain minimum time) and the time at which an input signal is applied are not significant. Such a flip-flop responds to *asynchronous* inputs.

A more sophisticated flip-flop incorporating two **AND** gates is shown in Fig. 13.24. Here an input is effective only when *enabled* by a **1** input at terminal E. In a digital system composed of many elements, it is usually necessary for the outputs of all elements to be synchronized. The synchronizing signal may come from a *clock*, and the enabling terminal is frequently designated **CLOCK** (**CK**). In a clocked system, transitions cannot run wild through a circuit; instead, changes occur in an orderly, one-step-at-a-time fashion. In addition to the synchronous inputs **R** and **S**, there may be asynchronous inputs to *clear* or *preset* the flip-flop.

The operation of a *clocked* RS flip-flop is illustrated by the typical waveforms of Fig. 13.24b. Initially output $\mathbf{Q = 0}$. If a **1** appears at **SET** when **ENABLE** goes to **1**,

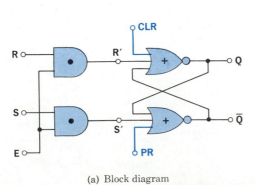

(a) Block diagram

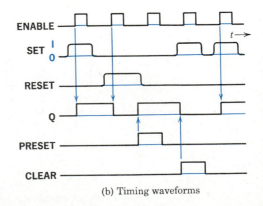

(b) Timing waveforms

Figure 13.24 A more sophisticated RS flip-flop.

the flip-flop is set with **Q** = **1**. At the next clock pulse, the presence of a **1** at **RESET** forces the output to **0**. At any time, a **1** at **PRESET** forces the output to **1**; a **1** input at the **CLEAR** terminal overrides other inputs and forces **Q** to **0**.

The Data Latch

The functional symbol for a simple RS flip-flop (without **PRESET** and **CLEAR**) is shown in Fig. 13.25a. One way to avoid the ambiguous state where **R** = **1** and **S** = **1**

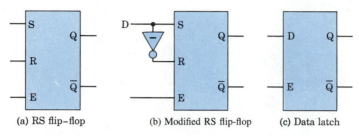

(a) RS flip–flop   (b) Modified RS flip-flop   (c) Data latch

Figure 13.25   Deriving a data latch from an RS flip-flop.

simultaneously is the circuit modification shown in Fig. 13.25b. By connecting an inverter between the R and S terminals and using only one input signal, the ambiguity is avoided and the number of terminals is reduced (an advantage in IC packages). When the **ENABLE** line is **HIGH**, the output **Q** follows the input **D**. In other words, when this *flip-flop* is enabled, the input data is transferred to the output line. After the **ENABLE** line goes **LOW**, no change in **Q** is possible, and the output is "latched" at the previous data value. This *data latch* is widely used as an element in digital systems; for example, a set of eight such latches could "remember" the eight digits representing a number or an instruction.

EXAMPLE 4

The enable and data inputs to a data latch are shown below. Predict the waveform of the output.

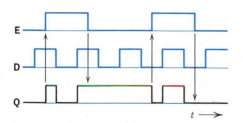

Figure 13.26   Typical data latch waveforms.

In the data latch, the output **Q** follows input **D** whenever enabled (**E** = **1**). When **E** goes to **0**, the output remains *latched* in the previous condition.

In Fig. 13.26, when **E** first goes **HIGH**, **D** = **1**; therefore, **Q** follows **D** and becomes **1**. As long as **E** is **HIGH**, **Q** follows any changes in **D**. When **E** goes **LOW**, **Q** = **D** = **1** and remains so. The output waveform is as shown.

The D Flip-Flop

In digital systems it is sometimes desirable to delay the transfer of data from input to output. For example, we may wish to maintain a present state at **Q** while we read in a new state that will be transferred to the output at the appropriate time. The D (for *delay*) flip-flop[†] shown in Fig. 13.27 is a refinement of the data latch incorporating a

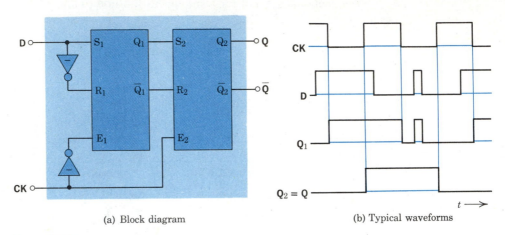

(a) Block diagram   (b) Typical waveforms

Figure 13.27   The D flip-flop.

second RS flip-flop. Here the data latch is enabled when the clock signal goes **LOW**, but the following RS flip-flop is enabled when **CLOCK** goes **HIGH**. In other words, Q_1 follows **D** whenever **CK** is **LOW**, but any change in the output of the combination $Q = Q_2$ is delayed until the next upward transition of **CK**. This is an *edge-triggered* flip-flop; Q_1 follows **D** while **CK** is **LOW**, then, on the leading edge of the clock pulse, the value of **D** is transferred to output **Q**. On the logic symbol of Fig. 13.28, the small triangle indicates an edge-triggered device.

Figure 13.28   Symbol for leading-edge-triggered D flip-flop.

Since the output can change only at the instant that the clock goes **HIGH**, the output can be synchronized with the outputs of other elements. Furthermore, a sudden spurious change in **D** similar to that shown in Fig. 13.27b will not affect the output. For proper operation of a practical device, the data input must be stable for a few nanoseconds before the device is clocked (the *set-up time*), and it must remain stable for a few nanoseconds after the clocking is initiated (the *hold time*).

[†]"Flip-flop" is a general term applied to the basic two-transistor element and to sophisticated, multigate IC devices.

The JK Flip-Flop

A widely used memory element is the JK flip-flop shown in Fig. 13.29. In its most common IC form, the output changes state on downward transitions of the clock pulse. The small circle on the symbol identifies this as a *trailing-edge-triggered*

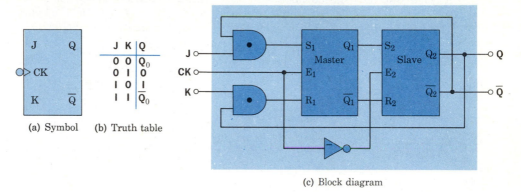

J	K	Q
0	0	Q_0
0	1	0
1	0	1
1	1	$\overline{Q_0}$

(a) Symbol   (b) Truth table

(c) Block diagram

Figure 13.29 The JK master-slave flip-flop.

flip-flop. The operation of this logic element is improved by employing a *master* flip-flop that is enabled on the upward transition of the clock pulse while the *slave* flip-flop is inactive. Then the slave is enabled on the downward transition and follows its master; that is, it takes on the state of the immobilized master.

In addition to avoiding the ambiguity referred to previously, this versatile device provides three different modes of response. Because of the feedback connections from output to input, the output of a JK flip-flop depends on the states of the inputs and the outputs at the instant the clock goes **LOW**. As indicated in the truth table, with **0** inputs at J and K, the clock has no effect, and the flip-flop remains in its present state Q_0. With unequal inputs, the unit behaves like an RS flip-flop. For **J = 1** and **K = 0**, the clock **SETS** the flip-flop to **Q = 1**; for **K = 1** and **J = 0**, the clock **RESETS** the flip-flop to **Q = 0**. (In other words, with **J ≠ K**, **Q = J** or **Q** follows **J**.) With **1** inputs at both J and K, the flip-flop toggles; that is, the output changes each time the clock goes **LOW**. (The operation of the JK flip-flop is revealed by a complete truth table constructed for the eight possible combinations of **J**, **K**, and Q_0; see Exercise 33.)

The sophistication of the JK response makes it useful in a variety of digital computer applications such as counters, arithmetic units, and registers. For greater flexibility, some versions include **PRESET** and **CLEAR** capabilities. In the unit shown in Fig. 13.30, the **PR** and **CLR** terminals are normally held **HIGH**. The small circles (inversion) indicate that if **PR** goes **LOW**, **Q** is forced to **1**; whereas if **CLR** goes **LOW**, **Q** is forced to **0**.

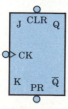

Figure 13.30 A JK flip-flop.

EXAMPLE 5

Connect a JK flip-flop to function as a data latch; that is, when it is **ENABLED**, the **DATA** is to be transferred to **Q** when the **CLOCK** goes **LOW**.

When **ENABLED**, **Q** should follow **J = D**, which requires **K ≠ J**.

When **DISABLED**, **Q** should remain "latched" in its present state, which requires **K = J = 0**. The truth table and the necessary connections are shown in Fig. 13.31.

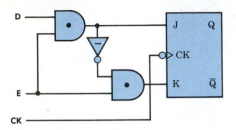

E	J	K
0	0	0
1	D	$\bar{J}$

Figure 13.31   A JK flip-flop connected as a data latch.

The T Flip-Flop

With the **J** and **K** inputs tied together and brought out to a single input terminal, the JK unit becomes a T or *toggle* flip-flop (Fig. 13.32). For **T = 0 (J = K = 0)**, the clock pulse has no effect on output **Q**. For **T = 1 (J = K = 1)**, the flip-flop toggles each time **CK** goes to **LOW**. The waveforms show that for **T** held **HIGH**, the output is a square wave of half the frequency of the clock; the device is a frequency divider. If the CK input responds to a sequence of events, the T flip-flop "divides by two."

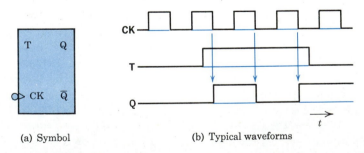

(a) Symbol   (b) Typical waveforms

Figure 13.32   A T flip-flop.

As we shall see in the next chapter, sets of flip-flops can be used to represent *binary numbers* in which each digit corresponds to the value of **Q (0 or 1)** of a flip-flop. A *register* is a set of flip-flops in which binary data can be stored. The same flip-flops can be reconnected to serve as a *counter* in which the number stored is the number of events being counted.

DIGITAL INTEGRATED CIRCUITS

The logic gates and memory elements previously described are available in IC form with significant advantages in small size, low power consumption, and low cost. Integrated circuits containing fewer than a dozen gates represent *small-scale integration* (SSI), while those with more than a hundred elements represent *large-scale integration* (LSI). In between are *medium-scale integration* (MSI) circuits. Digital ICs in common use range from 14-pin dual 4-input **NAND** gates (SSI) and 24-pin seven-segment lamp drivers (MSI) to 40-pin microprocessors incorporating more than 10,000 transistors (LSI). Currently available are *very large-scale integration* (VLSI) circuits incorporating more than 100,000 gates.

The active elements may be BJTs—the "bipolar" family of which TTL is the most common example—or MOSFETs (the MOS family) using either n-channel or p-channel devices in enhancement or depletion modes.

TTL. TTL circuits are superior to DTL units in most respects. The logic voltage swing is large (0.2 to 3.3 V), and they are relatively immune to noise. The fan-out (typically 10) is high, power consumption (10 mW/gate) is low, and the speed of operation (9 ns) is very high. These excellent characteristics have made TTL logic very popular, and TTL devices are available in great variety. Newer versions include the S series (high-speed Schottky), the LS series (low-power Schottky), and the ALS series (advanced LS). (See Fig. A.6 and the Lancaster reference at the end of this chapter.)

IIL. The necessity for isolating islands in fabricating chips (see p. 312) limits gate density in TTL technology. In the new *integrated injection logic* (I^2L), bipolar junction transistors are "merged" to provide interconnections without isolation regions. The high speed of bipolar elements is retained, and the increased density makes I^2L technology attractive for LSI applications.

ECL. ECL logic ic characterized by very high speed (typically 2-ns propagation delay) and large fan-out (16) but high power consumption (25 mW/gate) and low voltage swing (0.8 V).

MOS. The MOS inverter of Fig. 13.2c is extremely simple in appearance and in fabrication. An MOS gate requires much less "real estate" than a corresponding TTL unit, and therefore, the gate density on a silicon chip can be much higher; MOS is widely used in LSI. The high input resistance means low input currents and power consumption is low (typically 1 mW/gate). Furthermore, as fabrication techniques improve, each generation of MOS devices offers improved performance.

CMOS. In *complementary MOS* devices (CMOS), n- and p-channel MOSFETs are paired so that, during switching, an **ON** transistor is always available for rapid charging of the load capacitance, whereas the **OFF** transistor limits dc current consumption. The circuit complexity is increased, but power is consumed only during switching. The compromise reduces gate density but results in higher speed and very low power consumption (0.01 mW/gate), characteristics that are desirable in digital watches, for example.

SUMMARY

- Digital logic circuits employ discrete instead of continuous signals.
  The basic logic operations are: **AND**, **OR**, **NOT**, **NAND**, and **NOR**.

All operations can be achieved using only **NAND** gates or only **NOR** gates.
The validity of any logic statement can be demonstrated with a truth table.

- Logic circuits are classified in terms of the components (R, D, T) employed.
  Design factors include: cost, noise immunity, power consumption, signal degradation, fan-in, fan-out, speed, and reliability.

- Diodes and transistors are used as switches in electronic logic gates.
  Transistors are switched from cutoff to saturation by changes in i_B or v_{GS}.
  Integrated circuits permit high-performance gates at reasonable cost.

- Memory components store data processed by decision components.
  When instructed, a flip-flop must change rapidly from one distinct stable state to the other and clearly evidence the new state.
  The transistor flip-flop is an inexpensive, widely used binary storage device.

- The basic RS flip-flop is **SET** by **S = 1** and **RESET** by **R = 1**.
  When a data latch is enabled, output **Q** follows input **D**.
  In a D flip-flop, transfer of data from input to output is delayed.
  Operation of the versatile JK flip-flop is determined by inputs at J and K.
  The T flip-flop is toggled from one state to the other by successive triggers.
  Basic applications of flip-flops include counters and registers.

REFERENCES

1. A.P. Malvino and Donald P. Leach, *Digital Principles and Applications*, McGraw-Hill Book Co., New York, 2nd ed., 1975.

   An excellent introduction to digital systems with detailed discussions and many illustrative examples. Includes basic concepts and their practical applications in computers.

2. Victor Grinich and Horace Jackson, *Introduction to Integrated Circuits*, McGraw-Hill Book Co., New York, 1975.

   A good undergraduate treatment of IC design of digital and analog circuits.

3. John B. Peatman, *The Design of Digital Systems*, McGraw-Hill Book Co., New York, 1972.

   A well-written introduction to modern digital devices and systems and their engineering applications.

4. Don Lancaster, *TTL Cookbook*, Howard Sams and Co., Indianapolis, 1974.

   A practical guide to understanding and using TTL circuits. Includes construction hints, design techniques, and operating characteristics of popular gates, flip-flops, counters, registers, and memories.

   (Also see the Millman and Holt references at the end of Chapter 12 and the Horowitz and Hill reference at the end of Chapter 16.)

REVIEW QUESTIONS

1. Distinguish between analog and digital signals. Between gates and switches.
2. Why did the successful computer follow the invention of the transistor?
3. Explain how diodes and transistors function as controlled switches.
4. What is a "truth table"? How is it used?
5. Draw symbols for and distinguish between **AND**, **OR**, **NOR**, and **NAND** operations.
6. What is "**NOT** the complement of variable **A** inverted"?
7. Draw a **NAND** gate. Why are they used so widely?
8. How are electronic logic circuits classified? What is DTL? TTL? ECL? MOS? CMOS?
9. List the factors of importance in the practical design of logic circuits.
10. Show with a sketch how cutoff and saturation resistances are calculated.
11. What are the advantages of TTL gates?
12. What is "three-state" logic? How is it achieved? What is a "buffer"?
13. Draw a TTL **NOR** gate and explain its operation.
14. What is the principal advantage of ECL gates?
15. Explain the operation of an MOS gate. A CMOS gate.
16. Distinguish between decision and memory components; give an example of each.
17. What is a flip-flop? What is its function in a computer?
18. Explain the operation of a flip-flop consisting of two **NOR** gates.
19. Explain the operation of a transistor RS flip-flop.
20. Distinguish between a data latch and a D flip-flop.
21. Distinguish between RS, T, and JK flip-flops.
22. Explain how T flip-flops "count." How many are needed to count to ten?

EXERCISES

1. Devise a two-input **NAND** gate using relays (Fig. 13.1) and construct the truth table.
2. An electric light is to be controlled by three switches. The light is **ON** if both switches A and B are closed, or if switch C is closed by itself, or if all three switches are closed.
   (a) Draw up a truth table for this function.
   (b) Describe the operation of the light using the states of the switches A, B, and C and the words "OR" and "AND."
3. The "resistor logic" circuit of Fig. 13.33 has binary inputs of 0 and 4 V.
   (a) For $R = 1 \text{ k}\Omega$, construct a "truth table" showing actual voltages. If a detector with an adjustable threshold were available, what logic operations could be performed with this circuit?
   (b) At what threshold should the detector be set for **AND** operations?
4. Replace the resistors R in Fig. 13.33 by diodes to form a three-input diode **OR** gate. Construct the truth table of voltages.
5. How many rows are needed in the truth table of an n-input logic circuit?
6. Assuming 5-V inputs and a diode forward voltage drop of 0.7 V, construct a "truth table" showing actual voltages for the diode circuits of (a) Fig. 13.3 and (b) Fig. 13.4.
7. The three resistances in Fig. 13.5 are 1 kΩ each and the lower end of R_B is at -1 V. Calculate a few values and plot v_{BE} and v_{out} for v_{in} between 0 and 5 V.
8. Use **NOR** gates to form a two-input **AND** gate and construct the truth table as a check.
9. A portion of a table defining the performance of the basic gates is given:

Gate	Inputs	Output
AND	All inputs **1**	1
	Any one input **0**	0

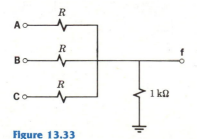

Figure 13.33

Complete the table for **NAND, OR, NOR,** and **NOT** gates.

10. The multiple-input logic gates commonly available are:

Inputs	2	3	4	5	8	> 8
NAND	x	x	x		x	x
NOR	x	x		x		
AND	x	x	x			
OR	x					

Only 2-input **OR** gates are available; however, 2-input **OR**s can be combined to form an n input **OR** circuit.
   (a) Design a 9-input **OR** gate using 2-input **OR** gates only.
   (b) The number of logic "levels" in part (a) introduces an undesirable delay. Design a 9-input **OR** gate using commonly available gates arranged so that signals pass through no more than two gates. (*Hint:* Use 3-input gates.)

11. The transistor of Fig. 12.15 is used in the switching circuit of Fig. 13.10a with $V_{CC} = 15$ V, $R_B = 5$ kΩ, and $R_C = 1$ kΩ.
   (a) Estimate β and α for this transistor.
   (b) Reproduce the i_C vs v_{CE} characteristics neglecting I_{CEO} and draw the load line.
   (c) For $v_i = 0.5$ V, *estimate* i_B, i_C, and v_o.
   (d) Repeat part (c) for $v_i = 1.7$ V and 2.8 V.
   (e) If the input (v_i) is "switched" from 0.5 V to 2.8 V, what are the corresponding output voltages?

12. When the transistor of Fig. 13.6b is conducting, $V_{BE} = 0.8$ V and $V_{CE} = 0.3$ V; $R = R_B = R_C = 2$ kΩ and $I_{CEO} = 10$ μA. The forward voltage drop on the diodes is 0.7 V. After several logic operations, the inputs at A are 1 and 2 V. Assuming 0 input at B, estimate the output voltages.

13. The transistor and diodes of Exercise 12 are used in Fig. 13.7b with $R_C = 2$ kΩ and $R_B = 20$ kΩ. Assuming a 5-V input at B, estimate the output for inputs of 1 and 2 V at A.

14. For a 2N3114 transistor switch (Fig. A-8) operating at $V_{CC} = 5$ V and $R_C = 100$ Ω, estimate the typical **ON** and **OFF** resistances at room temperature. Repeat for the worst possible conditions at 150° C. (*Note:* $h_{FE} = \beta$.)

15. In the silicon **NAND** gate of Fig. 13.12, $R_B = 2$ kΩ, $R_C = 5$ kΩ, and $\beta \cong 50$. The "load" across which V_o appears consists of the **A** inputs of four following DTL **NAND** gates.
   (a) For $V_A = 0.2$ V and $V_B = 0.2$ V, predict V_o.
   (b) For $V_A = 3$ V and $V_B = 4$ V, predict V_o.
   (c) Estimate the base and collector currents for conditions (a) and (b).

16. For the **NAND** gate of Fig. 13.7b, $R_B = 2$ kΩ, $R_C = 4.7$ kΩ, $\beta = 50$, and the current drawn at the output is negligible.
   (a) Sketch the i_B vs v_{BE} characteristic for the transistor and estimate the critical values of v_{BE} for the **ON** and **OFF** states.
   (b) Sketch the i_C vs v_{CE} characteristics and estimate the critical values of i_C and output voltage V_o.
   (c) Replace the active devices by dc circuit models and redraw the circuit. Assuming V_B is high, estimate the critical values of V_A for V_o **HIGH** and **LOW**, and sketch the transfer characteristic V_o vs V_A.

17. In Fig. 13.13b, R_2 (from base to emitter of T_2) is "large" and $R_B = 5.8$ kΩ.
   (a) For $V_A = 1$ V and $V_B = 4$ V, estimate V_P, V_{P2}, I_{B2}, and V_{out}.
   (b) Repeat for $V_A = V_B = 3.5$ V.

18. Assume T_3 in Fig. 13.14 is **ON** and estimate the voltage at the base of each transistor. Prove that T_4 is **OFF**.

19. For the circuit in Fig. 13.34:
   (a) Draw up a truth table for **F** in terms of **A** and **B**.
   (b) Identify the logic classification and the function performed.

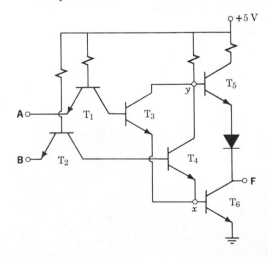

Figure 13.34

(c) For $V_A = 0.3$ and $V_B = 1.5$ V, predict V_x, V_y, and V_F.

(d) For $V_A = 2.3$ and $V_B = 1.5$ V, predict V_x, V_y, and V_F.

20. In the IC logic gate of Fig. 13.35, the threshold value of V_{BE} is 0.8 V and $V_{CE(sat)} = 0.2$ V. Predict voltages at x, y, and **F** and the condition (**ON**, **OFF**) of T_1 for:

(a) $V_A = V_B = 4$ V

(b) $V_A = 4$ V, $V_B = 0.2$ V

(c) $V_A = V_B = 0.2$ V

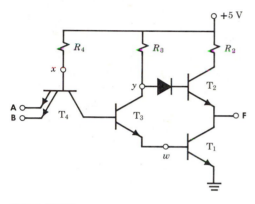

Figure 13.35

21. Classify the element in Fig. 13.35 in terms of components employed and logic function performed. Assuming $V_{BE} = 0.7$ V and $V_{CE(sat)} = 0.3$ V, complete the following table:

Condition			Approximate voltages					
T_3	T_2	T_1	F	w	y	x	A	B
—	—	ON	—	—	—	—	—	—
—	—	—	—	—	—	—	3.0	0.2

22. (a) In Fig 13.14, $V_C = 0$ and $V_A = V_B = 5$ V; what is V_{out}? If V_C is slowly increased to 5 V, what happens to V_{out}? At what value of V_C?

(b) With $V_C = 0$, $V_A = V_B$ is changed from 5 to 0 V. What happens to V_{out}? How is this behavior used in "three-state" logic?

23. In the circuit of Fig. 13.17a, T_3 serves as a "sink" for currents flowing from or to the inputs of subsequent logic gates "driven" by this gate. When the output ($\overline{AB}$) is **LOW**, the current *into* the output terminal is 2 mA; when the output is **HIGH**, the current *out of* terminal $\overline{AB}$ is 0.5 mA. As-

suming $V_{BE} = 0.7$ V and $V_{CE(sat)} = 0.3$ V, complete the following table ($B_1 =$ base of T_1):

Condition		Approximate voltages					
T_2	T_3	B_1	B_2	B_3	$\overline{AB}$	A	B
—	ON	—	—	—	—	—	—
—	—	—	—	—	—	0.3	3.3

24. Figure 13.36 incorporates two open-collector **NOT** gates and a three-state buffer (**ENABLED** when **LOW** in this case). Draw a truth table showing the states of **g** and **f** for inputs **A**, **B**, and **C**.

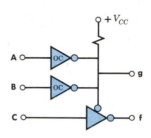

Figure 13.36

25. (a) Modify the circuit of Fig. 13.18 to create an **OR** gate where $V_o = $ **A** + **B**.

(b) Modify your circuit of part (a) to create a **NOR** gate.

26. Design, that is, draw the circuit diagram for, a 2-input **NAND** gate using MOS devices only.

27. Repeat Exercise 26 for a 2-input **NOR** gate.

28. Design a basic memory unit consisting of two **NAND** gates and describe its operation in a table similar to Table 13-1.

29. (a) Design a clocked RS flip-flop using four **NAND** gates.

(b) Modify the circuit to allow **PRESET** and **CLEAR** operations.

30. Draw the circuit diagram for a flip-flop using MOS devices only.

31. For the transistors of Fig. 13.23, $V_{CE(sat)} = 0.3$ V and $V_{CC} = 5$ V. Draw up a table showing the voltages at P_1 and P_2 for the following conditions: (a) T_1 initially conducting, (b) a negative pulse applied to S, (c) a positive pulse applied to S, (d) a negative pulse applied to R, (e) a positive pulse applied to R.

32. (a) Draw the block diagram and symbol for a D flip-flop that is "trailing-edge triggered," that is, the value of **D** is transferred to the output **Q** when **CK** goes **LOW**.
(b) Reproduce the waveforms for **CK** and **D** (Fig. 13.37) and show the resulting waveforms of Q_1 and of $Q_2 = Q$.

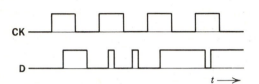

Figure 13.37

33. Reproduce the circuit of Fig. 13.29c. Draw up the truth table showing S_1, R_1, and **Q** (the *next* output state) for all values of **J**, **K**, and Q_0 (the *present* output state). Compare your result to the truth table of Fig. 13.29b.

34. A 500-Hz square wave is applied to the **CK** input of a T flip-flop with **T** held high.
(a) Draw 6 cycles of the square wave along with the corresponding flip-flop output.
(b) Draw a circuit showing two T flip-flops in tandem and the waveform of the output of T2.
(c) What operation is performed by the combination of flip-flops?

35. Analyze the memory element in Fig. 13.38. Draw up a truth table assuming the input is a series of pulses and **Q** is initially **0**. What function is performed by the **AND** gates?

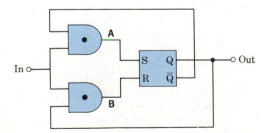

Figure 13.38

36. Two JK flip-flops that respond to downward transitions are connected in tandem (Fig. 13.39). For a 2-kHz square-wave input, determine the output.

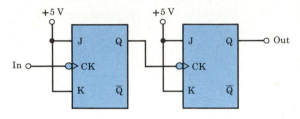

Figure 13.39

37. A JK flip-flop that responds to downward transitions is connected as in Fig. 13.40. For a 2-kHz square-wave input, determine the output.

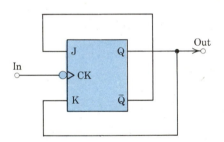

Figure 13.40

38. A CMOS digital device is fabricated as shown in Fig. 13.41. Draw a clearly labeled circuit diagram using standard symbols. If $|V_T|$ is 2.3 V for each unit, determine the outputs for inputs of 1 and 9 V. What logic function is performed?

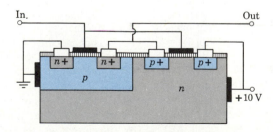

Figure 13.41

PROBLEMS

1. Devise a complete dc circuit model for a BJT in the switching mode, including the effect of $V_{CE(sat)}$.

2. A four-input DTL **NAND** gate similar to Fig. 13.12a "drives" eight similar **NAND** gates (a fan-out of 8). Assume $V_{CC} = 5$ V, $R_B = R_C = 5$ k$\Omega \pm 10\%$, $\beta = 30 \pm 40\%$, $V_{CE(sat)} = 0.3$ V, and $V_{junct} = 0.7 \pm 0.1$ V. Draw the circuit, define the "worst possible" operating condition, and predict the **LOW** output voltage under this condition.

3. In the DTL **NAND** gate of Fig. 13.12, $R_B = 2$ kΩ, $R_C = 4$ kΩ, and $\beta = 20$. For a "fan-out of 10," that is, the output V_o connected to the inputs of 10 similar gates, predict base and collector currents and output voltage V_o.

4. In the ECL gate of Fig. 13.42, $R_C = R_3 = 2$ kΩ and $R_E = R_4 = 5$ kΩ.

(a) For $V_A = V_B = 2.9$ V, estimate the voltages at P, E, and the Output terminal.

(b) Repeat part (a) for $V_A = V_B = 4.4$ V.

(c) Repeat part (a) for $V_A = 2.9$ V and $V_B = 4.4$ V and identify the function performed.

5. Two sensors are mounted on a half-white rotating disk as in Fig. 13.43. Sensor output is 5 V for white and 0 V for dark. Specify the digital element or elements to put in the black box so that the LED is **ON** for clockwise rotation. (*Hint:* Look at the waveforms.)

6. Four T flip-flops are connected in cascade to form a scale-of-sixteen counter. Draw a schematic diagram indicating the paths taken by pulses. Using feedback links to reintroduce counts into earlier stages, convert the unit into a decimal scalar, that is, a counter with one output pulse for every ten input pulses.

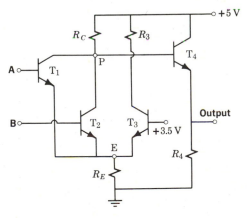

Figure 13.42 ECL gate.

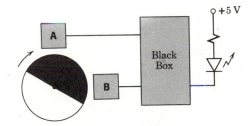

Figure 13.43

14

Digital Devices

Binary Numbers

Boolean Algebra

Logic Circuits

Registers and Counters

Memories

Information Processing

Digital information processing requires special electronic components. In Chapter 13 we saw how diodes and transistors can be used as switches in gates and flip-flops. The efficient design of digital circuits also requires a special numbering system and a special algebra. Our next step is to examine the binary number system and learn to apply logic theorems to binary relations. With this background we can analyze given logic circuits or synthesize new ones to realize desired logic functions.

Computers consist of large numbers of logic gates and memory elements organized to process data at high speed. Binary data or instructions are stored temporarily in *registers*. Binary *counters* are used in calculations and to keep track of computer operations. Instructions and data are stored at specified locations in *memory* and can be retrieved at will. With the understanding of registers, counters, memories, and computer organization gained here, we shall be prepared to study, in Chapter 15, the versatile electronic device called a *microprocessor*.

BINARY NUMBERS

In the decimal system a quantity is represented by the *value* and the *position* of a digit. The number 503.14 means

or
$$500 \quad + 0 \quad + 3 \quad + \tfrac{1}{10} \quad + \tfrac{4}{100}$$

$$5 \times 10^2 + 0 \times 10^1 + 3 \times 10^0 + 1 \times 10^{-1} + 4 \times 10^{-2}$$

In other words, 10 is the *base* and each position to the left or right of the decimal point corresponds to a *power* of 10. Perhaps it is unfortunate that we do not have 12 fingers, because in certain ways 12 would be a better base. In fact, such a base-12 or *duodecimal* system was used by the Babylonians, and we still use 12 in subdividing the foot, the year, and the clock face.

In representing data by **ON-OFF** switch position, there are only two possibilities and the corresponding numbers are **1** and **0**. In such a *binary* system the base is 2 and the total number of fingers on both hands is written **1010** since

$$1 \times 2^3 + 0 \times 2^2 + 1 \times 2^1 + 0 \times 2^0 = 8 + 0 + 2 + 0 = 10$$

In electronic logic circuits the numbers **1** and **0** usually correspond to two easily distinguished voltage levels specified by the circuit designer. For example, in TTL, **0** corresponds to a voltage near zero and **1** to a voltage near +5 V.

Number Conversion

Binary-to-Decimal Conversion. In a binary number, each position to the right or left of the "binary point" corresponds to a power of 2, and each power of two has a decimal equivalent.

To convert a binary number to its decimal equivalent, add the decimal equivalents of each position occupied by a **1**.

For example,

$$\textbf{110001} = 2^5 + 2^4 + 0 + 0 + 0 + 2^0 = 32 + 16 + 1 = 49$$

$$\textbf{101.01} = 2^2 + 0 + 2^0 + 0 + 2^{-2} = 4 + 1 + \tfrac{1}{4} = 5.25$$

Decimal-to-Binary Conversion. A decimal number can be converted into its binary equivalent by the inverse process, that is, by expressing the decimal number as a sum of powers of 2. An automatic, and more popular, method is the *double-dabble* process in which integers and decimals are handled separately.

To convert a decimal integer to its binary equivalent, progressively divide the decimal number by 2, noting the remainders; the remainders taken in reverse order form the binary equivalent.

To convert a decimal fraction to its binary equivalent, progressively multiply the fraction by 2, removing and noting the carries; the carries taken in forward order form the binary equivalent.

EXAMPLE 1

Convert decimal 28.375 into its binary equivalent.

Using the double-dabble method on the integer (a shorthand notation is shown at the left),

```
2 | 28
    |14       0   28 ÷ 2 = 14 with a remainder of 0
     |7       0   14 ÷ 2 =  7 with a remainder of 0
      |3      1    7 ÷ 2 =  3 with a remainder of 1
       |1     1    3 ÷ 2 =  1 with a remainder of 1
        0     1    1 ÷ 2 =  0 with a remainder of 1
```

The binary equivalent is **11100**.

Then converting the fraction,

$$0.375 \times 2 = 0.75 \text{ with a carry of } 0$$

$$0.75 \times 2 = 1.50 \text{ with a carry of } 1$$

$$0.50 \times 2 = 1.00 \text{ with a carry of } 1$$

The binary equivalent is **.011**.

28.375 is equivalent to binary **11100.011**.

Binary Arithmetic

Since the binary system uses the same concept of value and position of the digits as the decimal system, we expect the associated arithmetic to be similar but easier. (The binary multiplication table is very short.) In *addition*, we add column by column, carrying where necessary into higher position columns. In *subtraction,* we subtract column by column, borrowing where necessary from higher position columns. In subtracting a larger number from a smaller, we can subtract the smaller from the larger and change the sign just as we do with decimals.

EXAMPLE 2

Convert the given numbers to the other form and perform the indicated operations.

14	**1110**	13	**1101**	**1010**	10
+11	**+ 1011**	−10	**−1010**	**− 1101**	−13
25	**11001**	3	**0011**	**− 0011**	−3

In *multiplication,* we obtain partial products using the binary multiplication table (**0 × 0 = 0, 0 × 1 = 0, 1 × 0 = 0, 1 × 1 = 1**) and then add the partial products.[†] In *division,* we perform repeated subtractions just as in long division of decimals.

[†]Digital multiplying circuits are basically adding circuits.

EXAMPLE 3

Convert the given numbers to the other form and perform all the indicated operations.

(a)
$$\begin{array}{r} 14.5 \\ \times 1.25 \\ \hline 725 \\ 290 \\ 145 \\ \hline 18.125 \end{array}$$

(b)
$$\begin{array}{r} 101.1 \\ 11.1\overline{\smash{)}10011.01} \\ 111 \\ \hline 1010 \\ 111 \\ \hline 111 \\ 111 \\ \hline 0 \end{array}$$

After decimal-binary conversion, the operations are

(a)
$$\begin{array}{r} 1110.1 \\ \times\ 1.01 \\ \hline 11101 \\ 00000 \\ 11101 \\ \hline 10010.001 \end{array}$$

(b)
$$\begin{array}{r} 5.5 \\ 3.5\overline{\smash{)}19.25} \\ 175 \\ \hline 175 \\ 175 \\ \hline 0 \end{array}$$

Bits, Bytes, and Words

A single binary digit is called a "bit." All information in a digital system is represented by a sequence of bits. An 8-bit sequence is called a "byte"; a 4-bit sequence is a "nibble"; a 16-bit sequence is a "word." The number of bits in the data sequences processed by a given computer is a key characteristic. An *8-bit microprocessor* can receive, process, store, and transmit data or instructions in the form of bytes. Eight bits can be arranged in $2^8 = 256$ different combinations.

Other Notations

The number of years in a century can be written 100D or 100_{10}. In binary notation, this would be written $0 + 2^6 + 2^5 + 0 + 0 + 2^2 + 0 + 0 =$ **01100100B** or **01100100**$_2$; the suffix B or subscript 2 is used whenever necessary to avoid confusion.

Although 8-bit numbers are easy for computers, they are difficult for humans to deal with. In *octal* notation, a single decimal number from 0 to 7 is used to represent each group of three bits. As an octal number, **01100100B** would be written as **01 100 100→144Q = 144**$_8$. Three-digit octal numbers are easier to remember and easier to check than their 8-bit binary equivalents.

In the alternative notation most commonly used in microprocessor work, each group of four bits is represented by a single *hexadecimal* number. In "hex," **01100100B** would be written as **0110 0100→64H = 64**$_{16}$. Since four bits can take on sixteen different values, we supplement the ten decimal digits 0 . . . 9 with the letters, A, B, C, D, E, and F. For example, **11000011**$_2$**→ 1100 0011→C3**$_{16}$ and $255_{10} =$ **11111111**$_2$**→ 1111 1111→FF**$_{16}$. (See Table 14-1 on page 362. In this book, binary, octal, and hex numbers are in SANS SERIF type.)

Signed Magnitudes

In binary notation, an n-bit data word can represent the first 2^n nonnegative integers. To allow for both positive and negative numbers, the most significant bit (MSB) can be designated as the *sign bit* (**1** for negative numbers). The lower order bits then represent the *magnitude* of the number in "straight" binary notation. Although it

Table 14-1 Number Systems

Decimal	Binary	Hex	Octal
0	0000	0	00
1	0001	1	01
2	0010	2	02
3	0011	3	03
4	0100	4	04
5	0101	5	05
6	0110	6	06
7	0111	7	07
8	1000	8	10
9	1001	9	11
10	1010	A	12
11	1011	B	13
12	1100	C	14
13	1101	D	15
14	1110	E	16
15	1111	F	17

is used, this arrangement has two disadvantages: the number zero has two different representations, and two different arithmetic circuits are required to process positive and negative numbers.

Two's Complement Notation

A better notation for computers, one that is easily implemented in hardware, is based on the fact that adding the complement of a number is equivalent to subtracting the number. For example, in evaluating $9 - 3$, to subtract 3 from 9, we can "add the 10's complement" of 3 (i.e., $10 - 3 = 7$) to obtain $9 + 7 = 16 \rightarrow 6$ after discarding the final carry. In the decimal system, the 10's complement of a multidigit number is easily found by taking the 9's complement of each digit (by inspection) and then adding 1. In general, to subtract a two-digit number B from A we use the relationship

$$A - B = A + [100 - B] - 100 = A + [(99 - B) + 1] - 100 \quad (14\text{-}1)$$

where $(99 - B)$ is the 9's complement. (See Example 4.)

In the binary system, arithmetic is simplified if negative numbers are in *signed-2's complement* notation. In this notation, the MSB is the sign bit: **0** for plus, **1** for minus. To form the 2's complement of any number, positive or negative:

*Form the 1's complement by changing **1**s to **0**s and **0**s to **1**s.*
*Add **1**.*

If the result of an arithmetic operation has a **1** sign bit, it is a negative number in 2's complement notation; to obtain the true magnitude, subtract **1** and form the 1's complement.

EXAMPLE 4

(a) Obtain the 10's complement of 15 and 24.

Form the 9's complement of each digit, then add 1: $15 \to 84 + 1 = 85 \qquad 24 \to 75 + 1 = 76$.

(b) Represent -15 and -24 in 8-bit signed 2's complement notation.

Form the 1's complement of each digit, then add 1.

$$-15_{10} \to -1111 \to -00001111 \to 11110000 + 1 \to 11110001 \to 1\ 1110001$$

$$-24_{10} \to -11000 \to -00011000 \to 11100111 + 1 \to 11101000 \to 1\ 1101000$$

(c) Perform $24 - 15$ and $15 - 24$ directly and by complement notation.

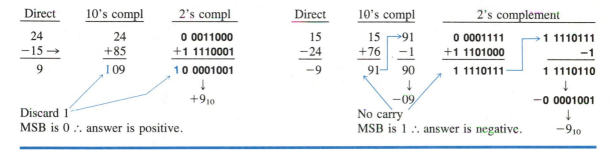

Binary-Coded Decimals (BCD)

For the convenience of humans, computer input/output devices may accept/provide decimals on the human side and binaries on the computer side. The conversion is simplified by *coding* each decimal digit, that is, replacing it by the 4-bit binary representation of the digit. For example, in the 8421 code $6_{10} \to \mathbf{0110}$, $3_{10} \to \mathbf{0011}$, and $363_{10} \to \mathbf{0011\ 0110\ 0011}$.

When a computer is to handle letters as well as numbers, an *alphanumeric code* is used. In the American Standard Code for Information Interchange (ASCII), seven bits are used to represent all the characters and punctuation marks on a teletypewriter keyboard plus some controls signals. (Note that $2^7 = 128$ combinations of 7 bits.) An eighth bit, the MSB, is a *parity bit* used in error detection. In the *even parity* convention, the MSB is set so that the number of **1**s in each ASCII character is even; the presence of an odd number of **1**s indicates an error.

BOOLEAN ALGEBRA

We see that binary arithmetic and decimal arithmetic are similar in many respects. In working with logic relations in digital form, we need a set of rules for symbolic manipulation that will enable us to simplify complex expressions and solve for unknowns; in other words, we need a "digital algebra." Nearly 100 years before the first digital computer, George Boole, an English mathematician (1815–1864) had formulated a basic set of rules governing the true-false statements of logic. Eighty-five years later (1938), Claude Shannon (at that time a graduate student at MIT) pointed out the usefulness of *Boolean algebra* in solving telephone switching problems and established the analysis of such problems on a firm mathematical basis. From our standpoint, Boolean algebra is valuable in manipulating binary variables in **OR**, **AND**, or **NOT** relations and in the analysis and design of all types of digital systems.

Boole's Theorems

The basic postulates are displayed in Tables 14-2 and 14-3. At first glance, some of the relations in Table 14-2 are startling. Properly read and interpreted, however,

Table 14-2 Boolean Postulates in 0 and 1

OR	AND	NOT
$0 + 0 = 0$	$0 \cdot 0 = 0$	$\bar{0} = 1$
$0 + 1 = 1$	$0 \cdot 1 = 0$	$\bar{1} = 0$
$1 + 0 = 1$	$1 \cdot 0 = 0$	
$1 + 1 = 1$	$1 \cdot 1 = 1$	

Table 14-3 Boolean Theorems in One Variable

OR	AND	NOT
$A + 0 = A$	$A \cdot 0 = 0$	$\bar{\bar{A}} = A$
$A + 1 = 1$	$A \cdot 1 = A$	
$A + A = A$	$A \cdot A = A$	
$A + \bar{A} = 1$	$A \cdot \bar{A} = 0$	

their validity is obvious. For example, "**1 OR 1**" (**1** + **1**) is just equivalent to **1**, and "**0 AND 1**" (**0** · **1**) is effectively **0**. In other words, for an **OR** gate with inputs of **1** at both terminals, the output is **1**, and for an **AND** gate with inputs of **0** and **1**, the output is **0**.

In general, the inputs and outputs of logic circuits are *variables*; that is, the signal may be present or absent (the statement is true or false) corresponding to the binary numbers **1** and **0**. The validity of the theorems in Table 14-3 may be reasoned out in terms of the corresponding logic circuit or demonstrated in a truth table showing all possible combinations of the input variables.

EXAMPLE 5

Demonstrate the theorems

$$A + \bar{A} = 1 \quad \text{and} \quad A \cdot 1 = A$$

by constructing truth tables.

Using the postulates, the truth tables are

A	$\bar{A}$	$A + \bar{A}$
0	1	1
1	0	1

A	$A \cdot 1$
0	0
1	1

Some of the more useful theorems in more than one variable are displayed in Table 14-4. The *commutation* rules indicate that the order of the variables in performing **OR** and **AND** operations is unimportant, just as in ordinary algebra. The *association* rules indicate that the order of the **OR** and **AND** operations is unimportant,

Table 14-4 Boolean Theorems in More Than One Variable

Commutation rules:	Association rules:	DeMorgan's theorems:
A + B = B + A	**A + (B + C) = (A + B) + C**	$\overline{A + B} = \overline{A} \cdot \overline{B}$
A · B = B · A	**A · (B · C) = (A · B) · C**	$\overline{A \cdot B} = \overline{A} + \overline{B}$
Absorption rules:	Distribution rules:	
A + (A · B) = A	**A · (B + C) = (A · B) + (A · C)**	
A · (A + B) = A	**A + (B · C) = (A + B) · (A + C)**	

just as in ordinary algebra. The second *distribution* rule indicates that variables or combinations of variables may be distributed in multiplication (not permitted in ordinary algebra) as well as in addition. The *absorption* rules are new and permit the elimination of redundant terms.

EXAMPLE 6

Derive the absorption rule

$$A + (A \cdot B) = A$$

using other basic theorems.

A	A · B	A + (A · B)
0	0	0
1	B	1

Factoring by the second distribution rule,

$$A + (A \cdot B) = (A + A) \cdot (A + B) = A \cdot (A + B)$$

we see that the two absorption forms are equivalent. Substituting **A · 1** for **A · A** (Table 14-3),

$$A \cdot (A + B) = A \cdot (1 + B) = A \cdot 1 = A$$

The truth table is shown at the left.

DeMorgan's Theorems

Augustus DeMorgan, a contemporary of George Boole, contributed two interesting and very useful theorems. These concepts are easily interpreted in terms of logic circuits. The first says that a **NOR** gate ($\overline{A + B}$) is equivalent to an **AND** gate with **NOT** circuits in the inputs ($\overline{A} \cdot \overline{B}$). The second says that a **NAND** gate ($\overline{A \cdot B}$) is equivalent to an **OR** gate with **NOT** circuits in the inputs ($\overline{A} + \overline{B}$). As generalized by Shannon, DeMorgan's theorems say:

> *To obtain the inverse of any Boolean function, invert all variables and replace all **OR**s by **AND**s and all **AND**s by **OR**s.*

Application of this rule is illustrated in Example 7 on page 366.

LOGIC CIRCUIT ANALYSIS

By comparing Example 7 with Example 1 in Chapter 13, we see that the theorems of Boolean algebra permit us to manipulate logic statements or functions directly, without setting up the truth tables. Furthermore, the use of Boolean algebra can lead to simpler logic statements that are easier to implement. This result is important when it is necessary to design a circuit to perform a specified logic function using the

EXAMPLE 7

Use DeMorgan's theorems to derive a combination of **NAND** gates equivalent to a two-input **OR** gate.

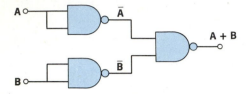

Figure 14.1 **OR** gate from **NAND** gates.

The desired function is $f = A + B$. Applying the general form of DeMorgan's theorems,

$$f = A + B = \overline{\overline{A} \cdot \overline{B}}$$

suggesting a **NAND** gate with **NOT** inputs.

Since $\overline{A \cdot A} = \overline{A}$, a **NAND** gate with the inputs tied together performs the **NOT** operation. The logic circuit is shown in Fig. 14.1.

available gates—only **NAND** gates, for example. DeMorgan's theorems are particularly helpful in finding **NAND** operations that are equivalent to other operations.

Up to this point we have been careful to retain the specific **AND** sign in $A \cdot B$ to focus attention on the *logic* interpretation. Henceforth, we shall use the simpler equivalent forms **AB** and **A(B)** whenever convenient. Also note that henceforth we shall follow convention and rely on the distinctive shapes of the logic symbols to identify their functions.

The *analysis* of a logic circuit consists in writing a logic statement expressing the overall operation performed in the circuit. This can be done in a straightforward manner, by starting at the input and tracing through the circuit, noting the function realized at each output. The resulting expression can be simplified or written in an alternative form using Boolean algebra. A truth table can then be constructed.

EXAMPLE 8

Analyze the logic circuit of Fig. 14.2.

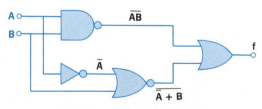

Figure 14.2 Logic circuit analysis.

Construct the truth table to demonstrate that this circuit could be replaced by a single **NAND** gate.

The suboutputs are noted on the diagram. The overall function can be simplified as follows:

$$f = \overline{AB} + \overline{\overline{A} + B}$$

$= (\overline{A} + \overline{B}) + A\overline{B}$	(DeMorgan's rule)
$= \overline{A} + \overline{B}(1 + A)$	(Distribution)
$= \overline{A} + \overline{B}$	$(1 + A = 1)$
$= \overline{AB}$	(DeMorgan's rule)

A	B	$\overline{AB}$	$\overline{\overline{A} + B}$	f
0	0	1	0	1
0	1	1	0	1
1	0	1	1	1
1	1	0	0	0

LOGIC CIRCUIT SYNTHESIS

One of the fascinating aspects of digital electronics is the construction of circuits that can perform simple mental processes at superhuman speeds. A typical digital computer can perform thousands of additions of 10-place numbers per second. The logic designer starts with a logic statement or truth table, converts the logic function into a convenient form, and then realizes the desired function by means of standard or special logic elements.

The Half-Adder

As an illustration, consider the process of addition. In adding two binary digits, the possible sums are as shown in Fig. 14.3a. Note that when **A = 1** and **B = 1**, the

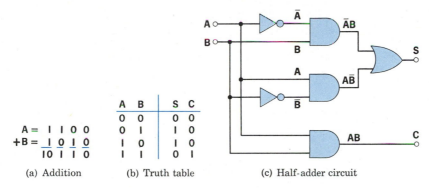

A	B	S	C
0	0	0	0
0	I	I	0
I	0	I	0
I	I	0	I

A = I I 0 0
+B = I 0 I 0
     10 I I 0

(a) Addition   (b) Truth table   (c) Half-adder circuit

Figure 14.3   Addition of two binary numbers.

sum in the first column is **0** and there is a *carry* of **1** to the next higher column. As indicated in the truth table, the half-adder must perform as follows: "**S** is **1** if **A** is **0 AND B** is **1**, **OR** if **A** is **1 AND B** is **0**; **C** is **1** if **A AND B** are **1**." In logic nomenclature, this becomes

$$S = \overline{A}B + A\overline{B} \quad \text{and} \quad C = AB \tag{14-2}$$

(A *full-adder* is capable of accepting the carry from the adjacent column.)

To *synthesize* a half-adder circuit, start with the outputs and work backward. Equation 14-2 indicates that the sum **S** is the output of an **OR** gate; the inputs are obtained from **AND** gates; inversion of **A** and **B** is necessary. Equation 14-2 also indicates that the carry **C** is simply the output of an **AND** gate. The corresponding logic circuit is shown in Fig. 14.3c.

There may be several different Boolean expressions for any given logic statement and some will lead to better circuit realizations than others. Algebraic manipulation of Eq. 14-2a yields

$$\overline{A}B + A\overline{B} = \overline{(A + \overline{B})(\overline{A} + B)} \qquad \text{(DeMorgan's theorems)}$$

$$= \overline{A\overline{A} + AB + \overline{A}\,\overline{B} + B\overline{B}} \qquad \text{(Multiplication)}$$

$$= \overline{\overline{A}\,\overline{B} + AB} \qquad (A\overline{A} = 0 \text{ and } B\overline{B} = 0)$$

$$= (A + B)(\overline{A} + \overline{B}) \qquad \text{(DeMorgan's theorems)}$$

$$= (A + B)\overline{A}\,\overline{B} \qquad \text{(DeMorgan's theorems)}$$

Looking again at the truth table (Fig. 14.3b), we see that another interpretation is: "**S** is **1** if (**A OR B**) is **1** **AND** (**A AND B**) is **NOT 1**." Therefore binary addition can be expressed as

$$S = (A + B)\overline{AB} \quad \text{and} \quad C = AB \tag{14-3}$$

The synthesis of this circuit, working backward from the output, is shown in Fig. 14.4. This circuit is better than that of Fig. 14.3c in that fewer logic elements are used

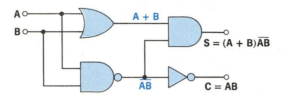

Figure 14.4 Another half-adder circuit.

and the longest path from input to output passes through fewer *levels*. The ability to optimize logic statements is an essential skill for the logic designer.

The Exclusive–OR Gate

The function $(A + B)\overline{AB}$ in Equation 14-3 is called the **Exclusive–OR** operation. As indicated by the truth table, it can be expressed as: "**A OR B** but **NOT** (**A AND B.**)" The alternate form (Eq. 14-2a) $\overline{A}B + A\overline{B}$ is called an "inequality comparator" because it provides an output of **1** if **A** and **B** are not equal.

EXAMPLE 9

Show that the inverse of the inequality comparator is an "equality comparator" and synthesize a suitable circuit.

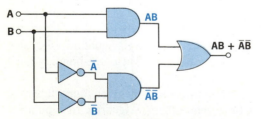

Figure 14.5 An equality comparator.

Algebraic manipulation of the inverse function using DeMorgan's theorems yields

$$\overline{\overline{A}B + A\overline{B}} = (A + \overline{B})(\overline{A} + B) = AB + \overline{A}\,\overline{B} \tag{14-4}$$

This is an equality comparator in that the output is **1** if **A** and **B** are equal. This function is highly useful in digital computer operation. Straightforward synthesis results in the circuit of Fig. 14.5.

The **Exclusive–OR** gate is used so frequently that it is represented by the special symbol $\oplus$ defined by

$$A \text{ XOR } B = A \oplus B = (A + B)\overline{AB} \tag{14-5}$$

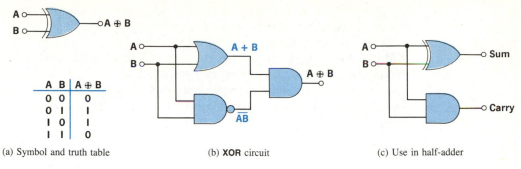

A B	A ⊕ B
0 0	0
0 1	1
1 0	1
1 1	0

(a) Symbol and truth table   (b) **XOR** circuit   (c) Use in half-adder

Figure 14.6   An **Exclusive-OR** circuit and its application.

One realization of the **Exclusive–OR** gate is shown in Fig. 14.6b. Assuming that this gate is available as a logic element, the half-adder takes on the simple form of Fig. 14.6c. As a further simplification we can treat the half-adder as a discrete logic element and represent it by a rectangular block labeled **HA**.

The Full-Adder

In adding two binary digits or *bits*, the half-adder performs the most elementary part of what may be highly sophisticated computation. To perform a complete addition, we need a *full-adder* capable of handling the carry input as well. The addition process is illustrated in Fig. 14.7a, where C is the carry from the preceding column. Each carry of **1** must be added to the two digits in the next column, so we need a logic circuit capable of combining three inputs. This operation can be realized using two half-adders and an **OR** gate. The values shown in Fig. 14.7b correspond to the third column of the addition. The operation of the full-adder can be seen more clearly by construction of a truth table. (Review Question 12.)

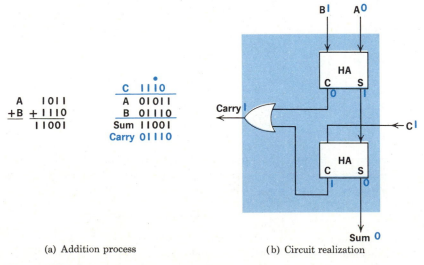

$$\begin{array}{r} A \quad 1011 \\ +B \ +1110 \\ \hline 11001 \end{array}$$

$$\begin{array}{l} \overset{\bullet}{C} \ \ 1110 \\ A \ \ 01011 \\ B \ \ 01110 \\ \hline Sum \ 11001 \\ Carry \ 01110 \end{array}$$

(a) Addition process   (b) Circuit realization

Figure 14.7   Design of a full-adder.

To perform the addition of "4-bit words," that is, numbers consisting of four binary digits, we need a half-adder for the first column and a full-adder for each additional column. In the *parallel binary adder,* the input to the half-adder consists of digits A_1 and B_1 from the first column. The input to the next unit, a full-adder, consists of carry C_1 from the first unit and digits A_2 and B_2 from the second column. The sum of the two numbers is represented by $C_4 S_4 S_3 S_2 S_1$. In subtraction, **NOT** gates provide the complement of the subtrahend (see Example 4, p. 363) and the process is reduced to addition.

A Design Procedure

A common problem in logic circuit design is to create a combination of gates to realize a desired function. The basic approach is to proceed from a statement of the function to a truth table and then to a Boolean expression of the function. The experienced logic designer then manipulates the Boolean expression into the simplest form. The realization of the final expression in terms of **AND**, **OR**, and **NOT** gates is straightforward.

For increased reliability on a spacecraft, triple sensing systems are used; no action is taken unless at least two of the three systems call for action. The truth table of the required *vote taker* is shown in Fig. 14.8a. Since the function is **YES (1)** only

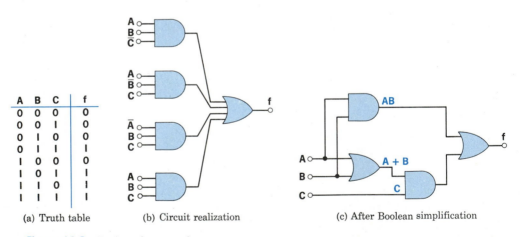

A	B	C	f
0	0	0	0
0	0	1	0
0	1	0	0
0	1	1	1
1	0	0	0
1	0	1	1
1	1	0	1
1	1	1	1

(a) Truth table   (b) Circuit realization   (c) After Boolean simplification

Figure 14.8 Design of a vote taker.

when a majority of the inputs are **YES**, the Boolean expression must contain a product term for each row of the truth table in which the function is **1**. Since the function is **YES** for any one *or* more of these rows, the Boolean expression is

$$f = AB\overline{C} + A\overline{B}C + \overline{A}BC + ABC \tag{14-6}$$

Assuming that the complement of each variable is available, as is true in most computers, the straightforward realization is a combination of four **AND** gates feeding an **OR** gate (Fig. 14.8b).

If the complements were not available, eight logic elements would be required and simplification of the circuit would be desirable. First, the Boolean function of Equation 14-6 can be expanded into

$$f = AB\overline{C} + A\overline{B}C + \overline{A}BC + ABC + ABC$$

since $ABC + ABC = ABC$. Factoring by the distribution rule yields

$$f = AB(\overline{C} + C) + C(AB + A\overline{B} + \overline{A}B)$$

Since $\overline{C} + C = 1$ and $AB + A\overline{B} + \overline{A}B = A + B$, the function becomes

$$f = AB + C(A + B)$$

This function requires only four logic elements (Fig. 14.8c).

MINIMIZATION BY MAPPING

In creating a logic circuit to realize a desired function, the designer seeks the optimum form. The criterion may be maximum speed (fewest logic levels) or minimum cost (fewest gate leads since the number of leads determines the cost of manufacture and the cost of assembly) or minimum design time (if only a few circuits are required). We have seen how Boolean algebra can be used in deriving simpler logic expressions. If the truth table is available or if the logic function is expressed as a "sum of products," the designer can go directly to the minimal expression by the mapping technique suggested by Maurice Karnaugh.

Karnaugh Maps

The *map* of the general logic function of three variables is shown in Fig. 14.9a. Each square in the map corresponds to one of the eight possible combinations of the three variables. The order of the columns is such that *combinations in adjacent squares differ only in the value of one variable*. Therefore, 2-square groups are independent of one variable; that is, $f_1 = \overline{A}B\overline{C} + \overline{A}BC = \overline{A}B$ and $f_2 = A\overline{B}C + \overline{A}\overline{B}C = \overline{A}C$. These relations are easy to derive using Boolean algebra, but they are *obvious by inspection* of the Karnaugh map.

The concept can be extended to groupings of four adjacent squares as shown in Fig. 14.9b, where the labels are omitted from the squares. The 4-square cluster outlined in color is independent of both **B** and **C** and the 4-square in-line group is independent of both **A** and **B**; that is, $f_3 = \overline{A}\overline{B}C + \overline{A}BC + ABC + A\overline{B}C = C$. Enlarging groups by overlapping simplifies the terms. Note that the map is "continuous" in that

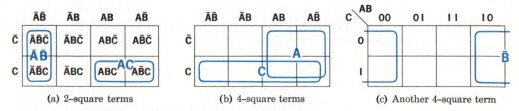

(a) 2–square terms   (b) 4–square terms   (c) Another 4–square term

Figure 14.9   Mapping a three-variable logic function.

the last column on the right is "adjacent" to the first column on the left. The 4-square group in Fig. 14.9c is just equal to $\overline{\mathbf{B}}$.

The standard labeling scheme for Karnaugh maps (Fig. 14.9c) is convenient for mapping from a truth table. Each square in the map corresponds to a row in the truth table. A specific logic function is *mapped* by placing a **1** in each square for which the function is **1**. When this has been done, possible simplifications are easily recognized.

EXAMPLE 10

Map the vote-taker function and simplify the circuit realization, if possible.

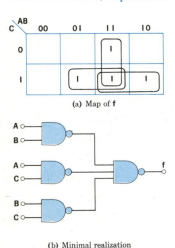

(a) Map of **f**

(b) Minimal realization

From the truth table of Fig. 14.8a, **1**s are placed in the squares corresponding to rows in the truth table for which the function is **1**, representing

$$f = \overline{\mathbf{A}}\mathbf{B}\mathbf{C} + \mathbf{A}\overline{\mathbf{B}}\mathbf{C} + \mathbf{A}\mathbf{B}\overline{\mathbf{C}} + \mathbf{A}\mathbf{B}\mathbf{C}$$

All the **1**s can be included in three overlapping 2-square groups; ∴ the complete function can be represented by

$$f = \mathbf{A}\mathbf{B} + \mathbf{A}\mathbf{C} + \mathbf{B}\mathbf{C}$$

By using DeMorgan's theorem, any "sum of products" can be converted to a "**NAND**ed product of **NAND**s." Here

$$f = \overline{\overline{\mathbf{A}\mathbf{B}} \cdot \overline{\mathbf{A}\mathbf{C}} \cdot \overline{\mathbf{B}\mathbf{C}}}$$

which can be synthesized using **NAND** gates only. Note that in the circuit realization of Fig. 14.10b the number of gate leads has been reduced to 9 input + 4 output = 13 from the 21 in Fig. 14.8b. This vote taker could be realized in a single SSI chip.

Figure 14.10 Simplification by Karnaugh mapping.

Mapping in Four Variables

The Karnaugh technique is even more valuable in simplifying functions in four variables. (For more than four variables, other techniques are usually more convenient.) As indicated in Fig. 14.11, 2-square groups are independent of one variable, 4-square groups are independent of two variables, and 8-square groups are independent of three variables. Note that the standard labeling scheme provides that adjacent rows differ by only one complement bar and the bottom row is *adjacent* to the top row. The four corner squares form a combination that is a little difficult to visualize.

There are detailed rules[†] for finding the minimal expression from a Karnaugh

---

[†]See pp. 69 ff. of the Peatman reference at the end of Ch. 13.

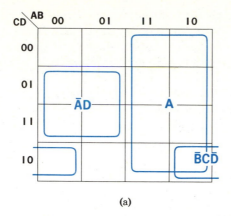

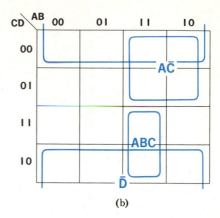

(a)   (b)

Figure 14.11   Examples of grouping on four-variable maps.

map, but some general guidelines will suffice for our purposes:

Include all 1s in groups of eight, four, two, or one.
Groups may overlap; larger groups result in simpler terms.
Of the possible combination of terms, select the simplest.

The technique is illustrated in Example 11.

EXAMPLE 11

Map the function

$$f = \overline{AB}(C + D) + \overline{A}C\overline{D} + AB\overline{C}D + A\overline{B}\overline{C}D$$

and obtain a minimal sum-of-products expression.

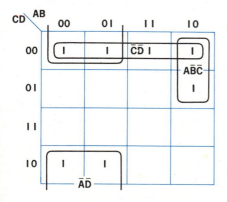

Figure 14.12   Four-variable simplification.

As an alternative to forming the truth table, let us map the function by considering the factors individually.

$\overline{AB}$ limits the first term to the **00, 01,** and **10** columns and $(\overline{C + D})$ corresponds to the first row, implying three 1s as shown in Fig. 14.12.

The $\overline{A}C\overline{D}$ term is independent of **B**, implying a 2-square group in the lower left-hand corner.

The four-variable terms imply 1s in the **1100** and **1001** squares.

All the 1s can be included in the two 4-square and one 2-square groups encircled. Therefore,

$$f = \overline{A}\overline{D} + \overline{C}\overline{D} + A\overline{B}\overline{C}$$

Other expressions are possible, but none will include fewer, simpler terms.

Two additional comments should be made. In some circuits certain combinations of inputs never occur; such *don't care* combinations may be mapped as **X**s and considered as either **0**s or **1**s, whichever provides the greatest simplification. In other circuits, the simplest realization results from implementing $\bar{f}$ as a sum of products and then inverting to obtain **f**.

The emphasis in this section is on synthesizing logic circuits from optimum arrangements of basic gates. This is the proper approach in designing complex circuits for mass production or in designing custom circuits requiring only a few gates. As we shall see, other approaches using standard IC packages are better where the cost of design time is an important factor.

REGISTERS

In addition to the logic circuits that process data, digital systems must include memory devices to store data and results. A flip-flop can store or "remember" one digit of a binary number, one bit. A *register* is an array of flip-flops that can temporarily store data or information in digital form. For example, the 8-bit registers in a microprocessor, composed of eight flip-flops in parallel, can handle 8-digit instructions or numbers. A great variety of registers is available in IC form.

Shift Registers

The sophistication of the response of the JK flip-flop makes it useful in computer applications. The *serial shift register* of Fig. 14.13 consists of four master-slave JK

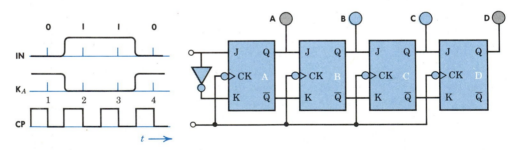

Figure 14.13 Entering **0110** into a 4-bit serial shift register.

flip-flops connected so that **J** ≠ **K**. At the trailing edge of each clock pulse, **Q** follows **J** in each flip-flop of the 4-bit register. The data are entered serially, that is, one bit at a time, and shifted right through the register at each clock pulse. (See Table 14-5.)

Table 14-5 Serial Shift Register

CP	IN	Q_A	Q_B	Q_C	Q_D
1	0 → 0				
2	1 → 1	0			
3	1 → 1	1	0		
4	0 → 0	1	1	0	

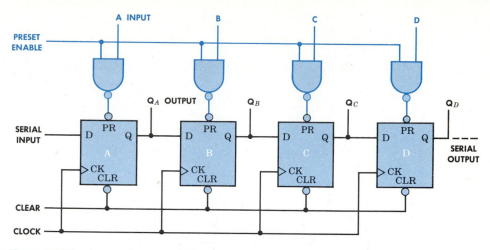

Figure 14.14 A general-purpose 4-bit shift register.

One application of such a register is as a *serial-to-parallel converter,* changing serial data to parallel form for processing all bits simultaneously. There is a single input line, but four lines are required for the parallel data output.

The shift register of Fig. 14.14 consists of D flip-flops with **CLEAR** and **PRESET** capabilities. It is similar to MSI TTL units available commercially. The symbols indicate that the flip-flops are cleared to **0** if **CLR** goes **LOW** while **PR** is inactive (**HIGH**) (clearing is independent of the clock level). On the positive-going edge of the clock signal, the input at **D** is transferred to **Q**. (See Fig. 13.27.)

Since both inputs and outputs are accessible, this unit can be used as a general purpose register. It can function as a 4-bit storage register (serial or parallel), as a serial-to-parallel converter (**SERIAL INPUT** to Q_A Q_B Q_C Q_D **OUTPUT**), or as a parallel-to-serial converter. For the last function, all stages are cleared to **0**, and the data to be loaded are applied to the **ABCD INPUTS**. A **HIGH** signal to **PRESET ENABLE** is **NAND**ed with any **1** input to send **PR LOW** setting **Q** to **1** in that stage (independent of the clock level). At the next upward transition of the clock pulse (unless the clock is inhibited), the data are shifted to the right; the value of Q_D is output and the value of Q_C is transferred to Q_D, etc. As we shall see, shifting bits one place to the right is equivalent to division by 2 in binary notation, and the shift operation is useful in computation.

A Practical Register—The 74173

The 74173 TTL unit is a versatile 4-bit register incorporating D flip-flops and three-state outputs for use in bus-organized systems. The "totem pole" outputs (R_4, T_4, D, and T_3 in Fig. 13.14) are capable of driving the bus lines directly. (Rated drive current is 16 mA.) As shown by the "pinout" in Fig. 14.15, the four data inputs (D) and the four data outputs (Q) use eight pins in the 16-pin package. Power supply (V_{CC}, GND), direct **CLEAR**, and **CLOCK** account for four pins. The remaining pins provide versatility.

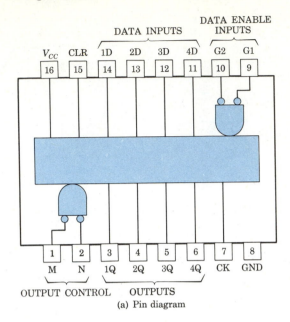

DATA INPUTS DATA ENABLE INPUTS

		INPUTS				OUTPUT Q
CLEAR	CLOCK	DATA ENABLE G1	G2	DATA D		
H	X	X	X	X		L
L	L	X	X	X		Q_0
L	↑	H	X	X		Q_0
L	↑	X	H	X		Q_0
L	↑	L	L	L		L
L	↑	L	L	H		H

H = high level (steady state)
L = low level (steady state)
↑ = low–to–high–level transition
X = irrelevant (any input including transitions)
Q_0 = the level of **Q** before the indicated steady–state input conditions were established.

OUTPUT CONTROL   OUTPUTS
(a) Pin diagram   (b) Function table

Figure 14.15   The 74173 4-bit register.

Access to the output control is provided by pins M and N. For **M + N = 0**, normal logic states are available; for **M + N = 1**, the three-state outputs are effectively disconnected from the bus. This means that the unit can be disconnected by a signal from either of two components of the system. Disabling of the outputs is independent of the clock level.

The entry of data into the flip-flops is also controlled. As indicated in the function table, when both **DATA ENABLE** inputs are **LOW** (**L**), data at the D inputs are loaded into their respective flip-flops on the next positive transition of the **CLOCK**.

These registers are used as bus buffer registers between data sources with inadequate drive or lacking three-state capability and the system bus. The typical propagation delay time is 23 ns, and the typical power consumption is 250 mW. They can accept clock frequencies up to 25 MHz.

COUNTERS

Flip-flops can be connected to function as electronic counters to count random events or to divide a frequency or to measure a parameter such as time, distance, or speed. Counters are used to keep track of operations in digital computers and in instrumentation. The JK master-slave flip-flop is widely used in counter design, but T and D flip-flops are also useful.

Divide-by-*n* Circuits

A divide-by-*n* counter produces one output pulse for *n* input pulses: *n* is called the "modulo" of the counter. As shown in Fig. 13.32, a toggle flip-flop divides by two; for T held **HIGH**, the output **Q** is the number of CK inputs divided by two. Two D

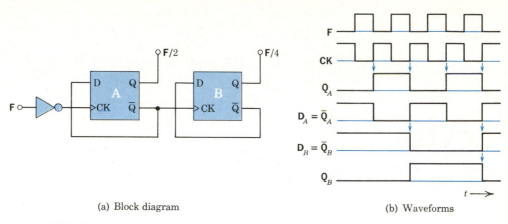

(a) Block diagram　　　　　　　　　(b) Waveforms

Figure 14.16　A divide-by-4 circuit using D flip-flops.

flip-flops can be used in the divide-by-four circuit of Fig. 14.16. With Q_A and Q_B cleared, when clock signal **F** goes **LOW**, **CK** goes **HIGH** and $D_A = \bar{Q}_A = 1$ is transferred to Q_A. The next time **F** goes **LOW**, Q_A changes to **LOW**; two events at **F** complete one cycle of Q_A or $Q_A = F/2$. When Q_A goes **LOW**, $\bar{Q}_A = CK_B$ goes **HIGH** and $D_B = \bar{Q}_B = 1$ is transferred to Q_B; four events at **F** complete one cycle of Q_B or $Q_B = F/4$.

Binary Ripple Counter

The counter of Fig. 14.17 consists of three JK flip-flops in cascade. With J and K held **HIGH** (the connections to +5 V are not shown), the flip-flops toggle at each downward transition of the pulse at CK. The lowest order bit Q_A changes state after each input pulse. The next bit Q_B changes state whenever Q_A goes **LOW** since Q_A supplies CK_B. Similarly, Q_C changes state whenever Q_B goes **LOW**. This is an *asynchronous, binary, modulo-8, ripple* counter; *asynchronous* because all flip-flops do not change at the same time; *binary* because it follows the binary number sequence with bit values of 2^0, 2^1, and 2^2; *modulo-8* because it counts through 8 distinct states; *ripple* because the changes in state ripple through the stages. The ripple effect is seen at the fourth input pulse where Q_A goes **LOW**, causing Q_B to go **LOW**, causing Q_C to go **HIGH**.

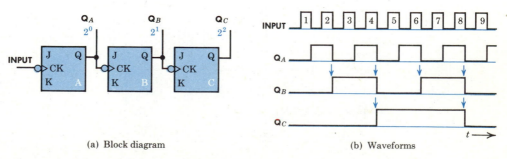

(a) Block diagram　　　　　　　　　(b) Waveforms

Figure 14.17　A three-stage binary ripple counter.

EXAMPLE 12

Design a modulo-5 binary counter using T flip-flops with **CLEAR** capability.

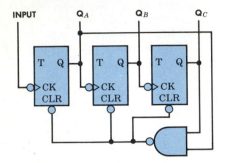

INPUT Q_A Q_B Q_C

Figure 14.18 A modulo-5 counter.

Three stages are required to count beyond 4. A modulo-5 counter must count up to 4 and then, on the fifth pulse, clear all flip-flops to 0. The sequence of states is

Count	Q_C	Q_B	Q_A	
0	0	0	0	
1	0	0	1	
2	0	1	0	
3	0	1	1	
4	1	0	0	
5	1	0	1	(Unstable)
	0	0	0	(Stable)

At the count of 5, the 1s at Q_A and Q_C can be **NAND**ed to generate a **CLEAR** signal as shown in Fig. 14.18.

Decade Counters

Counters that communicate with humans usually display results in the decimal system. However, flip-flops count in binary; therefore, binary numbers must be *coded* in decimal. To count to 10 in the 8421 code, four flip-flops are required. The usual way to get 10 distinct states is to modify a 4-bit binary counter so that it skips the last 6 states. It counts normally from 0 to 9, and then feedback logic is provided so that at the next count the decimal sequence is reset to zero. (See Exercise 29.)

Synchronous Counters

One disadvantage of ripple counters is the slow speed of operation caused by the long time required for changes in state to ripple through the flip-flops. Another problem is that the cumulative delays can cause temporary state combinations (and voltage spikes called *glitches*) that result in false counts. Both of these difficulties are avoided in *synchronous* counters in which all flip-flops change state at the same instant.

In the synchronous counter of Fig. 14.19, JK master-slave units are connected

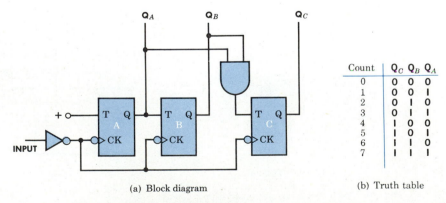

Q_A Q_B Q_C

Count	Q_C	Q_B	Q_A
0	0	0	0
1	0	0	1
2	0	1	0
3	0	1	1
4	1	0	0
5	1	0	1
6	1	1	0
7	1	1	1

(a) Block diagram

(b) Truth table

Figure 14.19 A synchronous modulo-8 counter.

(a) Pin diagram

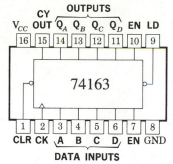

(b) Recommended operating conditions

		MIN	NOM	MAX	UNIT
Supply voltage, V_{CC}		4.75	5	5.25	V
High–level output current, I_{OH}				−800	μA
Low–level output current, I_{OL}				16	mA
Input clock frequency, f_{clock}		0		25	MHz
Width of clock pulse, $t_{w(clock)}$		25			ns
Width of clear pulse, $t_{w(clear)}$		20			ns
Setup time, t_{setup}	Data inputs A, B, C, D	15			ns
	Enable P	20			
	Load	25			
	Clear	20			
Hold time at any input, t_{hold}		0			ns

(c) Functional block diagram

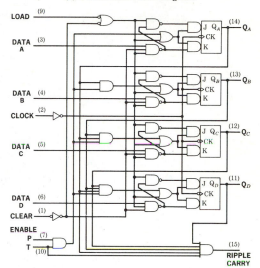

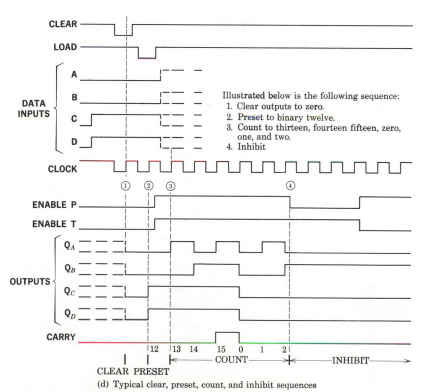

Illustrated below is the following sequence:
1. Clear outputs to zero.
2. Preset to binary twelve.
3. Count to thirteen, fourteen fifteen, zero, one, and two.
4. Inhibit

(d) Typical clear, preset, count, and inhibit sequences

Figure 14.20 The SN74163 synchronous 4-bit counter (Texas Instruments, Inc.).

as T (toggle) flip-flops. At each count, flip-flop *A* toggles, and the other flip-flops are clocked. The master-slave characteristic causes flip-flop *B* to toggle on the next count after Q_A becomes **1**, as called for in the truth table. Similarly, the **AND** gate causes flip-flop *C* to toggle on the next count after Q_A **AND** $Q_B = $ **1**. In general, synchronous counters are faster and more trouble-free.

A Practical Counter—The 74163

The 74163 (Fig. 14.20 on page 379) is a popular 4-bit counter using TTL technology. The technical features include:

4-bit binary—counts 0 to 15 and then resets; both P and T enable inputs must be **HIGH** to count.

Carry output—this permits cascading 74163 counters for higher counting capacity. The carry pin (15) provides a positive pulse that can enable successive stages.

Synchronous operation—flip-flops are clocked simultaneously so that, when enabled, outputs change simultaneously.

Buffered clock input—on the rising edge of the clock pulse, the four JK master-slave flip-flops are triggered.

Programmable—the outputs can be preset. When **LOAD** goes **LOW**, the values at the **DATA INPUTS** are set into the **OUTPUTS** after the next clock pulse.

Synchronous clear—When **CLEAR** is **LOW**, all flip-flop outputs are set **LOW** after the next clock pulse (regardless of inputs). This permits modifying the count length by decoding the maximum count desired using an external **NAND** gate connected to the **CLEAR** input.

The circuits are designed for operation in ambient temperatures from 0 to 70° C; typical power consumption is 325 mW.

EXAMPLE 13

Design a simple divide-by-12 counter using a 74163. Show the pin diagram and the external connections required.

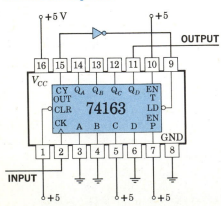

Figure 14.21 A divide-by-12 counter.

We could shorten the count length by decoding the outputs on the eleventh count (**1011**), using a 3-input **NAND** gate to drive the **CLEAR LOW** on the next count as explained earlier.

Another approach is shown in Fig. 14.21. To divide by 12, we prepare to load $16 - 12 = 4$ into the counter by setting **DATA INPUT C HIGH** and tying the other **DATA INPUTS LOW**.

In output state **1111**, the **CARRY** is inverted to set **LOAD LOW**; after the next clock pulse, DCBA = 0100 is loaded into the **OUTPUTS**. The counting sequence is

State: 8, 9, 10, 11, 12, 13, 14, 15, 4, 5, 6, 7, 8
Count: 0, 1, 2, 3, 4, 5, 6, 7, 8, 9, 10, 11, 0

On the 12th count, the output at Q_D goes **HIGH** in state 8D = **1000B**. This signal is used for the counter **OUTPUT**.

READ-AND-WRITE MEMORY

In a digital computer, instructions and numbers are stored in the *memory*, an organized arrangement of elements called *memory cells,* each capable of storing one bit of information. In *Read-And-Write Memory* (RAM), the computer can store ("write") data at any selected location ("address") and, at any subsequent time, retrieve ("read") the data. In *Read-Only-Memory* (ROM), data are initially and permanently stored (by the manufacturer or the user); the computer can read the data at any address, but it cannot alter the stored bits.

In Fig. 14.22a, the k address lines can designate $2^k = m$ words whose n bits are carried on the n parallel input and output lines. The 64 bits of the 4×16 ROM in Fig.

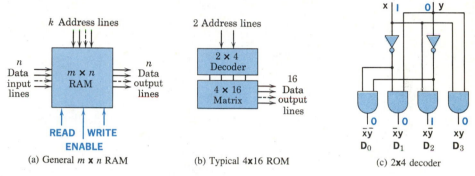

Figure 14.22 Examples of IC memories.

14.22b are stored in 64 memory cells arranged in $2^k = 2^2 = 4$ words of 16 bits each available at the 16 output lines. The address is coded as a k-bit binary number; a *decoder* translates the coded address and specifies one of the 2^k words. In the 2×4 decoder of Fig. 14.22c, the address **XY = 10** yields a **1** at D_2 ($X\overline{Y}$) and specifies that word 2 is to be read, that is, connected to the 16 output lines.

RAM

The Read-And-Write Memory of Fig. 14.22a includes an $m \times n$ matrix of memory cells. Each cell consists of a binary storage element and the associated control logic. In the simple cell shown in Fig. 14.23a, the cell is selected when **S** goes **HIGH**. With **R/$\overline{W}$ HIGH**, the output is enabled and the value of **Q** is read out on the output data line. With **R/$\overline{W}$ LOW**, the input is enabled to write in the input data value.

The organization of memory cells into an elementary RAM with a capacity of two 3-bit words is shown in Fig. 14.23b. The decoder activates one of the two word-select lines in accordance with the address. In the **WRITE** operation (**R/$\overline{W}$ LOW**), three bits of data are transferred from the input lines to the selected word. In the **READ** operation, the 3 bits of the selected word are transferred to the output lines via the **OR** gates. The outputs of the unselected words are all **LOW**.

Since the words in memory can be accessed in any order, this is a *random-access memory*. (Originally RAM meant random-access memory, but now it is interpreted as read-and-write memory.) In the 2×3 RAM shown, address selection is *linear;* that

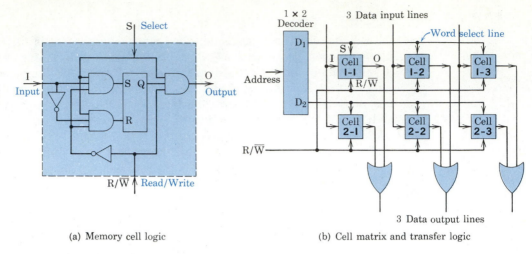

(a) Memory cell logic

(b) Cell matrix and transfer logic

Figure 14.23 Functional representation of a 2×3 RAM.

is, addressing is performed by activating one word-select line. In large memories, selection is *coincident;* that is, each cell is accessed by addressing an X select line to select the row and a Y select line to select the column. The intersection of the X and Y lines identifies one cell in a two-dimensional matrix. For example, a 1024-bit (1K) RAM could consist of 1024 memory cells arranged in a 32×32 matrix on the silicon chip. For operation as a 256×4 RAM, the 32 columns could be organized into 8 groups of 4 columns representing eight 4-bit words in each of 32 rows. The 32 row-select lines would be addressed by a 5-bit address ($2^5 = 32$), and the 8 column-select lines would be addressed by a 3-bit address ($2^3 = 8$).

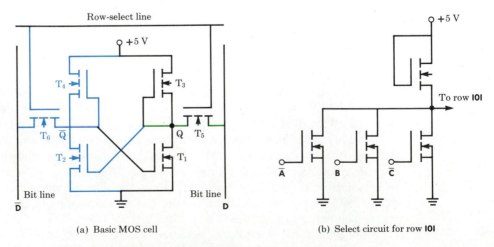

(a) Basic MOS cell

(b) Select circuit for row **101**

Figure 14.24 Typical MOS memory cell and row-select circuit.

MOS RAM

Monolithic semiconductor memory dominates the market for small, fast memory applications. The high packing density and low power consumption of MOS devices have led to their wide use; storage of 64K bits on a single IC chip is common.

The MOS memory cell of Fig. 14.24 contains 6 n-channel enhancement-mode MOSFETs. Transistor T_1 is the *driver* and T_3 the *load* for an inverter (Fig. 13.2c, p. 329) that is cross-coupled to a second inverter T_2-T_4 to form the storage flip-flop (Fig. 13.23). When T_2 is **ON** and T_1 is **OFF**, output **Q** is logic **1**. With the row-select line **LOW**, transistors T_5 and T_6 are **OFF**, and the cell is isolated from the bit lines. In a **READ** operation, when row-select goes **HIGH**, T_5 and T_6 provide coupling and **Q** appears on bit line **D**. A *sense amplifier* (similar to that shown in Fig. 14.25b) connected to the bit line provides buffering, and the proper logic level appears on the data output line. In a **WRITE** operation, the selected row of cells is connected to the bit lines, and **Q** is **SET** or **RESET** by a **1** placed on bit line **D** or $\overline{\textbf{D}}$ by the *write amplifier*. In this *static* RAM with *nondestructive* read-out, the state of the flip-flop persists as long as power is supplied to the chip, and it is not altered by the **READ** operation.

The row-select circuit of Fig. 14.24b is one of eight similar circuits that decode 3-bit addresses to select a row of MOS memory cells. The three select input pins are connected to buffer-inverters that generate **A**, $\overline{\textbf{A}}$, **B**, $\overline{\textbf{B}}$, **C**, and $\overline{\textbf{C}}$. The circuit is a **NOR** gate that performs the **AND** function on the complements of the inputs (see the truth table in Fig. 13.6a). For the address **101**, the **NOR** gate inputs are **0**, **0**, **0**; the driver transistor is **OFF**; the output is **HIGH**; and only one row-select line will go **HIGH**.

EXAMPLE 14

A 64-bit RAM organized to provide sixteen 4-bit words is to use circuitry similar to Fig. 14.23. The row-select circuits are similar to those in Fig. 14.24b. Specify the number of address lines, the number of **NOR** gate inputs in the row-select circuits, and the decoder connections for the fourth row.

To address 16 words requires 4 address lines since $2^k = m$ or $2^4 = 16$. Assuming linear selection, there will be 16 rows, and 4 input **NOR** gates are required to decode the addresses. Assuming the first 4 rows are designated **0000**, **0001**, **0010**, and **0011**, the fourth row-select inputs must be $\textbf{AB}\overline{\textbf{C}}\overline{\textbf{D}}$ so that all drivers are **OFF**.

For another address **0010**, the last transistor will be **ON**, the output will be **LOW**, and the fourth row will not be active.

Bipolar RAM

Bipolar read-and-write memories are extremely fast, but they are less compact and less energy efficient than MOS RAMs. They are TTL compatible and are used as small "scratch-pad" memories for data being processed.

The RAM cell of Fig. 14.25a employs two cross-coupled triple-emitter transistors in a flip-flop. One emitter of each transistor is connected to a bit line to provide for data transfer and storage. The other emitters are connected to address lines for coincident selection. The address lines are normally **LOW**, and the currents from all the conducting transistors in the matrix flow out along these lines.

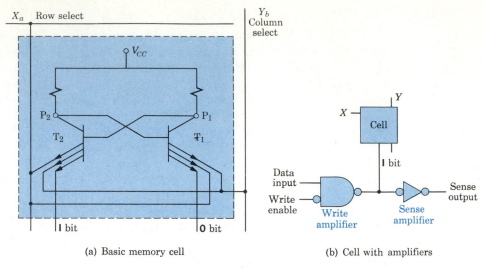

(a) Basic memory cell   (b) Cell with amplifiers

Figure 14.25   Bipolar RAM with coincident selection.

Assuming T_2 is **ON**, V_{P2} is **LOW**, T_1 is **OFF**, and V_{P1} is **HIGH**. (For example, with $V_{CC} = 3.5$ V and $V_X = V_Y = 0.3$ V, $V_{BE2} \cong 0.7$ V and $V_{CE(sat)} = 0.3$ V, $V_{P1} = V_X + V_{BE2} = 0.3 + 0.7 = 1.0$ V. Then $V_{P2} = V_X + V_{CE(sat)} = 0.3 + 0.3 = 0.6$ V; T_1 cannot be forward biased.) To address cell *a-b*, the corresponding address lines X_a and Y_b are taken **HIGH**. (All cells except the one addressed have at least one address line—X or Y—**LOW** and are unaffected.) As X_a and Y_b go **HIGH**, the connected emitters rise in potential, and the current in T_2 is transferred to the **1-bit** line (held at 0.5 V, say, slightly above logic **0**). This current activates the sense amplifier (Fig. 14.25),[†] and a logic **1** appears at the output. The **READ** operation has been performed.

The **WRITE** operation proceeds as follows. The particular address lines are taken **HIGH**, **WRITE ENABLE** goes **LOW**, and a **1**, say, is input to the write amplifier. The amplifier output takes the **1-bit** line **LOW**, and T_2 is turned **ON** (assuming it was **OFF**). The unaddressed cells on the same bit line are not affected since at least one emitter in each transistor is held **LOW** by an inactive address line. After **WRITE ENABLE** goes **HIGH**, the bit line returns to normal (0.5 V, say). When the address lines return to normal (0.3 V), the emitter current of T_2 is diverted to an address line, ready for any **READ** operation.

READ-ONLY MEMORY

In Read-Only Memory, binary data is physically and permanently stored by defining the state of the constituent memory cells. A set of input signals on the address lines is decoded to access a given set of cells whose states then appear on the output lines.

The function performed can be described in three ways. If the set of inputs is identified as an address and the output is the word (data or instructions) stored there, this is a "memory." If the set of inputs represents data coded in one form and the

---

[†]The sense amplifier is a transistor switch that is turned on when the **1-bit** line current enters the base.

outputs represent the same data coded in another form, this is a "code converter." If the input is identified as a set of binary variables (a binary *function*) and the corresponding output is identified as a related binary function, this is a "combinational logic circuit" that can replace a network of logic gates.

The same bipolar and MOS technologies are used in IC ROM as in RAM. In general, ROM is simpler than RAM since fewer control elements are necessary (Fig. 14.22b) and no provision is made for changing cell states. The essential components are: an address decoder, a matrix of memory cells, and appropriate buffers.

ROM Cells

The MOS memory cell shown in Fig. 14.26a consists of a single enhancement-mode transistor connected to word and bit lines plus a passive load R (actually another MOSFET). If this cell is selected, the word line goes **HIGH**, the conductivity of the

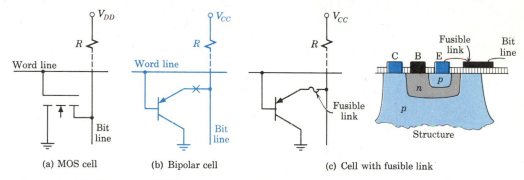

(a) MOS cell   (b) Bipolar cell   (c) Cell with fusible link

Figure 14.26 Elementary ROM cells.

n channel is enhanced, drain current flows, and the bit line is pulled **LOW**; a **0** bit is output. If a **1** is to be stored at this location in memory, this MOSFET is made inoperative by proper masking during chip fabrication. Instead of the normal thin oxide layer between gate and channel, a thick oxide layer is deposited, and the threshold voltage for conducting is excessively high. When the word line is raised, there is no pull down, and a **1** is output. The finished matrix consists of $m \times n$ similar transistors with a pattern of **1**s corresponding to the thick-oxide cells. This pattern is specified by the user ordering "custom-programmed" ROMs.

The bipolar memory cell of Fig. 14.26b operates in an analogous way. If this cell is selected, the word line goes **LOW**, the emitter-base junction is forward biased, and the bit line is pulled **LOW**; a **0** bit is output. If a **1** is to be stored at this location, the emitter connection at **X** is not completed during fabrication. The specifications of the customer are translated into a pattern of **0**s and **1**s represented by complete and incomplete transistors.

PROMs

The process just described is called *mask programming*; it can be done only by the manufacturer. Also available are *field programmable* memories called PROMs. The PROM cell of Fig. 14.26c contains a special conducting segment that can be melted by a high-current pulse after packaging. The PROM is fabricated with all **0**s; the user "programs" the unit by electrically changing appropriate **0**s to **1**s. This is an irreversible process; once fused, the links are destroyed. PROMs are economical in small quantities; the high cost of programming masks is justifiable only in large runs.

EPROMs

Erasable programmable read-only memories, called EPROMs for obvious reasons, can be programmed and erased repeatedly. They are particularly useful in the development of new products where the stored information is changed as the development proceeds. One common type employs MOS devices in which the silicon gate element "floats" in an insulating oxide layer. With the proper design, a high-voltage pulse can cause avalanche breakdown and injection of avalanche-created electrons across the very thin oxide layer into the floating gate. After the programming pulse is over, the trapped negative charge enhances the channel conductivity and creates a MOSFET representing, after buffering, a stored **1**. These EPROMs are packaged with transparent quartz covers. Irradiation of the floating gates with ultraviolet light for a few minutes dissipates the trapped charge and the memory is "erased."

A Practical EPROM—The 2732

The 2732 is a 4096-word by 8-bit erasable and electrically programmable ROM suited for digital system development where experimentation with memory pattern is

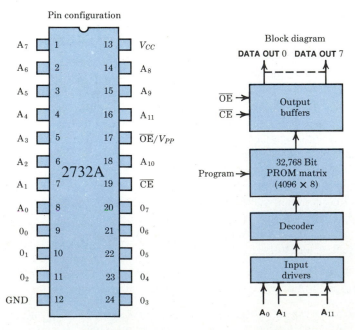

Figure 14.27   The Intel high-speed NMOS 2732A EPROM.

involved. The pinout and functional diagram for the 2732A (Intel's 250-ns access time version of the 2732) is shown in Fig. 14.27. On the 24-pin package, there are 2 power supply terminals, 12 address inputs ($A_0 - A_{11}$), 8 data outputs ($O_0 - O_7$), a chip enable pin ($\overline{CE}$), and a control/program pin (OE/V_{PP}).

As fabricated, all 32,768 bits are in the **1** state. Information is stored by selectively programming **0**s in the appropriate locations. In the programming process, each word is addressed in sequence, and a high-current pulse is applied to store the desired pattern in the 8 bits simultaneously. In practice, the complete bit pattern, called a "program," is loaded into a "PROM programmer." The microcomputer-controlled programmer runs through the addressing-pulsing sequence 32 times and verifies the stored program by checking against the loaded program—all in less than two minutes. To erase the program, the chip under the transparent quartz lid is exposed to high-intensity ultraviolet light for 15 to 20 minutes.

INFORMATION PROCESSING

In the typical digital system there are several sources of information and several destinations. Providing a transfer path for every possible combination of source and destination is usually not feasible.

Routing and Transfer

The practical approach is to provide a means for selecting the desired source and transferring its information to a common path or *bus* and providing a means for connecting the desired destination to the bus. This approach (Fig. 14.28) is called

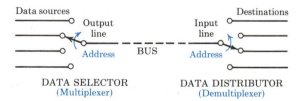

Figure 14.28   A bus-organized information transfer system.

"multiplexing." A multiplexer or *data selector* selects the desired source and places its information on the bus; a demultiplexer or *data distributor* transfers information on the bus to the selected destination.

Decoder/Demultiplexer/Data Distributor. The decoder of Fig. 14.22c translates the ROM address and places one of four 16-bit words on the ROM output lines. In general, decoders convert binary information from one coded form to another. The unit that controls the numerals displayed by a digital clock is a "BCD-to-seven-segment" decoder; decimal numbers coded in binary form are converted to provide readout of the corresponding seven-segment decimal figure. Also available are "BCD-to-decimal" decoders that drive one of ten indicator lamps in accordance with the BCD input address.

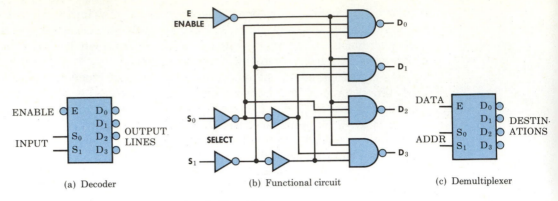

(a) Decoder   (b) Functional circuit   (c) Demultiplexer

Figure 14.29   A 2×4 line decoder/demultiplexer.

As shown in Fig. 14.29, the same unit can serve as a decoder or as a data distributor, depending on how the terminals are interpreted. When enabled by **E** going **LOW**, the decoder places a **0** on the **OUTPUT** line corresponding to the **INPUT** code; all other output lines remain **HIGH**. In Fig. 14.29c, **DATA** from the bus is applied to the **E** terminal of the demultiplexer and appears on the **DESTINATION** line selected by the **ADDRESS**.

A data distributor has one input line and many output lines. The TTL 75154 is a 4×16 line decoder/demultiplexer. In the 24-pin package, there are 4 address lines, 16 output lines, an enable input, a data input, and 2 power supply leads. It can decode a 4-bit address to send one of the 16 output lines **LOW**, or it can distribute input data to the addressed output line.

Multiplexer/Data Selector.   A multiplexer has many input lines and one output line. The multiplexer of Fig. 14.30 has four possible information sources; it chooses the one source whose address is on the select terminals. The information on the chosen

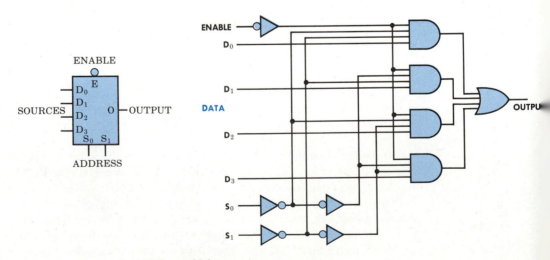

Figure 14.30   A 4×1 line multiplexer.

input line is transferred to the output line when the device is enabled. This is a "data selector" in that it selects a data source and transfers its data to the output. The TTL 74150 selects one of 16 data sources; the 74151 selects one of eight. The TTL 74153 is a dual 4 × 1 line multiplexer that can route 2-bit data from one of four sources to a 2-bit bus.

The basic logic circuit design approach outlined earlier proceeds from a function statement to a truth table and then to a logic expression that can be manipulated into the simplest form using Boolean algebra or Karnaugh mapping. As another approach, a ROM can replace a network of logic gates if the address input is identified as a set of binary variables and the stored words as the corresponding output function. In complex problems, the great saving in design time may justify the high cost of a ROM. In many problems of moderate complexity, the best approach is to save design time by using a data selector as a decision-making circuit. As shown in Example 15, logic circuit design is reduced to simply generating the truth table by placing **1**s and **0**s on the data selector inputs as required. Since the inputs can be easily changed, the data selector becomes, in effect, an EPROM.

EXAMPLE 15

Design a vote taker by using a 74151 1-of-8 data selector (Fig. 14.31).

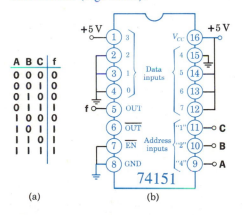

A	B	C	f
0	0	0	0
0	0	1	0
0	1	0	0
0	1	1	1
1	0	0	0
1	0	1	1
1	1	0	1
1	1	1	1

(a)   (b)

Figure 14.31   Data selector logic design.

The 74151 selects one of eight inputs and transfers the data on the selected input (or its complement) to the output. For normal operation, the enable pin must be **LOW**.

To synthesize a logic function, variables **A**, **B**, and **C** are applied to the address inputs. The data inputs, corresponding to rows of the truth table, are connected **LOW** or **HIGH** as required by the truth table of the vote taker. For **ABC = 011** → 3D, Data input 3 is connected to logic **1** → 5 V, etc. With **ENABLE** held **LOW**, an input of **011** on the address lines will result in f = 1; action will be taken in accord with the majority vote.

Note: A 1-of-8 data selector can generate any logic function of up to four variables by using the "folding" technique described on p. 141 of the Lancaster reference at the end of Chapter 13.

Computer Organization

An electronic digital computer is a vast assemblage of simple logic gates and memory devices organized to perform complex calculations at high speed by automatically processing information in the form of electrical pulses. The information consists of *data* and *instructions;* a complete set of instructions is called a *program*.

The functional organization of a computer is illustrated in Fig. 14.32 on page 390. Although the designs of individual computer components vary widely, all general-purpose computers follow the same general plan.

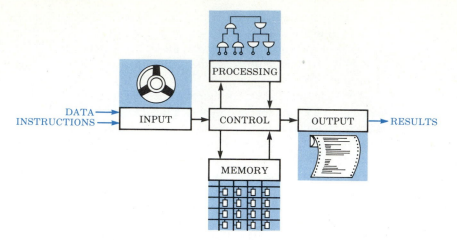

Figure 14.32 Functional components of a general-purpose digital computer.

Input

Instructions and data are entered into the machine through an input device that reads data on punched cards, magnetic tape, or magnetic disks and converts it to electrical waveforms. All communication from the operator to the machine is through the input device. Most commonly this is a magnetic tape reader or a disk drive.

Memory

Each instruction or datum is "stored" in the computer memory until it is "retrieved" for use. Flip-flops are used for temporary storage of data and instructions in active registers. The working computer memory usually consists of an array of ICs capable of storing millions of bits of information.

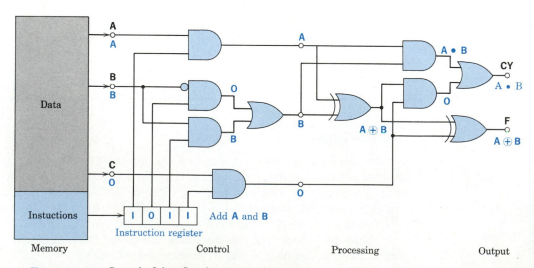

Figure 14.33 Control of data flow by program instructions.

Control

The flow of instructions, data, and results is governed by the control unit, which sets the various logic gates, feeds the numerical data, and provides the clock pulses that regulate the speed of all operations. Instructions are received from the input device and stored in one part of the memory; data are stored in another part. In accordance with the program, the control unit provides for a desired sequence of arithmetic operations and retrieves the numerical data from memory.

The operation of the control unit is illustrated in the elementary example of Fig. 14.33. Here stored numbers **A** and **B** are to be operated on in accordance with instructions in the form of a 4-bit word. The instruction is drawn from memory and placed in a register associated with a logic circuit of five gates. The instruction **1011**, as shown, sets the logic gates so that **A**, **B**, and **0** are available for processing. Other instructions set the logic gates for different operations (see Exercise 40).

Processing

The processing or *arithmetic* unit consists of logic gates arranged to perform addition, complementing, and incrementing, and the associated registers for temporary storage of data or results. One basic operation in computation is the addition of two binary numbers. In adding a column, the first number is taken from memory and placed in a register, the second number is added, a third is then added to the sum of the first two, and so on. In *parallel addition*, there is a separate adder for each bit and all orders are added in one operation. In *serial* addition, numbers are added one bit at a time by a single full-adder; the numbers and the sum are stored in shift registers. In Fig. 14.33, the processing logic is instructed to "Add **A** and **B**" and output the **SUM** and **CARRY**.

Output

The computer presents its results to the operator in display, print, or graphical form by means of an output device. Cathode-ray tubes are used for temporary display of results or plots. Modern line printers can convert electrical signals to typed characters at the rate of 20 lines per second. Electrostatic printing permits even higher speeds. Results on magnetic tape or disks are readily stored for reuse. Plotters are available for graphical presentation of calculations where this is desirable.

SUMMARY

- Conversions from binary to decimal and vice versa follow simple rules.
  Binary arithmetic follows the same rules as decimal arithmetic.
  Two's complement notation simplifies arithmetic with negative numbers.
- Boolean algebra is used to describe and to manipulate binary relations.
  Boolean theorems are summarized in Tables 14-2, -3, and -4.
  DeMorgan's theorems are particularly helpful in logic circuit synthesis.

- To analyze a logic circuit, trace through the circuit from input to output, noting the function obtained at each step.
  To synthesize a logic circuit, use Boolean algebra or Karnaugh mapping to simplify the function, and realize it with standard elements.

- Useful components and their logic functions include:

$$\text{Exclusive–OR gate:} \quad f = (A + B)\overline{AB} \quad (\text{or } f = A\overline{B} + \overline{A}B)$$

$$\text{Half-adder:} \quad S = (A + B)\overline{AB} \quad C = AB$$

$$\text{Equality comparator:} \quad f = AB + \overline{A}\,\overline{B}$$

- A register is an array of flip-flops that can store binary numbers.
  In general, data may be entered and extracted in serial or parallel form.
  Practical IC registers may provide **CLEAR**, **PRESET**, and **ENABLE** capability.

- An array of flip-flops may be connected to function as a digital counter.
  Synchronous counters are faster but more expensive than ripple counters.
  Practical IC counters provide versatile operation at low cost.

- In Read-And-Write Memory (RAM), the computer can store data at any location and retrieve it at any subsequent time; k lines can address 2^k words.
  RAM units consists of a matrix of memory cells and an address decoder.
  MOS devices offer high density at low power; bipolar devices are faster.

- In Read-Only Memory (ROM), data are permanently stored as cell states.
  ROMs can act as addressable memory, code converter, or decision logic.
  PROMs are field-programmable ROMs; EPROMs are erasable PROMs.

- In a bus-organized system, a multiplexer selects a source and puts its data on the bus; a demultiplexer transfers the data to a selected destination.

- A digital computer is a complex array of logic and memory elements organized to perform binary calculations at high speed.
  Computers include input, memory, control, processing, and output sections.

REFERENCES

1. Albert Malvino, *Digital Computer Electronics,* McGraw-Hill Book Co., New York, 1977, Chapters 6 and 7.
   A clear exposition of registers, counters, and memories as employed in microcomputers.

2. John Hilburn and Paul Julich, *Microcomputers/Microprocessors,* Prentice-Hall, Englewood Cliffs, N.J., 1976, Chapters 3 and 4.
   A good discussion of number systems and memories with microcomputer applications.

(Also see references at the end of Chapter 13.)

REVIEW QUESTIONS

1. Why is the binary system useful in electronic data processing?
2. How can decimal numbers be converted to binary? Binary to decimal? To octal?
3. Explain "2's complement" notation. What are its advantages?

4. What is a "binary coded decimal"? What is decimal 476 as a BCD in the 8421 code?
5. What is Boolean algebra and why is it useful in digital computers?
6. Which of the Boolean theorems in **0** and **1** contradict ordinary algebra?

7. Explain the association and distribution rules for Boolean algebra.

8. State DeMorgan's theorems in words. When are they useful?

9. How can a logic circuit be analyzed to obtain a logic statement?

10. How can a logic statement be synthesized to create a logic circuit?

11. What function is performed by a half-adder? An **Exclusive–OR** gate?

12. Explain how binary numbers are added in a parallel adder.

13. Explain how we can design a **NAND** circuit from a truth table.

14. Explain the principle behind the numbering of columns in a Karnaugh map.

15. How can a shift register serve as a serial-to-parallel converter?

16. What is meant by a "bus-organized" system?

17. Explain the operation of a binary ripple counter.

18. Distinguish between RAM and ROM and PROM.

19. How many address lines are required to address a 1-kilobyte memory?

20. Explain the **WRITE** operation in a MOS RAM.

21. Explain the **READ** operation in a bipolar ROM.

22. Differentiate between a multiplexer and a decoder.

EXERCISES

1. (a) Write in binary the following decimals: 4, 12, 23, 31.
   (b) Write in decimal the following binaries: **00011, 00101, 01010, 10101**.

2. (a) Write in binary the following decimals: 7, 13, 21, 39, 75.
   (b) Write in decimal the following binaries: **00111, 01100, 11110, 101011**.
   (c) Convert decimals 0.875 and 0.65 to binaries.

3. Given four numbers: (a) 165_{10}, (b) **11001101B**, (c) **207Q**, and (d) **D8**$_{16}$. Convert each to equivalent numbers in three different notations.

4. Repeat Exercise 3 for: (a) 95_{10}, (b) **10110101B**, (c) **126Q**, (d) **7E**$_{16}$.

5. Repeat Exercise 3 for: (a) 113_{10}, (b) **11001110B**, (c) **226Q**, (d) **C5**$_{16}$.

6. (a) Convert the number **247Q** to binary, decimal, hex, and BCD.
   (b) Convert the number **6B**$_{16}$ to binary, decimal, octal, and BCD.

7. Convert the BCD **10010011** to decimal, binary, and octal.

8. (a) Using signed 2's complement notation, express as 8-bit words the decimal numbers 45, 126, -30, and -48.
   (b) Show in binary notation the arithmetic operations: $126 - 45$, $45 + (-30)$, and $45 - 48$.
   (c) Write out a brief explanation of how subtraction is accomplished using 2's complement notation.

9. (a) Using signed 2's complement notation, express as 8-bit words the decimal numbers 25, 121, -17, and -96.

   (b) Show in binary notation the arithmetic operations: $121 - 25$, $25 + (-17)$, and $25 - 96$.
   (c) Write out a brief explanation of how subtraction is accomplished using 2's complement notation.

10. In general, what is the 2's complement of the 2's complement of binary number **A**?

11. Use truth tables to prove the following Boolean theorems:
    (a) **A + 1 = 1**
    (b) **AB + AC = A(B + C)**
    (c) **A + B = B + A**
    (d) **A(BC) = (AB)C**

12. Use Boolean theorems to prove the following identities:
    (a) $\mathbf{ABC + AB\overline{C} = AB}$
    (b) $\mathbf{A(\overline{A} + B) = AB}$
    (c) $\mathbf{AB + \overline{A}C = (A + C)(\overline{A} + B)}$
    (d) $\mathbf{(A + C)(A + D)(B + C)(B + D) = AB + CD}$
    (e) $\mathbf{(A + B)(\overline{A} + C) = AC + \overline{A}B}$

13. When writing equations for "Programmed Array Logic" circuits, complicated expressions must be broken down into simple "sums-of-products" (like Eq. 14-6).
    (a) Write the following expression as a sum-of-products. Show the Boolean theorems used during each step of simplification.

$$[(\mathbf{A \cdot B \cdot \overline{C}}) \cdot \overline{(\mathbf{A \cdot \overline{B} \cdot C})} + \mathbf{\overline{A}\,\overline{B}}] \cdot \mathbf{\overline{D}}$$

    (b) Invert the result from part (a), and factor it into a sum-of-products, showing theorems used.

14. The function **f = A + B** is to be realized using only **NAND** gates. Use DeMorgan's theorems to express **f** in terms of $\overline{C} \cdot \overline{D}$ where **C** and **D** can be expressed in terms of **A** and **B**. Draw the necessary logic circuit and check by constructing the truth table.

15. Analyze the logic circuit of Fig. 14.34 and determine **f** in terms of **A** and **B**. Simplify using Boolean algebra and check your result with a truth table.

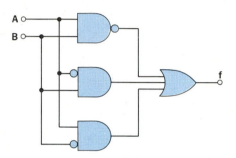

Figure 14.34

16. (a) What single gate is equivalent to the circuit of Fig. 14.35a? Check your answer using Boolean algebra.
    (b) Repeat part (a) for the circuit of Fig. 14.35b.
    (c) What theorem is illustrated in parts (a) and (b)?

17. The 8212 I/O (input/output) port consists of 8 data latches with noninverting three-state buffers as shown in Fig. 14.36. When this port (Device) is to be used (Selected) in the output Mode, the microprocessor places 8 bits of data on the **DI** lines and sends device select signal **DS = 1** and control signals **M = 1** and **S** (Strobe) = **1**. Assuming **S = 1** at all times:
    (a) Construct a truth table showing **M**, **DS**, **CK**, and **E**.
    (b) Explain the operation of this "output latch" when **M = 1**.
    (c) Explain the operation of this "gated buffer" when **M = 0**.

18. To provide a comparison of the most common logic operations, construct a table with inputs **A** and **B** and outputs **AND**, **NAND**, **OR**, **NOR**, **Exclusive–OR**, and **Equality Comparator**.

19. Synthesize logic circuits to realize the following functions as written:
    (a) $\overline{\overline{A + B} + A \cdot B} + \overline{A + B}$
    (b) $(\overline{A \cdot B} + A \cdot B) \cdot \overline{A + B}$
    (c) $(A + B) + \overline{AB} \cdot \overline{AB}$
    (d) $\overline{A + B} \cdot \overline{AB} + AB$

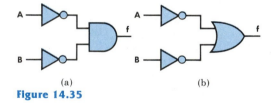

(a)   (b)

Figure 14.35

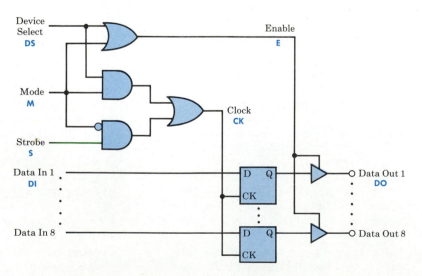

Figure 14.36 I/O port.

20. Given the logic function

$$f = \overline{AB + \overline{A}\,\overline{B}} + \overline{A}B$$

Assuming the complements are available, simplify the function using DeMorgan's theorem and synthesize it using the basic gates.

21. Map the following functions and find the minimal sum-of-products form:
(a) $ABC + \overline{A}\,\overline{B}\,\overline{C}$
(b) $AB + AC$
(c) $(A + \overline{C})(\overline{B} + \overline{C})$
(d) $ABC + BC + A\overline{B}C$

22. Evaluate the combination of the four corner squares of a four-variable Karnaugh map.

23. Map the following functions and find the minimal sum-of-products form:
(a) $ABC\overline{D} + A\overline{B}C + \overline{B}\,\overline{C}$
(b) $AB + \overline{A}\overline{B}CD + A\overline{B}C$
(c) $\overline{A}(C + D) + A\overline{B}C + A\overline{B}\,\overline{C}\,\overline{D}$

24. The array of delay flip-flops in Fig. 14.37 serves as a three-bit register.
(a) Describe the operation of this register.
(b) For the input waveforms shown, draw the waveform of Q_A, the first bit of output.

25. A 4-bit "serial-to-parallel converter" receives data one bit at a time (LSB first) and then transmits the four bits simultaneously.

(a) Using leading edge-triggered D flip-flops with active-low clear, design the converter; that is, draw a labeled connection diagram using standard flip-flop symbols.
(b) Assuming that converter outputs Q_A (LSB), Q_B, Q_C, and Q_D are initially **0**, show the output waveforms on the timing diagram of Fig. 14.38.

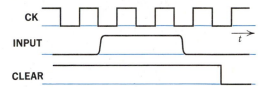

Figure 14.38

26. Design (i.e., draw a schematic diagram of) a binary counter that can count up to decimal 1000.

27. The "ring counter" of Fig. 14.39 is used in a computer to activate (**ENABLE**) several devices in turn.
(a) For input waveforms **X** and **CK** as shown, draw the output waveforms.
(b) In a truth table, show the output "data word" $Q_D Q_C Q_B Q_A$ after each downward transition of **CK**. Describe the operation of this counter.

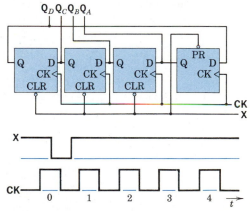

Figure 14.39 Ring counter.

28. (a) Define the operation of a leading edge-triggered D (delay) flip-flop.
(b) Construct a truth table for a synchronous divide-by-3 counter using two D flip-flops. Show "state," Q_1, Q_2, D_1, and D_2.
(c) Design the counter circuit so that Q_2 goes **HIGH** once in every 3 states.

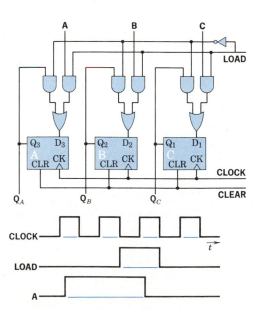

Figure 14.37 Three-bit register.

29. Design an 8421 BCD ripple counter as follows:

(a) Draw a block diagram of a 4-stage binary ripple counter using T flip-flops *ABCD* (**T** held **HIGH**).

(b) Show the truth table of flip-flop outputs for a decade counter that counts normally to decimal 9 and then resets to **0000**. How does it differ from the truth table for a binary counter in the tenth row?

(c) Modify the circuit to accomplish the following: On the eighth count, the change in state of the *D* (MSB) flip-flop is to disable the input to the *B* flip-flop so it will *not* change on count 10. (See part d.)

(d) Modify the circuit so that on the tenth count the *D* flip-flop will be reset to **0** by the output of the *A* flip-flop, without affecting the use of Q_C to toggle *D*.

(e) Check the operation of your decade counter by drawing **CK** and flip-flop waveforms.

30. In a certain application, you need a circuit that will start and stop counting clock pulses on command.

(a) Start by designing a 3-bit *synchronous* counter using positive edge-triggered JK flip-flops. Label the bits Q_A, Q_B, and Q_C, with Q_C the most significant bit.

(b) Now modify the circuit using **AND** gates so that it counts whenever the **COUNT ENABLE** lead is **HIGH**, stops counting when **CTE** goes **LOW**, and resumes counting from where it stopped when **CTE** goes **HIGH** again. Label bits as before.

31. Refer to the divide-by-12 counter of Fig. 14.21. Draw the waveforms of **INPUT** (clock), $Q_D Q_C Q_B Q_A$, **CARRY**, **LOAD**, and **OUTPUT** for one complete cycle of the counter. (See typical waveforms in Fig. 14.20.)

32. Assuming that each byte in RAM has a unique address:

(a) How many bytes can be specified by an address of 8 bits? Of 16 bits? Of *n* bits?

(b) How many memory cells are required in the RAMs of part (a)?

33. An array of eight memory cells is arranged in two rows and four columns. You are to design the addressing system consisting of a Row Decoder and a Column Decoder. Assume a row or column is selected when driven **HIGH** (logic **1**).

(a) Rows **0** and **1** are selected by Row Select (**RS**) values **0** and **1** and columns 0 to 3 similarly. Compose the truth tables for **RS** and for **CS$_1$** and **CS$_2$** (Column Select).

(b) Show the eight memory cells (lettered *A* through *D* and *E* through *H*). Design the decoders using **AND** and **NOT** gates only and show the decoder circuits on your diagram.

(c) Specify the address values that will select cell *F*.

(d) How many Row Select and Column Select lines would be required to address 1024 cells arranged in 16 rows and 64 columns?

34. Decimal numbers coded in binary are to drive the display of Fig. 14.40. (*Note:* The anodes of all LED segments are connected to +5 V; a segment is lit when its cathode is taken **LOW**, that is, logic **0**.) You are to design a "BCD-to-seven segment" code converter.

(a) Draw up a truth table showing decimal numbers, their binary codes, and the state of each segment.

(b) How many decoder input and output lines are required?

(c) Segment *b* should be lit for certain input numbers. Write the logic expression for segment *b* in terms of input variables **DCBA** (**A** is LSB).

(d) Simplify the logic expression and synthesize it using basic logic gates.

(e) Describe how a ROM could be used to perform the decoding function; specify the number of address and data lines. Draw a conclusion regarding the relative difficulty of ROM vs logic gate design.

Figure 14.40   Seven-segment decimal display.

35. Show how a ROM could be used to realize:

(a) The decoder of Fig. 14.29.

(b) The multiplexer of Fig. 14.30.

(c) The vote taker of Fig. 14.8.

36. Design (draw the circuit of) an 8-bit parallel-to-series data converter using a data selector and a clocked binary counter.
37. Using a 74151 data selector:
    (a) Design the sum circuit of the 1-bit full-adder of Fig. 14.7.
    (b) Design a 3-variable **Exclusive–OR** gate.
38. (a) Analyze the behavior of the circuit of Fig. 14.41 and draw up a truth table showing the output for values of the control lines **AB**.
    (b) Describe in words the function performed by this circuit.
    (c) Where in a computer would such a device be useful?

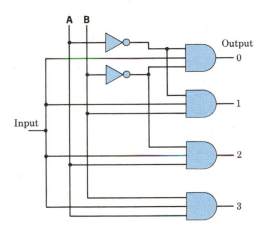

Figure 14.41

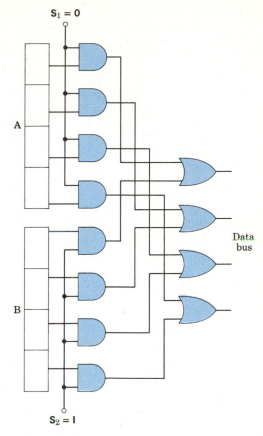

Figure 14.42

41. Refer to Fig. 14.33 and interpret the logic (ignore **CY**) instruction:
    (a) **1110**.
    (b) **1010**.

39. Four-bit registers A and B and a data bus are interconnected as shown in Fig. 14.42.
    (a) What operation is accomplished when $S_1 S_2$ is **01**?
    (b) What is the name and function of this computer component?
    (c) Where in a computer would such a component be used? For what purpose?
40. Refer to Fig. 14.33 and interpret the arithmetic instruction:
    (a) **1101** with **C = 1**.
    (b) **0101** with **C = 1**.

PROBLEMS

1. An electric light is to be controlled by three switches. The light is to be **ON** whenever switches A and B are in the same position; when A and B are in different positions, the light is to be controlled by switch C.
   (a) Draw up a truth table for this situation.
   (b) Represent the light function f in terms of **A**, **B**, and **C**.
   (c) Simplify the function and design a practical switching circuit.

2. One way of checking for errors in data processing is to "code" each word by attaching an extra bit so that the number of **1**s in each "coded" word is *even*, that is, the words all have *even parity*. Assuming that the probability of *two* errors in one word is *negligible*, the coded word can be checked for even parity at various points in the process. Assuming **Exclusive–OR** gates are available, design a *parity checker* for coded 4-bit words such that the output is **0** for even parity.

3. In a pocket calculator, each character of the 15-digit display consists of 7 light-emitting diodes arranged as shown in Fig. 14.40. For example, decimal "1" is displayed when segments *c* and *d* are lit.
   (a) Determine how each decimal from 0 to 9 could be represented and draw the corresponding segment display.
   (b) Draw up the truth table showing the state of each segment for the decimal numbers. (Assume "lit" equals logic **0**.)
   (c) If a four-stage binary counter similar to Fig. 14.17 is used to drive the display, segment *b* should be lit for certain combinations of outputs from flip-flops A through D. Write and simplify the logic expression governing segment *b*.

4. The "comparator," widely used in computers, has an output of **0** unless the inputs are identical. Synthesize, using **NAND** gates only, a circuit that "compares" the 2-digit binary numbers A_1A_2 and B_1B_2.

5. Devise a "half-subtractor."
   (a) Construct the truth table providing **A, B, Borrow**, and **Difference** columns.
   (b) Synthesize the circuit using only the five basic gates.
   (c) Synthesize the circuit assuming **Exclusive–OR** gates are available.

6. In the RAM memory cell shown in Fig. 14.43, $V_{CE(\text{sat})} = 0.3$ V and $V_{BE} \cong 0.7$ V. The word line can be switched from **LOW** (0.3 V) to **HIGH** (4 V). Currents in the data bit lines can be sensed. Assume initially that T_1 is **ON**.
   (a) With the word line **LOW**, what is voltage V_{P1}? V_{P2}? Is T_2 **ON** or **OFF**?
   (b) What is current I_D?
   (c) If now the word line is switched to **HIGH**, what is I_D? What memory function has been performed? What is the effect on T_1 of then returning the word line to **LOW**?
   (d) Now the word line and bit line *D* are simultaneously raised to **HIGH**. What is the effect on T_1? On T_2?
   (e) After returning word and bit lines to **LOW**, what memory function has been performed?

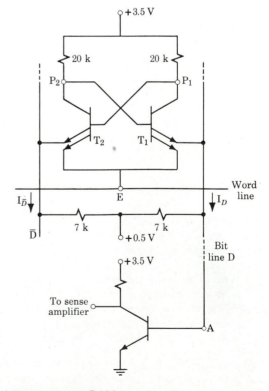

Figure 14.43   RAM memory cell.

15

Microprocessors

A Four-Bit Processor

A Stored-Program Computer

Microprocessor Programming

Practical Microprocessors

A Minimum Computer System

In 1969 Ted Hoff, an Intel Corporation engineer, was handed a difficult problem in LSI design. A Japanese calculator company wanted Intel to produce chips for an extremely complex logic design. Using his background in high density memories, Hoff came up with a bold new approach—put a tiny central processing unit (CPU) on one chip, store a program of instructions in memory on another, and use a third to move data in and out of the CPU. The resulting crude computer turned out to be too slow for its intended application, and it was rejected; but the idea was revolutionary, and it has changed the world in significant ways.

The age of electronic computers began in 1946 with ENIAC (Electronic Numerical Integrator And Computer), a 30-ton, 140-kW, 18,000-vacuum tube monster that could add two 12-digit numbers in 200 μs. In the 1950s, the vacuum tube was replaced by the transistor which, in turn, was replaced by the integrated circuit. The first microprocessor chip, a fourth-generation computer component about the size of three letters on this page and costing only a few dollars, just about matched the computing power of ENIAC. Its invention may well be the greatest technological development of our time.

The microprocessor is more than just a new device with applications in such diverse areas as automobile ignition and pollution control, traffic lights, microwave ovens, cash registers, entertainment electronics, medical diagnosis, and "smart" instruments. Its computing power, small size, and low cost have completely changed our approach to the design of all kinds of machines used in work and play, at home and in business, in the factory and in the laboratory. In particular, it has changed our philosophy of design in electronics; instead of creating an elaborate network of components to solve a data processing problem, we take a cheap, mass-produced "brain" with a flexible "memory" and "tell" it how to solve our problem.

In this brief introduction to a complex subject, we start with an elementary processor and learn how it operates and how it can be programmed. We gradually increase its complexity by adding capability and flexibility a little at a time until we are using a computer that is functionally equivalent to a commercial unit. With this background, we study a popular microprocessor and learn how to build and program a basic microcomputer.

A FOUR-BIT PROCESSOR

Let us start with a very simple data processor and see how it is organized to execute instructions. Later we shall add complexity in order to obtain more versatility.

An Elementary Processor

Some of the essential elements of a processor are shown in Fig. 15.1. Instructions in the form of 4-bit words placed in the *Instruction Register* cause the *Controller* to perform operations on the *Accumulator* register in response to signals received from the *Timer*. The Controller is a set of logic gates that guides the flow of binary data in response to the combination of bits in the Instruction Register (IR). (See Fig. 14.33.) The Accumulator is at the heart of the processor; arithmetic and logic operations are performed by the Controller on the contents of the Accumulator. Binary data represented by the *Input* switches can be transferred, on command, to the Accumulator; and the contents of the Accumulator can be transferred, on command, to the *Output* light display.

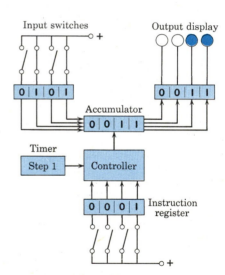

Figure 15.1 An elementary processor (after Cohn and Melsa).

The operation of the processor is illustrated in Fig. 15.2. Initially (Step 0), the Accumulator contains the binary word remaining after the last previous operation (say, **1111**), all lights in the display are **OFF** (**0000**), and the input switches are set for **0101**, for example. Note that each register always contains some 4-bit word; a register can never be "empty." The Timer is initially set at Step 0, and we assume that at each step of the Timer the Controller will execute the instruction in the IR. The first instruction is a 4-bit word placed in the IR; in this case, $2^4 = 16$ different instructions are possible by setting the four instruction switches. Let us define the instruction **0100** (in binary

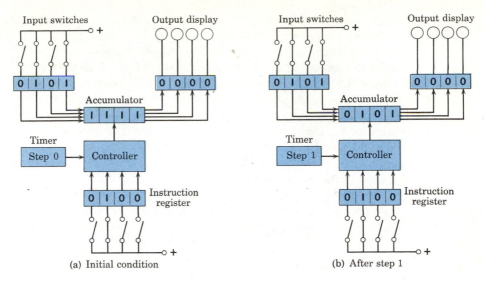

Figure 15.2 Operation of an elementary processor.

code) as **INPUT TO A**, meaning: "The data at the Input is moved to register A (the Accumulator)." At the first signal from the Timer (Step 1), the Controller "decodes" the instruction and enables the input to the A register; the Input word **0101** is transferred to the Accumulator (Fig. 15.2b).

To display the Input word at the Output, a second step is required. Let us define the instruction **0101** as **OUTPUT FROM A**, meaning: "The content of register A is moved to the Output." In Fig. 15.3a, instruction **0101** has been placed in the IR by resetting the instruction switches. At the next signal from the Timer (Step 2), the Controller decodes the instruction to enable the output of the A register, and the word **0101** is displayed at the Output (Fig. 15.3b). Note that the content of the Accumulator is unaffected by executing this instruction; this is a "nondestructive readout" of

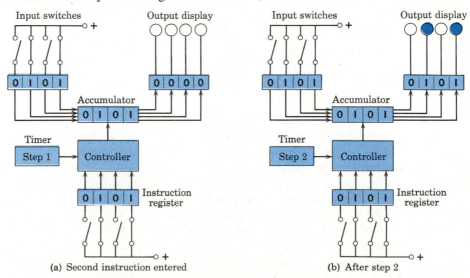

Figure 15.3 Operation of an elementary processor (continued).

information. Note also that the same binary word (**0101**) represents "data" in the Accumulator and an "instruction" in the IR.

In addition to transferring data in and out, the Controller is capable of manipulating the content of the Accumulator. One useful instruction (designated **0001** here) is **COMPLEMENT A**, meaning: "The content of register A is complemented; each zero bit becomes **1**, and each **1** becomes **0**." This instruction is useful in performing arithmetic operations.

Another useful instruction (designated **0010**) is **ROTATE A LEFT**, meaning: "Each bit in the Accumulator is shifted left one position; the low-order bit is set to the value shifted out of the high-order bit position." Similarly, the instruction **0011** can be defined as **ROTATE A RIGHT**, meaning: "Each bit in the Accumulator is shifted right one position; the high-order bit is set to the value shifted out of the low-order bit position." These instructions cause the Accumulator to act as a shift register with an end-around loop.

A Simple Program

In any microcomputer, the Controller must be told, in great detail, just what to do; in other words, it must be *programmed*. The programmer decides which tasks are to be performed in what order and defines the sequence in a written program. The program is a detailed list of specific instructions, selected from a fairly small repertoire of allowable instructions, each corresponding to an elementary step in the overall operation.

Suppose, for example, that: "The binary number at the Input is to be multiplied by two and the result displayed at the Output." Since in binary numbers a shift of one position left is equivalent to a multiplication by two, the programmer breaks the desired operation down into the following sequence and adds some explanatory "comment."

Step	CODE	INSTRUCTION	Comment
1	0100	INPUT TO A	Move data at Input to Accumulator.
2	0010	ROTATE A LEFT	Multiply content of A by two.
3	0101	OUTPUT FROM A	Move result in A to Output.

The execution of this program is illustrated in Fig. 15.4 where Input and Output are represented by registers. Here it is assumed that the Input number (limited to **0111** in this example) is **0101**, the initial content of the Accumulator is **xxxx**, and the initial setting of the Output is **yyyy** (irrelevant). The first instruction is placed in the IR and then, during Step 1, the Input number is moved to the Accumulator for manipulation. The second instruction is placed in the IR and then, during Step 2, the content of A is shifted left one position. The third instruction is placed in the IR and then, during Step 3, the Input number multiplied by two is transferred to the Output. Although this program is very simple the concepts involved are powerful and, when expanded and extrapolated, are capable of sophisticated information processing.

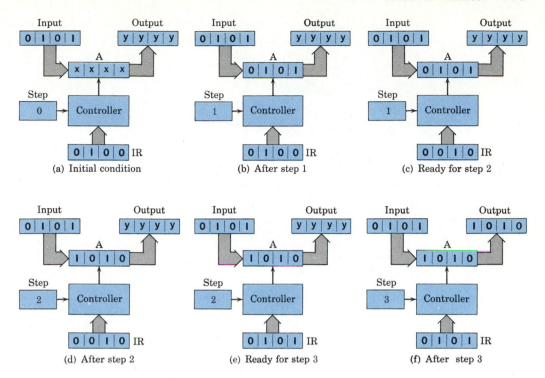

Figure 15.4   Step-by-step execution of a simple program.

A More Versatile Processor

In order to solve more challenging problems, we now increase the capability of our elementary processor by four significant additions.

Multiple I/O Ports. In general a processor can receive information or data from more than one input device and send results to more than one output device. Each device is connected to a separate "port" of the processor identified by number. Henceforth, our input/output (I/O) transfer instructions must include a second 4-bit word specifying the particular port involved. For example, let us define the two-word instruction **0100–0100** as **INPUT TO A FROM PORT 4**. Similarly, to transfer data from the Accumulator to Output Port 7, we will use **OUTPUT TO PORT 7 FROM A (0101–0111)**. Henceforth, our Controller will know that following the "operation code" ordering an I/O transfer, the next word identifies the specific port.

A Bidirectional Data Bus. To facilitate transfer of Accumulator data to and from multiple ports, we add a "bidirectional data bus," a common path (four lines in parallel) for the transmission of 4-bit words in either direction. The use of three-state buffers permits the physical connection of many devices to the common bus. (See Fig. 13.16.)

Some General Purpose Registers. As shown in Fig. 15.5, our more versatile (but still simplified) processor incorporates a data bus permitting information flow

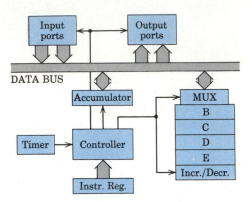

Figure 15.5 A more versatile processor.

between the Accumulator and I/O ports and also between the Accumulator and four general purpose RAM registers via a multiplexer (MUX). Registers B, C, D, and E are convenient locations for the storage of data used in computation or control. The content of each register can be incremented or decremented. Unlike the A register, however, these registers are not capable of data manipulation; for example, it is not possible to complement the content of the B register directly. Furthermore, only the A register can communicate with the I/O ports.

More Instructions. To take advantage of the additional registers, we expand the capability of the Controller to respond to instructions placed in the IR. The Instruction Set of Table 15-1 has been expanded to include three new types of instruction.

The **INCREMENT/DECREMENT** instructions are useful in counting; for example, register B (designated by the code number **0000**) could be used as a counter that would be incremented after each in a series of external events or internal program steps. The two-word instruction would be **0110–0000**.

The **LOAD** instructions permit transferring information between the Accumulator and one of the general purpose registers. For example, to transfer the content of the Accumulator to register D for temporary storage, the two-word instruction is **LOAD D FROM A**, coded as **1010–0010**. The second word of such a **LOAD** instruction is the code designation of the register specified as source or destination.

The **LOAD DATA** (sometimes called "move immediate") instructions allow data to be loaded directly into a register. For example, in counting it is sometimes desirable to start by loading the desired number of events (say, **0101**) into a register (say, B) and then count down to zero by decrementing the register. To accomplish the operation **LOAD B** with **0101** requires an operation code, a register specification, and the data; the required three-word instruction is coded as **1101–0000–0101**.

The logic instructions **AND**, **OR**, and **XOR** permit bit-by-bit operations on the content of the Accumulator and a register. In effect, four simultaneous logic operations are performed with one instruction; the results are stored in the Accumulator,

Table 15-1 Instruction Set for a 4-Bit Processor

CODE	INSTRUCTION	Meaning
0000	HALT	The processor is stopped; registers are unaffected.
0001	COMPLEMENT A	Content of A is complemented bit-by-bit.
0010	ROTATE A LEFT	Each bit in A is shifted left one position; low-order bit takes the value shifted out of high-order position.
0011	ROTATE A RIGHT	Each bit in A is shifted right one position; high-order bit takes the value shifted out of low-order position.
0100 pppp	INPUT TO A FROM p	The data at the specified port p is moved to the Accumulator.
0101 pppp	OUTPUT FROM A TO p	The content of A is moved to the specified port p without changing the content of A.
0110 rrrr	INCREMENT r	Content of register r is incremented by one. [Register codes: A(0111), B(0000), C(0001), D(0010), E(0011)]
0111 rrrr	DECREMENT r	Content of register r is decremented by one.
1000 rrrr	ADD TO A, r	Add to the content of A the content of register r without changing the content of r.
1001 rrrr	LOAD A FROM r	Transfer to the Accumulator the content of r without changing the content of r.
1010 rrrr	LOAD r FROM A	Transfer to register r the content of A without changing the content of A.
1011 rrrr	LOAD r FROM MEM	Transfer to r the content of the memory location whose address is in register E without changing MEM.
1100 rrrr	LOAD MEM FROM r	Transfer content of r to the memory location whose address is in register E without changing content of r.
1101 rrrr dddd	LOAD r WITH data	Load into register r the data d specified in the third word.
1110 LLRR	LOGIC OP A WITH R	Perform logic operation LL on content of A with content of register RR without changing content of RR. [Logic codes: OR(00), XOR(01), AND(11)] [Register codes: B(00), C(01), D(10), E(11)]
1111 cccc addr	JUMP IF cc TO addr	If condition specified in second word is met, jump to program instruction whose address is given in third word; if condition not met, continue sequentially. [Condition codes: UNcond(0000), Not Zero(1000), Zero(1001), No Carry(1010), Carry(1011), Positive(1110), Negative(1111)]

and the other register is unaffected. One simple application is in selecting the low-order bit, say, from a 4-bit number. A **0001** "mask" is moved into a register and **AND**ed with **A**; the result in A contains only the low-order bit. The instruction would be **1110–1110** to **AND (LL = 11)** the content of **D(RR = 10)** with **A**.

To provide some computation capability, the arithmetic **ADD** operation is included. By using this instruction, the content of a specified register is added to the content of the Accumulator, and the result is placed in A. As we have seen (p. 362), subtraction can be accomplished by **ADD**ing the 2's complement of the subtrahend. (*Note:* Table 15-1 contains other instructions that will be used after we learn how programs are stored in memory.)

EXAMPLE 1

Show the instructions necessary to subtract decimal 5 = **0101B** from the content of A.

CODE	INSTRUCTION	Comment
1010 0001	**LOAD** **C FROM A**	Temporarily store content of Accumulator in register C.
1101 0111 0101	**LOAD** **A WITH** **0101**	Move immediately decimal number 5 into Accumulator.
0001 0110 0111	**COMPLEMENT A** **INCREMENT** **A**	Form the 2's complement of decimal 5 in the Accumulator.
1000 0001	**ADD TO A**, **C**	Add initial content of Accumulator to 2's complement of subtrahend.

A STORED-PROGRAM COMPUTER

The processor becomes truly useful when we provide it with a memory into which we can write the program. The processor fetches the instructions from memory, decodes them to arrange the processing circuitry, executes the instructions, and then makes the results available at output ports. Using modern LSI technology, the instruction decoder, controller, accumulator, auxiliary registers, timing circuits, and I/O buffers can be fabricated on a single chip and mass produced at low cost. A customized digital system results when the mass-produced processor is provided with a tailor-made program stored in memory.

Program Storage

A memory is a set of locations where binary information can be stored. For example, a 64-bit memory could be organized into sixteen 4-bit words with each word location identified by a 4-bit address. The programmer stores the program instructions at successive locations in memory; this simplifies addressing because the next address is just the present address incremented by one.

In Fig. 15.6, a 16-word memory is shown with 4-bit address and data buses. The program used in Example 1 has been written into locations 0–9. If this were a field programmable memory (see PROM, p. 385) fabricated with all **0**s, the programmer would address each location and electrically change the **0**s to **1**s where appropriate.

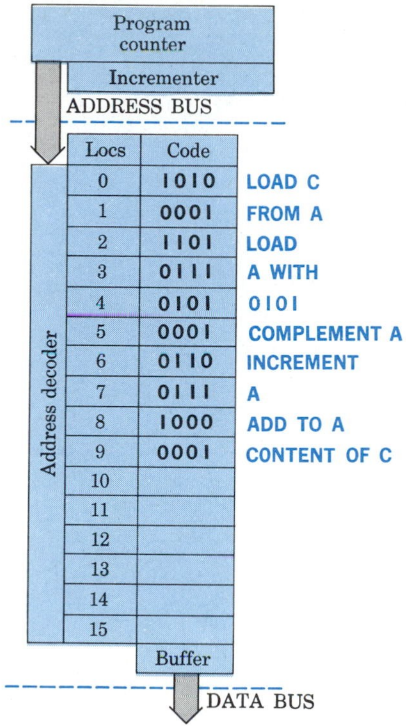

Locs	Code	
0	1010	LOAD C
1	0001	FROM A
2	1101	LOAD
3	0111	A WITH
4	0101	0101
5	0001	COMPLEMENT A
6	0110	INCREMENT
7	0111	A
8	1000	ADD TO A
9	0001	CONTENT OF C
10		
11		
12		
13		
14		
15		

Program counter / Incrementer / ADDRESS BUS / Address decoder / Buffer / DATA BUS

Figure 15.6 A program stored in ROM.

The Program Counter

The processor includes a register that contains the address of the next instruction. The processor updates this Program Counter (PC) by adding **1** to the counter content each time it fetches an instruction. We say that the PC always "points" to the next instruction. To execute the given program, the PC would be set to **0** and then incremented through nine addresses.

Timing

The activities of the processor are cyclical: fetching an instruction, transferring data, performing an operation, writing the result, fetching the next instruction, and so on. To execute an instruction in a few microseconds requires precise timing governed by a clock. In general, an interval called a "state" (equal to several clock periods) is necessary for the completion of a specific processing activity, and there are several states in a "machine cycle." The time required to fetch and completely execute an instruction is called an "instruction cycle" and includes one or more machine cycles.

The typical processor can transmit only one address during a machine cycle. The first machine cycle is always an instruction fetch; if execution requires another reference to memory or to an addressable I/O device, a second machine cycle is required.

EXAMPLE 2

Assuming the initial content of A is **1111**, show the content of the IR, A, C, and PC registers for the first two instruction cycles of the program in Example 1.

Locs	CODE	Instruction
0	1010	LOAD C
1	0001	FROM A
2	1101	LOAD
3	0111	A WITH
4	0101	0101

Instruction Cycle	IR	A	C	PC
1	xxxx	1111	yyyy	0000
	1010	1111	yyyy	0001
	0001	1111	1111	0010
2	1101	1111	1111	0011
	0111	1111	1111	0100
	0101	0101	1111	0101

A Basic Microcomputer

A stored-program computer is shown in Fig. 15.7. If the CPU components within the dashed line are fabricated on a single IC chip, this is a *microprocessor* (μP). The combination of microprocessor, IC memory devices, and I/O ports constitutes a *microcomputer* (μC).

In this more sophisticated system, a Program Counter, Incrementer, Address Bus, and external Memory are provided. Data and address buses are terminated in pins. Timing is performed by a Clock, and Timing and Control circuitry is separated from the Arithmetic/Logic Unit (ALU).

The ALU

In accordance with Control signals, the ALU performs arithmetic and logical operations on binary data. All ALUs contain an Adder that will combine the content of the Accumulator with a word on the internal data bus and place the result back on the data bus. The ALU also contains hardware that will efficiently perform incrementing, left or right bit shifts, and logic operations such as **AND, OR,** and **XOR**.

The ability of the ALU to "raise flags" signaling certain key conditions that occur in data processing greatly increases the versatility of the μP. A "conditional instruction" specifies an operation to be performed "only if" a certain condition has been satisfied; for example, if an addition results in a carry, the μP is diverted to a sequence of instructions to handle that contingency. Conditional instructions are extremely valuable in a stored-program computer because they provide decision-making capability.

In Fig. 15.7, the Flag register is set by the ALU in accordance with the result of the last instruction. Most μPs provide Zero, Carry, and Sign flags that indicate

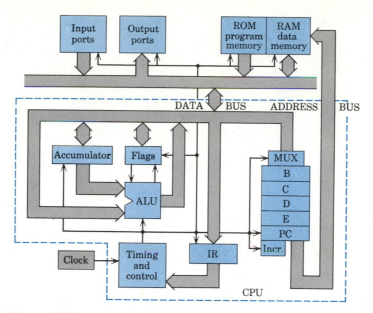

Figure 15.7 A basic microcomputer.

whether the result of the last instruction is zero (or not zero), generates a carry (or no carry), or is positive (or negative). For example, if the last instruction is an **ADD** and a carry results, "the Carry flag is set" meaning that a **1** is placed in the Carry bit of the flag register. During a conditional instruction, the status of the specified flag is tested to determine if the "branch" in the program is to be followed.

JUMP Instructions

So far we have considered only *sequential* programs that run through a series of instructions in order. **JUMP** instructions allow us to add great variety by altering the sequence in which stored instructions are executed. Since the Program Counter tells the μP where to get the next instruction, a **JUMP** instruction must change the content of the PC register. As shown in the Instruction Set of Table 15-1, the three-word **JUMP IF cc TO addr** instruction means: "If the condition specified in the second word is met, control is passed to the program instruction whose address is given in the third word." In effect, the binary word **addr** is forced into the PC. Unconditional jumps are useful in "looping" when we wish the μP to execute a part of the program over and over. Conditional jumps provide "intelligence"—the ability to make decisions and implement different programs depending on circumstances.

Data Memory

The μC of Fig. 15.7 can address either the ROM Program Memory or a RAM Data Memory. The RAM memory can function like a series of registers into which data can be written and from which data can be retrieved when needed. Our instruction set provides a flexible memory addressing scheme in which the address of the location to be accessed is stored internally in register E. To load the content of the Accumulator into memory location 6, the address is first placed in E by using the instruction **LOAD E WITH 0110**. Then the content of A is stored by using **LOAD MEM FROM A**. The **LOAD MEM** operation code (**1100**) transfers data from the specified register (A, here) to the memory location (**0110**) whose *address* is stored in E. This "indirect addressing" scheme is particularly convenient where we need to address the same memory location repeatedly or where we wish to address locations sequentially. To store or retrieve data words sequentially, the E register is simply incremented.

EXAMPLE 3

Write a program segment, starting at location **0000**, to do the following: Decrement and test the content of register B. If the content of B = **0**, store the content of C in data memory location **0110**; if not, decrement B again and repeat.

LOCS	CODE	INSTRUCTION	Comment
0000	1101	LOAD	Load into register E the
0001	0011	E WITH	specified memory
0010	0110	0110	location.
0011	0111	DECREMENT	Count by decrementing
0100	0000	B	register B.
0101	1111	JUMP	If decrement of B sets Zero
0110	1001	IF ZERO	flag, jump to "store C"
0111	1011	TO 1011	instruction.
1000	1111	JUMP	Loop back and decrement
1001	0000	UNCOND	again.
1010	0011	TO 0011	
1011	1100	LOAD MEM	Store content of C in
1100	0001	FROM C	memory at location
			whose address is in E.

Input/Output Operations

In performing its information processing function, the μP must "talk" to many devices: keyboards, sensors, disks, cathode-ray tubes, printers, and control mechanisms. Between each device and the digital system there must be an "interface" unit that transcribes parallel binary data to and from the "language" used by the device. In Fig. 15.8, we assume that each I/O device provides its own interface so that the I/O ports process binary data that is compatible with the μP signals.

A μP using the architecture shown in Fig. 15.8 would execute an **INPUT TO A FROM p** instruction by placing the port address **pppp** on the address bus and sending an **I/O READ** signal to the input *device decoder*. The 4×16 decoder ($2^4 = 16$) generates a *device select* signal to **ENABLE** the one addressed input port to place its data on the data bus. The μP addresses an I/O port and reads the content of the port buffer just as it addresses a memory location and reads its content.

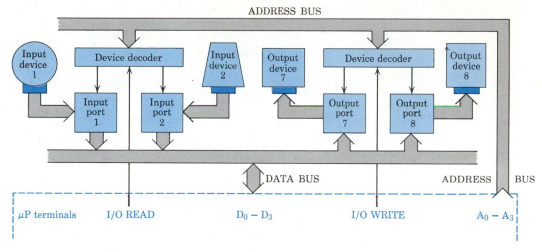

Figure 15.8 Input/output control connections.

MICROPROCESSOR PROGRAMMING

The ALU is capable of performing a very few operations at very high speed. Our Instruction Set indicates that the ALU in combination with various registers and data processors can be controlled by binary operation codes to perform 16 unique instructions. Following these instructions, our μP can transfer data inside the CPU, perform logical and arithmetic manipulations, make decisions based on manipulation results, and move data into and out of the μP.

The μP cannot, by itself, perform simple multiplication or figure percentage. It must be told in unambiguous terms, in complete detail, just what to do and in just what order. The physical components, connections, and logic circuits that decode and execute the operation codes constitute the computer *hardware*. The lists of specific instructions, selected from those allowed by the μP manufacturer and organized to control computer operations, are *software*.

Programming Technique

Programming is the art of directing the processor hardware to perform the elementary steps in solving a computation or control problem in an effective way. In writing a program (in general), the programmer:

- Defines the problem to be solved.
- Determines a feasible solution plan.
- Draws up a flow chart identifying the sequence of operations.
- Writes the program instruction by instruction.
- Checks out the program to see that it solves the problem.

Flow Charts

A flow chart is a block diagram representing the structure of the program. The essential symbols are rectangles (with one input and one output), used for processes, and diamonds (with two or more outputs), used at decision points. Lines connecting the blocks indicate the flow of information. The flow chart is a valuable aid in writing a program and explaining it to someone else. Flow charts visually break large, complex programs into smaller logical units and make it easier to make changes and corrections. With the program structure clearly defined in English, the flow chart can be converted directly into a set of instructions, using any programming language.

Programming technique can be illustrated using Example 3. In general terms, a delay is to be introduced into a μP-controlled operation.

Define the Problem. The programmer converts this to: "N instruction cycles after a certain event, the content of register C is to be stored in a specified memory location."

Determine a Solution Plan. One possibility is: "Use register B as a counter and assume an appropriate count has been loaded into B. Load the specified address (**0110**) into register E. Decrement and test register B. If the content of B = **0**, store the content of C in data memory location **0110**; if not, decrement B again and repeat the test."

Draw a Flow Chart. The solution plan is displayed graphically in Fig. 15.9. The process rectangles contain brief English expressions for the operations performed; the decision diamond contains the question: "Does Counter B = **0**?" The two outputs are labeled "Yes" and "No" for convenience. The other labels outside the blocks are also for convenience in identifying points in the program.

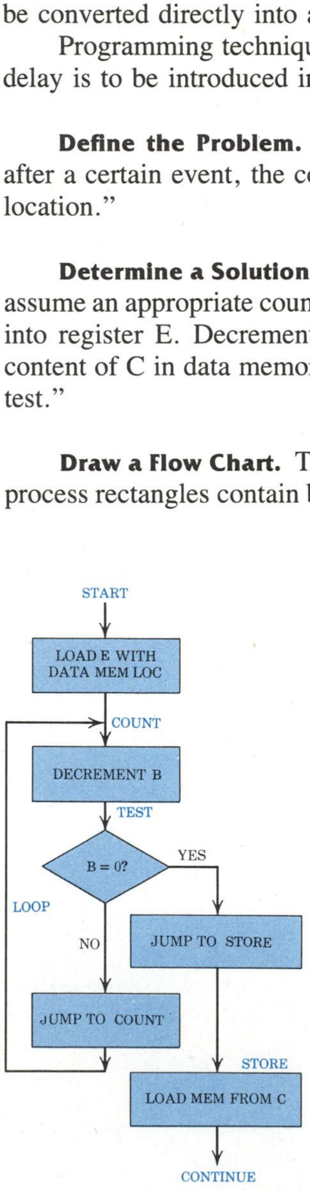

Write the Program. This can be done directly from the flow chart; the results should agree with Example 3. The decision alternatives, the repeating loop, and the **JUMP TO STORE** command, all seen clearly in the flow chart, are more obscure in the written program. The more elaborate flow chart of Example 4 includes two methods of counting and two types of jump commands. The next step would be to write the program.

Figure 15.9 Flow chart for Example 3.

EXAMPLE 4

A processor is to read in 3 integers from input port 4, store them in external RAM locations starting at **1000**, add the 3 integers, and send their sum to output port 5. Outline a solution plan and draw a flow chart. (Disregard overflows and memory limitations.)

Solution Plan
Use register D for sums.
Preset counter B to $16 - 3 = 13$ and count up.
Clear counter C and count up.
Use register E as a memory pointer.

Bring in integers one at a time and store them in sequential RAM locations.
Do three integers.
Fetch integers from memory one at a time and add them.
Do three integers.
Output the sum.

Flow Chart
The solution plan leads to the flow chart of Fig. 15.10.

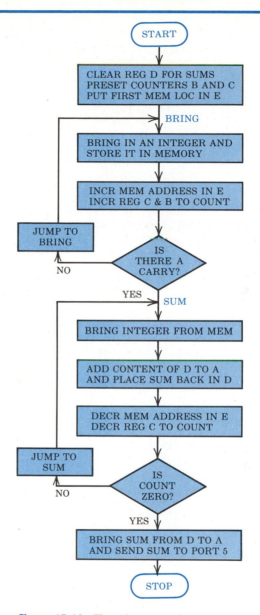

Figure 15.10 Flow chart example.

Check the Program. The smallest error may make a program completely ineffective. The identification and elimination of all errors is called "debugging." Working through the flow chart and checking the written program instruction by instruction are essential steps in debugging, but the only true check is an error-free execution by the processor.

After the program has been written and debugged, it must be physically entered into ROM program memory by electrically setting **1**s and **0**s into the ROM registers.

This could be done manually for Example 3 by entering the first 4-bit location (**0000**) into the address lines of a PROM and then introducing high currents in the Code pattern (**1101**). The second location would then be addressed and the bit pattern **0011** entered, and so on until 13 memory locations were filled. The program of instructions understandable to humans would then reside in ROM in a form understandable to the machine, that is, in *machine language*.

Program Development

Looking ahead to the use of real μPs with instruction sets of a hundred or so 8-bit instructions in practical applications employing programs requiring a thousand or so memory locations, we see some disadvantages of manual programming in machine language. We are going to need help in storing an error-free program in a reasonable time, and fortunately, it is available. The principal programming aids are:

Text Editor. A text editor is a computer program that allows the programmer to make changes, correct errors, and make substitutions easily in preparing, and displaying in human readable form, an error-free program.

Assembler. A symbolic assembler is a program that automatically translates instructions in symbolic form (**DECR B**) into machine language (**0111–0000**); it also keeps track of memory locations for the programmer.

System Monitor. The system monitor is a computer program that facilitates the overall μP programming process by performing intricate supervisory operations in response to simple instructions.

Development System. Text Editor, Assembler, and Monitor programs are controlled by a small computer called a Development System (Fig. 15.11). With a teletypewriter (TTY) keyboard for entering instructions and commands and a cathode-ray tube (CRT) for displaying them, the development of programs for new μP-based products is simplified.

PROM Programmer. This is a computer-controlled development tool that will read the final "software" program and automatically load it into a PROM.

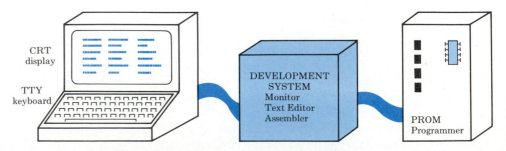

Figure 15.11 A microprocessor development system with peripherals.

Programming Languages

The set of instructions that directs any computer can be written in *machine language*, in *assembly language*, or in a *higher-level language*.

As we have seen, in machine language the "words" are sets of binary digits that can be stored in memory or moved in and out of registers along buses. Binary instruction words control sets of switches according to the architecture of the computer. In a given μP, all communication must be in the machine language fixed by the manufacturer in the fabrication process.

High-level languages such as ALGOL (algebraic language) or FORTRAN (formula translation language) simplify programming sophisticated data processing because a single symbol such as $\sqrt{}$ can represent a complex sequence of machine steps. Such high-level languages are translated into machine language by a *compiler*, a sophisticated computer program itself. High-level languages make programming easy, but they are inefficient in their use of program memory that is usually "costly" in a μC.

An assembly language uses short symbolic phrases that are understandable to people to represent specific binary instructions that are understood by the machine. The phrases, made up of alphanumeric symbols, are called *mnemonics* (memory aids). Because it is derived from English, assembly language simplifies the programmer's job, reduces the number of errors, and makes errors easier to find. Assembly language programs are efficient in their use of memory and, because they require fewer instructions, they execute faster than those written in high-level language. Assembly language is translated into machine language by an Assembler, essentially a computer-based dictionary. For a given μP, any user can devise his or her own set of mnemonics and the necessary Assembler.

Assembly Language Programming

The translation dictionary embodied in an Assembler is similar to the Code-Instruction columns of our Instruction Set for a 4-Bit Processor, but our instructions are much too long for efficient handling by a computer. If we are willing to use brief but unambiguous abbreviations and a strictly prescribed arrangement of symbols (syntax), we can greatly simplify assembling the program. For example, if we reserve "**IN**" for input operations, and if we agree to follow "**IN A**" by a comma and then the code number of the port, the instruction **INPUT TO A FROM port 2** becomes "**IN A,2.**" The role of the Assembler is indicated in Fig. 15.12.

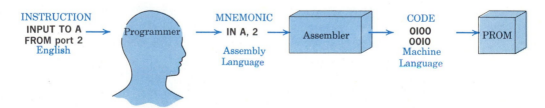

Figure 15.12   The role of the Assembler in programming.

The Assembler program includes a fixed dictionary of mnemonics and their machine language translations into binary code (Table 15-2). It also can "build" an address dictionary wherein **LABEL**s assigned by the programmer to certain instructions are identified by their addresses in memory. For example, in a **JUMP** instruction, the address of the next instruction must be specified. In the flow chart of Fig. 15.9, there is a jump to the address of the **DECREMENT B** instruction, but that address in memory is not known until after the program is assembled. However, if that instruction carries the **LABEL** "COUNT," the programmer can refer to it by **LABEL** and the Assembler will straighten it out after assembly is complete. (Labels contain no more than five characters.)

In assembly language syntax, each instruction occupies a single line. First comes the **LABEL** (if needed), then the **MNEMONIC** (always present), then any Comment the programmer finds useful in checking the program or explaining it to someone else. Sometimes the entity (data, complement, address) on which the instruction operates is identified separately as the **OPERAND** (2 in the **IN A,2** instruction); here we shall include operation and operand in the **MNEMONIC**.

The Assembler is well trained; for example, if the first entry in an instruction line is a character (in ASCII code), it is identified as the first character of a **LABEL**. If the first entry is a space, the Assembler assumes there is no **LABEL** and treats the first character as the beginning of a **MNEMONIC**. The beginning of a Comment is signaled by a special character (a semicolon here). In creating the machine language program, the Assembler completely ignores the Comment, although the Comment is carried in the display and will be printed in any printout of the assembled program.

Table 15-2 Instruction Mnemonics

CODE	MNEMONIC	Instruction	CODE	MNEMONIC	Instruction
0000	HALT	Stop processing	1010 rrrr	LOAD r,A	Load r from A
0001	COMPL A	Complement A	1011 rrrr	LOAD r,M	Load r from MEM
0010	ROTL A	Rotate A left	1100 rrrr	LOAD M,r	Load MEM from r
0011	ROTR A	Rotate A right			
0100 pppp	IN A,p	Input to A from p	1101 rrrr dddd	LDIM r,d	Load r with data
0101 pppp	OUT p,A	Output to p from A	1110 00RR	OR R,A	OR R with A
0110 rrrr	INCR r	Increment r	1110 01RR	XOR R,A	XOR R with A
0111 rrrr	DECR r	Decrement r	1110 11RR	AND R,A	AND R with A
1000 rrrr	ADD A, r	Add to A, r	1111 cccc addr	JP cc,addr	JUMP if cc to addr
1001 rrrr	LOAD A, r	Load A from r			

To provide the Assembler with needed information that it cannot deduce, special *assembler directives* are entered as mnemonics:

ORG addr—The program is to originate at address **addr** in program memory. (This information is needed to locate the labels.)

END —There are no more executable instructions forthcoming. (This tells the Assembler its task is completed.)

EXAMPLE 5

Assuming the program is to be stored in memory starting at location **0000**, show the program of Example 3 in assembly language syntax as it would be displayed after assembly.

LOCS	CODE	LABEL	MNEMONICS	Comment
			ORG 0000	;Start program at loc **0000**
0000	1101, 0011, 0110		LDIM E,0110	;Load E with spec mem loc **0110**
0011	0111, 0000	COUNT	DECR B	;Count by decrementing reg B
0101	1111, 1001, 1011		JP Z,STORE	;If decrement of B sets Zero flag, ;jump to store C instruction
1000	1111, 0000, 0011		JP UN,COUNT	;Loop back and decrement again
1011	1100, 0001	STORE	LOAD M,C	;Store C in mem at specified addr
			END	;No more instructions

In Example 5, note the following:

- Each instruction occupies a single line. (Comments may run over.)
- Only the first location for a multiword instruction is shown.
- The Assembler will "define" **COUNT** as **0011** and **STORE** as **1011**. If a new instruction is inserted between **JP Z,STORE** and **JP UN,COUNT**, the Assembler will automatically redefine **STORE**.
- Ordinarily no Comment is required for **ORG** and **END** directives.

PRACTICAL MICROPROCESSORS

The basic function of a μP is to perform arithmetic, logic, and control operations on data obtained from input devices; the results are made available to various output devices, all in accordance with a stored program of instructions. The 4-bit processor that we have studied in detail can perform all the basic functions, but it is severely limited in the size of data words handled (4-bit) and the number of instructions (16).

Practical Limitations

Ideally, a μP of negligible size and cost should accept unlimited amounts of data in any form, respond to programs of any length composed of instructions of infinite variety, deal with any number of peripheral devices, and perform all operations instantaneously without consumption of power. Although available μPs approximate these ideal characteristics, there are practical limitations within which μP manufacturers must work.

A major constraint on the design of LSI components is the limited number of terminals available for external connection. Because most industrial testers will not accept components with more than 40 (or 42) pins, common μPs are limited to 40 pins. The number of internal elements, now 10 to 20,000 transistors on a single chip, continues to climb, but the total number of terminals for all purposes is fixed. With a minimum of 2 pins for power supply and 2 for clocking, say, only 36 pins are available for data, address, and control buses (unless a multiplexing arrangement is used). The most common result of this limitation is a design providing 8 bidirectional data lines, 16 address lines, and 12 control lines (Fig. 15.13). A device with 8 data lines is said to have a "word size" of 8 bits or one byte. (See Example 6.)

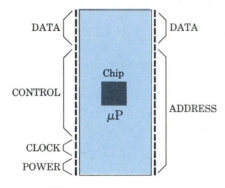

Figure 15.13   A 40-pin DIP package.

Because instructions are carried over the data bus, the instruction code of an 8-bit processor would allow a maximum of $2^8 = 256$ different instructions. For practical reasons, all bit combinations are not convenient, and the number of unique instructions is usually only one-third to one-half of the total possible.

EXAMPLE 6

(a) The Intel 8080 is an "8-bit microprocessor with 16 address lines." What is the range of signed integers that can be processed?

(a) In signed 2's-complement notation, 8-bit positive integers range from **00000000** to **01111111** or from 0 to 127D and negative integers range from **10000000** to **11111111** or from -128 to -1D. Therefore, integers from -128 to $+127$D can be processed.

(b) How many unique locations can be addressed?

(b) Since $2^{16} = 65,536$, the number of locations possible is 65,536.

The complexity of the operations performed by a μP depends on the density of circuit elements since this limits the number of internal registers and the extent of "invisible" internal microprograms that control ALU operations. In comparison to bipolar devices, MOS technology provides higher density but lower speed. The newer NMOS technology used in the Intel 8085, Motorola 6800, and Zilog Z80, for example, is faster than p-channel MOS because of the greater mobility of the electrons employed in conduction. Clock rates of 5 MHz permit typical instruction cycles of less than 1 μs.

A complete digital system including processor (μP), program memory (ROM), data memory (RAM), and input/output ports (I/O) is called a "microcomputer" (μC). In the most common computer configuration, the CPU is on one chip and the other components are on one or more additional chips. The great virtue of microprocessors is that mass-produced, low-cost hardware can be used to solve diverse control problems through individualized programs.

It is now possible to put all essential elements (except quartz crystal and power supply) on a single LSI chip—the Intel 8048, for example. This is truly "a computer on a chip," and it is an amazingly versatile computer considering its small size and low cost. Furthermore, since the program is internal, the lines ordinarily reserved for addressing are available for external communication. On the other hand, the chip area devoted to memory and I/O is unavailable for computation, and the ROM portion must be individually masked during fabrication. As a result of these limitations, the principal application of one-chip microcomputers is in simple control devices (requiring small memories) that are produced in very large numbers (justifying the high mask costs).

The Z80

Microprocessor technology is changing rapidly and devices described here are likely to become obsolete in a short time. However, at the time of its introduction (1976), the Zilog Z80 CPU was an 8-bit microprocessor of advanced design and it continues to provide great versatility and high speed in an elegantly simple configuration. It uses n-channel depletion-mode MOS technology (Fig. 12.5) that provides high element density and permits operation at 5 V with a 4-MHz clock. Clock signal and power supply use three pins, and the data bus and 16 address lines use another 24 pins, leaving 13 pins of the 40-pin DIP (Dual In-line Package) for a sophisticated set of control signals. The instruction set includes 158 distinct instructions that provide flexibility and computing power rivaling some fairly large (and expensive) minicomputers. The combination of simplicity and versatility makes the Z80 attractive for our purpose.

Registers. The architecture is similar to that of the basic CPU shown in Fig. 15.7. As shown in Fig. 15.14, there is an 8-bit Accumulator (A) that holds the result of arithmetic and logic operations and an 8-bit Flag register (F) that indicates specific operating conditions. The six general purpose registers B, C, D, E, H, and L are arranged so that they can be used as individual 8-bit registers or as 16-bit register pairs BC, DE, and HL. The HL pair can be used to store the high (H) and low (L) bytes

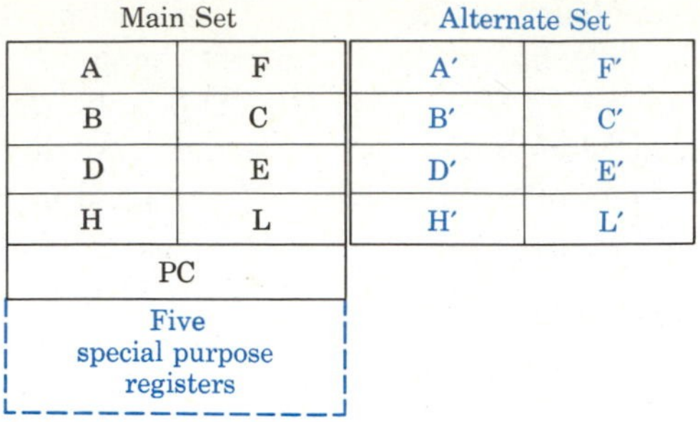

Main Set		Alternate Set	
A	F	A′	F′
B	C	B′	C′
D	E	D′	E′
H	L	H′	L′
PC			
Five special purpose registers			

Figure 15.14 Register set of the Z80.

of a 16-bit address in RAM memory for a transfer-to-or-from-memory instruction. The Program Counter (PC) is a 16-bit register that contains the address of the next instruction. The other five special-purpose registers provide increased programming and operating convenience; their use requires more background than is justified in this introduction.

Alternate Registers. The alternate set of Accumulator, Flag, and general purpose registers is particularly useful in systems where an I/O device can *interrupt* the program when it is ready to be serviced. For example, an output device such as a line printer may be slow in comparison to the μP controlling it. An efficient way to handle such a device is to supply it with one byte of information and then let the μP continue with other control or computation activities as defined by the program. When the device has finished printing a character, say, it sends an *interrupt* signal to the μP. At the next convenient time, the μP shifts operation to the alternate set of registers, goes through the appropriate *service routine* to supply to the printer the data for the next character, and then returns to its main program. Without this duplicate set, the contents of operating registers would have to be saved by proper storage in RAM so that the main program could continue where it left off at the interruption. Because the time-consuming "save" operation is avoided, the Z80 is known for its fast response to interrupts.

ALU. Arithmetic and logic instructions executed in the ALU include:

ADD	COMPARE	INCREMENT
SUBTRACT	SHIFT LEFT	DECREMENT
AND	SHIFT RIGHT	SET BIT
OR	ROTATE LEFT	RESET BIT
XOR	ROTATE RIGHT	TEST BIT

The **SUBTRACT, SHIFT**, and **ROTATE** operations are particularly useful in integer multiplication and division. The bit manipulations are useful in setting flags or external control conditions.

Pin Description. The Z80 "pinout" is shown in Fig. 15.15. Arrowheads differentiate among input, output, and bidirectional signals. A bar over a label indicates that the pin is active low. Included are the following:

> *Address Bus* (**A₀-A₁₅** [MSB]) is a 16-bit, three-state bus. I/O addressing uses the 8 lower bits to select up to 256 input or 256 output ports.
>
> *Data Bus* (**D₀-D₇**) is an 8-bit bidirectional, three-state bus.

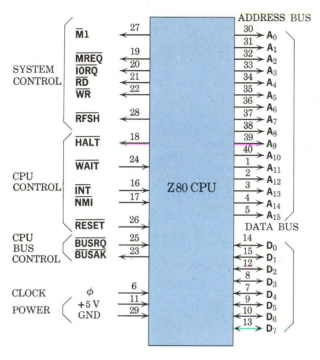

Figure 15.15 Z80 pin designation (Zilog, Inc.).

> *Machine Cycle 1* (**M1**) indicates that the CPU is in the op code fetch cycle.
>
> *Memory Request* (**MREQ**) indicates that the address bus holds a valid address for a memory read or memory write operation.
>
> *I/O Request* (**IORQ**) indicates that the address bus holds a valid address for an I/O read or write.
>
> *Read* (**RD**) indicates that the CPU wishes to read data. The device addressed uses this signal to put data on the data bus.
>
> *Write* (**WR**) indicates to memory or an I/O device that the data bus holds valid data to be stored.
>
> *Refresh* (**RFSH**) is used to refresh dynamic memories.
>
> *Wait* (**WAIT**) tells the CPU that the device addressed is not ready for a data transfer.
>
> *Halt* (**HALT**) indicates that the CPU is "stopped" (ignoring program instructions) and is awaiting an interrupt.
>
> *Interrupt Request* (**INT**) accepts the signal generated by an I/O device requiring service.

NonMaskable Interrupt (**NMI**) is a high priority interrupt to save program status in case of an "error," for example, a power failure; it forces the CPU to restart in a designated location.

Reset (**RESET**) forces the PC to zero and gets the CPU ready to start at location zero.

Bus Request (**BUSRQ**) asks the CPU to disconnect from data and address buses so that external devices can control these buses temporarily.

Bus Acknowledge (**BUSAK**) indicates that the CPU three-state data, address, and control elements are set to the high-impedance state, that is, disconnected.

EXAMPLE 7

Draw up a function table showing the Z80 output signal levels used in controlling read/write operations on memory and I/O devices.

The outputs are active low; therefore:

MREQ	IORQ	RD	WR	Function
LOW	HIGH	LOW	HIGH	Read from memory.
LOW	HIGH	HIGH	LOW	Write into memory.
HIGH	LOW	LOW	HIGH	Read from I/O.
HIGH	LOW	HIGH	LOW	Write into I/O.

Timing. In executing instructions, the Z80 steps through a series of elementary operations corresponding to *machine cycles*. Depending on its complexity, each machine cycle takes from three to six clock cycles, and complete instructions are executed in one to six machine cycles. (See Table 15-4.) Figure 15.16 shows how machine cycles make up a typical instruction cycle. The first machine cycle (**M1**) is always an instruction fetch during which the next instruction code is transferred to the instruction register. During T_1, **M1** goes **LOW**, the content of PC is placed on the address bus, then **MREQ** and **RD** go **LOW** to select the program memory (as in Example 7). During T_2, the memory responds by placing the instruction on the data bus, and

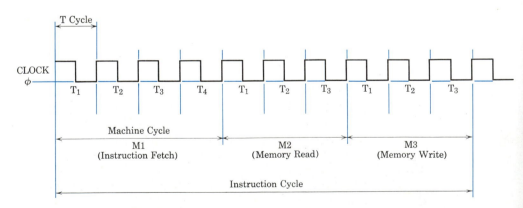

Figure 15.16 Example of Z80 timing.

then the μP loads the instruction into the IR. During T_3 and T_4, the PC is incremented, and the current instruction is decoded and executed (if no further reference to memory is required). A memory read or write generally requires three clock periods unless "wait states" are requested by the memory device with a $\overline{\text{WAIT}}$ signal. Other operations are handled in essentially the same way.

The Z80 Instruction Set

The Z80, like the Motorola 6800, evolved from the Intel 8080, a very popular 8-bit microprocessor. The Z80 can execute 158 instructions, including all 78 of the 8080 and all 16 of our 4-bit processor. For compatibility, the machine language codes for the 78 common instructions are identical in the Z80 and 8080; the mnemonics are different, however, and therefore different assemblers are required. For comparison, some examples are shown in Table 15-3.

Note that the concept of mnemonic and code developed for our simple processor is found in current microprocessors. The availability of 8 bits permits one-word instructions where two words were required with 4 bits. The **n** designates an 8-bit number. The **SSS** designates one of eight general-purpose "source" registers; a **DDD** code designates one of eight "destination" registers.

Table 15-3 Instruction Code Examples

Our 4-Bit Processor		Zilog Z80		Intel 8080	
MNEMONIC	CODE	MNEMONIC	CODE	MNEMONIC	CODE
HALT	0000	HALT	01110110	HLT	01110110
LOAD A,r	1001 rrrr	LD A,r	01111rrr	MOV A,r_2	01111sss
IN A,p	0100 pppp	IN A,(n)	11011011 ←n→	IN port	11011011 pppppppp

Note also that 8-bit codes are confusing and errors are easily made in working with them. For that reason, hexadecimal notation is used to represent instruction, data, and address codes. To convert a binary number to "hex," represent each 4-bit group by its hex equivalent. For example, the **HALT** code becomes **0111 0110 → 76**, and the address **0101101010001111** becomes **0101 1010 1000 1111 → 5A8F** in hex.

Table 15-4 on page 424 lists the mnemonics and codes for a basic set of Z80 instructions. The full list of 158 instructions, detailed timing diagrams, and suggestions for use of the more sophisticated Z80 capabilities are included in the *Z80-CPU Technical Manual* available from the manufacturer. For our purposes the abridged instruction set is more than adequate. It will permit writing a variety of programs to illustrate practical (but limited) application of microprocessors.

Table 15-4 Abridged Z80 Instruction Set

Four-Bit Processor			Z80			
Instruction	MNEMONIC	MNEMONIC	CODE[1]	Bytes	Clock Periods	Flags[2]
Stop processor	HALT	HALT	76	1	4	
*No operation		NOP	00	1	4	
Complement A	COMPL A	CPL	2F	1	4	
Rotate A left	ROTL A	RLCA	07	1	4	C ← prev MSB
Rotate A right	ROTR A	RRCA	0F	1	4	C ← prev LSB
Input to A from p	IN A,p	IN A,(n)	DB n	2	10	
Output to p from A	OUT p,A	OUT (n),A	D3 n	2	11	
Increment r	INCR r	INC r	00rrr100	1	4	Z,S ↕
Decrement r	DECR r	DEC r	00rrr101	1	4	Z,S ↕
Add to A,r	ADD A,r	ADD A,r	10000rrr	1	4	C,Z,S ↕
Load A from r	LOAD A,r	LD A,r	01111rrr	1	4	
Load r from A	LOAD r,A	LD r,A	01rrr111	1	4	
*Load r from r′		LD r,r′	01rrrr′r′	1	4	
Load r from MEM	LOAD r,M	LD r,(HL)	01rrr110	1	7	
Load MEM from r	LOAD M,r	LD (HL),r	01110rrr	1	7	
Load r with data	LDIM r,d	LD r,n	00rrr110 ←n→	2	7	
*Load HL with addr		LD HL,nn	21nn	3	10	
OR R with A	OR R,A	OR r	10110rrr	1	4	C←0; Z,S ↕
XOR R with A	XOR R,A	XOR r	10101rrr	1	4	C←0; Z,S ↕
AND R with A	AND R, A	AND r	10100rrr	1	4	C←0; Z,S ↕
*Jump uncond to addr	JP UN addr	JP nn	C3nn	3	10	
Jump if cc to addr	JP cc,addr	JP cc,nn	11ccc010 nn	3	10	
*Compare A with MEM [A − (HL)]		CP (HL)	BE	1	7	Z ↕ ; C set if A < (HL)

Notes:
*New or modified instruction.
[1] In hex where convenient; otherwise binary.
[2] C = Carry; Z = Zero; S = Sign = 1 (negative) if MSB = 1.
↕ = Flag is affected by the result of the operation.
n = 8-bit number.
Numbers in parentheses are "addresses."
The Z80 uses the lower 8 bits of a 16-bit address first.

Jump Conditions				Registers	
cc	ccc	Condition		r	rrr
NZ	000	Not Zero		B	000
Z	001	Zero		C	001
NC	010	No Carry		D	010
C	011	Carry		E	011
P	110	Sign Positive		H	100
M	111	Sign Negative		L	101
				A	111

A common application of microprocessors is processing a set of data. Figure 15.17 shows the flowchart for a typical program using the HL register pair of the Z80 as a memory location pointer. The actual program as displayed by a commercial assembler is reproduced in Table 15-5 on p. 426.

EXAMPLE 8

Assuming the program of Example 5 is to be stored in a **1 K** (1024-byte) PROM starting at location **256D = 100000000B = 0100H**, show the PROM contents and the program in Z80 assembly language syntax. (*Note:* The Z80 uses the lower 8 bits of a 16-bit address first.)

The programmer writes a specific mnemonic for each instruction, clearly identifying numbers in hex. The assembler converts the mnemonics to machine language and displays the code in hex, one byte per 8-bit location. After the assembler defines **COUNT** and **STORE**, the display is as shown in black below.

LOCS	CODE	LABEL	MNEMONICS	Bytes	Periods
			ORG 0100H		
0100	21 06 00		LD HL,0006H	3	10
0103	05	COUNT	DEC B	1	4
0104	CA 0A 01		JP Z,STORE	3	10
0107	C3 03 01		JP COUNT	3	10
010A	71	STORE	LD (HL),C	1	7
			END	11 Total	41

How long would the Z80 take to execute this 5-instruction segment?

At 4 MHz, the 41 clock periods represent

$$41 \div 4 \times 10^6 \cong 10 \ \mu s.$$

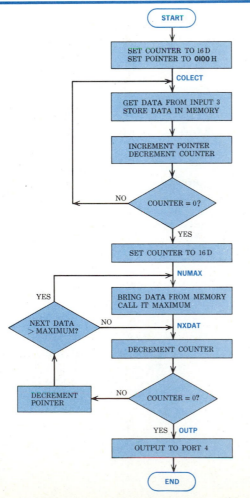

Figure 15.17 Flow chart for the program displayed in Table 15-5.

Table 15-5 Print-Out by a Commercial Assembler of a Z80 Program

DISPLAY LINE NO.			
	0001	;	CROMEMCO CDOS Z80 ASSEMBLER VERSION 02.12
	0002	;	
	0003	;	PROGRAMMING EXAMPLE
	0004	;	A COMMON APPLICATION OF MICROPROCESSORS IS PROCESSING A
	0005	;	SET OF DATA. THIS IS DONE BY COLLECTING THE DATA, STORING IT IN
	0006	;	AN "ARRAY", MANIPULATING THE DATA, AND OUTPUTTING IT. IN THE Z80
	0007	;	THIS IS FACILITATED BY USING REGISTER PAIR HL AS A MEMORY
	0008	;	LOCATION POINTER.
	0009	;	
	0010	;	PROCESSING A SET OF DATA
	0011	;	
	0012	;	WRITE A PROGRAM TO COLLECT SIXTEEN 8-BIT UNSIGNED DATA
	0013	;	NUMBERS (PRESUMED TO BE AVAILABLE AT PORT 3H), STORE THEM IN
	0014	;	MEMORY (STARTING AT LOCATION 0100H), SELECT THE LARGEST, AND
	0015	;	SEND IT TO PORT 4H.

MEM LOCS	OBJECT CODE		LABEL	MNEMONIC	COMMENT
		0016	**LABEL**	**MNEMONIC**	**COMMENT**
		0017		ORG 0000H	; START THE PROGRAM AT 0000H
0000	0E10	0018	BEGIN:	LD C,16D	; SET THE COUNTER TO 16 DECIMAL
0002	210001	0019		LD HL,0100H	; SET REGISTER PAIR HL TO LOC 0100H
		0020	;		
0005	DB03	0021	COLECT:	IN A,(3H)	; BRING DATA INTO ACCUMULATOR
0007	77	0022		LD (HL),A	; PUT IN MEM LOC POINTED TO BY HL
0008	2C	0023		INC L	; INCREMENT HL TO POINT TO NEXT LOC
0009	0D	0024		DEC C	; DECREMENT COUNTER.
000A	C20500	0025		JP NZ,COLECT	; CONT COLLECT IF C NOT EQUAL ZERO
		0026	;		
000D	0E10	0027		LD C,16D	; SET COUNTER TO 16 DECIMAL
000F	7E	0028	NUMAX:	LD A,(HL)	; BRING IN PROVISIONAL MAXIMUM
0010	0D	0029	NXDAT:	DEC C	; DECREMENT COUNTER
0011	CA1C00	0030		JP Z,OUTP	; JUMP TO OUTPUT COMMAND IF C ZERO
0014	2D	0031		DEC L	; DECR POINTER TO NEXT LOCATION
0015	BE	0032		CP (HL)	; IS NEXT DATA GREATER THAN MAX?
		0033			; THE CARRY FLAG IS SET IF (A) < (HL)
0016	DA0F00	0034		JP C,NUMAX	; IT IS, SO REPLACE PROVISIONAL MAX
0019	C31000	0035		JP NXDAT	; NO, KEEP LOOKING
		0036	;		
001C	D304	0037	OUTP:	OUT (4H),A	; OUTPUT THE TESTED MAXIMUM
001E	76	0038		HALT	; ALL DONE SO STOP
001F	(0000)	0039		END BEGIN	; TELL THE ASSEMBLER TO STOP

ERRORS 0

CROSS REFERENCE LISTING

BEGIN	0018	0039	*Notes:* Items in color have been added.
COLECT	0021	0025	Some lines are left blank to make it easier to read.
NUMAX	0028	0034	BEGIN identifies this program segment.
NXDAT	0029	0035	ERRORS refer to syntax errors.
OUTP	0037	0030	Items in "label dictionary" are listed at left.

A MINIMUM COMPUTER SYSTEM

In addition to the CPU, a working microcomputer must include clock signal, program memory, data memory, input port, output port, and power supply. For proper operation of the control system, some logic gates may also be required. We are familiar with all the components; now let us put them together in a practical Z80 microcomputer.

Clock

The only requirement is that the clock input be a square wave swinging from 0 to 5 V (nominal) at a frequency between 500 kHz and 4 MHz. At frequencies close to the maximum, the oscillator should be controlled by a quartz crystal to provide the necessary stability.

Program Memory

For flexibility let us choose a user-programmed memory such as the 2732 EPROM (see p. 386). This 4096-byte (4K) erasable/programmable memory (Fig. 15.18) is well suited to digital system experimentation. During programming, the

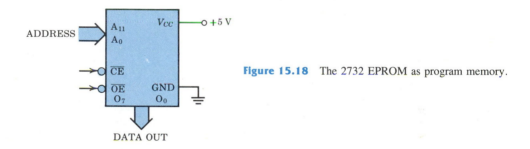

Figure 15.18   The 2732 EPROM as program memory.

$\overline{OE}/V_{PP}$ input is pulsed from a TTL level up to 21 V. In the read mode, a single 5-V power supply is required and all inputs are at TTL levels. By applying a **HIGH** signal to the $\overline{CE}$ input, the 2732A is placed in the standby mode and power is reduced by 75%.

The three-state data outputs of the 2732 allow us to physically connect the 8 Data Out pins to the data bus. When $\overline{CE}$ (Chip Enable) is taken **LOW**, this device is selected; when $\overline{OE}$ (Output Enable) is taken **LOW**, the outputs come out of the high-impedance state and place on the data bus the content of the location addressed by the 12 address lines A_0–A_{11}.

Data Memory

To store data and the results of intermediate calculations we need some read-and-write memory. For simplicity we choose a *static* RAM whose contents remain stable as long as power is available over a *dynamic* RAM, which requires "refreshing" every

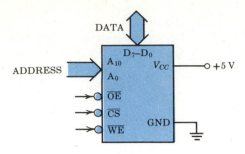

Figure 15.19 The 2116 static RAM as data memory.

millisecond or so but which is cheaper for large systems. The 2116 in Fig. 15.19 is an NMOS 2K (2048-word × 8-bit) static RAM that operates on a single 5-V supply and is TTL compatible.

During a **WRITE** cycle, the $\overline{OE}$ (Output Enable) signal is **HIGH** to place the Data pins in the input mode, signal $\overline{WE}$ (Write Enable) goes **LOW**, and the data is written into the location designated by A_0–A_{10} (Address Inputs). During a **READ** cycle, $\overline{OE}$ goes **LOW** to place the Data pins in the output mode, signal $\overline{WE}$ goes **HIGH**, and the data at the designated address is made available.

Input Port

The simplest input port is a register into which an input unit places data that, at the appropriate instant, is transferred to the data bus. The 8212 is a versatile input/output port consisting of an 8-bit data-latch register (Fig. 13.25) with three-state output buffers and control logic for operating in various modes. With the Mode terminal (**MD**) low and the Strobe terminal (**STB**) high (Fig. 15.20), the device operates as a "gated buffer." The outputs are in the high-impedance state until the device is *selected* by taking $\overline{DS}_1$ low **AND DS**$_2$ high. While the device is selected by the appropriate signal from the CPU, the input unit data are transferred to the data bus.

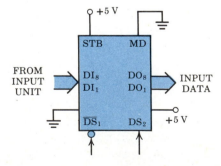

Figure 15.20 The 8212 as an input port.

Output Latch

A simple buffer is not satisfactory as an output port because the CPU operates so fast that output data are valid for only a few μs. In Fig. 15.21, with **MD** high (output

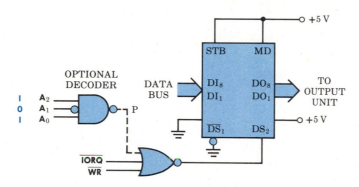

Figure 15.21 The 8212 as an output latch.

mode), the output buffers are enabled, and the device select signal ($\overline{\textbf{DS}}_1 \cdot \textbf{DS}_2$) provides the **CLOCK** signal that latches the data bus values into the output register of the 8212. While the 8212 is "selected," **CK** = **DS**$_2$ is high, and the output **Q** of each flip-flop follows the input **DI** value. When the 8212 is "unselected," **CK** goes low and **Q** remains at the latched value (see Exercise 17 in Chapter 14). The output unit can take its time processing or displaying the output data, which remain constant until the CPU again selects the output 8212.

Port Selection

The Z80 control outputs can be used to "notify" the input or output port that it has been selected. When the CPU is ready to write data at an output port, $\overline{\textbf{IORQ}}$ and $\overline{\textbf{WR}}$ go low. The output of the **NOR** gate in Fig. 15.21 goes high supplying **DS**$_2$ = **1** and selecting the output latch.

In a more flexible system with, say, eight output ports, port selection can be handled by **NAND** gate decoders. In the **OUT** (**n**),**A** instruction, the port address **n** is placed on the three lower-order lines of the address bus. In Fig. 15.21, only address **101** will take point P low and select this device. Each of the other seven ports would have its own unique decoder.

Power Supply

The Z80 requires a single 5-V supply capable of supplying 1.1 W at 5 V $\pm$ 5%. The 8212s and TTL logic gates also operate at 5 V and require less than 1 W each. The 2732 EPROM needs about 0.75 W in ROM operation and only 0.2 W in standby. The 2116 requires 0.2 W when active and only 0.025 W on standby. These power supply requirements are easily met.

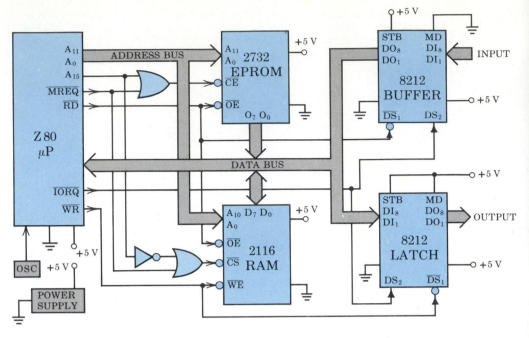

Figure 15.22 A minimum Z80-based microcomputer.

A Practical Microcomputer

The complete microcomputer shown in Fig. 15.22 is simple in layout and connection. Only a fraction of the capability of the components is used here and only part of the terminals are directly involved. In general, output terminals can be left open, but unused input terminals should be connected high (+5 V) or low (ground) depending on their characteristics.

In accordance with a 4-kilobyte program stored in the 2732 EPROM, the Z80 CPU will input 8-bit data through the 8212 buffer, use the 2116 2K RAM to process it following customized instructions, and make it available at the 8212 output latch for display or control. With a few dollars worth of chips, a few hours of construction time, and some imaginative programming, the designer can create a fast and powerful computer.

SUMMARY

- In a processor, binary instruction words placed in the Instruction Register are interpreted by the Controller, which then performs arithmetic and logic operations on the content of the Accumulator in response to signals from the Timer.
  The Controller has a limited repertoire of allowable instructions.
  A program is a detailed list of specific instructions, each corresponding to an elementary step in the overall operation.

- For greater versatility, the processor can be provided with a bidirectional Data Bus, multiple I/O ports, general purpose registers, more instructions, a Memory in which the program is stored, and a Program Counter that always contains the address of the next instruction.

- Conditional instructions, contingent upon certain key conditions that occur in data processing, provide decision-making capability when flags are raised. Jump instructions add variety by altering the sequence of instructions.
  Indirect addressing provides flexibility in storing or retrieving data from RAM.
  I/O ports are addressed and read or written into just as memory locations are.

- Programming is the art of directing the processor to perform the elementary steps in solving a computation or control problem in an effective way.
  Flow charts visually break large, complex programs into smaller logical units that are easier to follow and easier to change and correct.
  Program development is simplified by a Development System computer controlling Editor, Assembler, and System Monitor programs.

- Programs can be written in machine, assembly, or higher-level language.
  Within a μP, all communication is in machine language fixed at manufacture.
  Assembly language uses short symbolic mnemonics understandable to people to represent specific binary instructions that are understood by the machine.
  Assembly language instructions are translated into machine language by an Assembler that can also build a dictionary of Labels and their addresses in memory.

- A practical 40-pin, 8-bit μP may have 8 bidirectional data lines, 16 address lines, and 12 control lines with 4 pins available for power and clocking.
  A complete digital system including processor, program memory, data memory, input/output ports, and clock, is called a microcomputer.
  Clock rates of 5 MHz permit typical instruction cycles of less than 1 μs.

REFERENCES

1. Adam Osborne, *An Introduction to Microcomputers*, Vol. 0 *The Beginner's Book*, Vol. I *Basic Concepts*, Adam Osborne and Associates, Berkeley, California, 1976.
   A descriptive introduction with pictures and detailed examples of computer components and operations.
2. David Cohn and James Melsa, *A Step By Step Introduction To 8080 Microprocessor Systems*, Dilithium Press, Forest Grove, Oregon, 1977.
   A well-organized and clearly illustrated introduction to a popular system.
3. Rodnay Zaks, *Microprocessors: From Chips To Systems*, Sybex, Inc., Berkeley, California, 1977.
   A good introduction to microprocessor concepts and operations. Compares available devices and describes practical applications.
4. Harry Garland, *Introduction to Microprocessor System Design*, McGraw-Hill Book Co., 1979.
   A clearly written introduction to practical design; general concepts and specific applications.
   (Also see the Malvino and Hilburn references at the end of Chapter 14.)

REVIEW QUESTIONS

1. What is the role of the Instruction Register? Controller? Accumulator? Timer?

2. How does a processor know that a binary "word" is an instruction and not a number?

3. What is an instruction? A program?
4. Why are three-state buffers used in processor registers?
5. Describe two ways of using a register as a counter?
6. How can a mass-produced processor be used in a customized application?
7. Why are instructions stored sequentially?
8. Explain the phrase: "The PC always points to the next instruction."
9. What is an instruction fetch? Why does it come first?
10. List three different functions of the ALU.
11. Why are **JUMP** instructions useful?
12. Explain "indirect addressing"; "software."
13. Why are flow charts useful?

14. Differentiate between machine, assembly, and higher-level languages.
15. Differentiate between assembler and compiler.
16. How does an assembler treat **LABELS**?
17. How are assembler directives used?
18. How does a microcomputer differ from a microprocessor?
19. Write a one-sentence technical description of the Z80.
20. Define the 13 Z80 control signals.
21. How does the Z80 execute a **CPL** instruction? An **IN A,(n)** instruction?
22. Explain the following Z80 instructions: **RLCA**; **ADD A,r**; **LD r,r′**; **LD r,(HL)**; **CP (HL)**.
23. List the components of a minimum Z80-based computer and explain the function of each.

EXERCISES

1. The initial condition of an elementary processor (Fig. 15.4a) is defined by: Step = **0**, Input = **1110**, A = **1011**, Output = **0110**. The "program" consists of Step 1: **0011**, Step 2: **0001**, and Step 3: **0101**. Following Fig. 15.4, show the step-by-step execution of this program.
2. Repeat Exercise 1 for the 4-step program whose instruction codes are: **0001**, **0101**, **0100**, **0011**.
3. A simple controller for the elementary processor of Fig. 15.2 is shown in Fig. 15.23. Complete the wiring diagram so that the proper registers are **ENABLED** (with a **1**) when codes **0100** and **0101** are placed in the Instruction Register.

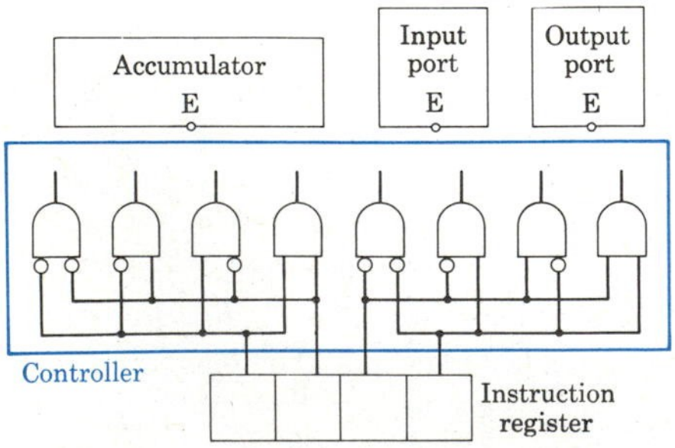

Figure 15.23

4. Write instructions (from Table 15-1) to perform the following operations. Show Code, Instruction, and Comment as in Example 1.

(a) Store data at input port 2 in register C.
(b) Complement the content of register D.
(c) Clear register D for use as a counter.
(d) "Mask off," that is, set to zero, the two MSB of a 4-bit word in the accumulator.
(e) Compare the contents of registers A and B; if they are equal, jump to program memory address **1010**. (*Hint:* Use **XOR**.)
(f) Add the numbers stored in data memory locations **1000** and **1001**, and store the sum in register B. (*Hint:* Use **LOAD r FROM MEM**.)

5. The Controller of our 4-bit processor needs an Instruction Decoder that will output a **1** to certain devices for each instruction code in the table below. Using **NOT** and **AND** gates, design and draw the logic circuit for such a decoder. Label the input and output terminals.

Instruction Code				Terminal Activated
I_4	I_3	I_2	I_1	
0	0	0	0	**HALT**
0	0	0	1	**COMPLEMENT A**
0	0	1	0	**ROTATE A LEFT**
1	1	1	1	**JUMP**

6. Write a program with comments for the 4-bit processor that:
(a) Increments register B without using the **INCR r** instruction.

(b) Decrements register B without using the **DECR r** instruction and justify including unique **INCR** and **DECR** instructions.

7. Write a program for the 4-bit processor that forms the 2's complement of the content of the A register without using the **COMPL A** instruction. (*Hint:* Use **XOR**.)

8. Using two- and three-word instructions complicates controller operation.
   (a) Explain why the complication occurs.
   (b) Suggest how a more sophisticated controller could be implemented.

9. Write instructions (from Table 15-1) to perform the following operations. Show Code, Instruction, and Comment as in Example 1.
   (a) Store the data at input port 3 in register D.
   (b) Complement the content of register C.
   (c) Clear register B for use as a counter.
   (d) "Mask off," that is, set to zero, the two LSB of a 4-bit word in the accumulator.
   (e) Compare the contents of registers C and A; if they are equal, jump to program memory address **1100**. (*Hint:* Use **XOR**.)
   (f) Add the numbers stored in data memory locations **1010** and **1011**, and store the sum in register E. (*Hint:* Use **LOAD r FROM MEM**.)

10. Decode the program below, add comments, and describe the function performed. Show the content of B and PC at the beginning of each instruction cycle.

LOCS	Code	LOCS	Code
0000	1101	0110	1001
0001	0000	0111	1011
0010	0011	1000	1111
0011	0111	1001	0000
0100	0000	1010	0011
0101	1111	1011	0000

11. (a) The programmer, by mistake, writes **0000** in location **1010** of Exercise 10. What is the effect of this "bug"? Show the content of B and PC for 12 instruction cycles.
    (b) Repeat part (a) for **0001** in location **1010**.
    (c) Comment on the effect of small errors in conditional instructions.

12. Repeat Exercise 9 for the following:
    (a) Complement the content of memory location **1111**.
    (b) Input a number from port 3, add it to the number stored in register C, and output the sum to port 2.
    (c) Subtract the number in register B from the number in register C and output the difference to port 4.
    (d) Fill data memory locations 0 to 15D with zeros.

13. Showing Code, Instruction (Table 15-1), and Comment, write a program segment to:
    (a) Simulate a "quad inverter" IC.
    (b) Multiply by 3 the number at port 4.
    (c) Reduce the number in register C by the number at port 5.
    (d) Substitute the MSB in the Accumulator for the MSB in register D.
    (e) Subtract the positive integer at port 2 from the positive integer at port 1 and output the difference at port 3.

14. Assuming the stored numbers in Exercise 4(f) are **0011** and **0100**, show the contents of the IR, A, E, B, and PC registers for each machine cycle of your addition "program." (See Example 2. Assume PC initially at **0000**.)

15. You are to write a program "routine" that will introduce a time delay of exactly 80 machine cycles. (Assume each word of instruction requires 1 machine cycle for execution.)
    (a) Outline, in general terms, a feasible solution based on Table 15-2.
    (b) Draw up a flow chart identifying the sequence of operations.
    (c) Write the program segment in assembly language syntax, showing memory locations and contents as in Example 5.

16. Repeat Exercise 15 for a program segment that will add the positive integer at input port 1 to the positive integer at port 2, and send the carry, if any, to port 4. (Assume register C initially contains **0001**.)

17. Write a Z80 program in assembly language syntax to execute the instructions of Example 4 (Fig. 15.10).

18. Write a Z80 program segment that will take a number (assumed less than 26D) from port 1, multiply it by 10, and send the product to port 2. (Use the **ROTATE** instruction.)

19. A "monostable multivibrator" (MM), when triggered, generates a single square pulse of given length. Prepare a flow chart and write a Z80 program to simulate an MM with a 1-s pulse length. Assume 1 clock period = 2 μs.

20. An alarm is to be sounded whenever the temperature sensed at input port 1 exceeds 120° C. The alarm is activated by placing the code number **99H** on output port 4. Write an appropriate Z80 program in assembly language syntax; the first instruction is to be in program memory location **0010H**.

21. One hundred 8-bit unsigned integers are stored in RAM at consecutive locations starting at **0001H**. Write a Z80 program in assembly language syntax to add the numbers five at a time, send each sum to output port 3, and stop after processing all the numbers (20 sums). Include flow chart.

22. Sixty-five signed numbers are stored in data memory locations **0100** to **0140H**. Prepare a flow chart and write a Z80 program that will instruct the processor to survey the list of numbers and tally the number of negative and positive entries in registers B and C, respectively.

23. One hundred 8-bit numbers reside in data memory at consecutive locations starting at **0101H**. The code number **FA** is known to occur at least once; you are to find the first occurrence. Prepare a flow chart and write a Z80 program (starting at program memory location **0010H**) that will search for the first occurrence of the code number **FA** and display the lower 8 bits of its address at output port 5.

24. At one point in a microcomputer operation, it is necessary to calculate 2^N where N is a positive integer available at port 1. The result is to be sent to port 2 unless the result is too large and causes an "overflow," in which case a 1 is to be displayed at port 3. After the calculation, the processor is to stop until instructed further. Draw a clearly labeled flow chart and write an efficient Z80 assembly language program to start in memory location **0020H**.

25. In one mode of operation, the Z80 can be "interrupted" and caused to perform a program segment called a "service routine," the final instruction of which is a **RETURN** to the main program.

   The pressure (in psi) in a closed vessel is 3.5 times the reading of a strain gauge (interfaced to port 4). Periodically the pressure is tested. Any time the pressure exceeds 65 psi, the positive pressure difference in psi is to be recorded on a strip chart (interfaced to port 8). Prepare a flow chart and write an efficient program from the interrupt to **RETURN** in assembly language syntax. (Disregard overflows. Let the microprocessor do the calculations.)

26. When a certain computer is turned on, it first determines the size of the RAM memory (assumed to be < 256 bytes) that is available to the CPU. Prepare a flow chart and write a Z80 program that will test each successive byte of memory, starting at **1000H**, by storing the number **10101010B** and then recalling it to see whether the number was stored correctly. The address of the last contiguous byte of working memory is to be stored in register pair HL.

27. (a) Using a decoder (Fig. 14.29) and appropriate logic, design a circuit that will select one of four output latches.

   (b) Use four additional 2 × 4 decoders in a circuit to select one of 16 output latches. (*Hint:* Use outputs of first decoder to enable the other decoders.)

CHAPTER

16

Operational Amplifiers

Inverting Circuits

Noninverting Circuits

Nonlinear Applications

Practical Considerations

The Analog Computer

I t was pointed out in Chapter 3 that the name *operational amplifier* was first applied to the amplifiers employed in analog computers to perform mathematical operations such as summing and integration. With sophisticated integrated-circuit amplifiers available for less than a dollar, the design of signal processing equipment has been radically altered. For linear or analog systems, the IC op amp plays the same basic building-block role as do the IC logic and memory elements in digital systems. In digital information processing systems, op amps are used to convert physical signals from and to the real world. Combinations of op amps and digital devices are widely used in instrumentation and control.

Our purpose here is to learn how these versatile amplifiers work and how they can be used in practical circuits. We start by reviewing ideal operational amplifiers and then look at some more circuits and applications. Next we consider some practical limitations and design aspects. Finally we examine the electronic analog computer to see how operational amplifiers can be arranged to simulate physical systems.

Ideal Operational Amplifiers

An op amp is a direct-coupled, high-gain voltage amplifier designed to amplify signals over a wide frequency range. Typically, it has two input terminals and one output terminal and a gain of at least 10^5 and is represented by the symbol in Fig.

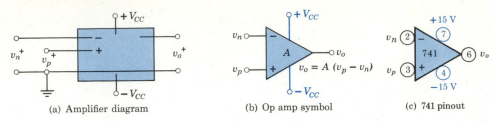

Figure 16.1 Operational amplifier diagram, symbol, and pinout.

16.1b. It is basically a *differential amplifier* responding to the difference in the voltages applied to the positive and negative input terminals. (Single-input op amps correspond to the special case where the + input is grounded.) It is normally used with external feedback networks that determine the function performed.

The characteristics of the ideal op amp are as follows:

Voltage gain $A = \infty$
Output voltage $v_o = 0$ when $v_n = v_p$
Bandwidth $BW = \infty$
Input impedance $Z_i = \infty$
Output impedance $Z_o = 0$

Although these are extreme specifications, commercially available units approach the ideal so closely that many practical circuits can be designed on the basis of these characteristics.

INVERTING CIRCUIT APPLICATIONS

In most cases, impedances Z_i and Z_o can be represented by resistances R_i and R_o as shown in the model of Fig. 16.2a. In the amplifier circuit of Fig. 16.2b, the input

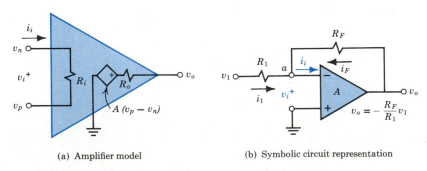

Figure 16.2 The basic inverting amplifier circuit.

signal is applied to the negative terminal and the positive terminal is grounded. The input voltage v_1 is applied in series with resistance R_1, and the output voltage v_o is fed back through resistance R_F. Because the op amp gain A is very large, $v_i = -v_o/A \cong 0$. Because R_i is very large, $i_i = v_i/R_i \cong 0$. For an ideal op amp, closely approximated by a commercial unit, $v_i = 0$ and $i_i = 0$; therefore, the sum of the currents into node a is

$$i_1 + i_F = \frac{v_1}{R_1} + \frac{v_o}{R_F} = 0 \tag{16-1}$$

and the gain of the inverting amplifier circuit is

$$\frac{v_o}{v_1} = A_F = -\frac{R_F}{R_1} \tag{16-2}$$

This basic relation has a variety of interesting interpretations.

Low-Pass Filter

A *passive filter* is a frequency-selective network consisting of passive resistors, inductors, and capacitors. (See Fig. 7.9.) Amplifiers can be used to realize *active filters* that avoid the use of bulky and expensive inductors and have other advantages. Low-cost op amps with desirable input and output properties are widely used in active filters.

As a simple example, consider Fig. 16.3 a. This is a form of the general inverting circuit with impedance elements $\mathbf{Z}_F$ and $\mathbf{Z}_1$ in the feedback network. Here $\mathbf{V}_1$ is a

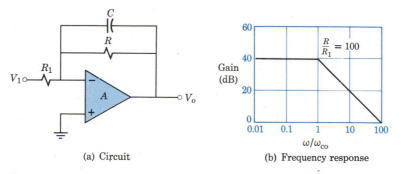

(a) Circuit (b) Frequency response

Figure 16.3 Simple low-pass filter.

sinusoidal signal of variable frequency or a combination of signals of various frequencies. The voltage gain (a transfer function here) is

$$\mathbf{A}_F = \frac{\mathbf{V}_o}{\mathbf{V}_1} = -\frac{\mathbf{Z}_F}{\mathbf{Z}_1} = -\frac{\dfrac{1}{(1/R) + j\omega C}}{R_1} = -\frac{R/R_1}{1 + j\omega RC} \tag{16-3}$$

The cutoff or half-power frequency is defined by $\omega_{co} RC = 1$, or

$$\omega_{co} = \frac{1}{RC} \tag{16-4}$$

The voltage gain decreases rapidly above ω_{co} and this serves as a *low-pass filter*.

The response curve of Fig. 16.3b, drawn for $R/R_1 = 100$, does not exhibit a sharp cutoff. More complicated circuits designed by more sophisticated methods and using several op amps can provide sharp high-, low-, or band-pass filtering. In designing active filters for high-frequency applications (above 1 MHz), the frequency response of the op amp itself must be taken into account.

Digital-to-Analog Converter

To translate a digital number to an analog signal, we need a digital-to-analog converter (DAC). Most commonly, the digital input is a binary word and the analog output is a voltage or current. For example, the 4-bit number

$$1100 = 1 \times 2^3 + 1 \times 2^2 + 0 \times 2^1 + 0 \times 2^0 = 8 + 4 + 0 + 0 = 12D \quad (16\text{-}5)$$

could be translated to an output signal amplitude of 12 V. The conversion involves a weighted sum corresponding to the output of the op amp summing circuit of Fig. 3.20.

For precise analog conversion, the bits must be represented by precise voltages, a condition that is *not* required in digital systems. Assuming that electronic switches providing precise outputs are incorporated into the digital interface register, the design of a simple DAC follows Example 1. More sophisticated circuits that provide input voltage conditioning, close voltage ratios, and fast operating speeds, are available in IC form at low cost.

EXAMPLE 1

Design a digital-to-analog converter to convert the binary number stored in a 7-bit register into a proportional voltage.

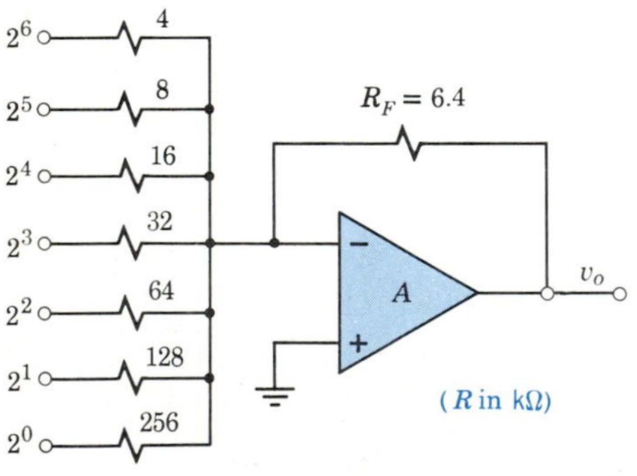

Figure 16.4 An op amp DAC.

For a summing amplifier (Eq. 3-22),

$$v_o = -\left(\frac{R_F}{R_1} v_1 + \frac{R_F}{R_2} v_2 + \cdots + \frac{R_F}{R_7} v_7 \right)$$

We assume that the register provides inputs at precisely 0 and +4 V and that the output in volts is to be equal to the decimal value $\times$ 0.1.

If only the most significant bit is set,

$$1000000B = 64D \rightarrow 6.4 \text{ V}$$

If we assume $R_F = 6.4$ kΩ, the output is

$$v_o = -\frac{R_F}{R_7} v_7 = -\frac{6.4}{R_7} 4 = -6.4 \text{ V}$$

or

$$R_7 = \frac{6.4 \times 4}{6.4} = 4 \text{ k}\Omega$$

Then $R_6 = 2 \times R_7 = 8$ kΩ, and $R_1 = 2^6 \times R_7 = 256$ kΩ as shown in Fig. 16.4.

Current/Voltage Converters

An example of op amp versatility is in driving a low-resistance load (a coil, for example) with a voltage source of high internal resistance. The *voltage-to-current converter* shown in Fig. 3.21 performs this function effectively if the load R_o can "float" with neither side grounded. If one side of the load must be grounded, the remainder of the circuit (including the op amp power supply) must be isolated from ground.

A good constant-current source for a grounded load can be made from an op amp and an external transistor. In Fig. 16.5a, reference voltage V_1 is obtained from a

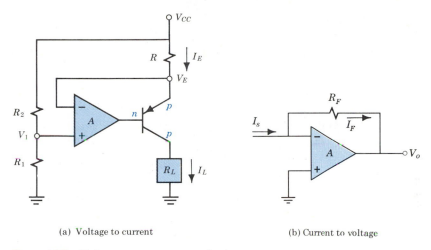

| (a) Voltage to current | (b) Current to voltage |

Figure 16.5 Voltage-current converter circuits.

voltage divider. Feedback requires that $V_E \cong V_1$; if V_E tends to fall below V_1, the forward bias on the base-emitter (np) junction decreases and emitter current I_E drops, bringing V_E back to V_1. In effect, a voltage $V_{CC} - V_1$ establishes an emitter current in the *pnp* transistor and current

$$I_L = I_C \cong I_E = \frac{V_{CC} - V_1}{R} \tag{16-6}$$

is maintained in a changing load R_L. The voltage source has been converted into a current source.

The inverse function is performed by the circuit of Fig. 16.5b. Here a given current I_s is introduced into the $-$ terminal of the op amp, setting up an equal current in the feedback resistor R_F. Since the amplifier input voltage v_i is practically zero, the output voltage is

$$V_o = -I_F R_F = -I_s R_F \tag{16-7}$$

and the source current has been converted into a voltage. The *current-to-voltage* converter is useful where a current is to be measured without introducing an undesired resistance ($R_{oF} \cong 0$) or where a current source, such as a photocell, has a high shunt resistance. (See Example 2 on page 440.)

Differential Amplifier

The *direct-coupled differential amplifier* shown in Fig. 16.7 is widely used in instrumentation. In the typical transducer, a physical variable such as a change in temperature or pressure is converted to an electrical signal that is then amplified to a useful level. The differential amplifier is advantageous because it discriminates against dc variations, drifts, and noise and responds only to the significant changes.

EXAMPLE 2

A photodiode generates $0.2 \, \mu A$ of current per μW of radiant power for a constant reverse-bias voltage. Design a circuit to measure radiant power with a voltmeter.

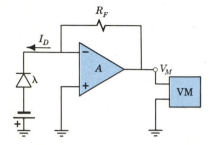

Figure 16.6 Photodiode light meter.

The photocurrent could be allowed to flow through a resistance to develop a measurable voltage, but this voltage would reduce the reverse bias and alter the photodiode characteristic.

Instead, we use the circuit of Fig. 16.6. Since $v_i \cong 0$, there is no bias voltage change with a change in I_D.

If $R_F = 5 \, k\Omega$ and the radiant power is $100 \, \mu W$, the voltmeter will read

$$V_M = I_D R_F = (100 \times 0.2) \times 10^{-6} \times 5 \times 10^3 = 100 \text{ mV}$$

and the scale factor of the "light meter" is $1 \, \mu W/mV$.

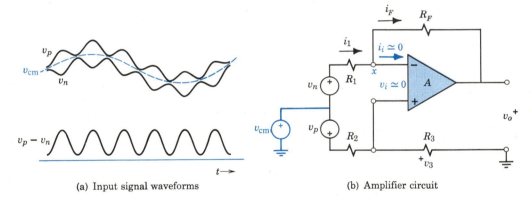

(a) Input signal waveforms   (b) Amplifier circuit

Figure 16.7 A direct-coupled differential amplifier.

In Fig. 16.7, a transducer output is represented by a *common-mode voltage* v_{cm} and a differential voltage $v_p - v_n$. Assuming an ideal op amp, $v_i = 0$ and the potential at node x is equal to the voltage across R_3, or

$$v_3 = (v_{cm} + v_p)\frac{R_3}{R_2 + R_3} = v_x \tag{16-8}$$

Since $i_i = 0$, the currents i_1 and i_F are equal, or

$$i_1 = \frac{v_{cm} + v_n - v_x}{R_1} = i_F = \frac{v_x - v_o}{R_F} \tag{16-9}$$

Combining these equations and simplifying yields

$$v_o = v_{cm}\left(\frac{R_3}{R_2 + R_3} - \frac{R_F}{R_1} + \frac{R_F}{R_1} \cdot \frac{R_3}{R_2 + R_3}\right)$$

$$+ v_p\left(\frac{R_3}{R_2 + R_3} + \frac{R_F}{R_1} \cdot \frac{R_3}{R_2 + R_3}\right) - v_n\left(\frac{R_F}{R_1}\right) \tag{16-10}$$

If we design the network so that $R_F/R_1 = R_3/R_2$, the coefficient of v_{cm} goes to zero, the *common-mode signal* is rejected, and the output is

$$v_o = -\frac{R_F}{R_1}(v_n - v_p) \tag{16-11}$$

The ability of a differential amplifier to discriminate against signals that affect both inputs in common and to amplify the small differences in input is a valuable characteristic.

NONINVERTING CIRCUIT APPLICATIONS

In Fig. 16.8, the input signal is applied to the noninverting + terminal, and a fraction of the output signal is fed back to the − terminal. Here R_1 and R_F constitute

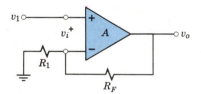

Figure 16.8 The noninverting amplifier circuit.

a voltage divider across the output voltage. For an ideal op amp with $v_i = 0$,

$$v_1 - \frac{R_1}{R_1 + R_F}v_o = v_i = 0 \tag{16-12}$$

and

$$A_F = \frac{v_o}{v_i} = \frac{R_1 + R_F}{R_1} \tag{16-13}$$

This basic noninverting amplifier has two distinctive characteristics. First, output signals are in phase with those at the input. Second, the input resistance is very high, approaching infinity in practical terms, and the output resistance is very low. This means that noninverting amplifiers do not "load" their sources and, in turn, they are not affected by their loads.

Voltage Follower

A useful special case of the noninverting circuit is shown in Fig. 16.9. Here $R_F = 0$ and $R_1 = \infty$ (open circuit). From Eq. 16-13, the circuit gain is now

$$A_F = \frac{v_o}{v_1} \cong \frac{R_1 + R_F}{R_1} = 1 \tag{16-14}$$

The output voltage is just equal to the input voltage and this is a *voltage follower,* so called because the potential at the v_o terminal "follows" the potential at the v_1 terminal.

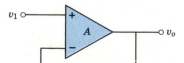

Figure 16.9 The voltage-follower circuit.

EXAMPLE 3

In the circuit of Fig. 16.10a, $R_s = 1$ kΩ and $R_L = 10$ kΩ. For the op amp, $A = 10^5$, $R_i = 100$ kΩ, and $R_o = 100$ Ω. For $v_o = 10$ V, calculate v_s and v_o/v_s and estimate the input resistance of the circuit.

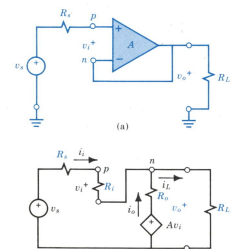

(a)

(b)

Figure 16.10 Voltage follower.

The circuit is redrawn in Fig. 16.10b to show R_i in series with the input and R_o in series with Av_i as in Fig. 16.2a. For $v_o = 10$ V,

$$i_L = \frac{v_o}{R_L} = \frac{10}{10^4} = 10^{-3} \text{ A}$$

Expecting i_i to be very small, $i_o \cong i_L$ and we write

$$Av_i = v_o + i_o R_o \cong v_o + i_L R_o$$

$$= 10 + 10^{-3} \times 10^2 = 10.1 \text{ V}$$

$$\therefore v_i = (Av_i)/A = 10.1 \times 10^{-5} \text{ V}$$

Hence $i_i = v_i/R_i = v_i/10^5 = 1.01 \times 10^{-9}$ A and the assumption regarding i_i is justified. Then

$$v_s = v_o + i_i(R_s + R_i)$$

$$= 10 + 1.01 \times 10^{-9}(1.01 \times 10^5)$$

$$= 10.0001 \text{ V}$$

$$A_F = v_o/v_s = 10/10.0001 = 0.99999$$

This is indeed a voltage follower with unity gain.

With feedback, the input resistance is

$$R_{iF} \cong \frac{v_s}{i_i} \cong \frac{10}{1.01 \times 10^{-9}} \cong 10^{10} \text{ }\Omega$$

a very high value.

Unity-Gain Buffer

Example 3 demonstrates that the gain of a voltage follower is almost exactly 1; its usefulness lies in its ability to isolate a high-resistance source from a low-resistance load. To provide this isolation, the isolating network should have a very high input resistance and a very low output resistance. In general, such an isolating network is called a "buffer." We cannot use the ideal op amp model to derive the input and output characteristics of such a *unity-gain buffer* because they depend on the nonideal properties of the op amp. Instead we use the model shown in Fig. 16.11b that includes the finite values of R_i and R_o for the op amp, and we proceed without the simplifying assumptions made in Example 3.

Input Resistance. Writing loop equations for the rearranged circuit,

$$v_1 - R_i i_1 - R_o(i_1 - i_2) - Av_1 + Av_o = 0 \tag{16-15}$$

$$v_1 - R_i i_1 - R_L i_2 = 0 \tag{16-16}$$

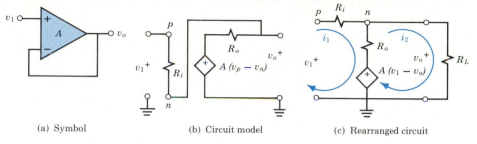

(a) Symbol   (b) Circuit model   (c) Rearranged circuit

Figure 16.11   Analysis of the unity-gain buffer circuit.

By inspection, $v_o = R_L i_2$ and, from Eq. 16-16, $i_2 = (v_1 - R_i i_1)/R_L$. Substituting these values in Eq. 16-15 yields

$$v_1 - R_i i_1 - R_o i_1 + R_o \frac{v_1 - R_i i_1}{R_L} - A v_1 + A R_L \frac{v_1 - R_i i_1}{R_L} = 0 \qquad (16\text{-}17)$$

Solving, the input resistance with feedback is

$$R_{iF} = \frac{v_1}{i_1} = \frac{R_i + R_o + (R_o/R_L)R_i + A R_i}{1 + R_o/R_L}$$

$$= \frac{R_L R_i + R_L R_o + R_o R_i + A R_L R_i}{R_L + R_o} \qquad (16\text{-}18)$$

For the practical case of A very large and $R_L \gg R_o$, this becomes

$$R_{iF} = \frac{(A + 1)R_i R_L}{R_L} + \frac{R_o(R_L + R_i)}{R_L} \cong A R_i \qquad (16\text{-}19)$$

A buffer using an op amp with $R_i = 100 \text{ k}\Omega$ and $A = 10^5$ would present an input resistance of $10,000 \text{ M}\Omega$ (as in Example 3).

   Output Resistance. Working from Fig. 16.11c, the output resistance can be obtained as v_{OC}/i_{SC}. For $i_2 = 0$,

$$v_{OC} = v_1 - R_i i_1 = v_1 - R_i \frac{v_1 - A v_1 + A v_{OC}}{R_i + R_o} \qquad (16\text{-}20)$$

Solving,

$$v_{OC} = v_1 \frac{R_i + R_o + (A - 1)R_i}{R_i + R_o + A R_i} = v_1 \frac{R_o + A R_i}{R_o + (1 + A)R_i} \qquad (16\text{-}21)$$

For the output shorted, $v_o = 0$ and

$$i_{SC} = \frac{v_1}{R_i} + \frac{A v_1}{R_o} = v_1 \frac{R_o + A R_i}{R_o R_i} \qquad (16\text{-}22)$$

For the practical case of A very large and $R_i \gg R_o$, these yield

$$R_{oF} = \frac{v_{OC}}{i_{SC}} = \frac{R_o R_i}{R_o + (1 + A)R_i} \cong \frac{R_o}{A} \qquad (16\text{-}23)$$

A buffer using an op amp with $R_o = 100 \ \Omega$ and $A = 10^5$ will present an output resistance of $0.001 \ \Omega$.

EXAMPLE 4

An instrumentation transducer is characterized by a voltage $V_T = 5$ V in series with a resistance $R_T = 2000$ Ω. It operates an indicator characterized by an input resistance of 100 Ω. Predict the voltage and power delivered to the indicator with and without the use of a buffer.

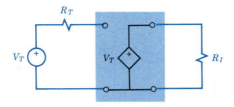

Figure 16.12 Buffer application.

Disregarding the buffer (Fig. 16.12), the voltage across the indicator would be

$$V_I = V_T \frac{R_I}{R_T + R_I} = 5 \frac{100}{2100} = 0.238 \text{ V}$$

and the power delivered would be

$$P_I = V_I^2 / R_I = (0.238)^2 / 100 = 0.566 \text{ mW}$$

With a unity-gain buffer, the output voltage follows the input. Since $R_i \rightarrow \infty$, no input current flows and, since $R_o \rightarrow 0$, there is no output voltage drop. Therefore, the model of the voltage follower is as shown. Now $V_I = V_T = 5$ V and the power is

$$P_I = V_I^2 / R_I = 5^2 / 100 = 250 \text{ mW}$$

The power gain achieved by inserting the buffer is

$$\frac{P_2}{P_1} = \frac{250}{0.566} = 442$$

Voltage Regulator

The simple Zener diode voltage regulator of Fig. 11.18 is designed to minimize the effects of supply voltage fluctuation (V_1) and load current variation (I_L) on load voltage V_L. By using a high-gain amplifier and negative feedback, much better regulation can be obtained.

In Fig. 16.13a, transistor T acts as a variable voltage dropping element in series with the load resistance R_L so that $V_{CE} = V_1 - V_L$. The feedback factor is $H =$

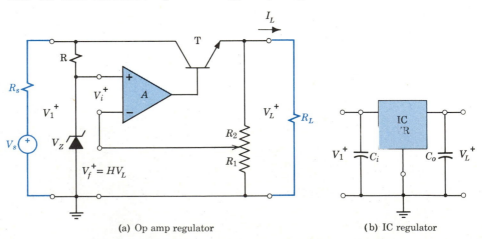

(a) Op amp regulator   (b) IC regulator

Figure 16.13 Voltage regulators with feedback.

$R_1/(R_1 + R_2)$ and the voltage fed back, $V_f = HV_L$, is applied to the inverting terminal of an op amp such as the 741. In effect, V_f is compared to a reference voltage obtained from the Zener diode; the error is amplified and used to control base current to the *pass element* T. If for any reason the load voltage tends to drop, V_f decreases slightly, V_i increases significantly, and the base current to T increases, increasing emitter current and stabilizing V_L.

The pass transistor must be able to dissipate the power $P_D = V_{CE}I_L$; an adequate heat sink is required. The variation of V_{BE} and V_Z with temperature is a principal limitation on voltage stability. Furthermore, the circuit of Fig. 16.13a provides no high-current protection; a short circuit may destroy T. These difficulties are overcome with the sophisticated circuitry available in IC form. Figure 16.13b shows a standard fixed-voltage monolithic regulator available at low cost. Capacitors C_i and C_o reduce inductive effects at the input and improve transient response at the output. Such regulators will maintain specified voltages within 0.05% against wide variations in input voltage and load current.

NONLINEAR APPLICATIONS

The ideal op amp is a linear device in that the output is directly proportional to the input for all values of input voltage. There are several important nonlinear applications of op amps; one of the simplest is the *comparator*.

Comparator

In Fig. 16.14a, if the input voltage v_1 is larger than the reference voltage V_R, the output voltage v_o is positive. Since the gain is very large, only a small voltage difference will drive the amplifier into *saturation* because the maximum output swing is limited by the supply voltage. (See Fig. 13.10.) The transfer characteristic indicates that a small decrease in v_i (a fraction of a millivolt) will drive the op amp from positive saturation to negative saturation (tens of volts).

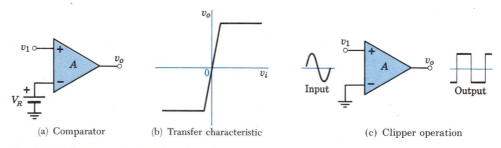

(a) Comparator   (b) Transfer characteristic   (c) Clipper operation

Figure 16.14 Nonlinear applications of an op amp.

If $V_R = 0$, this becomes a *zero-crossing comparator* (Fig. 16.14b). Such a comparator can be used to convert a sine wave into a square wave by the clipping action shown in Fig. 16.14c.

Digital Voltmeter

The availability of stable, high-gain amplifiers whose characteristics are known precisely has contributed greatly to the art of instrumentation. By the use of amplifiers, the power required from the system under observation is reduced; the power needed to drive the indicating or recording instrument is provided by the amplifier. In contrast to the traditional moving-coil instrument, an electronic voltmeter may present an input impedance of millions of ohms and follow signal variations at megahertz frequencies.

Digital voltmeters, combining flexible digital controls and precise amplifiers, offer many advantages over other voltmeters: greater speed, higher accuracy and resolution, reduced operator error, and compatibility with other digital equipment. The widely used *dual-slope technique* is illustrated in Fig. 16.15. The unknown input

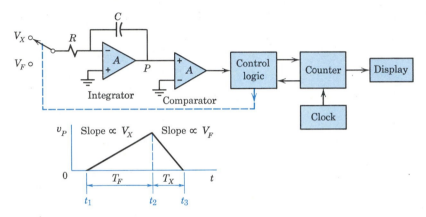

Figure 16.15 A digital voltmeter using dual-slope integration.

voltage V_X is applied to the integrator at time t_1, and capacitor C charges at a rate proportional to V_X for a fixed time T_F. At time t_2 (after a fixed number of clock pulses has been counted), the control logic switches the integrator input to fixed reference voltage V_F and C discharges at a rate proportional to V_F. The counter is reset at t_2 and counts until the comparator indicates that the integrator output voltage v_P has returned to zero at t_3. To transfer the same charge,

$$V_X T_F = V_F T_X \qquad \text{or} \qquad V_X = V_F(T_X/T_F) = kT_X$$

The count at t_3, proportional to the input voltage, is displayed as the measured voltage.

The integration process averages out noise and, by integrating over 1/60 s, 60-Hz hum is almost eliminated. Three to five readings are taken per second. By using the appropriate *signal conditioner,* currents, resistances, or ac voltages can be measured by the same instrument. Such a digital multimeter (DMM) may incorporate automatic range and polarity selection so that measurement is reduced to selecting a variable and pressing a button.

Analog-to-Digital Converter

The analog data obtained in measurements on a physical system (temperature, pressure, displacement, voltage) can be converted into digital form for processing in

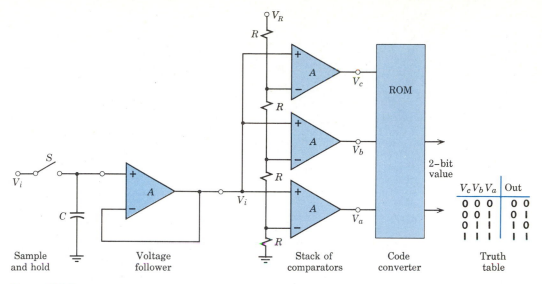

Figure 16.16 A parallel-comparator A/D converter.

various ways. One technique is illustrated in Fig. 16.16. Assuming that analog signal voltage V_i is available, the conversion is initiated by closing electronic switch S that charges capacitor C to V_i. The "sampled" input voltage is "held" after S is opened because the input resistance of the voltage follower is extremely high.

Each comparator in the stack of three compares V_i to a "threshold" voltage derived from a precision voltage divider across the fixed reference V_R. Each comparator whose threshold voltage is below the analog signal voltage is driven into saturation (Fig. 16.14b), and the output is limited to the supply voltage. For example if $V_i > \frac{1}{4}V_R$, the first comparator saturates, providing an output $V_a = V_{CC}$ taken as logic **1**. If $\frac{2}{4}V_R < V_i < \frac{3}{4}V_R$, the first and second comparators saturate, providing an input code **011** to the ROM. The ROM code converter considers this input as an address and outputs the stored digital number **10**. In a practical A/D converter providing 5-bit digital output, $2^5 - 1 = 31$ comparators are required. High-speed IC converters operating on 20-ns cycles are available at low cost.

Square-Wave Generator

A simple and inexpensive *square-wave generator* can be constructed with one op amp and a pair of back-to-back Zener diodes. In Fig. 16.17a, a capacitor C is charged through resistor R_F from output voltage v_o limited to $+V_Z$ or $-V_Z$ by the diodes in combination with R_S. The op amp compares v_1 to $\frac{1}{2}V_Z$ obtained from the voltage divider where $R_2 = R_3$ (in this case). When $v_i = \frac{1}{2}V_Z - v_1$ changes sign, v_o changes sign. Half of v_o is fed back (positive feedback) to the noninverting terminal to drive the op amp into saturation.

To see how the circuit works, assume that $R_2 = R_3$ and that $V_Z = 10$ V. At $t = 0^-$, v_1 is approaching -5 V. At $t = 0$, v_1 reaches -5.01 V, say, and v_i goes positive, sending the op amp toward positive saturation but limited to $+10$ V. Since v_1 is the voltage on C, it cannot suddenly change and, at $t = 0^+$, $v_1 \cong -5$ V. Since

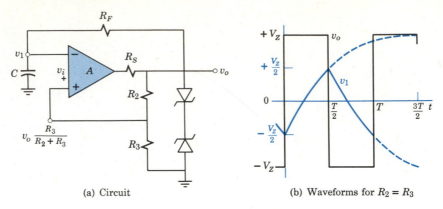

(a) Circuit   (b) Waveforms for $R_2 = R_3$

Figure 16.17   An elementary square-wave generator.

$v_o = +10$ V, the voltage now tending to force current through R_F is $v_o - v_1 = 10 - (-5) = 15$ V. The capacitor voltage will rise exponentially, or

$$v_1 = 15(1 - e^{-t/R_F C}) - 5 \qquad (16\text{-}24)$$

When v_1 exceeds $+\frac{1}{2}V_Z = +5$ V, the + input terminal is less positive than the −terminal, v_i changes sign, and v_o goes negative. Half of v_o is fed back to make v_i more negative, and v_o is driven rapidly and solidly to $-V_Z$.

In general, where $R_3/(R_2 + R_3) = H$ and $v_1 = -HV_Z$ at $t = 0$,

$$v_1 = (1 + H)V_Z(1 - e^{-t/R_F C}) - HV_Z \qquad (16\text{-}25)$$

for the first half-cycle. At $t = T/2$, $v_1 = +HV_Z$. Substituting these values in Eq. 16-25 and solving, the period is

$$T = 2R_F C \ln\frac{1 + H}{1 - H} \qquad (16\text{-}26)$$

Such a square-wave generator with symmetric Zener diodes and stable network elements will provide good square waves in the audiofrequency range.

EXAMPLE 5

Design a simple square-wave generator for operation at 200 Hz $(T = 0.005$ s) with an amplitude of ± 10 V.

Two 10-V diodes (1N4104) are connected back-to-back as in Fig. 16.17a. Assuming op amp operation at ± 15 V, approximately $15 - 10 = 5$ V is to be dropped across R_S. Therefore, for $I_Z = 1$ mA, say,

$$R_S = V/I_Z = 5/0.001 = 5000 \ \Omega$$

Selecting $C = 0.1 \ \mu\text{F}$, $R_2 = R_3 = 1 \ \text{M}\Omega$, $H = 0.5$.

$$R_F = \frac{T}{2C \ln 3} = \frac{5 \times 10^{-3}}{2 \times 10^{-7} \times 1.1} = 22,750 \cong 22 \ \text{k}\Omega$$

Triangular-Wave Generator

Circuits containing a few operational amplifiers can be used to generate pulse trains, ramps, sawtooths, sine waves, or almost any other waveform of interest. For example, consider the *triangular-wave generator* in Fig. 16.18. In response to

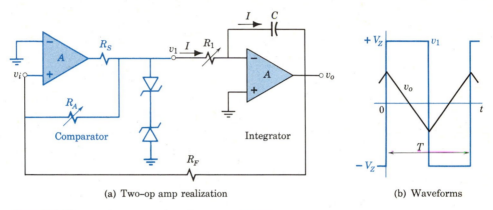

(a) Two–op amp realization   (b) Waveforms

Figure 16.18 An elementary triangular-wave generator.

changes in v_i, the comparator switches between positive and negative saturation with its output v_1 clamped at $+V_Z$ or $-V_Z$. Assume that $v_1 = +V_Z$ at $t = 0$. The current flowing into the integrator is $I = V_Z/R_1$, which also serves to charge capacitor C. The integrator output v_o is just the capacitor voltage, or

$$v_o = V_0 - \frac{1}{C} \int_0^t I \, dt = V_0 - \frac{I}{C} t \tag{16-27}$$

This is a negative-slope *ramp* voltage. A part of $v_o - v_1$ is fed back through R_F to the + terminal of the comparator. When v_i changes sign to negative, the comparator switches to negative saturation, v_1 switches to $-V_Z$, and constant current I is reversed. This initiates a positive-going ramp and v_o is triangular. The amplitude can be controlled by R_A, which adjusts the feedback factor, and the frequency can be adjusted by R_1, which controls the capacitor charging current.

PRACTICAL CONSIDERATIONS

The ideal op amp is characterized by infinite gain, infinite bandwidth, infinite input impedance, and zero output impedance. By the ingenious use of technological advances, design engineers provide at incredibly low cost commercial amplifiers that come close to the ideal.

A Four-Stage Op Amp

To provide the desired characteristics, several stages are required. In Fig. 16.19, the first stage is a *differential amplifier* with a double-ended output; this provides high input resistance, high gain for *difference* signals, and rejection of signals *common* to

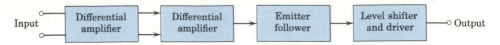

Figure 16.19 Block diagram of a four-stage operational amplifier.

both terminals. The second stage is a single-output differential amplifier that provides more gain and more discrimination. The third stage is an *emitter follower* to provide low output resistance and isolation of the amplifier from the load. The last stage is a combination *level shifter* and *driver* stage. The level shifting corrects for any dc offsets that have been introduced by bias networks or by component imbalances. The driver is a power amplifier to provide large output currents from a low output resistance.

The differential amplifier[†] is a versatile component of many electronic instruments and deserves further consideration. As shown in Fig. 16.20a, transistor T3 is

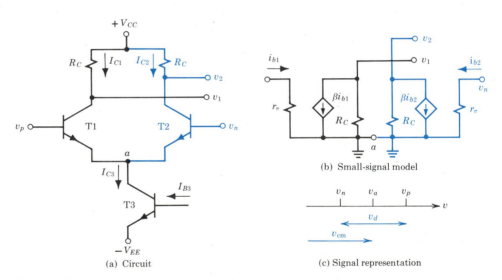

Figure 16.20 Analysis of a differential amplifier.

a constant-current source where $I_{C3} = \beta I_{B3}$ and I_{B3} is obtained from a biasing network. The remainder of the circuit is a symmetric amplifier of matched resistors and transistors; in integrated circuit form, the corresponding elements are fabricated simultaneously and are nearly identical in all respects. Qualitatively speaking, any increase in v_p increases I_{B1}, which increases I_{C1} and lowers v_1. Since $I_{C1} + I_{C2} = I_{C3} = $ constant, an increase in I_{C1} results in a corresponding decrease in I_{C2} and v_2 goes up. The output $v_o = v_2 - v_1$ responds to the *difference* in v_p and v_n. If both v_p and v_n increase due to a *common* signal, T1 and T2 would have to respond in identical fashion, but I_{C1} and I_{C2} are unaffected since their sum must be constant. Therefore, $v_2 = v_1$ and there is no output.

[†]Here we refer to an electronic circuit, a portion of an IC op amp; see the footnote on p. 70. The entire op amp can also function in a differential amplifier circuit (Fig. 16.7).

For a quantitative analysis, we consider the effect of *small changes* in V_p and V_n. Referring to Fig. 12.15a, we see that a small change in v_{BE} (from a steady value of, say, 0.7 V) produces a proportionate small change in base current Δi_B. For such a "small signal,"

$$\Delta i_{B1} = \frac{\Delta v_{BE}}{r_\pi} = \frac{v_p - v_a}{r_\pi}$$

where v_p and v_a are signal values and r_π is an effective input resistance (the reciprocal of the slope of the i_B-v_{BE} curve). Referring to Fig. 12.15b, the resulting small change in collector current will be $\Delta i_{C1} = \beta \Delta i_{B1}$. Summing the currents into node a, we write

$$\Sigma i = (1 + \beta)\Delta i_{B1} + (1 + \beta)\Delta i_{B2} = (1 + \beta)\frac{v_p - v_a}{r_\pi} + (1 + \beta)\frac{v_n - v_a}{r_\pi} = 0$$

$$(16\text{-}28)$$

Hence, $(v_p - v_a) + (v_n - v_a) = 0$, which yields

$$v_a = \frac{v_p + v_n}{2} \qquad (16\text{-}29)$$

or v_a takes on a value just halfway between v_p and v_n (Fig. 16.20c). Now let the input signals be described in terms of a common-mode signal v_{cm} and a difference signal v_d so that $v_p = v_{cm} + v_d/2$ and $v_n = v_{cm} - v_d/2$. Using the small-signal models,

$$v_1 = -\beta\frac{v_p - v_a}{r_\pi}R_C = -\frac{\beta R_C}{r_\pi}(v_{cm} - v_a + v_d/2)$$

$$v_2 = -\beta\frac{v_n - v_a}{r_\pi}R_C = -\frac{\beta R_C}{r_\pi}(v_{cm} - v_a - v_d/2)$$

and the output voltage is

$$v_o = v_2 - v_1 = \frac{\beta R_C}{r_\pi}v_d = \frac{\beta R_C}{r_\pi}(v_p - v_n) \qquad (16\text{-}30)$$

The common-mode signal is rejected and the difference signal is amplified.

Common-Mode Rejection Ratio

Ideally, a differential amplifier responds only to the difference voltage $v_p - v_n$ and not to the common-mode voltage v_{cm}. Practically, however, an op amp with an input v_{cm} produces some output v_{ocm}; the ratio

$$A_{cm} = \frac{v_{ocm}}{v_{cm}} \qquad (16\text{-}31)$$

is called the *common-mode gain*. The ratio of differential gain A to A_{cm} is called the common-mode rejection ratio, CMRR, where

$$\text{CMRR} = \frac{A}{A_{cm}} \qquad (16\text{-}32)$$

or, in decibels, CMRR(dB) $= 20 \log$ CMRR. Typical values run from 50 to 100 dB.

EXAMPLE 6

A 741 op amp has a gain $A = 2 \times 10^5$ and a CMRR of 90 dB. The input consists of a difference signal $v_p - v_n = 10 \ \mu V$ and a common-mode signal 100 times as large; that is, $v_p = 1005 \ \mu V$ and $v_n = 995 \ \mu V$. Determine A_{cm} and the output voltage.

By Eq. 16-32, the common-mode gain is

$$A_{cm} = A/10^{90/20} = 2 \times 10^5 \times 10^{-4.5}$$

$$= 2 \times 10^{0.5} = 6.3$$

By superposition, the output voltage is

$$v_o = A_d v_d + A_{cm} v_{cm}$$

$$= 2 \times 10^5 \times 10^{-5} + 6.3 \times 10^{-3}$$

$$= 2.0063 \text{ V}$$

Even in this extreme case, imperfect CMRR introduces only a 0.3% error in the output.

Input Offset Voltage

When both input terminals are grounded, the ideal op amp develops zero output voltage. Under the same conditions, a practical amplifier will have a finite output because of inevitable small imbalances in the components. By definition, the *input offset voltage* V_{OS} is the dc input voltage required to reduce the output to zero. A practical op amp can be modeled by an ideal op amp with an input voltage source equal to V_{OS}, typically a few millivolts (Fig. 16.21). For the configuration shown with

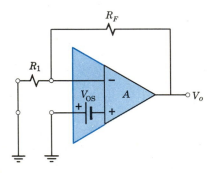

Figure 16.21 An op amp with input voltage offset.

$v_1 = 0$, V_{OS} is equivalent to a signal applied to the noninverting terminal. Therefore, the output voltage will be

$$V_o = \frac{R_1 + R_F}{R_1} V_{OS} \tag{16-33}$$

The offset voltage and its "drift" with changes in temperature and time is one of the important sources of error in op amp circuits. Op amps with appropriately low values of offset voltage must be selected or special nulling circuits must be employed. Many IC op amps provide connections to internal points for this purpose.

Bias Current Offset

The input terminals of a differential amplifier (Fig. 16.20a) carry the base bias currents for the transistors in the input stage. In an op amp circuit, these dc currents (typically a fraction of a μA) produce IR voltages that appear in the output as spurious signals. A practical op amp can be modeled by an ideal op amp with current generators in each lead (Fig. 16.22). In an IC op amp, bias currents I_{B1} and I_{B2} will be very nearly equal and their effects can be balanced out by the insertion of resistor R_2.

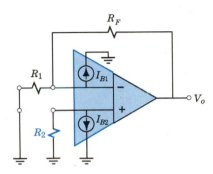

Figure 16.22   An op amp with bias current offset.

By superposition, the total effect can be predicted as the sum of the effects of each current generator considered separately. Since V_{R1} is essentially zero, all I_{B1} flows through R_F and its contribution to output voltage is

$$V_{o1} = I_{B1}R_F \tag{16-34}$$

The Thévenin equivalent of I_{B2} and R_2 is a signal $V_T = -I_{B2}R_2$ at the noninverting terminal; its contribution is

$$V_{o2} = \frac{R_1 + R_F}{R_1}(-I_{B2}R_2) \tag{16-35}$$

Hence the output offset due to bias currents is

$$V_o = V_{o1} + V_{o2} = I_{B1}R_F - I_{B2}R_2(1 + R_F/R_1) \tag{16-36}$$

To minimize the effects of the nearly equal bias currents, we equate their coefficients and solve for R_2, obtaining

$$R_2 = \frac{R_1R_F}{R_1 + R_F} \tag{16-37}$$

For this value of R_2, the remaining output offset is

$$V_o = (I_{B1} - I_{B2})R_F = I_{OS}R_F \tag{16-38}$$

where I_{OS} is the *input offset current*, which is usually much smaller than average bias current $(I_{B1} + I_{B2})/2$. As indicated in Example 7, input offset current is important in amplifier circuits with large resistance values. Input offset voltage is important in circuits with high gain.

EXAMPLE 7

An op amp with the following "typical" specifications:

Input bias current = 0.1 μA
Input offset current = 0.02 μA
Input offset voltage = 1 mV

is used in an inverting amplifier with $R_1 = 20$ kΩ and $R_F = 1$ MΩ.

Specify the proper value of balancing resistor R_2. Then predict the output offset due to bias current without R_2 and with R_2. Predict the output offset due to input offset voltage.

By Eq. 16-37,

$$R_2 = \frac{R_1 R_F}{R_1 + R_F} = \frac{(2 \times 100) \times 10^8}{(2 + 100) \times 10^4} = 19.6 \text{ k}\Omega$$

Without R_2, bias current I_{B1} produces

$$V_o = I_{B1}R_F = 1 \times 10^{-7} \times 1 \times 10^6 = 0.1 \text{ V}$$

With R_2, the input offset current produces

$$V_o = I_{OS}R_F = 2 \times 10^{-8} \times 1 \times 10^6 = 0.02 \text{ V}$$

By Eq. 16-33, the input offset voltage produces

$$V_o = V_{OS}(R_1 + R_F)/R_1$$

$$= 10^{-3}(1.02 \times 10^6)/2 \times 10^4 = 0.051 \text{ V}$$

Frequency Response

As we shall see in Chapter 19, the frequency response of an amplifier is defined by the *bandwidth*, the range of frequencies between the upper and lower *breakpoints* or "3-dB points" in a Bode plot (Fig. 9.12a).

Whereas the ideal op amp has an infinite bandwidth, the gain A of a practical op amp begins to drop off as the frequency increases. In many applications, however, the frequency response may be considered *flat* because the use of feedback greatly increases the bandwidth of the circuit.

Since time constant and frequency are reciprocal quantities, with feedback the bandwidth is increased by the same factor as the time constant is reduced, that is, by the factor $1 - GH$ (Eq. 9-8). For an op amp with $G = A = 10^5$ in a circuit where gain with feedback $A_F = G_F = G/(1 - GH) = 100$, the factor $1 - GH = G/G_F = 1000$. The bandwidth of the circuit will be 1000 times as great as the bandwidth of the op amp itself.

Stability

Actually, the high-frequency gain of an op amp is usually limited on purpose by the designer. At high frequencies, inherent capacitive effects introduce phase shift of the output signal with respect to the input. If this phase shift equals 180°, the feedback will be *positive,* and oscillation is possible if the loop gain AH is equal to 1 (see p. 233).

One way to ensure that the amplifier is free from oscillation, or *stable*, is to include an *RC* frequency compensation network. Such a network is essentially a low-pass filter (Fig. 16.3) and introduces a maximum of 90° phase shift and a *gain rolloff* of 20 dB per decade or 6 dB per octave. This gain characteristic is superimposed on the characteristic of the basic amplifier and reduces the gain to unity before a 180°

phase shift can occur. In an op amp, frequency compensation may be internal or connections may be brought out for external compensation.

Slew Rate

The bandwidth determines the ability of an operational amplifier to follow rapidly changing small signals. In addition, there are limitations on the rate at which the output can follow large signals. This results from the fact that only finite currents are available within the amplifier to charge the various internal capacitances. The maximum rate at which the output voltage can swing from most positive to most negative is the *slew rate* ρ, typically a few volts per microsecond. For example, for an op amp with a slew rate of 1 V/μs operating at $V_{CC} = \pm 15$ V, the switching time will be

$$T_{SW} \cong \frac{\text{supply voltage}}{\text{slew rate}} = \frac{30}{1} = 30 \ \mu s$$

Slew rate may be a limitation on frequency response to large sinusoids, but it is of primary importance in switching applications such as the clipper in Fig. 16.14c.

THE ANALOG COMPUTER

The modern electronic analog computer is a precision instrument. Its basic purpose is to predict the behavior of a physical system that can be described by a set of algebraic or differential equations. The programming procedure is to arrange the operational amplifiers to perform the operations indicated in the describing equations and provide a means for displaying the solution.

In addition to the op amps (identified by function in Fig. 16.23), the practical computer includes an assortment of precision resistors and capacitors, a function

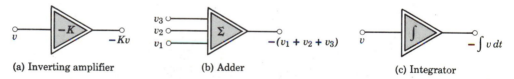

(a) Inverting amplifier   (b) Adder   (c) Integrator

Figure 16.23   Symbolic representation of operational amplifier functions.

generator to provide various inputs, means for introducing initial conditions, potentiometers for introducing adjustable constants, switches for controlling the operations, an oscilloscope or recorder for displaying the output, and a problem board for connecting the components in accordance with the program. In the hands of a skillful operator, the analog computer faithfully simulates the physical system, provides insight into the character of the system behavior, and permits the design engineer to evaluate the effect of changes in system parameters before an actual system is constructed.

Solution by Successive Integration

In one common application, the analog computer is used to solve linear integro-differential equations. To illustrate the approach, let us predict the behavior of the

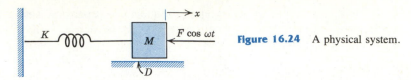

Figure 16.24 A physical system.

familiar physical system shown in Fig. 16.24. Assuming that mass is constant, that the spring is linear, and that friction force is directly proportional to velocity, the system is described by the linear differential equation

$$\Sigma f = 0 = -F \cos \omega t - M\frac{d^2x}{dt^2} - D\frac{dx}{dt} - \frac{1}{K}x \qquad (16\text{-}39)$$

and a set of initial conditions. The behavior of the system can be expressed in terms of displacement $x(t)$ or velocity $u(t)$ where $u = dx/dt$. We wish to display this behavior, so we proceed to program the computer to solve the equation.

The first step is to solve for the highest derivative. Anticipating the inversion present in the operational amplifier used, we write

$$\frac{d^2x}{dt^2} = -\left(\frac{F}{M}\cos \omega t + \frac{D}{M}\frac{dx}{dt} + \frac{1}{KM}x\right) \qquad (16\text{-}40)$$

To satisfy this equation, the mathematical operations required are: addition, integration, inversion, and multiplication by constants. The required addition and the two integrations are shown in Fig. 16.25; in each operation there is an inversion.

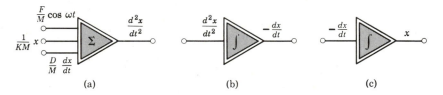

Figure 16.25 Operations needed in solving Eq. 16-40 by successive integration.

The next step is to arrange the computer elements to satisfy the equation. Knowing the required inputs to the adder, we can pick off the necessary signals (in the form of voltages) and introduce the indicated multiplication constants and inversions. Ignoring initial conditions for the moment, one possible program is outlined in Fig. 16.26.

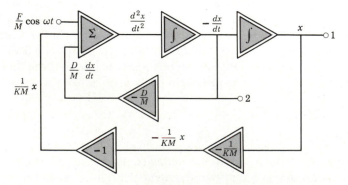

Figure 16.26 Analog computer program for Eq. 16-40.

Closing the circuit imposes the condition that the equation be satisfied. A properly synchronized cathode-ray tube connected at terminal 1 would display the displacement $x(t)$. The velocity $u(t)$ is available at terminal 2, but an inversion would be necessary to change the sign. Here six operational amplifiers are indicated; by shrewd use of amplifier capabilities, the solution can be accomplished with only three op amps.

EXAMPLE 8

For the physical system in Fig. 16.24, mass $M = 1$ kg, friction coefficient $D = 0.2$ N · s/m, and compliance $K = 2$ m/N. Devise an analog computer program using only three op amps to obtain the displacement and velocity for an applied force $f = F \cos \omega t$ N. (*Note:* Assume the op amps are single-input inverting.)

From Example 11 in Chapter 3, we know that one op amp can add and integrate the weighted sum. In this case, $M = 1$, $D/M = \frac{1}{5}$, and $1/KM = \frac{1}{2}$; after one integration, Eq. 16-40 becomes

$$\frac{dx}{dt} = -\int \left(F \cos \omega t + \frac{1}{5}\frac{dx}{dt} + \frac{1}{2}x \right) dt \quad (16\text{-}41)$$

A convenient reference is to let $RC = 1$ and specify resistances in megohms and capacitances in microfarads. For $C = 1$ μF and $R_1 = 1$ MΩ, a pure integration is performed; for $R_2 = 0.2$ MΩ, v_2 has a weighting of 5 in the result. On that basis, the appropriate values for Eq. 16-41 are shown in Fig. 16.27.

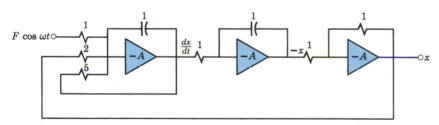

Figure 16.27   Analog computer program for solving Eq. 16-41. Resistance in MΩ and capacitance in μF.

The Practical Computer

In the practical computer there are provisions and limitations that we have not considered. One provision is for inserting a variety of initial conditions. How could we impose the condition that the initial displacement in Example 8 is $X_O = 10$ m? In these programs, displacement is represented by a voltage at a particular point in the circuit. A convenient place to put an initial voltage is on a capacitor. Figure 3.35a indicates that, since v_i is negligibly small because of the high gain A, the voltage across the 1-μF capacitor is just equal to v_o. An initial voltage of 10 V on the capacitor of the second integrator of Fig. 16.27 is analogous to an initial displacement of 10 m. The practical computer provides voltage sources and potentiometers (variable voltage dividers, see Fig. 16.28) to introduce initial displacements or velocities by closing the *reset* switch. To *compute* the switch is opened.

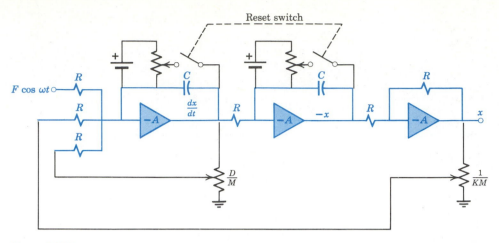

Figure 16.28 A practical computer with provision for introducing initial conditions.

Potentiometers also provide flexibility in the form of adjustable multiplying constants. Using calibrated potentiometers, the designer has a continuous range of parameter values at his fingertips. If the multiplying constants are greater than unity, the value of R in Fig. 16.28 can be changed to allow the use of potentiometers.

Another advantage of the analog computer is that the time scale can be changed at will. In Example 8, 1 s of computer time corresponds to 1 s of real time. To slow down an event that takes only a few milliseconds, we can arrange the *time scale* so that 1 ms of real time corresponds to 1 s of computer time. To change the time scale by a factor a, we can let the computer time variable $t_c = at$ and rewrite the differential equation. Increasing the integrator time constant RC by a factor a slows down the computer by the same factor. Changing the time scale changes the voltage levels produced by integration, and it may be necessary to modify the *amplitude scale*.

One limitation of the practical computer is that the amplifiers are linear over only a finite voltage range, usually around ± 10 V. To ensure accuracy and to avoid possible damage, the amplifier outputs should never exceed the rated maximum value. On the other hand, if the signal voltage is too low at any stage, it may be masked by the *noise* voltages that are present in any electronic equipment. (One reason for preferring integration over differentiation in analog computation is that random noise voltages may have large time derivatives, while their integrated values are usually zero.) Once the time-scale factor has been established, an amplitude-scale factor can be introduced to keep signal voltages as large as possible without exceeding the rated values.

The plot of velocity in Example 9 must be interpreted also because u_t is equal to $10u_c$. One possibility is to adjust the amplitude scale so that the output of the first integrator is $10 \, dx/dt_c$ instead of dx/dt as shown in Fig. 16.27. Since the force, velocity, and displacement are all represented by voltages, provision must be made to place these at proper levels; in complicated systems, amplitude scaling becomes difficult. Fortunately, most commercial computers have overload indicators on each amplifier; the variable to any amplifier that shows overloading can be rescaled.

EXAMPLE 9

For the physical system shown in Fig. 16.24, mass $M = 1$ kg, friction coefficient $D = 2 \text{ N} \cdot \text{s/m}$, and compliance $K = 0.0001$ m/N. Plan an analog computer program to obtain a record showing displacement and velocity for an applied force $f = -200 \cos 200t$ N.

Note: The available pen recorder will not follow variations above about 2 Hz.

With this very stiff spring the governing equation becomes

$$\frac{d^2x}{dt^2} + 2\frac{dx}{dt} + 10,000x = -200 \cos 200t$$

The frequency of the applied force is 200 rad/s or 32 Hz. Reasoning by analogy from Eq. 7-33, we see that the undamped natural frequency of the system is

$$\omega_n = \frac{1}{\sqrt{KM}} = \frac{1}{\sqrt{0.0001}} = 100 \text{ rad/s or 16 Hz}$$

Because the pen recorder available will not follow such rapid variations, we wish to expand the time scale by a factor of 10. Letting computer time $t_c = 10t$, $u = dx/dt = 10 \, dx/dt_c$ and $d^2x/dt^2 = 100 \, d^2x/dt_c^2$. In terms of computer time,

$$100\frac{d^2x}{dt_c^2} + 20\frac{dx}{dt_c} + 10,000x = -200 \cos 20t_c$$

or

$$\frac{d^2x}{dt_c^2} + 0.2\frac{dx}{dt_c} + 100x = -2 \cos 20t_c$$

The program for such a computation proceeds as in Example 8 (see Fig. 16.27). The resulting plot of displacement $x(t_c)$ must be interpreted in terms of the changed time variable.

An interesting application of the analog computer is as a *function generator*. A desired time function can be obtained as a solution to another equation. For example (see Problem 6), with the proper initial conditions the function $y = Y \cos \omega t$ can be obtained as the solution of the equation

$$\frac{d^2y}{dt^2} + \omega^2 y = 0$$

SUMMARY

- The availability of low-cost high-performance op amps has radically altered the design of electronic signal processing equipment.
- The ideal op amp has infinite gain, bandwidth, input impedance, and common-mode rejection ratio and zero output impedance.
  Negative feedback is used to obtain the desired operating characteristics.
- Op amps are used in inverting, noninverting, and nonlinear modes in a great variety of applications.
  Active filters with improved characteristics can be realized with op amps.
  Digital signals can be converted to analog signals and vice versa.

Voltage signals can be converted to current signals and vice versa.

The unity-gain buffer is used to isolate a source from a load.

High gain and negative feedback provide precise voltage regulation.

A digital voltmeter can be made from an integrator and a comparator.

Various waveforms can be generated by op amps operating nonlinearly.

■ The differential amplifier is a symmetric two-input amplifier that provides high differential gain and a high common-mode rejection ratio.

Practical limitations on op amp performance include: input offset voltage, bias current offset, frequency response, stability, and slew rate.

■ The analog computer is a flexible model of the system being studied; system behavior is represented by continuously varying quantities.

Programming consists in arranging the operational amplifiers to perform efficiently the operations indicated in the equations describing the system.

Integration is preferable to differentiation for practical reasons.

■ In a practical computer there are provisions for introducing initial conditions and for adjusting parameter values.

Changes in time and amplitude scale may be necessary.

REFERENCES

1. Roger Melen and Harry Garland, *Understanding IC Operational Amplifiers,* Howard W. Sams & Co., Indianapolis, 1971.

   An excellent nonmathematical introduction to IC op amps; how they are made, how they work, and how they are used. Practical applications.

2. Jacob Millman, *Microelectronics,* McGraw-Hill Book Co., New York, 1979, Chs. 15 and 16.

   A thorough treatment of op amp design and parameter measurement; discussion of many applications in analog and digital systems; description of some practical op amps available in IC form.

3. Jerry Eimbinder, ed., *Application Considerations for Linear Integrated Circuits,* Wiley-Interscience, New York, 1970.

   Practical advice for design engineers selecting and applying IC op amps. Specifications, general applications, and design considerations.

4. Jerald Graeme and Gene Tobey, eds., *Operational Amplifiers: Design and Application,* McGraw-Hill Book Co., New York, 1971.

   Thorough treatment of design and applications of IC op amps; prior circuit theory required. Appendix includes concise summary of basic concepts.

5. Paul Horowitz and Winfield Hill, *The Art of Electronics,* Cambridge University Press, Cambridge, 1980.

   Encyclopedic reference book emphasizing practical design techniques; linear and digital devices, their characteristics, and their application to instrumentation and signal processing.

REVIEW QUESTIONS

1. List the specifications of an ideal op amp.
2. Explain the operation of a low-pass filter.
3. Why must bit voltages be precise in a DAC?
4. In Fig. 16.5a, little current is drawn from the source; where does I_L come from?
5. Why is a differential amplifier used in the input stage of an op amp?
6. Distinguish between difference and common-mode signals.
7. What is a voltage follower? How does it work?
8. Explain the operation of an op amp voltage regulator.
9. What are the advantages of digital voltmeters?
10. What is the function of the ROM in the parallel-comparator ADC?

11. Explain the operation of the square-wave generator in Fig. 16.17.
12. Why does input bias current cause output voltage error?
13. Why is high-frequency gain intentionally reduced in an op amp?
14. What is the basic difference between digital and analog computers?

15. Why is the differentiator less useful than the integrator in computation?
16. Outline the procedure followed in programming an analog computer.
17. What would be needed to permit the solution of nonlinear equations?
18. What are the objectives of time scaling and amplitude scaling?

EXERCISES

1. Determine the error in Eq. 16-2 if $R_1 = R_F = 1$ MΩ, $R_i = 10$ MΩ, $A = 5 \times 10^4$, and R_o (the output impedance of the amplifier) is negligible.
2. An op amp has the following low-frequency parameters: $R_i = 1$ MΩ, $R_o = 100$ Ω, and $A = 10^5$. Stating any necessary assumptions, design an amplifier circuit with:
   (a) $A_F = -200$, $R_{iF} = 5$ kΩ.
   (b) $A_F = -30$, $R_{iF} = 20$ kΩ.
   (c) $A_F = +20$, $R_{iF} \geq 1$ MΩ.
   (d) $A_F = +50$, $R_{iF} \geq 1$ MΩ.
3. For inputs $v_1(t)$ and $v_2(t)$, devise a circuit to generate an output $v_o = +10 \int v_1 \, dt - 5v_2$. The input resistance at each input terminal is to be ≥ 100 kΩ.
4. Design a circuit, i.e., specify X and Y in Fig. 16.29, so that $v_o(t)$ is obtained from $v_1(t)$ in Fig. 16.30.

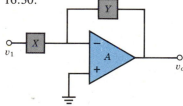

Figure 16.29

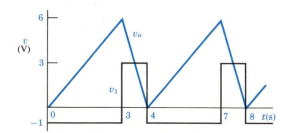

Figure 16.30

5. Design a low-pass filter with a dc gain of 46 dB and a cutoff frequency of 1 kHz.
6. Design a low-pass filter with a dc gain of 60 dB and a cutoff frequency of 100 Hz.

7. In Fig. 16.29, X consists of $R_1 = 10$ kΩ in series with $C_1 = 1$ μF and Y is $R_F = 100$ kΩ. If v_1 is a sinusoidal signal of variable frequency ω, derive an expression for V_o/V_1, plot $|V_o/V_1|$ in dB vs. ω, and identify the function performed.
8. For the amplifier circuit in Fig. 16.31:
   (a) Define v_o in terms of the given input.
   (b) If $v_1(t) = \sqrt{2} \, V_1 \cos \omega t$, define Z_F and describe the operation performed on sinusoidal inputs.

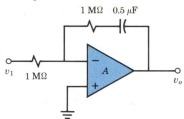

Figure 16.31

9. For the circuit of Fig. 16.29, where $X = L$ and $Y = R$, derive an expression for v_o in terms of v_1. What function is performed by the circuit?
10. Derive an expression for gain of the circuit of Fig. 16.32 as a function of frequency ω. For $R_1 = 10$ kΩ, $C_1 = 1$ μF, $C = 0.1$ μF, and $R = 100$ kΩ, determine the critical frequencies and sketch the gain as in Fig. 16.3b. What function is performed at low frequencies? At high frequencies?

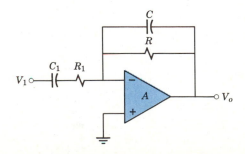

Figure 16.32

11. Design a digital-to-analog converter (DAC) that will convert the binary number stored in an 8-bit latch into a proportional *positive* voltage. The latch provides outputs at 0 and +4 V; the maximum output voltage is to be approximately 8 V. Use ideal op amps with $R_F = 1\ \text{k}\Omega$.

12. A voltmeter requires 1 mA to provide a full-scale deflection of 10 V. It is to be converted into an electronic voltmeter with an input resistance of at least 10 MΩ and ranges of 0 to 1 and 0 to 10 V.
    (a) Calculate the "input" resistance of the basic instrument.
    (b) Using an ideal op amp, design a circuit to provide the desired characteristics.
    (c) Predict the voltage indicated on the two voltmeters when connected, in turn, across a device represented by $V_T = 6$ V in series with $R_T = 500\ \Omega$. Draw a conclusion.

13. The current in a 100-Ω load R_L is to be proportional to $v_1 + 10v_2$, where v_1 and v_2 are sources with effective output resistances of over 10 kΩ. Using op amps of the type described in Exercise 2, design an effective circuit.

14. (a) Analyze the circuit of Fig. 16.33 and derive an expression for i_L in terms of the given quantities.
    (b) A source characterized by $v_s = 2$ V in series with $R_s = 1$ kΩ is connected to the v_1 terminal. If $R_1 = 100$ kΩ, $R_F = 1$ MΩ, and $R_2 = 10$ kΩ, determine i_L. What function is performed by this circuit and what are its virtues?

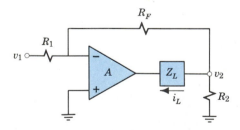

Figure 16.33

15. In Fig. 16.33, the input combination of v_1 in series with R_1 is replaced by i_1 in parallel with R_1. Stating any necessary assumptions, derive an expression for i_L in terms of i_1. Evaluate the effect of R_1 and Z_L on the operation and define the function performed.

16. Starting from basic principles and stating any assumptions:
    (a) Express v_x and v_y in terms of v_a, v_b, and v_o in Fig. 16.34 and derive an expression for v_o.
    (b) Define the function of this circuit.
    (c) For $R_1 = R_2 = 50$ kΩ, $v_a = -3 + \sin \omega t$, and $v_b = \sin \omega t + 2 \sin 2\omega t$, predict v_o and explain the virtue of this circuit.

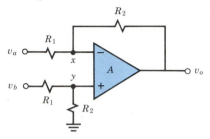

Figure 16.34

17. A cheap op amp with $A = 10^4$, $R_i = 10$ kΩ, and $R_o = 1$ kΩ is used in a unity-gain buffer. Predict R_{iF} and R_{oF}.

18. The op amp of Exercise 2 is used in the circuit of Fig. 16.5b. Derive a general expression for input resistance R_{iF} and evaluate it for the given op amp parameters and $R_F = 1$ kΩ.

19. The op amp circuit shown in Fig. 16.35 is suggested as a means for obtaining a steady output V_o for any value of V_s between 8 and 12 V, say.
    (a) Describe in words how this circuit might operate to provide a steady output voltage.
    (b) Stating any necessary assumptions, for $V_s = 10$ V, predict current I, voltage V_{CE}, and voltage V_o.
    (c) With V_s at 10 V, say, explain what would happen to the circuit variables if V_s increased to 12 V.

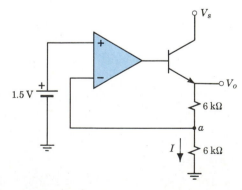

Figure 16.35

20. (a) Derive Eq. 16-26 for the square-wave generator.
  (b) For convenience, it is desired to have the period $T = 2R_F C$; specify the value of H.

21. In Fig. 16.18a, $V_Z = 10$ V, $R_S = 5$ kΩ, $R_A = R_F = 100$ kΩ, $R_1 = 100$ kΩ, and $C = 0.1$ μF. Assume that at $t = 0$, $v_1 = V_Z$, and $v_o = 0$ V; at what time t will v_i change sign? Determine the amplitude and the frequency of the triangular wave.

22. Determine the CMRR for the differential amplifier of Fig. 16.7 if:
  (a) $R_1 = R_2 = 10$ kΩ and $R_F = R_3 = 100$ kΩ.
  (b) $R_1 = 10$ kΩ, $R_2 = 9$ kΩ, and $R_F = R_3 = 100$ kΩ.

23. An IC differential amplifier (Fig. 16.20) operates with $I_{C3} = 200$ μA. Assuming $\beta = 100$, estimate the input resistance $R_i = (v_p - v_n)/i_{b1}$ at room temperature.

24. For the differential amplifier of Fig. 16.20, $\beta = 99$, $r_\pi = 20$ kΩ, $R_{C1} = 100$ kΩ, and $R_{C2} = 110$ kΩ. Estimate the differential gain, the common-mode gain, and the CMRR.

25. The op amp of Exercise 2 has an input bias current of 1 μA and an input offset current of 0.1 μA.
  (a) Design an amplifier with a gain of $+20$, assuming $R_F = 100$ kΩ.
  (b) Design an amplifier with a gain of -20, assuming $R_F = 100$ kΩ.

26. A 741 op amp has a low-frequency gain of 10^5 and a unity-gain frequency of 1 MHz.
  (a) Estimate the upper cutoff frequency and plot gain in dB versus frequency on a log scale.
  (b) Plot on the same graph the curve for an inverting amplifier with a low-frequency gain of -100 using the 741.

27. Draw block diagrams of computer programs to solve:
  (a) $\begin{cases} x - 3y = 5 \\ x + 2y = \sin \omega t \end{cases}$

  (b) $\begin{cases} x + 2y = t \\ x - 3y = 0 \end{cases}$

28. Draw block diagrams of efficient computer programs to solve:
  (a) $\dfrac{d^2 y}{dt^2} + 2y = 5 \sin \omega t$

  (b) $\dfrac{d^3 x}{dt^3} - 3\dfrac{d^2 x}{dt^2} + 2\dfrac{dx}{dt} + x = 0$

29. Draw an efficient computer program to solve:

$$2\frac{d^2 y}{dt^2} + 4\frac{dy}{dt} + y = 2$$

provided $y = 3$ and $dy/dt = 1$ at $t = 0$.

30. Write the equation whose analog is shown in Fig. 16.36 (C in μF, R in MΩ).

31. (a) Repeat Exercise 30 for Fig. 16.37.
  (b) What is $y(t)$?

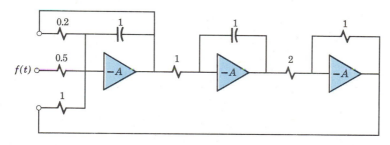

Figure 16.36

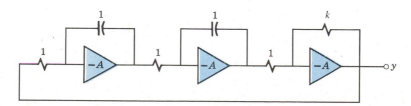

Figure 16.37

PROBLEMS

1. The output of a flowmeter is $v = Kq$, where q is in cm^3/s and $K = 20\,mV \cdot s/cm^3$. The effective output resistance of the flowmeter is $2000\,\Omega$. Design a circuit that will develop an output voltage $V_o = 10\,V$ (to trip a relay) after $200\,cm^3$ have passed the metering point.

2. A mistake is made in connecting up a voltage follower; v_1 is connected to the $-$ terminal and v_o is fed back to the $+$ terminal. Derive an expression for gain A_F and estimate R_i; comment on the usefulness of this circuit. (*Hint:* In the real amplifier of Fig. 16.20a, the gain may be less than 1 for extremely small signals.)

3. Derive a general expression for the output resistance of an op amp connected in the inverting circuit.

4. Derive a general expression for the input resistance of an op amp in the noninverting circuit.

5. An inverting amplifier ($R_1 = 1\,k\Omega$ and $R_F = 10\,k\Omega$) employs an op amp with an input offset voltage of 4 mV. The offset voltage is to be balanced out by introducing a null current I_N at the junction of R_1 and R_F in Fig. 16.21.
   (a) Determine the appropriate value of I_N.
   (b) If I_N is to be obtained from a 1-MΩ potentiometer across a ±15-V power supply, determine the setting of the potentiometer.

6. Devise a computer program to solve the equation

$$\frac{d^2y}{dt^2} + \omega^2 y = 0$$

and specify the initial conditions so that the function generated is $y = A \cos \omega t$.

17

Large-Signal Amplifiers

Practical Amplifiers

Biasing Circuits

Power Amplifiers

Other Amplifiers

The two major functions performed by electronic devices are switching and amplifying. We have studied the use of diodes and transistors in switching and other nonlinear applications, and we have learned to use nearly ideal op amps in a great variety of special circuits. Using the characteristics of two-port devices described in Chapter 12 and the circuit theory of Part I, we are now ready to study the analysis and design of various types of amplifiers.

The design of electronic amplifiers to meet critical performance and cost specifications requires a great deal of knowledge and judgment. Electrical engineers will gain the detailed knowledge in subsequent courses and the judgment from practical experience. Other engineers and scientists will be more concerned with the use of existing amplifiers; if they do any designing (in connection with instrumentation, for example), it will be under circumstances where optimum performance is not essential and a simplified approach is satisfactory.

In this chapter we discuss the factors that influence amplifier performance, illustrate the basic analysis techniques, and present methods for designing simple amplifiers and predicting their performance. Our approach is first to look at the practical considerations in amplifier operation and devise circuits for maintaining the proper operating conditions, then to analyze the performance of one type of power amplifier, and finally to describe briefly some other important types of amplifiers. In subsequent chapters we shall learn to design amplifiers where frequency response and operating characteristics must be considered.

PRACTICAL AMPLIFIERS

Amplifier Classification

There are many ways of describing amplifiers. A *single-stage* amplifier consists of one amplifying element and the associated circuitry; in general, several such elements are combined in a *multistage* amplifier. In a sound reproduction system, for example, the first stages are *small-signal* voltage (or current) amplifiers designed to amplify the output of a phonograph pickup, a few millivolts, up to a signal of several volts. The final stage is a *large-signal* or *power* amplifier that supplies sufficient power, several watts, to drive the loudspeaker.

Such amplifiers are called *audio* amplifiers if they amplify signals from, say, 30 to 15,000 Hz. In measuring structural vibrations, temperature variations, or the electrical currents generated within the human body, very low-frequency signals are encountered; to handle signals from zero frequency to a few cycles per second, *direct-coupled* amplifiers are used. In contrast, the *video* amplifier in a television receiver must amplify picture signals with components from 30 to 4,000,000 Hz.

A video amplifier is a *wide-band* amplifier that amplifies equally all frequencies over a broad frequency range. In contrast, a *radiofrequency* amplifier for the FM broadcast band (around 100 MHz) is *tuned* to select and amplify the signal from one station and to reject all others. In this chapter we are primarily interested in simple untuned audio power amplifiers.

Amplification and Distortion

The terms amplification and gain are used almost interchangeably. For sinusoidal signals or for a particular sinusoidal component of a periodic signal the voltage gain is

$$\mathbf{A}_V = \frac{\mathbf{V}_{\text{out}}}{\mathbf{V}_{\text{in}}} = A\,e^{j\theta} \tag{17-1}$$

where $\mathbf{A}_V$ is the complex ratio of two phasors. In a *linear amplifier* A and θ are independent of signal amplitude and frequency, and the output signal is a replica of the input signal.[†]

If there is *distortion* in the amplifier, the output is not a replica of the input. In Fig. 17.1, output is not proportional to input and there is *nonlinear* or *amplitude* distortion. In other words, A is not a simple constant. As a result of amplitude distortion, there are frequency components in the output that are not present in the input. Fourier analysis of the output would reveal the presence of a "second harmonic," a component of twice the frequency of the "fundamental" input signal.[‡] Amplitude distortion usually occurs when excessively large signals are applied to nonlinear elements such as transistors.

At the other extreme is distortion due to *noise*, random signals unrelated to the input. If the input signal is too small, the output consists primarily of noise and is not a replica of the input. The "snow" that appears on a television screen when only a weak

---

[†] If $\theta = k\omega$, a time lag is introduced, but there is no phase distortion.
[‡] The calculation of harmonic distortion is outlined in Chapter 18 of J. Millman and C. Halkias: *Integrated Electronics*, McGraw-Hill Book Co., New York, 1972.

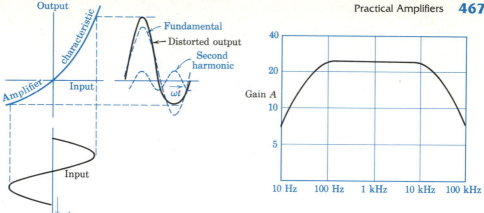

Figure 17.1 Amplitude distortion.

Figure 17.2 Frequency distortion.

signal is available is a visual representation of noise. One source of noise is the random thermal motion of electrons in the amplifier circuit elements. The *shot effect* of individual electrons crossing a junction is another source of noise. Noise is of greatest importance in input stages where signal levels are small; any noise introduced there is amplified by all subsequent stages. The *dynamic range* of any amplifier is bounded at one end by the level at which signals are obscured by noise and at the other by the level at which amplitude distortion becomes excessive.

In Fig. 17.2, the *frequency-response curve* of an audio amplifier indicates that there is *frequency distortion*; all frequencies (within a finite band) are not amplified equally. In other words, gain A in Eq. 17-1 is a function of frequency. A signal consisting of a fundamental at 1 kHz, a tenth harmonic at 10 kHz, and a hundredth harmonic at 100 kHz would have a different waveform after amplification. No amplifier is completely free from frequency distortion.

If θ is a function of frequency, the relative amplitudes of the signal components may be unchanged but the relative phase positions are shifted. As shown in Fig. 17.3, such *phase distortion* changes the shape of the output wave. The eye is sensitive to such distortion but the ear is not; a human ordinarily cannot distinguish between the two signals. On the other hand, the ear is quite sensitive to amplitude or frequency distortion.

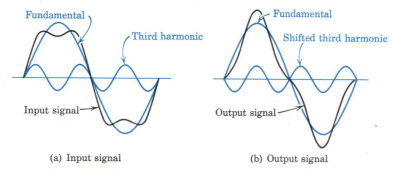

(a) Input signal

(b) Output signal

Figure 17.3 Phase distortion.

Phase and frequency distortion are caused by circuit elements such as capacitive and inductive reactances that are frequency dependent. Some transistor parameters also are frequency dependent. In the design of untuned or wide-band amplifiers special steps are taken to reduce the variation in gain with frequency.

Practical Considerations

Some of the practical considerations in amplifier design are illustrated in the simplified two-stage audio amplifier of Fig. 17.4.

Biasing. Transistors are maintained at the proper operating point by dc power supplies and biasing networks. In portable units, batteries are used; in other units, the readily available 60-Hz current is rectified and filtered to provide the necessary direct current. In Fig. 17.4, the battery V_{CC} supplies the collector voltages to transistors T1 and T2 and, through resistors R_B and R_D, the base-biasing currents. When several different bias voltages are required, they are obtained from a voltage divider or by series *dropping* resistors.

Coupling. The voltage generated in the phonograph pickup is coupled to T1 by a combination of R_{VC} and C_C. Variable resistor R_{VC} provides volume control by voltage-divider action. Transistor T1 is coupled to the second stage by means of collector resistor R_C and coupling capacitor C_C. Such an *RC* coupling circuit develops a useful signal output across R_C and transfers it to the input of the next stage, but dc voltages and currents are blocked by C_C.

The loudspeaker is coupled to T2 by transformer T. Transformer coupling is more expensive than *RC* coupling, but it is more efficient and it permits impedance matching for improved power transfer. The combination of R_D and C_D is a *decoupling filter* to prevent feedback of amplified signals to the low-level input stage. Such feedback is likely to occur when the common battery supplying several stages begins to age and develops an appreciable internal resistance.

Load Impedance. If possible, the load impedance of an untuned amplifier is made purely resistive to minimize the variation in gain with frequency. In tuned amplifiers, the load impedance is usually a parallel resonant circuit. For voltage amplifiers, large values of R_C are desirable; but large R_C requires large supply voltages, so the selected value is a compromise.

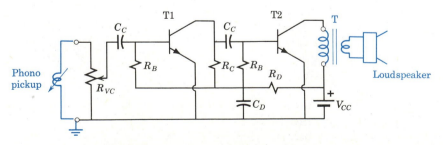

Figure 17.4 A simplified two-stage phonograph amplifier.

Input and Output Impedance. When several amplifier stages are connected in *cascade*, the output characteristics of one stage are influenced by the input characteristics of the next. For ac signals, R_C, the load resistance of T1, is effectively shunted by the biasing resistance R_B and also by the input resistance of T2. Ideally the input impedance of an amplifier stage should be high to minimize "loading" of the preceding stage, and the output impedance should be low for efficient power transfer. (See p. 69.)

Unintentional Elements. Figure 17.4 is a wiring diagram and shows components, not circuit elements. Even a short, straight conductor can store a small amount of energy in the form of a magnetic field or an electric field (between the conductor and ground or a metal chassis). The magnetic field effect is important only at extremely high frequencies and will be neglected here. The electric field effect can be represented by a *wiring capacitance*.

In the same way we represent the energy storage due to a difference in potential between the gate and source of a MOSFET by an *equivalent capacitance* (Fig. 18.10). At a transistor junction there is a charge separation and a potential difference across the depletion region (see Fig. 11.10); a change in junction potential causes a change in charge distribution ($q = Cv$), and this effect is represented by a *junction capacitance* C_j.

Bipolar transistors exhibit another charge storage effect. Under equilibrium conditions, the base must be electrically neutral. Some of the charge from majority carriers (holes in a *pnp* transistor) diffusing from the emitter into the base is neutralized by carriers (electrons in *pnp*) from the base connection. A change in emitter-base potential causes a transfer of charge into or out of the base; this effect is represented by a *diffusion capacitance* C_d. A circuit model used for analyzing amplifier performance at high frequencies must include these unintentional elements and, therefore, it will be quite different from the wiring diagram of Fig. 17.4.

BIASING CIRCUITS

The objective in biasing is to *establish* the proper operating point and to *maintain* it despite variations in temperature and variations among individual devices of the same type. Furthermore, this objective is to be achieved without adversely affecting the desired performance of the circuit. The biasing problem is difficult because of:

- The wide variations in device parameters expected in mass-produced transistors,
- The complicated interrelations among transistor variables, and
- The inherent sensitivity of semiconductor devices to temperature.

Because of its importance, much attention has been given to this problem and as a result many ingenious circuits are available for the designer's use.

FET Biasing

Field-effect transistors are used primarily as "small-signal" amplifiers because the parabolic transfer characteristic introduces amplitude distortion if the signals are large. (See Example 2 in Chapter 12.) If the excursions along the load line are small (Fig. 12.16b), the biasing requirements are not critical.

DE MOSFET. Biasing a depletion-or-enhancement MOSFET is extremely simple because of its unique transfer characteristic that allows a quiescent operating point at $V_{GS} = 0$. The transfer characteristic of a DE MOSFET (Fig. 12.5b) is repeated in Fig. 17.5b. To obtain this characteristic, the drain-source channel must be biased by a dc voltage that places v_{DS} in the normal operating region between turn-on and breakdown where drain current i_D is nearly independent of v_{DS}. The necessary circuit (Fig. 17.5c) contains a voltage supply V_{DD}, a drain resistor R_D across which signal voltages can be developed, and a gate resistor R_G that allows any charge that might build up on the highly insulated gate to "leak" off.

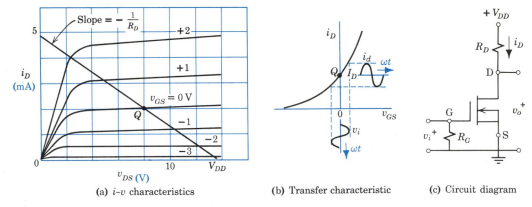

(a) i-v characteristics   (b) Transfer characteristic   (c) Circuit diagram

Figure 17.5 A DE MOSFET amplifier.

As indicated by the transfer characteristic, a small-signal voltage applied to the input produces a signal current at the output. The output voltage $v_o = V_{DD} - i_D R_D$ contains a signal component that is an amplified replica of the input voltage.

JFET. Biasing a depletion-mode JFET is complicated by the fact that positive values of v_{GS} are not permitted; ac signals must be superimposed on a steady negative dc value V_{GS}. (For nomenclature, refer to Table 17-1.) A possible Q point is shown in Fig. 17.6a and b. A further complication is that V_{GS} for a given I_D varies from sample to sample and with temperature.

Table 17-1 Transistor Nomenclature

	Field-Effect Transistors		Bipolar Transistors	
	Voltages	Currents	Voltages	Currents
Instantaneous total value	v_{GS}, v_{DS}	i_D	v_{EB}, v_{CB}	i_E, i_C
Instantaneous signal component	v_{gs}, v_{ds}	i_d	v_{eb}, v_{cb}	i_e, i_c
Quiescent or dc value	V_{GS}, V_{DS}	I_D	V_{EB}, V_{CB}	I_E, I_C
Effective (rms) value of signal	V_{gs}, V_{ds}	I_d	V_{eb}, V_{cb}	I_e, I_c
Supply voltage (magnitude)	V_{GG}, V_{DD}		V_{BB}, V_{CC}	

To provide the necessary negative value of V_{GS}, one possibility would be to use a negative source V_{GG}, but this would require a second battery. Another simple possibility is to place a resistor R_S between source and ground so that the source is at

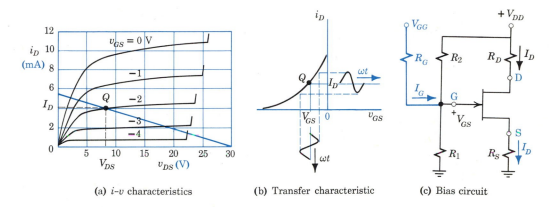

(a) i-v characteristics   (b) Transfer characteristic   (c) Bias circuit

Figure 17.6   Biasing a depletion-mode JFET.

a potential $+I_D R_S$ with respect to ground; if then the gate is at ground potential, $V_{GS} = -I_D R_S$ as desired. A more complicated but better biasing scheme is shown in Fig. 17.6c, where R_1 and R_2 form a voltage divider to place the gate at the desired potential with respect to ground. As a first step in the analysis, replace V_{DD}, R_1, and R_2 by the Thévenin equivalents

$$V_{GG} = \frac{R_1}{R_1 + R_2} V_{DD} \quad \text{and} \quad R_G = \frac{R_1 R_2}{R_1 + R_2} \tag{17-2}$$

To facilitate design, these equations are solved for the circuit values

$$R_2 = R_G \frac{V_{DD}}{V_{GG}} \quad R_1 = \frac{R_G R_2}{R_2 - R_G} \tag{17-3}$$

Summing voltages around the gate "loop" yields

$$\Sigma V = 0 = V_{GG} - I_G R_G - V_{GS} - I_D R_S \tag{17-4}$$

Disregarding the negligible gate current through a reverse-biased junction, the voltage equation reduces to

$$I_D R_S = V_{GG} - V_{GS} \quad \text{or} \quad R_S = \frac{V_{GG} - V_{GS}}{I_D} \tag{17-5}$$

For a specified Q point (I_D, V_{GS}) and chosen values of V_{GG} and R_G, the required values of R_S, R_1, and R_2 are easily calculated from Eqs. 17-3, 5. Furthermore, by making the stable quantity V_{GG} large compared to the variable quantity $|V_{GS}|$, the effect of any shift in V_{GS} is reduced.

EXAMPLE 1

The JFET of Fig. 17.6 is to be operated at a quiescent point defined by $I_D = 4$ mA, $V_{DS} = 8$ V, and $V_{GS} = -2$ V. Design an appropriate biasing circuit with $V_{DD} = 30$ V.

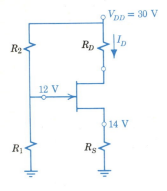

Figure 17.7   JFET bias design.

Using the voltage-divider circuit of Fig. 17.7, assume $V_{GG} = 12$ V so that $V_{GG} - V_{GS}$ is large compared to V_{GS} for stability, and assume $R_G = 6$ MΩ to keep the input resistance high. Then, by Eqs. 17-3 and 5,

$$R_2 = R_G \frac{V_{DD}}{V_{GG}} = 6\frac{30}{12} = 15 \text{ M}\Omega$$

$$R_1 = \frac{R_G R_2}{R_2 - R_G} = \frac{6 \times 15}{15 - 6} = 10 \text{ M}\Omega$$

$$R_S = \frac{V_{GG} - V_{GS}}{I_D} = \frac{12 - (-2)}{0.004} = 3.5 \text{ k}\Omega$$

Summing voltages around the drain "loop" yields

$$\Sigma V = 0 = V_{DD} - I_D R_D - V_{DS} - I_D R_S$$

For $V_{DS} = 8$ V,

$$R_D = (V_{DD} - V_{DS} - I_D R_S)/I_D$$

$$= (30 - 8 - 14)/0.004 = 2 \text{ k}\Omega$$

Note that a 20% shift in V_{GS} (to -2.4 V) would cause only a 3% change in $V_{GG} - V_{GS}$ and in I_D (Eq. 17-5).

Enhancement MOSFET. For an enhancement-only MOSFET (Fig. 12.3), normal operation requires a V_{GS} of the proper polarity to attract the holes or electrons necessary for conduction in the channel. If the voltage-divider bias circuit of Fig. 17.8b is used, V_{GG} is specified to provide the proper magnitude and polarity of $V_{GS} = V_{GG} - V_S = V_{GG} - I_D R_S$. (Some representative voltages are shown in color.)

The simpler bias circuit shown in Fig. 17.8c is satisfactory in many situations. Since the gate current is negligible, $V_{GS} = V_{DS}$ and operation is in the normal region for $|V_{GS}| > |V_T|$. (The design of such a bias circuit was illustrated in Example 1 of Chapter 12; the design values are shown in color in Fig. 17.8c.) In this *drain-feedback bias* circuit, any change in drain current due to changes in device or circuit parameters is *fed back* to the gate circuit in such a way as to compensate for the parameter change. For example, if I_D increases for any reason, V_{DS} decreases and the corresponding decrease in V_{GS} tends to decrease I_D. Feedback resistor R_G must be very large to minimize the degrading effect of *signal* current flowing from the output back to the input.

BJT Biasing

In normal operation of a bipolar transistor the emitter-base junction is forward biased and the collector-base junction is reverse biased. Since in the common-emitter configuration these bias voltages have the same polarity, a single battery can supply

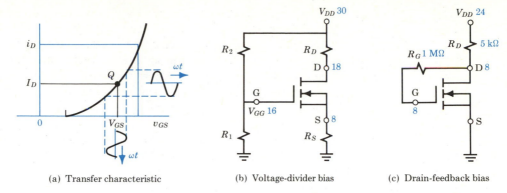

(a) Transfer characteristic   (b) Voltage-divider bias   (c) Drain-feedback bias

Figure 17.8   Biasing an n-channel enhancement-only MOSFET.

both through an appropriate circuit. Here again, however, the variation in parameters among mass-produced transistors and the inherent sensitivity of semiconductors to temperature complicate the biasing problem, particularly in the large-signal amplifiers to be discussed in the next section. After demonstrating that the simplest solution is unsatisfactory, we analyze one commonly used bias circuit and outline an approximate design procedure.

Fixed-Current Bias. One possibility is to provide the desired dc base current from V_{CC} as in Fig. 17.9. For a forward-biased emitter the junction voltage is a fraction

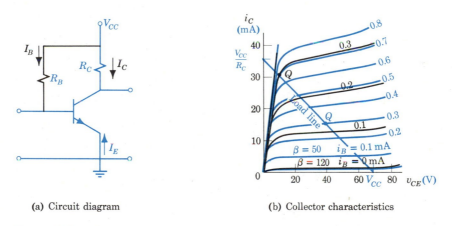

(a) Circuit diagram   (b) Collector characteristics

Figure 17.9   Fixed-current bias and effect of β change.

of a volt and may be neglected in comparison to V_{CC}. On this basis, the quiescent base current is $I_B \cong V_{CC}/R_B$ and the operating point would be determined if the collector characteristics were known precisely.

But the characteristics are not known precisely. For steady or dc values (see Table 17-1) Eq. 12.6 becomes

$$I_C = \frac{\alpha}{1-\alpha}I_B + \frac{I_{CBO}}{1-\alpha} = \beta I_B + (1+\beta)I_{CBO} \qquad (17\text{-}6)$$

Even if I_B is fixed, collector current I_C may vary widely with the large variations in β and I_{CBO} expected among mass-produced transistors. As shown in Fig. 17.9b, a value $I_B = 0.3$ mA that places the quiescent point in the linear region of a transistor with $\beta = 50$ (characteristics in color) pushes operation into the saturation region of another transistor of the same type with $\beta = 120$. Also, I_C will vary widely with temperature changes because β increases nearly linearly with temperature and I_{CBO} increases exponentially with temperature.

Thermal Runaway. The power loss in a transistor is primarily at the collector junction because the voltage there is high compared to the low voltage at the forward-biased emitter junction. If the collector current I_C increases, the power developed tends to raise the junction temperature. This causes an increase in I_{CBO} and β and a further increase in I_C, which tends to raise the temperature. In a transistor operating at high temperature (because of high ambient temperature or high developed power), a regenerative heating cycle may occur that will result in *thermal runaway* and possibly destruction of the transistor.

For equilibrium, the power developed in the transistor is equal to the power (heat) dissipated to the surroundings. The ability to dissipate heat by conduction is proportional to the difference between junction and ambient temperatures and inversely proportional to the thermal resistance of the conducting path. For a constant ambient temperature, the ability to dissipate heat increases as the junction temperature goes up. Equilibrium will be upset and thermal runaway may occur when the increase in power to be dissipated is greater than the increase in dissipation ability due to the increase in junction temperature. One way to avoid this cumulative effect is to cool the collector junction; power transistors utilize the heat conducting capability of the metal chassis on which they are mounted or special heat-radiating fins. Another way is to use an effective biasing circuit.

Self Bias. The ingenious self-biasing circuit of Fig. 17.10 decreases the effect of changes in β or temperature on the quiescent operating point. Its operation is based on the fact that the critical variable to be stabilized is the collector current rather than

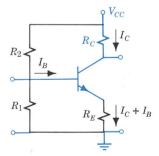

Figure 17.10 A self-biasing circuit.

the base current. The combination of R_1 and R_2 constitutes a voltage divider (as in Fig. 17.6c) to bring the base to the proper potential to forward bias the emitter junction. If I_C tends to increase, perhaps because of an increase in β due to a rise in temperature, the current $I_C + I_B$ in R_E increases, raising the potential of the emitter with respect to ground. This, in turn, reduces V_{BE}, the forward bias on the base-emitter junction, reduces the base current and, therefore, limits the increase in I_C. In other words, any increase in collector current is fed back to the base circuit and modifies the bias in such a way as to oppose a further increase in I_C.

Quantitative analysis of the circuit is simplified if the voltage divider (Fig. 17.11a) is replaced by its Thévenin equivalent where

$$V_{BB} = \frac{R_1}{R_1 + R_2} V_{CC} \quad \text{and} \quad R_B = \frac{R_1 R_2}{R_1 + R_2} \tag{17-7}$$

After replacing the transistor by a dc model derived from Eq. 17-6 (Fig. 17.11c),

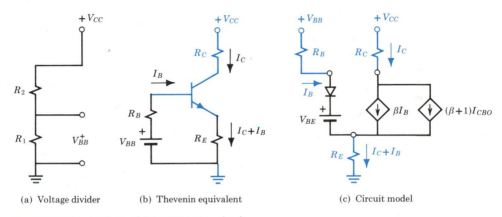

(a) Voltage divider   (b) Thevenin equivalent   (c) Circuit model

Figure 17.11   Analysis of the self-biasing circuit.

Kirchhoff's voltage law around the base loop yields

$$V_{BB} - I_B R_B - V_{BE} - (I_C + I_B)R_E = 0 \tag{17-8}$$

From Eq. 17-6, represented in the model by two current sources in parallel,

$$I_B = \frac{I_C}{\beta} - \frac{\beta + 1}{\beta} I_{CBO} \tag{17-9}$$

Substituting in Eq. 17-8 and solving,

$$I_C = \frac{V_{BB} - V_{BE} + \left(\dfrac{\beta + 1}{\beta}\right) I_{CBO}(R_B + R_E)}{R_E + \dfrac{R_B + R_E}{\beta}} \tag{17-10}$$

This is a general equation for collector current in a self-biasing circuit where V_{BE}, β, and I_{CBO} are device parameters that are known imprecisely and that are sensitive to temperature change. Typically, $|V_{BE}|$ decreases 2.5 mV/°C, β may vary by 6:1 and increases linearly with temperature, and I_{CBO} may vary by 10:1 and approximately doubles for every 10° C increase above 25° C.

EXAMPLE 2

The β of individual specimens of a silicon transistor varies from 30 to 180. If V_{BE} may vary from 0.5 to 0.9 V and I_{CBO} may vary from 1 to 10 nA, predict the extreme variation in I_C in a self-biasing circuit where $R_1 = 10$ kΩ, $R_2 = 90$ kΩ, $R_E = 2$ kΩ, $R_C = 15$ kΩ, and $V_{CC} = 28$ V.

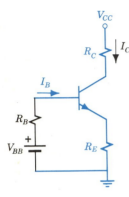

Figure 17.12 Stabilizing effect of a self-biasing circuit.

By Eqs. 17-7, the Thévenin equivalents are

$$R_B = \frac{R_1 R_2}{R_1 + R_2} = \frac{10 \times 90}{10 + 90} = 9 \text{ k}\Omega$$

$$V_{BB} = \frac{V_{CC} R_1}{R_1 + R_2} = \frac{28 \times 10}{10 + 90} = 2.8 \text{ V}$$

By Eq. 17-10, the collector current (Fig. 17.12) is

$$I_C = \frac{V_{BB} - V_{BE} + \left(\dfrac{\beta + 1}{\beta}\right) I_{CBO}(R_B + R_E)}{R_E + \dfrac{R_B + R_E}{\beta}}$$

For $\beta = 30$, $V_{BE} = 0.9$ V, and $I_{CBO} = 1$ nA (the "worst case"),

$$I_C = \frac{2.8 - 0.9 + \left(\dfrac{31}{30}\right) 10^{-9}(11{,}000)}{2000 + 11{,}000/30}$$

$$= \frac{1.9 + 0.00001}{2367} = 0.8 \text{ mA}$$

For $\beta = 180$, $V_{BE} = 0.5$ V, and $I_{CBO} = 10$ nA,

$$I_C = \frac{2.8 - 0.5 + \left(\dfrac{181}{180}\right) 10^{-8}(11{,}000)}{2000 + 11{,}000/180}$$

$$= \frac{2.3 + 0.0001}{2061} = 1.1 \text{ mA}$$

For these extreme variations, I_C shifts only 0.3 mA. Note that for a typical silicon transistor, $(1 + \beta) I_{CBO} = I_{CEO}$ is negligible in bias calculations.

Example 2 demonstrates that a properly designed self-biasing circuit is effective in stabilizing I_C despite variations in device parameters. Equation 17-10 reveals the design criteria; to stabilize operation, each term containing a variable parameter should be made "small" with respect to a circuit constant. Since the terms are to be "small," rough approximations are acceptable in evaluating the terms; for example, for $\beta \geq 20$, $(\beta + 1)/\beta \cong 1$, and for $R_B \geq 5R_E$, $R_B + R_E \cong R_B$. With these approximations in mind, the design criteria are:

To make I_C independent of I_{CBO}, make

$$\left(\frac{1 + \beta}{\beta}\right) I_{CBO} (R_B + R_E) \cong I_{CBO} R_B \ll V_{BB} - V_{BE} \qquad (17\text{-}11)$$

To make I_C independent of β, make

$$\frac{R_B + R_E}{\beta} \cong \frac{R_B}{\beta} \ll R_E \qquad (17\text{-}12)$$

To make I_C independent of V_{BE}, make

$$V_{BE} \ll V_{BB} \qquad (17\text{-}13)$$

The effect of temperature on the Q point can be predicted in terms of changes in V_{BE} and I_{CBO}. Assuming the circuit is designed to be independent of β (Eq. 17-12), the collector current is approximated by

$$I_C = \frac{V_{BB} - V_{BE} + I_{CBO}R_B}{R_E} \qquad (17\text{-}14)$$

Assuming V_{BB}, R_B, and R_E are constant (actually $\pm 10\%$ variations are to be expected), the *change* in I_C will be

$$\Delta I_C = \frac{-\Delta V_{BE} + \Delta I_{CBO}R_B}{R_E} \qquad (17\text{-}15)$$

The relative sensitivity of silicon and germanium is evaluated in Example 3.

EXAMPLE 3

A silicon transistor with $V_{BE} = 0.7$ V and $I_{CBO} = 10$ nA and a germanium transistor with $V_{BE} = 0.3$ V and $I_{CBO} = 5$ μA (typical room temperature values) are used in the circuit of Example 2. Predict the effect on collector current of a 50° increase in operating temperature. (Use the typical rules given at the bottom of p. 475.)

The change in V_{BE} will be

$$\Delta V_{BE} = (-2.5 \text{ mV/}°\text{C})50°\text{C} \cong -0.1 \text{ V}$$

I_{CBO} will change by the factor

$$2^{\Delta T/10} = 2^{50/10} = 2^5 = 32$$

and the new values will be

$$\text{Si:} I_{CBO} = 32 \times 10 = 320 \text{ nA}$$

$$\text{Ge:} I_{CBO} = 32 \times 5 = 160 \text{ } \mu\text{A}$$

By Eq. 17-15, the *change* in I_C will be

$$\text{Si:} \Delta I_C = \frac{+0.1 + 310 \times 10^{-9} \times 9 \times 10^3}{2 \times 10^3}$$

$$= \frac{0.1 + 0.003}{2 \times 10^3} \cong 0.05 \text{ mA}$$

$$\text{Ge:} \Delta I_C = \frac{+0.1 + 155 \times 10^{-6} \times 9 \times 10^3}{2 \times 10^3}$$

$$= \frac{0.1 + 1.4}{2 \times 10^3} = 0.75 \text{ mA}$$

Bias Design, Approximate Method

As demonstrated in Example 3, silicon transistors are relatively unaffected by temperature changes and the small change in collector current is primarily due to ΔV_{BE}. Germanium transistors are sensitive to temperature changes and the principal factor is ΔI_{CBO}.

In stabilizing against variations in β, Eq. 17-12 indicates that R_B should be small compared to βR_E. In practice, the ratio $\beta R_E / R_B$ is limited. R_B must not be too small because it appears directly across the input and tends to divert part of the signal current. Also, R_E must not be too large because part of the dc supply voltage V_{CC} appears across R_E (Fig. 17.13). As R_E is increased, less voltage is available for developing an output signal across R_C, and the collector circuit efficiency is reduced. (To prevent an undesirable ac voltage variation across R_E, it is customarily by-passed by a capacitor that offers low impedance at all signal frequencies. Coupling capacitors isolate the dc bias currents from the signal source and the following stage. See Fig. 17.16.)

The designer must select values of R_1, R_2, and R_E to provide optimum performance, but the criteria are different in every situation and skill and experience are essential to an optimum design. If large temperature changes are not expected and if less than optimum performance is acceptable, the following procedure is satisfactory:

1. Select an appropriate nominal operating point (I_C, I_B, and V_{CE}) from the manufacturer's data (see Fig. 17.9b).
2. Arbitrarily assume that $V_E = I_E R_E \cong I_C R_E \cong 3$ V, say, and solve for R_E.
3. Select V_{CC} and R_C.
   (a) If V_{CC} is specified, $R_C \cong (V_{CC} - V_{CE} - 3)/I_C$ (17-16)
   (b) If R_C is specified, $V_{CC} \cong 3 + V_{CE} + I_C R_C$ (17-17)
4. Arbitrarily select R_B equal to $\beta_{\min} R_E / 10$.   (17-18)
   (For $\beta = 50$, a 10% change in β will cause a 1% change in I_C.)
5. Calculate $V_{BB} = I_B R_B + V_{BE} + (I_B + I_C) R_E$.   (17-19)
   (Lacking other information, use $I_B = I_C / \beta$ and assume $V_{BE} \cong 0.7$ V for silicon or 0.3 V for germanium.)
6. Calculate R_2 and R_1 (see Eqs. 17-7) from

$$R_2 = R_B \frac{V_{CC}}{V_{BB}} \quad \text{and} \quad R_1 = \frac{R_2 R_B}{R_2 - R_B} \quad (17\text{-}20)$$

The exact sequence in which these steps are taken can be modified to fit a given situation. (See Example 4, p. 482.)

POWER AMPLIFIERS

To drive a loudspeaker or a recording instrument requires an appreciable amount of power. In analyzing or designing a *power amplifier*, large signals and the accompanying nonlinearity must be considered. Transistors capable of delivering high power output are expensive and therefore careful attention to efficient amplifier design is justified.

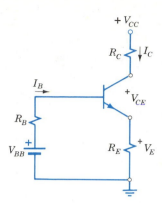

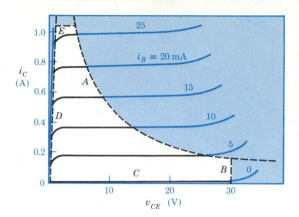

Figure 17.13 Bias design.

Figure 17.14 Permissible operating region of a transistor.

The basic design problem is to provide the desired power output with a stable circuit that uses the amplifying device efficiently and safely. Let us first consider the permissible operating region, see how operating point and load resistance are selected, derive expressions for power and efficiency, and then analyze a common type of audio-frequency amplifier that uses a power transistor.

Permissible Operating Region

An electronic amplifier must operate without introducing excessive distortion and without exceeding the voltage, current, and power limitations of the device. The bipolar junction transistor provides a large linear region and is commonly used where large signal excursions are required. The permissible operating region can be indicated on the output characteristics of the transistor as in Fig. 17.14. Line A represents the maximum allowable power dissipation P_D for the transistor; this is an hyperbola defined by $V_{CE}I_C = P_C = P_D$, where P_D is established by the manufacturer. Operation above this line may damage the device. Line B reflects the fact that at high collector voltages the avalanche effect causes a rapid increase in collector current and the curves become nonlinear.

For any electronic device there are regions of excessive nonlinearity that should be avoided. Line C bounds the region in which collector current is approaching zero and a further decrease in signal value (i_B) does not produce a corresponding decrease in output current. Line D bounds the saturation region in which a further increase in signal does not produce a corresponding increase in output current. Line E indicates an arbitrary limit within which the transistor manufacturer guarantees its specifications rather than a maximum allowable current.

Operating Point and Load Line

The purpose of the biasing arrangement is to locate and maintain operation in the permissible region. Within this limitation we wish to obtain maximum power output.

In general, the load line should lie below the maximum dissipation curve, and its slope (determined by the load resistance) should reflect a compromise between large signals and low distortion.

The factors influencing the choice of Q point and load resistance are shown in Fig. 17.15, where R_L is assumed to be in the collector circuit. Quiescent point Q_1 is well below the maximum dissipation curve and does not permit maximum voltage and

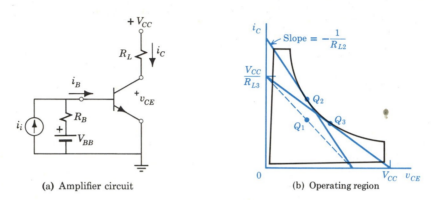

(a) Amplifier circuit   (b) Operating region

Figure 17.15   Location of the load line for a simple amplifier.

current swings. A second possibility is Q_2 where the load line for $R_{L2} < R_{L1}$ is drawn tangent to the hyperbola; this permits the same voltage swing as Q_1 and a larger current swing. The load line for $R_{L3} > R_{L1}$, tangent to the hyperbola at Q_3, permits the same current swing as Q_1 and a larger voltage swing. Operation at any point along the maximum dissipation curve between Q_2 and Q_3 will permit approximately the same signal power output. The choice will depend on practical factors such as the available supply voltage or the resistance of the load.

The practical amplifier of Fig. 17.16a includes some additional complexities. Since the capacitances are open circuits for dc bias currents, the operating point and *dc load line* are determined from the equivalent dc circuit where $R_{dc} = R_C + R_E$. For maximum voltage swings, operation should be near the midpoint of the load line or $V_{CC} \cong 2V_{CE}$. Since the quiescent value $V_{CE} = V_{CC} - I_C R_{dc}$,

$$R_{dc} = R_C + R_E \cong \frac{V_{CE}}{I_C} \qquad (17\text{-}21)$$

a typical design condition.

The ac signals "see" a different circuit. $R_B = R_1 \| R_2$ is chosen large enough so that most of the signal current flows into the base. C_E is an ac short circuit; therefore R_E is effectively removed. V_{CC} is a steady dc source and there can be no ac voltage across it; therefore the upper end of R_C is effectively grounded. C_{C2} is an ac short circuit, therefore R_C and R_L are effectively in parallel across the output voltage v_o, which is the signal component of v_{CE}. The ac load line defined by

$$R_{ac} = R_C \| R_L = \frac{R_C R_L}{R_C + R_L} \qquad (17\text{-}22)$$

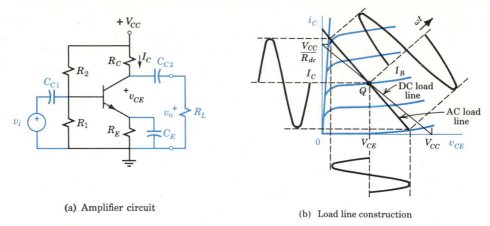

(a) Amplifier circuit

(b) Load line construction

Figure 17.16 Operation of a large-signal amplifier.

is shown in Fig. 17.16b. For this circuit $R_{dc} > R_{ac}$ and the optimum location of the Q point is at the midpoint of the ac load line. (The ac load line defines the relation between signal values and device characteristics; the dc load line just identifies the supply voltage V_{CC} required for operation at Q. See Example 4 on p. 482.)

Transformer Coupling

As illustrated in Example 4, the supply battery voltage V_{CC} must exceed the average collector voltage V_{CE} by the average voltage drop across the collector resistor $I_C R_C$. Furthermore, most of the signal power developed is dissipated in R_C instead of being transferred to R_L. The necessary supply voltage can be reduced and the power transfer improved by using a *transformer* to couple the output signal to the load.

As pointed out on p. 221, a transformer passes on ac signals but, since only *changing* magnetic flux induces secondary voltage, dc currents flow only in the low-resistance primary winding. With transformer coupling (Fig. 17.18), the supply voltage V_{CC} is approximately equal to V_{CE} instead of $V_{CE} + I_C R_C$; the dc power is correspondingly reduced. Also, by proper choice of the turn ratio a transformer permits impedance matching for maximum power transfer. For example, the low resistance of a loudspeaker (R_L) can be made to appear to a transistor as a much higher value (R_L'). For resistances, Eq. 8-28 becomes

$$R_L' = \left(\frac{N_1}{N_2}\right)^2 R_L \tag{17-23}$$

To gain these benefits, the output of an audiofrequency amplifier may be coupled to the load by a transformer. The performance of a well-designed transformer approaches the ideal over the major part of the audible range, but the output falls off at very low and very high frequencies. (See Fig. 21.20.)

Power and Efficiency

The calculation of output power and efficiency can be illustrated for the case of the simplified transformer-coupled transistor amplifier of Fig. 17.18. Here the chosen

EXAMPLE 4

The audiofrequency amplifier of Fig. 17.17 employs the silicon power transistor for which characteristics are shown. The manufacturer specifies $I_C(\text{max}) = 1$ A, $V_{CE}(\text{max}) = 50$ V, $P_D(\text{max}) = 5$ W (case at $25°$ C), and $\beta_{\text{min}} = 30$. Design the amplifier for maximum power to $R_L = 750\ \Omega$; an approximate design is satisfactory. (Assume capacitances are "large" and use "preferred values" of resistance.)

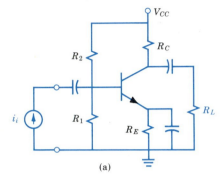

(a)

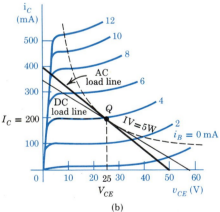

(b)

Figure 17.17 Amplifier design.

To select the operating point, the maximum dissipation line $I_C V_{CE} = 5$ W is sketched in as shown. The ac load line should be tangent to the hyperbola and there is a range of slopes with nearly equal output powers. The high-voltage region is judged to be slightly more linear than the high-current region, so the decision is made to use the full allowable voltage and let the current be limited by dissipation.

At $V_{CE} = V_{CEQ} = 50/2 = 25$ V and $I_C = 0.2$ A,

$$P_C = V_{CE} I_C = 25 \times 0.2 = 5\ \text{W}$$

and the corresponding ac load resistance is

$$R_{ac} = \frac{V_{pk}}{I_{pk}} = \frac{V_{CE}}{I_C} = \frac{25}{0.2} = 125\ \Omega$$

$$\therefore R_C = \frac{1}{\dfrac{1}{R_{ac}} - \dfrac{1}{R_L}} = \frac{1}{0.008 - 0.0013} = 150\ \Omega$$

To design the bias circuit, we assume a voltage $V_E = 3$ V; the emitter resistance is

$$R_E \cong \frac{V_E}{I_C} = \frac{3}{0.2} = 15\ \Omega$$

Then the necessary supply voltage is

$$V_{CC} = 3 + V_{CE} + I_C R_C = 3 + 25 + 30 = 58\ \text{V}$$

To minimize the effect of changes in β (Eq. 17-12), let

$$R_B = \frac{\beta_{\text{min}} R_E}{10} = \frac{30 \times 15}{10} = 45\ \Omega$$

Next, we read $I_B = 4$ mA, assume $V_{BE} = 0.7$ V for silicon, and by Eq. 17-19,

$$V_{BB} = I_B R_B + V_{BE} + (I_B + I_C) R_E$$

$$= 0.004 \times 45 + 0.7 + (0.204)\,15$$

$$= 0.18 + 0.7 + 3.06 \cong 4\ \text{V}$$

To complete the bias design, by Eqs. 17-20,

$$R_2 = R_B \frac{V_{CC}}{V_{BB}} = 45\,\frac{58}{4} = 652 \cong 680\ \Omega$$

$$R_1 = \frac{R_2 R_B}{R_2 - R_B} = \frac{680 \times 45}{680 - 45} = 48 \cong 47\ \Omega$$

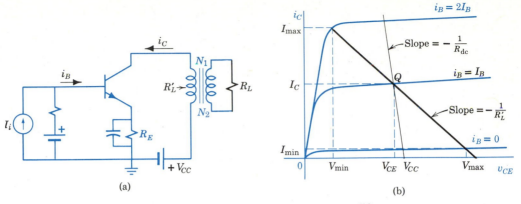

Figure 17.18 Simplified circuit of a transformer-coupled power amplifier.

Q point and desired signal swing define the ac load line whose slope is $-1/R'_L$. Once R'_L is determined, the turn ratio N_1/N_2 is selected to provide a match with the actual load resistance R_L. The dc load line has a slope of $-1/R_{dc}$; in this case R_{dc} is just R_E plus the small winding resistance of the transformer primary. Hence $I_C R_{dc} \ll V_{CE}$, and therefore $V_{CC} \cong V_{CE}$.

Assuming a sinusoidal signal current in a resistive load, the amplitude of the sinusoid is just one-half the difference between the maximum and minimum values of current, and the rms value is

$$I_c = \frac{I_{pk}}{\sqrt{2}} = \frac{1}{2\sqrt{2}}(I_{max} - I_{min}) \tag{17-24}$$

The output signal power is

$$P_o = I_c^2 R'_L = V_c I_c = \frac{V_c^2}{R'_L} \tag{17-25}$$

where V_c is the rms value of the signal voltage across R'_L.

With transformer coupling, the average dc power supplied by the collector battery (R_E is to be small and is neglected for the moment) is

$$P_{CC} = V_{CC} I_C \cong V_{CE} I_C \tag{17-26}$$

For sinusoidal signals, the average power dissipated in the transistor is

$$P_D = \frac{1}{2\pi} \int_0^{2\pi} v_{CE} i_C \, d(\omega t)$$

$$= \frac{1}{2\pi} \int_0^{2\pi} (V_{CE} - \sqrt{2} V_c \sin \omega t)(I_C + \sqrt{2} I_c \sin \omega t) \, d(\omega t)$$

$$= V_{CE} I_C - V_c I_c \tag{17-27}$$

This interesting result indicates that the power dissipated is the difference between a constant input $V_{CE} I_C$ and a variable output $V_c I_c$. The losses to be dissipated are low when the signal output is high, and the power to be dissipated is maximum in the quiescent condition with no signal applied.

The efficiency of the output circuit is defined as signal power output over dc power input or

$$\text{Efficiency} = \frac{P_o}{P_I} = \frac{V_c I_c}{V_{CE} I_C} \tag{17-28}$$

The theoretical limit for efficiency in this type of amplifier can be determined by considering the idealized current amplifier whose characteristics are shown in Fig.

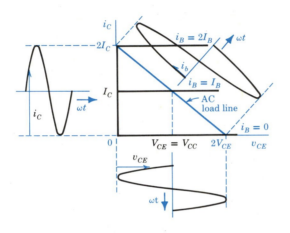

Figure 17.19 Operation of an idealized current amplifier.

17.19. A signal of amplitude equal to I_B produces a total current swing from $2I_C$ to zero and a total voltage swing from $2V_{CE}$ to zero. The ideal efficiency is

$$\frac{P_o}{P_I} = \frac{(V_{CE}/\sqrt{2})(I_C/\sqrt{2})}{V_{CE} I_C} \times 100 = 50\% \tag{17-29}$$

At maximum signal input, the actual efficiency of transistor amplifiers is around 40 to 45%. (See Example 5.) The *average* efficiency with variable signal input is usually much lower.

A small but significant amount of input power is required to "drive" the base of a large-signal amplifier. In Example 5, the rms value of base signal current is $4/\sqrt{2}$ mA and the necessary rms value of base-emitter signal voltage may be around $0.5/\sqrt{2}$ V. (The actual value can be predicted from I_C vs. V_{BE} characteristics provided by the manufacturer.) The required input power of approximately 1 mW is supplied by a small-signal amplifier of the type discussed in Chapter 19.

OTHER TYPES OF AMPLIFIERS

The emphasis in this chapter is on untuned power amplifiers. Some special forms of these amplifiers and some different types of amplifiers deserve mention.

EXAMPLE 5

Modify the design of Example 4 to provide transformer coupling for maximum power output to a load of 5 Ω, and predict the amplifier output and efficiency.

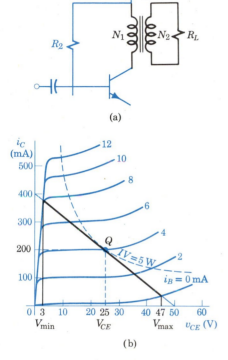

(a)

(b)

Figure 17.20 Amplifier performance.

The circuit modification is shown in Fig. 17.20a. The transformer should match the 5-Ω load to the 125-Ω load line and a commercial unit would be so labeled. The turn ratio should be (by Eq. 17-23)

$$N_1/N_2 = \sqrt{R'_L/R_L} = \sqrt{125/5} = 5$$

Now $I_C R_C \cong 0$ and the necessary supply voltage is

$$V_{CC} = V_{CE} + 3 = 25 + 3 = 28 \text{ V}$$

(R_1 and R_2 must be recalculated.)

To determine the amplifier performance, we note that the ac load line in Fig. 17.20b intersects the $i_B = 8$ mA line at $V_{min} \cong 3$ V, and the $i_B = 0$ line at $V_{max} = 47$ V; therefore the maximum undistorted voltage swing is

$$\frac{V_{max} - V_{min}}{2} = \frac{47 - 3}{2} = 22 \text{ V} = \sqrt{2}V_c$$

The power output is

$$P_o = \frac{V_c^2}{R'_L} = \frac{(22/\sqrt{2})^2}{125} = 1.94 \text{ W}$$

Neglecting R_E, the collector circuit efficiency is

$$\frac{P_o}{P_I} = \frac{1.94}{5} \times 100 \cong 39\%$$

If the emitter resistance loss is included, $V_{CC} = 28$ V and, assuming $I_E \cong I_C$, the efficiency is

$$\frac{P_o}{V_{CC}I_C} = \frac{1.94}{28 \times 0.2} \times 100 \cong 35\%$$

Class B and C Operation

If corresponding values of i_B and i_C are obtained from the ac load line of Fig. 17.16b and plotted as in Fig. 17.21a, the *transfer characteristic* is obtained. For distortion-free operation of untuned amplifiers the operating point is placed in the center of the linear part of the transfer characteristic and output current flows throughout the input-signal cycle. This condition is called *class A* operation and results in low distortion and also low efficiency because of the high value of dc current I_C with respect to the output signal current.

If the amplifier is biased to cutoff, output current flows only during the positive half-cycle and the output is badly distorted. The resulting *class B* operation is useful in a push-pull circuit (Fig. 17.22). The second transistor supplies the other half-cycle of output current and the even harmonic distortion is cancelled. The average value of collector current is much lower, and therefore, the dc power input is less and the efficiency is higher.

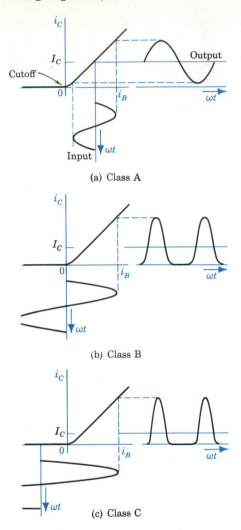

Figure 17.21 Amplifier operation modes.

If the amplifier is biased beyond cutoff, output current flows during only a small part of the cycle. With *class C* operation, the output is highly distorted, but with a tuned load impedance this presents no problem. (See Fig. 17.24.) The tuned circuit selects the fundamental component of the signal and rejects all others. A sinusoidal output is obtained and the efficiency is high because the average collector current is relatively low.

Push-Pull Amplifiers

Much of the distortion introduced in large-signal amplifiers can be eliminated by using two transistors in the *push-pull* circuit of Fig. 17.22. The input transformer T_1 receives a sinusoidal voltage from a low-level source. The signals applied to the two transistors are 180° out of phase, and the resulting collector currents are 180° out of phase. The output transformer T_2 delivers to the load a current that is proportional to

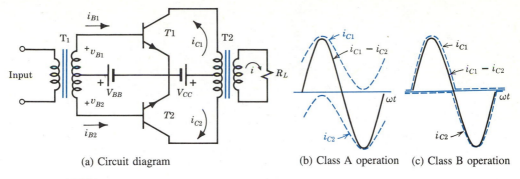

(a) Circuit diagram   (b) Class A operation   (c) Class B operation

Figure 17.22 Basic push-pull amplifier circuit.

the difference of the two collector currents. Any even-harmonic distortion components tend to cancel and the only distortion is that due to odd harmonics. Also the performance of transformer T_2 (a critical component) is improved because the dc components of i_{C1} and i_{C2} just cancel; magnetic core saturation and the accompanying nonlinearity are avoided.

The push-pull circuit is particularly useful for class B operation. Bias supply V_{BB} is set for the turn-on voltage, about 0.7 V for silicon transistors. (In practice, V_{BB} is obtained from a voltage divider across V_{CC} or from the voltage drop across a diode.) As v_{B1} goes positive, i_{B1} begins to flow and i_{C1} follows (Fig. 17.22c). Since v_{B2} is negative, transistor $T2$ is cut off. When v_{B1} goes negative, v_{B2} is positive and transistor $T2$ takes over. If the two transistors have identical characteristics, as they will have in integrated circuit form, even harmonics are cancelled and the result is a nearly pure sinusoidal output.

Following reasoning similar to that in the derivation of Eq. 17-29, the ideal class B efficiency can be calculated. For two transistors, $P_I = 2V_{CC}I_C$ and $I_C = I_{dc} = I_{max}/\pi$. Therefore,

$$\text{Efficiency} = \frac{P_o}{P_I} = \frac{V_c I_c}{2V_{CC}I_C} = \frac{(V_{max}/\sqrt{2})(I_{max}/\sqrt{2})}{2V_{max}(I_{max}/\pi)}$$

$$= \frac{\pi}{4} = 0.785 \quad \text{or} \quad 78.5\%$$

Since *losses* in a transistor must not exceed the rated heat dissipation, two transistors in class B can supply nearly six times the power output of one similar transistor in class A. (See Exercise 32.)

The expensive and bulky transformers can be eliminated by using an *npn* and a *pnp* transistor in the *complementary-symmetry* configuration. In Fig. 17.23a, the base-emitter voltage is nearly constant as long as the transistor is forward biased. If bias resistors R_1 and R_2 are chosen to place operation in the linear region, V_o will differ from V_i by a small constant value; in other words, the voltage at the emitter will *follow* the input voltage. (As we shall see in Chapter 19, the *emitter follower* has important advantages in small-signal amplifiers.)

If R_1 and R_2 are selected to bias the *npn* transistor to cutoff, V_o will follow V_i during the positive swing of the input signal and be zero during the negative half-cycle. While

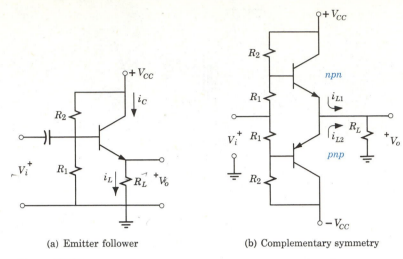

(a) Emitter follower   (b) Complementary symmetry

Figure 17.23 A class B complementary amplifier.

V_i is zero or negative, $i_C \cong i_L = 0$ and negligible power is dissipated. If a *pnp* transistor is added in the complementary connection of Fig. 17.23b, i_{L2} will flow during the negative half-cycle of V_i and the output of this push-pull amplifier will be linear. In IC versions of this circuit, resistances R_1 are replaced by diodes whose forward voltages "track" the base-emitter voltages of the transistors. The high efficiency of class B operation is obtained at low cost.

Tuned Amplifiers

The gain of a transistor amplifier depends on the load impedance. If a high-Q parallel resonant circuit is used for the load (Fig. 17.24), a very high resistance is presented at the resonant frequency and therefore the voltage gain is high. The

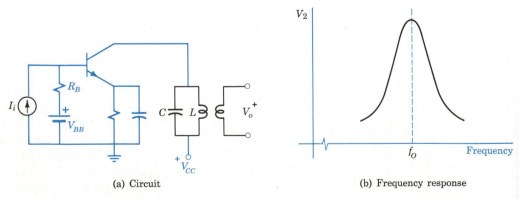

(a) Circuit   (b) Frequency response

Figure 17.24 A tuned amplifier.

frequency-response curve has the same shape as that of the resonant circuit. A narrow band of frequencies near resonance is amplified well, but signals removed from the resonant frequency are discriminated against.

The high selectivity of the load impedance eliminates nonlinear distortion; any harmonics in the input signal or in the collector current itself develop little voltage across the load impedance. As long as the collector current has a component at the resonant frequency, the output is nearly sinusoidal. With distortion eliminated, high efficiency is achieved by operating the transistor in a nonlinear region.

SUMMARY

- An amplifier is a device for raising the level of a signal voltage, current, or power. For sinusoidal signals, the voltage gain is

$$\mathbf{A} = \frac{\mathbf{V}_{\text{out}}}{\mathbf{V}_{\text{in}}} = A\,e^{j\theta}$$

  If A and θ are constant (or if $\theta = k\omega$), the amplification is linear and there is no distortion.

- In practical amplifiers, power supplies and biasing networks must maintain operation at the proper point, coupling circuits must transfer signals from one stage to the next without excessive discrimination, and load impedances must provide desired output without requiring excessive supply voltages.

- For FETs used as small-signal amplifiers, biasing is not critical and various circuits are available.
  For BJTs used as large-signal amplifiers, biasing by a combination of emitter resistor and voltage divider works satisfactorily.
  For noncritical design, an approximate bias procedure is available.

- The permissible operating region of a transistor is defined by maximum allowable current, voltage, power, and distortion.
  For large-signal class A operation, the midpoint of a properly located ac load line is the optimum quiescent point.

- Transformer coupling reduces the required supply voltage, isolates the signal output, permits impedance matching, and improves efficiency.
  In an idealized transformer-coupled, class A current amplifier:

$$\text{Input power (dc)} = P_I = V_{CC}I_C = V_{CE}I_C$$

$$\text{Output power (max)} = P_o = V_cI_c = \tfrac{1}{2}V_{CE}I_C = \tfrac{1}{2}P_I$$

- Class B and C amplifiers operate efficiently because output current is cut off during ineffective portions of the signal cycle.
  The push-pull amplifier reduces distortion and improves efficiency.
  The tuned amplifier uses a resonant load circuit to obtain high selectivity and low distortion.

REFERENCES

1. Jacob Millman, *Microelectronics*, McGraw-Hill Book Co., New York, 1979, Chapters 11 and 18.

2. Charles Holt, *Electronic Circuits*, John Wiley and Sons, New York, 1978, Chapters 15 and 16.

3. Donald Schilling and Charles Belove, *Electronic Circuits: Discrete and Integrated*, McGraw-Hill Book Co., New York, 1979, Chapters 4 and 5.

REVIEW QUESTIONS

1. Distinguish between tuned and wide-band amplifiers and give an application of each (other than those in the text).
2. Distinguish between frequency, phase, and amplitude distortion. Explain the effect of each in a hi-fi system.
3. What determines the dynamic range of an amplifier?
4. What is the purpose of biasing in a MOSFET? In a BJT?
5. What are the relative merits of RC and transformer coupling?
6. Explain the practical considerations involved in the selection of values for: R_1, R_2, R_D, R_S in Fig. 17.8b.
7. Explain the operation of drain-feedback bias.
8. Why is fixed-current transistor bias unsatisfactory?
9. Explain thermal runaway.
10. Explain the function of each circuit element in Fig. 17.25.
11. Explain how the circuit of Fig. 17.25 stabilizes I_C if V_{CE} changes.
12. What are the practical limits on R_1 and R_3 in Fig. 17.25?
13. How are the quiescent point and load line related to the permissible operating region?

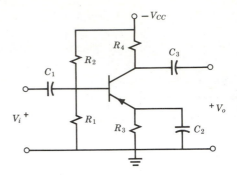

Figure 17.25

14. Sketch a set of transistor characteristics and indicate regions of high amplitude distortion.
15. Explain the statement: "A power transistor runs cool when the power output is high."
16. How does a transformer improve amplifier efficiency?
17. Why is impedance matching desirable?
18. Explain how push-pull amplifiers work. Where are they used?
19. How are "tuned amplifiers" tuned? Where are they used?
20. Explain why class B operation is more efficient than class A.

EXERCISES

1. Explain by means of sketches the differences between amplitude, frequency, and phase distortion. How would each be evidenced in a hi-fi system?
2. A MOSFET with the characteristics of Fig. 17.5a is used as a voltage amplifier with $V_{DD} = 20$ V and $R_D = 6$ kΩ.
   (a) Draw a "zero-bias" circuit and predict the quiescent values I_D and V_{DS}.
   (b) Sketch the transfer characteristic for a "zero-bias" circuit.
   (c) Estimate from the transfer characteristic the output signal current i_d for an input voltage $v_i = 0.5 \sin \omega t$ V.
   (d) Estimate the voltage gain of the amplifier.
3. Redraw the circuit of Fig. 17.5c modified by inserting resistance R_S between the S terminal and ground.

   (a) Derive an equation for i_D as $f(v_{DS})$ for the modified circuit.
   (b) The MOSFET of Fig. 17.5a is to operate at $I_D = 1$ mA and $V_{DS} = 8$ V; for $R_D = 5$ kΩ, specify R_S and V_{DD}. Check your answers by drawing the load line.
4. A JFET for which $I_{DSS} = 8$ mA and $V_p = -6$ V is used in the circuit of Fig. 17.26 where $V_{DD} = 32$ V, $V_{GG} = 10$ V, $R_D = 3$ kΩ, $R_S = 2.5$ kΩ, and $R_G = 10$ MΩ.
   (a) Estimate I_D by assuming $V_{GS} = 0$ and $I_G = 0$.
   (b) Using the estimated value of I_D, estimate V_{GS} by Eq. 12-1 and recalculate I_D from the circuit. Draw a conclusion about the method in part (a).

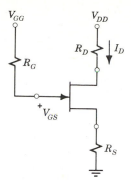

Figure 17.26

5. The bias circuit for a JFET is similar to Fig. 17.7 with $V_{DD} = 24$ V, $R_1 = 4$ MΩ, and $R_2 = 8$ MΩ. You are to complete the design to provide quiescent operation at $I_D = 2$ mA, $V_{DS} = 7$ V, and $V_{GS} = -3$ V.
   (a) Draw a circuit suitable for completing the bias design.
   (b) Use your circuit to specify R_S and R_D.

6. In the circuit of Fig. 17.7, $V_{DD} = 24$ V, $R_D = 3$ kΩ, $R_S = 5$ kΩ, $R_1 = 6$ MΩ, and $R_2 = 12$ MΩ.
   (a) Assume $V_{GS} = -2$ V and predict the quiescent values I_D and V_{DS}.
   (b) Assume V_{GS} is actually -1 V and repeat part (a).
   (c) Draw a conclusion about the stability of this bias scheme.

7. A JFET for which $I_{DSS} = 5$ mA and $V_{GS(OFF)} = -6$ V is to be operated at a Q point defined by $I_D = 3$ mA and $V_{DS} = 10$ V. Assuming $R_{in} \geq 10$ MΩ and $R_D = 3$ kΩ, design an appropriate bias circuit.

8. For the MOSFET of Fig. 17.8, the threshold voltage is 2 V and the drain current is 8 mA at a gate-source voltage of 6 V.
   (a) Sketch the transfer characteristic and predict the drain current at a quiescent point defined by $V_{GS} = 4$ V and $V_{DS} = 10$ V.
   (b) Design the bias circuit for $V_{DD} = 24$ V; i.e., specify R_2 and R_D.

9. The enhancement MOSFET of Fig. 12.3 is to be operated in the normal region at $I_D = 2$ mA. Draw a drain-feedback bias circuit and specify V_{DD} assuming $R_D = 7$ kΩ.

10. The data sheet of an enhancement MOSFET specifies $V_T = -3$ V and $I_D = -4$ mA for $V_{GS} = V_{DS} = -8$ V.

(a) Determine V_{GS} for $I_D = -2$ mA.
(b) Specify R_D for drain-feedback bias at $I_D = -2$ mA with $V_{DD} = 16$ V.

11. In the circuit of Fig. 17.9a, $I_B = 0.1$ mA, I_{CBO} for the germanium transistor is 5 μA, and $β$ ranges from 50 to 200.
   (a) Predict the corresponding range of values of I_C.
   (b) If $β = 150$ and if I_{CBO} quadruples as a result of temperature increase, predict the change in I_C.

12. A silicon transistor with a nominal $β = 100$ is to operate with $I_C \cong 2$ mA and $V_{CE} = 6$ V in the circuit of Fig. 17.9.
   (a) For $V_{CC} = 12$ V, specify R_C and R_B.
   (b) If the $β$ is actually 40, estimate the actual value of I_C for the value of R_B specified in part (a).

13. The transistor of Exercise 12 is to operate under the same quiescent conditions in the circuit of Fig. 17.10.
   (a) For $R_C = 3$ kΩ and $R_E = 0.5$ kΩ, specify V_{CC}.
   (b) For $R_1 = 12$ kΩ and $R_2 = 68$ kΩ, calculate V_{BB} and R_B.
   (c) If the $β$ is actually 40, estimate the actual value of I_C.
   (d) Compare the performance of the circuit with that in Exercise 12.

14. The one-stage audio amplifier shown in Fig. 17.25 employs a silicon transistor with a nominal $β$ of 100 ($β_{min} = 70$).
   (a) Replace the BJT by a suitable model, consider the effect of the capacitors at dc, and draw a new circuit appropriate for bias analysis.
   (b) Use your circuit to explain how this bias scheme "stabilizes the collector current against changes in $β$ and V_{BE}."
   (c) Given $V_{CC} = 15$ V and $R_3 = 2$ kΩ, use your circuit to specify R_1, R_2, and R_4 for quiescent operation at $I_C = -1.25$ mA and V_{CE} at -5 V. State any assumptions.

15. In Fig. 17.16a, $V_{CC} = 10$ V, $R_1 = 58$ kΩ, $R_2 = 142$ kΩ, $R_C = 8$ kΩ, and $R_E = 3.9$ kΩ. Drawing an appropriate circuit model and stating any assumptions, estimate dc current I_C and voltage V_{CE}.

16. In the self-bias circuit (Fig. 17.10) $R_1 = 70$ kΩ, $R_2 = 140$ kΩ, $R_E = 3.9$ kΩ, and $V_{CC} = 15$ V. Estimate I_C and specify resistor R_C to establish

the quiescent voltage across the silicon transistor (V_{CE}) at 5 V.

17. A silicon transistor whose β may vary from 50 to 200 (nominal $\beta = 100$) is to be operated at $I_C = 2$ mA and $V_{CE} = 5$ V. For this amplifier application, $R_C = 2$ kΩ.
    (a) Using the approximate procedure, design an appropriate bias network and draw a labeled circuit diagram.
    (b) Determine the maximum variation in I_C expected for the given variation in β.

18. A germanium transistor with a nominal β of 60 is to be operated at $I_C = 0.5$ mA and $V_{CE} = 4$ V with the available $V_{CC} = 10$ V.
    (a) Using the approximate procedure, design an appropriate self-biasing network and draw a labeled circuit diagram.
    (b) If the actual β varies from 20 to 100, predict the maximum variation in I_C.

19. The data sheet for a 2N3114 specifies a maximum power dissipation of 5 W with the case at 25° C and a "derating factor" of 28.6 mW/°C above 25° C to maintain a maximum junction temperature of 200° C.
    (a) Explain the term "derating factor" in terms of energy transformations within the transistor and between the transistor and the case.
    (b) If during operation the case temperature of the 2N3114 may rise to 75° C, estimate the allowable power dissipation.

20. The collector dissipation for the transistor of Fig. 17.20b is to be limited to 3 W, maximum I_C is 1 A, and maximum V_{CE} is to be 60 V.
    (a) Reproduce the characteristics and sketch in the permissible operating region.
    (b) For operation at $V_{CE} = 20$ V, specify a suitable R_{ac} and the nominal I_B for maximum power output.

21. A silicon transistor with a β of 50 and a maximum allowable collector dissipation of 10 W is to be used in the circuit of Fig. 17.15a. The maximum collector current is 2 A and the maximum collector voltage is 50 V.
    (a) Making and stating any desirable simplifying assumptions, sketch the permissible operating region for this transistor.
    (b) Estimate the maximum possible signal power output in R_L and the efficiency.

22. The silicon transistor in the circuit of Fig. 17.16a has a nominal β of 100; $V_{CC} = 15$ V, $R_1 = 10$ kΩ, and $R_2 = 30$ kΩ.

(a) Specify R_E and R_C to put the Q point at $V_{CE} = 6$ V and $I_C = 2$ mA.
(b) Sketch the i_C vs. v_{CE} characteristics and draw the dc load line.
(c) For $R_L = 3$ kΩ, draw the ac load line and estimate the output voltage v_o for an input current $i_b = 10 \sin \omega t$ μA.

23. Repeat part (b) of Exercise 21 assuming that R_L is replaced by an ideal transformer providing matched coupling to $R_L = 750$ Ω. Specify the turn ratio of the transformer. Compare the results to those for Exercise 21, and draw a conclusion.

24. Specify the turn ratio of a transformer to provide impedance matching between a 3600-Ω source and:
    (a) A 16-Ω speaker.
    (b) Four 16-Ω speakers operated in parallel.

25. A transistor is rated at $I_C(\max) = 1$ A, $V_{CE}(\max) = 60$ V, $P_D(\max) = 3$ W. Estimate the maximum signal power output for an amplifier using this transistor in class A operation.

26. In Example 4 the Q point is at $V_{CE} = 25$ V and $I_{CE} = 200$ mA.
    (a) For no input signal, estimate the collector power supplied by the battery, the power dissipated in R_C and R_E, the power dissipated at the collector of the transistor, and the power delivered to R_L.
    (b) Repeat for an input signal $i_b = 2 \cos \omega t$ mA.
    (c) Define "efficiency" for this power amplifier and calculate it.
    (d) Compare the power output and efficiency to the results of Example 5.

27. The permissible operating region of a power transistor is defined by $P_D = 5$ W, $I_C(\max) = 1$ A, $V_{CE}(\max) = 100$ V, and $V_{CE}(\min) = 2$ V.
    (a) Sketch the transistor characteristics and select an appropriate quiescent point for operation in the circuit of Fig. 17.15a.
    (b) Specify R_L for nearly maximum power output.
    (c) Calculate total dc power in, maximum signal power out, and overall efficiency.
    (d) If the actual load is 10 Ω, specify the appropriate transformer turn ratio and recalculate the overall efficiency of the amplifier.

28. The transistor of Fig. 17.14 is to operate with an effective load resistance of 45 Ω.

(a) Select an optimum operating point and specify I_C and V_{CC}.

(b) Show the Q point on a sketch of the characteristics and estimate the maximum symmetric swing in collector current and collector voltage.

(c) Estimate the rms value of input current required from the driver.

(d) Assuming that a transformer matches the load to $R_L' = 45\ \Omega$, calculate the dc power in, the maximum signal power out, and the overall efficiency.

29. The transistor of Fig. 17.14 is used in the circuit of Fig. 17.18 where R_L is a 4-Ω loudspeaker, $N_1/N_2 = 3$, and $R_E = 5\ \Omega$. The Q point is to be at $I_C = 0.4$ A and $V_{CE} = 10$ V.

(a) Reproduce the characteristics and draw the load line.

(b) Specify the battery voltage and estimate the input signal current (rms) required.

(c) Estimate the maximum current and power delivered to the loudspeaker.

30. A transistor rated at $I_C(\text{max}) = 1.6$ A, $V_{CE}(\text{max}) = 50$ V, $V_{CE(\text{sat})} \cong 0$ V, $I_{CEO} \cong 0$, and $P_D(\text{max}) = 10$ W is used in the circuit of Fig. 17.15a.

(a) Draw, approximately to scale, the permissible operating region.

(b) For $R_L = 100\ \Omega$, select the "best" quiescent point, explaining your reasoning.

(c) Predict the average collector dissipation, the maximum power output to R_L, and the collector circuit efficiency.

(d) If your Q point does not take full advantage of the capability of the transistor, rearrange the circuit to provide maximum possible power to the same 100-Ω load and repeat part (c).

31. If two transistors of the type described in Exercise 25 are used in a class B push-pull amplifier, estimate the maximum signal power output and draw a conclusion.

32. For class B operation:

(a) Show the operation of an amplifier on the idealized characteristics of Fig. 17.19.

(b) Demonstrate that two transistors in class B can supply nearly six times the power of one similar transistor in class A.

33. In the complementary amplifier of Fig. 17.23, $V_{CC} = 24$ V and $R_L = 8\ \Omega$. Estimate the maximum value of V_i for proper operation. Neglecting the effect of the constant base-emitter drop, estimate the average (dc) collector current and the signal power delivered to R_L for the maximum allowable signal input.

PROBLEMS

1. An inexperienced designer uses the transistor of Example 2 in the circuit of Fig. 17.9a with $V_{CC} = 20$ V and $R_C = 5$ kΩ.

(a) Specify R_B for a nominal $I_C = 1$ mA.

(b) Predict the extreme variation in I_C for this circuit, compare it to the variation in Example 2, and draw a conclusion regarding the effectiveness of this method.

2. A JFET (somewhat like Fig. 17.6a) and a BJT (somewhat like Fig. 17.9b) are connected in the *current-source bias* circuit of Fig. 17.27. Estimate I_D and V_{DS}. (*Hint:* Estimate the voltages at points 1, 2, and 3 in turn.) Evaluate the stability of this bias circuit.

3. A 2N3114 transistor (Fig. A8) is to be operated as a single-stage audiofrequency amplifier at $I_C = 2$ mA and $V_{CE} = 40$ V with $R_C = 10$ kΩ.

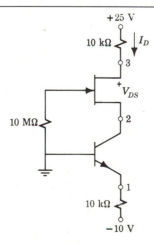

Figure 17.27

Draw an appropriate wiring diagram and, stating all assumptions, design the bias network.

4. The circuit of Fig. 17.28 is suggested as a simple means for obtaining improved stability.

   (a) Explain qualitatively what happens if I_C tends to rise as a result of an increase in I_{CBO} or β.

   (b) Derive an approximate expression for I_C in terms of I_{CBO} and β (somewhat like Eq. 17-10).

   (c) Under what conditions (i.e., for what relation between R_C and R_B) will I_C be insensitive to changes in β?

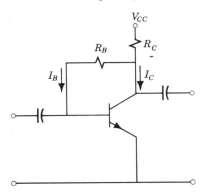

Figure 17.28

5. Prove that the point at which a load line is tangent to the hyperbola $P_D = V_{CE}I_{CE}$ is the midpoint of the load line.

6. A transformer-coupled audio amplifier is to supply 1 W to a 16-Ω loudspeaker. Available is a transistor whose permissible operating range is de-fined by $P_D = 4$ W, $V_{CE}(\text{max}) = 80$ V, $V_{CE}(\text{min}) = 2$ V at $I_C = 200$ mA, and $I_C(\text{max}) = 1$ A. Stating all assumptions, design the amplifier (draw the wiring diagram and specify the quiescent point, bias resistances, supply voltage, capacitor values, and transformer turn ratio) and estimate the total dc power requirement.

7. Demonstrate that the maximum theoretical efficiency for a class A amplifier with a resistive load in the collector circuit (not transformer coupled) is 25%.

8. The circuit of Fig. 17.29 has two outputs labeled 1 and 2. Predict I_C and V_{CE} for $v_i = 0$. Predict and sketch the output voltages for $v_i = V_m \sin \omega t$. What function does this circuit perform? Where would it be useful?

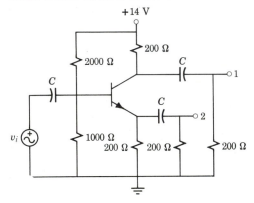

Figure 17.29

18

Small-Signal Models

Large and Small Signals

Diodes

Field-Effect Transistors

Bipolar Junction Transistors

The use of models to represent complicated phenomena or devices or systems is common practice throughout engineering and science. On the basis of experimental observations, the scientist may propose a tentative model that is then tested under various conditions. The engineer uses models to simplify the analysis of a known device or system or to predict the behavior of a proposed design. These models may be physical or mathematical in form; a linear model amenable to mathematical analysis is particularly desirable.

To predict the behavior of diodes and transistors in switching and biasing applications, we have used circuit models composed of ideal circuit elements. Working from characteristic curves, we devised relatively simple models to represent complicated physical devices under specified conditions. In an extension of that approach, we are now going to derive linear circuit models that are used to predict the behavior of semiconductor devices in practical applications over a wide range of frequencies. The character of the model depends on the accuracy required; more sophisticated models are used in computer-aided design for optimum performance. The character of the model also depends on the size of the signals expected.

LARGE AND SMALL SIGNALS

The circuit models derived in Chapters 11 and 13 are based on piecewise linearization of the characteristic curves and are useful for predicting the behavior of diodes and transistors under static conditions or where relatively large swings in voltage occur. A different approach is necessary if we are interested in the response to small signals superimposed upon dc values.

A Mathematical Model

Consider the circuit of Fig. 18.1 in which two sinusoidal signals $v_M(t)$ and $v_C(t)$ are superimposed on a bias voltage V_0 and the total is applied to a diode circuit. Using

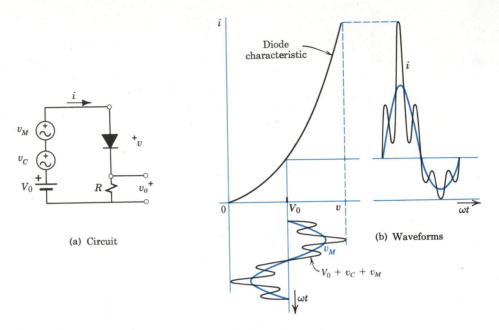

(a) Circuit   (b) Waveforms

Figure 18.1   Effect of nonlinearity on the sum of two signals.

the analytical technique of Chapter 2, we can represent a portion of the diode characteristic by a *mathematical model* described by the power series

$$i = a_0 + a_1v + a_2v^2 \tag{18-1}$$

If the voltage drop across R is relatively small, the full voltage

$$v = V_0 + v_M + v_C = V_0 + V_M \sin \omega_M t + V_C \sin \omega_C t \tag{18-2}$$

appears across the diode and

$$
\begin{aligned}
i = {}&a_0 + a_1V_0 + a_1V_M \sin \omega_M t + a_1V_C \sin \omega_C t \\
&+ a_2V_0^2 + a_2V_M^2 \sin^2 \omega_M t + a_2V_C^2 \sin^2 \omega_C t \\
&+ 2a_2V_0V_M \sin \omega_M t + 2a_2V_0V_C \sin \omega_C t \\
&+ 2a_2V_MV_C \sin \omega_M t \sin \omega_C t
\end{aligned}
$$

Substituting for $\sin^2 \omega t$ and $\sin \omega_M t \sin \omega_C t$ and rearranging terms,

$$
\begin{aligned}
i = {}&a_0 + a_1V_0 + a_2V_0^2 + \tfrac{1}{2}a_2V_M^2 + \tfrac{1}{2}a_2V_C^2 \\
&+ (a_1 + 2a_2V_0)V_M \sin \omega_M t + (a_1 + 2a_2V_0)V_C \sin \omega_C t \\
&- \tfrac{1}{2}a_2V_M^2 \cos 2\omega_M t - \tfrac{1}{2}a_2V_C^2 \cos 2\omega_C t \\
&+ a_2V_MV_C \cos (\omega_C - \omega_M)t - a_2V_MV_C \cos (\omega_C + \omega_M)t
\end{aligned}
\tag{18-3}
$$

The output voltage v_o is directly proportional to i and contains the same eleven components. The first five terms constitute a dc component dependent on the amplitude of the input signals, as expected in a nonlinear device. The next two terms are replicas of the input signals resulting from the linear term of Eq. 18-1. Then there are two terms representing second harmonics generated by the nonlinearity. The last two terms are of particular interest in communication.

Amplitude Modulation

In a typical communication or control system, a high-frequency easily propagated signal serves as a *carrier* and information is superimposed on the carrier in the process called *modulation*. In amplitude modulation (AM), the amplitude of the carrier $v_C = V_C \sin \omega_C t$ is varied by the modulating signal $v_M = V_M \sin \omega_M t$. The equation of the amplitude-modulated wave is

$$v_{AM} = (1 + m \sin \omega_M t)V_C \sin \omega_C t \qquad (18\text{-}4)$$

where m is the *degree of modulation* $= V_M/V_C$, approximately 80% in Fig. 18.2.

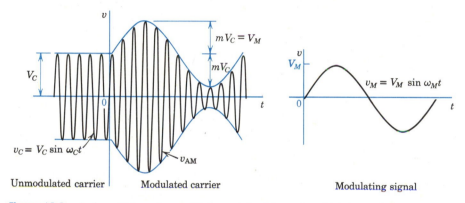

Figure 18.2   A sinusoidal carrier amplitude-modulated by a sinusoidal signal.

Expanding Eq. 18-4,

$$\begin{aligned}
v_{AM} &= V_C \sin \omega_C t + mV_C \sin \omega_M t \sin \omega_C t \\
&= V_C \sin \omega_C t + \tfrac{1}{2}mV_C \cos (\omega_C - \omega_M)t - \tfrac{1}{2}mV_C \cos (\omega_C + \omega_M)t
\end{aligned} \qquad (18\text{-}5)$$

This equation reveals that the amplitude-modulated wave consists of three sinusoidal components of constant amplitude. The carrier and the upper and lower *side frequencies* are shown in the frequency spectrum of Fig. 18.3a. In general the modulating signal is a complicated wave containing several components and the result is a carrier plus upper and lower *sidebands*. If the 910-kHz carrier of an AM station is modulated by frequencies from 100 to 4500 Hz, the frequency spectrum is as shown in Fig. 18.3b. The station is assigned a *channel* 10 kHz wide to accommodate side frequencies from 0 to 5000 Hz. The amplifiers in AM transmitters and receivers are designed to amplify well over a 10-kHz bandwidth and to discriminate against frequencies outside this range. In comparison, channels for frequency modulation (FM)

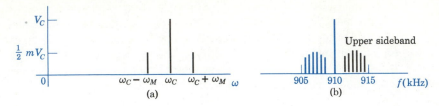

Figure 18.3 Frequency spectra of AM waves.

broadcast are 200 kHz wide and permit lower noise and higher frequency response for greater fidelity. For satisfactory television transmission, sidebands approximately 6 MHz wide are required.

EXAMPLE 1

A carrier signal $v_C = 100 \sin 10^7 t$ V is to be modulated by a signal $v_M = 100 \sin 10^4 t$ V, using a device with a characteristic defined by $i = v + 0.003v^2$ mA. The resulting current is filtered by a circuit that passes all frequencies from 9.9 to 10.1 Mrad/s and rejects all others. Describe the output signal.

Comparing the device characteristic to Eq. 18-1, $a_0 = 0$, $a_1 = 1$, and $a_2 = 0.003$. Only the seventh, tenth, and eleventh components in Eq. 18-3 have frequencies within the passband of the filter.

The magnitudes of the carrier and side-frequency components are

$$a_1 V_C = 100 \text{ mA} \quad \text{and} \quad a_2 V_M V_C = 30 \text{ mA}$$

The output, an amplitude-modulated signal, is

$$i = 100 \sin 10^7 t - 30 \cos 1.001 \times 10^7 t$$
$$+ 30 \cos 0.999 \times 10^7 t \text{ mA}$$

Comparing the second term with the corresponding term in Eq. 18-5,

$$\tfrac{1}{2} m I_C = \tfrac{1}{2} m \times 100 = 30 \quad \text{or} \quad m = \frac{60}{100}$$

and the degree of modulation is 60%.

Effect of Signal Size

As indicated in Example 1, a diode can be used as a *modulator*. The same nonlinear device will also serve as a *detector* to recover the information from the three radiofrequency signals or as a *mixer* to obtain the difference or "beat frequency" between two sinusoidal signals. (See Problems 1 and 3.) We know that the diode can also be used as a switch or as a rectifier. In all these functions, the signal excursions extend into the nonlinear regions of the diode characteristics. Obviously, the function actually performed is determined by the relative size of the coefficients of the various components in the output. These in turn depend on the factors a_0, a_1, and a_2 in the device characteristic, the bias voltage V_0, and the signal amplitudes V_M and V_C.

For linear applications, where the distorting effects of nonlinearity are to be avoided, the bias voltage is selected to place operation on a portion of the characteristic where factor a_2 is relatively small. Then the signal amplitudes are kept small. Since the coefficients of the unwanted components vary as the square or the product of the signal amplitudes, any device can be considered to be linear if the signals are sufficiently small. By definition,

A "small signal" is one to which a given device responds in a linear fashion.

DIODES

If we are interested in the response of a diode to small signals superimposed upon dc values, a special circuit model is necessary.

Dynamic Resistance

As shown in Fig. 18.4, a small signal voltage v superimposed on a steady voltage V_{dc} produces a corresponding signal current i. The current-voltage relation for small

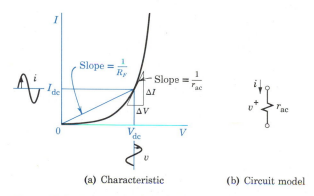

(a) Characteristic   (b) Circuit model

Figure 18.4   Dynamic resistance of a diode.

signals is defined by the slope of the curve at the operating point specified by V_{dc}. The *dynamic resistance* or *ac resistance* is defined as

$$r_{ac} = \frac{dV}{dI} \cong \frac{\Delta V}{\Delta I} \text{ in ohms} \tag{18-6}$$

In general, the dynamic resistance r_{ac} will differ appreciably from the static forward resistance defined as $R_F = V_{dc}/I_{dc}$.

The value of r_{ac} can be predicted for a theoretical diode for which

$$I = I_s(e^{eV/kT} - 1)$$

as in Eq. 11-21. Taking the derivative,

$$\frac{dI}{dV} = \frac{e}{kT}I_s e^{eV/kT} = \frac{e}{kT}(I + I_s) \cong \frac{eI}{kT} \tag{18-7}$$

Hence, for this theoretical diode,

$$r_{ac} = \frac{dV}{dI} = \frac{v}{i} = \frac{kT}{eI} = \frac{kT/e}{I} \tag{18-8}$$

The dynamic resistance is directly proportional to absolute temperature T and inversely proportional to dc diode current I. At room temperature, $T = 293$ K $\cong 300$ K, and $kT/e \cong 0.025$ V $= 25$ mV. Then

$$r_{ac} = \frac{0.025}{I} = \frac{25}{I(\text{in mA})} \Omega \tag{18-9}$$

This provides a simple means for estimating r_{ac} in practical diodes; however, the ohmic "body resistance" of the semiconductor (in series with the junction) may be the dominant factor at large currents.

FIELD-EFFECT TRANSISTORS

In the elementary amplifier of Fig. 18.5a, the input is a signal voltage v_i superimposed on the gate-source bias voltage V_{GG}. The signal voltage $v_{gs} = v_i$ modulates the width of the conducting channel and produces a signal component i_d of the drain current i_D. (For nomenclature, see Table 17-1 on p. 470.) The signal current develops across resistor R_L an amplified version of the input voltage.

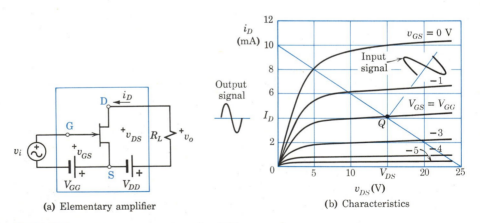

(a) Elementary amplifier (b) Characteristics

Figure 18.5   Small-signal operation of an FET.

The quantitative behavior of a small-signal amplifier could be determined graphically as we did in the case of large-signal BJT amplifiers, but this approach is not good for two reasons. First, the work is laborious and the results lack the generality needed in optimizing a design. Second, individual devices of a given type vary widely in their characteristics and the curves provided by the manufacturer are only general indications of how that type behaves on the average. A better approach is to devise a convenient model that will accurately predict device performance within limitations. If we limit our consideration to small signals superimposed on dc values, we are justified in assuming linear relations among signals in devices that are highly nonlinear in their large-signal behavior. On this basis, the complicated physical device inside the colored rectangle of Fig. 18.5a is to be replaced, insofar as small signals are concerned, by a linear circuit model.

Two-Port Parameters

A diode is a two-terminal device or one-port, and its circuit model also has a single pair of terminals. Transistors are two-ports and can be represented as in Fig.

Figure 18.6 The general two-port.

18.6. In general, of the four variables identified only two are independent, and the behavior can be defined by a pair of equations such as

$$\begin{cases} v_1 = f_1(i_1, i_2) \\ v_2 = f_2(i_1, i_2) \end{cases} \quad \text{or} \quad \begin{cases} i_1 = f_3(v_1, v_2) \\ i_2 = f_4(v_1, v_2) \end{cases} \tag{18-10}$$

For linear networks, the first pair of equations would define a set of impedances as parameters[†] and the second pair would define a set of admittances as parameters. Another acceptable formulation is

$$\begin{cases} v_1 = f_5(i_1, v_2) \\ i_2 = f_6(i_1, v_2) \end{cases} \tag{18-11}$$

For linear networks, the first equation of this pair would define an impedance and a voltage factor, and the second equation would define a current factor and an admittance. This mixture is called a set of *hybrid parameters*, and it is a particularly convenient set to use for electronic devices.

Evaluation of Parameters

In selecting an appropriate formulation for the field-effect transistor, we note that the input current is very small because in the JFET the gate-source junction is reverse biased and in the MOSFET the gate is insulated. Also, we note that the FET is a voltage-controlled device (Fig. 18.5b). Therefore, we write (Eq. 18-10b),

$$\begin{cases} i_G = 0 \\ i_D = f(v_{GS}, v_{DS}) \end{cases} \tag{18-12}$$

[†]This is more easily seen if the first pair of Eqs. 18-10 is rewritten:

$$\begin{cases} V_1 = Z_{11}I_1 + Z_{12}I_2 \\ V_2 = Z_{21}I_1 + Z_{22}I_2 \end{cases} \quad \text{for the linear sinusoidal case.}$$

Since $i_G = 0$ at all times, only the second equation is significant. Because we are interested in small or "differential" changes in the variables, we express the total differential of the drain current as

$$di_D = \frac{\partial i_D}{\partial v_{GS}} \, dv_{GS} + \frac{\partial i_D}{\partial v_{DS}} \, dv_{DS} \tag{18-13}$$

Equation 18-13 says that a change in gate voltage dv_{GS} and a change in drain voltage dv_{DS} both contribute to a change in drain current di_D. The effect of the partial contributions is determined by the coefficients $\partial i_D/\partial v_{GS}$ and $\partial i_D/\partial v_{DS}$; these are the "partial derivatives" of the drain current with respect to gate voltage and drain voltage.

If the drain-source voltage is held constant, $dv_{DS} = 0$ and the partial derivative with respect to gate-source voltage is equal to the total derivative or

$$\frac{\partial i_D}{\partial v_{GS}} = \frac{di_D}{dv_{GS}}\bigg|_{v_{DS}\,=\,k} = g_m \tag{18-14}$$

where g_m is the *transconductance* in siemens, so called because it is the ratio of a differential current at the output to the corresponding differential voltage at the input. Typical values of g_m lie between 500 and 10,000 μS.

A second FET parameter can be defined by holding the gate-source voltage constant so that $dv_{GS} = 0$ and

$$\frac{\partial i_D}{\partial v_{DS}} = \frac{di_D}{dv_{DS}}\bigg|_{v_{GS}\,=\,k} = \frac{1}{r_d} \tag{18-15}$$

where r_d is the *dynamic drain resistance* and is just equal to the reciprocal of the slope of a line of constant v_{GS}. Typical values of r_d lie between 20 and 500 kΩ. (Instead of r_d, the *output conductance* $g_{os} = g_d = 1/r_d$ may be specified.)

For the JFET or the DE MOSFET, the drain current in the normal operating region is (Eq. 12-1)

$$i_D = I_{DSS}\left(1 - \frac{v_{GS}}{V_p}\right)^2 \tag{18-16}$$

At a dc bias voltage V_{GS}, the transconductance is

$$g_m = \frac{di_D}{dv_{GS}} = -\frac{2I_{DSS}}{V_p}\left(1 - \frac{V_{GS}}{V_p}\right)$$

The transconductance evaluated at $V_{GS} = 0$ is

$$g_{mo} = -\frac{2I_{DSS}}{V_p}\left(1 - \frac{0}{V_p}\right) = -\frac{2I_{DSS}}{V_p} \tag{18-17}$$

sometimes called y_{fs}, the *forward transadmittance*. In general,

$$g_m = g_{mo}\left(1 - \frac{V_{GS}}{V_p}\right) = g_{mo}\sqrt{\frac{I_D}{I_{DSS}}} \tag{18-18}$$

Since the manufacturer usually gives the value of g_{mo} (or y_{fs}) along with I_{DSS}, Eq. 18-18 can be used to calculate g_m for any operating value of I_D or V_{GS} (after solving for V_p from Eq. 18-17). For E MOSFETs, the data sheet usually includes a curve of g_m vs. I_D.

EXAMPLE 2

Estimate the model parameters of the FET of Fig. 18.5b for operation at a Q point defined by $v_{DS} = 15$ V and $v_{GS} = -2$ V. Use Δi and Δv to approximate di and dv.

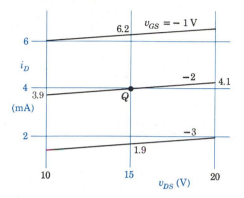

Figure 18.7   Evaluating parameters.

Assuming $V_p \cong -6$ V and $I_{DSS} \cong 10$ mA, use Eq. 18-17 to calculate the value of g_m at $V_{GS} = -2$ V and compare to the value obtained graphically.

Taking increments from the given operating point (Fig. 18.7) and reading the corresponding values yields

$$g_m = \left.\frac{di_D}{dv_{GS}}\right|_{15\,\text{V}} \cong \frac{\Delta i_D}{\Delta v_{GS}}$$

$$= \frac{(6.2 - 1.9)10^{-3}}{-1 + 3} = 2150\ \mu\text{S}$$

$$r_d = \left.\frac{1}{\dfrac{di_D}{dv_{DS}}}\right|_{-2\,\text{V}} \cong \frac{\Delta v_{DS}}{\Delta i_D}$$

$$= \frac{20 - 10}{(4.1 - 3.9)10^{-3}} = 50\,\text{k}\Omega$$

$$g_m = -\frac{2I_{DSS}}{V_p}\left(1 - \frac{V_{GS}}{V_p}\right)$$

$$= -\frac{2 \times 10^{-2}}{-6}\left(1 - \frac{-2}{-6}\right) = 2200\ \mu\text{S}$$

approximately the same as the measured value.

Small-Signal Model

On the diagram of the elementary amplifier in Fig. 18.8a, v_{GS}, v_{DS}, and i_D are total instantaneous quantities representing small ac variations superimposed on steady or dc values. The dc values fix the quiescent point and determine the values of the FET parameters; the ac values correspond to the signals, which are of primary interest. We wish to devise a small-signal model that will hold only for ac values and that will enable us to predict the performance of the FET in an ac application such as this elementary amplifier.

For differential changes, Eq. 18-13 states that

$$di_D = \frac{\partial i_D}{\partial v_{GS}}\,dv_{GS} + \frac{\partial i_D}{\partial v_{DS}}\,dv_{DS} \tag{18-19a}$$

Replacing the differentials by small signals and using the FET parameters, Eq. 18-19a becomes

$$i_d = g_m v_{gs} + \frac{1}{r_d}\,v_{ds} \tag{18-19b}$$

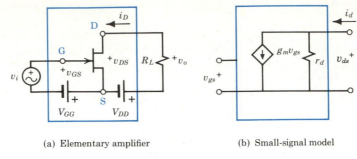

(a) Elementary amplifier   (b) Small-signal model

Figure 18.8 Derivation of a small-signal model of an FET.

This relation must be satisfied by the ac circuit if it is to be a functional model. Equation 18-19b says that the drain-signal current consists of two parts. In the corresponding parallel circuit, one branch contains a controlled-current source $g_m v_{gs}$ directly proportional to the input signal voltage, and the other branch carries a current v_{ds}/r_d directly proportional to output voltage. The circuit of Fig. 18.8b is the logical result; it is the circuit model of Eq. 18.19b.

The functional relationship holds equally well for effective values of sinusoidal quantities (or to the corresponding phasors) and Eq. 18-19b becomes

$$I_d = g_m V_{gs} + \frac{1}{r_d} V_{ds} \qquad (18\text{-}19c)$$

EXAMPLE 3

Predict the voltage gain of the amplifier of Fig. 18.8a if $R_L = 5$ kΩ.

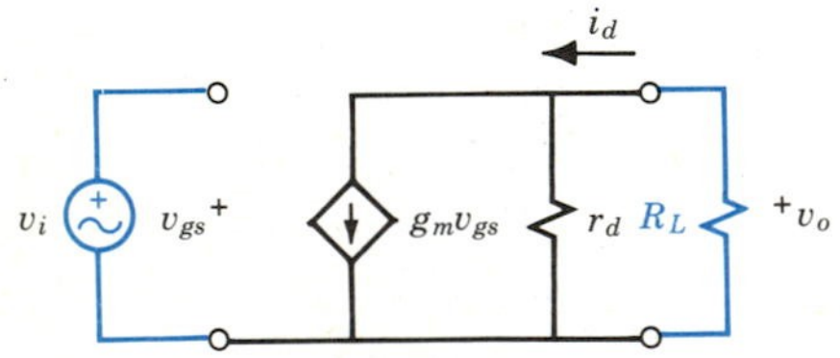

Figure 18.9 Voltage gain calculation.

The amplifier is replaced by the circuit model of Fig. 18.9 where $g_m = 2200$ μS and $r_d = 50$ kΩ.

Since $r_d \gg R_L$,

$$v_o = -g_m v_i (R_L \| r_d) \cong -g_m v_i R_L$$

and the voltage gain is

$$\frac{v_o}{v_i} = -g_m R_L = -2200 \times 10^{-6} \times 5 \times 10^3 \cong -11$$

The minus sign indicates an inversion or phase reversal of the output signal.

The use of models in the analysis of FET amplifiers is illustrated in Example 3. One conclusion is that for practical values of R_L (limited by available supply voltage V_{DD}), voltage gains are small; much larger gains can be obtained with BJT amplifiers. (See Example 4.) Note that r_d typically is so large compared to practical values of R_L that reasonably good predictions can be made with a model consisting of controlled-current source $g_m v_{gs}$ alone.

High-Frequency Model

At high frequencies consideration must be given to the charging currents associated with various capacitive effects. In Fig. 18.10a, the small-signal model has been modified to take into account capacitances inherent in field-effect transistors. C_{gs} and

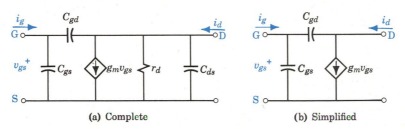

(a) Complete   (b) Simplified

Figure 18.10   High-frequency FET models.

C_{gd} represent the capacitances between gate and source and gate and drain; in a JFET these arise from the reverse-biased *pn* junctions and will be of the order of 1 to 10 pF. C_{ds} represents the drain-to-source capacitance and will be small, say, 1 to 5 pF. As predicted by this model, the voltage gain of an FET amplifier drops off as frequency increases. A quantitative analysis of high-frequency performance is given in Chapter 19; the simplified model of Fig. 18.10b is usually satisfactory.

BIPOLAR JUNCTION TRANSISTORS

Figure 12.16 shows the graphical analysis of an elementary large-signal amplifier. For small signals, a linear circuit model is much more convenient in both analysis and design. As with the FET, our objective is a simple model that will predict the performance of a properly biased transistor (BJT) insofar as small ac variations are concerned.

An Idealized Model (Common-Base)

The i-v characteristics of a typical transistor are displayed in Fig. 12.12. The general relations can be expressed in terms of voltages or currents as the independent variables. We have the choice of whether to deal with impedances, admittances, or a hybrid combination in defining parameters. If we choose the last, the general relations for a common-base transistor are

$$\begin{cases} v_{EB} = f_1(i_E, v_{CB}) \\ i_C = f_2(i_E, v_{CB}) \end{cases} \tag{18-20}$$

Before deriving a general model, let us reason out a circuit model from the

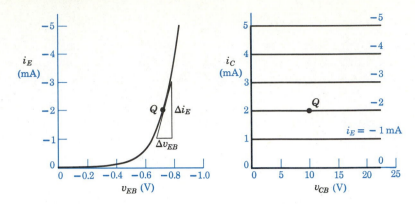

Figure 18.11 Idealized transistor characteristics (CB).

idealized v-i characteristics of Fig. 18.11. The emitter characteristics are those of a theoretical diode and the collector characteristics indicate a constant α and negligible I_{CBO}. For operation at a quiescent point Q defined by $I_E = -2$ mA and $V_{CB} = 10$ V, a small change Δv_{EB} produces a small change Δi_E that, in turn, produces a small change Δi_C (for nomenclature, see Table 17-1 on p. 470). The small-signal behavior of this idealized transistor is defined by

$$\begin{cases} v_{eb} = r_e i_e \\ i_c = -\alpha i_e \end{cases} \tag{18-21}$$

where

$$r_e = \frac{\Delta v_{EB}}{\Delta i_E} = \frac{v_{eb}}{i_e} = \text{the dynamic junction resistance} \tag{18-22}$$

$$-\alpha = \frac{\Delta i_C}{\Delta i_E} = \frac{i_c}{i_e} = \text{the forward current-transfer ratio} \tag{18-23}$$

Since $i_b = -(i_e + i_c)$ by Kirchhoff's current law, the circuit representation of these relations is shown in Fig. 18.12b. The emitter-base junction of a transistor behaves like a junction diode and r_e is given by Eq. 18-8. At room temperature,

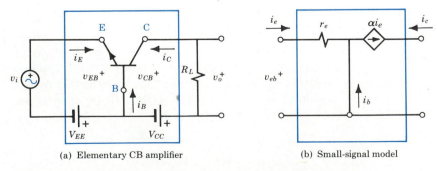

(a) Elementary CB amplifier   (b) Small-signal model

Figure 18.12 Derivation of a small-signal model of a transistor (CB).

$$r_e = \frac{0.025}{I_E} = \frac{25}{I_E(\text{in mA})} \; \Omega \qquad (18\text{-}24)$$

This gives a simple, accurate means for estimating r_e in practical transistors.

EXAMPLE 4

(a) A transistor with $\alpha = 0.98$ is to operate at room temperature with $I_E = 0.5$ mA. Estimate the dynamic emitter resistance.

By Eq. 18-24,

$$r_e = \frac{25}{I_E(\text{in mA})} = \frac{25}{0.5} = 50 \; \Omega$$

(b) If operation increases the junction temperature by 60 K, predict the new emitter resistance.

For this 20% increase in junction temperature T,

$$kT/e \cong 25(300 + 60)/300 = 30 \text{ mV}$$

and

$$r_e' = 30/0.5 = 60 \; \Omega$$

(c) The transistor is to operate at room temperature in the amplifier circuit of Fig. 18.12a where $R_L = 5$ kΩ. For an output signal $v_o = 2$ V, determine the signal input current i_i and the voltage gain.

For calculations involving small signals, the transistor is replaced by the model of Fig. 18.12b.

In Fig. 18.13, $i_e = i_i$ and $v_o = \alpha i_e R_L = \alpha i_i R_L$. The necessary signal current is

$$i_i = \frac{v_o}{\alpha R_L} = \frac{2}{0.98 \times 5000} = 410 \; \mu\text{A}$$

The input voltage required is

$$v_i = r_e i_i = 50 \times 410 \times 10^{-6} = 0.0205 \text{ V}$$

The voltage gain is

$$\frac{v_o}{v_i} = \frac{2}{0.0205} = 98$$

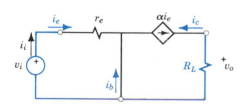

Figure 18.13 Application of CB model.

The Common-Emitter Configuration

One advantage of a model is that it may provide new insight into the behavior of the actual device. Insofar as small signals are concerned, the linear model of Fig. 18.13 can replace the actual physical transistor and its dc supplies. If the transistor is reconnected as shown in Example 5 on p. 508, the base-emitter terminals become the input port and the collector-emitter terminals become the output port. To represent this common-emitter configuration, the model is merely rearranged, but the device performance is dramatically changed by this rearrangement.

The possibility of a current gain as well as a voltage gain is an important advantage of the common-emitter amplifier. Because this configuration is used so extensively, we are justified in deriving a special model. Although the model of Fig.

18.15a could be used, it is not convenient because the controlled source is not a function of the input current; also, the character of the current gain is obscured in this model. A better form is that of Fig. 18.15b, where β and r_π are to be determined. As a first step, we note that $i_b + i_e = \alpha i_e$. Solving,

$$i_e = -\frac{i_b}{1 - \alpha} \quad \text{or} \quad \alpha i_e = -\frac{\alpha}{1 - \alpha} i_b = -\beta i_b \quad (18\text{-}25)$$

EXAMPLE 5

The transistor of Example 4 is reconnected as a common-emitter amplifier. Determine the signal input current i_i now required.

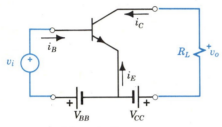

(a) Schematic diagram

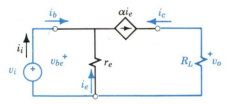

(b) Circuit model

Figure 18.14 Small-signal circuit model reconnected in the CE configuration.

As shown in Fig. 18.14b,

$$i_e = -i_b + \alpha i_e \quad \text{or} \quad i_e = \frac{-i_b}{1 - \alpha}$$

Then

$$v_o = \alpha i_e R_L = -\frac{\alpha}{1 - \alpha} i_b R_L$$

For the same 5-kΩ load resistance and the same 2-V output, the input signal is

$$i_i = i_b = -\frac{(1 - \alpha)v_o}{\alpha R_L} = -\frac{0.02 \times 2}{0.98 \times 5000} = -8.2 \ \mu A$$

In comparison to the 410 μA required in the common-base amplifier, only 8.2 μA is required here.

There is an effective current gain of

$$\frac{410}{-8.2} \cong -50$$

The voltage gain is

$$\frac{v_o}{v_i} = \frac{\alpha i_e R_L}{-i_e r_e} = -\frac{\alpha R_L}{r_e} = -98$$

The negative signs indicate a 180° phase reversal of the output signal with respect to the input.

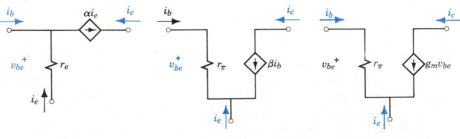

(a) Generator controlled by emitter current

(b) Generator controlled by base current

(c) Generator controlled by base-emitter voltage

Figure 18.15 Circuit models for an idealized transistor (CE).

where β is now the small-signal current gain (*CE*). For most purposes, $\beta_{ac} \cong \beta_{dc}$.

The current generator in Fig. 18.15b is controlled by the input current i_b. Now the character of the current gain is more clear. The effect of a variation in base current (a signal input to the base) appears in the collector circuit as a controlled source whose magnitude is $\alpha/(1 - \alpha)$ times as great. For a typical transistor with $\alpha = 0.98$, $\beta = 0.98/0.02 = 49$.

The new relation between the input voltage and current is obtained by noting that

$$v_{be} = -v_{eb} = -r_e i_e = +r_e \frac{i_b}{1 - \alpha} \tag{18-26}$$

The input resistance for the common-emitter configuration is therefore

$$r_\pi = \frac{v_{be}}{i_b} = \frac{r_e}{1 - \alpha} = (\beta + 1) r_e \tag{18-27}$$

Since $I_C = \alpha I_E$, Eq. 18.24 can be used to obtain

$$r_\pi = \frac{1}{1 - \alpha} r_e = \frac{\alpha}{1 - \alpha} \frac{25}{I_C} = \beta \frac{25}{I_C (\text{in mA})} \Omega \tag{18-28}$$

The "r_π-β model" of Fig. 18.15b represents the small-signal behavior of an idealized transistor in the common-emitter configuration. (It could have been deduced directly from an idealization of the curves of Fig. 12.15 where $I_B \cong I_C/\beta$.) Alterna-

---

EXAMPLE 6

An amplifier employing a transistor with $\beta = 50$ and operating at $I_C = 0.5$ mA in the common-emitter configuration is to deliver a signal current of 0.2 mA rms to a load resistance of 5 kΩ. Estimate the signal current gain and voltage gain.

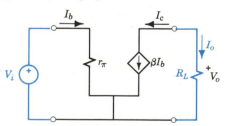

Figure 18.16   Application of the CE transistor model.

For this estimate, the r_π-β model of Fig. 18.16 is satisfactory. Since $I_o = -I_c = -\beta I_b$, the signal current gain is

$$\frac{I_o}{I_b} = -\beta = -50$$

The necessary input signal current (rms) is

$$I_b = \frac{-I_o}{\beta} = -\frac{0.2 \times 10^{-3}}{50} = -4 \times 10^{-6} = -4 \ \mu A$$

By Eq. 18-28 the input resistance is

$$r_\pi = \beta r_e = \beta \frac{25}{I_C} = 50 \times \frac{25}{0.5} = 2500 \ \Omega$$

$$\therefore V_i = I_b r_\pi = -4 \times 10^{-6} \times 2.5 \times 10^3 = -10 \, \text{mV}$$

The voltage gain is

$$\frac{V_o}{V_i} = \frac{+I_o R_L}{V_i} = -\frac{2 \times 10^{-4} \times 5000}{10 \times 10^{-3}} = -100$$

tively, the output current can be defined in terms of an input voltage (Fig. 18.15c). Since $v_{be} = r_\pi i_b \cong \beta r_e i_b$,

$$\beta i_b = \frac{1}{r_e} v_{be} = g_m v_{be} \tag{18-29}$$

where g_m, the small-signal transconductance for the BJT, is given by

$$g_m = \frac{1}{r_e} = 40 I_C \tag{18-30}$$

Under certain conditions the common-base arrangement is superior, but where current amplification as well as voltage amplification is important, the common-emitter amplifier possesses obvious advantages. (See Example 6.) Under other circumstances (discussed in Chapter 19), the common-collector configuration is advantageous.

The Hybrid-π Model

For more precise predictions of transistor behavior, more sophisticated models are necessary. In Fig. 18.17, the *base-spreading resistance r_b* takes into account the

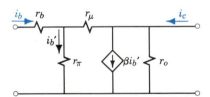

Figure 18.17   The more precise hybrid-π circuit model.

ohmic resistance of the long, thin base region. Typically, r_b varies from 50 to 150 Ω; lacking other information, a value of 100 Ω can be assumed with little error if r_π is relatively high. The resistance r_μ accounts for a very small current due to changes in collector-base voltage; this effect is negligibly small in most cases and r_μ will be ignored in our work. The resistance r_o provides for the small change in i_C with v_{CE} for given values of i_B (see Fig. 12.15).

In this model, proposed by L. J. Giacoletto, the circuit elements are arranged in a π-configuration known as the hybrid-π model.[†] It is particularly useful, when

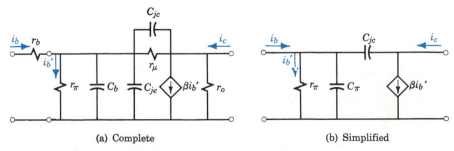

(a) Complete   (b) Simplified

Figure 18.18   High-frequency hybrid-π models.

[†]It is customary to use the *internal* base-emitter voltage $v'_{be} = i'_b r_\pi$ as the control variable here.

properly modified, in predicting the high-frequency behavior of transistors. Junction capacitances C_{je} and C_{jc} and capacitance C_b accounting for base-charging current can be connected in parallel with r_π and r_μ as shown in the high-frequency hybrid-π model of Fig. 18.18a. If r_b and r_o are neglected and r_μ ignored, and if C_b and C_{je} are combined into C_π, the simplified high-frequency model of Fig. 18.18b results. This model will be discussed in more detail in connection with amplifier performance in Chapter 19. (Note that the r_π-β model is a simple hybrid-π for moderate frequencies.)

The General h-Parameter Model

For any configuration and any small-signal application, a general linear model can be used to represent the transistor and predict its performance. Hybrid parameters are most commonly used and are most frequently supplied by manufacturers. The chief advantages of h-parameters are the ease with which they can be determined in the laboratory and the facility with which they can be handled in circuit calculations.

Choosing input current i_I and output voltage v_O as the independent variables and rewriting Eq. 18-11 in conventional transistor notation,

$$\begin{cases} v_I = f_5(i_I, v_O) \\ i_O = f_6(i_I, v_O) \end{cases} \tag{18-31}$$

For the common-emitter configuration, for example, v_I is the base-emitter voltage and i_O is the collector current.

Because of the nonlinearity of these functions, no simple model can be formulated to predict total voltages and currents. Since we are interested in small-signal behavior, let us consider differential changes in the variables. Taking the total differential, these become

$$\begin{cases} dv_I = \dfrac{\partial v_I}{\partial i_I} di_I + \dfrac{\partial v_I}{\partial v_O} dv_O \\[3mm] di_O = \dfrac{\partial i_O}{\partial i_I} di_I + \dfrac{\partial i_O}{\partial v_O} dv_O \end{cases} \tag{18-32}$$

Equation 18-32a says that a change in input current di_I and a change in output voltage dv_O both contribute to a change in input voltage dv_I. The effectiveness of the partial contributions is dependent on the coefficients $\partial v_I/\partial i_I$ and $\partial v_I/\partial v_O$; these partial derivatives of input voltage with respect to input current and output voltage and the corresponding partial derivatives of output current are the h-parameters of a small-signal circuit model.

Noting that the differential quantities di_i and dv_O correspond to small signals i_i and v_o, Eqs. 18-32 can be written as

$$\begin{cases} v_i = \dfrac{\partial v_I}{\partial i_I} i_i + \dfrac{\partial v_I}{\partial v_O} v_o = h_i i_i + h_r v_o \\[3mm] i_o = \dfrac{\partial i_O}{\partial i_I} i_i + \dfrac{\partial i_O}{\partial v_O} v_o = h_f i_i + h_o v_o \end{cases} \tag{18-33}$$

where h_i = input impedance with output short-circuited (ohms),
$\quad\ h_f$ = forward transfer current ratio with output short-circuited,
$\quad\ h_r$ = reverse transfer voltage ratio with input open-circuited,
$\quad\ h_o$ = output admittance with input open-circuited (siemens).

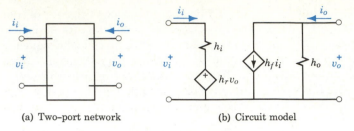

(a) Two–port network   (b) Circuit model

Figure 18.19 General h-parameter representation.

The first of Eqs. 18-33 says that the input voltage is the sum of two components. This indicates a series combination of an impedance drop $h_i i_i$ and a controlled voltage source $h_r v_o$ directly proportional to the output voltage. The second equation indicates a parallel combination of a controlled current source $h_f i_i$ and an admittance current $h_o v_o$. The h-parameter model of Fig. 18.19b follows from this line of reasoning.

The general model is applicable to any transistor configuration. In a specific case, the parameters are identified by a second subscript b, e, or c, depending on whether the base, emitter, or collector is the common element.

EXAMPLE 7

The transistor of Example 6 has the following common-emitter h-parameters: $h_{ie}=2500\ \Omega$, $h_{re} = 4 \times 10^{-4}$, $h_{fe} = 50$, and $h_{oe} = 10\ \mu S$ as shown in Fig. 18.20.

For $R_L = 5\ k\Omega$ and $I_o = 0.2$ mA rms, estimate the current gain and voltage gain of the amplifier and compare with the results of Example 6.

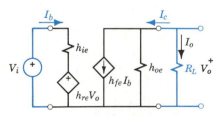

Figure 18.20 Application of general h-parameters in the CE configuration.

Applying the current-divider principle,

$$I_o = -I_c = -\frac{1/R_L}{h_{oe} + 1/R_L} h_{fe} I_b$$

The current gain is

$$\frac{I_o}{I_b} = -\frac{h_{fe}/R_L}{h_{oe} + 1/R_L} = -\frac{50/5000}{(10 + 200)10^{-6}} = -47.6$$

The required input signal is

$$I_b = I_o/(-47.6) = 200/(-47.6) = -4.2\ \mu A$$

In the input loop,

$$V_i = I_b h_{ie} + h_{re} V_o$$
$$= -4.2 \times 10^{-6} \times 2500 + 4 \times 10^{-4} \times 1$$
$$= (-10.5 + 0.4)10^{-3} = -10.1\ mV$$

Since $V_o = I_o R_L = 0.2 \times 10^{-3} \times 5 \times 10^3 = 1$ V,

$$\frac{V_o}{V_i} = \frac{1}{-0.0101} = -99$$

The simpler model introduces only small errors.

Determination of *h*-Parameters

The *h*-parameters are small-signal values implying small ac variations about operating points defined by dc values of voltage and current. They could be determined from families of characteristic curves. In actual practice, however, the parameters are determined experimentally by measuring the ac voltages or currents that result from ac signals introduced at the appropriate locations. For example, to determine h_{fe}, a small ac voltage is applied between base and emitter, and the resulting ac currents in the base lead and the collector lead are measured (the collector circuit must be effectively short-circuited to ac). Then h_{fe} is the ratio of collector current to base current and is just equal to β. Similar measurements permit the determination of the other small-signal parameters. Note that h_{ie} is $r_{\pi} + r_b \cong r_{\pi}$ and $h_{ib} \cong r_e$.

Since transistor parameters vary widely with operating point, measurements are made under standard conditions. Usually, values are for a frequency of 1000 Hz at room temperature (25° C) with an emitter current of 1 mA and a collector-base voltage of 5 V. Transistor manufacturers publish typical, minimum, and maximum values of the parameters along with average static characteristics. Typical values are shown in Table 18-1. An engineer designing a transistor amplifier, for example, must arrange the circuit so that the performance of the amplifier is within specifications despite expected variations in transistor parameters. (In Chapter 19 we see how feedback can be used to stabilize performance.)

Table 18-1 Typical Transistor Parameters

	2N1613	2N3114	2N699B			2N2222A	
			Min.	Typical	Max.	Min.	Max.
h_{ie} (kΩ)	2.2	1.5		2.8		2	4
h_{re} ($\times 10^{-4}$)	3.6	1.5		3.5			8
h_{fe}	55	50	35	70	100	50	300
h_{oe} (μS)	12.5	5.3		11		5	35
h_{ib} (Ω)	27	27	20	27	30		
h_{rb} ($\times 10^{-4}$)	0.7	0.25		0.5	1.25		
h_{fb}	-0.98	-0.981		-0.987			
h_{ob} (μS)	0.16	0.09	0.1	0.12	0.5		

Comparison of BJTs with FETs

The much higher values of g_m permit much higher voltage gains with the BJT as compared to the FET. On the other hand, the much higher values of input resistance for the FET permit lower "loading" of the source as compared to the BJT. The more linear transfer characteristic of the BJT permits larger signal swings without distortion; the square-law transfer characteristic of the FET is advantageous in modulation and mixing. In integrated circuit form, the MOSFET is smaller and simpler and therefore cheaper than the BJT.

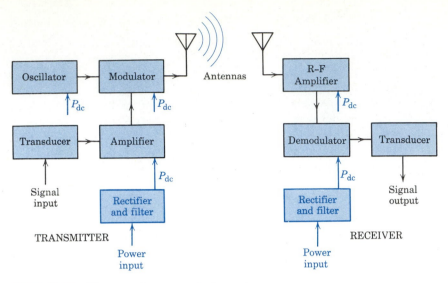

Figure 18.21 Elements of a communication system.

A Communication System

We now have in our repertoire of electronic devices all the essential elements of a telemetry or communication system (Fig. 18.21).

In a radio broadcasting system the signal input is an audible variation in air pressure on a microphone or a mechanical vibration of a phonograph needle. The transducer, a microphone or pickup, converts the acoustic vibration into an electrical signal that, when amplified, is available for modulating. The radiofrequency carrier signal is generated in the oscillator and combined with the audiofrequency signal in the modulator. Usually the modulator stage also performs amplification so that a high-level amplitude-modulated signal is fed into the transmitting antenna.

Energy is radiated from the antenna in the form of an electromagnetic wave, part of which is intercepted by the receiving antenna. The selected carrier and sidebands are amplified and fed into the demodulator. The detector output, usually amplified further, is converted by the transducer (a loudspeaker) into a replica of the original input signal. The power required may be obtained from 60-Hz ac lines, converted into dc power by rectifier-filter combinations, and supplied to each stage of the communication system.

SUMMARY

- The function performed by a diode depends on bias voltage and signal amplitude. A "small signal" is one to which a given device responds in a linear fashion.
- A modulator superimposes information on a carrier by modifying the carrier amplitude, frequency, or phase.

  An AM wave consisting of carrier and sidebands is described by

  $$v_{AM} = V_C \sin \omega_C t - \frac{mV_C}{2} \cos(\omega_C + \omega_M)t + \frac{mV_C}{2} \cos(\omega_C - \omega_M)t$$

  A demodulator recovers the original signal from the modulated wave.

- The small-signal performance of complicated and highly nonlinear electronic devices can be predicted using relatively simple linear models.

  For small signals superimposed on steady values, the significant diode parameter is the dynamic resistance $r_{ac} = dv/di$, the reciprocal of the slope of the i-v characteristic at the operating point.

- To derive the small-signal circuit model of a nonlinear two-port:
  1. Write the governing equation in terms of the chosen variables.
  2. Take the total differential to obtain the equations for small variations.
  3. Note that differentials can be approximated by small signals, and write the corresponding equations for small signals.
  4. Interpret the equations in terms of an electrical circuit with appropriate parameters, including passive elements and controlled sources.

- For a field-effect transistor, the input current is negligible and the small-signal model consists of transconductance g_m and drain resistance r_d (Fig. 18.22).

  For a junction transistor, the general model is complicated; at moderate frequencies the hybrid-parameter CE model is as shown in Fig. 18.23.

  In many BJT applications, a satisfactory model consists simply of an input resistance r_π and a controlled-current source βI_b (Fig. 18.24).

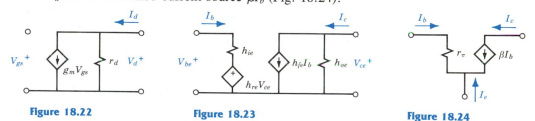

Figure 18.22   **Figure 18.23**   **Figure 18.24**

- Because individual transistors vary widely from the norm, it is customary to use parameter value ranges published by the manufacturer, minimizing the effects of parameter variations by proper circuit design.

REVIEW QUESTIONS

1. Explain the differences between V/I, $\Delta V/\Delta I$, dV/dI, and $\partial V/\partial I$.

2. Explain the difference between static resistance and dynamic resistance.

3. Sketch and label the spectrum of frequencies transmitted by your favorite AM station during a typical music program.

4. Explain how a nonlinear device produces amplitude modulation. Demodulation.

5. Why is a high frequency needed for a radio broadcast carrier?

6. How many telephone conversations (two-way) can be carried on a telephone cable that transmits frequencies up to 8 MHz? How many television programs?

7. Why is the ac model of an FET simpler than an equally precise model of a BJT?

8. In analytic geometry, what does $z = f(x, y)$ represent in general? Could $i_b = f(v_c, v_b)$ be similarly represented? What is the graphical interpretation of $\partial z/\partial y$?

9. Sketch a set of FET characteristics and define, graphically, the two parameters.

10. What are the criteria for selecting the quiescent point in an amplifier?

11. Does the circuit of Fig. 18.8b contain any information not in Eq. 18-19b? Why use the circuit?

12. What is meant by: "The results of a graphical analysis of performance from characteristic curves lack generality"?

13. Why are no dc supplies such as batteries shown in the linear models? What is the ac impedance of a battery?
14. Explain how a current gain is obtained in a transistor amplifier.
15. Define the h-parameters graphically in terms of common-base characteristic curves. Why are h-parameters not calculated this way?

EXERCISES

1. A voltage $v = V_M \sin \omega_M t + V_C \sin \omega_C t$ is applied to a device whose mathematical model is $i = a_0 + a_2 v^2$.
   (a) Determine the device current.
   (b) Identify the frequency components of current.
2. The amplitude of a 740-kHz signal is varied sinusoidally from zero to $2V_C$ at a rate of 4000 times per second. Sketch the resulting modulated wave and list the frequencies of the components appearing in the transmitted wave.
3. An amplitude-modulated wave is described by the equation

   $$v = 5\cos 2.05 \times 10^6 t + 10 \sin 2.1 \times 10^6 t - 25\cos 2.15 \times 10^6 t$$

   Identify the carrier frequency (rad/s) and the frequencies of any modulating signals.
4. A signal is described by the equation

   $$v = (8 + 5\sin 1000t + 2\sin 2000t)\sin 10^5 t$$

   List the frequencies (rad/s) present in the signal and show them on a frequency spectrum. Identify the carrier and the upper and lower sidebands.
5. Channel 4, a typical TV channel, extends from 66 to 72 MHz. A ruby laser operates in the red region with a wavelength of 7500 Å. If the useful bandwidth of a laser beam is equal to 0.2% of its carrier frequency, how many TV channels could be carried on a single laser beam?
6. Two signals $a = \sin \omega t$ and $b = 5\sin 20\omega t$ are multiplied together.
   (a) Show on a frequency spectrum all the components of the product.
   (b) If a voltage equal to the product is fed into a device for which $i_o = v + 0.2v^2$, show on a frequency spectrum all the components of the output current.
7. A signal $v = A\sin 100{,}000t + B\sin 103{,}000t$ is applied to the input of an FET whose output is defined by $i_o = av^2$.

   (a) List the frequencies of the components appearing in the output.
   (b) Show the components on a frequency spectrum.
   (c) Under what conditions does this represent demodulation, or the separation of an audible signal from a carrier?
   (d) Sketch a filter circuit to pass only the audible component to load R_L.
8. Derive a general expression for the dynamic resistance of a device for which $i = v + 0.01 v^2$ A and evaluate r_{ac} at $V = 10$ V. Predict $i(t)$ for $v(t) = 10 + \sin \omega t$ V by adding the responses to the two components. Why does "superposition" work here?
9. A silicon diode is characterized by $I_s = 10$ nA. Predict the dc forward resistance and the dynamic resistance at:
   (a) $T = 25°$ C and current = 0.1 mA.
   (b) $T = 25°$ C and current = 10 mA.
   (c) $T = 100°$ C and current = 10 mA.
10. Starting from the second pair of Eqs. 18-10, and assuming a linear two-port network, derive a circuit model with admittance parameters (Y_{11}, etc.).
11. Starting from the first pair of Eqs. 18-10, and assuming a nonlinear two-port device, derive a small-signal model with impedance parameters (z_{11}, etc.).
12. A semiconductor two-port is described by the equations

    $$\begin{cases} i_1 = 10^{-7} v_1 \\ i_2 = 2 \times 10^{-3} v_1 + 10^{-5} v_2 \end{cases}$$

    Devise, draw, and give parameter values for a circuit model of this device.
13. Represent the MOSFET of Fig. 12.4c by a small-signal model and estimate the parameters at a point defined by $v_{GS} = 0$ and $v_{DS} = 8$ V.
14. For a JFET with $g_{mo} = 4000$ μS and $V_p = -4$ V, predict g_m at:
    (a) $V_{GS} = -1$ V; (b) $I_D = 2$ mA.

15. A DE MOSFET with $I_{DSS} = 5$ mA, $g_{mo} = 5000$ μS and a negligibly large r_d is used in the circuit of Fig. 17.5c. If $V_{DD} = 25$ V and $R_D = 3$ kΩ, predict the voltage gain v_o/v_i.

16. For a DE MOSFET with $I_{DSS} = 10$ mA and $g_{mo} = 5000$ μS, predict g_m at:
(a) $I_D = 5$ mA; (b) $V_{GS} = -1$ V.

17. The MOSFET of Exercise 16 ($r_d = 50$ kΩ) is used in the circuit of Fig. 18.8a with $R_L = 5$ kΩ and $V_{GG} = 2$ V.
(a) Specify I_D and V_{DD} for operation at $V_{GS} = -2$ V and $V_{DS} = 8$ V.
(b) Predict the voltage gain.

18. For the MOSFET of Exercise 13, specify the value of load resistance to provide a voltage gain of 10.

19. An E MOSFET with a transconductance of 3000 μS and a dynamic drain resistance of 100 kΩ is used in the circuit of Fig. 18.5a where $v_i = 0.5 \sin \omega t$ and $R_L = 5$ kΩ. Draw an appropriate circuit and predict output voltage v_o.

20. The JFET of Exercise 14 is used in the circuit of Fig. 18.8a with $V_{GG} = -1$ V, $V_{DD} = 30$ V, and $R_L = 4$ kΩ. Predict the quiescent drain current I_D and the input voltage V_i(rms) to produce an output voltage $V_o = 30$ mV.

21. At high frequencies a semiconductor two-port is defined by the equations:

$$\begin{cases} i_1 = 10^{-8}\, dv_1/dt - 10^{-9}\, dv_{21}/dt \\ i_2 = 2 \times 10^{-3}\, v_1 + 10^{-9}\, dv_{21}/dt \end{cases}$$

Devise, draw, and give parameter values for a circuit model of this device.

22. A semiconductor two-port is described by the equations

$$\begin{cases} i_1 = 10^{-6}\, e^{40v_1} \\ i_2 = -0.98i_1 + 10^{-6}\, v_2 \end{cases}$$

Devise a small-signal model for this device at an operating point defined by $i_1 = 2$ mA and $v_2 = -5$ V.

23. A two-port electronic device is defined by the following equations

$$\begin{cases} i_1 = 0.1v_1 + 0.1\, dv_1/dt \\ v_2 = 5v_1 + 20i_2 \end{cases}$$

where currents are in amperes and voltages in volts. Devise a suitable circuit model and label the circuit elements with the appropriate values.

24. A transistor with an α of 0.99 is used in the circuit of Fig. 18.12.

(a) Is this an *npn* or a *pnp* transistor?
(b) Is it biased properly for normal operation? If not, make the appropriate changes in battery polarity.
(c) Redraw the circuit using a simple small-signal model for an idealized transistor to replace the portion within the colored lines.
(d) Derive a general expression for voltage gain $A_V = v_o/v_i$.
(e) Evaluate A_V for operation at $I_E = 0.5$ mA and $R_L = 2$ kΩ.

25. The transistor of Exercise 24 is reconnected in the circuit of Fig. 18.14a. Use the same model and repeat parts (c), (d), and (e).

26. Compare the voltage gains, current gains, and power gains (i.e., the product of voltage gain times current gain) for the amplifiers of Exercises 24 and 25 and draw a conclusion.

27. A silicon BJT with $\beta = 40$ and $I_{CEO} = 10$ nA is connected in the circuit of Fig. 17.15a where $V_{CC} = 20$ V, $R_B = 25$ kΩ, $V_{BB} = 3.2$ V, and $R_C = 2$ kΩ.

(a) Write the theoretical equations for $i_B = f(v_{BE})$ and $i_C = f(i_B)$.
(b) Sketch approximately to scale the base and collector characteristics for $0 < i_B < 0.25$ mA.
(c) Replace the transistor by an approximate dc model based on simplification of the curves of part (b).
(d) Redraw the circuit and use your model to predict quiescent values of I_B, I_C, and V_{CE}.
(e) Replace the transistor by a simple small-signal model, neglect large R_B, and estimate the signal components of i_C and v_{CE} for $i_i = 0.1 \sin \omega t$ mA.

28. In the elementary amplifier of Fig. 18.14, at the quiescent point $I_C = 1.5$ mA and $V_{CE} = 4.5$ V.

(a) Redraw the circuit for small-signal analysis using the r_π-β model with $\beta = 60$.
(b) Specify R_L for a voltage gain $|v_o/v_i| = 180$.
(c) Specify the battery voltage V_{CC}.

29. The 2N3114 parameters in Table 18-1 are for $I_C = 1$ mA.

(a) Estimate r_π, r_b, and β for $I_C = 1$ mA.
(b) Assuming r_b and β are constant, estimate r_π and h_{ie} for $I_C = 0.2$ mA.

30. A transistor has the following parameters known to $\pm 10\%$ accuracy: $r_b = 100$ Ω, $r_\pi = 2500$ Ω,

$r_\mu = 20\ M\Omega$, $\beta = 100$, and $r_o = 200\ k\Omega$. Predict the voltage gain of the elementary amplifier of Fig. 18.14 with $R_L = 2\ k\Omega$:

(a) Using the circuit model of Fig. 18.17.

(b) Using the circuit model of Fig. 18.15.

(c) Compare the results of parts (a) and (b) and draw a conclusion.

31. A 2N3114 transistor operating at 25° C and $I_E = 1$ mA is used in an elementary amplifier to provide 2 V rms across $R_L = 5\ k\Omega$. The input source can be represented by V_s in series with $R_s = 1\ k\Omega$.

(a) Using the general h-parameter model, specify V_s.

(b) Using the r_π-β model, repeat part (a) and draw a conclusion.

32. For certain parameters, the manufacturer of the 2N699B specifies guaranteed minimum and maximum values (Table 18-1). Determine the corresponding range of values of r_π. Is high precision justifiable in calculations using these parameters?

33. A 2N1613 is to be used ($I_C = 1$ mA, $V_{CE} = 5$ V) in a common-emitter amplifier to provide 1.5 V rms output across a load resistance R_L. The available open-circuit signal is 10 mV from a source with an internal resistance of 500 Ω.

(a) Draw and label the circuit with an h-parameter model.

(b) Simplify the model where appropriate to permit an approximate analysis.

(c) Specify the necessary load resistance and the collector supply voltage V_{CC}.

34. Compare the voltage gain possibilities of an FET rated at $I_{DSS} = 4$ mA and $g_{mo} = 5000\ \mu S$ and a BJT rated at $\beta = 100$ at $V_{CE} = 5$ V, both operating at $I_D = I_C = 1$ mA. Predict the actual voltage gains with $R_C = R_D = 5\ k\Omega$ assuming the voltage source has an output resistance of 2500 Ω, and draw a conclusion.

PROBLEMS

1. When a nonlinear device is used as a detector, the important factor is the second-degree term in Eq. 18-1. Prove that an FET whose characteristic is defined by $i = av^2$ in combination with a suitable filter will recover the modulating signal from an amplitude-modulated wave.

2. The diode-capacitor circuit of Fig. 3.28 can be used as a detector for AM signals. Assume that the input voltage v is the modulated carrier of Fig. 18.2 and that time constant $R_L C$ is large compared to the period T_C but small compared to T_M.

(a) Sketch v_L, the output signal.

(b) If the "load," an earphone, includes an inductive reactance that is large at the carrier frequency, sketch the output current i_L.

3. A *mixer* is widely used in radio and TV receivers to obtain a signal with a frequency equal to the *difference* in the frequencies of two input signals. Because of its square-law transfer characteristics an FET makes an ideal mixer. Demonstrate that if the sum of two signals at frequencies ω_x and ω_y is applied to the gate-source terminals of a DE MOSFET, the output current will contain a component at frequency $\omega_x - \omega_y$. What are the other components of the output and how could they be eliminated?

4. The "Darlington pair" shown in Fig. 18.25 is frequently packaged as a single component or made available as an integrated circuit. Assuming identical transistors (β, r_π) and stating any other necessary assumptions, predict the current gain of the pair and the input and output resistances of the pair in the given circuit. Draw a conclusion about the advantages of such a component.

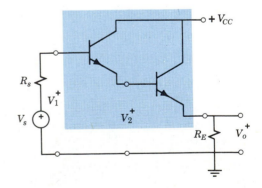

Figure 18.25

5. A transistor amplifier is terminated in a load resistance R_L. Using the general h-parameter circuit model (Fig. 18.19b), derive an expression for voltage gain v_o/v_i.

6. A transistor amplifier is terminated in a load resistance R_L. Using the general h-parameter circuit model (Fig. 18.19b), derive an expression for input resistance v_i/i_i.

7. In a particular transistor application (Fig. 18.26), the usual load resistance is omitted and an emitter resistance R_E is inserted. V_{BB} and V_{CC} are adjusted to bring operation to an appropriate quiescent point. Replace the transistor with an r_π-β model and derive expressions for: (a) v_E in terms of v_i (voltage gain), and (b) input resistance $r_i = v_i/i_b$. Assuming $\beta = 50$, $r_\pi = 1200\ \Omega$, and $R_E = 2000\ \Omega$, evaluate r_i and compare with the

input resistance of an ordinary common-emitter amplifier (Fig. 18.14a).

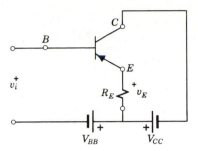

Figure 18.26

19

Small-Signal Amplifiers

RC-Coupled Amplifiers

Frequency Response

Multistage Amplifiers

Feedback Applications

Followers

To amplify the signal from a strain gage or a microphone to a power level adequate to drive a recorder or a loudspeaker usually requires several stages of amplification. If the final amplifier is a power transistor, appreciable driving current is needed and several stages of current amplification may be required. The low-level voltage or current amplification is provided by "small-signal" amplifiers that respond linearly.

In a small-signal amplifier, the input signals are small in comparison to the dc bias and the resulting output swings are small in comparison to the quiescent operating values of voltage and current. Under small-signal operation, bias is not critical and amplitude distortion is easily avoided. Although graphical analysis is possible, the linear models developed in Chapter 18 are much more convenient to use.

We have already used the linear models to predict the performance of single-stage amplifiers at moderate frequencies. Now we are going to investigate the gain of untuned amplifiers as a function of frequency and consider the effect of cascading several stages. Field-effect and bipolar transistors with *RC* coupling are commonly employed for small-signal amplification in the frequency range from a few cycles per second to several megahertz, and therefore they are emphasized in this discussion. Although BJTs and FETs operate on different physical bases, their external behavior is quite similar and the same approach is used in predicting their frequency responses.

The overall performance of systems can be greatly improved by the use of feedback; the gain of amplifiers can be made much greater, or more stable, or less sensitive to change, or less dependent on frequency. We conclude this chapter by investigating the application of feedback concepts to amplifiers.

RC-COUPLED AMPLIFIERS

Each stage of a multistage amplifier consists of an electronic two-port, a biasing network, and the coupling circuits. Commonly the output of a transistor is coupled to the next stage by means of a load resistor and a coupling capacitor.

Frequency-Response Curve

The voltage gain of one stage of an *RC*-coupled amplifier is shown in Fig. 19.1. The gain is relatively constant over the *midfrequency range*, but it falls off at lower

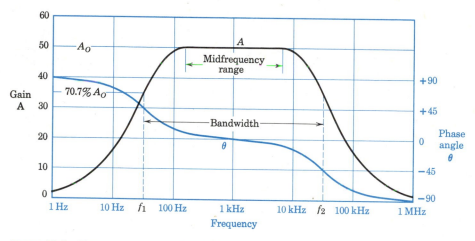

Figure 19.1 Frequency response of an *RC*-coupled single-stage amplifier.

frequencies and at higher frequencies. In ordinary amplifiers, the frequency-response curve is symmetric if frequency is plotted on a logarithmic scale as shown. The *bandwidth* is defined by lower and upper *cutoff frequencies* or *half-power frequencies* f_1 and f_2, just as in the frequency response of a resonant circuit. At the half-power frequencies, the response is 70.7% of the midfrequency gain, or $A_1 = A_2 = A_0/\sqrt{2}$, *and the phase angle* $\theta = \pm 45°$.

A Generalized Amplifier

In the general amplifier circuit of Fig. 19.2, the input signals come from a "source," maybe a phono pickup or a previous amplifier stage, that is characterized by an open-circuit voltage $\mathbf{V}_s$ and Thévenin equivalent impedance $\mathbf{Z}_s$. The input signals consist of sinusoids of various amplitudes and frequencies so variables are shown as effective-value phasors. The "one-stage" amplifier is characterized by in-

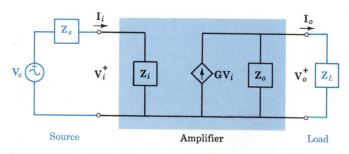

Figure 19.2 A generalized one-stage amplifier circuit.

put and output impedances $\mathbf{Z}_i$ and $\mathbf{Z}_o$ and a controlled source (controlled by input current or voltage) with gain $\mathbf{A} = \mathbf{V}_o/\mathbf{V}_i$ where $\mathbf{Z}_i$, $\mathbf{Z}_o$, and $\mathbf{A}$ are complex functions of frequency. The "load" of the amplifier may be a transducer or a following amplifier stage. Given the components of the input signal and the characteristics of the source and load, we can predict the output if we know the frequency response of the amplifier.

Typical small-signal amplifier circuits are shown in Fig. 19.3; these are "RC-coupled amplifiers" because of coupling capacitors C_{C1} and C_{C2} and their associated resistances. Our approach is to replace these complicated circuits by linear

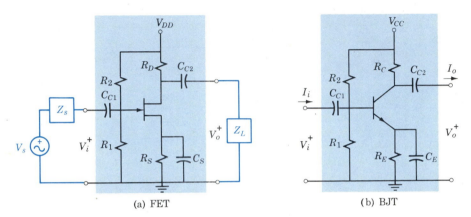

(a) FET   (b) BJT

Figure 19.3 Single-stage RC-coupled amplifiers.

models that respond in the same way to ac signals. We assume that the transistors are operating at proper quiescent points and the bias voltages and currents are of no concern. Bias resistors R_S and R_E are assumed to be properly bypassed by C_S and C_E so they do not appear in the models of Fig. 19.4. Since batteries are short circuits to ac signals, the upper terminal of R_D or R_C is grounded through the battery. To input signals, R_1 and R_2 are effectively in parallel, replaced by $R_1 \| R_2 = R_G$ or R_B. The batteries are omitted. On the other hand, unintentional elements such as junction capacitances (C_j) and wiring capacitances (C_W), which do not appear on the wiring diagram, must be shown in the amplifier models. Reflecting the most common situation, the source impedance is simply R_s and the load is represented by C_L and R_L in parallel.

Small-Signal Amplifier Models

The device models are those derived in Chapter 18. (See Figs. 18.10b and 18.18b.) For the FET, C_{gs} and C_{gd} are gate-source and gate-drain capacitances and g_m is the transconductance; r_d has been omitted since usually $r_d \gg R_D$. For the BJT, C_π is the effective input capacitance representing junction and diffusion capacitances, r_π is the small-signal input resistance, and C_{jc} is the collector-base junction capacitance. Mutual resistance r_μ is ignored and output resistance r_o has been omitted since $r_o \gg R_C$. Base-spreading resistance r_b has been neglected on the assumption that

$r_b \ll R_s + r_\pi$; therefore, $V'_{be} = V_{be}$. The current source can be considered to be controlled by the base current or the base-emitter voltage since $g_m V_{be} = \beta I'_b$ where g_m is the transconductance and β is the small-signal short-circuit current gain.

The determination of the values of individual elements in Fig. 19.4 is quite difficult; the specific values depend upon such factors as quiescent voltage and current,

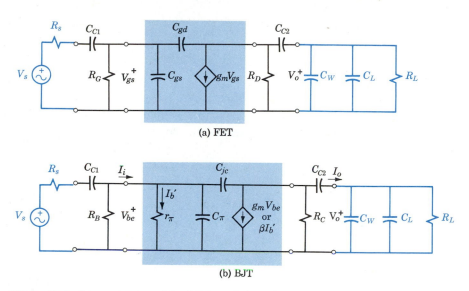

(a) FET

(b) BJT

Figure 19.4 Linear circuit models of *RC*-coupled amplifiers.

operating temperature, characteristics of the following stage, and configuration of the physical circuit. The important thing from our viewpoint here is that an actual amplifier can be replaced, insofar as small signals are concerned, by a linear circuit model that can be analyzed by methods we have already mastered. Fortunately, the models for FET and BJT are similar in form and the same analysis can be used for both.

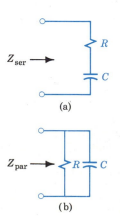

(a)

(b)

Figure 19.5 *Z* of *RC* combinations.

These linear models hold fairly well for frequencies from a few cycles per second to several megahertz. It would be possible to perform a general analysis on the circuits as they stand, but it is much more convenient, and more instructive, to consider some simplifying approximations. We recall that for series and parallel combinations of resistance R and capacitance C (Fig. 19.5),

$$Z_{ser} = \sqrt{R^2 + (1/\omega C)^2} = R\sqrt{1 + (1/\omega CR)^2} \quad (19\text{-}1)$$

$$Z_{par} = \frac{1}{\sqrt{(1/R)^2 + (\omega C)^2}} = \frac{R}{\sqrt{1 + (\omega CR)^2}} \quad (19\text{-}2)$$

We note that in Eq. 19-1 if $\omega CR \geq 10$, $Z_{ser} \cong R$ within 0.5%. Also, in Eq. 19-2, if $\omega CR \leq 0.1$, $Z_{par} \cong R$.

The fact that gain is independent of frequency over the midfrequency range (Fig. 19.1) indicates that there is a range of frequencies over which the capacitive effects are negligible. This is the case in practical amplifiers (see Exercise 5), and it leads to another important conclusion. Since ωC increases with frequency, series capacitances become important only at frequencies lower than the midfrequencies and parallel capacitances become important only at frequencies higher than the midfrequencies.

FREQUENCY RESPONSE

The analysis of series and parallel RC combinations leads to the convenient conclusion that the frequency response of an RC amplifier can be divided into three regions. In the midfrequency range, the capacitances can be neglected. In the low-frequency range, the series capacitances C_C must be considered. In the high-frequency range, the parallel capacitances must be considered.

Midfrequency Gain

In a properly designed untuned amplifier, there is a range of frequencies over which the general models of Fig. 19.4 can be replaced by the purely resistive circuits of Fig. 19.6.

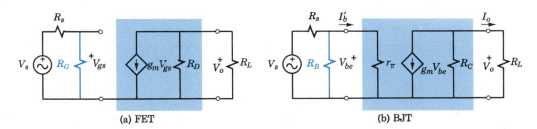

(a) FET   (b) BJT

Figure 19.6   Midfrequency models of amplifiers.

FET. Since $R_G \gg R_s$, R_G can be omitted in most cases. Where the parallel combination of R_D and R_L is represented by R_o, the output voltage is

$$V_o = -g_m V_{gs}(R_D \| R_L) = -g_m V_s R_o \qquad (19\text{-}3)$$

and the midfrequency voltage gain is

$$A_{VO} = \frac{V_o}{V_s} = -g_m R_o \qquad (19\text{-}4)$$

where subscripts V and O are for "voltage" and "midfrequency." The negative sign indicates a 180° phase reversal of the output signal with respect to the input. (See Example 1.)

BJT. For proper stability R_B should be small compared to βR_E (Eq.17-12), whereas for negligible loading of the input circuit R_B should be large compared to r_π.

EXAMPLE 1

A JFET with $g_m = 2000$ μS, $C_{gs} = 20$ pF, and $C_{gd} = 2$ pF is used in an RC-coupled amplifier with $R_G = 1$ MΩ, $R_S = 2$ kΩ, and $R_D = 5$ kΩ. Signal source resistance $R_s = 1$ kΩ and load resistance $R_L = 5$ kΩ. Coupling capacitance $C_{C1} = C_{C2} = 1$ μF and $C_S = 20$ μF; the load capacitance (including wiring) $C_L = 20$ pF. Predict the voltage gain V_o/V_s at $\omega = 10^4$ rad/s ($f = 1590$ Hz).

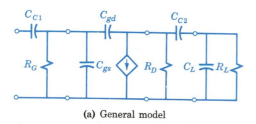

(a) General model

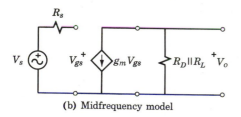

(b) Midfrequency model

Figure 19.7 Midfrequency simplification.

At the given frequency,

$$1/\omega C_S = 1/10^4 \times 20 \times 10^{-6} = 5 \ \Omega$$

Therefore, $R_S = 2000$ Ω is effectively shorted and the model is as shown in Fig. 19.7a.

Since C_{C1} and C_{C2} are effectively in series with R_G and R_L, 1 MΩ and 5 kΩ respectively, the reactances

$$1/\omega C_C = 1/10^4 \times 10^{-6} = 100 \ \Omega$$

are negligibly small (Eq. 19-1).

With C_{C1} and C_{C2} omitted, C_L is in parallel with $R_o = R_D \| R_L = 2500$ Ω and C_{gs} is in series with R_s. The reactances

$$1/\omega C_L = 1/\omega C_{gs} = 1/10^4 \times 20 \times 10^{-12} = 5 \ \text{M}\Omega$$

are so large that C_L and C_{gs} can be omitted.

The reactance

$$1/\omega C_{gd} = 1/10^4 \times 2 \times 10^{-12} = 50 \ \text{M}\Omega$$

is so large that a negligible current will flow in C_{gd}.

In other words, 1590 Hz is in the "midfrequency" range of this amplifier and the model of Fig. 19.7b is appropriate. The midfrequency gain (Eq. 19-4) is

$$A_{VO} = -g_m R_o = -2000 \times 10^{-6} \times 2500 = -5$$

Conclusion: Working from the simplified model, calculation of the midfrequency gain is straightforward.

Here we assume that the ideal condition $r_\pi \ll R_B \ll \beta R_E$ is achieved so the circuit is stable *and* simple to analyze. (See Exercise 6.) If R_B is neglected in Fig. 19.6b, r_π and R_s constitute a voltage divider. Where the parallel combination of R_C and R_L is represented by R_o, the output voltage is

$$V_o = -g_m V_{be}(R_C \| R_L) = -g_m \frac{r_\pi}{r_\pi + R_s} V_s R_o = -\beta \frac{V_s}{r_\pi + R_s} R_o \qquad (19\text{-}5)$$

and the midfrequency voltage gain is

$$A_{VO} = \frac{V_o}{V_s} = -\frac{r_\pi}{r_\pi + R_s} g_m R_o = -\beta \frac{R_o}{r_\pi + R_s} \qquad (19\text{-}6)$$

Performance of these one-stage amplifiers is influenced by the characteristics of the source (R_s) and the load (R_L). In the multistage amplifiers discussed later, the source and the load for each stage are other similar stages.

Low-Frequency Response

Below the midfrequencies, the susceptances of the parallel capacitances are negligibly small, but the reactances of the coupling capacitances C_C become increasingly important.

Coupling Capacitors. The general models can be reduced to the simpler low-frequency models shown in Fig. 19.8, where R_G has been retained because it is

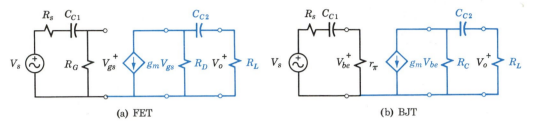

(a) FET   (b) BJT

Figure 19.8   Low-frequency models of amplifiers.

significant but R_B has been omitted since it is large compared to r_π. With this simplification, the FET and BJT models are analogous and one analysis will serve for both.

The input and output circuits are "high-pass filters" (see Fig. 7.10) of slightly different form. The general circuit (Fig. 19.9a) can be represented by the Thévenin

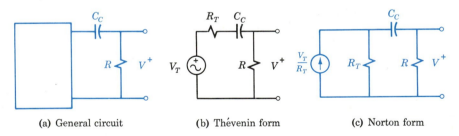

(a) General circuit   (b) Thévenin form   (c) Norton form

Figure 19.9   High-pass filter circuits.

form as in the input circuit or by the Norton form as in the output circuit; the two forms are equivalent. For midfrequencies where C_C is effectively shorted, the output of the Thévenin circuit is

$$V = V_0 = \frac{R}{R + R_T} V_T \tag{19-7}$$

At low frequencies where C_C is significant, the output of the Thévenin circuit is

$$\mathbf{V}_L = \frac{R}{R + R_T - j\dfrac{1}{\omega C_C}}\mathbf{V}_T = \frac{\dfrac{R}{R + R_T}\mathbf{V}_T}{1 - j\dfrac{1}{\omega C_C (R + R_T)}} = \frac{V_0}{1 - j\dfrac{1}{\omega C_C (R + R_T)}} \tag{19-8}$$

after dividing through by $R + R_T$. In words, the low-frequency output V_L of either filter circuit is related to the midfrequency output V_O by a complex factor dependent on frequency and an RC product. As frequency is decreased, a larger fraction of voltage V_T appears across C_C and the voltage V at the output is reduced. The *cutoff* or *half-power frequency* where $|V_L/V_O| = 1/\sqrt{2}$ is defined by

$$\omega_{co} = \frac{1}{(R + R_T)C_C} \tag{19-9}$$

The behavior of the FET or BJT amplifiers in Fig. 19.8 at low frequencies can now be predicted. At a frequency defined by

$$\omega_{11} = \frac{1}{C_{C1}(R_s + R_G)} \quad \text{or} \quad \omega_{11} = \frac{1}{C_{C1}(R_s + r_\pi)} \tag{19-10}$$

the input voltage V_{gs} or V_{be} will be down to 70% of V_O and the possible amplifier gain will be correspondingly reduced. At a frequency defined by

$$\omega_{12} = \frac{1}{C_{C2}(R_D + R_L)} \quad \text{or} \quad \omega_{12} = \frac{1}{C_{C2}(R_C + R_L)} \tag{19-11}$$

the output voltage V_o will be down to 70% of $g_m V_{gs} R_o$ or $g_m V_{be} R_o$ and the amplifier gain will be correspondingly reduced. The overall low-frequency voltage gain for the FET amplifier can be expressed as

$$\mathbf{A}_L = \mathbf{A}_O \cdot \frac{1}{1 - j\dfrac{1}{\omega C_{C1}(R_s + R_G)}} \cdot \frac{1}{1 - j\dfrac{1}{\omega C_{C2}(R_D + R_L)}} \tag{19-12}$$

or, for either FET or BJT, the *relative gain* at low frequencies is

$$\frac{\mathbf{A}_L}{\mathbf{A}_O} = \frac{1}{1 - j\omega_{11}/\omega} \cdot \frac{1}{1 - j\omega_{12}/\omega} \tag{19-13}$$

To predict the behavior of a given circuit, determine ω_{11} and ω_{12}; the higher of the two determines the cutoff frequency (unless they are very close together; see Exercise 14). To design a circuit for a specified behavior, calculate C_{C1} and C_{C2} from Eqs. 19-10 and 11, specify the larger at its calculated value, and make the smaller several times its calculated value.

Bypass Capacitors. The foregoing analysis of low-frequency response assumes that C_S (or C_E) effectively bypasses R_S (or R_E) at the lowest frequency of interest. In practice, it turns out that C_S must be much larger than C_{C1} and C_{C2} and, in most cases, C_S is the important element in determining the lower cutoff frequency. The practical approach is to assume that the coupling capacitors are still effective at the frequencies where C_S becomes critical. With this assumption, the low-frequency FET amplifier model is as shown in Fig. 19.10a.

At low frequencies, the voltage gain is reduced because the current $g_m V_{gs}$ flowing through Z_S (the parallel combination of R_S and C_S) develops a voltage that subtracts

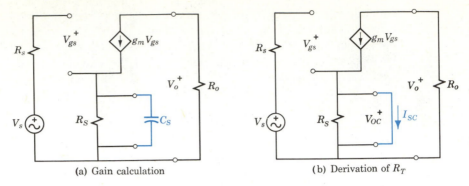

(a) Gain calculation   (b) Derivation of R_T

Figure 19.10 Low-frequency amplifier model where C_S is critical.

from the signal voltage V_s. In the input loop, $\mathbf{V}_{gs} = \mathbf{V}_s - g_m \mathbf{V}_{gs} \mathbf{Z}_S$, or the gate voltage is reduced to

$$\mathbf{V}_{gs} = \frac{\mathbf{V}_s}{1 + g_m \mathbf{Z}_S} \tag{19-14}$$

and the voltage gain V_o/V_s is reduced by the factor $1 + g_m \mathbf{Z}_S$. The lower cutoff frequency occurs when $1 + g_m \mathbf{Z}_S$ has a magnitude of $\sqrt{2}$. Since $\mathbf{Z}_S$ is a complex quantity, there is no simple expression for the proper value of C_S. Instead we shall use a useful technique to determine an approximate expression.[†]

With R_S unbypassed, the gain is low (but not zero, because Z_S has a maximum value of R_S). With R_S perfectly bypassed, the gain is $A_V = g_m R_o$. Therefore, let us specify C_S to be large enough so that at f_1 the reactance $1/\omega_1 C_S$ is equal to the *effective resistance* bypassed. We define the effective resistance as the Thévenin equivalent resistance looking in at the terminals of R_S across which C_S is connected. Letting $Z_S = R_S$ in Eq. 19-14,

$$V_{OC} = g_m V_{gs} R_S = g_m \frac{V_s}{1 + g_m R_S} R_S = \frac{g_m R_S}{1 + g_m R_S} V_s \tag{19-15}$$

Since $I_{SC} = g_m V_s$ with R_S shorted, the Thévenin equivalent is

$$R_T = \frac{V_{OC}}{I_{SC}} = \frac{R_S}{1 + g_m R_S} = \frac{1}{g_m + 1/R_S} = R_S \| 1/g_m \tag{19-16}$$

or the effective resistance is the parallel combination of R_S and $1/g_m$. Therefore the design criterion (see Example 2) is

$$\omega_1 = 2\pi f_1 = \frac{1}{C_S R_T} = \frac{g_m + 1/R_S}{C_S} \tag{19-17}$$

In the BJT amplifier, C_E is the important element in determining the lower cutoff frequency. Here again an exact analysis is complicated and an approximate expression is satisfactory for our purposes. The approach employed in establishing the design criterion for C_S in the FET amplifier would work, but using a different approach may

---

[†]Suggested by Malvino; a general expression is given on p. 449 of Schilling and Belove. See References at the end of Chapter 17.

EXAMPLE 2

Specify C_S for a lower cutoff frequency of 20 Hz in the circuit of Fig. 19.10 where $g_m = 2000\ \mu S$ and $R_S = 2\ k\Omega$. Show that the voltage gain at 20 Hz is down to approximately 70% of its midfrequency value.

By Eq. 19-17,

$$C_S = \frac{g_m + 1/R_S}{2\pi f_1} = \frac{0.002 + 0.0005}{2\pi \times 20} \cong 20\ \mu F$$

For this value of C_S and $f = 20$ Hz,

$$g_m \mathbf{Z}_S = \frac{g_m}{(1/R_S) + j\omega C} = \frac{0.002}{0.0005 + j0.0025}$$

$$= 0.784 \underline{/-78.7^\circ}$$

$$1 + g_m \mathbf{Z}_S = 1 + (0.15 - j0.77) = 1.15 - j0.77$$

$$= 1.38 \underline{/33^\circ}$$

The magnitude of $1 + g_m \mathbf{Z}_S$ is $1.38 \cong \sqrt{2} = 1.41$; therefore f_1 is approximately the lower cutoff frequency and Eq. 19-17 is justified.

be more enlightening. As before, we assume that the coupling capacitors are effective at the frequency where C_E becomes critical and draw the low-frequency amplifier model as shown in Fig. 19.11a.

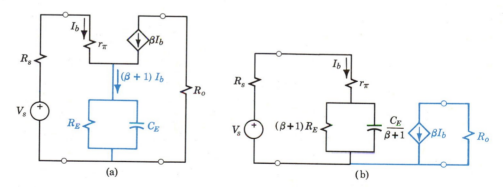

(a)   (b)

Figure 19.11   The low-frequency model where C_E is critical.

The current $(\beta + 1)\mathbf{I}_b$ flowing through $\mathbf{Z}_E$ (the parallel combination of R_E and C_E) develops a voltage that, in effect, opposes the signal voltage $\mathbf{V}_s$. Summing the voltages around the left-hand loop and solving for the current,

$$\mathbf{I}_b = \frac{\mathbf{V}_s}{R_s + r_\pi + (\beta + 1)\mathbf{Z}_E} \tag{19-18a}$$

where

$$(\beta + 1)\mathbf{Z}_E = \frac{\beta + 1}{\dfrac{1}{R_E} + j\omega C_E} = \frac{1}{\dfrac{1}{(\beta + 1)R_E} + j\omega \dfrac{C_E}{\beta + 1}} \tag{19-18b}$$

Equations 19-18a and b are satisfied by the circuit model of Fig. 19.11b. If we neglect the current through the relatively large resistance $(\beta + 1)R_E$,

$$I_b = \frac{V_s}{R_s + r_\pi - j\dfrac{\beta + 1}{\omega C_E}} \tag{19-19}$$

If the degrading effect of $\mathbf{Z}_E$ is to be small, the reactance of $C_E/(\beta + 1)$ must be small compared to the resistances with which it is in series. Current I_b will drop to 70.7% of its midfrequency value and the gain will drop to 70% of its midfrequency value at frequency ω_1 where

$$\frac{1}{\omega_1 C_E/(\beta + 1)} = R_s + r_\pi \tag{19-20}$$

Therefore, the design criterion is

$$\omega_1 = 2\pi f_1 = \frac{\beta + 1}{C_E(R_s + r_\pi)} \tag{19-21}$$

In practice, C_E is specified to provide the desired low-frequency response and then C_{C1} and C_{C2} are selected to be several times the values indicated by Eqs. 19-10 and 11. Note that the cutoff frequency is just a convenient measure of the low-frequency response of the amplifier and does not imply that no amplification occurs below this frequency.

EXAMPLE 3

In the circuit of Fig. 19.12, $R_s = 2$ kΩ, $R_E = 1$ kΩ, and $C_E = 50$ μF. R_1 and R_2 are very large. For $r_\pi = 1.5$ kΩ and $\beta = 50$, estimate the lower cutoff frequency and check the validity of the assumption made in deriving Eq. 19-20. For $R_C = 5$ kΩ and $R_L = 10$ kΩ, specify appropriate values for C_{C1} and C_{C2}.

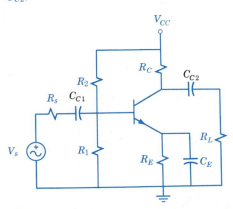

Figure 19.12   Low-frequency response.

By Eq. 19-21 (see Fig. 19.11),

$$f_1 = \frac{\omega_1}{2\pi} = \frac{\beta + 1}{2\pi C_E(R_s + r_\pi)}$$

$$= \frac{50 + 1}{2\pi \times 5 \times 10^{-5}(2 + 1.5)10^3} = 46 \text{ Hz}$$

The admittance of $(\beta + 1)R_E$ is

$$\frac{1}{(\beta + 1)R_E} = \frac{1}{51 \times 10^3} \cong 2 \times 10^{-5} \text{ S}$$

which when squared (see Eq. 19-2) is negligibly small in comparison to the square of

$$\frac{\omega_1 C_E}{\beta + 1} = \frac{1}{R_s + r_\pi} = \frac{1}{3.5 \times 10^3} = 29 \times 10^{-5} \text{ S}$$

For the same value of f_1, Eqs. 19-10 and 11 indicate

$$C'_{C1} = \frac{1}{2\pi f_1(R_s + r_\pi)} = \frac{1}{2\pi \times 46(3500)} \cong 1 \ \mu\text{F}$$

Therefore, specify a value of, say, $5C'_{C1} = 5 \ \mu$F.

$$C'_{C2} = \frac{1}{2\pi f_1(R_C + R_L)} = \frac{1}{2\pi \times 46(15,000)} \cong 0.23 \ \mu\text{F}$$

Therefore, specify a value of, say, $5C'_{C2} = 1 \ \mu$F.

High-Frequency Response

Above the midfrequencies, the reactances of coupling and bypass capacitances are negligibly small, but the susceptances of the other capacitances become increasingly important. Under these conditions, the high-frequency versions of the amplifier circuit models are as shown in Fig. 19.13, where $R_o = R_D \| R_L$ or $R_C \| R_L$ and

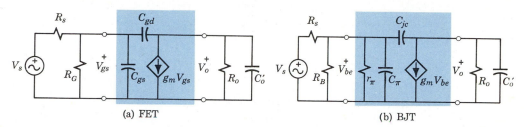

(a) FET (b) BJT

Figure 19.13 High-frequency models of amplifiers.

$C_o' = C_W + C_L$ (wiring plus lead capacitance). The analysis of these analogous circuits is complicated by the presence of C_{gd} and C_{jc} that provide negative feedback from the output to the input and therefore tend to reduce the gain at high frequencies.

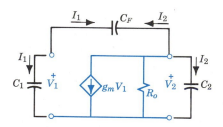

Figure 19.14 Miller capacitances.

As a first step, let us simplify the circuit using a technique first suggested by J. M. Miller. In Fig. 19.14, the voltage across feedback capacitance C_F causing current I_1 is

$$V_1 - V_2 = V_1 - (-g_m R_o)V_1 = V_1(1 + g_m R_o) = V_1(1 + A) \qquad (19\text{-}22)$$

where $V_2/V_1 = -A = -g_m R_o$ is the gain of this circuit. This current could be accounted for in the input circuit by a capacitance that is $(1 + A)$ times as large under a voltage only $1/(1 + A)$ times as large, i.e., by a capacitance

$$C_1 = C_F(1 + A) = C_F(1 + g_m R_o) \qquad (19\text{-}23)$$

under a voltage V_1. Following the same reasoning, the current I_2 could be accounted for by a capacitance

$$C_2 = C_F(1 + 1/A) = C_F(1 + 1/g_m R_o) \cong C_F \qquad (19\text{-}24)$$

under a voltage V_2 since $V_2 - V_1 = V_2(1 + 1/A) \cong V_2$. When these substitutions are made and the "Miller capacitances" are combined with C_{gs} and C_o', respectively, the simplified circuit model is as shown in Fig. 19.15b where $C_{eq} = C_{gs} + C_{gd}(1 + A)$ and $C_o = C_{gd} + C_W + C_L$.

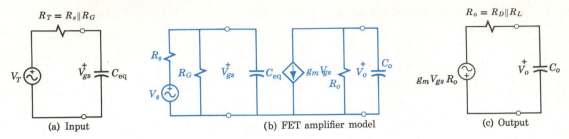

(a) Input (b) FET amplifier model (c) Output

Figure 19.15 Low-pass filter circuits.

FET. In the Thévenin equivalent (Fig. 19.15a), $R_T = R_s \| R_G \cong R_s$ since $R_G \gg R_s$, in general. Now the input circuit has the form of a *low-pass filter*. At midfrequencies, $V_{gs} = V_T \cong V_s$. At high frequencies where C_{eq} is significant,

$$V_{gs} = \frac{1/j\omega C_{eq}}{(R_s \| R_G) + 1/j\omega C_{eq}} V_s = \frac{V_s}{1 + j\omega C_{eq}(R_s \| R_G)} \tag{19-25}$$

As frequency is increased, a smaller fraction of the voltage V_s appears across C_{eq} and voltage V_{gs} is reduced. At the upper cutoff frequency defined by

$$\omega_{21} = \frac{1}{C_{eq}(R_s \| R_G)} \tag{19-26}$$

the input voltage V_{gs} will be down to 70% of V_s and the possible amplifier gain will be correspondingly reduced. Reasoning by analogy, we can say that at an upper cutoff frequency

$$\omega_{22} = \frac{1}{C_o(R_D \| R_L)} \tag{19-27}$$

the output voltage V_o will be down to 70% of $g_m V_{gs} R_o$ and the amplifier gain will be correspondingly reduced. The overall high-frequency gain for the FET amplifier can be expressed as

$$\mathbf{A}_H = \mathbf{A}_O \cdot \frac{1}{1 + j\omega C_{eq}(R_s \| R_G)} \cdot \frac{1}{1 + j\omega C_o(R_D \| R_L)} \tag{19-28}$$

or the relative gain at high frequencies is

$$\frac{\mathbf{A}_H}{\mathbf{A}_O} = \frac{1}{1 + j\omega/\omega_{21}} \cdot \frac{1}{1 + j\omega/\omega_{22}} \tag{19-29}$$

To predict the behavior of a given circuit, determine ω_{21} and ω_{22} from Eqs. 19-26 and 27; the lower of the two determines the upper cutoff frequency (unless they are very close together). In designing a circuit for a specified behavior, the designer does not have freedom in specifying C_{eq} and C_o since these are determined by device parameters, operating conditions, and gain. If the predicted performance is not satisfactory, one possibility is to reduce R_D, which reduces gain (and therefore C_{eq}) and increases ω_{21} and ω_{22}.

EXAMPLE 4

In Example 1, an FET ($g_m = 2000$ μS, $C_{gs} = 20$ pF, and $C_{gd} = 2$ pF) was used in an RC-coupled amplifier with $R_s = 1$ kΩ, $R_G = 1$ MΩ, $R_D = 5$ kΩ, and $R_L = 5$ kΩ to give a midfrequency gain of $-g_m R_o = -5$. If $C_W + C_L \cong 4 + 16 = 20$ pF, predict the upper cutoff frequency.

The equivalent input and output capacitances are

$$C_{eq} = C_{gs} + C_{gd}(1 + g_m R_o) = 20 + 2(6) = 32 \text{ pF}$$

$$C_o = C_W + C_L + C_{gd} = 4 + 16 + 2 = 22 \text{ pF}$$

The equivalent input and output resistances are

$$R_s \| R_G = 10^3 \times 10^6/(10^3 + 10^6) = 999 \cong 1000 \ \Omega$$

$$R_D \| R_L = 5000 \times 5000/(5 + 5) \times 10^3 = 2500 \ \Omega$$

The individual cutoff frequencies are

$$\omega_{21} = 1/C_{eq}(R_s \| R_G) = 1/32 \times 10^{-12} \times 10^3$$

$$= 31.3 \text{ Mrad/s}$$

$$\omega_{22} = 1/C_o(R_D \| R_L) = 1/22 \times 10^{-12} \times 2500$$

$$= 18 \text{ Mrad/s}$$

The upper cutoff frequency of the amplifier is

$$18/2\pi \cong 2.9 \text{ MHz}$$

The manufacturer's data sheets usually give values that are easy to measure. The "reverse transfer capacitance" $C_{rss} = C_{gd}$; the "input capacitance with output shorted" $C_{iss} = C_{gs} + C_{gd}$. Therefore, $C_{gs} = C_{iss} - C_{rss}$. We have not considered C_{ds}, which is usually negligible. (See Table 19-1.)

Table 19-1 Typical FET Parameters

		2N4416 n-channel JFET Min/Max	3N163 p-channel E MOSFET Min/Max
Static parameters			
$V_{gs(OFF)}$	Gate-source cutoff voltage	$-2.5/-6.0$ V	
$V_{GS(th)} = V_T$	Gate-source threshold voltage		$-2/-5$ V
I_{DSS}	Saturation drain current	$5/15$ mA	
$I_{DS(ON)}$	ON drain current		$-5/-30$ mA
Small-signal parameters at 1 kHz			
$g_{fs} = g_{mo}$	Forward transconductance	$4500/7500$	$2000/4000$ μS
$g_{oss} = 1/r_d$	Output conductance	50	250 μS
$C_{rss} = C_{gd}$	Reverse transfer capacitance	0.8	0.7 pF
$C_{iss} = C_{gs} + C_{gd}$	Input capacitance	4	2.5 pF
$C_{oss} = C_{ds}$	Output capacitance	2	3 pF

BJT. Comparing Figs. 19.13a and b and reasoning by analogy, the upper cutoff frequency for a BJT amplifier is defined by

$$\omega_{21} = \frac{1}{C_{eq}(R_s \| r_\pi)} \quad \text{or} \quad \omega_{22} = \frac{1}{C_o(R_C \| R_L)} \tag{19-30}$$

where $C_{eq} = C_\pi + C_{jc}(1 + g_m R_o)$ and $C_o = C_{jc} + C_W + C_L$ (Fig. 19.15b). (More precisely, the equivalent input resistance is $R_s \| r_\pi \| R_B$, but in a well-designed amplifier $R_B \gg r_\pi$.) The overall high-frequency gain can be expressed as

$$\mathbf{A}_H = \mathbf{A}_O \cdot \frac{1}{1 + j\omega C_{eq}(R_s \| r_\pi)} \cdot \frac{1}{1 + j\omega C_o(R_C \| R_L)} \tag{19-31}$$

and the relative gain at high frequencies is

$$\frac{\mathbf{A}_H}{\mathbf{A}_O} = \frac{1}{1 + j\omega/\omega_{21}} \cdot \frac{1}{1 + j\omega/\omega_{22}} \tag{19-32}$$

The upper cutoff frequency is determined by the lower of ω_{21} and ω_{22} just as in the case of the FET amplifier.

Analysis of the high frequency response of transistors is complicated by the fact that *lumped* circuit models can only approximate effects that are, in fact, *distributed* throughout a finite region in the transistor. For example, finite times are required for signals in the form of changes in charge density to be propagated across the base of a bipolar transistor. At higher frequencies the current gain I_c/I_b begins to decrease in magnitude and a phase lag appears. In the hybrid-π model of Fig. 19.16a, these charging effects are represented by capacitance C_π.

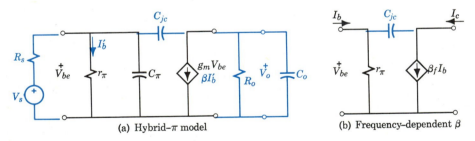

(a) Hybrid-π model   (b) Frequency-dependent β

Figure 19.16   High-frequency BJT models.

Another approach is to represent the transistor by parameters that are frequency dependent. We define β as the short-circuit current gain at low and medium frequencies and let $I_c/I_b = \beta_f$, a function of frequency. At the *beta cutoff frequency* f_β, current gain β_f is just equal to 70.7% of the low-frequency value. The significant relation between current phasors (Fig. 19.16b) is

$$\frac{\mathbf{I}_c}{\mathbf{I}_b} = \frac{\beta}{1 + j(f/f_\beta)} = \frac{\beta}{1 + j(\omega/\omega_\beta)} \tag{19-33}$$

and the magnitude of the high-frequency current gain is

$$\beta_f = \frac{\beta}{\sqrt{1 + (f/f_\beta)^2}} \tag{19-34}$$

where f_β may be specified by the manufacturer. At very high frequencies, $\beta_f = \beta/(f/f_\beta) = \beta f_\beta/f$ or current gain is inversely proportional to frequency. Alternatively, therefore, the manufacturer may specify the frequency

$$f_T = \beta f_\beta \tag{19-35}$$

at which the current gain β_f would fall to unity, or it may state the value of β_f at a specified frequency f.

To avoid working with frequency-dependent parameters, we can use constant circuit elements with the appropriate frequency response. In Fig. 19.16a, the fall-off in short-circuit current gain is accounted for by C_π alone because, when V_o is shorted, gain A in Eq. 19-23 is zero. Therefore,

$$f_T = \beta f_\beta = \frac{\beta}{2\pi C_\pi r_\pi} = \frac{1}{2\pi C_\pi r_e} \tag{19-36}$$

since $r_\pi = \beta r_e$. The determination of parameter values from manufacturer's specifications is illustrated in Example 5.

EXAMPLE 5

The manufacturer of the 2N3114 specifies $h_{ib} = r_e = 27\ \Omega$ and $h_{fe} = \beta = 50$ at 1 kHz (Appendix 8) and $\beta = 2.7$ at 20 MHz. Determine f_T, f_β, and C_π.

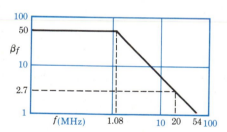

Figure 19.17 High-frequency response.

The 2.7 value is for β_f at $f = 20$ MHz. By Eq. 19-35, the current gain becomes unity at frequency $f_T = \beta f_\beta = \beta_f \times f = 2.7 \times 20$ MHz $= 54$ MHz and the beta cutoff frequency in Fig. 19.17 is

$$f_\beta = \frac{f_T}{\beta} = \frac{54\text{ MHz}}{50} = 1.08\text{ MHz}$$

Since the common-base emitter resistance $h_{ib} = r_e$, by Eq. 19-36,

$$C_\pi = \frac{1}{2\pi f_T r_e} = \frac{1}{2\pi \times 54 \times 10^6 \times 27} = 109\text{ pF}$$

For ordinary transistors used in audiofrequency amplifiers, the decrease in β is unimportant; for special high-frequency planar transistors, the effect may be unimportant up to many megahertz. In amplifiers employing high-frequency transistors, the upper cutoff frequency may be determined by the effective shunting capacitance C_o, due to wiring and other parasitic effects, as indicated by Eq. 19-30. If the voltage gain is large, the collector-base junction capacitance C_{jc} becomes critical. (The practical value of this parameter is the common-base output capacitance designated C_{ob} or C_{cb} on the data sheet.) In practice, the engineer determines which is the limiting factor and bases the amplifier design on a careful analysis of that factor.

MULTISTAGE AMPLIFIERS

To achieve the desired voltage or current gain and the necessary frequency response, several stages of amplification may be required.

Cascading

The stages are usually connected in *cascade;* the output of one stage is connected to the input of the next. Where the preceding and following stages are similar, $R_L = r_\pi$ (Fig. 19.6b) and the midfrequency *per stage* voltage gain for the BJT is

$$A_{VO} = \frac{V_o}{V_{be}} = \frac{-g_m V_{be} R_o}{V_{be}} = -g_m R_o \tag{19-37}$$

just as for the FET. For the BJT the midfrequency *per stage* current gain is

$$A_{IO} = \frac{I_o}{I_b} = \frac{V_o/r_\pi}{V_{be}/r_\pi} = \frac{V_o}{V_{be}} = -g_m R_o = -\frac{\beta}{r_\pi} R_o \tag{19-38}$$

Since R_o is determined in part by r_π and is always less than r_π, the current gain per stage is always less than β.

By defining a single stage to include the input characteristics of the next amplifying element, the combined effect of cascaded stages is easy to predict. In Fig. 19.18,

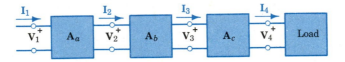

Figure 19.18   Block diagram of amplifiers in cascade.

the individual stages are represented by rectangular blocks labeled with the individual voltage gains. The overall voltage gain is

$$A = \frac{V_4}{V_1} = \frac{V_2}{V_1} \frac{V_3}{V_2} \frac{V_4}{V_3} = A_a A_b A_c \tag{19-39}$$

or

$$A \underline{/\theta} = A_a A_b A_c \underline{/\theta_a + \theta_b + \theta_c} \tag{19-40}$$

In words, the overall voltage gain is the product of the gain amplitudes and the sum of the phase shifts; both A and θ are functions of frequency. A similar statement would describe the overall current gain.

Gain in Decibels

The calculation of overall gain in multistage amplifiers is simplified by using decibels. As indicated in Eq. 9-27,

$$\text{Gain in decibels} = 20 \log \frac{V_2}{V_1} \tag{19-41}$$

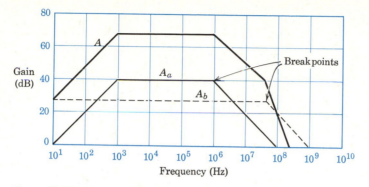

Figure 19.19 Frequency response expressed in decibels.

The advantage of the decibel unit here is that when response is plotted in dB the overall response curves of a multistage amplifier can be obtained by adding the individual response curves. In Fig. 19.19, the response curves for two stages a and b are plotted in dB. From Eqs. 19-13 and 19-32 it can be deduced that for $f \ll f_1$ relative gain is proportional to frequency, and for $f \gg f_2$ relative gain is inversely proportional to frequency. On log-log scales, these relations are straight lines with slopes of 20 dB per decade and the response of an RC-coupled amplifier can be approximated by curve A_a. (Actually, the gain is down 3 dB at the *breakpoint frequencies*.) A second amplifier with better low-frequency response is approximated by curve A_b. The overall gain of the combination is the sum of the two curves, and the straight-line approximation is easily drawn. For a single-stage amplifier with two high-frequency breakpoints, the frequency response would resemble curve A of Fig. 19.19.

Gain-Bandwidth Product

The *bandwidth* of an amplifier is a useful design criterion; an audio amplifier needs a bandwidth of about 20 kHz whereas a video amplifier requires a bandwidth of about 4 MHz. For an untuned amplifier the bandwidth is defined as BW $= \omega_2 - \omega_1 \cong \omega_2$ since ω_1 is small compared to ω_2. For a single-stage (Fig. 19.20), the gain is

$$\mathbf{A} = \frac{\mathbf{V}_o}{\mathbf{V}_{gs}} = \frac{g_m}{\dfrac{1}{R_o} + j\omega C_{eq}} = \frac{g_m R_o}{1 + j\omega R_o C_{eq}} = \frac{A_O}{1 + j\omega R_o C_{eq}} \qquad (19\text{-}42)$$

Where $\omega_2 = 1/R_o C_{eq}$, any increase in bandwidth by decreasing R_o results in a decrease in midfrequency gain. The product of gain magnitude and bandwidth is

$$A_O \omega_2 = g_m R_o \frac{1}{R_o C_{eq}} = \frac{g_m}{C_{eq}} \le \frac{g_m}{C_{gs}} \qquad (19\text{-}43)$$

since $C_{eq} \cong C_{gs}$ if the Miller effect is not large. Because g_m and C_{gs} are device parameters:

The gain-bandwidth product is a constant for a specific FET.

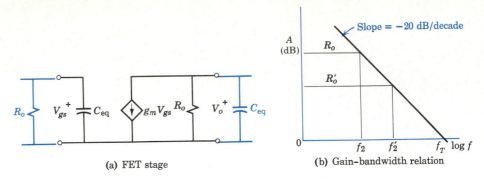

(a) FET stage   (b) Gain–bandwidth relation

Figure 19.20   Gain-bandwidth relation for an FET amplifier.

For the FET of Example 4, $g_m/C_{gs} = 2 \times 10^{-3}/20 \times 10^{-12} = 10^8$ rad/s $\cong$ 15 MHz. With this FET a gain of 5 permits a bandwidth of less than 3 MHz (2.9 MHz in Example 4 where $R_s = 1$ kΩ); if a bandwidth of 5 MHz is required, the maximum possible gain is 3.

     A bipolar transistor has a similar *figure of merit*, but it is usually expressed in a different way. If the high-frequency response is limited by the beta cutoff, the bandwidth is approximately f_β and the maximum possible gain is β, so the *gain-bandwidth product* is

$$A_o f_2 \leq \beta f_\beta = f_T \qquad (19\text{-}44)$$

The frequency f_T at which $\beta_f = 1$ typically ranges from 100 kHz for alloy junction power transistors to 10 GHz for high-frequency transistors.

EXAMPLE 6

The specifications for a certain silicon transistor indicate a minimum h_{fe} of 40 and a typical f_T of 8 MHz. Is this transistor suitable as a radiofrequency amplifier at 1 MHz?

By Eq. 19-35 at $f = 1$ MHz,

$$\beta_f = \frac{\beta f_\beta}{f} = \frac{f_T}{f} = \frac{8}{1} = 8$$

By Eq. 19-44

$$f_\beta = \frac{f_T}{\beta} = \frac{f_T}{h_{fe}} = \frac{8 \text{ MHz}}{40} = 0.2 \text{ MHz}$$

Although some current gain is possible at 1MHz, the beta cutoff is only 0.2 MHz and another transistor should be selected.

     Example 7 illustrates the complete design of a single-stage small-signal amplifer to meet frequency-response specifications.

EXAMPLE 7

Design a single-stage amplifier with a voltage gain of 34 dB, flat within 3 dB from 46 Hz to 200 KHz. Source and load resistances are $R_s = 2$ kΩ and $R_L = 10$ kΩ.

Use a 2N3114 with $h_{fe} = \beta = 50$ (1 kHZ) and $h_{ie} = 1.5$ kΩ $= r_b + r_\pi \cong r_\pi$ at $I_C = 1$ mA and $V_{CE} = 5$ V; $C_{ob} = C_{jc} = 6$ pF and $f_T = 54$ MHz.

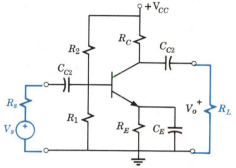

(a) Circuit diagram

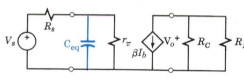

(b) Small–signal model

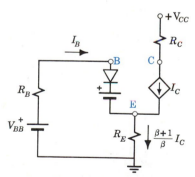

(c) DC bias design model

Figure 19.21   BJT amplifier design.

Replace the circuit diagram by the small-signal model of Fig. 19.21b, assuming $R_B \gg r_\pi$.

At midfrequencies,

$$A_{VO} = \left|\frac{V_o}{V_s}\right| = \frac{\beta I_b (R_C \| R_L)}{I_b (R_s + r_\pi)} = 10^{dB/20} \cong 50$$

Solving,

$$R_o = R_C \| R_L = \frac{(R_s + r_\pi)A_{VO}}{\beta} = \frac{(2 + 1.5)50}{50} = 3.5 \text{ k}\Omega$$

Hence

$$R_C = \frac{1}{(1/R_o) - (1/R_L)} = \frac{10^3}{0.286 - 0.1} = 5.38 \text{ k}\Omega$$

Assuming high-frequency response is limited by C_{eq} with $C_\pi = 109$ pF from Example 5,

$$C_{eq} = C_\pi + (1 + \beta R_o/r_\pi)C_{jc}$$

$$= 109 + (1 + 50 \times 3.5/1.5)6 = 815 \text{ pF}$$

$$f_2 = \frac{1}{2\pi C_{eq}(R_s \| r_\pi)} = \frac{1}{2\pi \times 815 \times 10^{-12} \times 860} = 228 \text{ kHz}$$

Therefore, the high-frequency response is acceptable.

Assuming bias is not critical (Fig. 19.21c), let

$$R_E = \frac{3}{I_E} \cong \frac{3}{I_C} = \frac{3}{0.001} = 3 \text{ k}\Omega$$

$$R_B \cong \frac{\beta R_E}{10} = \frac{50 \times 3}{10} = 15 \text{ k}\Omega$$

Then

$$V_{CC} = V_E + V_{CE} + I_C R_C = 3 + 5 + 1(5.38) \cong 13 \text{ V}$$

$$V_{BB} = R_B I_C/\beta + V_{BE} + V_E = 15 \times 1/50 + 0.7 + 3 = 4 \text{ V}$$

$$R_2 = R_B V_{CC}/V_{BB} = 15 \times 13/4 = 48.8 \to 47 \text{ k}\Omega$$

$$R_1 = R_B R_2/(R_2 - R_B) = 15 \times 47/32 = 22.03 \to 22 \text{ k}\Omega$$

For low-frequency response, use the results of Example 3 where $C_E = 50$ μF, $C_{C1} = 5$ μF, and $C_{C2} = 1$ μF.

FEEDBACK AMPLIFIERS

The concept of feedback was introduced in Chapter 9 and was used in Chapter 17 to stabilize the operating points of field-effect and bipolar transistors. The combination of amplifier gain and positive or negative feedback can be used to perform some very valuable functions.

Positive Feedback

By using the rules of block diagram algebra, all linear feedback networks can be reduced to the basic system of Fig. 19.22a, for which the gain is $\mathbf{G}_F = \mathbf{G}/(1 - \mathbf{GH})$. The behavior of the system can be predicted by examining the denominator $1 - \mathbf{GH}$.

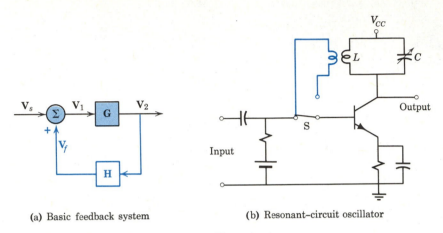

(a) Basic feedback system   (b) Resonant–circuit oscillator

Figure 19.22 Positive feedback in an oscillator circuit.

For **GH** positive but less than unity, the denominator is less than unity and the system gain is greater than the forward gain **G**. This condition of *positive feedback* was used in *regenerative receivers* to provide high gain in the days when electronic amplifiers were poor and signals were weak. Operation with positive feedback becomes less stable as *GH* approaches unity, so regenerative receivers required constant adjustment.

If the value of **GH** approaches $1 + j0$, the denominator approaches zero and the overall gain increases without limit. An amplifier of infinite gain can provide an output with no input; such a device is called an *oscillator* or signal generator and is an important part of many electronic systems.

Oscillators

When a circuit is designed to oscillate,[†] special provision is made to feed back a portion of the output signal in the correct phase and amplitude to make the **GH** product just equal to $1 + j0$. By making **GH** frequency dependent, oscillation can be obtained at a single desired frequency. Either **G** or **H** can be the frequency-dependent element of the feedback system of Fig. 19.22a.

The oscillator of Fig. 19.22b is basically an amplifier with a parallel resonant circuit for a load impedance. Gain is a maximum for signals at the resonant frequency of the collector circuit. This is called a *tuned* amplifier because the frequency for maximum amplification can be selected by varying the capacitance. (Such an amplifier is used in a radio receiver to amplify the signals transmitted by a selected station while

---

[†]The fact that stable operation occurs in the nonlinear region of the device characteristic complicates the design and analysis of oscillators.

rejecting all others.) To create a "self-excited amplifier" or oscillator, switch S is shifted (and the "Input" terminals are shorted) and part of the voltage across the tuned circuit is fed back into the input. In this case, **G** is a sensitive function of frequency.

For audiofrequencies, the values of L and C required for a resonant circuit are impractically large. The *phase-shift oscillator* of Fig. 19.23 avoids this requirement and is fairly common for the generation of audiofrequency signals. The output of the FET amplifier appears across R_D. The first RC section constitutes a phase-shifting network in that the voltage across R is from 0° to 90° ahead of the voltage across R_D depending on the frequency. For a particular frequency, the three RC sections produce a total phase shift of 180°; the additional 180° required for oscillation is inherent in the gate-drain voltage relations in an FET. In this case, **H** is the frequency-sensitive element of the feedback system.

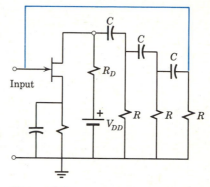

Figure 19.23 An FET phase-shift oscillator.

To provide good waveform, an oscillator should operate in the linear region of the amplifier. In the *resistance-capacitance tuned oscillator* of Fig. 19.24a, widely used as a laboratory instrument, the necessary phase reversal for positive feedback is provided by the two-stage amplifier. The input to the first stage is obtained from the voltage divider $Z_1/(Z_1 + Z_2)$ across the output of the second stage. Voltage V_{in} is in phase with V_{out} for $\omega = 1/\sqrt{R_1 R_2 C_1 C_2}$ (the principle of the Wien bridge) and oscillation occurs. In the usual case, $R_1 = R_2 = R$ and $C_1 = C_2 = C$ and the oscillator

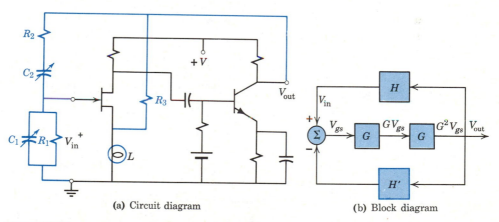

(a) Circuit diagram   (b) Block diagram

Figure 19.24 A resistance-capacitance tuned oscillator.

frequency is $\omega_o = 1/RC$. As the oscillations build up, the current through R_3 and incandescent lamp L increases, raising the temperature of the filament. The voltage developed across this nonlinear resistance subtracts from the input voltage and limits the amplitude of the oscillations. The negative feedback loop is shown in Fig. 19.24b, where H' is a nonlinear function of output voltage.

NEGATIVE FEEDBACK APPLICATIONS

Negative feedback, where the amplitude and phase of the signal fed back are such that the overall gain of the amplifier is less than the forward gain, may be used to advantage in amplifier design. Since additional voltage or current gain can be obtained easily with transistors, the loss in gain is not important if significant advantages are obtained. Amplifier performance can be improved in the following ways:

1. Gain can be made practically independent of device parameters; gain can be made insensitive to temperature changes, aging of components, and variations from typical values.
2. Operating point can be made practically independent of device parameters, and it can be made insensitive to temperature changes or aging of components.
3. Gain can be made practically independent of reactive elements and therefore insensitive to frequency.
4. Gain can be made selective to discriminate against distortion or noise.
5. The input and output impedances of an amplifier can be improved greatly.

Stabilizing Gain

For precise instrumentation or long-term operation, amplifier performance must be predictable and stable. However, amplifier components such as transistors and IC elements vary widely in their critical parameters. By employing negative feedback and sacrificing gain (which can be made up readily), stability can be improved.

EXAMPLE 8

The midfrequency gain of a small-signal amplifier is, by Eq. 19-6,

$$A_{VO} = -\beta \frac{R_o}{r_\pi + R_s} \cong -100$$

In a certain mass-produced item, it is expected that random variations in parameters and the effects of temperature variations might reduce the nominal gain from 100 to as low as 50.

Investigate the stabilizing effect of using two identical stages and a feedback loop with $H = -0.0099$.

For two stages, the forward gain is $G_N = (-100)^2 = 10,000$. With feedback, the nominal gain is

$$G_{FN} = \frac{G_N}{1 - G_N H} = \frac{10,000}{1 - 10^4(-0.0099)} = 100$$

If the actual gain per stage drops to $G_A = (-50)^2 = 2500$, the amplifier gain is

$$G_{FA} = \frac{G_A}{1 - G_A H} = \frac{2500}{1 - 2500(-0.0099)} = 97.1$$

With feedback (and an additional stage), the change in gain is reduced to less than 3%.

Stabilizing Operation

The self-biasing circuit for decreasing the effect of changes in β or temperature on the operating point of a transistor is essentially a feedback device. As indicated in Fig. 19.25a, if I_C tends to increase, the current $I_C + I_B$ in R_E increases, raising the potential of the emitter with respect to ground. This, in turn, reduces the forward bias, reduces the base current and, therefore, limits the increase in I_C. In other words, an increase in I_C is fed back in such a way as to oppose the change that produced it.

EXAMPLE 9

Represent the response of the self-biasing circuit of Fig. 19.25a by a block diagram and derive the "stability factor" $\Delta I_C / \Delta \beta$.

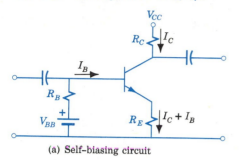

(a) Self-biasing circuit

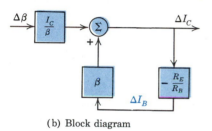

(b) Block diagram

Figure 19.25 Bias stabilization.

Where I_{CBO} is negligible (as in silicon transistors), $I_C = \beta I_B$ and

$$\Delta I_C = \Delta\beta I_B + \beta\Delta I_B = \Delta\beta(I_C/\beta) + \beta\Delta I_B \quad (19\text{-}45)$$

With fixed-current bias ($\Delta I_B = 0$), changes in β produce proportionate changes in I_C.

With self-bias (Fig. 19.25a),

$$V_B - I_B R_B - V_{BE} - (I_C + I_B)R_E = 0$$

Assuming V_{BE} is constant, for small changes this becomes

$$-\Delta I_B R_B - \Delta I_C R_E - \Delta I_B R_E = 0$$

or

$$\Delta I_B = -\Delta I_C R_E/(R_B + R_E) \cong -\Delta I_C R_E/R_B \quad (19\text{-}46)$$

where R_B is large compared to R_E.

Equations 19-45 and 46 are realized in Fig. 19.25b. For this closed-loop system $G = 1$, and

$$\frac{\Delta I_C}{\Delta\beta} = \frac{I_C/\beta}{1 + \beta R_E/R_B} \qquad \text{or} \qquad \frac{\Delta I_C}{I_C} = \frac{\Delta\beta/\beta}{1 + \beta R_E/R_B}$$

For $\beta R_E = 10 R_B$ (Eq. 17-18), an 11% change in β results in only a 1% change in I_C.

Improving Frequency Response

No amplifier amplifies equally well at all frequencies; at the upper cutoff frequency, for example, the voltage gain is down to 70.7% of the midfrequency value. By sacrificing gain, however, the upper cutoff frequency can be raised. In the practical case, gain can be made insensitive to frequency over the range of interest by using negative feedback.

Using system notation, the gain **G** is frequency dependent because of storage effects in the electronic device and reactive elements in the circuit. The feedback factor **H,** however, can be obtained through purely resistive elements. If **H** is obtained by means of a voltage divider tap on a precision noninductive resistor, then **H** = H, a constant. If **GH** is very large at the frequency of interest, the basic gain equation becomes

$$\mathbf{G}_F = \frac{\mathbf{G}}{1 - \mathbf{GH}} \cong \frac{\mathbf{G}}{-\mathbf{GH}} = -\frac{1}{H} \qquad (19\text{-}47)$$

and the gain is independent of frequency, temperature, or aging.

In Example 10, the midfrequency gain is decreased by a factor of $1 - GH$, but the bandwidth is increased by the same factor. As is usual in such cases, we trade gain for bandwidth.

EXAMPLE 10

An amplifier whose high-frequency response is given by $\mathbf{G} = A_O/[1 + j(\omega/\omega_2)]$ is used in the basic feedback circuit (Fig. 19.22a). If $A_O = 1000$, $\omega_2 = 10^4$ rad/s, and $H = -0.009$, derive an expression for high-frequency response with feedback and predict the new upper cutoff frequency ω_{2F}.

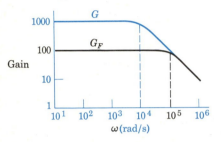

Figure 19.26   Gain-bandwidth tradeoff.

The response is shown in Fig. 19.26. With feedback,

$$\mathbf{G_F} = \frac{\mathbf{G}}{1 - \mathbf{GH}} = \frac{\dfrac{1000}{1 + j(\omega/\omega_2)}}{1 + \dfrac{9}{1 + j(\omega/\omega_2)}}$$

$$= \frac{1000}{1 + j\left(\dfrac{\omega}{\omega_2}\right) + 9} = \frac{100}{1 + j\left(\dfrac{\omega}{10\omega_2}\right)} \quad (19\text{-}48)$$

In the midfrequency range, ω is small compared to $10\omega_2$ and

$$G_F \cong 100$$

At the new cutoff frequency, gain will be down by a factor of $\sqrt{2}$; therefore, $\omega_{2F}/10\omega_2 = 1$, or

$$\omega_{2F} = 10\omega_2 = 10(10^4) = 10^5 \text{ rad/s}$$

Reducing Distortion

Distortion-free amplification is obtained easily at low signal levels. In the last power amplifier stage, however, efficient performance usually introduces nonlinear distortion because of the large signal swings. A common type of distortion results in the introduction of a second harmonic as shown in Fig. 17.1. (Also see Eq. 18-3.)

A nonlinear amplifier can be represented by the combination of linear blocks and rms signals shown in Fig. 19.27a with switch S open. The distortion component V_D is introduced in the output with no corresponding component in the input V_i. Without feedback, the output signal is

$$V_o = GV_i + V_D \quad (19\text{-}49)$$

and the distortion in percent of the fundamental is $(V_D/GV_i) \times 100$. Closing the switch S introduces feedback. In Fig. 19.27b, using the corollary of rule 4 on p. 231, the distortion component V_D has been shifted ahead of G by inserting the factor $1/G$. Now

$$V_o = \frac{G(V_i + V_D/G)}{1 - GH} = \frac{G}{1 - GH}V_i + \frac{V_D}{1 - GH}$$

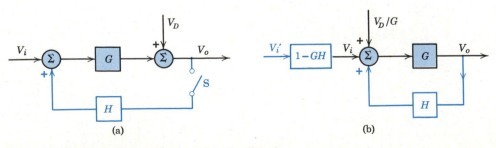

Figure 19.27   Block diagram representation of a distortion component.

If another stage of low-level, distortionless amplification with a gain of $1 - GH$ is added, the new output is

$$V_o = GV_i' + \frac{V_D}{1 - GH} \qquad (19\text{-}50)$$

EXAMPLE 11

The output of a power amplifier contains a 10-V, 1000-Hz fundamental and "20% second-harmonic distortion." The gain $G = 100\underline{/0°}$. Predict the effect of feedback on the distortion if $H = -0.19$.

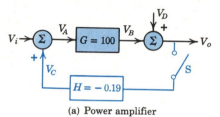

(a) Power amplifier

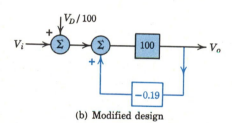

(b) Modified design

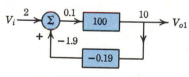

(c) 1000-Hz signal

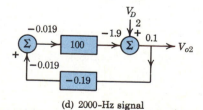

(d) 2000-Hz signal

Figure 19.28 Reducing distortion in a power amplifier by using negative feedback.

The block diagram for the system is as shown in Fig. 19.28a. Blocks G and H are linear models so the distortion component V_D is introduced as an extraneous signal not present in the input to the amplifier. By definition, 20% second-harmonic distortion indicates the presence of a 2000-Hz signal of magnitude

$$V_D = 0.20V_{B1} = 0.20 \times 10 = 2 \text{ V}$$

Applying rule 4 (p. 231), the summing point for V_D is shifted in front of G and the block diagram is that of Fig. 19.28b. With feedback, the gain is

$$G_F = \frac{G}{1 - GH} = \frac{100}{1 + 19} = 5$$

To retain the original power output, the input signal must be increased to

$$V_i = \frac{V_{o1}}{G_F} = \frac{10}{5} = 2 \text{ V at 1000 Hz}$$

The output is now G_F times the equivalent input or

$$V_o = G_F(V_i + V_D/100)$$

At 1000 Hz, $V_{o1} = G_F V_i = 5 \times 2 = 10 \text{ V}$

At 2000 Hz, $V_{o2} = G_F V_D/100 = 5 \times 0.02 = 0.1 \text{ V}$

The voltage distributions for fundamental and second-harmonic signals are shown in Figs. 19.28c and d. Since the blocks are linear, the superposition principle is applicable and the two signals can be handled separately.

The percentage of second-harmonic distortion is now

$$\frac{V_{o2}}{V_{o1}} \times 100 = \frac{0.1}{10} \times 100 = 1\% \text{ (a tolerable amount)}$$

The original amplifier required an input signal of $V_{o1}/100 = 0.1$ V; therefore, additional, distortionless, voltage gain of 2 V/0.1 V = 20 is necessary, but this is easily obtainable at these low levels.

As shown in Example 11, the relative amplitude of the distortion component compared to the desired component will be reduced by the factor $1 - GH$, which may be 10 to 100 or so.

Improving Impedance Characteristics

To minimize the "loading" effect of an electronic device such as an amplifier on a signal source, the input impedance of the device should be high. An FET with insulated gate is very good in this respect, but a BJT is not. Also, to provide efficient power transfer to the load, the output impedance of the device should be small. Neither the FET nor the BJT is optimum in this respect. By using negative feedback the input and output impedances can be dramatically improved. A particularly useful form of feedback circuit is the FET *source follower* or its BJT counterpart, the *emitter follower*.

FOLLOWERS

In the circuit of Fig. 19.29a, the load resistor R_L is in series with the emitter and the collector is tied to V_{CC}, an ac ground. The output voltage[†] $v_o = v_E$ differs from the base voltage v_B by the base emitter voltage $V_{BE} \cong 0.7$ V. Even for large swings of

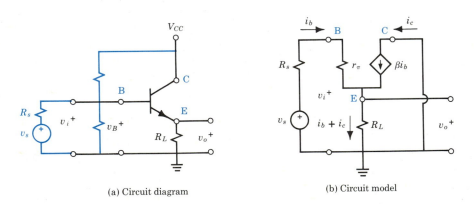

(a) Circuit diagram   (b) Circuit model

Figure 19.29   The emitter follower, a common-collector amplifier with feedback.

v_B, the voltage of the emitter with respect to ground "follows"; hence the name *emitter follower*.

[†]In this discussion we use lowercase letters to represent small signals. The relations hold equally well for effective values of sinusoidal signals.

The Emitter Follower

To examine the impedance characteristics of the emitter follower at moderate frequencies, we replace the transistor by its small-signal model (Fig. 19.29b). Summing voltages around the input loop and solving for the input current,

$$i_b = \frac{v_s}{R_s + r_\pi + (\beta + 1)R_L} \tag{19-51}$$

The output voltage is

$$v_o = (i_b + i_c)R_L = (\beta + 1)i_bR_L = \frac{(\beta + 1)R_Lv_s}{R_s + r_\pi + (\beta + 1)R_L} \tag{19-52}$$

and the voltage gain in this feedback circuit is

$$G_F = \frac{v_o}{v_s} = \frac{(\beta + 1)R_L}{R_s + r_\pi + (\beta + 1)R_L} \cong 1 \tag{19-53}$$

If β is large (50 to 150) and if R_L is of the same order of magnitude as $(R_s + r_\pi)$, then G_F approaches unity, but it can never exceed unity.

The virtue of an emitter follower is not in providing voltage gain. It does provide a current gain of $\beta + 1$ and a corresponding power gain, but its greatest usefulness is in providing impedance transformation. From Fig. 19.29b,

$$v_i = i_br_\pi + (i_b + i_c)R_L = i_br_\pi + (\beta + 1)i_bR_L \tag{19-54}$$

By definition, the input resistance is

$$R_i = \frac{v_i}{i_b} = r_\pi + (\beta + 1)R_L \cong (\beta + 1)R_L \tag{19-55}$$

The effective output resistance can be evaluated from open- and short-circuit conditions. With the output terminals open-circuited,

$$v_{OC} = v_o \cong v_s$$

since the gain is approximately unity (Eq. 19-53). With the output terminals shorted (shorting R_L), $i_b = v_s/(R_s + r_\pi)$ and the short-circuit current is

$$i_{SC} = i_b + i_c = (\beta + 1)i_b = \frac{\beta + 1}{R_s + r_\pi}v_s$$

Therefore,

$$R_o = \frac{v_{OC}}{i_{SC}} = \frac{R_s + r_\pi}{\beta + 1} \tag{19-56}$$

From Eqs. 19-55 and 19-56, it is clear that the emitter follower acts as an impedance transformer with a ratio of transformation approximately equal to $\beta + 1$. With a load resistance R_L, the resistance seen at the input terminals is $(\beta + 1)$ times as great. With the input circuit resistance $R_s + r_\pi$, the effective output resistance is only $1/(\beta + 1)$ times as large.

EXAMPLE 12

A 2N1893 transistor ($r_\pi = 2.8$ kΩ, $\beta = 70$) is used as an amplifier with $R_s = 1.2$ kΩ and $R_L = 4$ kΩ. Compare the input and output resistances in the common-emitter (CE) and common-collector (CC) configurations.

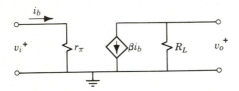

Figure 19.30   Calculation of input and output resistances.

Using the simplified circuit of Fig. 19.30, the input resistance in the common-emitter configuration is

$$R_i = \frac{v_i}{i_b} = r_\pi = 2.8 \text{ k}\Omega$$

For the common-collector circuit of Fig. 19.29a, the input resistance (Eq. 19-55) is

$$R_i = r_\pi + (\beta + 1)R_L = 2.8 + (70 + 1)4 = 287 \text{ k}\Omega$$

The CC input resistance is over 100 times as great.

In the CE amplifier with $R_L = 4$ kΩ, $R_o = 4000$ Ω. For the CC amplifier,

$$R_o = \frac{R_s + r_\pi}{\beta + 1} = \frac{1200 + 2800}{70 + 1} = 56 \text{ }\Omega$$

The output resistance is reduced by a factor of 71.

The Source Follower

The common-drain amplifier or "source follower" is similar to the emitter follower in that it provides improved (lower) output impedance in return for reduced (essentially unity) gain. The input impedance, already high for an FET, can be made even higher by using the bias arrangement of Fig. 19.31 where $R_S = R_1 + R_2$. The quiescent operating point is defined by

$$V_{DS} = V_{DD} - I_D(R_1 + R_2) \tag{19-57}$$

where V_{DS} is made approximately half of V_{DD} to put operation near the midpoint of the load line. The gate-source bias,

$$V_{GS} = -I_D R_1 \tag{19-58}$$

is typically a few volts so that $R_1 \ll R_2$ and $R_S \cong R_2$.

To evaluate the performance of the source follower, we replace the FET by its small-signal model (Fig. 19.31b). In this circuit, $v_{gs} = v_i - v_o$; in feedback notation, $H = 1$ and $G = g_m R_S$. Neglecting the current in the very large R_G and any output current, $i_d = g_m v_{gs} = g_m(v_i - v_o) = v_o/R_S$. Solving,

$$v_o = \frac{g_m R_S v_i}{1 + g_m R_S} \tag{19-59}$$

and the voltage gain is

$$G_F = \frac{v_o}{v_i} = \frac{g_m R_S}{1 + g_m R_S} \tag{19-60}$$

which approaches unity in the typical case where $g_m R_S \gg 1$.

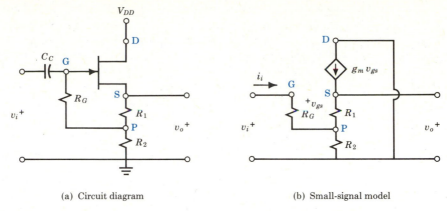

(a) Circuit diagram   (b) Small-signal model

Figure 19.31   The source follower or common-drain amplifier.

The effective output resistance can be evaluated from open- and short-circuited conditions. Since the open-circuit gain is approximately unity,

$$v_{OC} = v_o \cong v_i$$

With the output terminals shorted (shorting R_S),

$$i_{SC} = g_m v_{gs} = g_m(v_i - 0) = g_m v_i$$

Therefore, the effective output resistance is

$$R_o = \frac{v_{OC}}{i_{SC}} \cong \frac{v_i}{g_m v_i} \cong \frac{1}{g_m} \tag{19-61}$$

For a typical g_m of 2000 μS, the output resistance is about 500 Ω as compared to a common-source value (R_D) of several thousand ohms.

The higher input resistance resulting from the bias circuit of Fig. 19.31 can be predicted as follows. The ac input resistance is the ratio of an input voltage to the corresponding input current. Neglecting the effect of the very small current i_i on the voltage across R_2, the voltage at point P is

$$v_P = \frac{R_2}{R_1 + R_2} v_o = \frac{R_2}{R_1 + R_2} G_F v_i \tag{19-62}$$

Then the effective input resistance is

$$R_i = \frac{v_i}{i_i} = \frac{v_i}{(v_i - v_P)/R_G} = \frac{R_G}{1 - \dfrac{R_2}{R_1 + R_2} G_F} \tag{19-63}$$

If, say, $R_2 = 20 R_1$ and $G_F \cong 1$, $R_i = 21 R_G$; the input resistance is many megohms and the source follower presents a negligible load to the signal source. This characteristic makes it desirable as the input stage of an oscilloscope or a multistage instrumentation amplifier. (*Note:* At signal frequencies, the input capacitance $C_i \cong C_{gd}$ in parallel with R_i may determine the input impedance. See Problem 13.)

A source follower with very high input impedance followed by an emitter follower with very low output impedance provides an excellent buffer. In integrated circuit form where active devices are cheap, sophisticated circuits provide nearly perfect buffering at low cost.[†]

SUMMARY

- Linear models are used to predict the performance of amplifiers in which signals are much smaller than bias values.
  The frequency-response curves of all RC-coupled amplifiers are similar in shape.

- For one stage of RC-coupled voltage (or current) amplification,

  Midfrequency gain is $\qquad A_O = -g_m R_o \, [\text{FET}] \qquad$ or $\qquad -\beta R_o/(r_\pi + R_s) \, [\text{BJT}]$

  Low-frequency response is $\mathbf{A}_L = A_O \dfrac{1}{1 - j(\omega_{11}/\omega)} \cdot \dfrac{1}{1 - j(\omega_{12}/\omega)}$

  High-frequency response is $\mathbf{A}_H = A_O \dfrac{1}{1 + j(\omega/\omega_{21})} \cdot \dfrac{1}{1 + j(\omega/\omega_{22})}$

  Lower and upper cutoff frequencies ω_1 and ω_2 are determined by the critical combinations of R and C.

- For two stages of amplification in cascade, the overall gain is

  $$A \underline{/\theta} = A_1 A_2 \underline{/\theta_1 + \theta_2}$$

  Gain in dB $= 10 \log (P_2/P_1) \qquad$ or $\qquad 20 \log (V_2/V_1)$.

- The gain-bandwidth product is a measure of amplifier capability.

  $$\text{FET: } A_O \omega_2 \le g_m/C_{gs} \qquad \text{BJT: } A_O f_2 \le \beta f_\beta = f_T$$

- An oscillator is a self-excited amplifier in which a portion of the output signal is fed back in the correct phase and amplitude to make $\mathbf{GH} = 1 + j0$.
  By making $\mathbf{G}$ or $\mathbf{H}$ frequency dependent, oscillation is at a single frequency.

- The benefits of negative feedback outweigh the loss in gain:
  Gain can be made insensitive to device parameters.
  Operating point can be made insensitive to changes.
  Gain can be made insensitive to frequency.
  Noise and distortion can be discriminated against.
  Input and output impedances can be improved.

- In emitter (source) followers, the entire output is fed back into the input.
  Gain is less than unity but input and output impedances are optimized.
  The emitter follower acts as an impedance transformer with a ratio of approximately $\beta + 1$.
  The source follower offers similar advantages.

[†]In the equivalent circuit of the 741 (Fig. A9), transistors Q_{1-6} constitute a two-stage differential amplifier (Fig. 16-20). Q_1 and Q_2 are input buffers; Q_3 and Q_4 are common-base amplifiers with active loads Q_5 and Q_6. Transistor Q_6 also serves as an inverting amplifier driving the base of Q_{16}. Transistors Q_{8-13} provide bias voltages and temperature compensation. Q_{16} is an emitter follower driving a class B complementary amplifier (Fig. 17.23).

REVIEW QUESTIONS

1. What determines whether a signal is "small" or "large"?
2. Distinguish between midfrequency range and bandwidth.
3. In Fig. 19.3, explain the effects of eliminating C_S and halving C_{C1}.
4. Draw a wiring diagram of one stage of an RC-coupled amplifier using an E MOSFET and draw a linear circuit model that is valid for frequencies from 20 Hz to 200 kHz.
5. For the circuit model of the preceding question, state reasonable assumptions and derive circuit models valid for low, medium, and high frequencies.
6. Define in words and in symbols the cutoff frequencies.
7. Differentiate between coupling and bypass capacitors.
8. What limits the low-frequency response of an FET? Of a BJT?
9. What limits the high-frequency response of an FET? Of a BJT?
10. Why is emphasis placed on voltage gain of an FET amplifier and current gain of a BJT amplifier?
11. Explain the Miller technique for transforming feedback circuits. Why is it useful?
12. Compare FET and BJT amplifiers on the following characteristics: input impedance, voltage gain, current gain, and linearity.
13. What is meant by "cascading"? Where is it used?
14. What are the advantages of the decibel as a unit?
15. Define the "gain-bandwidth product." Why is it useful?
16. Is oscillation possible in an ordinary two-stage RC-coupled audio amplifier? Explain.
17. A portable high-gain amplifier tends to become oscillatory as the batteries age. It is known that the internal resistance of batteries increases with age. Explain the connection between these phenomena.
18. Explain the expression "trading gain for bandwidth."
19. Draw a block diagram to represent a "noisy" amplifier in which noise (perhaps an objectionable hum) appears in the output without a corresponding input. Suggest a means for reducing such noise.
20. Explain how an amplifier can be made insensitive to the effect of increased temperatures on transistor parameters.
21. Define the impedance characteristics of an ideal amplifier and explain your answer.
22. Why is a common-drain amplifier called a source follower?
23. Cite two virtues of the source follower.
24. Explain how an emitter follower functions as an impedance transformer. Where in a "hi-fi" set would this property be valuable?

EXERCISES

1. A BJT amplifier has a midfrequency gain of 200 with lower and upper cutoff frequencies at 10 Hz and 100 kHz, respectively. Draw the frequency-response curve showing gain and phase angle vs. frequency. What is the bandwidth of this amplifier?

2. In the generalized amplifier of Fig. 19.2, $Z_s = R_s = 2$ kΩ, $Z_i = R_i = 18$ kΩ, $G = 0.005$, $Z_o = R_o = 20$ kΩ.
   (a) Assuming Z_L is a pure resistance R_L, derive an expression for voltage gain V_o/V_s.
   (b) For $R_L = 20$ kΩ, evaluate the voltage gain.
   (c) Repeat part (a) assuming Z_L consists of $R_L = R_o$ in parallel with C_L.

3. In the generalized amplifier of Fig. 19.2, $Z_s \cong R_s = 2$ kΩ, Z_i consists of $R_i = 1$ MΩ in parallel with $C_i = 100$ pF, $G = 0.01 / 180°$, and $Z_o \cong R_o = 10$ kΩ.
   (a) Assuming Z_L is a pure resistance R_L, specify R_L for a voltage gain $|V_o/V_s| = 50$ at $\omega = 10$ krad/s.
   (b) Predict the voltage gain (magnitude and phase angle) at $\omega = 6.67$ Mrad/s.

4. In the FET amplifier of Fig. 19.3a, Z_s and Z_L are pure resistances. Assuming that the frequency is so low that wiring capacitance and the internal capacitances of the transistor can be neglected, draw a small-signal circuit model of the amplifier.

5. A silicon transistor ($\beta = 70$, $C_\pi = 20$ pF, $C_{jc} = 5$ pF, and $r_\pi = 2.8$ kΩ) is used in the circuit of Fig. 19.3b with $R_C = 10$ kΩ, $C_{C1} = C_{C2} = 5$ μF, $R_E = 2$ kΩ, $C_E = 50$ μF, $R_1 = 10$ kΩ, and $R_2 = 100$ kΩ. It is estimated that $C_L + C_W = 50$ pF. List the assumptions made in deriving the circuit of Fig. 19.6b and check their validity at $f = 5000$ Hz.

6. If $r_\pi \cong R_s$ and $R_B \cong 10r_\pi$ are known within $\pm 10\%$ in Fig. 19.6b, calculate the error made in predicting A_V by neglecting R_B. Draw a conclusion.

7. An FET for which $g_m = 2$ mS, $r_d = 200$ kΩ, $C_{gs} = 10$ pF, and $C_{gd} \cong 0$ is used in the circuit of Fig. 19.32 with $R_s = 1$ kΩ, $R_1 \| R_2 = R_G = 0.1$ MΩ, $R_D = 5$ kΩ, $R_S = 2$ kΩ, and R_L "very large," $C_{C1} = C_{C2} = 0.1$ μF, and C_S "large"; $C_W + C_L$ is estimated to be 10 pF.
   (a) Draw and label a linear circuit model valid for a wide range of frequencies, stating any useful assumptions.
   (b) Draw and label a simplified circuit model for medium frequencies, and predict the mid-frequency gain.

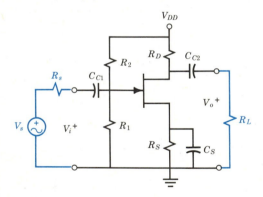

Figure 19.32

8. An enhancement MOSFET ($g_m = 5$ mS) is used in an RC-coupled amplifier.
   (a) Draw a circuit diagram similar to Fig. 19.32, with $R_1 = R_2 = 50$ MΩ, $R_S = 2.5$ kΩ, and $V_{DD} = 30$ V.
   (b) Draw a circuit model valid for moderate frequencies, and specify R_D for an open-circuit voltage gain $|V_o/V_i| = 40$.
   (c) For the specified R_D, with $R_s = 1$ kΩ and $R_L = 20$ kΩ connected across V_o, predict V_o/V_s.

9. In the circuit of Fig. 19.33, $R_1 = 20$ kΩ,

$R_2 = 80$ kΩ, $R_E = 2$ kΩ, and $V_{CC} = 10$ V. For this silicon transistor, $\beta = 100$.
   (a) Estimate I_C (by neglecting $I_B R_B$) and calculate r_π.
   (b) Specify R_C for a midfrequency voltage gain $|V_o/V_i| = 150$.
   (c) For the specified R_C and $R_s = 1$ kΩ, predict V_o/V_s.
   (d) For the specified R_C and R_s, predict V_o/V_s with $R_L = 10$ kΩ connected across V_o.

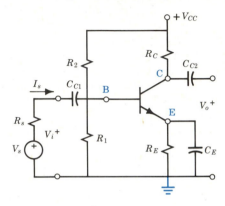

Figure 19.33

10. In the circuit of Fig. 19.33, $R_1 = 10$ kΩ, $R_2 = 90$ kΩ, $R_C = 10$ kΩ, and $I_C = 1.25$ mA. For this transistor $\beta = 50$.
    (a) Replace the transistor by an appropriate linear model and redraw the circuit to permit the prediction of the small-signal performance at moderate frequencies. (Show terminals B, C, and E on your diagram.)
    (b) Predict the midfrequency current gain I_{b2}/I_{b1} if the following stage employs a similar transistor under similar conditions.
    (c) If $R_s = 1$ kΩ, predict the overall voltage gain V_2/V_s under the conditions of part (b).

11. In the circuit of Fig. 19.33, $R_1 = 20$ kΩ, $R_2 = 60$ kΩ, $I_C = 2$ mA, and $\beta = 80$. The source is a phono pickup represented by $V_s = 10$ mV (rms) in series with $R_s = 500$ Ω. The "load" is another stage whose input resistance is 10 kΩ. Complete the design so that V_o is 1.6 V (rms) at midfrequencies.
    (a) Draw an appropriate small-signal model.
    (b) Specify the value of R_C.
    (c) Specify values of R_E and V_{CC}.

12. A silicon transistor ($\beta \cong 100$) is used in the amplifier of Fig. 19.33, where $R_s = 5$ kΩ, $R_1 = 58$ kΩ, $R_2 = 142$ kΩ, $R_C = 8$ kΩ, and

$R_E = 3.9$ kΩ. The capacitors are "large."
(a) Stating any necessary assumptions, draw a circuit appropriate for dc analysis and estimate the collector current for $V_{CC} = 10$ V.
(b) Redraw the circuit for small-signal analysis and specify resistance R_L (across V_o) for a midfrequency gain $|V_o/V_s| \cong 50$.

13. The amplifier of Exercise 12 is to have a lower cutoff frequency of 20 Hz. Assuming $I_C = 0.5$ mA and C_E and C_{C2} are "large," draw an appropriate circuit model and specify C_{C1}.

14. For a certain RC-coupled amplifier, $\omega_{11} = \omega_{12} = 100$ rad/s. What is ω_1 in this case?

15. A single-stage FET RC-coupled amplifier used in a mechanical engineering experiment must have a gain at 1 Hz equal to 90% of its midfrequency gain.
(a) What lower cutoff frequency is required?
(b) If $g_m = 2$ mS, $R_D = 10$ kΩ, and if the input resistance to the following stage is 1 MΩ, what value of C_{C2} is required?
(c) If the gain is halved by reducing R_D, is the required value of C_{C2} reduced?

16. For a certain n-channel JFET, the manufacturer specifies $g_m = 4000$ μS, and maximum values of $C_{iss} = 7$ pF and $C_{rss} = 3$ pF. This JFET is used in the amplifier of Fig. 19.3a with $R_s = 1$ kΩ, $R_1 = 1$ MΩ, $R_2 = 1.5$ MΩ, $C_{C1} = C_{C2} = 0.1$ μF, $R_D = 5$ kΩ, $R_S = 10$ kΩ, $C_S = 50$ μF, and $R_L = 1$ MΩ.
(a) Draw an appropriate small-signal model and predict the midfrequency gain V_o/V_s.
(b) Draw an appropriate small-signal model and predict the lower cutoff frequency.

17. In the amplifier of Exercise 10, $R_E = 1$ kΩ, $C_E = 50$ μF, and $C_{C1} = 5$ μF. C_{C2} is "large."
(a) Calculate the "lower cutoff frequencies" as defined by C_E and C_{C1} separately.
(b) What is the lower cutoff frequency for the amplifier?
(c) Specify an appropriate value for C_{C2}, explaining your reasoning.

18. In the amplifier of Exercise 10, $R_E = 1$ kΩ.
(a) Specify the value of C_E for a lower cutoff frequency of 50 Hz.
(b) Specify appropriate values of C_{C1} and C_{C2}.

19. In the circuit of Fig. 19.33, $R_s = 2$ kΩ, $R_E = 2$ kΩ, $R_C = 6$ kΩ, and R_L (across V_o) = 2 kΩ; $C_{C1} = C_{C2} = 5$ μF and $C_E = 100$ μF. The BJT with $\beta \cong 100$ operates at $I_C = 1.25$ mA and $V_{CE} = 5$ V. Stating any necessary assumptions, predict the midfrequency response and the lower

cutoff frequency. Draw and use appropriate circuit models.

20. For a certain silicon transistor in the common-emitter configuration, the small-signal current relations are:
$$\begin{cases} i_b = 0.004v_{be} + 10^{-10}\,dv_{be}/dt \\ i_c = 0.02v_{be} + 2 \times 10^{-5}v_{ce} \end{cases}$$
(a) Draw and label an appropriate circuit model.
(b) What determines the frequency response of this device?
(c) Determine the frequency at which the base signal voltage (and therefore the output current) is only 70.7% of its low-frequency value for a given input current.

21. The circuit model for an amplifier is shown in Fig. 19.34. Predict the upper and lower cutoff frequencies (rad/s).

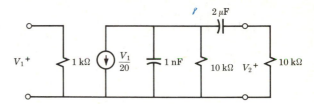

Figure 19.34

22. For the FET amplifier of Exercise 7:
(a) Draw and label a simplified circuit model valid for low frequencies and predict ω_1 and f_1.
(b) Draw and label a simplified circuit model valid for high frequencies and predict ω_2 and f_2.
(c) For this amplifier, define "midfrequency" and sketch the frequency response curve. (Plot three key points on the A_V vs. f curve; for a log plot on ordinary graph paper let unit distance correspond to a factor of 10 in frequency.)

23. For a certain RC-coupled amplifier, $\omega_{21} = \omega_{22} = 100$ krad/s. What is ω_2 in this case?

24. For the JFET amplifier of Exercise 16, draw an appropriate small-signal model and predict the upper cutoff frequency, assuming about 4 pF of wiring capacitance in the drain circuit.

25. For the 2N699B (see Table 18-1), beta is down to unity at 70 MHz.
(a) Devise an appropriate small-signal model and evaluate the parameters.

(b) Estimate the upper cutoff frequency of the current gain in an amplifier using the 2N699B, stating any assumptions made.

26. In an amplifier using the transistor of Exercise 25, $R_B = R_1 \| R_2$ is negligibly large, wiring capacitance is negligibly small, $C_{jc} = 13$ pF, and $R_o = 2$ kΩ. Predict the midfrequency voltage gain and the upper cutoff frequency.

27. An E MOSFET ($g_m = 3000$ μS, $C_{gs} = 6$ pF, $C_{gd} = 2$ pF) is used in the circuit of Fig.19.32 with $R_s = 2$ kΩ, $R_1 = 1.25$ MΩ, $R_2 = 5$ MΩ, $R_D = 10$ kΩ, $R_S = 4$ kΩ, and $R_L = 10$ kΩ. $C_{C1} = 0.2$ μF, C_S is "large," and C_o is negligible.
(a) Draw and label the wiring diagram.
(b) Use circuit models to specify C_{C2} for $f_1 = 20$ Hz.
(c) Use a circuit model to predict f_2.

28. A single-stage transistor amplifier has a midfrequency current gain of 40 and a lower cutoff frequency of 60 Hz. Predict the corresponding values for an amplifier consisting of three such stages in cascade.

29. For $A = 1/(1 + j\omega RC)$, explain why a log-log plot of $A(\omega)$ above $\omega_2 = 1/RC$ is a straight line of slope -20 dB per decade.

30. The output of a single-stage of transistor amplification is 2 V (rms) across a load resistance of 10 kΩ. The input is 0.5 mV (rms) across an input resistance of 1 kΩ.
(a) Express the power gain in decibels.
(b) Express the voltage gain and current gain in decibels.

31. A telephone system using cable with a power loss of 1.5 dB/km requires a minimum signal of 1 μW.
(a) If amplifiers are located 50 km apart, what amplifier output power is required?
(b) Assuming 10% overall amplifier efficiency, what total (dc) input power is required for New York to Los Angeles transmission?
(c) Compare the result of part (b) to the answer for part (c) of Exercise 24 in Chapter 9.

32. An amplifier consists of the following stages: Stage 1, midfrequency power gain $= 10$ dB, $f_1 = 0$ Hz, $f_2 = 3$ MHz. Stage 2, midfrequency power gain $= 30$ dB, $f_1 = 10$ Hz, $f_2 = 1$ MHz. Stage 3, midfrequency power gain $= 20$ dB, $f_1 = 100$ Hz, $f_2 = 300$ kHz.
(a) Plot the individual and overall response curves.
(b) If the midfrequency output is to be 2 W, what input power is required?

33. Two similar BJTs ($\beta = 50$, $I_C = 0.5$ mA) are used in the untuned amplifier of Fig. 19.35. You are to complete the design by specifying R_L for an overall gain $V_o/V_s \cong 1000$ at 2000 Hz.
(a) Stating any simplifying assumptions, draw and label an appropriate small-signal model of the amplifier.
(b) Use your model to derive a general expression for V_o/V_s and then specify R_L.
(c) Explaining your reasoning, predict the lower cutoff frequency for the amplifier.
(d) Plot the individual stage and overall frequency responses (up to 10 kHz) on a Bode plot similar to Fig. 19.19.

34. A transistor with a nominal $\beta = 50$ has a gain-bandwidth product $f_T = 30$ MHz. Describe, quantitatively, the high-frequency characteristics of this transistor. What is the current gain in dB at 30 MHz? At 2 MHz?

35. Each stage of a transistor amplifier is to provide a current gain of approximately 20 with $f_2 = 5$ MHz. For a $\beta = 80$, what minimum beta cutoff frequency is required?

36. A multistage amplifier is to provide a voltage gain of 1000 with an upper cutoff frequency of 200 kHz. Available is an FET for which $g_m = 2$ mS, $C_{rss} = 2$ pF, and $C_{iss} = 30$ pF. How many stages are required?

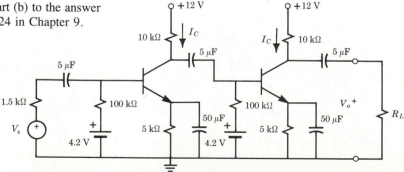

Figure 19.35

37. For the circuit of Fig. 19.36, draw an ac circuit model.

   (a) Note the polarities of ac voltages and determine the voltage feedback factor H.

   (b) Draw a block diagram for this circuit.

   (c) Derive an expression for overall mid-frequency gain from the circuit model, and from the block diagram.

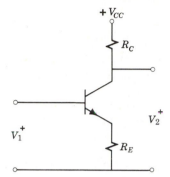

Figure 19.36

38. Repeat Exercise 37 for the circuit of Fig. 19.37, identifying *current* feedback.

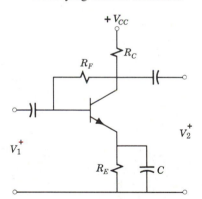

Figure 19.37

39. A feedback system is represented by Fig. 19.22a with $\mathbf{H} = 0.005\,\underline{/0°}$. Define the conditions necessary for oscillation.

40. For the circuit of Fig. 19.22b:

   (a) Determine the frequency of oscillation in words and symbols.

   (b) If $C = 200$ pF and $L = 25$ mH, estimate the oscillation frequency.

41. In the phase-shift oscillator of Fig. 19.23, $R = 1$ MΩ and $R_D = 10$ kΩ.

   (a) Estimate the approximate value of C for oscillation at 20 Hz; state any simplifying assumptions made.

   (b) If, instead, oscillation at 20 Hz is to be determined by a resonant circuit with $L = 1$H, approximately what value of C is required?

Is this a reasonable value for a variable capacitor?

42. For frequencies below $\omega = 2000$ rad/s, the gain of an amplifier is described by $\mathbf{G} = 0.1\,\omega\underline{/180°}$. For oscillation at $\omega = 1500$ rad/s, what value of $\mathbf{H}$ is required?

43. A common-emitter small-signal amplifier (Fig. 19.6b) uses a 2N699B with $I_C = 1$ mA and $V_{CE} = 5$ V. $R_C = 3$ kΩ and $R_B = 30$ kΩ.

   (a) Through what range can the *per stage* current gain vary without departing from the manufacturer's specifications (min to max)?

   (b) Introduce current feedback with $H = -0.2$ and recalculate the range of current gain.

   (c) How could the loss in current gain due to the introduction of negative feedback be made up?

44. Consider the basic feedback system of Fig. 9.3. Assuming small changes in G, derive a general expression for the fractional change in overall gain with feedback (dG_F/G_F) in terms of the change in forward gain (dG/G) and the values of G and H.

45. As the ambient temperature increases, input resistance h_{ie} increases more rapidly than $h_{fe}(\beta)$. As a result, for a certain transistor amplifier with a voltage gain of 200, a temperature increase of 50° C results in a 10% decrease in gain. You are to use feedback to reduce the net effect on overall gain to 2%.

   (a) Draw a block diagram of the amplifier, labeling V_1 and V_2.

   (b) Add feedback and specify the necessary feedback factor.

46. A cheap amplifier ordered from a catalog is supposed to have a voltage gain of 100, but when delivered proves to have a gain of 70. In fact, when first turned on it has a gain of only 50, which rises to 70 over the course of a minute or two. A second amplifier of the same type shows an initial gain of 90, rising to 110 during warmup.

   Design and diagram an arrangement to obtain a fairly stable gain of 100 by cascading the two amplifiers. Specify the feedback factor. Predict the percentage change in gain of your amplifier system during warmup.

47. Replace the transistor in Fig. 19.25a by an appropriate model and redraw the circuit for dc analysis with $V_{BB} = 2.9$ V, $R_B = 46$ kΩ, $R_E = 4$ kΩ, $R_C = 8$ kΩ, and $V_{CC} = 10$ V. β is around 100.

(a) Derive a literal expression for I_C and estimate the value.

(b) Estimate the change in I_C for a 30% increase in β.

(c) Remove R_E, adjust the value of V_{BB} to provide the same I_B, and repeat part (b).

(d) Explain in words the stabilizing effect of R_E.

48. The high-frequency response of an amplifier is defined by $A = A_O/(1 + j\omega/\omega_2)$ where $A_O = 100$ and $\omega_2 = 10$ krad/s.

(a) Plot a couple of points and sketch a graph of A/A_O versus ω on a logarithmic scale.

(b) Assume that a feedback loop with $H = -0.04$ is provided and repeat part (a). Label this graph "With feedback."

49. The low-frequency response of an amplifier is defined by $A = A_O/(1 - j\omega_1/\omega)$ where $\omega_1 = 200$ rad/s and $A_O = 200$.

(a) Plot a couple of points and sketch a graph of A/A_O versus ω with ω on a logarithmic scale.

(b) Assume that a feedback loop with $H = -0.045$ is provided and repeat part (a). Label this graph "With feedback."

50. An amplifier has a midfrequency gain of 100, but the gain is only 50 at 10 Hz. Devise a feedback system and specify a real value of H so that the gain at 10 Hz is within 5% of the new midfrequency gain.

51. In Fig. 19.28a, consider V_D to be the input and V_o the output and redraw the block diagram. Calculate the gain V_o/V_D for this system and interpret the result.

52. An audiofrequency amplifier with a voltage gain of 200 has an output of 5 V, but superimposed on the output is an annoying 60-Hz hum of magnitude 0.5 V.

(a) Devise a feedback system and specify H so that with the same 5-V output, the hum component is reduced to 0.025 V.

(b) Draw separate block diagrams for the audio signal and the hum component, showing the actual voltages at each point in the system.

53. A class A power amplifier (see Fig. 17.21a) using a 2N3114 is to supply 70 V (rms) of signal across a 10,000-Ω load resistance.

(a) How much ac power is supplied? To get this much power, the transistor must be driven into the nonlinear regions. It is estimated that the major element of distortion is a 7-V (rms) second harmonic. Express this as percent distortion.

(b) The source for the amplifier is a phono pickup with an output of 6 mV and a source resistance of 1500 Ω. How many amplifier stages (operating at $I_C = 1$ mA) of what current gain will be required to drive the final amplifier?

(c) It is suggested that the distortion can be reduced by providing feedback around the power amplifier. Specify the feedback factor necessary to reduce the distortion to 1%.

(d) To maintain the required output voltage, how must the intermediate amplifier be modified?

54. An emitter follower employs a 2N699B at $I_C = 1$ mA supplied by a source for which $V_s = 2$ V and $R_s = 2$ kΩ. For an emitter resistance $R_L = 1$ kΩ, predict the output voltage and the input and output resistances.

55. A device with a high output resistance ($R_s = 50$ kΩ) is to be coupled to a load resistance R_L by means of a 2N3114 transistor.

(a) Devise a circuit and specify R_L so that the input resistance seen by the source is approximately 50 kΩ.

(b) Predict the output resistance of the transistor under these circumstances.

56. The input to a device can be represented by $R = 100$ kΩ in parallel with $C = 100$ pF. It is fed by an FET for which $g_m = 2$ mS. Predict the upper cutoff frequency if the FET is used as:

(a) An ordinary amplifier with $R_D = 5$ kΩ.

(b) A source follower with $R_S = 5$ kΩ.

57. Modify the circuit of Fig. 19.31a so that V_i is applied (through C_C) between the gate and source terminals. Draw the small-signal model and express V_o/V_i in terms of the circuit parameters. Use this amplifier as one block in a block diagram representation of Eq. 19-60. What is the feedback factor H?

PROBLEMS

1. A practical JFET amplifier circuit is shown in Fig. 19.38.

(a) Redraw the circuit for dc analysis, assume

$V_{DS} = 10$ V, and estimate the bias voltage V_{GS} and quiescent current I_D.

(b) Redraw the circuit for ac analysis and evaluate the parameters given that $g_{mo} = 8$ mS and $I_{DSS} = 8$ mA.

(c) Predict the voltage gain.

(d) Draw one cycle of a sinusoidal $v_i(\omega t)$ and, on the same graph, show the voltages with respect to ground at points 1, 2, 3, and 4.

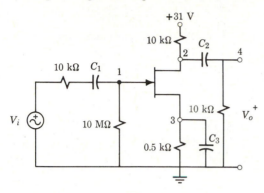

Figure 19.38

2. An elementary transistor amplifier is to be used as a preamplifier for a "hi-fi" set. The output of the phonograph pickup (low internal impedance) is only 2 mV, but 0.4 V is required as the input to the power amplifier that drives the loudspeaker. "Design" the preamplifier by specifying a suitable transistor and collector current and the appropriate load resistor and supply voltage.

3. An inexperienced designer omits the source by-pass capacitor C_S in Fig. 19.32. As a first step in determining the effect of the omission at mid-frequencies, replace the FET by a small-signal model and redraw the circuit. Derive an expression for V_o/V_i and evaluate the effect on voltage gain if $R_S = 5(1/g_m)$. What advantage would such a circuit offer?

4. Derive the criterion for C_E (Eq. 19-21) by using the approach followed in deriving the criterion for C_S (Eq. 19-17).

(a) Define the condition for ω_1.

(b) Determine the Thévenin equivalent resistance across which C_E is connected (Fig. 19.12).

(c) Express ω_1 in terms of the circuit parameters.

5. An FET amplifier similar to Fig. 19.3a with $R_D = 5$ kΩ is built and tested. The manufacturer specifies $g_m = 4000$ μS, $C_{rss} = 5$ pF, and $C_{iss} = 25$ pF for the FET. When V_o/V_i is observed on a CRO with an input capacitance of 30 pF, the midfrequency voltage gain is ob-

served to be 18 and the gain is down to 12.7 at 1 MHz. Analyze these experimental results and draw a conclusion.

6. A small-signal wide-band amplifier is to provide a voltage gain of 40 with a flat response from 20 Hz to approximately 100 kHz. Available is a 2N699B transistor (see Table 18-1, p. 513) with $\beta_f = 3.5$ at 20 MHz and $C_{jc} = 13$ pF. The resistance of the source is $R_s = 700$ Ω. Stating all assumptions, design the amplifier (draw the wiring diagram and specify the components and supply voltage).

7. The E MOSFET of Fig. 12.6 is used as an amplifier with V_s in series with R_s applied from gate to ground and V_o taken between drain and ground.

(a) Replace the MOSFET by a small-signal model and use nodal analysis to evaluate $A = |V_o/V_s|$ for $R_s = 10$ kΩ, $R_G = 1$ MΩ, $g_m = 2$ mS, and $R_D = 10$ kΩ.

(b) Derive an expression for input resistance of the amplifier and evaluate R_i.

(c) Replace R_G by its Miller equivalent, repeat parts (a) and (b), and draw a conclusion.

8. In the circuit of Fig. 18.8a, the input source consists of V_i in series with $R_i = 10$ kΩ and $R_L = 5$ kΩ. The JFET can be represented by the model of Fig. 18.10 with $g_m = 5$ mS, $r_d = 100$ kΩ, $C_{gs} = C_{gd} = 10$ pF, and $C_{ds} = 1$ pF.

(a) Draw a small-signal model of the amplifier and evaluate the capacitive reactances at $f = 10$ kHz.

(b) Redraw the circuit model neglecting any capacitances that have a small effect compared to the associated resistances and calculate the voltage gain V_o/V_i.

(c) Recalculate the reactances at $f = 1$ MHz and redraw the amplifier circuit as in part (b). Assume $V_o = 50$ mV (rms) and calculate the necessary voltage V_{gs} across C_{gs} and the voltage across C_{gd}. Calculate the currents through C_{gs} and C_{ds} and the necessary voltage V_i (rms).

(d) Calculate the voltage gain at 1 MHz and draw a conclusion.

9. An amplifier with a gain $G_1 = 200$ is subject to a 20% change in gain due to part replacement. Design an amplifier (in block diagram form) that provides the same overall gain but in which similar part replacement will produce a change in gain of only 0.1%. Use amplifier G_1 as part of your design.

10. Ten watts of audio power is to be supplied to a 250-Ω load.
    (a) What output voltage (rms) and current are required? For this voltage swing, the transistor must be driven into nonlinear regions resulting in distortion in the form of a 5-V (rms) second harmonic. Express this as "% distortion."
    (b) The available phono pickup has an output of 1 mV. If the final power amplifier has a voltage gain of 50, specify the number and gain of the necessary voltage amplifiers.
    (c) Show how distortion can be reduced by providing feedback around the final amplifier and specify the feedback factor to reduce the distortion to 1%. Specify the changes necessary in the voltage amplifiers to maintain the same output voltage.

11. An amplifier is represented by the general model of Fig. 19.2 with Rs in place of Zs. Reproduce the circuit and add negative voltage feedback with a feedback factor H. Derive expressions for G_F, R_{oF}, and R_{iF}, the new gain, output resistance, and input resistance. Draw a general conclusion.

12. An innovative experimenter disconnects C_E in Fig. 19.33 and finds the midfrequency voltage gain V_o/V_i greatly reduced. He expects that there may be some offsetting improvement and asks you to analyze the performance. Derive an expression for V_o/V_i and predict the effect of disconnecting C_E if $\beta = 100$, $R_C = 5$ kΩ, $R_E = 1$ kΩ, and $I_C = 1$ mA. Investigate the new input impedance and draw a conclusion.

13. A JFET for which $I_{DSS} = 5$ mA, $V_p = -3$ V, $C_{gs} = 20$ pF, and $C_{gd} = 2$ pF is used as a source follower in the circuit of Fig. 19.31a with $R_G = 5$ MΩ and $V_{DD} = 20$ V. For quiescent operation at $I_D = 2$ mA and $V_{DS} = 10$ V, design the circuit—specify R_1 and R_2. Predict the voltage gain, the output resistance, and the input impedance at 10 kHz.

14. The "hybrid" circuit shown in Fig. 19.39 provides a very high input impedance and a very low output impedance.
    (a) Describe the operation of the circuit in a qualitative way.
    (b) Replace the active devices by simple small-signal models and redraw the circuit for convenient analysis.
    (c) For $R_1 = 10$ MΩ, $g_{m1} = 3 \times 10^{-3}$ S, $I_C = 1$ mA, $\beta_2 = 100$, and $R_3 = 3$ kΩ, predict the voltage gain V_3/V_1, R_i, and R_o.

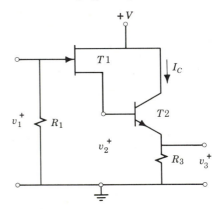

Figure 19.39

Electromechanics

CHAPTER **20**

Magnetic Fields and Circuits

Energy Conversion

Magnetic Fields

Magnetic Circuits

In the course of civilization, human beings strive to become free of the restrictions imposed by the physical world. Our success in overcoming natural limitations and thereby gaining control of our environment is directly related to our ability to control small and large amounts of energy. For example, we have extended our senses by developing systems that enable us to communicate throughout the solar system and devising memories that enable us to store millions of bits of data in tiny chips. In Part II of this book we were concerned primarily with electronic devices for controlling and processing information; consideration of the associated energy was secondary.

The civilization process has also been dependent upon our ability to work far beyond the limitations of our own muscles. Long ago humans learned to harness animals to supply power, and then we succeeded in applying the power of the winds and rivers to our purposes. Only recently have we learned to use the energy stored in fuels to develop mechanical power and, even more recently, electrical power. Part III is concerned primarily with devices for converting energy to and from electrical form.

ENERGY CONVERSION

We start with a survey of present energy sources and a look at some interesting new conversion devices that hold promise for the future.

Energy Sources

The important sources of energy include solar radiation, fossil fuels, and nuclear reaction; geothermal energy is an interesting speculation.

Solar. Each day there comes to the Earth from the Sun in the form of direct solar radiation an amount of energy equivalent to many years of consumption at our present rate. Summer radiation intensity is of the order of 1 kilowatt on each square meter; if efficient conversion methods were available, a small fraction of the Earth's surface could supply our current needs. Solar energy is used to a limited extent in low-temperature applications for water and space heating, and some use is made of direct radiation in powering remote installations and space laboratories. Giant power-generating satellites have been proposed, but at present the indirect effects of solar radiation are more significant.

Indirect Solar. Some of the radiation causes evaporation of sea water, which returns to the Earth as rain; a hydroelectric plant in the mountains receives its energy from the Sun indirectly. Wind (a first derivative of solar radiation) is a renewable, but somewhat undependable, resource that may be a significant energy source in the next century. Another part of the radiation causes the growth of plants and animals that are consumed by humans as fuel for their muscles. One interesting future possibility is to grow crops that efficiently produce glucose that can be converted to alcohol, a versatile fuel.

Fossil Fuels. The fossil fuels represent the cumulative effect of millions of years of irradiation of living organisms and their subsequent transformation into coal and petroleum. As indicated in Fig. 20.1, since 1945 oil and gas have been our major energy sources. Low production cost and flexibility accelerated the rise in importance of oil. More recently, new pipeline technology and our interest in clean air have made gas an attractive fuel. Coal is our most plentiful indigenous energy source; estimates of the recoverable reserves indicate several hundred years of supply. Production, transportation, and combustion of coal must be done carefully, however, if coal is to be an acceptable alternative to our dwindling supply of petroleum. Improved tech-

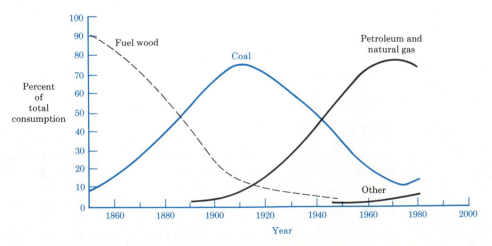

Figure 20.1 Shifts in energy sources (U.S Bureau of Mines).

niques for land restoration, emission control, and conversion of solid coal to clean gas or liquid indicate that coal may once again become our leading energy source. Currently about 91% of our energy comes from fossil fuels (coal 19%, natural gas 26%, and petroleum 46%) with the remainder from hydro and nuclear plants.

Nuclear. In chemical reactions, energy is either absorbed or released. When a molecule of hydrogen or carbon is burned, the energy released is a few electron-volts. In contrast, the energy released in nuclear *fission* is many million electron-volts per atom. In terms of energy released per unit mass, uranium fission "fuel" is nearly 3 million times as effective as coal. The total energy stored in fissionable material in the Earth's crust appears to be of the same order of magnitude as the supply of fossil fuel. For the last 20 years, nuclear has been "the fuel of the future." In 1982, about 5% of our total energy (the equivalent of 2 million barrels of oil per day) came from nuclear plants. Units now under construction or on order would double the existing capacity in the next decade. Everyday operation of plants here and abroad, particularly in France, Japan, and Russia, demonstrates the practicality of nuclear energy, but political and environmental considerations have slowed its development in the United States.

Geothermal. The heat of the Earth's interior is a potential energy source. Natural steam or steam produced by circulating water through fractured rocks can contribute significantly to our energy resources if the difficult environmental and technical problems (noise and air pollution, corrosion) can be solved.

Energy and Entropy

In a general consideration of energy conversion there are two fundamental principles that were originally formulated in connection with thermodynamics. The *first law* of thermodynamics states that energy can be converted from one form to another, but it can be neither created nor destroyed. In applying this so-called "law of conservation of energy," matter is included as a form of energy. One application is the *heat balance*, a detailed expression of the fact that all the energy inputs to a system must equal the sum of all the energy outputs plus any increase in energy storage. We shall make use of this approach in analyzing the performance of electromechanical devices.

The *second law* of thermodynamics says that no device, actual or ideal, can both continuously and completely convert heat into work; some of the heat is unavailable and must be rejected. For example, no engine could extract heat from sea water and convert it to work with no effect other than the cooling of the water. This so-called "law of degradation of energy" stipulates what transformations of energy are possible, whereas the first law governs the energy relations in a possible transformation. The unavailability of energy is measured by a property called *entropy*, which has some of the characteristics of probability; a uniform distribution of energy corresponds to a high entropy. We say that the energy of the universe remains constant, while the entropy tends toward a maximum. One of the primary activities of the engineer is directing the inevitable degradation of energy so that in the process some useful result is obtained.

An important characteristic of energy is whether it has an *ordered* or *disordered* form. The energy of an electric current in an inductor or the energy of a rotating flywheel is said to be ordered. In contrast, the random thermal energy of electrons or atoms in a solid is said to be disordered. Energy conversion from an ordered form to a disordered form, as in heating by an electric current or friction braking of a flywheel, can be achieved with 100% efficiency. The efficiency of energy conversion between ordered forms, as in an electric generator, can *approach* 100%. However, the efficiency of conversion from disordered to ordered form, as in a steam turbine (thermal to mechanical), is limited by the requirement that some energy be rejected; the maximum conceivable efficiency of such a conversion is $(T_1 - T_2)/T_1$ where T_1 is the highest temperature (K) in the cycle and T_2 is the lowest naturally available temperature. This concept of a limiting efficiency for heat engines was presented in 1824 by Sadi Carnot, a 28-year old French engineer.

Electrical Energy Generation

Since chemical energy is more ordered than thermal energy, higher conversion efficiencies are to be expected in generators employing chemical reaction.

Voltaic Cells. In a *voltaic cell*, two dissimilar electrodes are separated by an electrolyte in which conduction takes place by the motion of positive and negative ions. The chemical reactions at the electrode–electrolyte surfaces provide the energy for continuous current production. In a *primary* cell, such as the ordinary dry cell or mercury battery, the energy conversion is accompanied by irreversible changes in cell composition and the cell has a limited life. In a *secondary* cell, such as the lead-acid storage battery, the chemical reactions are reversible. During discharge, lead sulphate and water are formed; the water dilutes the sulfuric acid electrolyte and the specific gravity of the electrolyte is an indication of the state of charge. The battery is charged by sending a current from an external source through the electrolyte in the opposite direction. The reactions are reversed, lead and lead dioxide are formed at the negative and positive plates, respectively, and the battery is restored to its original condition.

Fuel Cells. A device that continuously converts the chemical energy in a fuel directly into electrical energy is very attractive. By avoiding the intermediate thermal energy stage, the cost and complexity of thermal-to-mechanical-to-electrical conversion apparatus are eliminated and the Carnot efficiency limitation is removed. The *fuel cell* is such a device; already it has been used to supply power in space vehicles and it shows great promise for the future.

The operation of a hydrogen–oxygen fuel cell of the type used in the Gemini spacecraft is illustrated in Fig. 20.2. The cell consists of two chambers and two porous electrodes separated by an electrolyte. Hydrogen supplied to the upper chamber diffuses through electrode A and, in the presence of a catalyst, reacts with the electrolyte to form positive ions and free electrons. The ions migrate through the electrolyte to electrode B. The electrons pass through the external load circuit and then to electrode B where they combine with oxygen and the ions to form water. The important point is that the electrons are forced to do useful work before the reaction is completed. In a properly designed cell, most of the energy that would appear as heat in a combustion reaction is available as electrical energy.

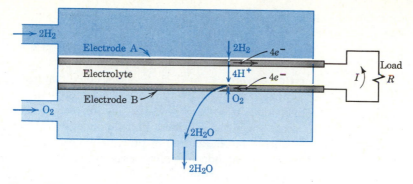

Figure 20.2 Operation of a hydrogen–oxygen fuel cell.

Solar Cells. Knowing the tremendous energy available in the form of solar radiation, we can appreciate the possibilities of a device that converts light energy directly into electrical energy. The semiconductor *solar cell* operates with fair efficiency, has an unlimited life, and has a high power capacity per unit weight. Already it is an important source of power for long-term satellites, and continued improvement in operating characteristics will make it competitive in other applications.

The operation can be explained in terms of our knowledge of *pn* junction diodes. As a result of the diffusion of majority carriers near the junction (see Figs. 11.10 and 11.11) a potential barrier is created. If the junction region is irradiated with photons possessing sufficient energy, electron-hole pairs are created and the density of minority carriers is greatly increased. Practically all the minority carriers drift across the junction and contribute a component of current I_p in the same direction as the I_s due to thermally generated carriers. For high light intensity and low bias voltage, the net current I in Eq. 11-21 is negative and the solar cell is *charging* the bias battery. Such a *pn* junction diode is a source of electrical energy. Alternatively, the solar cell may supply a load R where $V = (-I)R$.

A practical solar cell is constructed so that the junction is exposed to the light. As shown in Fig. 20.3b, a very thin layer of *p*-type material is created by diffusion of acceptor atoms into a heavily doped *n*-type silicon wafer. Electrical contact is provided by a thin translucent nickel plating (not shown) and a solder contact over the

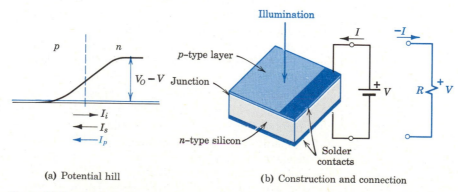

(a) Potential hill   (b) Construction and connection

Figure 20.3 Solar cell operation and construction.

bottom and along one edge of the top. Because the current and voltage output of a single cell are small, cells are usually connected in series and parallel to form solar batteries.

Thermoelectric Converters. In 1821, Seebeck noted that heat applied to a junction of dissimilar metals could produce a small current in a closed circuit. The efficiency of conversion was so low, however, that his *thermocouples* were applied only for measurement purposes. Advances in solid-state theory and technology have led to higher efficiencies and the design of practical thermoelectric converters.

The operation of a semiconductor converter is similar to that of a solar cell in that energy is added to generate electron-hole pairs near a *pn* junction. In Fig. 20.4a, the

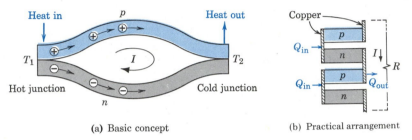

(a) Basic concept   (b) Practical arrangement

Figure 20.4   Semiconductor thermoelectric converters.

hot junction is maintained at temperature T_1 by an input of thermal energy. Holes generated near the junction drift across the junction into the *p* region, and electrons drift into the *n* region. At the cold junction, the effect is less and therefore there is a net flow of current consisting of electrons in the *n* material and holes in the *p* material.

Note that, if minority carriers drift across a *pn* junction, they must gain energy and this energy is provided by the heat source. If minority carriers are *forced* through junction 1 in the direction shown in Fig. 20.4a, junction 1 is *cooled*. This is the *Peltier effect* and permits the use of a *pn* junction as a heater or a cooler, depending on the direction of current flow. In this case, electrical energy is used to provide a *heat pump* effect.

Figure 20.4b shows a more practical form of converter arrangement. Because good thermoelectric materials are usually poor thermal conductors, the active elements are kept short and the heat input is by way of a good thermal and electrical conductor such as copper. A great advantage of the thermoelectric converter is that any form of thermal energy can be used. However, since the input energy is disordered, the efficiency of this device can only approach the Carnot ideal $(T_1 - T_2)/T_1$.

Thermionic Converters. In the *thermionic* converter, heat supplied to the cathode (emitter) provides the electrons with more than enough energy to overcome the cathode work function. The electrons escape with sufficient energy to move to the anode (collector) against a potential difference that includes an output voltage across a load. Filling the cathode–anode space with slow-moving positive ions neutralizes the negative "space charge" created by the emitted electrons and also improves the effective work functions of the cathode and anode surfaces.

The operation of a vapor-filled thermionic converter is shown diagrammatically in Fig. 20.5a.[†] The heat supplied to the cathode produces a great quantity of electrons with energies corresponding to the potential at point 1 in Fig. 20.5b. Just outside the emitter surface negative space charge is high and an appreciable part of the escape energy is lost. A much smaller potential drop (from V_2 to V_3) occurs across the balance

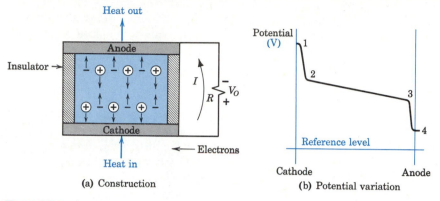

(a) Construction (b) Potential variation

Figure 20.5 A vapor-filled thermionic converter.

of the neutralized cathode–anode space. A free electron just outside the anode gives up energy corresponding to $V_3 - V_4$ upon entering the anode lattice structure. The energy available for useful work in an external load corresponds to potential V_4.

The thermionic converter is a high-temperature, low-voltage, high-current device. Cathodes operate efficiently at temperatures around 2000 K and this matches the delivery temperature of energy from a solar furnace or a nuclear reactor. A promising application is as the first stage in a nuclear power plant; the heat rejected at the anode could be used by a conventional steam turbine and the overall plant efficiency would be increased.

Electromechanical Energy Conversion

Electrical energy is still generated in the process used by Edison in his first central station in 1882. Fossil fuel is burned to obtain heat to convert water into steam to drive mechanical engines to force electrical conductors through a magnetic field to generate voltage and current. An atomic power plant differs only in that the heat is obtained from nuclear reaction. In a hydroelectric plant the electrical generator is driven by a water turbine instead of a steam turbine.

At present practically all electrical energy is *generated* by devices employing magnetic fields, and most electrical energy is *consumed* in devices employing magnetic fields in the conversion of electrical to mechanical energy. Most electromechanical devices could more properly be called *electromagnetomechanical* devices because magnetic fields provide the essential coupling in the energy-conversion process. The predominance of magnetic coupling is due to the high energy densities obtainable with commonly available magnetic materials; high energy density results in high power capacity per unit volume of machine.

[†]See V. C. Wilson, "Thermionic Power Generation," *IEEE Spectrum,* May 1964, p. 75.

We have already worked with magnetic fields in two different situations. We defined inductance as a measure of the ability of a circuit component to store energy in a magnetic field (Chapter 1) and employed the resulting v-i characteristic in analyzing circuits exhibiting this property. Also, we defined magnetic flux density in terms of the force on a moving charge and used this concept in studying the motion of electrons in a uniform magnetic field (Chapter 10). In the next four chapters we are going to study the operating principles of some important energy conversion devices in order to learn to predict their performance. We need a quantitative understanding of magnetic fields and the magnetic circuits employed to establish them.

MAGNETIC FIELDS

The behavior of an electric circuit can be completely described in terms of the voltage and current at various points along the path constituting the circuit. In contrast, it is characteristic of fields that they are distributed throughout a region and must be defined in terms of two or three dimensions. What is a magnetic field? About all we can say is that it is a region of space with some very useful properties. Does it really exist? From our standpoint it is a convenient concept for describing and predicting the behavior of devices that do exist.

Magnetic Flux and Flux Density

Magnetic fields are created by electric charge in motion and, in turn, the strength of magnetic fields is measured by the force exerted on a moving charge. In vector notation, the defining equation (Eq. 1-6) is

$$\mathbf{f} = q\mathbf{u} \times \mathbf{B} \qquad (20\text{-}1)$$

A *magnetic flux density B* of 1 tesla "exists" when a charge q of 1 coulomb moving normal to the field with a velocity u of 1 meter per second experiences a force f of 1 newton. If $\mathbf{u}$ is at an angle θ with respect to $\mathbf{B}$ (see Fig. 10.3), the direction of $\mathbf{f}$ is normal to the plane containing $\mathbf{u}$ and $\mathbf{B}$ and the magnitude of $\mathbf{f}$ is $quB \sin \theta$.

The summation obtained by integrating flux density over an area (Fig. 20.6a) is *magnetic flux ϕ* in webers defined by

$$\phi = \int \mathbf{B} \cdot d\mathbf{A} \qquad (20\text{-}2)$$

In describing a magnetic field, we represent magnetic flux by *lines of magnetic force* or just *lines*. The lines are drawn tangent to the flux density vector at any point in the field (Fig. 20.6b). The lines are close together where the flux density is high. An important fact is that the amount of magnetic flux leaving any closed surface is just equal to the amount entering; in other words, magnetic flux lines are continuous. This fact can be expressed mathematically by the equation

$$\oint \mathbf{B} \cdot d\mathbf{A} = 0 \qquad (20\text{-}3)$$

where the symbol $\oint$ indicates integration over a closed surface. In Fig. 20.6c the lines close within the magnet.

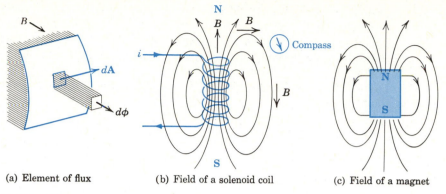

(a) Element of flux   (b) Field of a solenoid coil   (c) Field of a magnet

Figure 20.6 Two-dimensional representation of magnetic fields.

Fields Due to Currents

In establishing magnetic fields, we are interested in arranging the motion of charge to achieve a maximum effect. Copper conductors wound in compact coils provide an effective arrangement in many situations. To determine the magnetic effect of charges moving in conductors in various configurations, we first consider the effect of a current i flowing in a short element of conductor ds (Fig. 20.7). If all the charge dq in the element moves a distance ds in time dt, the velocity u is ds/dt or $ds = u\ dt$. Then

$$i\ ds = \frac{dq}{dt}(u\ dt) = dq\ u \tag{20-4}$$

and we see that current i in element ds is equivalent to charge dq moving at velocity u.

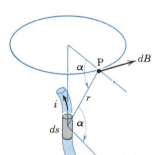

Figure 20.7 Magnetic field due to a current.

Experiments (first performed by the ingenious French physicist André Ampère in 1820) indicate that the contribution of current i in element ds to the magnetic flux density dB at point P in a homogeneous material is proportional to the current i and the cosine of angle α and inversely proportional to the square of the distance r. In MKS units,

$$dB = \mu\frac{i\ ds\ \cos\ \alpha}{4\pi r^2} \tag{20-5}$$

The factor μ is a property of the material surrounding the conductor and is called the *permeability* in webers per ampere-meter or the equivalent henrys per meter. The direction of the dB vector is tangent to the circle with its center on the extension of ds and passing through the point P as shown. A convenient rule is:

> *If the conductor is grasped in the right hand with the thumb extending in the direction of i, the fingers curl in the direction of B.*

Magnetic Field Intensity

The permeability of free space is $\mu_o = 4\pi \times 10^{-7}$ henrys per meter; the ratio of the permeability of any substance to that of free space is called the *relative permeability μ_r*, a dimensionless number. The relative permeability for most materials is near unity; the permeabilities of air and copper, for example, are practically the same as that of free space. The relative permeabilities of the *ferromagnetic* materials (iron, cobalt, nickel, and their alloys) may be in the hundreds of thousands. In other words, the magnetic flux density produced by a given current in a coil wound on a ferromagnetic core may be several thousand times as great as the flux density produced in air by the same current and coil.

To eliminate the effect of the medium, it is convenient to define the *magnetic field intensity H* where

$$\mathbf{H} = \frac{\mathbf{B}}{\mu} \tag{20-6}$$

in amperes per meter. The magnetic field intensity is a measure of the *tendency* of a moving charge to produce flux density; the actual value of B produced depends on the permeability of the medium.

Ferromagnetism

The spin of an orbital electron constitutes charge in motion, and therefore magnetic effects occur on an atomic level. Since in the atoms of most materials these electron spins are cancelled out by other electron spins, there is no net effect. In ferromagnetic materials, however, there are unbalanced electron spins and also a tendency for neighboring atoms to align themselves, so that their magnetic effects all add up.[†] In a specimen of unmagnetized ferromagnetic material, there are small *domains* in which all the atoms are aligned. Each domain (on the order of a thousandth of an inch in extent) is a region of intense magnetization. However, the domains are randomly oriented (Fig. 20.8a) and the specimen exhibits no net external magnetic field.

If an external field is applied, there is a tendency for the tiny magnets to align with the applied magnetic field or *polarize* just as a compass needle tends to align itself with the Earth's field. At low values of field intensity H (region 1 in Fig. 20.8b),

[†]See J. M. Ham and G. R. Slemon, *Scientific Basis of Electrical Engineering*, John Wiley & Sons, New York, 1961.

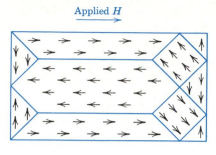

Applied *H*

(a) Domains in an unmagnetized specimen

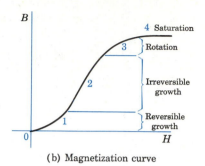

(b) Magnetization curve

Figure 20.8 Magnetization of an iron specimen.

domains nearly aligned with the applied field grow at the expense of adjacent, less favorably aligned domains in an elastic (reversible) process. This results in an increase in flux density B over that expected in free space. As H is increased (region 2), the direction of magnetization of misaligned domains switches, in an irreversible process, and this contributes to a rapid increase in B. At higher values of H (region 3), the directions of magnetization rotate until the contributions of all domains are aligned with the applied field. A further increase in field intensity produces no further effect within the ferromagnetic material and the material is said to be *saturated* (region 4). Commercial magnetic steels (usually called "iron") tend to saturate at flux densities of 1 to 2 teslas. For very high values of H the slope of the B–H curve approaches μ_o.

The flux density in a ferromagnetic material is the sum of the effects due to the applied field intensity H and the *magnetic polarization M* produced within the material. This relation can be expressed by the equation

$$B = \mu_o(H + M) \tag{20-7}$$

This can be rewritten as

$$B = \mu_o\left(1 + \frac{M}{H}\right)H = \mu_o\mu_r H = \mu H \tag{20-8}$$

From Fig. 20.8b it is clear that M/H is not a constant and therefore μ_r, the relative permeability, is not a constant. Because μ_r is greatly affected by rolling, stamping, and other material processing and because it depends on the previous history of magnetization (see Fig. 21.4), μ_r is seldom known precisely. However, for many calculations an average, constant value of μ_r can be assumed.

Field Around a Long Straight Conductor

Equation 20-5 expresses the contribution dB at point P by a current i in element ds. To determine the field around a conductor carrying current, we can obtain the total effect by integration. For the infinitely long straight conductor of Fig. 20.9, the

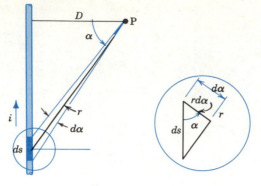

Figure 20.9 Field around a long conductor.

distance from the active element to point P is given by $r = D/\cos \alpha$. Then, considering differentials,

$$ds \cos \alpha = r \, d\alpha$$

and

$$\frac{ds \cos \alpha}{r^2} = \frac{r \, d\alpha}{r^2} = \frac{d\alpha}{r} = \frac{\cos \alpha \, d\alpha}{D}$$

Substituting in Eq. 20-5 and integrating,

$$B = \frac{\mu i}{4\pi} \int_{-\infty}^{+\infty} \frac{\cos \alpha}{r^2} \, ds = \frac{\mu i}{4\pi D} \int_{-\pi/2}^{+\pi/2} \cos \alpha \, d\alpha = \frac{\mu i}{4\pi D} \left[\sin \alpha \right]_{-\pi/2}^{+\pi/2} = \frac{\mu i}{2\pi D} \quad (20\text{-}9)$$

In words, flux density B is directly proportional to current i and varies inversely as the distance D.

EXAMPLE 1

Derive an expression for the flux density at the center of a compact N-turn circular coil.

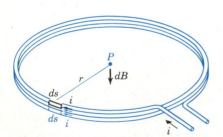

Figure 20.10 Flux density produced by a compact circular coil.

For the configuration of Fig. 20.10, $\alpha = 0$, $\cos \alpha = 1$, and Ampere's law becomes

$$dB = \frac{\mu i \, ds}{4\pi r^2}$$

The total flux density for a full turn is

$$B = \int dB = \frac{\mu i}{4\pi r^2} \int_0^{2\pi r} ds = \frac{\mu i (2\pi r)}{4\pi r^2} = \frac{\mu i}{2r}$$

For a compact N-turn coil, there are N similarly oriented elements carrying current i.

$$\therefore B = N \frac{\mu i}{4\pi r^2} \int_0^{2\pi r} ds = \frac{\mu N i}{2r} \quad (20\text{-}10)$$

An important principle can be demonstrated by considering the field intensity around the long straight conductor of Fig. 20.9. From Eqs. 20-6 and 20-9,

$$H = \frac{B}{\mu} = \frac{i}{2\pi D}$$

Along a circle of radius D, field intensity is constant and the line integral is

$$\oint \mathbf{H} \cdot d\mathbf{l} = \frac{i}{2\pi D} \cdot 2\pi D = i \qquad (20\text{-}11)$$

Although the relation is derived here for a special case, it has the following general interpretation.

The line integral of field intensity along any closed path is just equal to the current linked.

The quantity $\int \mathbf{H} \cdot d\mathbf{l}$ is called the *magnetomotive force* (mmf) because in magnetic circuits it plays a role analogous to the electromotive force in electric circuits. This principle, sometimes called the *mmf law*, is very useful in determining the fields due to currents in conductors of various configurations.

Field Produced by a Toroidal Coil

One conductor configuration for creating a strong magnetic field consists of many turns of wire wound on a cylindrical form; such a *solenoid* (Fig. 20.6) may have an air core or, for greater flux density, an iron core. Another common form is the *toroid* of Fig. 20.11 in which the wire is wound on a doughnut-shaped core. Let us use the mmf law to investigate the field in the vicinity of the toroidal coil.

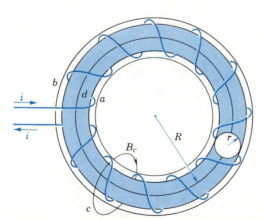

Figure 20.11　A toroidal coil.

Since closed path a links no current, the mmf along the path is zero; since $\int dl$ is finite, net H must be zero. We conclude that there is no field intensity and no magnetic flux in the direction of path a. Following the same reasoning, what conclusion do you reach regarding path b?

Closed path c links current i once and, therefore, an mmf of i A exists. The net current flow is clockwise around the toroid. Applying the right-hand rule, the resulting

H and $B = \mu_o H$ are in the direction of the arrow. A small amount of magnetic flux goes through the hole and into the paper. The flux density at any point in the interior or exterior could be calculated by using Ampere's law (Eq. 20-5).

If there are N turns on the toroid, closed path d links the current i N times. We know that the magnetic field is due to charge in motion. Insofar as path d is concerned, there are N conducting paths in parallel, each carrying a current $i = q/t$. We conclude that the total effect is as if a current Ni were linked once. By Eq. 20-11 the magneto-motive force in ampere-turns[†] is

$$\oint \mathbf{H} \cdot d\mathbf{l} = H(2\pi R) = Ni$$

and

$$B = \mu H = \frac{\mu}{2\pi R} Ni \qquad (20\text{-}12)$$

If the toroid has a circular cross section with a radius r, the core area is πr^2. If r is small compared to R, the flux density can be assumed to be uniform at the value given by Eq. 20-12 where R is the mean radius. On the basis of this assumption,

$$\phi = BA = \frac{\mu}{2\pi R} Ni(\pi r^2) = \frac{\mu r^2}{2R} Ni \qquad (20\text{-}13)$$

EXAMPLE 2

A coil consists of 1000 turns wound on a toroidal core with $R = 6$ cm and $r = 1$ cm. To establish a total magnetic flux of 0.2 mWb in a nonmagnetic core, what current is required? Repeat for an iron core with a relative permeability of 2000.

For a nonmagnetic core $\mu = \mu_o$. By Eq. 20-13,

$$i = \frac{2R\phi}{\mu r^2 N} = \frac{2 \times 6 \times 10^{-2} \times 2 \times 10^{-4}}{4\pi \times 10^{-7} \times 10^{-4} \times 10^3} \cong 190 \text{ A}$$

For an iron core, $\mu = \mu_o \mu_r = 2000\mu_o$, and

$$i = \frac{190}{2000} = 0.095 \text{ A} = 95 \text{ mA}$$

Note that the flux density is

$$B = \frac{\phi}{A} = \frac{2 \times 10^{-4}}{\pi \times 10^{-4}} \cong 0.64 \text{ T}$$

which is well below the saturation value for commercial "iron."

Letting $l = 2\pi R$ represent the length of the core and $\mathscr{F} = Ni$ the effective mmf, Eq. 20-13 can be rewritten as

$$\phi = \mu \frac{A}{l} \mathscr{F} \qquad (20\text{-}14)$$

[†]Since the number of turns is a dimensionless quantity, the *unit* of mmf is the *ampere*. However, we shall use "ampere-turns" (abbreviated A·t) to emphasize the significance of the number of turns in the usual case of a multiturn coil.

Does the form of this equation look familiar? Is there an analogy between this equation and Eq. 11-3? Do you see any possibility of considering a magnetic field as a "circuit"?

Summary of Magnetic Field Relations

Before leaving the subject of magnetic fields, let us list six important relations:

$$\mathbf{f} = q\mathbf{u} \times \mathbf{B} \qquad\qquad \int \mathbf{B} \cdot d\mathbf{A} = \phi$$

$$dB = \frac{\mu i \, ds \cos \alpha}{4\pi r^2} \qquad\qquad \oint \mathbf{B} \cdot d\mathbf{A} = 0$$

$$\mathbf{B} = \mu \mathbf{H} \qquad\qquad \oint \mathbf{H} \cdot d\mathbf{l} = Ni$$

Three of these relations are definitions. Can you identify them? The other three relations are based on experimental observations. Can you write out statements of the principles involved? Although many electric terms, such as voltage, current, and watt, are familiar because they are used in everyday activities, this is not true of magnetic terms. You should not go beyond this point without being sure that you know the precise names and the units of each of the variables in these six relations.

MAGNETIC CIRCUITS

A magnetic field is usually a means to an end rather than an end in itself. We wish to establish regions of intense magnetic flux density because of the effect of such fields on moving charges. The action of the field can produce beam deflection in an oscilloscope, torque in a motor, or voltage in a generator. By an optimum arrangement of coil and ferromagnetic material the engineer provides the necessary magnetic field at the lowest cost or with a minimum weight.

The Magnetic Circuit Concept

In the electric circuit of Fig. 20.12a, current I is proportional to voltage V and the constant of proportionality is the conductance $G = 1/R$ (Eq. 11-4); the conductance depends on the geometry of the conducting path and the property called conductivity. Equation 20-14 indicates that flux ϕ is proportional to magnetomotive force $\mathcal{F}$

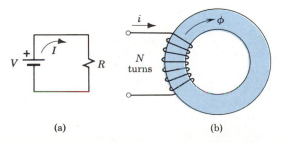

(a)   (b)

Figure 20.12   Electric and magnetic circuits.

and the constant of proportionality depends on the geometry of the magnetic path and a property called permeability. For Fig. 20.12 with mean length l and area A, the analogy is emphasized if we write the relations as follows:

$$\frac{V}{I} = R = \frac{1}{\sigma}\frac{l}{A} \qquad \frac{\mathscr{F}}{\phi} = \mathscr{R} = \frac{1}{\mu}\frac{l}{A} \qquad (20\text{-}15)$$

The ratio of mmf $\mathscr{F}$ to flux ϕ is $\mathscr{R}$, the *reluctance* in ampere-turns per weber. In many practical situations the flux density is uniform and a circuit approach, using a calculated reluctance, is possible.

EXAMPLE 3

Use a circuit approach to calculate the current required to establish the flux of 0.2 mWb in the iron core of Example 2.

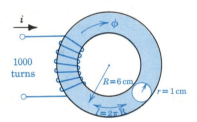

Figure 20.13 A magnetic circuit.

First the reluctance of the path in Fig. 20.13 is calculated. By Eq. 20-15,

$$\mathscr{R} = \frac{1}{\mu}\frac{l}{A} = \frac{1}{2000 \times 4\pi \times 10^{-7}}\frac{2\pi \times 0.06}{\pi(10^{-2})^2}$$

$$= 4.75 \times 10^5 \text{ A·t/Wb}$$

Then

$$\mathscr{F} = Ni = \phi\mathscr{R} = 2 \times 10^{-4} \times 4.75 \times 10^5 = 95 \text{ A·t}$$

and the current required in a 1000-turn coil is

$$i = \frac{\mathscr{F}}{N} = \frac{95 \text{ A·t}}{1000 \text{ t}} = 0.095 \text{ A} = 95 \text{ mA}$$

It must be emphasized that nothing "flows" in a magnetic circuit. Another difference is that in ferromagnetic materials permeability varies widely with flux density, whereas in most conductors conductivity is independent of current density within the normal operating range. In spite of these differences, the magnetic circuit concept is very useful. The analogous quantities are shown in Table 20-1.

Table 20-1 Electric and Magnetic Circuit Analogies

Electric		Magnetic	
Current density	J	Magnetic flux density	B
Current	I	Magnetic flux	ϕ
Electric field intensity	$\mathcal{E}$	Magnetic field intensity	H
Voltage	V	Magnetomotive force	$\mathscr{F}$
Conductivity	σ	Permeability	μ
Resistance	R	Reluctance	$\mathscr{R}$

$$\mathcal{E}l = V = IR = \frac{J}{\sigma}l \qquad\qquad Hl = \mathscr{F} = \phi\mathscr{R} = \frac{B}{\mu}l$$

Magnetic Circuit Calculations

The circuit approach is particularly useful if the magnetic field is confined to paths of simple geometry and if the flux density is uniform within each component of the path. In the relay of Fig. 20.14a, the coil wound on a *core* 1 establishes a magnetic

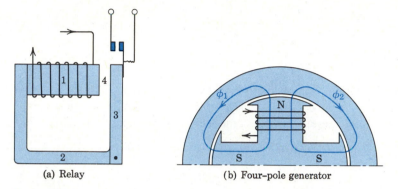

(a) Relay   (b) Four–pole generator

Figure 20.14 Devices incorporating series and parallel magnetic circuits.

flux that is largely confined to a path consisting of a fixed iron *yoke* 2, a movable iron *armature* 3, and an *air gap* 4. Half of a four-pole generator is shown in Fig. 20.14b. Flux ϕ_1 follows a path consisting of a pole structure N, an air gap, a yoke, another air gap, and a pole structure S. In the complete generator, there are four similar magnetic circuits.

Reasoning by analogy from electric circuits, we conclude that:
For magnetic circuit elements in series,

$$\phi_1 = \phi_2 = \cdots = \phi_n \qquad \text{and} \qquad \mathscr{F} = \mathscr{F}_1 + \mathscr{F}_2 + \cdots \mathscr{F}_n \qquad (20\text{-}16)$$

For magnetic circuit elements in parallel,

$$\phi = \phi_1 + \phi_2 + \cdots + \phi_n \qquad \text{and} \qquad \mathscr{F}_1 = \mathscr{F}_2 = \cdots = \mathscr{F}_n \qquad (20\text{-}17)$$

In magnetic circuits the "source" of mmf is usually a coil carrying a current. (It could be a permanent magnet.) The "applied" mmf is just equal to the sum of the mmf "drops" across the elements of a series magnetic circuit. The mmf "drops" across parallel elements are equal; the total flux is the sum of the fluxes in parallel elements.

Series Magnetic Circuits

A common problem is to find the current required to establish a given flux distribution. The procedure is as follows:

1. Analyze the magnetic circuit into a combination of elements in which the flux density is approximately uniform.
2. Determine the flux density $B = \phi/A$ in each element and the corresponding field intensity H.
3. Calculate the mmf drops $\mathscr{F} = Hl$ in each element.
4. Calculate the total mmf and the required current.

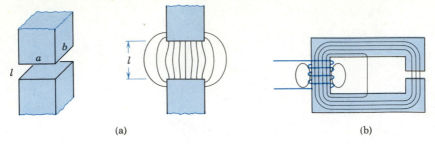

(a)   (b)

Figure 20.15 Air gap fringing and leakage flux.

The calculation of the mmf drop across an air gap is straightforward if it is assumed that the flux density is the same as that in the adjoining iron elements. Actually, there is always some *fringing* of magnetic flux (Fig. 20.15a) and the flux

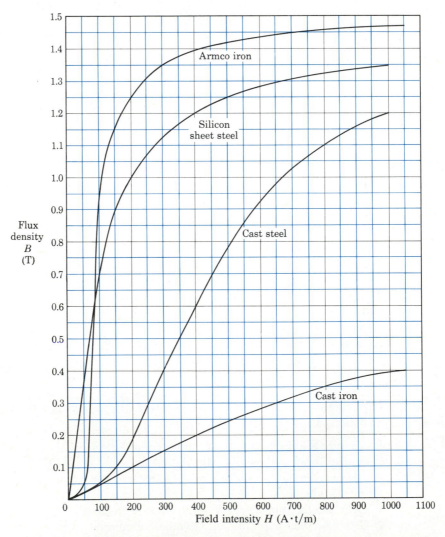

Figure 20.16 Typical magnetization curves for common magnetic materials.

density in the gap is lower than in the adjacent iron. One approximate rule for accounting for fringing in short gaps is to increase each dimension (*a* and *b*) by the length of the gap. However, neglecting fringing gives a conservatively high value for the mmf drop and simplifies the calculation. Some flux produced by a coil may return through a short air path (Fig. 20.15b) and never reach the air gap. This *leakage flux* is usually small if the permeability of the iron is high. As a result of flux leakage, a higher mmf is required to establish a given flux in the air gap.

The mmf drop across an iron section can be calculated by assuming a value of relative permeability or by working from average data in the form of *magnetization curves*. Curves similar to those in Fig. 20.16 are available from manufacturers of magnetic materials. It must be emphasized that these are typical curves and results based on such data are not precise. In fact, variations in values of magnetic properties of 3 to 5% are to be expected and there is no point in attempting more precise calculations.

EXAMPLE 4

Given the magnetic circuit shown in Fig. 20.17 with 500 turns wound on each leg, find the current required to establish a flux of 4 mWb across the 0.1-cm air gaps.

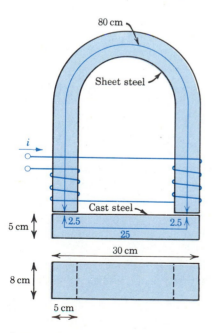

Figure 20.17   A series magnetic circuit.

Following the procedure outlined, we consider the magnetic circuit to consist of two iron elements and two air gaps in series. We assume that fringing is negligible and that the effective length of the cast steel element is $25 + 2.5 + 2.5 = 30$ cm; the increased flux density at the inner corner is partially offset by the decreased flux density at the outer corner.

Sheet steel: $B = \dfrac{\phi}{A} = \dfrac{4 \times 10^{-3} \text{ Wb}}{0.05 \times 0.08 \text{ m}^2} = 1 \text{ T}$

From Fig. 20.16, $H = 200$ A·t/m and
$$\mathscr{F} = Hl = 200 \times 0.8 \cong 160 \text{ A·t}$$

Cast steel: $B = \dfrac{\phi}{A} = \dfrac{4 \times 10^{-3} \text{ Wb}}{0.05 \times 0.08 \text{ m}^2} = 1 \text{ T}$

From Fig. 20.16, $H = 670$ A·t/m and
$$\mathscr{F} = Hl = 670 \times 0.3 \cong 200 \text{ A·t}$$

Air gaps: $\mathscr{F} = 2Hl = 2\dfrac{B}{\mu}l = \dfrac{2\phi l}{A\mu_o}$

$$= \frac{2(0.004)(0.001)}{0.004(4\pi \times 10^{-7})} = 1590 \text{ A·t}$$

By Eq. 20-16, the total mmf required is $\overline{1950 \text{ A·t}}$. Since the mmfs of the two coils are in the same direction, the effective number of turns is 1000 and the current required is

$$i = \frac{\mathscr{F}}{N} = \frac{1950}{1000} = 1.95 \text{ A}$$

Note that in Example 4 the high permeability of the iron results in a small mmf drop even though the path length is great compared to the length of the air gap. Since the mmf drop across the air gap is the predominant factor, as a first approximation the other mmf drops can be neglected. As a second approximation, the iron elements can be taken into account using average values of permeability.

EXAMPLE 5

Repeat Example 4, using the first approximation described in the preceding paragraph.

From Fig. 20.16, the average permeabilities for $B \le 1$ T are:

Sheet steel:

$$\mu_r = \frac{B}{\mu_o H} \cong \frac{1}{4\pi \times 10^{-7} \times 200} \cong 4000$$

Cast steel:

$$\mu_r = \frac{B}{\mu_o H} \cong \frac{1}{4\pi \times 10^{-7} \times 700} \cong 1000$$

Repeat Example 4, using the second approximation.

Assuming all the reluctance is in the air gap,

$$i = \frac{\mathscr{F}}{N} = \frac{1590}{1000} \cong 1.6 \text{ A}$$

For a series magnetic circuit with constant μ,

$$\mathscr{R} = \mathscr{R}_a + \mathscr{R}_{ss} + \mathscr{R}_{cs} = \frac{1}{\mu_o A}\left(\frac{l_a}{\mu_a} + \frac{l_{ss}}{\mu_{ss}} + \frac{l_{cs}}{\mu_{cs}}\right)$$

$$= \frac{1}{4\pi \times 10^{-7}(0.004)}\left(\frac{0.002}{1} + \frac{0.8}{4000} + \frac{0.3}{1000}\right)$$

$$\cong 5 \times 10^5 \text{ A·t/Wb}$$

Therefore,

$$i = \frac{\mathscr{F}}{N} = \frac{\phi\mathscr{R}}{N} \cong \frac{4 \times 10^{-3} \times 5 \times 10^5}{1000} \cong 2 \text{ A}$$

We conclude that the method to be used in a given problem depends on the data available and the precision required.

Series Circuit with Given Mmf

In Example 4 the problem is to find the mmf required for a given flux distribution. Knowing the flux density in each element of the circuit, we can determine the corresponding field intensity and mmf. Finding the precise flux distribution resulting from a given mmf is a more difficult problem because of the nonlinearity of the magnetization curves. (The approximate methods used in Example 5 assume linearity and work equally well for both classes of problems.)

The methods of analysis outlined in Chapter 2 are applicable to nonlinear magnetic circuits. The method of piecewise linearization (see Fig. 2.26a) is essentially the approach used in Example 5, where we assumed average values of relative permeability. If there is a single nonlinear element, a load line can be constructed as in Fig. 2.27. (If a magnetic circuit consists of an iron element, an air gap, and a multiturn coil, what are the analogs of V_T, R_T, and R_N?) An effective approach is first to obtain a rough approximation by assuming all the reluctance is in the air gap and then to proceed by "educated" trial and error.

EXAMPLE 6

The core of cast steel in Fig. 20.18 has a cross section of 10 cm^2 and an average length of 35 cm with a 1-mm air gap. It is wound with 200 turns of wire carrying a current of 3 A. Determine the total flux across the gap.

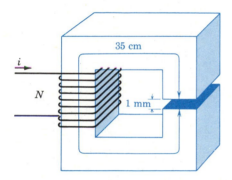

Figure 20.18 Magnetic flux calculation.

As a first approximation, assume that all the mmf is across the air gap. Then

$$B = \mu H = \frac{\mu_o \mathcal{F}}{l} = \frac{\mu_o Ni}{l} = \frac{4\pi \times 10^{-7} \times 200 \times 3}{10^{-3}}$$

$$= 0.75 \text{ T}$$

The actual value of B is less than this value because of the mmf drop in the cast steel. Assuming $B = 0.6$ T,

$$\mathcal{F}_a = H_a l_a = \frac{B l_a}{\mu_o} = \frac{0.6 \times 10^{-3}}{4\pi \times 10^{-7}} = 478 \text{ A} \cdot \text{t}$$

From Fig. 20.16, $H_{cs} \cong 400$ A·t/m and

$$\mathcal{F}_{cs} = H_{cs} l_{cs} = 400 \times 0.35 = 140 \text{ A} \cdot \text{t}.$$

The total mmf required is $478 + 140 = 618$ A·t. This is within 3% of the $3(200) = 600$ A·t available and, therefore, $B = 0.6$ T is an acceptable value. The total flux is $\phi = BA = 0.6 \times 10^{-3} = 0.6$ mWb.

Note: If greater precision were required, a second guess would be that $B_2/B_1 = \mathcal{F}_2/\mathcal{F}_1 = 600/618$.

Parallel Magnetic Circuits

The analysis of magnetic circuits containing parallel elements is based on analogy with the corresponding electric circuits. The same mmf (voltage) exists across elements in parallel and the total flux (current) is the sum of the fluxes (currents) in the parallel elements. The procedure is illustrated in Example 7 on page 582.

EXAMPLE 7

A flux of 3.6×10^{-4} Wb is to be established in the center leg of the sheet steel core shown in Fig. 20.19 (all dimensions in cm). Find the necessary current in the 300-turn coil.

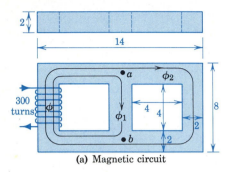

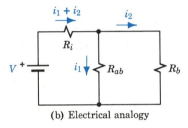

(b) Electrical analogy

Figure 20.19   Parallel circuit calculations.

Possible flux paths have been sketched in to indicate the general distribution of flux. Assuming that flux density between points a and b is approximately uniform, the flux density in the center leg is

$$B_1 = \frac{\phi_1}{A_1} = \frac{3.6 \times 10^{-4}}{0.02 \times 0.02} = 0.9 \text{ T}$$

From Fig. 20.16, for $B_1 = 0.9$ T, $H_1 = 150$ A·t/m and

$$\mathcal{F}_{ab} = H_1 l_1 \cong 150 \times 0.06 = 9 \text{ A·t}$$

Then

$$H_2 = \frac{\mathcal{F}_{ab}}{l_2} \cong \frac{9}{(6 + 6 + 6)10^{-2}} = \frac{9}{0.18} = 50 \text{ A·t/m}$$

From Fig. 20.16, for $H_2 = 50$ A·t/m, $B_2 \cong 0.35$ T and

$$\phi_2 = B_2 A_2 = 0.35 \times 4 \times 10^{-4} = 1.4 \times 10^{-4} \text{ Wb}$$

To the left of ab, the total flux is

$$\phi = \phi_1 + \phi_2 = (3.6 + 1.4)10^{-4} = 5 \times 10^{-4} \text{ Wb}$$

and

$$B = \frac{\phi}{A} = \frac{5 \times 10^{-4}}{4 \times 10^{-4}} = 1.25 \text{ T}$$

From Fig. 20.16, for $B = 1.25$ T, $H = 500$ A·t/m and

$$\mathcal{F} = Hl = 500 \times 0.18 = 90 \text{ A·t}$$

The required current in the 300-turn coil is

$$i = \frac{\mathcal{F} + \mathcal{F}_{ab}}{N} = \frac{90 + 9}{300} = 0.33 \text{ A}$$

SUMMARY

- The future course of civilization depends on our ability to improve the efficiency with which existing energy resources are used and to develop new sources of energy.

- The second law of thermodynamics indicates what energy transformations are possible; the first law governs the energy relations in a possible energy transformation. Fuel cells, solar cells, thermoelectric converters, and thermionic converters show promise for the future. However, most electrical energy is still generated by traditional thermal–mechanical–magneto–electrical devices.

■ Magnetic flux density **B** is defined in terms of the force exerted on a moving charge by the equation $\mathbf{f} = q\mathbf{u} \times \mathbf{B}$.
The contribution of an element ds carrying a current i is $dB = \mu i\, ds \cos \alpha/4\pi r^2$.
Magnetic flux $\phi = \int \mathbf{B} \cdot d\mathbf{A}$ Wb; flux lines are continuous or $\oint \mathbf{B} \cdot d\mathbf{A} = 0$.

■ The permeability of free space is $\mu_o = 4\pi \times 10^{-7}$ henrys per meter.
The relative permeability of a material is $\mu_r = \mu/\mu_o$.
Magnetic field intensity is defined by $\mathbf{H} = \mathbf{B}/\mu$ amperes per meter (or A·t/m).
The line integral of magnetic field intensity along any closed path is just equal to the current linked or $\oint \mathbf{H} \cdot d\mathbf{l} = i$ A (or A·t).

■ Ferromagnetism is due to the magnetic effects of unbalanced electron spins and the behavior of domains under the action of external magnetic fields.
The resulting magnetic polarization M is defined by $B = \mu_o(H + M) = \mu_o\mu_r H$.
Magnetization curves provide B–H data for magnetic materials.

■ Analogies between magnetic and electric quantities provide the basis for a circuit approach to magnetic field problems.

$$\text{Series:} \quad \phi_1 = \phi_2 = \cdots = \phi_n \quad \text{and} \quad \mathscr{F} = Ni = \Sigma Hl$$
$$\text{Parallel:} \quad \mathscr{F}_1 = \mathscr{F}_2 = \cdots = \mathscr{F}_n \quad \text{and} \quad \phi = \Sigma BA$$

■ To find the current required to establish a given magnetic flux:

1. Subdivide the circuit into uniform elements.
2. Determine B in each element and the corresponding H.
3. Calculate the mmf drops $\mathscr{F} = Hl$ in each element.
4. Calculate the total mmf and the required current.

■ To find magnetic flux for a given current, use graphical analysis or trial and error. Magnetic calculations are never precise; in a practical device with an air gap the mmf drop in the iron may be negligible.

REVIEW QUESTIONS

1. What are the advantages of solar, hydro, wind, coal, and nuclear energy sources? The disadvantages?
2. Why is magnetic circuit theory important in the study of modern energy conversion?
3. What is a magnetic field? What are its properties?
4. How could you determine experimentally if a magnetic field is present?
5. What are lines of magnetic force? How are they useful?
6. What is the flux density at the center of a long, straight, copper conductor carrying current i? A similar iron conductor?
7. Why is iron magnetic and aluminum is not?
8. Why does relative permeability vary with flux density?
9. Sketch the magnetic flux distribution around a long straight conductor.
10. Explain Equation 20.11 for a permanent magnet.
11. Given an $i\, ds$ element, what factors determine dB at point P?
12. Sketch analogous magnetic and electric circuits and list four analogous terms. How are these circuits basically different?
13. List five magnetic field quantities with their symbols and units.
14. In magnetic circuits, what corresponds to Kirchhoff's laws?
15. Explain the statement: "Applied mmf equals the sum of the mmf drops."
16. Under what circumstances can all the mmf be assumed across the air gap?
17. Given the current in a coil wound on a magnetic core with an air gap, outline the procedure for finding total flux.

EXERCISES

1. Assuming that energy use has doubled every 20 years, compare the energy to be used in the next two decades to the total energy used up to this time.
2. The heating value of a pound of coal is about 10^7 J. Assuming a conversion efficiency of 35%, estimate the amount of coal required to light a 100-W bulb 6 hr/day for a year.
3. Sketch a conducting element of length l carrying a current i consisting of carriers of charge q and velocity u. Taking into account area A, carrier density n, and total charge Q, demonstrate that $Qu = il$.
4. A current i in the short conducting element shown in Fig. 20.20 produces a flux density B_1 at point 1. Determine the magnitude and the direction of the flux density vectors at points 2, 3, and 4.

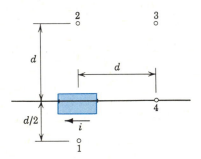

Figure 20.20

5. A circular coil of radius R carrying a current I lies in the x-y plane with its center at the origin. Derive an expression for the flux density B as a function of distance along the z axis.
6. A 10-turn compact circular coil 10 cm in diameter is to provide a flux density of 0.5 mT at its center. Specify the required current.
7. Demonstrate that the units of permeability are henrys per meter.
8. In a certain material the flux density is 1.0 T. Calculate the magnetic field intensity and the magnetic polarization if the relative permeability of the material is: (a) 1.00002 and (b) 2500.
9. The long straight conductor in Fig. 20.21 carries a current I in air. Determine the magnitude and direction of the:
   (a) Magnetic flux density at point A.
   (b) Force on element dl carrying I_2.
   (c) Force on element ds carrying I_3.

(d) Force on a similar element ds carrying I_4 into the paper.

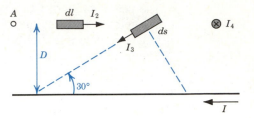

Figure 20.21

10. A straight conductor carrying a current I passes through the center of a square 1 cm on a side, in a plane parallel to two sides of the square and at an angle of 30° with respect to the normal to the square. Determine the total mmf around the square.
11. A straight conductor 1 m long and 2 mm in diameter carries a current of 10 A. Determine and plot the magnitude of the magnetic flux density in the air around the conductor.
12. One commonly used memory element for digital computers is a small toroidal core in which the direction of magnetization distinguishes between 0 and 1. If the core (Fig. 20.11) is of *ferrite*, for which $\mu_r = 1200$, and $R = 0.2$ cm and $r = 0.05$ cm, estimate the magnetomotive force required to establish a flux density of 1 mT.
13. The long solenoid coil of Fig. 20.22 is one way of getting a uniform flux density over a region. Recognizing this solenoid as a special form of the toroid of Fig. 20.11, the analysis is similar.
   (a) Evaluate $\int H\,dl$ along paths a and b.
   (b) Assuming an infinitely long solenoid, evaluate $\int H\,dl$ along path dL and express B in that region.

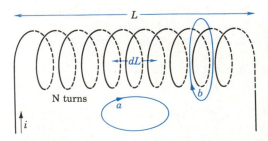

Figure 20.22

(c) For an "air-core" solenoid of length $L = 20$ cm, $r = 1$ cm, and $N = 100$ turns, predict the flux density and total flux across the median section for a current of 2 A.

14. Iron with a relative permeability of 1500 is formed into a toroidal core with $R = 12$ cm and $r = 1$ cm. Design a coil to provide a flux density of 0.8 T in the core.

15. Refer to the six relations on page 575.
(a) Which are *definitions*, and what is defined in each?
(b) Write out in words the principles expressed in the *experimental* relations.

16. (a) Derive an expression for the equivalent reluctance $\mathcal{R}_{EQ}$ of a magnetic circuit consisting of reluctances $\mathcal{R}_a$, $\mathcal{R}_b$, and $\mathcal{R}_c$ in series.
(b) Repeat part (a) for reluctances in parallel.

17. Calculate the average value of relative permeability for flux densities up to $B = 0.4$ T for:
(a) cast iron, (b) cast steel, (c) silicon sheet steel, and (d) Armco iron.

18. (a) Calculate about five values and plot a graph of relative permeability versus flux density for cast steel.
(b) Repeat for Armco iron.

19. For experiments in space propulsion, engineers developed a 6-inch diameter solenoid providing a flux density of 10.7 T across a 1-inch bore by using a superconducting magnet immersed in helium. Current is initiated by a 6-V battery and then continues without degradation. Estimate the current required in a 1000-turn coil to establish this flux density across 10 cm of air gap.

20. (a) A symmetric core of silicon sheet steel is similar to that in Fig. 20.23 but with no air gap. If $a = c = 5$ cm and $b = d = 25$ cm, find the total magnetic flux produced by a current of 2.5 A in the 200-turn coil.
(b) Repeat for a core of cast steel.

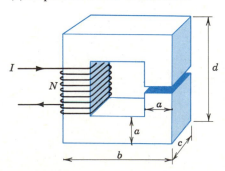

Figure 20.23

21. For the Armco iron core of Fig. 20.23, $a = 5$ cm, $c = 10$ cm, $b = d = 20$ cm, and $N = 300$. Predict the current required to establish a flux of 5 mWb:
(a) Across a 4-mm air gap.
(b) If there is no air gap.

22. For flux densities below 0.8 T, a magnetic material has a relative permeability of 5600; it saturates at 1.4 T.
(a) Sketch, approximately to scale, the B–H curve.
(b) The material is used in the core of Fig. 20.23 where $a = c = 10$ cm and $b = d = 35$ cm; the 400-turn coil has a resistance of 4 Ω. Predict the dc voltage required to establish a flux of 8 mWb across the 5-mm air gap.

23. A symmetric core similar to that in Fig. 20.23, but with no air gap, is to carry a total flux of 1.6 mWb. Dimensions are $a = c = 4$ cm, and $b = d = 20$ cm.
(a) Estimate the current required in a 100-turn coil if the core is Armco iron.
(b) Repeat for a cast steel core.
(c) Repeat for a cast iron core.

24. The core of Fig. 20.23 is Armco iron with $a = c = 10$ cm, and $b = d = 40$ cm; the current in the 300-turn coil is 10 A.
(a) Estimate the allowable air gap length for a magnetic flux of 12 mWb.
(b) Estimate the error in percent that would result from neglecting the mmf drop in the iron.

25. Armco iron is used in the core of Fig. 20.23 with $a = c = 10$ cm and $b = d = 50$ cm. You are to estimate the total flux across the 3-mm air gap when the current is 1.5 A in the 2000-turn coil.
(a) What assumption could be made to obtain an approximate value of B? Make such an assumption and calculate B.
(b) Is the actual value larger or smaller than this B?
(c) Assume a value of B and calculate the corresponding I.
(d) Compare this value of I with the given current. If they agree within 5%, the problem is solved; if not, repeat parts (c) and (d).

26. Silicon sheet steel is used in the core of Fig. 20.23 with $a = 4$ cm, $b = 20$ cm, $c = 5$ cm, and $d = 25$ cm. Predict the flux density in the 2-mm gap with a current of 5.5 A in the 400-turn coil.

27. Silicon sheet steel is used in the core of Fig. 20.24 with $a = 20$ cm, $b = 40$ cm, $c = 120$ cm, and $d = 60$ cm. $N_1 = N_2 = 500$ turns. Estimate the value of $I_1 = I_2$ to produce a total magnetic flux of 100 mWb across the 5-mm air gap. (*Hint*: Can symmetry be used to simplify this problem?)

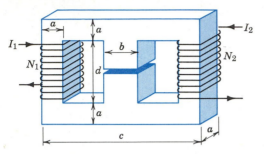

Figure 20.24

28. A single 500-turn coil on the center leg of the core of Exercise 27 replaces the two coils on the outer legs. Without detailed calculation, predict the required current I in terms of I_1. Which is the more economic design?

29. Cast steel is used in the core of Fig. 20.24 with $a = 5$ cm, $b = 8$ cm, $c = 40$ cm, and $d = 20$ cm. $N_1 = N_2 = 450$ turns; $I_1 = 5$ A and $I_2 = -5$ A.
(a) Draw the analogous electric circuit.
(b) Sketch the flux distribution showing lines close together where flux density is high, etc.
(c) Estimate the flux density in the 2-mm air gap.

PROBLEMS

1. The major home appliances and some "typical" energy consumptions are: water heater 15 kWh/day, air-conditioner 2 kW, refrigerator-freezer 4 kWh/day, dishwasher 0.6 kWh/cycle, range and oven 3 kWh/day, washer and dryer 3 kWh/cycle, and color TV 200 W. Estimate the annual energy cost for a home in your community.

2. In a large power plant the steam input to the turbine is at 1000° F and the condenser-cooling water from a lake is at 50° F. The overall plant efficiency is 40%. If improved engineering would permit using steam at 1100° F, what overall efficiency would be expected?

3. Estimate the fraction of the total area of your city to be covered by solar cells if the total electrical load were to be supplied by solar radiation.

4. Starting from Ampere's law (Eq. 20-5), derive an expression for the flux density at the center of a square coil L meters on a side carrying a current I. Repeat for a compact coil of N turns.

5. One way to obtain a nearly uniform magnetic field in air is to use two compact circular N-turn coils of radius R spaced a distance R apart along their common axis.
(a) Using the result of Exercise 5, derive an expression for the flux density B at the geometric center of the combination.

(b) For 100-turn coils of 10-cm radius carrying 5 A, calculate B at the geometric center of the combination and at the midpoint of one coil and draw a conclusion.

6. A core of silicon sheet steel has the general shape of Fig. 20.23. Stating all assumptions, determine a ratio of effective length of iron l_i to effective length of air l_a to define the conditions under which the mmf drop in the iron can be neglected without introducing an error greater than 5% in the flux calculations.

7. Design the winding for a high permeability silicon steel core in the shape of Fig. 20.23 to provide a flux density of 1 T across the 5 mm air gap ($a = 5$ cm and $c = 10$ cm). Assume that copper windings occupy half the area of the "window" in the core, the remainder being taken up with insulation and air space, and that the allowable current density in copper is 200 A/cm².
(a) Determine the area of the window and specify the dimensions of the core.
(b) Estimate the mean turn length and specify the wire resistance per meter for operation at 120 V dc.
(c) Select an appropriate wire size from a table of copper wire gages.

8. The core of Fig. 20.23 is silicon sheet steel with $a = c = 5$ cm and $b = d = 30$ cm. The 400-turn coil carries a current of 15 A. Estimate the total flux across the 5-mm air gap by using a "load line" drawn on a graph of ϕ vs. $\mathscr{F}$ for the core.

9. A core similar to that in Fig. 20.23 but without an air gap is placed around the neck of a TV picture tube to provide magnetic deflection. A second coil identical to that shown is wound on the right-hand leg; it carries the same current I in such a direction as to produce a *counterclockwise* mmf. Sketch the resulting magnetic field and describe qualitatively the deflection obtained.

21

Transformers

AC Excitation of Magnetic Circuits

Transformer Operation

Linear Circuit Models

Transformer Performance

In 1831 Michael Faraday applied a voltage to one coil and observed a voltage across a second coil wound on the same iron core. Fifty-three years later at an exhibition in Turin, Italy, the English engineers Gaulard and Gibbs demonstrated an "electric system supplied by inductors," using "secondary generators" with open iron cores.[†] Three young Hungarian engineers, Déri, Bláthy, and Zipernowsky, who visited the Turin exhibition, recognized the disadvantages of the open core and started work on an improved version. Five weeks later they shipped their "Transformer No. 1," a 1400-W model with a closed core of iron wires. Seventy-five of their transformers were used to supply the 1,067 Edison lamps that illuminated the 1885 Budapest exhibition. In one observer's words: "A new system of distribution has been inaugurated which . . . bids fair to mark an epoch. . . . It becomes possible to conduct the current from one central station to many consumers, even over very great distances." That prediction came true and the transformer is a key energy-conversion device in modern power-distribution systems.

The transformer transfers electrical energy from one circuit to another by means of a magnetic field that links both circuits. There are three reasons for investigating its behavior at this point in our studies: The transformer illustrates the use of magnetic circuits in energy conversion, it provides a good example of voltages induced by a changing magnetic field, and transformer action is the basis of operation of induction motors and other important electromechanical devices. First we consider magnetic circuits with sinusoidal flux variation, then we investigate the voltage and current relations in coils linking the same magnetic circuit. Next we treat the transformer as a two-port and derive a circuit model. Finally we use the model to predict transformer performance.

[†]Halacsy-von Fuchs: "Transformer Invented 75 Years Ago," *Electrical Engineering*, June 1961, p. 404.

AC EXCITATION OF MAGNETIC CIRCUITS

When a steady or dc′ voltage exists across a coil linking a magnetic core, the current that flows is limited by the *resistance* of the winding. The voltage RI developed across the resistance is just equal to the applied voltage V, and the device of Fig. 21.1a can be represented by the circuit model of Fig. 21.1b. The magnetic flux ϕ established in the core is determined by the magnetizing force H in ampere-turns per meter, the properties of the core as defined by a B–H curve, and the cross-sectional area A of the core.

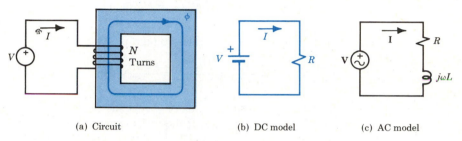

(a) Circuit   (b) DC model   (c) AC model

Figure 21.1   Magnetic circuit excited by dc and ac voltages.

In contrast, when an ac voltage exists across the same coil, the current is limited by the *impedance* of the winding. The voltage drop due to resistance is usually small, and the voltage drop due to inductive reactance is approximately equal to the applied voltage. The magnetic flux in the core is just that required to produce an induced voltage approximately equal to the applied voltage. In the circuit model of Fig. 21.1c, this effect is represented by an inductance L. Originally we described inductance as a measure of energy stored in a magnetic field, but we defined it as a circuit quantity in terms of its v-i characteristic. Now we are in a position to consider inductance in terms of the magnetic field that it represents.

Induced Voltage and Inductance

On the basis of his experiments, Faraday correctly concluded that the voltage induced in a multiturn coil linking a changing magnetic field is proportional to the number of turns N and to the time rate of change of flux ϕ or

$$v = N\frac{d\phi}{dt} = \frac{d\lambda}{dt} \tag{21-1}$$

where λ (lambda) is the number of *flux linkages* in weber-turns. If, in a 10-turn coil, 8 turns link a flux of 1 Wb and 2 turns link a flux of only 0.9 Wb, the total flux linkage is 8×1 plus 2×0.9 or 9.8 Wb · t. (Here again the number of turns is a dimensionless quantity, but we carry the term in the unit as a reminder of its physical significance.)

The polarity of the induced voltage can be determined by *Lenz's law*:

The induced voltage is always in such a direction as to tend to oppose the change in flux linkage that produces it.

Consider, for example, the case of a coil linked by a *decreasing* flux. If a closed path is provided, the current caused by the induced voltage will be in such a direction as to produce *additional* flux.

In general, the relation between flux and current is nonlinear; however, in many practical situations linearity can be assumed with negligible error. If the flux in a magnetic circuit varies sinusoidally so that

$$\phi = \Phi_m \sin \omega t$$

the induced voltage in an N-turn coil is

$$v = N\frac{d\phi}{dt} = N\omega\Phi_m \cos \omega t \tag{21-2}$$

The effective value of this sinusoidal voltage is

$$V = \frac{V_m}{\sqrt{2}} = \frac{N\omega\Phi_m}{\sqrt{2}} = \frac{2\pi}{\sqrt{2}}Nf\Phi_m = 4.44Nf\Phi_m \tag{21-3}$$

For ac excitation, the resulting flux Φ_m depends on the frequency as well as the magnitude of the applied voltage. If the resistance drop is neglected, the current that flows is just that required to establish the flux specified by Eq. 21-3.

By the definition of inductance, the voltage developed is

$$v = L\frac{di}{dt} \tag{21-4}$$

Combining this definition with Faraday's law,

$$L\frac{di}{dt} = N\frac{d\phi}{dt} = \frac{d\lambda}{dt} \qquad \text{or} \qquad L = N\frac{d\phi}{di} = \frac{d\lambda}{di} \tag{21-5}$$

For a linear magnetic circuit, ϕ is proportional to i or

$$L = \frac{N\phi}{i} = \frac{\lambda}{i} \tag{21-6}$$

and another interpretation of this circuit parameter is:

Inductance is the number of flux linkages per unit of current.

In Example 1 we see the important factors in designing an inductance. If many turns are wound on a high-permeability core, the flux linkage per ampere is high and the inductance is correspondingly great.

Energy Storage in a Magnetic Field

Another interpretation of inductance is as a measure of the ability of a circuit component to store energy in a magnetic field (Eq. 1-15). This is an important concept because the magnetic field is frequently a coupling medium in transforming energy from one form to another. How can the stored energy be calculated?

EXAMPLE 1

A current of 2 A in a coil produces a magnetic field described by the flux pattern shown in Fig. 21.2 where each line represents 1 mWb of flux. Determine the flux linkage created and the inductance of the coil.

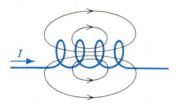

Figure 21.2 Flux linkages.

Since 2 coil turns link 4 mWb and the other 2 turns link only 2 mWb, the total flux linkage is

$$\lambda = 2 \times 4 \times 10^{-3} + 2 \times 2 \times 10^{-3} = 12 \text{ mWb} \cdot \text{t}$$

By Eq. 21-6 the inductance is

$$L = \frac{\lambda}{i} = \frac{12 \times 10^{-3}}{2} = 6 \times 10^{-3} \text{ H} = 6 \text{ mH}$$

In the toroidal core of Fig. 21.3a the mmf is $\mathcal{F} = Ni = Hl$. Assuming that the core is "thin" ($r \ll R$), flux density B and magnetizing force H are uniform across the area A and the total flux is $\Phi = BA$. Starting with an unmagnetized core, we can store

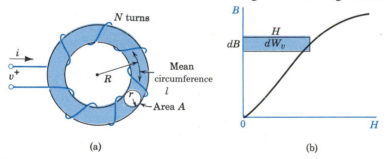

Figure 21.3 Energy storage in a magnetic field.

energy by building up a current and creating a magnetic field. The magnetic field energy stored comes from the electric circuit, and the electrical input is

$$W = \int_0^t vi \, dt = \int_0^t N \frac{d\phi}{dt} \cdot i \, dt = \int_0^\phi Ni \, d\phi \tag{21-7}$$

after changing variables. But $Ni = \mathcal{F} = Hl$ and $d\phi = A \, dB$, therefore

$$Ni \, d\phi = (Hl)(A \, dB) = (lA)H \, dB$$

where lA is the volume of the magnetic core. The energy stored per unit volume, i.e., the *energy density*, is

$$W_V = \frac{W}{lA} = \int_0^B H \, dB \tag{21-8}$$

which can be interpreted as the area between the magnetization curve and the B axis (Fig. 21.3b). Assuming linear magnetic characteristics, i.e., constant permeability,

$B = \mu H$ and Eq. 21-8 can be integrated to yield

$$W_V = \int_0^B H \, dB = \int_0^B \frac{B}{\mu} \, dB = \frac{1}{2}\frac{B^2}{\mu} \tag{21-9}$$

or

$$W_V = \int_0^B H \, dB = \int_0^H \mu H \, dH = \frac{1}{2}\mu H^2 \tag{21-10}$$

Either of these expressions can be used to calculate the energy density in any part of a magnetic field.

EXAMPLE 2

An N-turn coil (similar to Fig. 21.3a) is wound on a thin toroidal core of relative permeability μ_r, cross section A, and length l_i with an air gap of length l_a. For a flux density B in the iron, compare the energy densities and the total energies in the iron and in the air.

If the relative permeability is 2000 and if the length of the air gap is 1% of the length of the iron, estimate the fraction of the total energy that is stored in the air.

Assuming uniform flux distribution and neglecting fringing and leakage flux, the flux density in the air is the same as in the iron. By Eq. 21-9,

$$\frac{W_{Va}}{W_{Vi}} = \frac{B^2/\mu_o}{B^2/\mu_o\mu_r} = \mu_r$$

or the energy density in the air in a series magnetic circuit is μ_r times as great as that in the iron.

The ratio of the total energies is

$$\frac{W_a}{W_i} = \mu_r\frac{l_aA_a}{l_iA_i} = \mu_r\frac{l_a}{l_i} = 2000 \times 0.01 = 20$$

Therefore, more than 95% of the energy stored is in the air. In many practical problems all the energy stored can be assumed to be in the air gap.

Hysteresis

In magnetizing a ferromagnetic material by reorientation of the domains, most of the effects are irreversible (inelastic). When the external field is removed, the magnetic material does not return to its original state. If an iron specimen is saturated, point 1 in Fig. 21.4a, and then the field is removed ($H = 0$), the magnetic condition

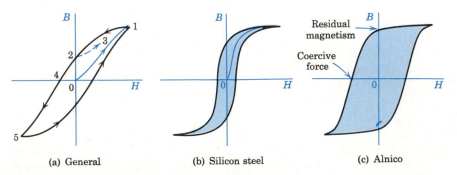

      (a) General   (b) Silicon steel   (c) Alnico

Figure 21.4   Hysteresis loops.

follows the line from 1 to 2. The ordinate at point 2 is called the *residual magnetism*. If a positive H is again applied, the condition follows the path 2 to 3. A negative magnetizing force (the *coercive force*) is required to bring the flux density to zero as at point 4. A large negative H produces saturation in the opposite direction (point 5). Reversing the magnetizing force causes the magnetic condition to follow the path 5 to 1. If the magnetizing force is due to an alternating current, the *hysteresis loop* is traced out once each cycle.

Just as a metal strip gets warm when it is repeatedly flexed, magnetic material gets warm when it is cyclically magnetized. In both cases, the energy that appears as an increase in temperature is due to inelastic action. The energy input to an initially unmagnetized sample (of unit volume) is the area between the curve 0–1 and the B axis; the energy returned is the area between the curve 1–2 and the B axis (Fig. 21.5). The difference in these areas is the energy converted to heat in an irreversible process. Following this line of reasoning for a complete cycle, we see that the area of the hysteresis loop is just equal to the energy lost per cycle. For silicon steel the loop is thin (Fig. 21.4b) and the *hysteresis loss* is small; for a "permanent magnet" the coercive force is large and the loop is fat (Fig. 21.4c). The area of the loop increases nonlinearly with maximum flux density. An empirical formula developed by Steinmetz for commercial magnetic steels gives the hysteresis power loss in watts as

$$P_h = K_h f B_m^n \tag{21-11}$$

where the constant K_h and exponent n vary with the core material; n is often assumed to be 1.6.

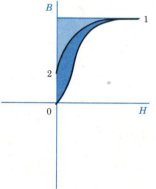

Figure 21.5 Hysteresis loss.

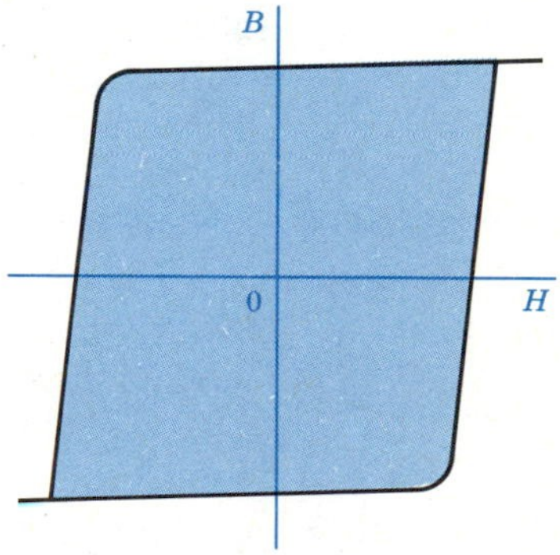

Figure 21.6 Square hysteresis loop.

In some magnetic materials the hysteresis loop is nearly rectangular (Fig. 21.6). In such *square-loop* materials the slope of the sides of the B–H curve is large and a small change in H can *switch* the core from nearly saturated in one direction to nearly saturated in the other. A tiny core and coil can serve as a magnetic flip-flop in switching circuits or as a storage element in a computer.

Eddy Currents

In early electric generators nearly three-fourths of the mechanical input appeared as heat in the magnetic circuit. A small part of this energy loss was due to hysteresis, but the major part was due to the fact that a changing magnetic flux induces voltages

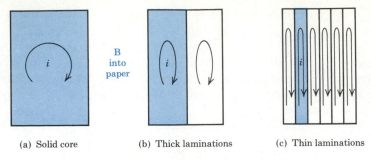

(a) Solid core   (b) Thick laminations   (c) Thin laminations

Figure 21.7   Eddy currents and the effect of core lamination.

in the core material itself. In a conducting iron core, induced voltages cause localized *eddy currents* and the resulting $i^2R = v^2/R$ power appears as heat. As shown in Fig. 21.7a, a changing flux (directed into the paper) induces a net current within the core material. The power loss can be reduced by decreasing v and increasing R. If, instead of a solid iron core, thin *laminations* are used (Fig. 21.7c), the effective induced voltage is decreased and the resistance of the effective path is increased. (What effect does laminating the core have on eddy-current path length? On cross-sectional area?) The laminations, perhaps 0.02 in. thick, are electrically insulated from each other by a thin varnish or just the scale produced in heating and rolling.

For a given core, the eddy-current power loss is given by[†]

$$P_e = K_e f^2 B_m^2 \tag{21-12}$$

Since induced voltage is proportional to fB_m (Eq. 21-2) and the loss varies as the square of the voltage, we expect the power loss to vary as $f^2B_m^2$. (See Example 3.) The constant K_e depends on the conductivity of the core material and the square of the thickness of the laminations.

TRANSFORMER OPERATION

The transformer is an electromagnetic energy converter whose operation can be explained in terms of the behavior of a magnetic circuit excited by an alternating current. In its most common form, the transformer consists of two (or more) multiturn coils wound on the same magnetic core but insulated from it. A changing voltage applied to the input or *primary* coil causes a changing current to flow, thus creating a changing magnetic flux in the core. Because of the changing flux, voltage is induced in the output or *secondary* coil. No electrical connection between input and output is necessary; the transformer may be used to insulate one circuit from another while permitting an exchange of energy between them. Since only changing currents are transformed, an output circuit can be isolated from a direct-current component of the input.

By adjusting the ratio of the turns on the two coils, we can obtain a voltage "step up" or "step down." The same device can be used to obtain a current step up or step down. In addition to transforming voltage or current, a transformer may be used to transform impedance to obtain maximum power transfer through impedance matching. Because these functions are performed efficiently and precisely, the transformer is an important energy conversion device and careful study of its behavior is justified.

[†]See p. 392 of W. H. Timbie, V. Bush, and G. Hoadley, *Principles of Electrical Engineering*, 4th ed., John Wiley & Sons, New York, 1951.

EXAMPLE 3

An iron-core inductor (Fig. 21.8) is designed to operate at 120 V at 60 Hz. Estimate the effect on hysteresis and eddy-current losses of operating at 150 V at 50 Hz.

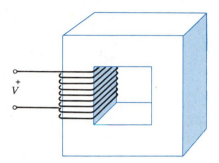

Figure 21.8   Calculation of core loss in an iron-core inductor.

Neglecting the IR drop, the induced voltage must be equal to the applied voltage, and by Eq. 21-3, the flux density must be

$$B_m = k\frac{V}{f}$$

where the constant k includes the number of turns and the core area. The ratio of the new B'_m to the old B_m is

$$\frac{B'_m}{B_m} = \frac{V'}{V} \cdot \frac{f}{f'} = \frac{150}{120} \cdot \frac{60}{50} = \frac{3}{2}$$

By Eq. 21-11 (assuming $n = 1.6$),

$$\frac{P'_h}{P_h} = \frac{f'}{f} \cdot \left(\frac{B'_m}{B_m}\right)^{1.6} = \left(\frac{5}{6}\right)\left(\frac{3}{2}\right)^{1.6} = 1.59$$

By Eq. 21-12,

$$\frac{P'_e}{P_e} = \left(\frac{f'}{f}\right)^2\left(\frac{B'_m}{B_m}\right)^2 = \left(\frac{5}{6}\right)^2\left(\frac{3}{2}\right)^2 = 1.56$$

The operation of electromagnetic devices can be greatly affected by relatively small changes in operating conditions. (How is I' related to I?)

Construction

In the basic configuration, two coils are wound on a common core. For power applications in the 25- to 400-Hz frequency range, close coupling is desired and the coils are intimately wound on a highly permeable closed iron core (Fig. 21.9a). Special design refinements make iron-core transformers useful over the audio-frequency range (20 to 20,000 Hz). For high frequencies (hundreds of kilohertz) and loose coupling, coils may be wound on a powdered iron "slug" (Fig. 21.9b) or with an air core.

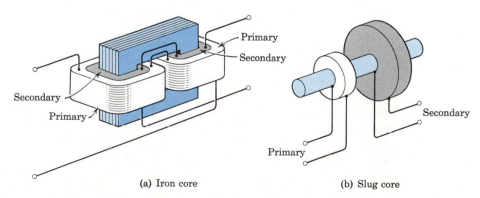

(a) Iron core

(b) Slug core

Figure 21.9   Two transformers.

To minimize resistance losses, the coils are usually wound with high-conductivity copper. In power and audio transformers, the cores are of high-permeability steel selected for low hysteresis loss and laminated to minimize eddy-current loss. The sum of hysteresis and eddy-current power is called *core loss* or *iron loss* in contrast to the *copper loss* due to the I^2R power in the windings.

Voltage Relations

In practice, the two coils are placed close together so that they link nearly the same flux. For clarity, in Fig. 21.10a the primary and secondary are shown on separate legs of the core, but it is assumed that both coils link the same flux. The voltage

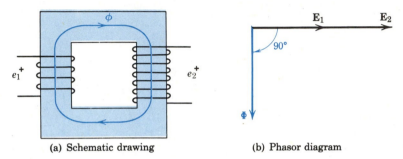

(a) Schematic drawing (b) Phasor diagram

Figure 21.10 Transformer operation.

induced by the changing flux we call an *electromotive force* (emf) represented by the symbol e. In general, the induced emf e is different from the terminal voltage v.

Assuming a sinusoidal variation in magnetic flux of the form $\phi = \Phi_m \sin \omega t$, the induced emfs are

$$e_1 = N_1 \frac{d\phi}{dt} = N_1 \omega \Phi_m \cos \omega t = \sqrt{2} E_1 \cos \omega t \qquad (21\text{-}13a)$$

and

$$e_2 = N_2 \frac{d\phi}{dt} = N_2 \omega \Phi_m \cos \omega t = \sqrt{2} E_2 \cos \omega t \qquad (21\text{-}13b)$$

where E_1 and E_2 are the effective values of the sinusoidal emfs. Therefore,

$$\frac{e_2}{e_1} = \frac{E_2}{E_1} = \frac{N_2}{N_1} \qquad (21\text{-}14)$$

or the emf ratio is just equal to the *turn ratio*. The relation between emf and magnetic flux phasors is shown in Fig. 21.10b. Since $\sin \omega t = \cos(\omega t - 90°)$, the flux phasor lags behind the emf phasor by 90°.

In practical transformers, the terminal voltages differ only slightly from the induced emfs and the terminal voltage ratio is approximately equal to the turn ratio. As an illustration, there may be a transformer on a power pole on your rear property line to transform the voltage from an efficient transmission value of 4000 V to a safe working value of 240 V. (Approximately what turn ratio is required?)

Exciting Current

With the secondary open-circuited, the emf E_2 appears across the terminals, V_2 is just equal to E_2, and V_1 is approximately equal to E_1. The current that flows in the primary when the secondary is open-circuited is called the *exciting current*. As shown in Fig. 21.11, the exciting current $\mathbf{I}_E$ consists of two parts. One component, the

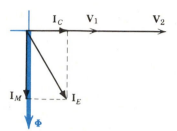

Figure 21.11 Components of the exciting current.

magnetizing current $\mathbf{I}_M$, establishes the necessary flux to satisfy Eq. 21-13a and can be calculated when the properties and dimensions of the iron core are known. As expected, $\mathbf{I}_M$ is in phase with the flux $\mathbf{\Phi}$ that it produces. The other component, the *core loss current* $\mathbf{I}_C$, represents the power dissipated in hysteresis and eddy-current loss. As expected, $\mathbf{I}_C$ is in phase with voltage $\mathbf{V}_1$ so that the product is power.

In an efficient transformer the core loss is small and $\mathbf{I}_E$ is approximately equal to $\mathbf{I}_M$. Strictly speaking, the magnetizing current is not sinusoidal for sinusoidal voltage and flux because of the nonlinear relation between $H = Ni/l$ and $B = \Phi/A$. However, for many practical transformers the magnetizing current is relatively small, and it is convenient to assume that i_M is sinusoidal and can be represented by phasor $\mathbf{I}_M$.

Current Relations

To develop the relation between primary and secondary currents under load conditions, let us assume (Fig. 21.12) that $\mathbf{V}_2 = \mathbf{E}_2$, $\mathbf{V}_1 = \mathbf{E}_1$, and $\mathbf{I}_E = \mathbf{I}_M$. When load $\mathbf{Z}_2$ is connected across the secondary by closing switch S, a current $\mathbf{I}_2 = \mathbf{V}_2/\mathbf{Z}_2$ flows. (Here θ_2 is assumed to be about 30° and the turn ratio $N_2/N_1 \cong 2$.) The load current $\mathbf{I}_2$ produces an mmf $N_2\mathbf{I}_2$ that *tends* to oppose the magnetic flux that produces it. (Using the right-hand rule, check the directions of the mmfs produced by positive currents in the primary and secondary coils.) But the flux cannot change if $\mathbf{E}_1$ is to equal $\mathbf{V}_1$ (Eq. 21-3), and therefore additional primary current $\mathbf{I}_1'$ must flow. To maintain the core flux, the new net mmf must equal the mmf due to $\mathbf{I}_M$ alone or

$$N_1\mathbf{I}_M - N_2\mathbf{I}_2 + N_1\mathbf{I}_1' = N_1\mathbf{I}_M \qquad (21\text{-}15)$$

This requires that

$$N_2\mathbf{I}_2 = N_1\mathbf{I}_1'$$

and

$$\frac{\mathbf{I}_1'}{\mathbf{I}_2} = \frac{N_2}{N_1} \qquad (21\text{-}16)$$

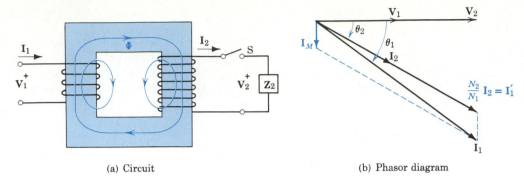

(a) Circuit   (b) Phasor diagram

Figure 21.12   Current relations in a loaded transformer.

We conclude that a secondary current $\mathbf{I}_2$ causes a component of primary current $\mathbf{I}_1'$; these currents are in phase and their ratio is just equal to the turn ratio N_2/N_1. (The ratio I_2/I_1' is the *reciprocal* of the turn ratio.)

Strictly speaking, some of the mmf N_2I_2 goes to produce the leakage flux shown in Fig. 21.12a; also, all the flux produced by the primary does not link the secondary. In a well-designed transformer, however, the leakage flux is small and there is little error in Eqs. 21-14 and 21-16.

The effect of a load across the output of a transformer is transferred to the source supplying the input. Because this is a difficult concept to grasp, let us describe the process again in different words. With the secondary open (the transformer unloaded), the input current is just the exciting current that establishes the necessary magnetic flux (so that $\mathbf{E}_1 \cong \mathbf{V}_1$) and supplies the core losses. The no-load power input is essentially the power dissipated in the core. For a given $\mathbf{V}_2$, the secondary current $\mathbf{I}_2$ that flows when a load is connected is determined in magnitude and angle by $\mathbf{Z}_2$. To maintain the magnetic flux at the required value, the mmf $N_2\mathbf{I}_2$ produced by the secondary current must be offset by an equal and opposite mmf $N_1\mathbf{I}_1'$ produced by additional primary current $\mathbf{I}_1'$. To the source supplying the transformer, the increased primary current represents an increase in apparent power. The total power input to the transformer is $P_1 = V_1I_1 \cos \theta_1$, where $\mathbf{I}_1$ is the phasor sum of the exciting current and the load current transferred to the primary. (See Example 4.)

LINEAR CIRCUIT MODELS

In deriving the voltage and current relations for a transformer and in solving Example 4, we made many simplifying assumptions. With just a little more effort we can derive a model that provides much greater precision. Furthermore, in spite of the nonlinearity of the magnetic circuit and the unpredictability of the hysteresis effect, precise results over a wide range of operating conditions can be obtained using a linear model. Finally, and this is important to engineers, the model parameters for a given transformer can be determined by two simple tests that can be performed with readily available instruments.

EXAMPLE 4

The laminated silicon steel core of a transformer (Fig. 21.13) has a mean length of 0.6 m and a cross section of 0.005 m². There are 150 turns on the primary winding (actually split between two "legs" of the core as in Fig. 21.9) and 450 turns on the secondary. The input is 200 V rms at 60 Hz. Estimate the primary current with the secondary open and with a resistance load of 120 Ω connected across the secondary.

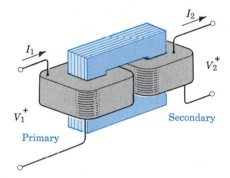

Figure 21.13   Transformer calculations.

The necessary maximum flux is, by Eq. 21-13a,

$$\Phi_m = \frac{\sqrt{2}E_1}{N_1\omega} = \frac{\sqrt{2} \times 200}{150 \times 2\pi \times 60} = 5 \times 10^{-3} \text{ Wb}$$

The maximum flux density is

$$B_m = \frac{\Phi_m}{A} = \frac{5 \times 10^{-3}}{5 \times 10^{-3}} = 1 \text{ T}$$

For silicon steel (Fig. 20.16) at this density, $H = 200 \text{ A} \cdot \text{t/m}$. The magnetizing current (rms value) is

$$I_M = \frac{1}{\sqrt{2}} \frac{\mathcal{F}}{N_1} = \frac{Hl}{\sqrt{2}N_1} = \frac{200 \times 0.6}{\sqrt{2} \times 150} \cong 0.6 \text{ A}$$

Since silicon steel has low hysteresis loss and the laminated core has low eddy-current loss, as an estimate we assume that $I_E = I_M = 0.6$ A.

Assuming that the voltage ratio is equal to the turn ratio,

$$V_2 = \frac{N_2}{N_1}V_1 = \frac{450}{150} \times 200 = 600 \text{ V}$$

With the resistance load connected,

$$I_2 = \frac{V_2}{R_2} = \frac{600}{120} = 5 \text{ A}$$

The component of primary current due to the load is

$$I_1' = \frac{N_2}{N_1}I_2 = \frac{450}{150} \times 5 = 15 \text{ A}$$

Since the exciting current is small and is to be added at nearly right angles to I_1' (see Fig. 21.12b), it is neglected and the primary current with load $\cong$ 15 A.

The Transformer as a Two-Port

In deriving a linear circuit model, we use the approach that worked well with transistors and incorporate our understanding of transformer operation. In wiring diagrams, an iron-core transformer is represented by the schematic symbol in Fig. 21.14a. In the two-port representation of Fig. 21.14b, we follow convention and show $\mathbf{I}_2$ as the current *out* because the transformer is considered to be a source supplying a load connected to the output port. Since transformers work only with changing cur-

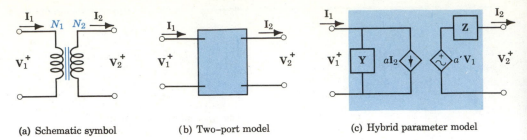

(a) Schematic symbol   (b) Two–port model   (c) Hybrid parameter model

Figure 21.14   Transformer representation.

rents, usually sinusoidal, we think in terms of phasor quantities and complex immittances.

In deriving a circuit model, we can choose any two independent variables and the choice determines the nature of the model parameters. Let us choose V_1 and I_2 (see Eq. 18-11) and write

$$\begin{cases} I_1 = f_1(V_1, I_2) = YV_1 + aI_2 & (21\text{-}17) \\ V_2 = f_2(V_1, I_2) = a'V_1 - ZI_2 & (21\text{-}18) \end{cases}$$

In these relations we recognize four hybrid parameters that are constant in a linear model. The first equation indicates that I_1 has two components and suggests a parallel circuit. The second equation indicates that V_2 has two components and suggests a series circuit. (The minus sign arises because I_2 is defined as the current *out*.) The result is shown in Fig. 21.14c; note that there is no electrical connection between input and output.

What is the physical nature of each of the four parameters? With the secondary open circuited, $I_2 = 0$ and I_1 is just equal to I_E, the exciting current. By Eq. 21-17, with $I_2 = 0$,

$$Y = \frac{I_E}{V_1} = \frac{I_C}{V_1} - j\frac{I_M}{V_1} = G + jB \qquad (21\text{-}19)$$

where I_C = the core-loss current (Fig. 21.15),
    I_M = the magnetizing current,
    G = a conductance accounting for power loss, and
    B = an inductive (negative) susceptance accounting for energy storage, *not* flux density.

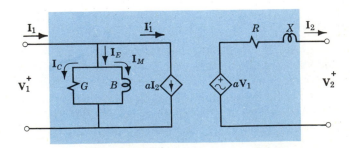

Figure 21.15   Hybrid parameter model of a transformer.

Since the open-circuit secondary voltage is approximately the primary voltage times the turn ratio, Eq. 21-18 indicates that parameter a' is given by

$$a' = \frac{N_2}{N_1} \qquad (21\text{-}20)$$

With a current $\mathbf{I}_2$ flowing in the secondary, we expect (Eq. 21-16) a component $\mathbf{I}_1'$ equal to $(N_2/N_1)\mathbf{I}_2$ to appear in the primary. Therefore,

$$a = \frac{N_2}{N_1} = a' \qquad (21\text{-}21)$$

With the transformer loaded, i.e., $\mathbf{I}_2 \neq 0$, the power lost in the resistance of the windings and the energy stored in the leakage fields (Fig. 21.12a) become appreciable. These effects are accounted for by the impedance

$$\mathbf{Z} = R + j\omega L = R + jX \qquad (21\text{-}22)$$

where R = an equivalent resistance including the effects of both windings,
L = an equivalent leakage inductance, and
X = an equivalent reactance.

Because of $\mathbf{Z}$, the output voltage changes with variations in load current.

Open- and Short-Circuit Tests

The parameters in the linear circuit model are easily and accurately determined by an ingenious laboratory procedure. With the secondary open circuited, $\mathbf{I}_2 = 0$, $\mathbf{I}_1 = \mathbf{I}_E$, and the transformer model is as shown in Fig. 21.16a. Voltmeter VM$_2$ has a very high resistance and appears to be an open circuit. Wattmeter WM measures the total power input, and VM$_1$ and AM$_1$ measure the primary voltage and exciting current, respectively.

In this *open-circuit test* the exciting current flows in the primary windings. However, I_E is small and, since power varies as the square of the current, the power lost in the windings is negligible and WM indicates only the core loss. Therefore, the parameters are

$$G = \frac{P_{OC}}{V_{1O}^2} \qquad\qquad Y = \frac{I_{1O}}{V_{1O}} \qquad (21\text{-}23)$$

$$B = -\sqrt{Y^2 - G^2} \qquad a = \frac{V_{2O}}{V_{1O}} \qquad (21\text{-}24)$$

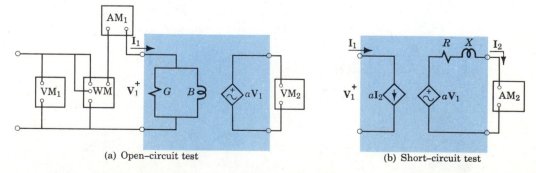

(a) Open–circuit test   (b) Short–circuit test

Figure 21.16 Instrumentation for determining model parameters.

(*Note:* V_{1O} is the primary voltage with the secondary open-circuited.) Since the core loss is dependent on B_m and f (see Example 3), and therefore dependent on V_1, the open-circuit test is performed at rated voltage and frequency. (Transformer ratings are discussed on p. 606.)

In contrast, the *short-circuit test* is performed at rated current. With the secondary short-circuited (ammeter AM_2 has a very low resistance and appears to be a short circuit), the transformer model is as shown in Fig. 21.16b. Since I_2 is limited only by the small internal impedance $R + jX$, the primary voltage required for rated

---

EXAMPLE 5

The primary of a transformer is rated at 10 A and 1000 V. On open circuit, instruments connected as in Fig. 21.16a indicate: $V_1 = 1000$ V, $V_2 = 500$ V, $I_1 = 0.42$ A, $P_{OC} = 100$ W. On short circuit, the readings are: $I_1 = 10$ A, $V_1 = 126$ V, $P_{SC} = 400$ W. Determine the hybrid parameters, predict the output voltage across a load impedance $Z_L = 19 + j12$ Ω, and draw a phasor diagram.

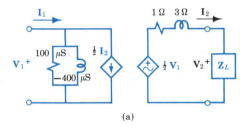

(a)

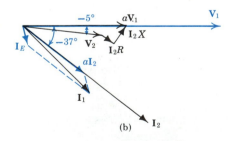

(b)

Figure 21.17 Transformer performance.

The parameters are determined from the test data. From open-circuit data,

$$G = \frac{P_{OC}}{V_{1O}^2} = \frac{100}{(1000)^2} = 100 \times 10^{-6} = 100 \ \mu S$$

$$Y = \frac{I_{1O}}{V_{1O}} = \frac{0.42}{1000} = 420 \ \mu S$$

$$B = -\sqrt{Y^2 - G^2}$$

$$= -\sqrt{(420)^2 - (100)^2} \times 10^{-6} \cong -400 \ \mu S$$

$$a = V_{2O}/V_{1O} = 500/1000 = 1/2$$

From short-circuit data,

$$R = \frac{a^2 P_{SC}}{I_{1S}^2} = \frac{400}{4 \times 10^2} = 1 \ \Omega$$

$$Z = a^2 \frac{V_{1S}}{I_{1S}} = \frac{126}{4 \times 10} = 3.15 \ \Omega$$

$$X = \sqrt{Z^2 - R^2} = \sqrt{(3.15)^2 - (1)^2} \cong 3 \ \Omega$$

The labeled circuit model and specified load are shown in Fig. 21.17a. The secondary current under load is

$$\mathbf{I}_2 = \frac{\frac{1}{2}\mathbf{V}_1}{\mathbf{Z} + \mathbf{Z}_L} = \frac{\frac{1}{2} \times 1000\,\underline{/0°}}{(1 + j3) + (19 + j12)}$$

$$= 20\,\underline{/-37°} \text{ A}$$

and the output voltage is

$$\mathbf{V}_2 = \mathbf{I}_2\mathbf{Z}_L = (20\,\underline{/-37°})(22.5\,\underline{/+32°}) = 450\,\underline{/-5°} \text{ V}$$

Note that under this load, the terminal voltage ratio is less than the turn ratio.

The complete phasor diagram is shown in Fig. 21.17b with $I_E = I_{1O}$ exaggerated in scale. Note that $a\mathbf{V}_1$ is equal to $\mathbf{V}_2 + \mathbf{I}_2\mathbf{Z}$.

current is very small. At this low voltage B_m is very small, the core loss is very small, and the exciting admittance $\mathbf{Y}$ is omitted from the model.

With the instruments of Fig. 21.16a applied to the short-circuited model, wattmeter WM indicates only the copper loss. Since $P_{SC} = I_{2S}^2 R = (I_{1S}/a)^2 R$,

$$R = \frac{a^2 P_{SC}}{I_{1S}^2} \tag{21-25}$$

Then

$$Z = \frac{aV_{1S}}{I_{2S}} = \frac{aV_{1S}}{I_{1S}/a} = a^2 \frac{V_{1S}}{I_{1S}} \tag{21-26}$$

and

$$X = \sqrt{Z^2 - R^2} = \sqrt{(a^2 V_{1S}/I_{1S})^2 - R^2} \tag{21-27}$$

Ammeter AM_2 is not essential, but the ratio I_{1S}/I_{2S} provides a check on the value of a determined from V_{2O}/V_{1O} in the open-circuit test.

There are two main reasons for representing a device by a linear model. First, the model permits general analysis and the drawing of general conclusions. Second, once the parameters are determined, the behavior of the device under various operating conditions can be predicted as in Example 5.

Other Models

The hybrid parameter model of Fig. 21.18a is convenient to work with; the parameters are easily determined and calculations are straightforward. Since an im-

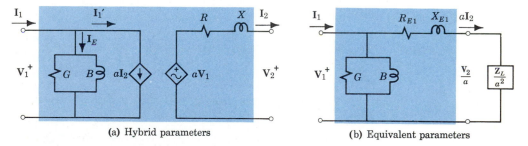

(a) Hybrid parameters   (b) Equivalent parameters

Figure 21.18 A model with parameter values referred to the primary.

pedance $\mathbf{Z}$ in the secondary circuit has the same effect as an impedance $\mathbf{Z}/a^2$ in the primary, the model can be converted to the alternate form of Fig. 21.18b. Here

$$\frac{R}{a^2} = R_{E1} \quad \text{and} \quad \frac{X}{a^2} = X_{E1} \tag{21-28}$$

where R_{E1} and X_{E1} are equivalent resistance and equivalent reactance *referred to the primary*. Similarly, V_2/a is the secondary voltage referred to the primary and $\mathbf{Z}_L/a^2$ is the load impedance referred to the primary. By using these parameters, a transformer problem is reduced to a simple parallel circuit problem.

A defect of the model of Fig. 21.18 is that it indicates that the exciting current is independent of the load current; this is not quite true. The induced emf $\mathbf{E}_1$ differs from the primary voltage $\mathbf{V}_1$ by the amount of the voltage drop across the resistance and leakage reactance of the primary winding. As the load current increases, the

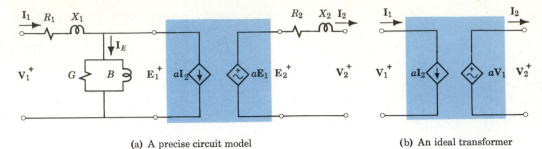

(a) A precise circuit model   (b) An ideal transformer

Figure 21.19   Conventional transformer models.

primary current increases, the voltage drop increases, and the necessary emf $\mathbf{E}_1$ decreases slightly. At high load currents, $\mathbf{E}_1$ is less than at no load, the necessary flux $\mathbf{\Phi}$ is less, and therefore $\mathbf{I}_E$ is less. If the portion of $\mathbf{Z}$ due to the primary resistance and leakage reactance is placed as shown in the conventional model of Fig. 21.19a, the reduction of $\mathbf{E}_1$ and $\mathbf{I}_E$ with increasing $\mathbf{I}_1$ is accounted for automatically. A second improvement is that here $a = \mathbf{E}_2/\mathbf{E}_1$, which is just equal to the turn ratio and is constant. Because of practical design considerations, in commercial transformers the resistances and leakage reactances divide so that $a^2R_1 \cong R_2 = \frac{1}{2}R$ and $a^2X_1 \cong X_2 = \frac{1}{2}X$ (see Problem 4). In performing the short-circuit test, the impedance seen at the primary terminals is

$$\frac{\mathbf{V}_{1S}}{\mathbf{I}_{1S}} = \left(R_1 + \frac{R_2}{a^2} \right) + j\left(X_1 + \frac{X_2}{a^2} \right) \tag{21-29}$$

EXAMPLE 6

Using the data of Example 5, determine the parameters for the more precise model of Fig. 21.19.

The values of G, B, and a are as previously calculated. Assuming the previously calculated resistance R and leakage reactance X divide in the usual way,

$$R_2 = (\tfrac{1}{2})R = \tfrac{1}{2}(1) = 0.5 \ \Omega$$

$$R_1 = \frac{1}{a^2}R_2 = 4R_2 = 4 \times 0.5 = 2.0 \ \Omega$$

$$X_2 = \tfrac{1}{2}X = \tfrac{1}{2}(3) = 1.5 \ \Omega$$

$$X_1 = \frac{1}{a^2}X_2 = 4X_2 = 4 \times 1.5 = 6 \ \Omega$$

The portion within the shaded area in Fig. 21.19b is called an *ideal transformer* because it provides for the transformation of voltage, current, and impedance without any of the imperfections of a real transformer. A real transformer approaches the ideal as the conductivity of the windings approaches infinity, the leakage flux approaches zero, the permeability of the core approaches infinity, and the core loss approaches zero. The ideal transformer is sometimes represented by adding the label "ideal" to the symbol in Fig. 21.14a, but its character is more clearly indicated by the symbolic representation in Fig. 21.19b.

Selecting a Model

The ideal transformer and the precise model represent the extremes of complexity ordinarily employed in power transformer analysis. In a given situation, the best model to use is the one that gives the required degree of precision with a minimum of effort; the hybrid parameter model of Fig. 21.18a is one useful compromise. If the transformer is heavily loaded, the exciting admittance Y may be of negligible importance. If voltage variation is of greater interest than power dissipation, the resistive elements (R and G) can be neglected and only the relatively larger reactive elements (X and B) retained. If core loss is unimportant, let $I_E = I_M$ or $G = 0$. (See Exercise 24.)

Models for Communications Transformers

Power transformers are ordinarily operated at fixed frequencies and the circuit model reactances are constant. The input and output transformers used in audio-frequency applications (for example, in connecting a power transistor to a loud-speaker) must handle signals from, say, 20 to 15,000 Hz. Because of the wide variation in reactances, the circuit model applicable at low frequencies is quite different from that appropriate at high frequencies. In addition the *distributed capacitance* of the windings, unimportant in most power transformer applications, becomes significant at high frequencies.

A convenient model for an output transformer is shown in Fig. 21.20a. This is derived from the model of Fig. 21.18b by neglecting G, replacing frequency-

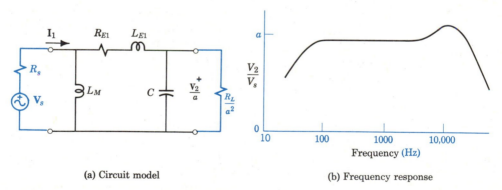

(a) Circuit model   (b) Frequency response

Figure 21.20   An audiofrequency transformer.

dependent B by L_M, and including a lumped shunt capacitance C to represent all the distributed capacitive effects. Laboratory measurements on the actual transformer indicate a response curve like that shown in Fig. 21.20b.

At moderate frequencies, corresponding to the midfrequency range of an amplifier, the admittances of the shunt paths represented by L_M and C and the reactance of the series element L_{E1} are negligible. If R_s and R_{E1} are small, the midfrequency voltage gain is just the turn ratio a. At low frequencies, the admittance $Y = 1/\omega L_M$ of the magnetizing circuit is high, part of the input current flows through this shunt path, and the output is reduced. At high frequencies, the effect of L_M is negligible, but the effects of L_{E1} and C are important. At the series resonant frequency of L_{E1} and C, the voltage gain may be greater than a, but at higher frequencies the gain drops sharply. For quantitative analysis the general model of Fig. 21.20a can be replaced by three simplified models that are applicable over restricted frequency ranges (see Exercise 29).

TRANSFORMER PERFORMANCE

For power applications the important external characteristics are voltage ratio, output power, efficiency, and the voltage variation with load. These may be obtained from manufacturer's specifications, from experimental measurements, or from calculations based on a circuit model.

Nameplate Ratings

The manufacturer of an electrical machine usually indicates on the *nameplate* the normal operating conditions.[†] A typical nameplate might read: "Transformer, 4400:220 V, 60 Hz, 10 kVA." The design voltages of the two windings are 4400 V and 220 V rms and the turn ratio is 1:20; either side may be the primary. At a frequency of 60 Hz, the design voltages bring operation near the knee of the magnetization curve and exciting current and core losses are not excessive.

The actual primary voltage will be that necessary to provide rated secondary voltage under rated load. On large transformers, taps on the windings allow small adjustments in the turn ratio. Using either side as a secondary, the rated *output* of 10 kVA ("full" load) can be maintained continuously without excessive heating (and the consequent deterioration of the winding insulation). Because the heating is dependent on the square of the current, the output is rated in apparent power (kVA) rather than in power (kW). Supplying a zero power-factor load, a transformer can be operating at rated output while delivering zero power.

The cross section of the iron core is determined by the operating voltage and frequency. The cross section of the copper conductor is determined by the operating current. Knowing the effect of these factors, an engineer can *rerate* a device under changed operating conditions.

When electrical isolation between primary and secondary is not required, an *autotransformer* offers a low-cost, high-efficiency alternative. In an autotransformer, all the turns link the same flux, but some of the turns carry both primary and secondary currents. For example, if 120 V is applied to 240 turns of a 300-turn winding, a

---

[†]See the section on Root-Mean-Square Rating on p. 68.

voltage of 150 V becomes available across the total. However, only a part of the output current flows through the 240-turn section. (See Exercise 13.)

Efficiency

Efficiency is by definition the ratio of output power to input power. A convenient form is

$$\text{Efficiency} = \frac{\text{output}}{\text{input}} = \frac{\text{output}}{\text{output} + \text{losses}} = 1 - \frac{\text{losses}}{\text{input}} \qquad (21\text{-}30)$$

Using the hybrid parameter model of Fig. 21.15, the output is $V_2 I_2 \cos \theta_2$ and the copper loss is $I_2^2 R$. The iron loss is independent of load current and is given by $V_1^2 G$ which is just equal to P_{OC} (Eq. 21-23).

By increasing the amount of copper, the $I^2 R$ loss is reduced; by increasing the amount (or quality) of iron, the eddy-current and hysteresis losses are reduced. The optimum design is based on economic considerations as well as technical factors. A *distribution* transformer that is always connected to the line should be designed with low iron losses. A *power* transformer that is only connected when in actual use usually has higher iron losses and lower copper losses than a comparable distribution transformer. It can be shown that for many engineering devices the maximum efficiency occurs when the variable losses are equal to the fixed losses (see Problem 5). In an *instrument* transformer that converts high voltages or high currents to lower values that are more easily measured, the important characteristic is a constant ratio of transformation.

Voltage Regulation

To hold the speed of a motor or the voltage of a generator constant under varying load we employ a *regulator*, and the variation in speed, or voltage, is expressed by the *regulation*. By definition the regulation is

$$\text{Regulation} = \frac{\text{no-load value} - \text{full-load value}}{\text{full-load value}} \qquad (21\text{-}31)$$

The voltage regulation of a transformer, then, is the change in output voltage (produced by a change in load current from zero to rated value) divided by the rated voltage. The regulation may be either positive or negative and is usually expressed in percent. (See Example 7 on p. 608.)

The voltage regulation is defined for voltage *magnitudes*, but usually it is calculated from a phasor diagram based on a circuit model. The primary cause of voltage drop with load is leakage reactance and if high precision is not required, a simplified model can be used. Note that by this convention the regulation is positive if the voltage decreases with increasing load.

EXAMPLE 7

The 10-kVA, 60-Hz, 1000/500-V transformer of Example 5 (P_{OC} = 100 W and P_{SC} = 400 W) supplies a 0.5 leading pf load. Predict the voltage regulation and the full-load efficiency.

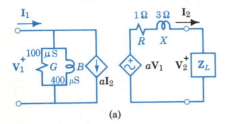

(a)

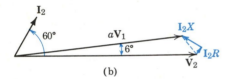

(b)

Figure 21.21 Voltage regulation and efficiency.

The model of Fig. 21.17 is repeated in Fig. 21.21. The internal impedance of the transformer is

$$\mathbf{Z} = R + jX = 1 + j3 = 3.16\underline{/+71.6°}\ \Omega$$

The rating is for *output* at 500 V and 10 kVA; therefore, V_2 = 500 V and I_2 = 10,000/500 = 20 A.

Under full load at 0.5 leading pf, the voltage drop across the internal impedance of the transformer is

$$\mathbf{V}_Z = \mathbf{I}_2\mathbf{Z} = 20\underline{/60°} \times 3.16\underline{/71.6°} = 63.2\underline{/131.6°}\ V$$

The phasor diagram (Fig. 21.21b) is drawn with $\mathbf{V}_2$ as a reference; it indicates that the no-load voltage $a\mathbf{V}_1$ is

$$\mathbf{V}_2 + \mathbf{I}_2\mathbf{Z} = (500 + j0) + (-42 + j47) \cong 460\underline{/6°}\ V$$

The voltage regulation (Eq. 21-31) is

$$\text{VR} = \frac{aV_1 - V_2}{V_2} = \frac{460 - 500}{500} = -0.08 \text{ or } -8\%$$

The negative VR means a rise in output voltage due to leading pf current in the leakage reactance X.

To determine efficiency, we note that the iron loss is given by P_{OC} = 100 W and the copper loss is $I_2^2 R$ = $(20)^2(1)$ = 400 W. The power output is $V_2 I_2 \cos\theta$ = 500 × 20 × 0.5 = 5000 W. By Eq. 21-30,

$$\text{Efficiency} = \frac{5000}{5000 + 100 + 400} \cong 0.91 \text{ or } 91\%$$

Even at this poor pf the efficiency is quite high.

SUMMARY

- With ac excitation of a coil, the magnetic flux in the core is just that required to induce a voltage approximately equal to the applied voltage.
  For a linear magnetic circuit, the induced voltage is

$$v = N\frac{d\phi}{dt} = \frac{d\lambda}{dt} = L\frac{di}{dt}$$

For a flux $\phi = \Phi_m \sin 2\pi ft$, $V(\text{rms})$ = $4.44Nf\Phi_m$.

- The energy density (energy per unit volume) in a magnetic field is

$$W_V = \int_0^B H\ dB \cong \frac{1}{2}\frac{B^2}{\mu} \cong \frac{1}{2}\mu H^2$$

In many problems all the energy stored can be assumed to be in the air gap.

■ In ferromagnetic materials subjected to alternating mmfs, the iron losses are due to hysteresis and eddy currents, and

$$P_i = P_h + P_e = K_h f B_m^n + K_e f^2 B_m^2$$

■ Transformers convert ac power from one voltage, current, or impedance level to another level. In an ideal transformer

$$\frac{V_2}{V_1} = \frac{N_2}{N_1} = a \qquad \frac{I_2}{I_1} = \frac{N_1}{N_2} \qquad \frac{Z_2}{Z_1} = \left(\frac{N_2}{N_1}\right)^2$$

In a practical transformer these relations are modified because of winding resistance, leakage reactance, and exciting current.

The exciting current includes magnetizing and core loss components.

■ Transformer performance can be predicted on the basis of a linear circuit model of appropriate complexity and precision.

Open- and short-circuit tests determine parameter values.

Distributed capacitance is important at high frequencies.

■ Transformers can operate at rated voltage and current continuously.

$$\text{Efficiency} = \frac{\text{output}}{\text{output} + \text{iron loss} + \text{copper loss}}$$

$$\text{Regulation} = \frac{\text{no-load value} - \text{full-load value}}{\text{full-load value}}$$

REVIEW QUESTIONS

1. Why were iron wires used in the Déri transformers?
2. What are the advantages of a closed core over an open core?
3. What determines the magnetic flux in a core excited by a direct voltage? An alternating voltage?
4. What is the effect on inductance of doubling the number of turns on a given core?
5. Describe an experimental method for measuring the energy stored in a given magnetic core.
6. Under what circumstances can all the energy stored in a magnetic circuit be assumed to be in the air gap?
7. What are eddy currents? How are they minimized?
8. Sketch a core with two windings, one connected by means of switch S_1 to a voltage source V_1 and the other connected by means of switch S_2 to a resistance R_2.
   (a) Explain the electric and magnetic behavior when S_1 is closed.
   (b) Explain the behavior when S_2 is then closed.

9. List four functions of a transformer.
10. How is induced emf different from terminal voltage?
11. What assumptions are made in deriving the model in Fig. 21.14b?
12. Compare the hybrid parameter models of the transistor and the transformer.
13. Can a 240:120-V transformer be used to step up 120 V to 240 V? To step up 240 V to 480 V? Why?
14. What is the distinction between exciting and magnetizing currents?
15. Outline the procedure and draw the circuit connections in the experimental determination of circuit model parameters.
16. Draw from memory three circuit models of increasing precision.
17. What is meant by "rated kVA" of a transformer? Why not "rated kW"?
18. Define the speed regulation of an electric motor driving a saw.

EXERCISES

1. In Example 6 of Chapter 20, the 3-A current is provided from a battery through a switch. If the switch is suddenly opened and the flux is reduced to zero in 10 μs, predict the voltage across the switch.

2. A current i in the N-turn coil of Fig. 21.22 establishes magnetic fluxes ϕ_1 and ϕ_2 in time t. Predict the average voltage V appearing across terminals ab.

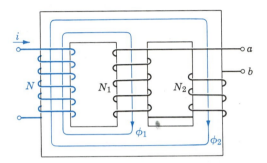

Figure 21.22

3. Compute the inductances of the toroidal coil of Example 2 in Chapter 20.

4. An inductor consists of N turns wound on a magnetic circuit of reluctance $\mathcal{R}$. Derive an expression for inductance L.

5. An N-turn inductance coil is wound on a high-permeability core of cross section A and length l_i with an air gap length l_a.
   (a) Stating any assumptions, derive an approximate expression for the inductance.
   (b) For $N = 800$ t, $A = 1$ cm^2, $l_i = 10$ cm, $l_a = 1$ mm, and $I = 10$ mA, predict the inductance.
   (c) If the core of part (b) is cast steel and $I = 2$ A, check your assumptions and predict the inductance.

6. An inductance coil for 120-V, 60-Hz operation is to be wound on an Armco iron core 4 cm × 3 cm in cross section and 30 cm in mean length. The maximum allowable flux density in the core ($\mu_r \cong 4000$) is 1.2 T.
   (a) How many turns are required?
   (b) What exciting current flows?

7. (a) Repeat Exercise 6(b) for the case of a 3-mm air gap in the core with the same N.
   (b) Compare the results for the two cases and draw a conclusion.

8. Sketch hysteresis loops that would be "ideal" for a transformer and for a computer memory element.

9. The hysteresis loss is 200 W and the eddy-current loss is 200 W when an iron-core inductance coil is operated at 120 V and 60 Hz. Estimate the losses for operation at: (a) 150 V and 60 Hz, (b) 120 V and 50 Hz, (c) 150 V and 72 Hz.

10. Redesign the coil of Exercise 9 (i.e., specify the necessary change in the number of turns) to reduce the total core loss to 300 W at 120 V and 60 Hz.

11. For the conditions of Example 3 on p. 595, predict the ratio I'/I. Comment.

12. Locate a transformer of any type or size and examine it carefully. Draw a sketch with approximate dimensions and list the purpose, manufacturer, approximate weight, and voltage and current specifications if available.

13. A coil of N turns is wound on a magnetic core of negligible reluctance. A tap (connection) is made N_2 turns from the end to form an *auto-transformer*, and a resistance R is connected across the N_2 turns. A voltage V (rms) is applied across the entire winding of N turns.
    (a) Sketch the circuit and predict the current I_R in the resistance and the input current I.
    (b) For $N = 1.25 N_2$, predict I and the current in the N_2 turns. Explain the operation.

14. The input to a power supply transformer (300 primary turns) is to be 120 V at 60 Hz. One secondary provides 120 mA at 500 V and another provides 2 A at 6 V (at unity pf).
    (a) Estimate the number of turns required on the secondaries.
    (b) Estimate the full-load primary current.

15. The core in Fig. 20.23 is made of silicon sheet steel with an air gap 2 mm long; $a = 4$ cm, $c = 5$ cm, and $b = d = 20$ cm. The 300-turn coil has a resistance of 2 Ω. Estimate the current for:
    (a) A steady magnetic flux of 2.5 mWb across the air gap.
    (b) An applied voltage of 120 V dc.
    (c) An applied voltage of 120 V at 60 Hz.
    (d) An applied voltage of 120 V at 40 Hz.

16. In a "clamp-on" ammeter, the core (similar to Fig. 20.23) is hinged so that it can be opened at the air gap and closed around any conductor car-

rying a current. The coil ($N = 20$ turns) supplies current to an ammeter. Sketch the connection and estimate the current flowing in the encircled conductor when the ammeter reads 1.5 A. Explain your reasoning.

17. A 500-W resistive load at 125 V is to be supplied from a 500-V, 60-Hz source using a transformer wound on an Armco iron core 10 cm² in cross section and 35 cm long.
    (a) If the maximum allowable flux density is 1.2 T, approximately how many primary and secondary turns are required?
    (b) Approximately what is the magnetizing current?
    (c) Approximately what are the full-load primary and secondary currents?

18. In Fig. 21.23, $v = V_m \cos 2000t$ V, $R_s = 100\,\Omega$, $L = 10$ mH, $C = 1\ \mu$F, and $R = 25\ \Omega$. Design (i.e., specify the turn ratios $N_2/N_1 = a$ and b for) the transformers to provide maximum power transfer to the load R.

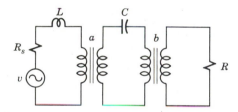

Figure 21.23

19. A 250:1000-V, 60-Hz transformer is tested. Open-circuit data are: $V_1 = 250$ V, $I_1 = 0.51$ A, $W_1 = 31.25$ W, $V_2 = 1000$ V. Short-circuit data are: $V_1 = 7.9$ V, $I_1 = 20$ A, $W_1 = 50$ W. Draw and determine the values for a hybrid parameter model.

20. The transformer of Exercise 19 is turned around and tested with power input to the 1000-V windings.
    (a) Predict the instrument readings for open- and short-circuit tests.
    (b) Draw and determine values for the hybrid parameter model of the transformer in this configuration.

21. The high-voltage winding of the transformer in Exercise 19 is connected to a 200-Ω resistance. Determine:
    (a) The actual primary voltage required to develop 1000 V across the resistance.
    (b) The actual primary current and show all voltages and currents on a phasor diagram.

22. Repeat Exercise 21 for a load impedance $\mathbf{Z}_L = 120 + j160\ \Omega$.

23. A transformer is rated at 100 kVA, 4000:1000 V, 60 Hz. The equivalent resistance and reactance referred to the primary are 3 Ω and 8 Ω, respectively. The exciting circuit parameters referred to the primary are $G = 0.1$ mS and $B = -0.4$ mS.
    (a) Draw and label an appropriate circuit model and predict the actual primary voltage required to supply $V_L = 1000$ V to a load $Z_L = 8 + j6\ \Omega$.
    (b) Determine the parameter values for the circuit model of Fig. 21.15.
    (c) Estimate the parameter values for the circuit model of Fig. 21.19a.

24. A transformer is labeled: 40 kVA, 4000:400 V, 60 Hz. The no-load primary current is 1 A. With the secondary short-circuited, the input to the primary is 1500 W at 250 V and 10 A.
    (a) Draw an appropriate circuit model, *stating* any necessary assumption.
    (b) Predict the primary current when a load $\mathbf{Z}_L = 3.85 - j0.2\ \Omega$ is connected across the 400-V secondary.
    (c) Repeat part (b) using the ideal transformer model and draw a conclusion.

25. For a certain 50-kVA, 400:2000-V, 60-Hz transformer, the model parameters (Fig. 21.19) are: $R_1 = 0.02\ \Omega$, $X_1 = 0.06\ \Omega$, $G = 2$ mS, $B = -6$ mS, $R_2 = 0.5\ \Omega$, and $X_2 = 1.5\ \Omega$.
    (a) Simplify the circuit model by neglecting the core loss, neglecting the voltage drop across $R_1 + jX_1$ due to the exciting current, and referring all quantities to the primary.
    (b) Using the simplified model, predict the no-load current and the primary current with a load $\mathbf{Z}_L = 100 + j75\ \Omega$.
    (c) For $\mathbf{Z}_L = 100 + j75\ \Omega$, predict the efficiency $= P_L/P_I$.

26. For the transformer of Exercise 25:
    (a) Simplify the circuit model as in part (a) by referring all quantities to the *secondary*.
    (b) Repeat parts (b) and (c) for $\mathbf{Z}_L = 59 - j48\ \Omega$.

27. A transformer is to be used to couple a 10-Ω loudspeaker to a transistor amplifier with a mid-frequency output resistance of 4000 Ω. Specify the turn ratio and express the power gain (in decibels) obtained by using the transformer instead of connecting the loudspeaker directly across the amplifier.

28. Four 20-W, 16-Ω loudspeakers connected in parallel are to be supplied by an amplifier with a 500-Ω output resistance. Specify the turn ratio of the output transformer and the output current of the amplifier.

29. Represent the transformer of Exercise 27 at low, moderate, and high frequencies by appropriately simplified circuit models.

30. The nameplate of a transformer with four separate coils reads: 10 kVA, 60 Hz, 460/230:230/ 115 V.
    (a) Interpret this nameplate and sketch the connection to provide a step down from 460 to 115 V.
    (b) What are the current ratings of the individual coils?

31. At a hydroelectric power plant the main transformer is a bank of three 66,667-kVA, 18,000: 220,000-V units connected in Δ which steps up the generated voltage for transmission at 220 kV. Sketch the transformer connection and calculate the current ratings of the primary and secondary windings.

32. A 5000:500-V, 100-kVA, 60-Hz transformer has a no-load input of 2 kW. $R_1 = 5\ \Omega, X_1 = 10\ \Omega$, $R_2 = 0.05\ \Omega$, and $X_2 = 0.1\ \Omega$.
    (a) Draw an appropriate circuit model.
    (b) If the transformer delivers 100 A at 500 V to a 0.8 leading power factor load, predict the efficiency.
    (c) For a 100-kVA, 0.4 lagging pf load, predict the efficiency.
    (d) Predict the voltage regulation for full load at 0.6 lagging pf.
    (e) Predict the voltage regulation for full load at 0.6 leading pf.

33. A 20-kVA, 4000:200-V transformer has the following parameters: $G = 30\ \mu S, B = -50\ \mu S$, $R = 0.06\ \Omega$, and $X = 0.08\ \Omega$ (Fig. 21.15).
    (a) Predict the efficiency at full load.
    (b) Repeat for a 0.4-pf lagging load drawing rated current.

34. For the transformer and load of Exercise 23(a), predict:
    (a) The efficiency.
    (b) The voltage regulation.

35. For a 300-kVA, 12,000:4000-V, 60-Hz transformer, open-circuit test readings are: 4000 V,

1.1 A, 2000 W, and short-circuit test readings are: 300 V, 25 A, and 2800 W.
    (a) On which "side" of the transformer were the O-C and S-C tests made?
    (b) Predict the efficiency for a resistance load of 300 kW.
    (c) Predict the efficiency for a 300-kVA load at 0.6 pf lagging.
    (d) Outline the procedure you would follow to predict the voltage regulation under the condition of part (c).
    (e) Carry out the calculations of part (d).

36. For the transformer of Exercise 33:
    (a) Draw a large-scale phasor diagram for a unity-pf load ("full load").
    (b) Predict the voltage regulation.

37. For the transformer of Example 7 on p.608, repeat the calculations of voltage regulation and efficiency for a 0.5 lagging pf.

38. A 100-kVA distribution transformer has a maximum efficiency of 96% at 70% of rated load. It operates at 70% load 8 hr/day and at 35% load 8 hr/day and at no load 8 hr/day. If electrical energy is 1¢/kWh and annual fixed charges (interest, depreciation, etc.) are 20% of purchase price, about how much extra cost would be justified for a similar 98% efficient transformer?

39. The transformer of Exercise 19 is rated at 5 kVA. Predict the voltage regulation and efficiency under full load at unity pf.

40. A transformer is labeled: 20 kVA, 120:240 V, 60 Hz. It is assumed to be well designed with optimum use of wire size and core size. (*Note:* Any practical insulation will stand 500 V.) You are asked if it can operate satisfactorily under the following conditions:

	Frequency (Hz)	Primary (V)	Secondary (V)
(a)	60	120	240
(b)	60	60	120
(c)	60	240	480
(d)	120	120	240
(e)	120	240	480
(f)	30	120	240
(g)	30	60	120

PROBLEMS

1. A low-resistance 50-turn coil is wound on a silicon sheet steel core with cross section of 20 cm² and mean length 50 cm. Determine the coil voltage for: (a) A current $i = 1 \cos 400t$ A, and (b) A current $i = 3 + 1 \cos 400t$ A, and compare the results. (c) Sketch a graph of inductive reactance versus dc current for this "saturable core reactor."

2. Predict the voltage ratio V_2/V_1 for the unusual transformer in Fig. 21.24 where $N_2 = N_1$.

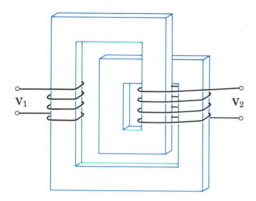

Figure 21.24

3. Design a transformer to operate on regular house current and supply 2.4 kW to a 60-Ω resistor. Available is a laminated silicon steel core similar to Fig. 21.8a; the cross section is 30 cm² and the mean length is 60 cm. State all assumptions.

(a) Estimate the maximum allowable flux density without an excessive value of H.
(b) Specify the number of turns for the primary.
(c) Estimate the rms value of the no-load primary current.
(d) Specify the number of turns on the secondary.
(e) Predict the full-load primary current.
(f) What size wire should be used? (Consult a handbook.)

4. Show that for symmetrically arranged coils and equal current density in primary and secondary windings, $R_2 = a^2 R_1$ in Fig. 21.19.

5. (a) Prove that in a transformer maximum efficiency occurs when the variable losses are equal to the fixed losses.
   (b) A transformer supplies 400 kVA to a resistance load with a core loss of 7.5 kW and a copper loss of 4.8 kW. Predict the maximum efficiency and the power output at which it occurs.

6. For most economic operation, the transformers of Exercise 31 are to reach their maximum efficiency of 98% at three-quarters of full load at unity pf. Specify the critical design parameters (Fig. 21.19) to meet this requirement.

7. Design (i.e., specify the parameters of) a 50-kVA, 4000:250-V, 60-Hz transformer to have a full-load efficiency of 98% and a voltage regulation of approximately 2%.

8. For the transformer of Exercise 35, specify the resistance load for which the efficiency will be maximum and predict this maximum efficiency.

22

Principles of Electromechanics

Translational Transducers

Rotational Transducers

Moving-Iron Transducers

In electromechanical energy conversion the coupling may be through the medium of an electric field or a magnetic field. Practical devices employ a variety of physical phenomena. The $q\mathcal{E}$ force on a charge in an electric field is used in a cathode-ray tube for electron acceleration, in a condenser microphone for acoustoelectric conversion, and in proposed space craft for propulsion. The *piezoelectric* effect, whereby certain crystalline materials distort under the action of electric fields, is used in crystal pickups and microphones. The *electrostriction* effect, whereby dielectric materials change dimension in the presence of electric fields, is used in vibration instruments. The analogous deformation of ferromagnetic materials, called *magnetostriction*, is used in supersonic underwater signaling. At the present time, these phenomena are limited in application to rather special situations.

In contrast, the generation of an electromotive force in a moving conductor in a magnetic field and the reverse process of development of a mechanical force by sending a current through a similar conductor find very wide application in instrumentation, automation, and power generation. These electro*magneto*mechanical phenomena are emphasized in this book, but the techniques developed for their analysis are useful in analyzing the behavior of devices employing other phenomena as well.

The purpose of this chapter is to present basic principles and to show their application to elementary forms of practical devices. Familiar techniques are used to derive linear models for devices employing translation and rotation. New methods are introduced for analyzing the behavior of iron elements moving in magnetic fields. The emphasis here is on principles and techniques useful in subsequent chapters for predicting the behavior of practical machines.

TRANSLATIONAL TRANSDUCERS

A device for changing energy from electrical to mechanical form or vice versa is called an electromechanical *transducer*. Usually such a transducer can operate as a *motor* converting electrical energy to mechanical, or as a *generator* converting mechanical energy to electrical. Let us consider how a conductor moving in a magnetic field performs these functions.

Electromotive and Mechanical Forces

In Fig. 22.1 a conductor of length l is moving at velocity u normal to a magnetic field of density B directed into the paper. Each charged particle in the conductor experiences a force

$$\mathbf{f}_q = q(\mathbf{u} \times \mathbf{B}) \tag{22-1}$$

The force on a positive charge q is in the direction of advance of a right-hand screw if $\mathbf{u}$ is imagined to rotate into $\mathbf{B}$. By this rule, positive charges are forced upward and

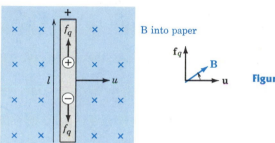

Figure 22.1 Moving-conductor transducer.

negative charges downward, and the net result (whether this is a metal or a semiconductor) is a difference of potential across the conductor. What is the magnitude of the emf generated?

The magnitude of the emf e can be determined by noting that charge separation continues until the force of the electric field produced is just equal to the force of the magnetic field causing it. Under equilibrium, the electric force per unit charge (or voltage gradient) is just equal to the magnetic force per unit charge or

$$\frac{\text{Electric force}}{\text{Charge}} = \mathcal{E} = \frac{e}{l} = \frac{\text{magnetic force}}{\text{charge}} = \frac{quB}{q} = Bu \tag{22-2}$$

For this case where $\mathbf{u}$, $\mathbf{l}$, and $\mathbf{B}$ are mutually perpendicular, the emf is

$$e = Blu \tag{22-3}$$

Since lu is the area "swept" per unit time, one interpretation of Eq. 22-3 is that the emf is just equal to the rate at which magnetic flux is being "cut" by the conductor. In general, the emf, a scalar, is given by the vector expression

$$e = \mathbf{B} \cdot (\mathbf{l} \times \mathbf{u}) = Blu \cos \alpha \sin \beta \tag{22-4}$$

where α is the angle from **B** to the normal to the plane containing **l** and **u**, and β is the angle from **l** to **u**. In most engineering devices the elements are arranged in the optimum orientation with $\alpha = 0°$ and $\beta = 90°$, and Eq. 22-3 applies.

If a path is provided for current flow (Fig. 22.2), there is a component of charge velocity u' in the direction of positive current and the charges experience a force f_d

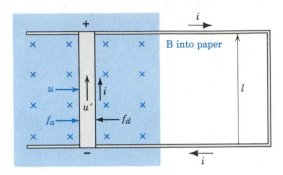

Figure 22.2   A conductor moving on rails.

that is transmitted to the atoms in the lattice of the conductor and therefore to the conductor itself. Since $qu' = il$ (see Eq. 20-4), the *developed force* is

$$\mathbf{f}_d = i(\mathbf{l} \times \mathbf{B}) \tag{22-5}$$

directed to the left. In general, $f_d = Bli \sin \gamma$ where γ (gamma) is the angle from **l** to **B**. For **l** perpendicular to **B**, the usual case, $\gamma = 90°$ and the magnitude of the developed force is

$$f_d = Bli \tag{22-6}$$

To *cause* motion in the direction of u, there must be an *applied* force f_a as shown, equal and opposite to the developed force f_d.

Note that the current i produces a magnetic field and this must be considered in some practical machines; here we assume that it is negligibly small compared to B. Note also that the conductor of length l moving at velocity u sweeps out an area $dA = lu\,dt$ in time dt. For uniform B, the rate of change of flux linked by the one-turn closed path in Fig. 22.2 is

$$\frac{d\lambda}{dt} = \frac{d\phi}{dt} = B\frac{dA}{dt} = Blu = e \tag{22-7}$$

As Faraday would have predicted, in this configuration the emf is just equal to the rate of change of flux linkage.

Bilateral Energy Conversion

With a path provided for current flow the moving conductor is a source of electrical energy, and

$$\text{Electrical power} = P_E = ei = Blu\,i \tag{22-8}$$

The necessary mechanical input is positive since the applied force is in the direction of the velocity, or

$$\text{Mechanical power} = P_M = f_a u = Bli\, u \tag{22-9}$$

As expected from the principle of energy conservation, the mechanical power in newton-meters per second is identically equal to the electrical power in watts.

The generated emf can be used to charge a battery as shown in Fig. 22.3. In this case

$$i = \frac{e - V}{R} \tag{22-10}$$

For $e = Blu > V$, current i is positive and the transducer is a *generator*. At a lower velocity, $e < V$, current i is negative, and the developed force is reversed. The conductor is forced to the right, the mechanical power input is negative, and the transducer is acting as a *motor*. The moving conductor transducer is said to be *bilateral*; the energy flow may proceed in either direction.

The Transducer as a Two-Port

Here, just as in the transformer, the magnetic field provides the coupling between input and output ports. Using familiar techniques, we should be able to derive a circuit model for this two-port device. The fact that mechanical force is dependent on a variable in the electrical portion of the system and electromotive force is dependent on a variable in the mechanical portion suggests the use of controlled sources. One possibility is shown in Fig. 22.3c.

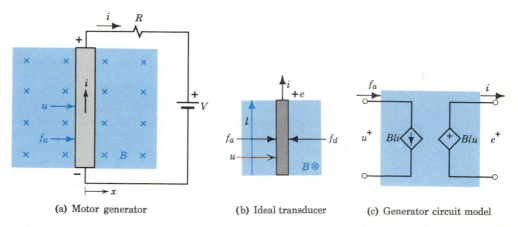

(a) Motor generator   (b) Ideal transducer   (c) Generator circuit model

Figure 22.3 A translational transducer.

The ideal transducer is sometimes called an "electromechanical ideal transformer." In this representation, force is analogous to current and velocity is analogous to voltage (see Table 5-1 and Fig. 21.19b). The term Bl corresponds to the turn ratio of the transformer. The polarity of the *force source Bli* is consistent with the reference direction of i in the *generator* representation shown.

A real transducer can be represented by accounting for the mass M of the conductor and the coefficient of sliding friction D. In a *motor*, the developed force

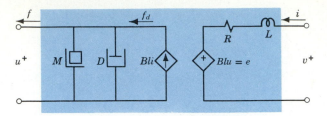

Figure 22.4 Model of a real transducer (motor).

must overcome the inertia force and the friction force; the governing equation for the mechanical side would be

$$f + M\frac{du}{dt} + Du = f_d = Bli \tag{22-11}$$

Part of the electrical input is lost in resistance or stored in inductance. Considering the terminal voltage v, the resistance R and inductance L of the conductor, the governing equation for the electrical side would be

$$v - L\frac{di}{dt} - Ri = e = Blu \tag{22-12}$$

What circuit configuration would satisfy this set of equations?

The symbols shown in Fig. 22.4 and Table 22-1 are conventional for representing M and D in mechanical circuits. Application of d'Alembert's principle (p. 38) to the parallel mechanical circuit yields Eq. 22-11; application of Kirchhoff's voltage law to the series electrical circuit yields Eq. 22-12. We conclude that this circuit model represents the electromechanical system described by the equations. This model is analogous to the hybrid parameter model proposed for a real transformer (Fig. 21.15). (What corresponds to the core loss? To the exciting current?) Note that the ideal transducer represents the *reversible* portion of the energy conversion occurring in a real transducer.

Table 22-1 Conventional Symbols for Mechanical Circuits

Element	Resistance (mechanical)	Mass	Compliance	Velocity source	Force source
Unit	$\dfrac{\text{Newton-second}}{\text{meter}}$	Kilogram	$\dfrac{\text{Meter}}{\text{Newton}}$	$\dfrac{\text{Meter}}{\text{Second}}$	Newton
Symbol				u_s	f_s
Characteristic	$f = Du$	$f = M\dfrac{du}{dt}$	$f = \dfrac{1}{K}\int u\, dt$	$u = u_s$	$f = f_s$

The Dynamic Transducer

The *moving-coil* or *dynamic* transducer is a practical form of the translational transducer. It is used in such devices as loudspeakers, disk recording heads, and phonograph pickups. The essential elements of a dynamic loudspeaker are shown in Fig. 22.5. The permanent magnet establishes an intense radial field in the annular air

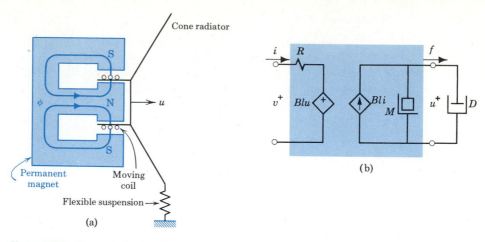

Figure 22.5 Dynamic loudspeaker construction and a simplified circuit model.

gap (shown in section). The moving coil consists of a few turns of fine wire wound on a light form and supported in the air gap. The coil form is attached to a light, stiff cone supported by a flexible suspension that keeps the cone and coil in place without restricting the axial motion.

An alternating current in the so-called *voice* coil develops a force that is transmitted to the cone. The moving cone produces air pressure waves and acoustic energy is radiated. In the simplified circuit model of Fig. 22.5b, the inductance of the moving coil and the effect of the cone suspension have been neglected. The elements M and D represent inertia effects and the loading effect of the air. At low frequencies the cone moves as a rigid piston, and the electrical energy is efficiently converted into sound waves if the cone is sufficiently large. At higher frequencies the effective mass of the cone presents a large mechanical admittance and the velocity is reduced.[†] In some high-fidelity systems the low-frequency signals are reproduced by a speaker with a large cone, the *woofer*, and the high-frequency signals are reproduced by the smaller *tweeter*. A *cross-over* network separates the sound into two bands; the signals below 1000 Hz, say, are filtered out by a frequency-selective network and sent to the woofer.

EXAMPLE 1

In a dynamic loudspeaker, a magnetic flux density of 0.6 T exists across the annular air gap of mean radius 1 cm, length 2 mm, and depth 1 cm (in the direction of u). A maximum force of 0.3 N is to be developed by a current $i = 0.2 \cos \omega t$ A.

Estimate the mmf required to establish the design flux density, and specify the number of turns on the voice coil.

Assuming all the reluctance is in the air gap, by Eq. 20-15,

$$\mathcal{F} = \frac{1}{\mu_o} B l_a = \frac{0.6 \times 2 \times 10^{-3}}{4\pi \times 10^{-7}} = 955 \text{ A·t}$$

By Eq. 22-6, for current $I_m = 0.2$ A in an N-turn coil, the developed force is $f_d = NBlI_m = NB(2\pi r)I_m$ or

$$N = \frac{f_d}{2\pi r B I_m} = \frac{0.3}{2\pi \times 0.01 \times 0.6 \times 0.2} = 39.8 \cong 40 \text{ turns}$$

[†]As the frequency increases, the cone radius becomes a larger fraction of a wavelength; therefore, the energy radiated remains high until the cone "breaks up" and no longer moves as a piston.

The Dynamic Pickup

The dynamic loudspeaker mechanism can also be used for a microphone. In fact, in some intercommunication systems the same device serves as microphone and speaker. Sound waves striking the cone cause motion of the coil in the magnetic field. In a well-designed microphone, the emf produced is an accurate replica of the sound signal for frequencies from 40 to 10,000 Hz.

In the dynamic phonograph pickup (Fig. 22.6), the vertical movement of the needle is transmitted to a tiny coil supported in the field of a permanent magnet. The small voltage generated is then amplified in a *preamplifier* that raises the signal above the *hum* level and also corrects the frequency distortion inherent in this type of pickup. The hum is due to voltages induced in the cable from pickup to amplifier by changing magnetic fields set up around transformers and other ac components. The frequency distortion is due to the fact that this type of pickup is *velocity sensitive*, as illustrated in Example 2.

EXAMPLE 2

A dynamic phonograph pickup (Fig. 22.6) consists of a 20-turn coil (length of each turn = 1 cm) moving normal to a field of $B = 0.2$ T. If the maximum allowable recording amplitude is 0.02 mm, predict the output voltage at 1000 Hz and at 100 Hz.

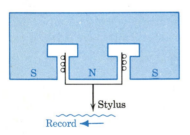

Figure 22.6   A dynamic phonograph pickup for vertical recordings.

The effective length of conductor in the moving coil is $l = $ 20 turns × 1 cm/turn = 20 cm = 0.2 m. For a sinusoidal displacement of amplitude 2×10^{-5} m,

$$x(t) = A \sin \omega t = 2 \times 10^{-5} \sin \omega t$$

and the velocity at 1000 Hz is dx/dt or

$$u = A\omega \cos \omega t = 2 \times 10^{-5} \times 2\pi \times 10^3 \cos \omega t \ \text{m/s}$$

Neglecting internal impedance, the output voltage is

$$v = e = Blu = 0.2 \times 0.2 \times 4\pi \times 10^{-2} \cos \omega t \ \text{V}$$

The rms value of output voltage at 1000 Hz is

$$V = \frac{0.2 \times 0.2 \times 4\pi \times 10^{-2}}{\sqrt{2}} \cong 0.0035 \ \text{V} = 3.5 \ \text{mV}$$

At 100 Hz the output voltage for the same recording amplitude is only $\frac{1}{10}$ as great. Since the amplitude is limited by the distance between grooves, some frequency compensation in the preamplifier is desirable.

ROTATIONAL TRANSDUCERS

In practice, *translational* transducers are limited to applications involving vibratory motion with relatively low displacement. To develop appreciable voltages, velocities must be high and this means relatively high frequencies. In contrast, an electromechanical transducer employing *rotation* of conductors in a magnetic field can produce large low-frequency or direct voltages (or torques) in easily constructed configurations.

The d'Arsonval Mechanism

The mechanism of the ordinary dc ammeter was invented more than one-hundred years ago, but it is still one of our best current-measuring devices. The d'Arsonval movement is widely used in instrumentation, and it is a good illustration of a simple rotational transducer.

As shown in Fig. 22.7, a moving coil is wound on a rectangular form that is supported on jeweled bearings in the air gap between the poles of a permanent magnet

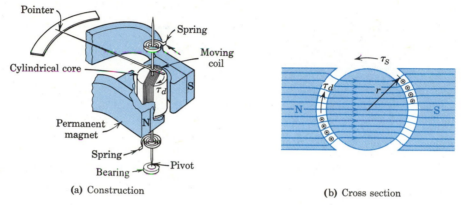

(a) Construction   (b) Cross section

Figure 22.7   The d'Arsonval meter movement.

and a stationary iron core. Current flowing in the coil produces forces that tend to rotate the coil assembly. The current is conducted to and from the coil by two fine springs that also supply a restoring torque to oppose rotation of the coil. A lightweight pointer attached to the assembly indicates the angular deflection on a calibrated scale.

Analysis of the behavior of the mechanism is simplified if we stipulate that rotation is restricted, by means of mechanical stops, to the region of uniform radial magnetic field. If there are N turns or $2N$ conductors of length l normal to the magnetic field at a radius r, the torque τ_d (tau) developed by a current i is

$$\tau_d = f_d \cdot r = B(2Nl)i \cdot r = 2NBlri = k_M i \qquad (22\text{-}13)$$

The combined effect of the two spiral springs is a restoring torque τ_S that is proportional to angular deflection, or

$$\tau_S = \frac{\theta}{K_R} \qquad (22\text{-}14)$$

where K_R is the rotational compliance in radians (or degrees) per newton-meter. Under equilibrium conditions, the restoring torque is just equal to the developed torque, and

$$\theta = 2NBlrK_R \cdot i = k_\theta i \qquad (22\text{-}15)$$

The deflection is directly proportional to the current in the moving coil, and the meter scale is linear.

EXAMPLE 3

A d'Arsonval ammeter (Fig. 22.7) has a rectangular coil form with length $l = 1.25$ cm and width $w = 2r = 2$ cm. Each spiral spring has a compliance $K_R = 16 \times 10^7$ °/N·m. A uniform flux density of 0.4 T is supplied by the permanent magnet. Determine the number of turns so that full-scale deflection of 100° is obtained with a current of 1 mA.

Noting that two springs are only half as compliant as one and then solving Eq. 22-15,

$$N = \frac{\theta}{2Blri(K_R/2)}$$

$$= \frac{100}{2 \times 0.4 \times 0.0125 \times 0.01 \times 0.001 \times 8 \times 10^7}$$

$$= 12.5 \text{ turns}$$

(Is it easy to construct a coil with an extra half turn?)

Meter Applications

The d'Arsonval moving-coil mechanism is a versatile device for displaying electrical currents or other physical variables that are directly related to currents.

The Ammeter. Current in a branch is measured by an ammeter connected in series with the branch. The movement in Example 3 provides full-scale deflection with a current of 1 mA. How could it be used in a 0- to 5-A ammeter? In the connection of Fig. 22.8a, some of the meter current is *shunted* through R_S and only a part of the

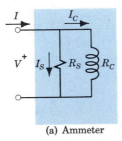

(a) Ammeter

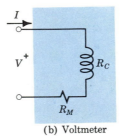

(b) Voltmeter

Figure 22.8   DC meter connections.

current to be measured flows through the moving coil. If the coil resistance R_C is 20 Ω, then full-scale deflection requires a voltage of

$$V = I_C R_C = 0.001 \times 20 = 0.020 \text{ V}$$

For full-scale deflection with a meter current of 5 A, the current in the *shunt* R_S is $5 - 0.001 = 4.999$ A. The resistance of the shunt should be

$$R_S = \frac{V}{I_S} = \frac{V}{I - I_C} = \frac{0.02}{4.999} \cong 0.004 \text{ Ω} \qquad (22\text{-}16)$$

A sturdy and precisely adjusted resistance may be mounted inside the meter case or be available for external connection. A circuit for a multirange ammeter is shown in Fig. 22.9. Four shunts are available to give four ammeter ranges; a make-before-break selector switch selects the appropriate range.

The Voltmeter. Voltage across a branch is measured by a voltmeter in parallel with the branch. Could the movement of Example 3 be used in a voltmeter? The d'Arsonval mechanism is essentially a current-measuring instrument, but it can be used as a voltmeter by adding a series resistance R_M as in Fig. 22.8b. For a 0- to 300-V voltmeter, full scale deflection is obtained with 1 mA in the moving coil and a voltage drop of 0.02 V; the remainder of the voltage must appear across the *multiplier* resistance. Therefore,

$$R_M = \frac{V_M}{I_C} = \frac{V - V_C}{I_C} = \frac{300 - 0.02}{0.001} \cong 300,000 \ \Omega \qquad (22\text{-}17)$$

A circuit for a multirange voltmeter is shown in Fig. 22.9b. Four multiplier resistances

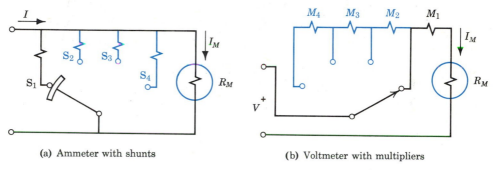

(a) Ammeter with shunts

(b) Voltmeter with multipliers

Figure 22.9 Multirange instrument connections.

are available to provide four voltmeter ranges; their values can be calculated from the equations

$$\frac{V_1}{I_M} = R_M + M_1 \qquad \frac{V_2}{I_M} = R_M + M_1 + M_2 \qquad (22\text{-}18)$$

and so forth.

The Ohmmeter. A useful application of the d'Arsonval movement is in measuring the dc resistance of passive circuits. A simple ohmmeter is shown in Fig. 22.10. The battery (usually 1.5 V), a fixed resistance R_1, and a variable "Ohms Adjust" resistor R_A are mounted within the ohmmeter case. With the test prods shorted, R_A is varied until the voltmeter reads full scale (usually 1 V). When the test prods are across an unknown resistance R_X, the voltage read is the voltage across R_1. Assuming R_M is very large compared to R_1,

$$V_M = \frac{R_1}{R_1 + R_X} \times \text{full scale} \qquad (22\text{-}19)$$

Full-scale deflection corresponds to 0 resistance; half-scale deflection is obtained when $R_X = R_1$. Very high resistance values are crowded into the lower part of the scale. Since

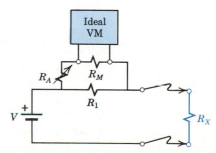

Figure 22.10 A simple ohmmeter.

the voltmeter reading changes most rapidly with a given percentage change in resistance when $R_X = R_1$, midscale readings are most precise. A selector switch is used to change scales by selecting values of R_1; typical ranges are for midscale readings of 10 Ω, 1 kΩ, and 100 kΩ. Practical ohmmeters use a more complicated circuit but the general principle of resistance measurement is the same.

The Multimeter. Since the same d'Arsonval movement can serve for current, voltage, and resistance measurements, an effective arrangement is to include shunts, multipliers, and battery in the same case along with the necessary selector switch. Such a combination volt-ohm-milliammeter (VOM) is called a *multimeter*. In addition to dc quantities, ac voltages can be measured by using a rectifier circuit.

Rectifier Instruments. Semiconductor diodes can be used to convert alternating current to direct current, which is measured by a d'Arsonval movement. In the full-wave bridge circuit (Fig. 22.11), an alternating current at the terminals produces

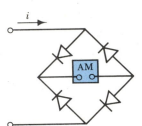

Figure 22.11 A rectifier-type instrument.

a unidirectional current through the moving coil and the deflection is proportional to the average current. These instruments are calibrated on sinusoidal waves to read rms values. Since the rms value of the sinusoid is $0.707/0.636 = 1.11$ times the average value, the instrument indicates 1.11 times the average value of the rectified waveform of any applied current. It is accurate for sinusoids only; for any other waveform a correction must be applied. (See Exercise 15.)

Meter Response

In addition to providing a convenient method of solution, a model should give insight into the general behavior of a device. Let us see what we can learn from the general model of a d'Arsonval movement. Neglecting the small inductance of the moving coil and the friction in the jeweled bearings, the model is as shown in Fig. 22.12a. For rotation, the appropriate mechanical variables are torque τ and angular velocity ω. The element J represents the rotational inertia of the coil assembly and K_R is the combined spring compliance. The mechanical proportionality constant $k_M = \tau_d/i$ is defined by Eq. 22-13. The electrical proportionality constant k_E is defined as

$$k_E = \frac{e}{\omega} = \frac{2NBlu}{\omega} = 2NBlr = k_M \qquad (22\text{-}20)$$

since $u = r\omega$; we see that in this electromechanical transducer also there is a simple "turn ratio."

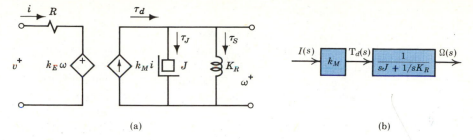

Figure 22.12 Circuit model of a d'Arsonval movement.

Suddenly connecting the instrument in a line corresponds to applying a step current to the circuit model. What is the dynamic response of the meter? By using the turn-ratio concept, the rotational admittances on the mechanical side could be referred to the electrical side. In Fig. 21.18, transformer parameters were referred to the primary; here turn ratio a is equal to $k_E = k_M = 2NBlr$. An alternative approach is to use the transfer-function concept of Chapter 9. For exponentials, the transfer functions for the "electromechanical transformer" and the mechanism are

$$\tau_d = 2NBlr\; i = k_M i \qquad \text{or} \qquad \frac{\mathrm{T}_d(s)}{I(s)} = k_M \tag{22-21}$$

$$\tau_d = J\frac{d\omega}{dt} + \frac{1}{K_R}\int \omega\, dt \qquad \text{or} \qquad \frac{\mathrm{T}_d(s)}{\Omega(s)} = sJ + \frac{1}{sK_R}$$

Therefore, the transfer function of the movement is

$$\frac{\Omega(s)}{I(s)} = \frac{\mathrm{T}_d(s)}{I(s)} \cdot \frac{\Omega(s)}{\mathrm{T}_d(s)} = \frac{k_M}{sJ + 1/sK_R} = \frac{sk_M K_R}{s^2 J K_R + 1} \tag{22-22}$$

Because the transfer function indicates purely imaginary poles, we expect an undamped sinusoidal response.

If the only mechanical resistance is that due to bearing friction, the needle will indeed oscillate widely. How could the long wait for a steady reading be reduced? There are two commonly used methods for introducing the necessary damping. One is to mount a large but light vane on the coil assembly and introduce air-friction damping. Another is to make the coil form of a conducting material, such as aluminum; the coil form then becomes a shorted turn, and coil motion induces an emf that results in energy dissipation. (How would this effect be represented in the circuit model of Fig. 22.12? How is the admittance function altered? See Exercise 16.)

In a properly designed meter movement with optimum damping, the needle quickly reaches a steady reading. In the steady state, $\omega = 0$, $e = k_E\omega = 0$, and $\tau_J = J\, d\omega/dt = 0$. The model then indicates that

$$\theta = \int \omega\, dt = K_R \tau_S = K_R \tau_d = K_R k_M \cdot i \tag{22-23}$$

which agrees with Eq. 22-15. The dynamic response of electromechanical transducers is treated in more detail in Chapter 25.

The Elementary Dynamo

Rotation of the coil in a d'Arsonval mechanism is limited to something less than 180°. If we are to realize the advantages of this configuration, we must permit continuous rotation. One possibility is to mount a rectangular coil on a shaft and connect the ends of the coil to conducting *slip rings* attached to the shaft but insulated from it as in the elementary *dynamo* of Fig. 22.13. As the N-turn coil rotates at angular velocity ω in a uniform magnetic field B, the generated emf e is connected to the external circuit by fixed *brushes* sliding on the rotating slip rings.

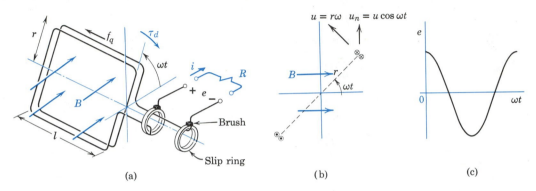

(a)   (b)   (c)

Figure 22.13   Elementary dynamo construction and operation.

Electromotive Force. The polarity of the generated emf is determined by considering the force on positive charges in the moving conductors. The right-hand rule indicates that the force f_q on positive charges in the upper conductors in Fig. 22.13b is into the paper; therefore, the front slip ring is negative at the instant shown. (What is the direction of the force in the lower conductors?) The magnitude of the emf can be determined by any of the three expressions employed so far. For $2N$ conductors of length l moving at velocity u in a magnetic field of density B, the emf (Eq. 22-4) is

$$e = 2NBlu \cos \alpha \sin \beta$$

Where $\beta = 90°$ and $2rl = $ area A,

$$e = 2NBlr\omega \cos \omega t = NBA\omega \cos \omega t \tag{22-24}$$

The rate at which flux is being cut by $2N$ conductors (Eq. 22-3) is

$$e = 2NBlu_n = 2NBlr\omega \cos \omega t = NBA\omega \cos \omega t \tag{22-25}$$

The rate of change of flux linkage (Eq. 22-7) is

$$e = \frac{d\lambda}{dt} = NB\frac{d(A \sin \omega t)}{dt} = NBA\omega \cos \omega t \tag{22-26}$$

Equation 22-26 shows that, in a uniform magnetic field, the emf is determined by the area A of the coil and not the particular shape or dimensions. The same emf or torque is developed by a circular coil, say, of the same area. Equations 22-13 and 20 were derived for rectangular coils in a uniform field; a more general formulation is

$$k_M = \frac{\tau_d}{i} = k_E = \frac{e}{\omega} = NBA \qquad (22\text{-}27)$$

Torque. In the d'Arsonval movement there is a uniform radial field and Eq. 22-27 indicates that

$$e = k_E \omega = NBA\omega \qquad (22\text{-}28)$$

Comparing Eqs. 22-24 and 27 and reasoning by analogy with Eq. 22-13, we anticipate that in the elementary dynamo

$$\tau_d = k_M i \cos \omega t = NBAi \cos \omega t \qquad (22\text{-}29)$$

The direction of this torque depends on the direction of the current (Example 4). If the

---

EXAMPLE 4

The elementary dynamo of Fig. 22.13 is driven at constant angular velocity by a mechanical source. The output terminals are connected to a load resistance R. Determine the electrical power supplied and the mechanical power required.

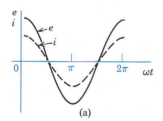

(a)

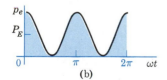

(b)

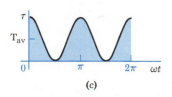

(c)

Figure 22.14 Power and torque relations in an elementary dynamo.

Assuming that the resistance and inductive reactance of the coil are small compared to R, all the emf appears across R and the current is

$$i = \frac{e}{R} = \frac{NBA\omega \cos \omega t}{R} = I_m \cos \omega t$$

The average electrical power is

$$P_E = \left(\frac{I_M}{\sqrt{2}}\right)^2 R = \frac{(NBA\omega)^2}{2R}$$

Assuming that the friction torque is negligibly small, the applied torque must be equal and opposite to the developed torque; its magnitude (see Fig. 22.14c) is

$$\tau_a = NBA \cos \omega t \cdot i = NBA \cos \omega t \frac{NBA\omega \cos \omega t}{R}$$

The average mechanical power is the product of the average torque times the angular velocity of rotation or

$$P_M = \mathrm{T}_{av}\omega = \frac{1}{2}\frac{\omega(NBA)^2}{R}\omega = \frac{(NBA\omega)^2}{2R} = P_E$$

Under these idealizing assumptions, all the mechanical power is converted to electrical power.

coil is rotating in the direction shown in Fig. 22.13a and if a resistance R is connected across the terminals, the current will flow as shown and the right-hand rule indicates a clockwise torque. This is a *developed* torque that must be overcome by the mechanical power source. If an electrical power source is used, however, this torque is available for doing mechanical work. Example 4 demonstrates that the ideal elementary dynamo is another bilateral electromechanical energy converter.

A General Torque Equation

As preparation for the analysis of a variety of rotating machines we need a general expression for torque. A single, simple relation that would apply to ac or dc machines employing electro- or permanent magnets in various configurations would be very convenient. Let us take another look at the elementary dynamo and try to abstract the essential idea.

As shown in Fig. 22.15b, a current i_2 in the upper conductor of length l in a field of uniform magnetic flux density B_1 develops a force

$$\mathbf{f} = i_2(\mathbf{l} \times \mathbf{B}_1)$$

and the torque developed by an N-turn coil of width $2r$ and area $2rl$ is

$$\tau_d = 2NB_1 l i_2 \cdot r \sin \delta = B_1 A \cdot N i_2 \cdot \sin \delta$$

$$= k_a B_1 i_2 \sin \delta \tag{22-30a}$$

where δ (delta) is the angle between $\mathbf{B}_1$ and the normal to the plane of the coil. But what is $N i_2$? This is the mmf of the coil, and it establishes a flux density $\mathbf{B}_2$ in the direction shown. An alternative form of the torque expression is

$$\tau_d = k_b B_1 B_2 \sin \delta \tag{22-30b}$$

In words, the torque can be considered as due to the *interaction of an electric current and a magnetic field* or to the *interaction of two fields*. But B_1 itself is just a man-

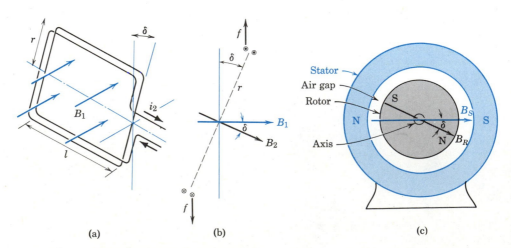

(a)   (b)   (c)

Figure 22.15 Torque on a coil in a magnetic field.

ifestation of a current i_1, either in a coil or in the atoms of a magnet, so the torque can also be considered as due to the *interaction of two currents*. A third formulation of the basic relation is

$$\tau_d = k_c i_1 i_2 \sin \delta \qquad\qquad (22\text{-}30c)$$

The angle δ is called the *torque angle* or *power angle*. The torque increases with $\sin \delta$ and reaches a maximum at $\delta = 90°$. Figure 22.15c symbolizes the general rotating machine. The stationary portion or *stator* sets up the field B_S and the rotating portion or *rotor* sets up the field B_R. Either or both of these fields may be established by electric currents in a practical machine. For a given current (or quantity of copper) and a given flux (or quantity of iron), the machine designer realizes the maximum torque if he arranges for δ to be $90°$. For continuous torque, $\sin \delta$ must not change sign. These are critical considerations and dictate the mechanical arrangement and electrical connection of most practical rotating machines.

Rotating Machines

A rotating electromechanical converter has several important advantages. Torques and velocities can be constant instead of pulsating, and high velocities are possible without the high accelerating and decelerating forces inherent in translational devices. High velocities permit high voltages and, therefore, higher electrical power per unit weight. Because frictional losses increase rapidly with velocity, the bearings are placed on shafts of small radius (and low linear velocity) and the active conductors are at a much larger radius.

AC Machines. A sinusoidal emf is generated when the coil of an elementary dynamo is turned at constant angular velocity in a uniform magnetic field. A more effective arrangement of copper and iron is shown in Fig. 22.16. Direct current is supplied to the field windings on the rotor through brushes and slip rings. The small air gap minimizes the magnetizing current required for a given flux density. The ac emf ($e = d\lambda/dt$) is generated in conductors connected in series and placed in slots in the stator. Having the high-power conductors stationary minimizes the problems of

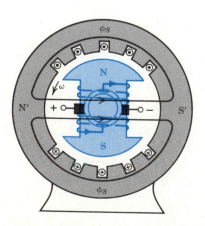

Figure 22.16 An alternator.

insulation, centrifugal force, and electrical connection. By distributing the turns along the periphery of the stator and shaping the air gap properly, the output voltage can be made to approximate a sinusoid. Most of the world's electrical power is generated in more elaborate versions of this basic alternator.

Under what conditions can the alternator function as a motor? If the field is energized to establish B_R and a direct current is supplied to the stator to establish B_S, the rotor will turn until $\delta = 0$ and then stop. When the fields are aligned, $\sin \delta = 0$ and there is no torque; this is a position of equilibrium. But motoring action is possible if we can upset the equilibrium in just the right way. Assume that at a particular instant the current is out of the upper conductors and into the lower conductors in Fig. 22.16. This creates a stator field with poles N' and S' as shown and $\delta = 90°$. The developed torque tends to turn the rotor clockwise (in the $-\omega$ direction). Because of inertia the rotor will rotate past the position of $\delta = 0$. If at that instant the current in the stator is reversed, the new torque is also clockwise and the rotor continues to turn. The stator current can be reversed by a mechanical switch, or an alternating current of just the right frequency can be used. The latter arrangement is the basis of the *synchronous motor* treated in more detail in Chapter 24.

DC Machines. To obtain unidirectional torque from direct currents, we use a rotating mechanical switch called a *split-ring commutator*. The same device permits the generation of direct currents in a rotating machine. In dc machines, the *field* winding is stationary and the assembly of conductors and commutators, called the *armature*, rotates. The operation of the commutator is indicated in Fig. 22.17. At the

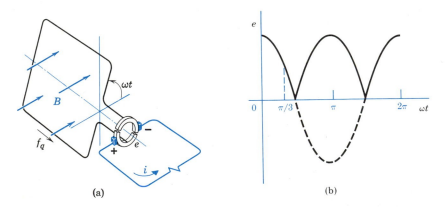

Figure 22.17 Commutator operation and generated voltage waveform.

instant shown, $\omega t = \pi/3$ and the coil emf is decreasing. At $\omega t = \pi/2$, the emf is zero and the coil is shorted by the brushes. An instant later the lower conductor begins to move up across the flux and the generated emf is reversed, but the lower conductor has automatically been switched to the negative brush and the terminal polarity is unchanged. The commutator could be called "a synchronous full-wave rectifier."

In the more practical form of Fig. 22.18a, the armature conductors are embedded in a cylindrical iron rotor. The stationary field windings establish a high and nearly uniform flux density in the small air gap. The emf generated in a single conductor (number 1) is shown in Fig. 22.18b; instant ωt_a corresponds to the left-hand figure with

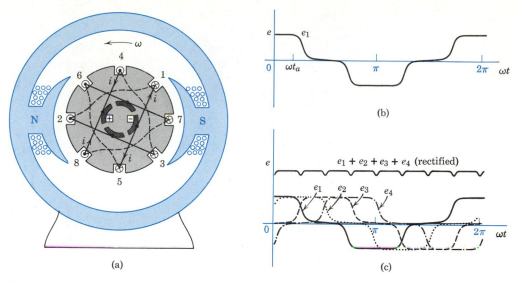

Figure 22.18 Two-pole dc generator construction and voltage waveforms.

conductor 1 just leaving the S pole. Starting at the negative brush (drawn on the inside of the commutator but actually riding on the outside) and going around the circuit, we see that conductors 1, 2, 3, and 4 are connected in series. The sum of the individual emfs appears at the positive brush. Going in the other direction from the negative brush, we see that the other four conductors are connected in series. In this winding there are two *parallel paths*. As the armature rotates, carrying the segmented commutator with it, conductor 1 is switched out of one path and conductor 8 takes its place.

In a practical machine there are many conductors and many commutator segments. At any instant many coils are connected in series between brushes and a few coils are short-circuited (by the brushes) as they are being switched from one circuit to another. As each coil is switched, there is a momentary drop in terminal voltage, resulting in what is called commutator *ripple*, but this is small compared to the total effect of many coils in series.

As indicated in Fig. 22.18a, the current is into the paper in all conductors moving upward, and out of the paper in all conductors moving downward. The net result of all the currents is a flux density distribution that can be represented by a vector directed upward. As the armature rotates, this flux density *remains fixed in space* at the optimum torque angle $\delta = 90°$. The resulting developed torque is clockwise and, in a generator, must be overcome by the mechanical power source. If direct current from an electrical source is introduced at the brushes, a constant unidirectional torque is developed and the machine functions as a motor.

The two-pole commutator machine resembles the d'Arsonval mechanism in many ways. As would be expected from Eq. 22-20, for constant flux density in the air gap

$$K_M = \frac{T}{I} = K_E = \frac{E}{\Omega} \tag{22-31}$$

where the capital letters indicate direct or average values of torque T, armature current I, emf E, and angular velocity Ω. If total flux per pole Φ is a variable (by controlling

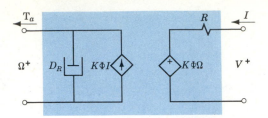

Figure 22.19 Steady-state model of a commutator machine (motor).

field current and, thereby, flux density), the expressions for torque and emf are

$$T = K\Phi I \quad \text{and} \quad E = K\Phi\Omega \quad (22\text{-}32)$$

The steady-state model of a commutator machine is shown in Fig. 22.19. In the steady state, energy storage terms are not significant and only the rotational mechanical friction D_R and electrical resistance R are considered. The value of K depends on the machine construction and the electrical connection, as illustrated in Example 5.

EXAMPLE 5

In a dc machine (Fig. 22.20), the armature is wound on a laminated iron cylinder 15 cm long and 15 cm in diameter. The N and S pole faces are 15 cm long (into the paper) and 10 cm along the circumference. The average flux density in the air gap under the pole faces is 1 T. If there are 80 conductors in series between the brushes and the machine is turning at $n = 1500$ rpm, predict the no-load terminal voltage.

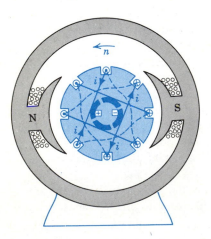

Figure 22.20 Voltage calculation.

The dc voltage can be predicted in any of the three ways expressed in Eqs. 22-24, 25, and 26. In this case the rate of flux cutting is a convenient approach. For N_s conductors (*not* coils) in series, the average emf is

$$E = N_s \frac{\Delta\Phi}{\Delta t} \frac{\text{Wb}}{\text{s}} = N_s\Phi \frac{\text{Wb}}{\text{pole}} p \frac{\text{poles}}{\text{rev}} \frac{n}{60} \frac{\text{rev}}{\text{s}} = \frac{N_s\Phi pn}{60} \text{ V}$$

$$(22\text{-}33)$$

Here the flux per pole is

$$\Phi = BA = 1 \times 0.15 \times 0.1 = 0.015 \text{ Wb}$$

Therefore,

$$E = 80 \times 0.015 \times 2 \times \frac{1500}{60} = 60 \text{ V}$$

Equation 22-33 can be derived directly from Eq. 22-25 by taking into account the number of poles and recognizing that the dc voltage is just the average of a rectified sinusoid. The value of K in Eq. 22-32 can be derived from Eq. 22-33 by noting that $\Omega = 2\pi n/60$. Solving,

$$K = \frac{E}{\Phi\Omega} = \frac{N_s p}{2\pi} \quad (22\text{-}34)$$

The number of conductors in series is conveniently determined from $N_s = Z/p'$ where Z is the total number of conductors and p' is the number of parallel paths. In one common type of winding, $p' = p$.

MOVING-IRON TRANSDUCERS

Most of the devices described so far employ a moving conductor in a magnetic field. The elementary alternator of Fig. 22.16 differs in that the iron rotor, which could be a permanent magnet, moves past the stationary conductors, but this is essentially just a convenient method of obtaining relative motion of the conductors with respect to the field. Another type of electromagnetomechanical energy converter employs an initially unmagnetized iron member moving in the direction of a magnetic field. Such *moving-iron* transducers cannot be analyzed conveniently in terms of forces on currents; a different approach is required.

In the simple door chime of Fig. 22.21a, part of an iron core or plunger is within the solenoid. When a current i flows in the solenoid, the plunger is accelerated upward

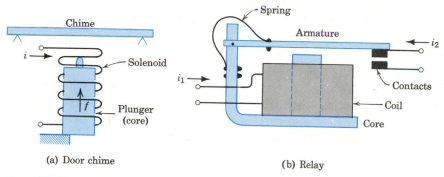

(a) Door chime

(b) Relay

Figure 22.21 Common examples of moving-iron transducers.

and strikes the chime; to design such a chime we need to know the force on the core as a function of position. A relay is a control device whereby the current in circuit 2 can be turned on or off by a current in circuit 1. In Fig. 22.21b, a current i_1 energizes the coil and moves the armature against the restraining force of the spring. The movable contact meets the fixed contact and circuit 2 is closed. Mechanical stops (not shown) limit the armature travel. Relay specifications include the *pickup* or closing current and the *dropout* or opening current. We shall learn how these currents can be predicted.

The *iron-vane* mechanism of Fig. 22.22a is another form of the moving-iron transducer. Current in a multiturn coil magnetizes two soft-iron elements. Since the polarities are the same and like poles repel, the movable vane rotates with respect to the fixed vane. The developed torque, opposed by a spring, is always in the same direction; therefore, this is a good mechanism for measuring alternating currents. In a *reluctance pickup* (Fig. 22.22b), a vibratory motion of the small iron armature changes the air gap and, therefore, the reluctance of the magnetic circuit. The flux established by the permanent magnet changes and a voltage is induced in the pickup coil. We need to know whether the generated emf is proportional to the displacement, velocity, or acceleration of the armature.

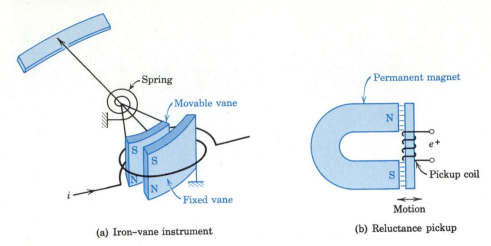

(a) Iron–vane instrument   (b) Reluctance pickup

Figure 22.22   Other moving-iron transducers.

Virtual Work Method

One approach to the quantitative analysis of an electromechanical transducer employs the principle of *virtual work*. We are interested in determining a force; therefore, we assume that a small displacement takes place, calculate the work that would be done in such a *virtual* displacement, and then determine the actual force.

If the plunger of Fig. 22.21a were to move an infinitesimal vertical distance dx, the work done on the plunger would be $f\,dx$. Where does this energy come from? It must come from the electrical input. Does all the electrical input appear as mechanical work? No, some of it appears as heat from electrical resistance and mechanical friction, and some of it may be stored in the magnetic field. Let us minimize i^2R by using high-conductivity wire and minimize mechanical friction by proper design. Then practically all the electrical input appears as either mechanical work or as an increase in energy stored in the magnetic field.

In a linear magnetic system, the energy stored is $\frac{1}{2}Li^2$; if the current is constant at a value I, the change in energy stored is $\frac{1}{2}I^2\,dL$. (Does the inductance change as the core moves? Yes, because the reluctance changes.) The electrical energy input is

$$dw = p\,dt = ei\,dt = (N\,d\Phi/dt)I\,dt = I\,d(N\Phi)$$

where $d(N\Phi)$ is the change in flux linkage associated with a virtual displacement dx. But $N\Phi = LI$ (Eq. 21-6) if I is constant and $dw = I\,d(N\Phi) = I^2\,dL$. Expressing the idea in the previous paragraph, we write

$$\text{Electrical input} = \text{work done} + \text{increase in energy stored} \qquad (22\text{-}35)$$

or

$$I^2\,dL = f\,dx + \tfrac{1}{2}I^2\,dL$$

Solving,

$$f = \frac{I^2\,dL - \tfrac{1}{2}I^2\,dL}{dx} = \tfrac{1}{2}I^2\frac{dL}{dx} \qquad (22\text{-}36)$$

This is a relation of great simplicity and, it turns out, of considerable generality.

The force is directly proportional to the rate of change of inductance and is in the direction associated with an increase in inductance (since I^2 is always positive).

Reasoning by analogy, what result would you anticipate for the torque in a rotational transducer? A similar analysis yields

$$\tau = \tfrac{1}{2}I^2\frac{dL}{d\theta} \tag{22-37}$$

Another basic conclusion is that in such a linear magnetic system, half the electrical energy input is converted to mechanical energy and the other half appears as an increase in the energy stored in the magnetic field.

EXAMPLE 6

A magnetic circuit is completed through a soft-iron rotor as shown in Fig. 22.23. Derive an expression for torque as a function of angular position.

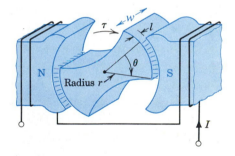

Figure 22.23   Reluctance torque.

To simplify the analysis, let us assume:

1. All the reluctance of the magnetic circuit is in the air gaps of length l.
2. There is no fringing so the effective area of each gap is the area of overlap $= A = r\theta w$.

The total reluctance for two gaps in series is

$$\mathcal{R} = \frac{2l}{\mu_o A} = \frac{2l}{\mu_o r w\theta}$$

For an N-turn coil, the inductance is

$$L = \frac{N\Phi}{I} = \frac{N\mathcal{F}}{I\mathcal{R}} = \frac{N^2 I}{I\mathcal{R}} = \frac{N^2 \mu_o r w\theta}{2l}$$

By Eq. 22-37 the torque is

$$\tau = \tfrac{1}{2}I^2\frac{dL}{d\theta} = \frac{\mu_o N^2 I^2 rw}{4l}$$

or the torque is independent of θ under the assumed conditions. Actually, there is some positive torque for negative values of θ. (Why?) Also note that there is no torque when overlap is complete because $dL/d\theta$ drops to zero.

The *reluctance torque* evaluated in Example 6 is important in practical machines. Small electric-clock motors operate on this principle. Some medium-size synchronous motors are pulled into synchronism by reluctance torque, and in large machines the reluctance effect contributes a significant part of the total torque.

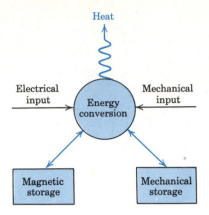

Figure 22.24 An energy conversion system.

Energy Balance Method

Equation 22-35 is an abbreviated *energy balance* based on the law of conservation of energy. In general, the energy added to a system goes to increase the energy stored within the system or appears as an output from the system (Fig. 22.24). Some of the chemical energy of the fuel supplied to a diesel-electric system is stored in the kinetic energy of the flywheel, or as thermal energy evidenced by an increase in temperature, and the remainder appears as useful electrical output or unavailable heat radiated or carried away in the exhaust. In dealing with electromagnetomechanical converters, it is convenient to write the energy balance as

$$\underset{\text{input}}{\text{Mechanical}} + \underset{\text{input}}{\text{electrical}} = \underset{\text{storage}}{\text{mechanical}} + \underset{\text{storage}}{\text{magnetic}} + \text{heat} \qquad (22\text{-}38)$$

This equation applies to all types of converters. In a generator, mechanical input is positive and electrical input is negative. In a d'Arsonval movement, electrical input is positive, mechanical input is zero, and mechanical energy is stored in the springs. During current changes, the kinetic energy of the moving coil may be significant. In a relay, the change in energy stored in the air gap just balances the mechanical output (see Eq. 22-39). Also, in every real device some energy is converted into heat in irreversible processes. Copper loss, hysteresis and eddy-current losses, and bearing and air-friction losses must all be considered. The energy-balance approach is particularly useful in situations in which overall effects are more easily handled than detailed phenomena[†] (See Example 7).

The Electromagnet

A versatile example of the moving-iron transducer is the electromagnet shown in Fig. 22.26. An electric current, either alternating or direct, develops a pull on the

[†]See Chapters 4 and 5 of H.H. Skilling, *Electromechanics*, John Wiley & Sons, New York, 1962. This excellent book is out of print but there may be a copy in your school library.

EXAMPLE 7

A commutator machine is rated at 5 kW, 250 V, 2000 rpm. The armature resistance R_A is 1 Ω. Driven from the electrical end at 2000 rpm, the no-load power input to the armature is $I_A = 1.2$ A at 250 V with the field winding ($R_F = 250$ Ω) excited by $I_F = 1$ A. Predict the efficiency of this machine operating as a 5-kW generator.

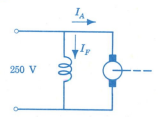

(a) Wiring diagram for no-load test

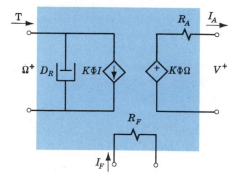

(b) Steady-state generator model

Figure 22.25   Generator efficiency.

In Fig. 22.25, input power $I_F^2 R_F = 1^2 \times 250 = 250$ W is required to provide the necessary magnetic flux. This is power lost and it appears as heat.

In no-load steady-state operation there is no output and no change in energy storage; therefore, the armature input of 1.2 A at 250 V = 300 W is all loss. A small part ($I_A^2 R_A = 1.2^2 \times 1 = 1.44$ W) is copper loss, but most of the input power goes to supply air, bearing, and brush friction and the eddy-current and hysteresis losses associated with flux changes in the rotating armature core. All these losses are dependent on speed, but they are nearly independent of load.

At full-load of 5 kW, $I_A = P/V = 5000/250 = 20$ A and the armature copper loss is

$$I_A^2 R_A = 20^2 \times 1 = 400 \text{ W}$$

With the generator driven from the mechanical end, the rotational losses can be assumed to be equal to the no-load armature input and are represented by D_R in the model. Then the energy balance (Eq. 22-39) is

$$\begin{matrix} \text{Mech.} \\ \text{input} \end{matrix} + \begin{bmatrix} \text{field} \\ \text{input} \end{bmatrix} - \begin{matrix} \text{elec.} \\ \text{output} \end{matrix} = 0 + 0 + \begin{bmatrix} \text{field} \\ \text{loss} \end{bmatrix} + \begin{matrix} \text{arm.} \\ \text{loss} \end{matrix} + \begin{matrix} \text{rotat.} \\ \text{loss} \end{matrix}$$

$$\text{Mech. input} + [250 - 5000] = [250 + 400 + 300]$$

Solving,

$$\text{Mech. input} = 5000 + 400 + 300 = 5700 \text{ W}$$

(Field input is all loss and appears on both sides.)

$$\text{Efficiency} = \frac{\text{output}}{\text{input}} = \frac{\text{elect. output}}{\text{mech. input} + \text{elect. input}}$$

$$= \frac{5000}{5700 + 250} = 0.84 \text{ or } 84\%$$

---

pivoted armature and raises the mass M a distance dl. Motion of the armature may be used to close or open contacts, to open or close valves, or to engage mechanisms. The important design criterion is the force developed for a given current.

Considering the shaded region as an electromechanical system, we see that there is electrical energy input and mechanical energy output, we suspect that there is energy stored, and we know that there is energy lost irreversibly. In terms of the energy-balance equation, the mechanical input for a virtual displacement dl is $-(f\,dl)$ since the work is actually done *by* the system. If during this infinitesimal displacement we

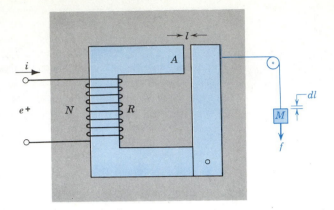

Figure 22.26 An electromagnet with a pivoted armature.

adjust the current so that the magnetic flux remains constant, $N \, d\phi/dt$ is zero and there is no induced emf in the coil. Then the electrical input is just $i^2R \, dt$, where dt is the time required for displacement dl. (How are these conditions different from those imposed in deriving Eq. 22-36?)

If the motion of the armature is horizontal, there is no change in potential energy stored; and if the motion is slow, there is no change in kinetic energy stored. The energy stored in the magnetic field does change. From Eq. 21-9 we know that the energy per unit volume is $\frac{1}{2}B^2/\mu$. If ϕ and B are constant, there is no change in the energy stored in the iron, but the energy stored in the air gap *decreases* by $(\frac{1}{2}B^2/\mu_o)A \, dl$, where A is the area of the gap if fringing is neglected.

The final term in the energy balance is primarily the $i^2R \, dt$ loss in the coil winding. With a frictionless pivot there is no mechanical friction loss, and with no change in ϕ there is no eddy-current or hysteresis loss. (Was this a convenient condition?) Equation 22-38 becomes

$$-f \, dl + i^2R \, dt = 0 - \frac{1}{2}\frac{B^2}{\mu_o}A \, dl + i^2R \, dt \qquad (22\text{-}39)$$

Actually, current i is a variable, but since the same term appears on both sides of the equation its evaluation is unimportant. Solving for the developed force in newtons,

$$f = \frac{B^2A}{2\mu_o} \qquad (22\text{-}40)$$

This result is general and does not require that ϕ (and B) be constant (see Exercise 31). Since the flux density possible in practical materials is limited by saturation, this relation places an upper limit on the force developed by an electromagnet of a given cross section.

To determine the relation between force and current, we assume that the permeability of the iron is so high that all the mmf in the magnetic circuit appears across the reluctance of the air gap. Then the flux density is

$$B = \frac{\phi}{A} = \frac{\mathscr{F}}{\mathscr{R}_a A} = \frac{Ni}{(l/\mu_o A)A} = \frac{\mu_o Ni}{l}$$

and Eq. 22-40 becomes

$$f = \frac{\mu_o N^2 i^2 A}{2l^2}$$

(22-41)

where μ_o = the permeability of free space = $4\pi \times 10^{-7}$ H/m,
N = the number of turns on the coil,
i = the current in amperes,
A = the area of the air gap in meters squared, and
l = the length of the gap in meters.

Note that the force increases as the armature moves and the air gap length decreases. For very small air gaps the assumptions must be reexamined.

EXAMPLE 8

In the relay shown in Fig. 22.27, the contacts are held open by a spring exerting a force of 0.2 N. The gap length is 4 mm when the contacts are open and 1 mm when closed. The coil of 5000 turns is wound on a core 1 cm² in cross section. Predict the pickup and dropout currents for this relay.

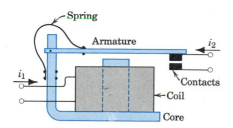

Figure 22.27 Relay performance.

To simplify the solution, we assume:

1. All reluctance is in a uniform air gap.
2. Fringing is negligible so that $A = 10^{-4}$ m².
3. Spring force is constant and acts at a distance equal to half the distance from the pivot to the air gap.
4. Friction and inertia effects are negligible.

For pickup or closing, the gap length is 4×10^{-3} m and the required force is $\frac{1}{2} \times 0.2$ N = 0.1 N. Solving Eq. 22-41 for current,

$$i = \sqrt{\frac{2l^2 f}{\mu_o N^2 A}} = \sqrt{\frac{2 \times 16 \times 10^{-6} \times 0.1}{4\pi \times 10^{-7} \times 25 \times 10^6 \times 10^{-4}}}$$

$$\cong 0.032 \text{ A or } 32 \text{ mA}$$

Since current is directly proportional to gap length, the opening or dropout current is approximately $32/4 = 8$ mA.

To check the adequacy of the core area and the validity of the reluctance assumption, we note that

$$B = \frac{\mu_o N i}{l} = \frac{4\pi \times 10^{-7} \times 5 \times 10^3 \times 32 \times 10^{-3}}{4 \times 10^{-3}}$$

$$= 0.05 \text{ T, a very conservative value}$$

SUMMARY

■ A conductor moving in a magnetic field generates an emf $e = \mathbf{B} \cdot (\mathbf{l} \times \mathbf{u}) = Blu \cos \alpha \sin \beta$

and develops a force $\mathbf{f} = i(\mathbf{l} \times \mathbf{B})$ where $f = Bli \sin \gamma$

The translational transducer is a bilateral energy converter that can be represented by a two-port model with $P_E = ei = P_M = fu = Bliu$

The dynamic transducer used in loudspeakers and pickups consists of a coil in translation in a transverse magnetic field.

■ The d'Arsonval mechanism widely used in instrumentation consists of a coil in rotation in a radial magnetic field; the coil generates an emf $e = NBA\omega$ and develops a torque $\tau = NBAi$.

■ A coil rotating at constant speed in a uniform magnetic field generates an emf $e = NBA\omega \cos \omega t$ and develops a torque $\tau = NBAi \cos \omega t$.
The emf can be evaluated in terms of the velocity of moving conductors, the rate at which flux is cut, or the rate of change of flux linkage.
The torque can be evaluated in terms of the interaction of: two electric currents, two magnetic fields, or a current and a field.

■ A useful form of the general torque equation is $\tau = kB_R B_S \sin \delta$.

■ A practical alternator, with direct current supplied to the rotor, generates a sinusoidal emf in the stator.
The alternator can operate as a motor if the rotor turns at synchronous speed and alternating current is supplied to the stator.

■ The commutator permits the generation of direct current and the production of unidirectional torque; as the rotor turns, conductors are automatically switched to the proper circuit.

$$\frac{T}{I} = \frac{E}{\Omega} = K\Phi = \frac{N_s p}{2\pi}\phi$$

■ The method of virtual work and the energy-balance concept are particularly useful in analyzing moving-iron transducers; in general,

$$f = \tfrac{1}{2}I^2\frac{dL}{dx} \quad \text{and} \quad \tau = \tfrac{1}{2}I^2\frac{dL}{d\theta}$$

■ The force developed by an electromagnet is approximately

$$f = \frac{B^2 A}{2\mu_o} = \frac{\mu_o N^2 i^2 A}{2l^2}$$

REVIEW QUESTIONS

1. List four basically different reversible electromechanical phenomena.
2. On a transducer, terminal A is positive with respect to terminal B. If an ammeter indicates current is flowing *out* of terminal B, is this transducer acting as a motor or as a generator? Explain.
3. Explain how, in a moving current-carrying conductor in a magnetic field, positive charges are forced in one direction and the conductor in another.
4. Outline the steps in deriving an expression for emf generated in a moving conductor.
5. In Fig. 22.4, what corresponds to the turn ratio in a transformer?
6. Draw and label a mechanical circuit model to represent a device governed by the equation $Bli = M\, du/dt + Du + \int u\, dt/K$.

7. An advertisement says: "The superior tone quality is due to the highly compliant suspension and massive magnet structure of the speaker." Analyze this statement. What does it mean? Are you impressed?
8. A loudspeaker has two coils; one consists of a few turns in an air gap and the other consists of many turns on an iron core. What function is served by each?
9. What frequency-response characteristic is desirable in the preamplifier following a dynamic phonograph pickup?
10. Why is the current to a d'Arsonval instrument carried by the spiral springs? What is the maximum possible deflection?
11. Write an expression for sensitivity (°/A) for a d'Arsonval mechanism.

12. How can the same instrument measure voltage and current?

13. Explain the concept of a "turn ratio" in a rotating coil transducer.

14. How is the circuit of Fig. 22.12b altered when the movement is damped?

15. Write three different expressions for the emf generated in a coil moving with respect to a permanent magnet.

16. Write an expression for torque in an ideal dynamo that does not include any magnetic factors. Explain the energy basis.

17. What is the torque angle? How does it affect machine design?

18. What advantages do rotational devices have over translational devices?

19. What condition is necessary if a machine with direct current in the rotor and alternating current in the stator is to develop useful torque?

20. Draw a sketch of an actual commutator (automobile generator or starter motor) showing segments, insulation, and brushes.

21. Explain the operation of a commutator in a dc motor.

22. Are there eddy-current or hysteresis losses in a dc generator? Why?

23. Given a bar of soft iron in a magnetic field, is the torque in such a direction that it will increase or decrease the stored magnetic energy?

24. What is the meaning of "virtual"? What is "virtual work?"

25. Explain the operation of a reluctance clock motor.

26. Make an energy balance for a dc generator that is accelerating. Clearly identify each term.

27. How would the energy balance (Eq. 22-39) differ for current held constant?

28. How is the lifting force of an electromagnet affected by a slight reduction in pole-face area?

EXERCISES

1. Derive Eq. 22-2 by equating the work done *by* an external agency to the work done *on* the moving charges.

2. In Fig. 22.28, a long, straight conductor carries a current I in air.
   (a) Express the magnetic flux density and the magnetic field intensity at point A and indicate their directions.
   (b) Express the force on an electron b moving with velocity U as shown and indicate its direction.
   (c) Repeat for element c of length dl carrying current I_2.
   (d) Repeat for element d.
   (e) Repeat for a similar element e carrying current into the paper.

3. Assume that a magnetic field of density $B = 0.1$ T is emerging vertically from your desk surface. Holding a pencil 15 cm in length inclined at an angle of 30° from the desk, you move your hand laterally at a speed of 60 cm/s to the right. Predict the magnitude and polarity of the voltage generated in the pencil lead.

4. In Fig. 22.29, a conducting bar moves with constant speed u (so that its position is $x = ut$) in a region of linearly increasing magnetic flux density into the paper. A resistance R is connected between the high conductivity rails. Determine:
   (a) $i(t)$, the current in the bar.
   (b) $f(t)$, the force on the bar.

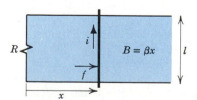

Figure 22.29

5. In the transducer of Fig. 22.2, the resistance of the rails is negligible compared to the resistance R of the conductor of mass M which slides with

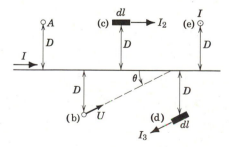

Figure 22.28

friction coefficient D on the rails. The conductor is initially at rest and a constant force f_a is applied at time $t = 0$.
(a) Draw an appropriate circuit model.
(b) Determine the steady-state velocity U.

6. An electromechanical two-port (Fig. 22.30) consists of a bar of resistance R and mass M restrained by a spring of compliance K in a field of flux density B. Write the governing equations for the electrical and mechanical sides, and devise an appropriate circuit model.

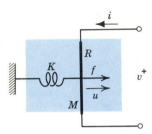

Figure 22.30

7. An electric rail "car" of mass M is driven by a conducting axle of length l and friction coefficient D moving normal to a component B of the Earth's magnetic field. From the electrified rails, the axle receives current I.
(a) Sketch the system and draw an appropriate circuit model.
(b) Use your model to derive an expression for velocity $u(t)$ if the brake is suddenly released, and evaluate the final velocity U_f.
(c) If $B = 10^{-4}$ T, $M = 1000$ kg, $l = 2$ m and $D = 0.05$ N·s/m, calculate the required axle current I for a peak velocity of 36 km/h. Comment on the feasibility of this scheme.

8. In a dynamic loudspeaker similar to Fig. 22.5a, a magnetic flux density of 0.6 T is available across the annular air gap of radius 1 cm, length 2 mm, and depth 1 cm (in the direction of u). A maximum force of 0.8 oz is to be developed by a current $i = 0.2 \cos \omega t$ A. Estimate the required mmf to provide the design flux density, and specify the number of turns on the voice coil.

9. A dynamic phonograph pickup is similar to the speaker in Fig. 22.5. The lateral displacement of the needle is transmitted directly to the 20-turn coil that vibrates in a field of $B = 0.1$ T. Each turn of the coil is 3 cm long.
(a) How many grooves per centimeter are there on a typical 33-rpm record? (What is the playing time?)
(b) If approximately 80% of the space per groove is available for recording, what is the maximum peak-to-peak displacement allowable?
(c) What is the maximum voltage (rms value) generated in this pickup at 100 Hz? At 10,000 Hz?
(d) For a uniform response (output voltage versus frequency), what provisions in regard to amplitude of vibration should be made in recording?

10. An ammeter similar to Fig. 22.7 is to deflect "full-scale" (2 rad) with a current of 1 mA in the square coil, 2 cm on a side. A flux density of 0.4 T is available from a permanent magnet, and the compliance of each spring is 10^5 rad/N·m. Complete the design by specifying the remaining parameter.

11. A d'Arsonval movement provides full-scale deflection with a current of 5 mA or an applied voltage of 30 mV. Specify the shunt and multiplier to provide ranges of 0 to 1 A and 0 to 50 V.

12. Repeat Exercise 11 for ranges of 0 to 5 A and 0 to 300 V.

13. Using a 100-μA movement with a 10,000-Ω internal resistance:
(a) Design a multirange dc milliammeter with ranges of 1, 10, and 100 mA.
(b) Design a multirange voltmeter with ranges of 3, 10, 50, and 150 V.

14. Design an ohmmeter, using the movement of Exercise 13 with a half-scale reading of:
(a) 50 Ω (b) 5000 Ω.

15. A d'Arsonval meter and a rectifier instrument (calibrated in rms values) are used to measure the current in Fig. 22.31.

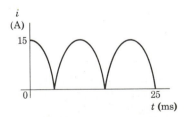

Figure 22.31

(a) Predict the reading on each instrument.
(b) Repeat for the voltage in Fig. 22.32.

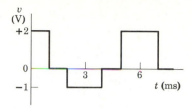

Figure 22.32

16. The circuit model of an undamped meter movement is shown in Fig. 22.12a. One method of introducing damping is to make the coil form of a conducting material such as aluminum.
    (a) How does this introduce the desired damping?
    (b) Redraw Fig. 22.12a to include the effect of the coil form.
    (c) Redraw Fig. 22.12b and derive the transfer function for the parallel portion of the circuit.
    (d) Define the condition for critical damping of the mechanical portion of the movement.

17. An elementary dynamo (Fig. 22.13a) consists of an N-turn coil of length l and radius r meters turning at n rpm in a field of B teslas. A resistance R is connected across the output.
    (a) Express the generated voltage $e(t)$.
    (b) For an applied torque $\tau_a = T_m \cos \omega t$, determine the output current and the mechanical power input.

18. Given the general equation $\tau_d = kB_1 i_2 \sin \delta$, define τ_d and δ. Identify B_1, i_2, and δ for a dc motor and for an ac generator. Evaluate δ.

19. A dc generator similar to Fig. 22.18, but with four stator poles, turns at 2000 rpm. Each pole face (at the air gap) is 10 cm × 12 cm and $B = 1$ T. Between brushes there are 75 turns effectively in series, and each turn has a total length of 42 cm.
    (a) Estimate the emf generated.
    (b) For an output current of 50 A, estimate the torque developed and the mechanical power input.

20. In the dc generator of Exercise 19, the effective air gap is 3 mm and there are 500 turns per field pole (see Fig. 20.14b). Stating any necessary assumptions, estimate the values of k_a and k_c in Equation 22-30.

21. (a) If continuous torque is to be developed in the machine in Fig. 22.18a, what can you say about the field established by the rotor?
    (b) The machine of Fig. 22.16 is to be used as a motor. DC current in the rotor will establish a magnetic field defined by N and S rotating at speed ω. For continuous torque, what can you say about the magnetic field N′–S′ in the stator?
    (c) The stator of the machine in Fig. 22.16 provides a moving field rotating at speed ω. If the rotor is replaced by the rotor in Fig. 22.18a, what can you say about the field in the rotor if continuous torque is to be developed?

22. An alternator similar to Fig. 22.16 but with 6 poles on the rotor is driven at 1200 rpm.
    (a) Sketch the alternator, labeling the poles, and predict the output frequency.
    (b) With no output current, the required torque is 5 N·m; with an output current of 20 A, the required torque is 45 N·m. Neglecting stator resistance, estimate the generated emf.

23. The alternator of Fig. 22.16 generates an emf $e = 120\sqrt{2} \cos 2\pi 60t$ V.
    (a) Calculate the speed in rpm.
    (b) Predict the new emf if the speed is increased by 10%.
    (c) At the original speed, predict the emf if the current in the rotor winding is decreased by 10%.
    (d) If 10 additional slots are cut in the unslotted portion of the stator and another identical winding installed, predict the emf $e_2(t)$ generated in the new winding under the original conditions.

24. In Fig. 22.18b, the emf generated at 1200 rpm by a single conductor has an average value of 4 V.
    (a) Estimate the total flux per pole and the maximum emf.
    (b) Predict the dc output voltage (no load) if each path consists of 60 conductors in series.
    (c) If a load current of 10 A flows, predict the electrical power generated and the counter torque developed.

25. The field winding of a 10-kW, 6-pole, 220-V dc generator produces 10 mWb of magnetic flux per pole. There are 72 armature slots and 2 conductors are placed in each slot. If the coils are connected internally so that there are 2 parallel paths (i.e., half the conductors are connected in series), estimate the speed of rotation for rated voltage at no load.

26. A commutator machine with an effective armature resistance of 1 Ω requires an input current of 2 A at 200 V just to turn the rotor at rated speed

of 2000 rpm (no load). Draw an appropriate circuit model and predict the mechanical power and torque output for an input of 20 A at 200 V.

27. A magnetically soft iron bar (low residual magnetism) is mounted on an axis (into the paper) between the poles of a magnet (Fig. 22.33).
    (a) As the bar is rotated through 360°, how does the magnetic polarity of the bar change?
    (b) At what positions is the torque zero? Maximum?
    (c) What position does the bar take if left to swing freely?
    (d) At what position is the total energy stored in the system a minimum?

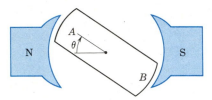

Figure 22.33

28. In the ac ammeter of Fig. 3.12, the total inductance with all coils aligned is 10 mH and, with the rotating coil L_R reversed, 6 mH. Assuming that $L = L_S + L_R + 2M \cos \theta$ (where $\theta = 0°$ when the coils are aligned), estimate the torque developed by a current of 2 A for $\theta = -90°$.

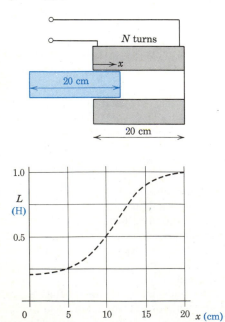

Figure 22.34

29. Figure 22.34 shows the variation of inductance for a solenoid and plunger.
    (a) Specify the required current for a maximum pull of 20 N.
    (b) Plot a curve of pull vs. plunger position.
30. In the reluctance pickup of Fig. 22.22b, the permanent magnet provides a total mmf $\mathcal{F}$ across the two air gaps of area A and length l. With the N-turn coil open-circuited and the armature moving at velocity u, derive expressions for:
    (a) The induced emf e.
    (b) The necessary applied force f_a.
31. In an electromagnet (Fig. 22.26), the current i is held constant. Stating any necessary assumptions, write the energy-balance equation for a small motion dl, and derive Eq. 22-41.
32. (a) Starting with Eq. 22-40 and carefully explaining your reasoning as you go along, derive Eq. 22-36.
    (b) Check Eq. 22-41 dimensionally.
33. A magnetic contactor used to close circuits carrying heavy currents is similar to Fig. 22.26 with contacts mounted on the pivoted armature. The cross section of the core is 8 cm² and the length is 25 cm. Specify the current required in a 200-turn coil to develop a force of 20 N across a 1-cm gap.
34. With a current of 5 A in the actuating coil of the contactor of Exercise 33, estimate the force required to open the contacts when the air gap is zero.
35. The dimensions of the electromagnet in Fig. 22.35 are in cm; the depth of the silicon steel core and armature is 5 cm.
    (a) Design a coil (specify N and i) to be placed on the center leg to provide a total pull on the armature (supported by springs) of 60 N at an air gap length $l = 1$ cm.
    (b) For the design values of N and i, estimate the force when the armature is in contact with the core.
    (c) Using the results of parts (a) and (b) and your knowledge of magnetic circuits, sketch a graph of force f as a function of gap length l.

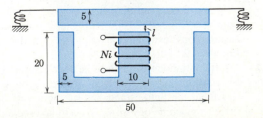

Figure 22.35

PROBLEMS

1. In a simplified form of magnetohydrodynamic generator (Fig. 22.36), the electrodes are parallel and of length $L = 30$ cm. A field of $B = 1$ T exists in the channel which has a rectangular cross section 10 cm by 20 cm. The plasma has a resistivity $\rho = 10^{-5}$ $\Omega \cdot$ m and an average velocity $U = 10$ m/s.
   (a) Explain why an emf should be developed across the electrodes.
   (b) Derive a literal expression for the electrode voltage indicated by a high-resistance voltmeter.
   (c) Estimate the maximum power obtainable from this generator under the given conditions.
   (d) What nonelectrical measurements would reveal this power?

2. The device sketched in Fig. 22.36 is proposed to pump a hot, corrosive liquid (such as molten sodium in a nuclear reactor).
   (a) What electrical or magnetic properties must the liquid possess?
   (b) For positive current I, how would the device work?
   (c) For $B = 1$ T across the 10-cm $\times$ 30-cm pipe, what current would be required for a pressure rise of 5 N/cm^2 (approximately 6 psi)?

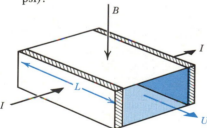

Figure 22.36

3. A d'Arsonval instrument cannot be employed to measure rapidly changing currents. Investigate the response of such an instrument to alternating voltages of the form $v = V_m \cos \omega t$. Work in terms of phasors, using θ for angular displacement and $\dot{\theta}$ for angular velocity to distinguish from frequency ω. Neglect the inductance of the moving coil and the frictional resistance of the bearings.
   (a) Draw an appropriate circuit model, clearly labeled, and derive an expression for the input impedance.

(b) Sketch the variation in impedance magnitude Z as a function of frequency ω, labeling significant values of Z and ω.
   (c) Derive an expression for the ratio θ/V and sketch a graph of the ratio θ_m/V_m as a function of ω.

4. The movement of a 50-mV (full-scale) d'Arsonval meter is characterized by: $R = 50\,\Omega$, $J = 10^{-8}$ N $\cdot$ m $\cdot$ s^2/rad, and for each spring $K_R = 2 \times 10^6$ rad/N $\cdot$ m.
   (a) Determine the frequency of the undamped oscillation and specify an appropriate value for D_R for reasonable damping.
   (b) Determine and sketch the response $\theta(t)$ of the damped movement to a suddenly applied voltage $v = 25$ mV.

5. Prepare a chart summarizing the characteristics of some of the devices studied so far. Rule a page into 4 columns headed Device, Input, Output, and Function. Provide 10 rows for the following devices: Amplifier, Rectifier, Modulator, Transformer, Loudspeaker, Pickup, Ammeter, Relay, Commutator Motor, and Alternator. Complete the chart by filling in the blanks with brief statements (10 words or less).

6. You are to derive an expression for the average torque developed by the motor of Fig. 22.37. The reluctance of the magnetic circuit is a minimum at $\theta = 0°$. Assume that the rotor shape is such that $\mathcal{R} = \mathcal{R}_0 - \mathcal{R}_a \cos 2\theta$. (Does this look reasonable?) If the impedance of the winding is negligibly small, $v = e$ and a sinusoidal voltage produces a sinusoidal magnetic flux $\phi = \Phi_m \cos \omega t$. Since unidirectional torque is produced only if the speed of rotation is synchronized with the alternation of the flux, let $\theta = \omega t - \delta$. What is θ when the flux is max-

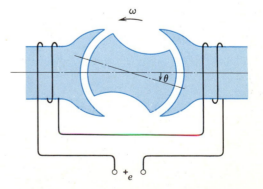

Figure 22.37

imum? If the instantaneous torque is represented by the sum of components of various frequencies, which component corresponds to average torque?

7. The capacitance of a parallel plate capacitor is $C = \epsilon A/x$ where ϵ = permittivity, A = area (of one plate), and x = spacing.

   (a) Using the method of virtual work (*Hint:* hold V constant), derive an expression for the force between the plates as a function of voltage V, capacitance C, and spacing x.

   (b) Eliminate C from the expression for force derived in part (a) and interpret physically the sign of the result.

   (c) Repeat part (a) holding charge Q constant and allowing x to change slowly.

8. In the magnetic clutch of Fig. 22.38, the dimensions are: $a = b = d = 1$ cm, $c = 3$ cm. The depth of the silicon steel core is 2 cm and the clutch plate is circular.

   (a) For an effective air gap of 0.2 mm and a friction coefficient of 0.9, estimate the torque that can be transmitted.

   (b) Estimate the horsepower that can be transmitted at 1000 rpm.

   (c) Estimate the required number of ampere-turns on the actuating coil.

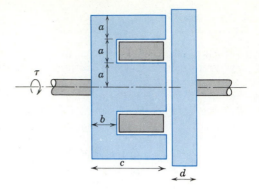

Figure 22.38

9. Design an electromagnet for lifting compressed automobiles in a junk yard.

   (a) Select a suitable material and a shape that will provide two lifting surfaces with rectangular blocks of compressed metal.

   (b) Assume a maximum acceleration of 16.1 ft/s^2 and estimate the necessary lifting force in newtons.

   (c) Allowing for paint and surface irregularities, specify the pole face area required and the necessary ampere-turns.

CHAPTER 23

Direct-Current Machines

Generators

Motors

The rotating machine is preeminent as a means for converting great amounts of electromechanical energy. The design of efficient and economical machines is a highly developed art, demanding specialized training and extensive experience; the number of engineers who *design* electrical machines is small. In contrast, engineers in all branches of engineering *use* electrical machines. This chapter and the next are concerned with the use of existing machines in steady-state applications under normal conditions of operation.

Our purpose here is to provide the background necessary to select machines to meet general requirements and to predict the behavior of the machines selected. Electrical machines vary widely in operating characteristics, power requirements, flexibility, ruggedness, and cost; therefore, the features of only the most common types of machines are described in detail. Prediction of the precise behavior of a given machine is quite difficult. By making simplifying assumptions, however, we can derive relatively simple models that give satisfactory results under most conditions. Being aware of the effect of the assumptions made, we can (with more training) modify the results if greater precision is required.

From Chapter 22 we know the principles of emf generation and torque development that apply to all rotating electromagnetic devices. Now we are going to investigate the construction features and operating characteristics of the most important types of machines. Our procedure, as with transistors and transformers, is to derive circuit models and use these to predict steady-state performance. Later (in Chapter 25) we shall study the dynamic behavior of machines as preparation for applications in control systems.

Classification of Electrical Machines

The rotating machine consists essentially of axial conductors moving in a magnetic field established in a cylindrical gap between two iron cores, but in practice it takes on a great variety of forms. One basis of classification is in terms of the current flowing in the stator and rotor windings. On this basis:

- Direct-current machines—direct current in both stator and rotor.
- Synchronous machines—alternating current in one, direct in the other.
- Induction machines—alternating current in both stator and rotor.

The construction of an induction motor is relatively simple, but the prediction of performance is quite complicated and a sophisticated model is required. In contrast, the construction of a dc motor is complicated but a simple model is quite satisfactory. Therefore, we start our study with dc machines.

DIRECT-CURRENT GENERATORS

The stator or *field* of a dc generator or motor (Fig. 23.1b) consists of an even number of magnetic poles (alternating N and S around the circumference) excited by direct current flowing in the field windings. The rotor or *armature* consists of a

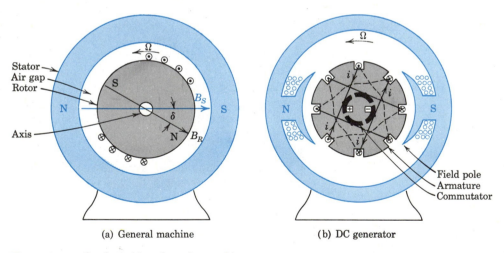

(a) General machine   (b) DC generator

Figure 23.1   Configurations of rotating machines.

cylindrical iron core carrying the active conductors embedded in slots and connected to the segments of the *commutator*. Direct current is carried to and from the armature by stationary *brushes* (drawn on the inside of the commutator but actually riding on the outside). The commutator automatically switches the conductors so that the external current from a generator or the torque from a motor is steady and unidirectional (see Fig. 22.18 and the accompanying explanation). The location of the brushes ensures that the torque angle δ is 90°.

Basic Relations

The emf generated in each conductor can be calculated from $e = Blu$ and the total emf is determined by the number of conductors in series at any time. The torque developed in any conductor can be calculated from $\tau_d = Blir$ and the total torque is the summation of the individual contributions. For a given machine in the steady state, the basic relations are

$$E = K\Phi\Omega \quad \text{and} \quad T_d = K\Phi I_A \tag{23-1}$$

where E = the generated emf in volts,
    Φ = the air gap flux per pole in webers,
    Ω = the angular velocity in radians per second,
    T_d = the developed torque in newton-meters,
    I_A = the armature current in amperes, and
    K = a constant for the given machine (see Eq. 22-34).

Equations 23-1 apply *at the air gap*; the terminal voltage differs from the generated emf by the armature resistance voltage drop and the shaft torque differs from the developed torque by the mechanical resistance torque.

The dc machine is a bilateral energy converter so that at the air gap the developed mechanical power is just equal to the generated electrical power or

$$T_d\Omega = EI_A \tag{23-2}$$

The air gap power represents only the reversible portion of the electromechanical energy conversion. Losses due to irreversible energy transformations occur in all practical machines.

Circuit Model of a Generator

For a dc generator in the steady state, the circuit model of Fig. 23.2 is applicable. Under steady-state conditions, currents and velocities are constant (or changing very

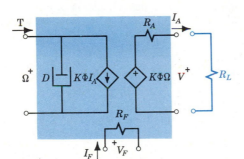

Figure 23.2 Circuit model for a generator.

slowly) so that there are no changes in mechanical or magnetic energy storage to consider. This means that the rotational inertia and electrical inductance terms are omitted from the model. The field resistance R_F must be considered in efficiency calculations.

In the linear model, the mechanical "conductance" term D represents all rotational losses ($P_M = \Omega^2 D$ analogous to $P_E = V^2 G$). It can be evaluated by measuring the no-load power input when the machine is operated at rated speed Ω and rated voltage E (and, therefore, rated air gap flux Φ). Included in the rotational losses are: bearing and brush friction, air friction or *windage*, and core losses due to hysteresis and eddy currents in regions of changing magnetic flux. (Where is there changing flux in a dc machine? In the armature core where B alternates and in the pole faces where B fluctuates due to changes in reluctance as each slot passes.)

The electrical resistance R_A is an effective resistance that takes into account the dc resistance of the armature winding, the effect of nonuniform current distribution in the armature conductors, the brush-contact loss, and the resistance of any auxiliary windings in the armature circuit. It also could include *stray-load* loss that arises from the distortion of flux and current distribution under heavy load. Here we assume that R_A is a fixed, measurable value.

In the model of Fig. 23.2 we assume that D, R_A, R_F, and K are constants; therefore, this is a linear model. To minimize errors, we should determine the constants under rated conditions; for small variations from a chosen operating point, linearity can be assumed with little error. The *external characteristic* is defined by

$$V = E - I_A R_A = K\Phi\Omega - I_A R_A \tag{23-3}$$

Under certain conditions we may choose to assume that the magnetic circuit also is linear. Then the magnetic flux is directly proportional to field current and

$$V = K' I_F \Omega - I_A R_A \tag{23-4}$$

EXAMPLE 1

A dc generator is rated at 10 kW, 200 V, and 50 A at 1000 rpm. $R_A = 0.4\ \Omega$ and $R_F = 80\ \Omega$. Predict the no-load voltage at 1000 rpm and the full-load voltage at 800 rpm if the field current is kept constant.

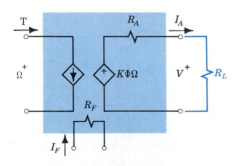

Figure 23.3 Generator model.

For constant-speed operation (resulting from a governor on the driving engine), the mechanical portion of the circuit model is not significant in the calculation of electrical quantities. From the model of Fig. 23.3 we see that the terminal voltage is

$$V = K\Phi\Omega - I_A R_A = E - I_A R_A$$

At no load, $I_A = 0$ and $V = V_{nl} = E$. Therefore,

$$V_{nl} = E = V_{fl} + I_A R_A = 200 + 50 \times 0.4 = 220\text{ V}$$

If the field current is constant, the generated emf E is directly proportional to speed; therefore, at 800 rpm

$$E' = E \times \frac{800}{1000} = 220 \times \frac{800}{1000} = 176\text{ V}$$

The terminal voltage at 800 rpm under full load (50 A) is

$$V_{fl}' = E' - I_A R_A = 176 - 50 \times 0.4 = 156\text{ V}$$

To keep field current constant requires a "separate" dc source. The external characteristic is shown in Fig. 23.6b.

Field Excitation

One basis for classifying dc machines is according to the arrangement for field excitation. In general, motors and generators may be *separately excited* or *self-excited*. The conventional symbols for field and armature are shown in Fig. 23.4a; here

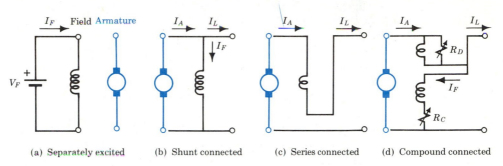

(a) Separately excited   (b) Shunt connected   (c) Series connected   (d) Compound connected

Figure 23.4   Wiring diagrams showing possible excitation connections for a generator.

the field is separately excited by connecting it to an independent source. A field winding for *shunt* connection directly across the armature usually consists of many turns of fine wire, and I_F is a few percent of I_A. A field winding for *series* connection consists of a few turns of heavy wire since it carries the entire armature current. Series generators are unsatisfactory for most applications, but series motors are widely used.

In *compound*-connected machines, all field poles have both shunt and series turns. The total mmf is either the sum or difference of $N_{sh}I_F$ and $N_{ser}I_A$ as desired. The relative influence of the shunt windings can be adjusted by means of a variable resistor or *control rheostat R_C* in the shunt field circuit. The series field mmf is a function of load current, but its relative effect can be reduced by connecting a low-resistance *diverter R_D* across the series field. This low-resistance path carries part of the armature current around the series winding.

The relation between generated emf and field current is defined by a *magnetization curve* (Fig. 23.5). Since E is proportional to Φ and I_F is proportional to $\mathcal{F}$, this is similar to a Φ–$\mathcal{F}$ curve for a magnetic circuit with an air gap. It is easily obtained experimentally by measuring no-load voltage as a function of field current at a constant speed. Curves for any other speed are obtained by direct proportion. Because of *residual magnetism*, a small emf is generated even with zero field current. Over a wide range the curve is nearly linear. At high values of I_F corresponding to high values of H, *saturation* of the iron core has an appreciable effect.

The Self-Excited Shunt Generator

The magnetization curve is independent of the source of I_F. In a separately excited generator, I_F is controlled by the operator and, within limits, the no-load voltage can be set to any value. In a self-excited generator, the no-load voltage is determined by the magnetization curve and the total resistance of the field circuit. (In Fig. 23.6b, V_R and I_R are the *rated* values of terminal voltage and current.) If the very

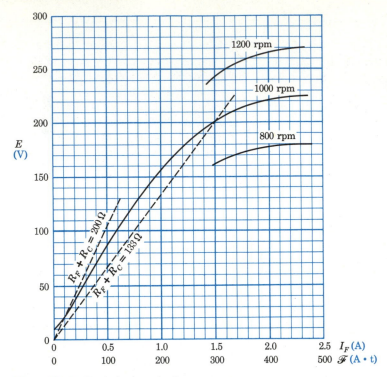

Figure 23.5 Typical magnetization curve.

small $I_A R_A$ drop is neglected, the generated emf $E = K\Phi\Omega$ appears across R_F plus R_C, the resistance of the control rheostat, and

$$E = (R_F + R_C)I_F \tag{23-5}$$

This represents a straight line with a slope $R_F + R_C$. But E is also a function of I_F as defined by the magnetization curve. The intersection of these curves is the simultaneous solution of the two relations. (See Example 2.)

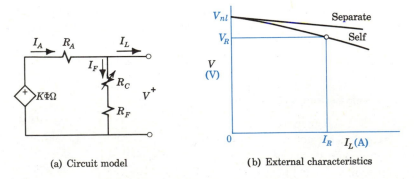

(a) Circuit model

(b) External characteristics

Figure 23.6 A self-excited generator.

EXAMPLE 2

The dc generator of Example 1 is rated at 10 kW, 200 V, and 50 A at 1000 rpm. $R_A = 0.4 \; \Omega$ and $R_F = 80 \; \Omega$. The magnetization curve is shown in Fig. 23.5. Predict the values of R_C to yield a no-load voltage of 200 V and a full-load voltage of 200 V.

For a speed of 1000 rpm, $E = V_{nl} = 200$ V when $I_F = 1.5$ A; therefore, solving Eq. 23-5 yields

$$R_C = \frac{E}{I_F} - R_F = \frac{V_{nl}}{I_F} - R_F = \frac{200}{1.5} - 80 = 53 \; \Omega$$

For a full-load voltage of 200 V, the generated emf (see Example 1) must be 220 V. Figure 23.5 indicates a field current of 2 A; therefore,

$$R_C = \frac{V_{fl}}{I_F} - R_F = \frac{200}{2} - 80 = 20 \; \Omega$$

The process of voltage buildup in a self-excited generator can be explained in terms of Fig. 23.5. With the generator rotating, a small voltage (about 10 V here) exists due to residual magnetism. When the field circuit is closed, this voltage causes a small field current that increases the flux that increases the voltage, etc. This cumulative process continues until the stable operating point is reached; at this point the voltage generated is just that required to produce the current to establish the necessary flux. No further increase in voltage is possible, except by lowering the resistance of the field circuit. Note that if the field circuit resistance is increased slightly, to 200 Ω in Example 2, the no-load terminal voltage is drastically reduced. (What would happen if the polarity of the residual voltage were opposite to that assumed here? When the field switch is closed, the current would *decrease* the flux, decreasing the voltage and preventing voltage buildup.)

The Compound Generator

The variation in terminal voltage with a change in load current is usually undesirable. As indicated in Fig. 23.6b, the drop in voltage in a separately excited generator is primarily due to $I_A R_A$. In a self-excited generator the drop in terminal voltage decreases the field current and reduces the generated voltage still further. In contrast, the excitation provided by a series winding increases with armature current. A compound generator with the shunt and series windings properly proportioned produces a full-load voltage just equal to the no-load value; such a machine is said to be *flat-compounded*. (See Example 3 on p. 654.)

Under certain circumstances the "drooping" characteristic of the differentially compounded generator may be desirable (Fig. 23.7b). The degree of compounding can be reduced by use of a diverter. Under other circumstances a rise in voltage with load may be desirable and the machine is *over-compounded*.

Armature Reaction

In deriving the basic relations for a conductor moving in a magnetic field, we assumed that the field due to current in the conductor itself was negligibly small. In a loaded machine, this is no longer true. As shown in Fig. 23.8a, the armature current

EXAMPLE 3

The generator of Examples 1 and 2 is to be operated with the shunt field (200 turns per pole) separately excited.

Design a series winding (Fig. 23.7a) to provide flat compounding, and predict the full-load terminal voltage if the series winding is incorrectly connected.

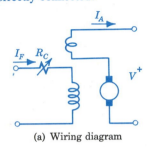

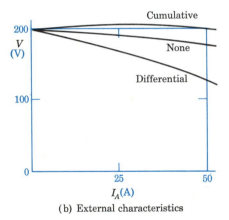

(a) Wiring diagram

(b) External characteristics

Figure 23.7 A compound generator.

We assume that the resistance of the series winding is negligible. From Example 2 (or reading from Fig. 23.5), the required mmf is:

For 200 V at no load,

$$NI_F = 200 \times 1.5 = 300 \text{ A} \cdot \text{t/pole}$$

For 200 V at full load,

$$NI_F = 200 \times 2 = 400 \text{ A} \cdot \text{t/pole}$$

The difference of $400 - 300 = 100$ A·t is to be supplied by an armature current of 50 A; therefore, the series winding should provide

$$\frac{100 \text{ A} \cdot \text{t/pole}}{50 \text{ A}} = 2 \text{ turns/pole}$$

Knowing the number of turns per pole on the shunt winding used in determining the magnetization curve, we can convert the horizontal scale of Fig. 23.5 to mmf. In *cumulative* compounding the mmfs of the shunt and series windings add. In *differential* compounding the series winding is connected so that the series mmf is in opposition to that produced by the shunt field.

For differential compounding with $I_F = 1.5$ A and $I_A = 50$ A, the net mmf per pole is

$$200 \times 1.5 - 2 \times 50 = 300 - 100 = 200 \text{ A} \cdot \text{t}$$

This corresponds to $I_F = 1$ A (in the 200-turn shunt winding) and the magnetization curve indicates $E \cong 156$ V. Then

$$V = E - I_A R_A = 156 - 50 \times 0.4 = 136 \text{ V}$$

The external characteristics of the generator are as shown.

establishes an appreciable magnetic field B_q directed at right angles or *in quadrature* with respect to the main field. The result of this *armature reaction* is a distortion of the flux distribution (Fig. 23.8b) with two practical results.

In the first place, commutation is affected adversely because the point of zero generated emf is shifted (in the direction of rotation in a generator). One practical arrangement is to place conductors carrying armature current in the faces of the main poles. Such *compensating windings* oppose the armature mmf and reduce the distortion of the flux distribution. Most machines also have small auxiliary field windings carrying the load current and placed midway between the main poles. Such *commutating poles* aid in reversing the current in the conductors as they undergo commutation by providing the necessary voltage impulse ($v \Delta t = L \Delta i$).

A second effect of armature reaction is a reduction in generated emf due to a reduction in the average flux density. Because the magnetic circuit is nonlinear, the increase in flux density in regions of high mmf is less than the decrease in regions of

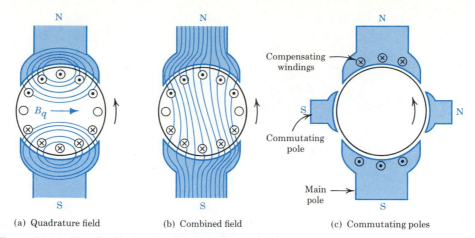

(a) Quadrature field   (b) Combined field   (c) Commutating poles

Figure 23.8   Flux distribution showing armature reaction in a generator.

low mmf. The total flux per pole is reduced and therefore the emf. This effect contributes to the curvature in the external characteristics in Fig. 23.7b.

Armature reaction is of great importance in the design and performance of large machines, and certain devices such as the automobile generator voltage regulator depend on armature reaction for their operation. In most of our problems, however, the additional complexity of considering armature reaction is not justified and we shall neglect its effect.

DIRECT-CURRENT MOTORS

If electrical power is supplied to a dc machine, it can operate as a motor. The essential difference between a motor and a generator is in the direction of armature-current flow. The same circuit model is applicable to a motor, or it can be turned around as in Fig. 23.9b. DC motors are classified as shunt, series, or compound,

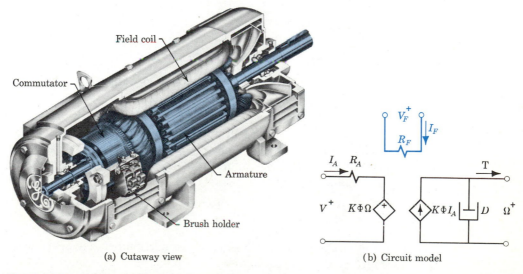

(a) Cutaway view   (b) Circuit model

Figure 23.9   A dc motor (Courtesy General Electric Company).

according to the method of field connection. Since torque depends on field flux and armature current, the operating characteristics of the various types differ widely.

The Shunt Motor

The basic relations (Eqs. 23-1) apply to the shunt motor, but it is usually more convenient to work with speed in rpm. Letting $\Omega = n \times 2\pi/60$, the basic equations are

$$E = K\Phi\frac{2\pi n}{60} = kn\Phi \qquad \text{and} \qquad T_d = K\Phi I_A \tag{23-6}$$

where n is the speed in rpm and k is a new constant.

On the basis of Fig. 23.9b, the voltage equation for a motor becomes

$$V = E + I_A R_A = kn\Phi + I_A R_A \tag{23-7}$$

The speed characteristic is revealed by solving Eq. 23-7 so that

$$n = \frac{V - I_A R_A}{k\Phi} \tag{23-8}$$

With Constant Voltage. In Fig. 23.10, the field circuit is connected directly across the supply line; the armature voltage is constant and the field current is constant for any setting of the rheostat R_C. If the effect of armature reaction is neglected, Φ

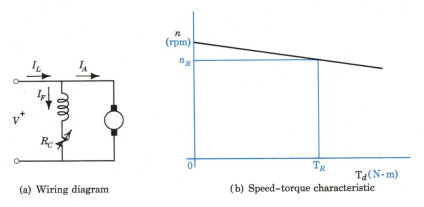

(a) Wiring diagram   (b) Speed–torque characteristic

Figure 23.10 A shunt motor with constant applied voltage.

remains constant. Since even at full load $I_A R_A$ is a small percentage of V, the shunt motor is essentially a constant-speed machine. At any given load, however, the speed is readily adjusted, within limits, by changing the field current with a control rheostat.

A specific expression for speed as a function of developed torque is obtained by solving the torque equation for I_A and substituting in Eq. 23-8 to yield

$$n = \frac{V}{k\Phi} - \frac{R_A}{kK\Phi^2}T_d = n_{nl} - m\,T_d \tag{23-9}$$

Equation 23-9 indicates a straight-line relation between speed and developed torque (Fig. 23.10b).

EXAMPLE 4

The dc machine of the previous examples is to be used as a shunt motor (Fig. 23.11). No-load tests indicate that the rotational losses at rated speed are 1100 W. For a terminal voltage of 220 V, specify the field current for a full-load (I_A = 50 A) speed of 1000 rpm, and estimate the no-load speed.

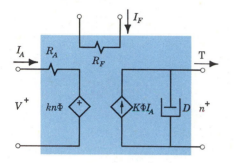

Figure 23.11   Shunt motor calculations.

Solving Eq. 23-7, for I_A = 50 A and R_A = 0.4 Ω,

$$E = kn\Phi = V - I_A R_A = 220 - 50 \times 0.4 = 200 \text{ V}$$

The magnetization curve at a full-load speed of 1000 rpm (Fig. 23.5) indicates I_F = 1.5 A for E = 200 V.

To supply the rotational losses, the armature input must be 1100 W (plus negligibly small $I_A^2 R_A$ loss). Therefore, at no load,

$$I_A = \frac{P_{nl}}{V} = \frac{1100}{220} = 5 \text{ A}$$

The no-load speed could be determined by solving Eq. 23-8, using the value of $k\Phi$ determined from full-load data. A simpler approach is to use the proportionality between E and n which leads to

$$n_{nl} = \frac{E_{nl}}{E_{fl}} \times n_{fl} = \frac{(V - I_A R_A)_{nl}}{(V - I_A R_A)_{fl}} \times n_{fl}$$

$$= \frac{218}{200} \times 1000 = 1090 \text{ rpm}$$

With Adjustable Voltage.   When the shunt motor is operating at constant voltage, torque is proportional to field flux whereas speed is inversely proportional to flux; reducing the flux to increase the speed makes less torque available. Greater flexibility is possible if the armature and field voltages are obtained from separate power supplies. As indicated by Eq. 23-8, speed can be controlled by armature voltage V or field voltage V_F (actually Φ). By holding Φ constant, the torque capability can be maintained (Eq. 23-6) while the speed can be effectively controlled over a wide range by adjusting the armature voltage. This arrangement is commonly used in industrial applications of large motors.

Just how does a machine adjust to an increased load? As torque is applied to the shaft the motor slows down and the emf decreases. A small decrease in emf results in a large increase in I_A (Eq. 23-7) and a large increase in developed torque. The motor slows down just enough to develop the required torque. If the shaft load is reduced, the motor speeds up slightly and the armature current (representing power input) is reduced.

As indicated by Example 4, the variation in speed of a constant-voltage shunt motor under load is quite small. The *speed regulation*, analogous to the voltage regulation of a transformer or generator, is defined as

$$SR = \frac{n_{nl} - n_{fl}}{n_{fl}} \tag{23-10}$$

For the shunt motor of Example 4, the speed regulation is 9%. Typical values for shunt motors are in the range from 5 to 12% and justify classifying the shunt motor as a constant-speed machine.

The Series Motor

If a field winding consisting of a few turns of heavy wire is connected in series with a dc armature (Fig. 23.12), the resulting *series motor* has characteristics quite different from those of a shunt motor. The basic equations (Eq. 23-6) still apply, but now magnetic flux is dependent on load current.

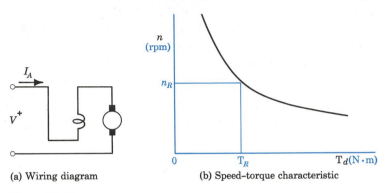

(a) Wiring diagram   (b) Speed–torque characteristic

Figure 23.12   A series motor.

To simplify the analysis, we shall assume that the series motor is operating on the linear portion of the magnetization curve and $\Phi = k_1 I_A$. Then the emf and torque equations become

$$E = kn\Phi = knk_1 I_A = k_E n I_A \tag{23-11a}$$

and

$$T_d = K\Phi I_A = Kk_1 I_A^2 = k_T I_A^2 \tag{23-11b}$$

Equation 23-11b indicates that the torque developed by a series motor is proportional to the square of the armature current. (How does this compare with the performance of a shunt motor?) To determine the speed-torque characteristic, we note that

$$V = E + I_A(R_A + R_{\text{ser}}) = (k_E n + R_T)I_A \tag{23-12}$$

where $R_T = R_A + R_{\text{ser}}$, the total resistance of the armature circuit. Solving Eq. 23-12 for I_A and substituting in Eq. 23-11b,

$$T_d = k_T \frac{V^2}{(k_E n + R_T)^2} \tag{23-13}$$

For values of n such that $k_E n \gg R_T$, Eq. 23-13 indicates that the developed torque is inversely proportional to the square of the speed. This relation is displayed in Fig. 23.12b. The series motor is *not* a constant-speed machine; this is not necessarily a disadvantage, but it does limit the application of series motors. In Example 5, if the torque drops to approximately 30% of the full-load value, the machine speed doubles; to prevent excessive speeds, series motors are mechanically coupled to their loads at all times.

EXAMPLE 5

(a) The dc machine of the previous examples is to be operated as a series motor rated at 220 V and 50 A at 1000 rpm. The resistance of the series winding is estimated to be 0.2 Ω. Design the series winding.

At full load the emf must be

$$E = V - I_A(R_A + R_{\text{ser}}) = 220 - 50(0.4 + 0.2) = 190 \text{ V}$$

The magnetization curve (Fig. 23.5) indicates a required mmf of $NI = 200 \times 1.4 = 280$ A·t. Then the series winding must provide

$$N_{\text{ser}} = \frac{NI}{I_A} = \frac{280 \text{ A·t}}{50 \text{ A}} \cong 5\frac{1}{2} \text{ turns per pole}$$

If this is not convenient, 6 turns can be wound per pole and a diverter used to ensure 280 A·t.

(b) Predict the torque developed at rated speed and at twice rated speed.

To determine torque we must evaluate two machine constants. We assume operation on the linear portion of the magnetization curve and use the linear model of Fig. 23.13. From Eq. 23-6, using rated values,

$$K\Phi = \frac{60E}{2\pi n} = \frac{60 \times 190}{2\pi \times 1000} = \frac{5.7}{\pi}$$

and the rated value of developed torque is

$$T_R = K\Phi I_A = \frac{5.7}{\pi} \times 50 \cong 91 \text{ N·m}$$

From Eq. 23-11a,

$$k_E = \frac{E}{nI_A} = \frac{190}{1000 \times 50} = 0.0038$$

Equation 23-13 indicates that at twice rated speed,

$$T_2 = T_R \frac{(k_E n + R_T)_R^2}{(k_E n + R_T)_2^2} = 91 \left(\frac{3.8 + 0.6}{7.6 + 0.6} \right)^2 \cong 26 \text{ N·m}$$

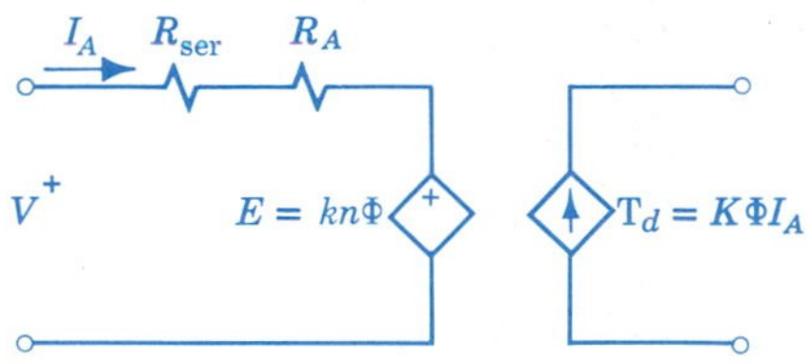

Figure 23.13 Series motor performance calculations using a circuit model.

Starting Direct-Current Motors

At starting the speed is zero, the emf of a dc machine is zero, and the applied voltage appears across the armature resistance (see Eq. 23-7 or Fig. 23.9). To prevent excessive armature current, the starting voltage must be reduced. If the armature voltage is adjustable, the motor is smoothly accelerated by increasing the voltage from zero. Automatic control to limit the starting current to 1.5 or 2 times the rated current is often used. If power is supplied at constant voltage, a starting resistance is connected in series with the armature; the resistance is shorted out in one or several steps as the motor comes up to speed. For high starting torque in either case, full voltage should be applied to the field of a shunt motor.

In a series motor, the field current is equal to the armature current. Since the torque varies as the square of the current (neglecting saturation), for a starting current higher than rated value the starting torque of a series motor is greater than that of a comparable shunt motor.

Special *starting boxes* for dc motors provide a variable starting resistance and protection against loss of voltage and loss of field current. (See Exercise 18.) Reversing either the armature or the field connection, but not both, changes the direction of rotation.

Speed Control

The shunt motor offers two important virtues: ease in adjusting speed and flexibility in matching the speed-torque characteristics of the load. To match the constant-torque characteristic of many mechanical loads, a motor can be operated with rated field and armature currents and adjustable armature voltage; developed torque is then independent of speed. Other loads require high torque at low speed and low torque at high speed; such "constant-horsepower" characteristics can be matched by using constant armature voltage and varying the field current. Because of magnetic circuit saturation in a practical machine, the range of speed control by field current is limited. The lowest speed corresponds to the maximum field current when $R_C = 0$. (What would happen if the field circuit were opened accidentally?) Large machines may employ both methods of speed control by using sophisticated regulators on both armature and field circuits.

The speed of a series motor is dependent on the load and is not so easily controlled. Under heavy torque loads, the series motor slows down, the emf decreases, and the armature current increases until sufficient torque is developed. An adjustable voltage supply or a series resistor can be used to vary the speed. (What would happen if the mechanical load were disconnected accidentally?) The high starting torque of series motors makes them useful in locomotives, trolley buses, cranes, and hoists.

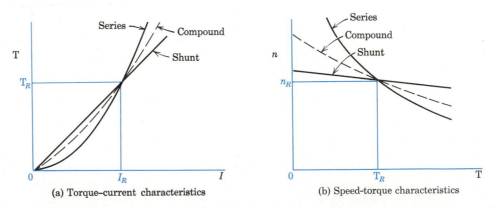

Figure 23.14   Operating characteristics of dc motors.

The *compound motor* has a shunt and a series field connected so that their mmfs add. The operating characteristics are intermediate between those of shunt and series motors (see Fig. 23.14). By adjusting the relative strength of the shunt and series fields, a motor can be designed to provide a desired compromise.

Can a dc motor operate on ac? Since torque is proportional to the product of flux and armature current, if field and armature currents reverse simultaneously, the torque will be pulsating but unidirectional. A *universal motor* is a series-wound commutator-type motor in which the entire magnetic circuit is laminated to minimize eddy-current and hysteresis losses. (A shunt winding is not used because its high inductance would introduce a large phase angle between I_F and I_A and reduce the average torque and the efficiency.) Operating at high speeds, up to 15,000 rpm, the universal motor provides high-power output for a given weight. These compact motors are ideal for portable, high-speed appliances such as hand drills, food mixers, and vacuum cleaners.

Adjustable Voltage Supply

To realize all the virtues of a dc motor, an adjustable voltage power supply is required.

Ward–Leonard System. One possibility is to use an ac motor driving a dc generator and an *exciter* (to supply field excitation current). In Fig. 23.15, the motor-generator set converts ac power to adjustable voltage dc power. With control rheostat

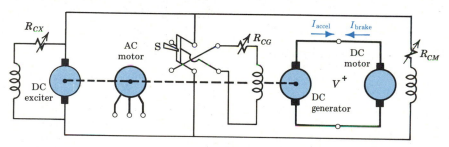

Figure 23.15   Ward–Leonard system adjustable voltage supply.

R_{CM} set for rated field current, dc motor voltage V is controlled by the generator field current through R_{CG}. Motor speed can be controlled smoothly from standstill to rated speed in either direction; reversing the direction of generator field current reverses the direction of motor rotation. Speeds above the rated speed can be obtained by reducing the motor field current. *Regenerative braking* of the load is possible by reducing the generator field current, thereby reducing V below the motor emf $E = kn\Phi$ and reversing the power flow; the generator acts as a motor driving the ac machine as a generator feeding the rotational energy of the load back into the ac line. Regenerative braking saves energy while providing smooth control.

Thyristor Rectifier System. Electronic rectifiers using SCRs offer several significant advantages over motor-generator sets. SCRs are more efficient, longer lived, lighter in weight, and more easily maintained; they do not, however, offer the inherent regenerative braking capability of a motor-generator set.

The half-wave SCR supply of Fig. 23.16 is suitable for driving small dc motors. The dc field current is provided through diode D_1 from the ac supply; when the ac

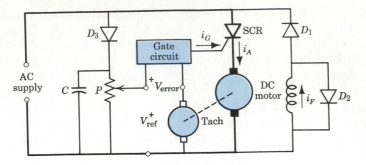

Figure 23.16 Half-wave SCR adjustable voltage supply.

polarity reverses cutting off D_1, field current continues to flow through *free-wheeling diode* D_2. (The inductance of the field itself provides some filtering action.) Capacitor C is charged through D_3 and an adjustable reference voltage V_{ref} is obtained at the speed-controlling potentiometer P. Whenever the tachometer voltage differs from the reference voltage, i.e., whenever the motor speed differs from the reference speed, the resulting error voltage shifts the firing angle of the gate circuit and thereby adjusts the conduction angle of the SCR that supplies armature current (see also Fig. 12.28).

Performance of DC Motors

Previous examples have illustrated the approach to calculating speed and torque. To compute output and efficiency, the mechanical and electrical losses must be taken into account. The parameters in the circuit model may be computed by using rather elaborate formulas, or they may be measured in the laboratory. A no-load test of a rotating machine is similar to the open-circuit test of a transformer and yields analogous data. Full-load tests may be performed using a *dynamometer* to measure the shaft torque.

The *power-flow diagram* in Fig. 23.17 is an aid in visualizing the energy conversion processes in a machine. The power-flow diagram, the energy balance, and the circuit model are closely related. Under dynamic conditions additional terms must be introduced to account for changes in magnetic and mechanical energy storage. The midpoint of the diagram corresponds to the air gap with iron losses entered on the mechanical side.

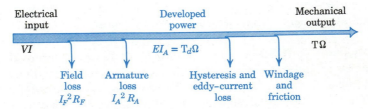

Figure 23.17 Power-flow diagram for a dc motor.

EXAMPLE 6

The dc machine of Example 4 is rated as a shunt motor at 220 V, 50 A, at 1000 rpm. $R_F = 80 \ \Omega$ and $R_A = 0.4 \ \Omega$. At rated speed, the field current is 1.5 A and the rotational losses are 1100 W. Determine the output under rated conditions and predict the full-load efficiency. (Do not charge $I_F^2 R_C$ losses to the motor.)

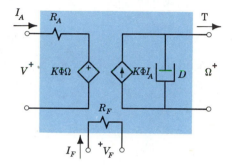

Figure 23.18 Shunt motor model.

The circuit model of Fig. 23.18 is applicable. At rated current and voltage,

$$E = V - I_A R_A = 220 - 50 \times 0.4 = 200 \text{ V}$$

By Eq. 23-6,

$$K\Phi = \frac{60E}{2\pi n} = \frac{60 \times 200}{2\pi \times 1000} = \frac{6}{\pi}$$

and the rated value of developed torque is

$$T_d = K\Phi I_A = \frac{6}{\pi} \times 50 = \frac{300}{\pi} = 95.5 \text{ N·m}$$

The developed mechanical power at the air gap is

$$P_M = T_d\Omega = K\Phi I_A \Omega = \frac{300}{\pi} \frac{2000\pi}{60} = 10,000 \text{ N·m/s}$$

Alternatively, the developed mechanical power is equal to the electrical power at the air gap or

$$P_M = P_E = EI_A = 200 \times 50 = 10,000 \text{ W or N·m/s}$$

Subtracting the rotational losses, the power available at the shaft is

$$P = 10,000 - 1100 = 8900 \text{ W or } \frac{8900}{746} = 11.9 \text{ hp}$$

The overall efficiency is

$$\frac{\text{Output}}{\text{Input}} = \frac{\text{shaft output}}{VI_A + I_F^2 R_F} = \frac{8900}{220 \times 50 + 1.5^2 \times 80}$$

$$= \frac{8900}{11,180} \cong 0.8 \text{ or } 80\%$$

Although dc motors are less common than ac motors, they are ideally suited to certain applications. The commutator makes the dc machine relatively high in first cost and in maintenance, and where there is moisture, dust, or explosive vapor, the enclosure must be selected with care. However, the dc motor provides easily adjusted speed, high efficiency, and great flexibility. DC motors range in size from tiny units turning fans in electronic apparatus to 10,000-hp motors driving the rolls in steel mills. Other applications include ship propulsion, fork lifts, punch presses, and battery-powered vehicles.

SUMMARY

■ In a direct-current machine, the basic steady-state relations are:

$$E = K\Phi\Omega \quad \text{and} \quad T_d = K\Phi I$$

■ In a dc generator, the external characteristics are defined by:

$$V = E - I_A R_A \quad \text{and} \quad T_d = T_a - \Omega D$$

Generators may be separately excited or self-excited.
The magnetization curve defines E as a function of I_F.

■ Self-excited voltage buildup stabilizes where the field-resistance line intersects the magnetization curve.
In a flat-compounded generator, the increase in series field mmf $N_{ser}I_A$ offsets the drop in armature voltage $R_A I_A$.
Armature reaction results in a distortion of flux distribution.

■ A dc motor differs from a generator only in the direction of I_A:

$$E = kn\Phi = V - I_A R_A \qquad T_d = K\Phi I_A = T_a + D\Omega$$

	Shunt Motor	*Series Motor*
Flux:	Constant but adjustable	Varies with I_A
Emf:	$E = kn\Phi$	$E = k_E n I_A$
Torque:	$T_d = K\Phi I_A$	$T_d = k_T I_A^2$
Speed:	Nearly constant	Widely varying
	$n = \dfrac{V - I_A R_A}{k\Phi}$	$T_d = k_T \dfrac{V^2}{(k_E n + R_T)^2}$
	$n = n_{nl} - k'T_d$	$\dfrac{n}{n_R} \cong \left(\dfrac{T_{dR}}{T_d}\right)^{1/2}$
Speed Control:	Efficient over wide range, by varying V_A and I_F	Some control by varying V_A or inserting R_{ser}
Starting Torque:	Fair; $\dfrac{T}{T_R} = \dfrac{I_A}{I_{AR}}$	Good; $\dfrac{T}{T_R} = \left(\dfrac{I_A}{I_{AR}}\right)^2$

In starting a dc motor, $E = 0$ at $n = 0$; therefore, it must be started with low V or with series R to limit I_A and $I_A^2 R_A$ heating.

Speed regulation of a motor is defined as $SR = \dfrac{n_{nl} - n_{fl}}{n_{fl}}$.

Characteristics of compound motors are intermediate between those of shunt and series motors.
Universal motors are specially designed series motors that will operate on ac or dc.

REVIEW QUESTIONS

1. Explain the statement: "The air gap power represents the reversible portion of electromechanical energy conversion."
2. Outline a laboratory procedure for determining D in Fig. 23.1.
3. Why and where are there eddy-current and hysteresis losses in a dc machine?
4. Given the rated voltage of a generator, estimate the no-load voltage.
5. On a compound generator, how could you distinguish between the series and shunt field windings by inspection?
6. Draw a wiring diagram of a compound generator showing control rheostat and diverter; what is the function of each?
7. What would happen if the polarity of the residual magnetism in a self-excited generator were reversed?
8. Explain the operation of a flat-compounded generator.
9. Why is an under-compounded generator more stable than an over-compounded machine?

Sketch their V-I characteristics.
10. What is armature reaction and why is it of concern to designers?
11. Sketch speed-torque curves for series, shunt, and compound motors.
12. What happens if the field connection of a fully loaded shunt motor is accidentally broken? Repeat for a lightly loaded motor.
13. How does a shunt motor react (electrically) to increased load?
14. How does a series motor react (electrically) to increased load?
15. In starting a shunt motor, should V_F be maximum or minimum? Why?
16. In comparison to I_F control of speed, why is V_A control more expensive and R_A control less efficient?
17. What are the advantages of an SCR voltage supply over an m-g set?
18. How is the output voltage of an SCR supply adjusted?

EXERCISES

1. In a dc generator similar to that in Fig. 23.1b, there are 64 slots on the armature, each containing two conductors that are always in series.
   (a) How many commutator segments are required?
   (b) How many conductors are in series between the brushes?
   (c) If the flux per pole is 0.15 Wb, what speed is required (rpm) to generate an emf $E = 240$ V?
2. A separately excited dc generator is rated at 250 V, 100 A, 2000 rpm; $R_A = 0.2\ \Omega$, $R_F = 125\ \Omega$.
   (a) Draw an appropriate linear model for steady-state, constant-speed operation.
   (b) Predict the no-load terminal voltage ($I_A = 0$).
   (c) Predict the full-load terminal voltage ($I_A = 100$ A) if the field current is decreased 10%.
3. A dc generator operated at speed Ω with constant flux per pole develops an emf $E = k\Omega$. The rotational inertia of the armature is J and the mechanical friction coefficient is D_R. The resistance of the armature circuit is R and the inductance is L.
   (a) Draw an appropriate circuit model showing

input torque T and output current I.
   (b) Use the circuit model to predict the steady-state speed at input torque T when supplying a load resistance R_L.
   (c) To maintain rated speed of 1200 rpm at no-load ($I = 0$), an input torque of 2 N·m is required. Use this information to evaluate a machine parameter in SI units.
4. A 20-kW, 250-V, 1200-rpm generator is driven at a constant speed of 1200 rpm and the field current is adjusted to provide an output voltage of 250 V at a load current of 80 A. When the load is disconnected, the output voltage rises to 262 V.
   (a) Evaluate the effective armature resistance.
   (b) The no-load driving power is 800 W. Evaluate the D factor in Fig. 23.2 for this generator.
   (c) Predict the driving torque required under full load.
   (d) If the speed is increased to 1500 rpm and the field current reduced by 10%, predict the new terminal voltage and the new no-load driving power.

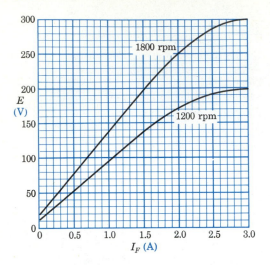

Figure 23.19 Magnetization curve.

5. The magnetization curve of Fig. 23.19 describes a separately excited dc generator.
   (a) Predict the no-load voltage at 1200 rpm with $I_F = 2$ A and at 2100 rpm with $I_F = 1$ A.
   (b) Specify the field current for a no-load voltage of 150 V at 1800 rpm and at 1500 rpm.
6. A separately excited dc generator is rated at 250 V, 100 A, 2000 rpm with $I_F = 4$ A. The no-load voltage is observed to be 260 V. Predict the output voltage at $I_L = 100$ A when driven at 1500 rpm with $I_F = 5$ A. State assumptions and use a circuit model.
7. A self-excited dc generator, known to be in good mechanical condition, fails to generate anywhere near rated voltage when driven by a motor. List four possible causes for this unsatisfactory operation and indicate the steps you would take to eliminate each.
8. A self-excited dc generator operates at 1800 rpm (see Fig. 23.19). Predict the no-load voltage if the total resistance of the field circuit is 120 Ω. Repeat for operation at 1200 rpm.
9. A compound generator with $R_A = 1$ Ω and $R_{ser} = 0.2$ Ω has 2 turns/pole on the series field and 300 turns/pole on the shunt field. The no-load voltage is 200 V with separate excitation at $I_F = 1$ A. Stating any necessary assumptions:
   (a) Estimate the terminal voltage at $I_A = 30$ A with the same separate excitation.
   (b) Repeat part (a) for compound excitation and describe the degree of compounding.
10. A 1500-rpm dc shunt generator with $R_A = 0.5$ Ω generates a no-load voltage of 240 V with $I_F =$

2 A in the 200-turn/pole shunt winding. Specify the number of turns per pole of series winding (of negligible resistance) to provide "flat-compounding" under full load of $I_L = 50$ A.

11. A self-excited dc generator operates at 1800 rpm (Fig. 23.19).
    (a) Predict the no-load voltage if the total resistance of the shunt field circuit (300 turns/pole) is 125 Ω.
    (b) If $R_A = 0.5$ Ω, estimate the shunt field current for a full-load ($I_A = 60$ A) voltage equal to the no-load voltage of part (a).
    (c) If there are 2 turns/pole on the series field (negligible resistance), describe the degree of compounding.

12. Predict how I_A and speed n of a typical dc shunt motor would be affected by the following changes in operating conditions:
    (a) Keeping I_F and T constant and halving V.
    (b) Keeping I_F and P constant and halving V.
    (c) Keeping V and T constant and doubling I_F.
    (d) Keeping P constant and halving I_F and V. (*Note:* T = load torque, V = terminal voltage, and P = power output.)

13. A dc shunt motor is rated at 1500 rpm, 50 hp output with a total line current of 180 A at 250 V. At no load, rated speed requires $I_A = 15$ A and $I_F = 3$ A.
    (a) Draw a clearly labeled circuit model appropriate for predicting motor performance. Stating any necessary assumptions, evaluate the parameters insofar as possible.
    (b) With the motor adjusted to provide rated speed at rated load at 250 V, the mechanical load is reduced until the total line current is 140 A. Predict the new speed and estimate the hp output.

14. In Fig. 23.20, V = 200 V, auxiliary resistor R = 100 Ω, and $R_F = 100$ Ω; R_A is negligibly small.
    (a) Determine the relation between speed and torque for this special machine.
    (b) If the speed is 1200 rpm, predict the developed torque at I = 1.5 A.

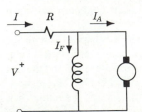

Figure 23.20

15. A 100-hp, 700-V series motor is rated at 120 A at 1200 rpm ($R_A = 0.25$ Ω, $R_{ser} = 0.1$ Ω). For $I_A = 60$ A, predict the speed and the power output by:
    (a) Making simplifying assumptions for an approximate answer.
    (b) Using the derived relations (Eqs. 23-12 and 23-13). Draw a conclusion.

16. State how I_A and n (see Exercise 12) of a typical dc series motor would be affected (approximately) by halving the terminal voltage if:
    (a) T remains constant.
    (b) P remains constant.
    (c) T varies as the square of n.

17. A 50-hp, 550-V series motor is rated at 74 A at 750 rpm ($R_A = 0.35$ Ω and $R_{ser} = 0.15$ Ω). Stating any simplifying assumptions:
    (a) Draw and label a circuit model appropriate for predicting performance.
    (b) Predict the armature current if the load torque is twice the rated value.
    (c) Predict the speed in part (b).
    (d) Estimate the current drawn and the torque developed when this motor is running at 1000 rpm.

18. Using the starting box of Fig. 23.21, a shunt motor is started by closing switch S and moving contact lever L to the first button on R_1. As the motor speeds up, the lever is advanced; at the last button it is held in position by magnet M against the pull of a spring.
    (a) Explain the functions of R_1, R_2, and M in terms of desired starting conditions for a dc motor.
    (b) Design a starting box (i.e., specify the value of R_1 in Fig. 23.21) for the shunt motor of Exercise 13 to limit the starting armature current to twice the rated current.

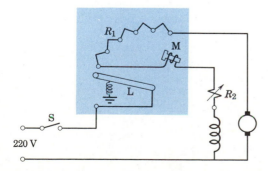

Figure 23.21   Starting box.

19. The output of a paper mill is to be wound on a 10-in. diameter core at a constant speed of 115 ft/s with a tension of 120 lb. The finished roll is 30 in. in diameter. Specify the motor to drive the windup roll by hp, type, and method of speed control.

20. A dc shunt motor is rated at 10 hp, 200 V, 1200 rpm. $R_F = 100$ Ω and $R_A = 0.4$ Ω. *Stating any necessary assumptions:*
    (a) Estimate the no-load speed.
    (b) Estimate the no-load speed with a 50-Ω control rheostat in the field circuit.
    (c) With the control rheostat of part (b) removed and with a 2-Ω control resistance in series with the armature, estimate the starting current and torque and predict the running speed at $I_A = 40$ A.

21. The torque required by a mixer is defined by $T = 3 \times 10^{-5}n^2$ N·m where n is speed in rpm. Available are a shunt motor ($R_A = 0.4$ Ω, $R_F = 240$ Ω) and a series motor ($R_T = 0.6$ Ω), each rated at 10 hp and 1200 rpm at 240 V.
    (a) Sketch the speed-torque curves of the two motors and the mixer on the same graph.
    (b) Stating any necessary assumptions, predict the operating speeds of the two motors.

22. A 20-hp, 240-V, 1200-rpm shunt motor with $R_A = 0.2$ Ω follows the magnetization curve of Fig. 23.5.
    (a) Estimate the field current for full load ($I_A = 70$ A).
    (b) While running at full load, the field current is quickly increased to 1.7 A. Predict the new I_A and the initial torque available for regenerative braking.

23. A small "personal" electric car for urban use requires full power to accelerate to full speed of 30 km/h in 10 s, but only 30% power to cruise at full speed. A typical cycle consists of acceleration, cruising for 1 minute, and deceleration to a stop. Estimate the percentage increase in operating range possible if regenerative braking at 60% efficiency is used.

24. The ac supply voltage in Fig. 23.16 is $v_s = V_m \sin \omega t$ and the gate circuit fires the SCR at $\omega t = 60°$. On the same time axis:
    (a) Sketch the waveforms of v_s and i_A, taking into account the significant inductance of the armature.
    (b) Show i_G and i_F waveforms, assuming the time constant of the field winding is approximately equal to one cycle of the supply.

(c) Show the waveform of v_A. (*Hint*: How is v_A related to e_A during the time the SCR is OFF?)

25. Given a self-excited shunt generator:
 (a) Draw an appropriate steady-state circuit model for this machine.
 (b) Draw a clearly labeled power-flow diagram with input on the left.

26. Calculate the rated efficiency of the series motor of Exercise 15.

27. Calculate the rated efficiency and predict the speed regulation and no-load armature current of the shunt motor of Exercise 13.

28. A 20-hp, 250-V, 1500-rpm shunt motor with $R_A = 0.2\ \Omega$ requires a no-load total input of 8 A.
 (a) Predict the full-load efficiency.
 (b) Predict the speed regulation.

29. A shunt motor with automatic controls is designed for a "base speed" of 1000 rpm at which it delivers 100% torque and 100% horsepower at full armature voltage and full field current. Above or below base speed, control is by adjusting I_F or V_A. Plot torque and horsepower (for rated I_A) vs. speed from 0 to 2000 rpm. Label the regions controlled by I_F and V_A.

PROBLEMS

1. A shunt generator is rated at 100 V at 1 A with $I_F = 0.02$ A; another is rated at 100 V at 25 A with $I_F = 1$ A. Show how these could be connected as a two-stage power amplifier. Estimate the dc power gain possible.

2. The magnetization curve for a 1800-rpm shunt generator is shown in Fig. 23.19. Machine constants are: $R_A = 0.5\ \Omega$, $L_A = 0.5$ H, $R_F = 120\ \Omega$, and $L_F = 30$ H. For self-excited operation:
 (a) Estimate the steady-state no-load terminal voltage.
 (b) Predict the time required for the voltage to build up to 80% of its ultimate no-load value. (*Hint*: What voltage is available to cause di/dt?)

3. A 100-hp, 700-V shunt motor with $R_A = 0.3\ \Omega$ and $R_F = 250\ \Omega$ operates at 2000 rpm with a control rheostat of 50 Ω in the field circuit. Predict the no-load speed with and without the 50-Ω rheostat.

4. A series motor ($R_T = 0.25\ \Omega$) is rated at 25 hp at 1500 rpm with 85 A at 250 V. Calculate the rated output torque and predict the current, speed, and hp output if the load torque is twice the rated value.

5. Six SCRs are used in the three-phase adjustable voltage motor supply in Fig. 23.22.
 (a) Sketch the waveforms of the supply voltages. (See Fig. 7.26.)
 (b) Under what circumstances will motor current flow through SCRs 1 and 6? 2 and 6?
 (c) Describe the operation of the circuit.
 (d) For a firing angle of 90°, sketch the wave-

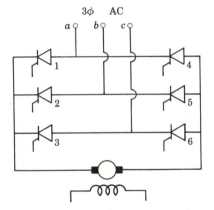

Figure 23.22

form of the voltage across the motor.
 (e) Repeat part (d) for $\theta = 120°$.

6. Predict the steady-state performance of a machine labeled: DC SHUNT MOTOR, 50 hp, 1500 rpm, 250 V, 180 A. At no load, rated speed requires $I_F = 3$ A and $I_A = 15$ A.
 (a) Draw a clearly labeled schematic wiring diagram, an appropriate circuit model, and a power-flow diagram.
 (b) Mechanical load is applied to the shaft until the total line current is 143 A. Predict the speed, developed torque, mechanical power output in hp, and the overall efficiency.

7. The electric car of Exercise 23 weighs 1100 lb including driver. Estimate the required motor rating in hp and the operating range with a 12-V, 100-A·h battery.

24

Alternating-Current Machines

Alternators

Synchronous Motors

Induction Motor Operation

Induction Motor Performance

In Chapter 23 we applied our knowledge of electromechanics to practical machines carrying direct currents in the stator and rotor windings. We considered the operating principles of these dc machines and derived simple circuit models for use in predicting their performance under changing conditions.

Now we are going to follow the same approach in studying synchronous machines and induction motors. In the induction motor, alternating current is applied to the stator and alternating currents are induced in the rotor by transformer action. In the synchronous machine, direct current is supplied to the rotor and alternating current flows in the stator. The same machine can function as *alternator* or *motor;* the only difference is in the direction of energy flow.

ALTERNATORS

In the alternator of Fig. 24.1, current is transferred to the rotor by stationary brushes riding on sliprings. Because the required field power is much less than that associated with the armature, placing the field on the rotor is the most practical arrangement. In other designs transformer action is used to transfer alternating current to the rotor and there convert it to direct current, using semiconductor diodes mounted on the rotor. This arrangement eliminates the necessity for any sliding contacts and greatly reduces maintenance problems.

As pointed out in connection with Fig. 7.26, the emf generated in each turn of the 4-turn coil shown is slightly out of phase with the emf generated in its neighbor because it experiences maximum flux density at a different instant. By proper distribution of the turns and shaping of the field pole, the total emf can be made to approach

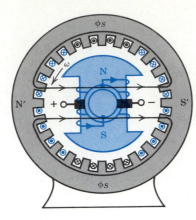

Figure 24.1 An elementary, two-pole synchronous machine.

a sinusoid. Two other similar windings (in color) occupy the remainder of the periphery of the stator and the result is a balanced set of three sinusoids that can be connected in Y or Δ (see Figs. 7.27 and 7.29).

Basic Relations

In this discussion the emphasis is on three-phase alternators although two-phase machines are sometimes used, particularly in control systems. For convenience we assume that all machines are Y-connected and balanced, and we work in terms of *one phase*; the phase voltage is the line-to-neutral voltage ($V_p = V_l/\sqrt{3}$) and the phase torque is just one-third of the total torque. Voltages and currents are sinusoidal and are represented by phasors. Assuming linear magnetic circuits greatly simplifies the mathematics and leads to results that are quite satisfactory for our purposes.

On the basis of these assumptions the basic relations for generated emf and developed torque (Eqs. 22-24, 30b) are

$$E = kn\Phi_R \qquad \text{and} \qquad T_d = k_T\Phi_R\Phi_S \sin \delta \qquad (24\text{-}1)$$

where E and Φ are rms values, n is speed in rpm, and T_d is an average value. Rotor flux Φ_R is due to dc field current in the rotor stator, flux Φ_S is due to ac current flowing in the stator, and δ is the torque angle. In generator operation, the frequency of the emf is directly related to the speed of rotation. A complete cycle of emf is generated as a pair of magnetic poles pass a given coil; therefore, the frequency in hertz is

$$f = \frac{n}{60}\frac{p}{2} \qquad (24\text{-}2)$$

In motor operation, to develop unidirectional torque the rotor must turn at *synchronous speed n_s* given by

$$n_s = \frac{120f}{p} \qquad (24\text{-}3)$$

It is convenient to remember that for 60-Hz operation the highest synchronous speed is 3600 rpm, obtained with a 2-pole machine ($p = 2$).

Circuit Model of an Alternator

For steady-state operation, the circuit model for the synchronous machine is as shown in Fig. 24.2. (It is quite similar to that for the dc machine in Fig. 23.2.) The phasors **E**, **I**, and **V** are rms values of per-phase quantities; emf E is defined by a

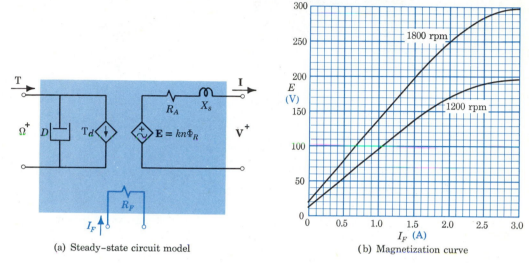

(a) Steady–state circuit model   (b) Magnetization curve

Figure 24.2   Representation of an alternator.

magnetization curve. The *armature resistance* R_A is an effective resistance that represents copper losses and takes into account nonuniform distribution of current in the armature conductors. The *synchronous reactance* X_s represents the effect of armature reaction and leakage inductance. Armature reaction results in a component of flux that induces a voltage in quadrature with the current and therefore can be represented by an inductive reactance. The effect of stator leakage flux that does not react with the rotor is also represented by an inductive reactance. In our simplified approach, these two effects are combined into a single reactance X_s that is assumed to be constant.[†]

The assumption regarding X_s holds fairly well for a machine with a uniform air gap. In such a *cylindrical-rotor* machine, the field windings are embedded in slots in an otherwise smooth rotor. The assumption is inaccurate for the more typical *salient-pole* machine shown in Fig. 24.1. In either case, the machine constants can be determined by open- and short-circuit tests just as for the transformer. In all but the smallest machines, the synchronous reactance is much larger than the armature resistance; therefore, we neglect R_A except in calculating losses.

Alternator Characteristics

In considering the electrical characteristics of an alternator driven at constant speed, the simplified model of Fig. 24.3a is satisfactory. Note that R_A is neglected and

---

[†]For a more complete discussion of synchronous reactance, see Chapter 3 of George McPherson, *An Introduction to Electrical Machines and Transformers,* John Wiley & Sons, New York, 1981.

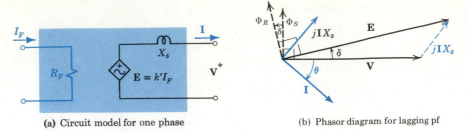

(a) Circuit model for one phase

(b) Phasor diagram for lagging pf

Figure 24.3 Electrical characteristics of an alternator.

$E = k'I_F$ for a linear magnetic circuit. The external characteristic (per phase) is defined by

$$V = E - I(R_A + jX_s) \cong E - jIX_s \qquad (24\text{-}4)$$

or

$$E = V + jIX_s \qquad (24\text{-}5)$$

The corresponding phasor diagram is shown in Fig. 24.3b. Phasor **V** is taken as a horizontal reference and, assuming an inductive load, **I** is drawn at a lagging angle θ. The jIX_s voltage drop is 90° ahead of **I**, and **E** is the sum of **V** and jIX_s. Since **E** is directly determined by rotor flux Φ_R and **V** bears a similar relation to air gap flux Φ_S linking the stator, the angle between **E** and **V** is just equal to δ, the angle between Φ_R and Φ_S.

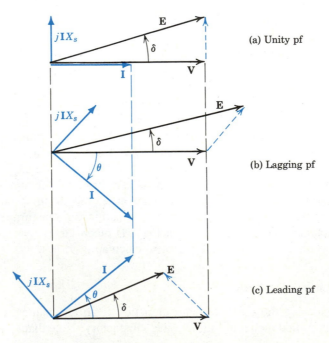

(a) Unity pf

(b) Lagging pf

(c) Leading pf

Figure 24.4 Effect of load power factor on the operation of an alternator supplying constant power $P = VI \cos \theta$.

The behavior of the alternator under various power factor conditions is illustrated in Fig. 24.4. We assume that the alternator is supplying a constant power $P = VI \cos \theta$, at a fixed voltage and frequency, as the power factor is changed. At unity pf, E and V are approximately equal in magnitude. For lagging-pf operation, E is considerably larger than V; therefore, I_F must be increased above the value required for a unity-pf load. For leading-pf operation, E is considerably smaller than V and less field current is required. A physical explanation for this behavior is that the leading current in the stator produces a component of mmf (by armature reaction) that aids the mmf established by the field current; therefore, less I_F and E are required for the same V. Lagging-pf operation is considered in Example 1.

EXAMPLE 1

A three-phase 6-pole alternator is rated at 10 kVA, 220 V, at 60 Hz. Synchronous reactance is $X_s = 3 \, \Omega$. The no-load terminal voltage (rms) follows the magnetization curve of Fig. 24.2b. Determine the rated speed and predict the field current required for full-load operation at 0.8 lagging pf.

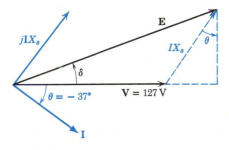

Figure 24.5 Alternator performance.

The rated speed is given by (Eq. 24-3)

$$n_s = \frac{120f}{p} = \frac{120 \times 60}{6} = 1200 \text{ rpm}$$

Assuming Y connection, the phase voltage is

$$V_p = \frac{220}{\sqrt{3}} = 127 \text{ V}$$

the rated phase current is

$$I = \frac{P_A/3}{V_p} = \frac{10,000/3}{127} = 26.2 \text{ A}$$

and the voltage drop across the synchronous reactance is

$$IX_s = 26.2 \times 3 = 78.6 \text{ V}$$

The phasor diagram for 0.8 lagging-pf operation is shown in Fig. 24.5. The necessary value of emf is

$$\mathbf{E} = \mathbf{V} + j\mathbf{I}X_s = \mathbf{V} + j(I \cos \theta + jI \sin \theta)X_s$$

$$= \mathbf{V} - IX_s \sin \theta + jIX_s \cos \theta$$

$$= 127 + 47.2 + j62.9 \cong 184 \underline{/20°} \text{ V}$$

An rms voltage of 184 V at 1200 rpm requires 2.3 A.

The Rotating Field

For unidirectional torque, $\sin \delta$ must not change sign (Eq. 24-1) and therefore the stator field must rotate at the same speed as the rotor field. How can a rotating field be produced by stationary coils? This concept is fundamental to the operation of synchronous and induction motors. It can be explained in terms of the mmf contributed by the three balanced currents in Fig. 24.6a. The instantaneous values of the currents are shown in (b). Positive currents in the concentrated windings produce mmfs in the

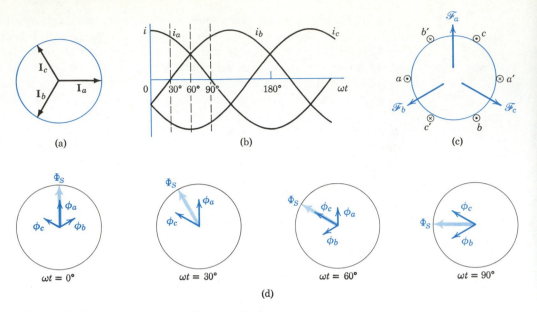

Figure 24.6 Production of a rotating stator field.

directions indicated in (c). At the instant $\omega t = 0°$, say, i_a is a positive maximum and i_b and i_c are negative and one-half maximum. Assuming a linear magnetic circuit, the principle of superposition applies, and the flux contributions of the three currents are as shown in (d). The stator flux Φ_S is the vector sum of the three contributions. At $\omega t = 30°$, the relative magnitudes of the three currents have changed and the position of the resulting stator flux has shifted 30°. At $\omega t = 60°$, the current magnitudes have changed and the resultant flux vector has shifted another 30°. We say that a rotating magnetic field exists because the position of maximum flux rotates at synchronous speed.

It was pointed out in Chapter 7 that in a balanced three-phase circuit, power is constant rather than pulsating, as in a single-phase circuit. We should have expected that an alternator supplying constant electrical power at a constant speed would develop a constant torque. For constant torque, δ must remain constant or the stator flux must move in synchronism with the rotor flux. Figure 24.6 is a demonstration of a situation that we could have anticipated.

SYNCHRONOUS MOTORS

Another feature of synchronous machines, which should come as no surprise, is that by changing the direction of power flow the alternator becomes a motor. In a dc machine, the direction of current flow is reversed by changing the field current and thereby changing the generated emf. But Fig. 24.4 indicates that in a synchronous machine there may be large changes in **E** with no change at all in the power delivered by the alternator. How can such a machine be made to deliver torque to a load?

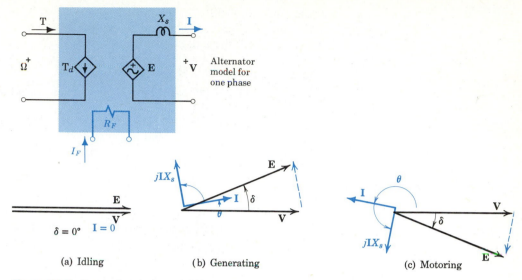

Figure 24.7 Power flow in a synchronous machine using the alternator model.

Motor Characteristics

Consider an alternator driven at synchronous speed by a gasoline engine. The output terminals of the alternator are connected to the three-phase line of the local power company. Referring to the model in Fig. 24.7, let **V** be one phase voltage and let I_F be adjusted so that **E** = **V**. Under this condition, **I** = 0 and there is no power transfer between the alternator and the line; the alternator is idling. Now let us open the throttle of the gasoline engine and attempt to accelerate the rotor. As the rotor is driven ahead of the stator field, there is a voltage difference between **E** and **V**, and current **I** flows to satisfy Eq. 24-5. (Note that the magnitude of **E** does not change.) The electrical power output must be matched by a corresponding mechanical input or

$$P_E = 3VI \cos \theta = P_M = 3\mathrm{T}_d \Omega \qquad (24\text{-}6)$$

where $\mathrm{T}_d = k_\mathrm{T} \Phi_R \Phi_S \sin \delta$ as before. The increase in developed torque associated with angle δ equals the increase in applied torque and the average speed of the engine-alternator combination does not change. (The forward shift in δ accompanying mechanical input is clearly visible if the rotor is illuminated by a stroboscopic light.)

If the engine throttle is returned to the idle position, the torque angle returns to zero and the rotor continues to turn at synchronous speed. If the throttle is closed, the rotor *tends* to slow down; the engine is now, in effect, an air compressor acting as a load on the synchronous machine. As the rotor shifts back through an angle $\delta^\dagger$ (Fig. 24.7c), the emf **E** is retarded in phase with respect to **V** and a voltage difference exists across X_s. The resulting current **I** is nearly 180° out of phase with **V**. The electrical *output VI* cos θ is *negative*, or the machine is absorbing electrical power from the line and delivering mechanical power to the shaft. It is operating as a *synchronous motor*.

[†]Note that δ is in "electrical degrees" where there are 180° electrical between adjacent magnetic poles. In a 4-pole machine, one revolution corresponds to $(p/2) \times 360 = 720$ electrical degrees.

A more convenient model for a synchronous motor is shown in Fig. 24.8. Here the reference direction of I has been changed and in-phase current represents positive electrical input. In general, the input power per phase is

$$P = VI \cos \theta \qquad (24\text{-}7)$$

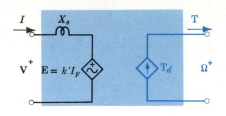

The voltage relation for a motor is

$$\mathbf{V} = \mathbf{E} + j\mathbf{I}X_s \qquad (24\text{-}8)$$

and the current is

$$\mathbf{I} = \frac{\mathbf{V} - \mathbf{E}}{jX_s} \qquad (24\text{-}9)$$

Figure 24.8 Circuit model for a synchronous motor.

if the armature resistance is neglected.

On the basis of this simple model, as interpreted by these equations, the behavior of synchronous motors under various operating conditions can be predicted.

EXAMPLE 2

The alternator of Example 1 ($X_s = 3\ \Omega$) is rated at 10 kVA, 220 V, and 26.2 A at 1200 rpm. Predict the torque angle and the field current for unity-pf operations as a motor at rated load.

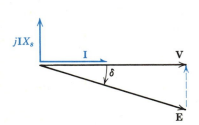

Figure 24.9 Unity-pf operation.

The phasor diagram for unity-pf operation is shown in Fig. 24.9. A slight increase in $\mathbf{E}$ over the idling value (Fig. 24.7a) is required. Solving Eq. 24-8,

$$\mathbf{E} = \mathbf{V} - j\mathbf{I}X_s = 127 - j26.2 \times 3 \cong 150\underline{/-32°}\ \text{V}$$

and the torque angle δ is 32°.

Referring again to the magnetization curve of Fig. 24.2b, we note that an emf of 150 V at 1200 rpm requires a field current of 1.65 A.

Power Relations

Since torque varies as the sine of the angle δ, there is a maximum possible developed torque and power. If the load on a motor is increased continuously, the rotor field drops further and further behind the stator field until this maximum is reached. Any further increase in load torque pulls the rotor into a position of less than maximum developed torque, and the rotor falls out of step and out of synchronism. At less than synchronous speed the developed torque is first in one direction and then in the other, and the machine comes to a shuddering halt. Some "loads" may overhaul the motor and drive it as an alternator at greater than synchronous speed, pulling it out of synchronism.

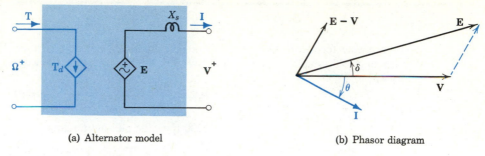

(a) Alternator model   (b) Phasor diagram

Figure 24.10 Derivation of developed power in an alternator.

What happens if the input to an alternator is increased continuously? An approximate expression for the maximum electrical power developed in an alternator can be derived by using the simplified model and phasor diagram of Fig. 24.10. By Eq. 24-5, assuming $V\underline{/0°}$ as a reference,

$$\mathbf{I} = \frac{\mathbf{E} - \mathbf{V}}{jX_s} = -j\left[\frac{E\underline{/\delta}}{X_s} - \frac{V\underline{/0°}}{X_s}\right] = -j\left[\frac{E(\cos\delta + j\sin\delta)}{X_s} - \frac{V}{X_s}\right]$$

The real part of $\mathbf{I}$ is $(E/X_s)\sin\delta$ which is equal to $I\cos\theta$. Then the electrical power developed per phase is $VI\cos\theta$ or

$$P = \frac{VE}{X_s}\sin\delta \tag{24-10}$$

The electrical power at the air gap is just equal to the developed mechanical power, which is the product of developed torque and speed. Therefore,

$$T_d = \frac{P}{\Omega} = \frac{VE}{\Omega_s X_s}\sin\delta \tag{24-11}$$

In a motor the electrical power developed is negative, which is consistent with the negative value of δ associated with mechanical power output (Fig. 24.7c). The black line in Fig. 24.11 is the power angle curve based on Eq. 24-10 and holds for cylindrical-rotor machines. Most synchronous motors have salient poles and the reluctance torque contribution is appreciable. It can be shown (see Problem 6 in Chapter 22) that the reluctance torque varies as $\sin 2\delta$; the colored line in Fig. 24.11 shows the variation expected in salient-pole machines. The advantages of salient-pole construction are a higher maximum or *pull-out* torque and a smaller torque angle for a given load in the normal operating region (below $\delta \cong 30°$).

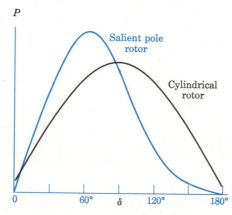

Figure 24.11 Power angle curves for synchronous motors.

EXAMPLE 3

Predict the pull-out torque for the 10-kVA, 200-V, 1200-rpm motor of Example 2 operating at $E = 150$ V.

Assuming cylindrical-rotor construction, Eq. 24-10 applies. For $\delta = 90°$, the maximum total power is

$$P_{max} = \frac{3VE}{X_s} = \frac{3 \times 127 \times 150}{3} \cong 19{,}000 \text{ W or N·m/s}$$

At 1200 rpm, the pull-out torque is

$$T_{max} = \frac{P_{max}}{\Omega_s} = \frac{19{,}000 \times 60}{2\pi \times 1200} \cong 150 \text{ N·m}$$

Note that the maximum power is nearly twice the rated value.

Reactive Power Relations

A unique feature of the synchronous motor is its ability to carry shaft load and draw a leading current. In an industrial plant operating at a lagging power factor, use of some relatively expensive synchronous motors may be justified by the savings resulting from improved power factor. (See the discussion preceding Example 3 in Chapter 7.)

The effect of varying field current on the power factor of an unloaded motor is displayed by the phasor diagrams in Fig. 24.12. Because the motor carries no shaft load, the torque angle is zero; for normal excitation $E = V$ and the motor draws no current. If, as in Fig. 24.12b, the field current is increased, E exceeds V and the resulting current is determined by Eq. 24-9. Such an *overexcited* machine draws a leading current; i.e., the sinusoidal current reaches its maximum 90° ahead of the sinusoidal terminal voltage. The power drawn is still zero since $VI \cos \theta = 0$, but the

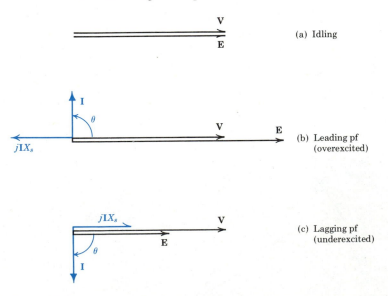

Figure 24.12 Effect of excitation on the power factor of an unloaded synchronous motor ($\delta = 0°$).

reactive power. $P_X = VI \sin \theta$ is appreciable. These conditions are characteristic of a capacitor (or condenser) and therefore a rotating machine of this type is called a *synchronous condenser*.

EXAMPLE 4

A three-phase induction furnace (Fig. 24.13a) draws 750 kVA at 0.6 lagging pf. A 1000-kVA synchronous motor is available. If the overall pf of the combination is to be unity, estimate the mechanical load that can be carried by the motor.

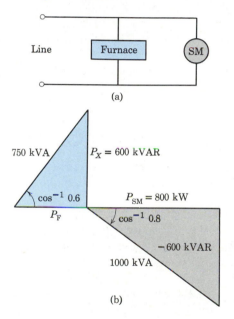

(a)

(b)

Figure 24.13 Power factor correction.

The reactive power of the furnace is

$$P_X = 750 \sin (\cos^{-1} 0.6) = 600 \text{ kVAR}$$

as shown in Fig. 24.13b. By operating at leading pf, the synchronous motor can supply this reactive power. If losses are neglected, the available power of the synchronous motor is

$$P_{SM} = \sqrt{P_A{}^2 - P_X{}^2} = \sqrt{1000^2 - 600^2} = 800 \text{ kW}$$

At the cost of a small increase in copper loss in the motor field windings, the necessary reactive power is supplied to the furnace and 80% of the motor rating is available for any mechanical load.

If the field current is below normal, E is less than V and a lagging current flows (Fig. 24.12c). Such an *underexcited* machine draws a positive reactive power and could be called a *synchronous inductor;* however, underexcited machines do not offer advantages comparable to those of the synchronous condenser. The phasor diagrams in Fig. 24.12 are drawn for a motor with no shaft load; in general, there is shaft load and **I** has an in-phase component.

In dc machines, the direction of power flow is determined by the relative magnitude of E and V and can be controlled by I_F. In a synchronous machine, changing the relative magnitude of E affects only the reactive power flow. To change the direction of power flow requires a change in the torque angle δ and this is accomplished by changing the mechanical power supplied to the shaft. Opening the throttle of the driving engine advances the rotor and increases δ, and electrical power is produced. Loading the shaft retards the rotor and increases δ in a negative direction, and mechanical power is produced.

Performance of Synchronous Motors

Because torque is approximately proportional to angle δ (a springlike characteristic), and because rotor inertia is high and bearing friction is low, the response of a synchronous motor to a sudden change in load tends to be oscillatory. The resulting wide swings in δ, superimposed on a steady synchronous speed, are similar to the oscillations of the undamped d'Arsonval movement (see p. 625). The motor is said to be "hunting" for an equilibrium condition. To eliminate this instability, short-circuited *damping windings* (similar to the squirrel cage shown in Fig. 24.16) are imbedded in each pole face. While δ is changing, there is a change in flux linkage, voltage is induced in the short-circuited turns, energy is dissipated, and the oscillations are damped.

A synchronous motor has no inherent starting torque; however, if a motor with damping windings is connected to the ac line with the field circuit open, strong currents are *induced* in the stationary windings. These currents react with the rotating stator field to develop sufficient torque to bring the rotor nearly to synchronous speed. Applying dc field current then provides "pull-in torque" and the rotor locks into step with the rotating stator field. Most synchronous motors are started by this induction motor action, which is described in the next section. Commercial designs provide starting torques of the order of 120% of rated and pull-out torques of the order of 175% of rated.

The synchronous motor is a constant-speed device with a relatively high efficiency and a controllable power factor. The ability to improve plant power factor while carrying shaft load offsets the higher cost. Another feature is that by increasing the number of poles, low constant speeds are available without gearing. Synchronous motors are ideal for constant-speed applications where high starting torque is not required. They are found in all types of industrial plants and are used to drive compressors, pumps, crushers, and other loads up to several thousand horsepower. Their range of application is greatly expanded with the adjustable frequency drive described on p. 692.

Stepper Motors

To position a shaft or a work piece precisely, we can use a special form of synchronous motor driven by digital signals. The rotor of the *stepper motor* in Fig. 24.14 is a permanent magnet, although soft-iron rotors developing reluctance torque (Fig. 22.21) can be used. The four stator windings receive electrical pulses from a microprocessor-controlled power unit. If winding C is excited, the rotor assumes the position corresponding to $\theta = 0°$. If windings B and C are excited simultaneously, the developed torque drives the rotor to equilibrium at $\theta = -45°$. Increasing the number of stator poles reduces the angle per step. A sequence of pulses from the control unit can step a valve to a specified opening, locate the pen of an X-Y plotter, or control the position of a robot arm.

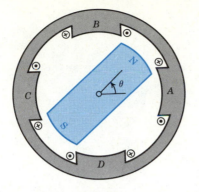

Figure 24.14 Stepper motor.

INDUCTION MOTOR OPERATION

In any rotating machine, we can conceive of torque as a result of the interaction of a current and a magnetic field, and we can write $T_d = k_a B_1 i_2 \sin \delta$ (Eq. 22-30a). In a dc machine, direct current is supplied to the stator to establish B_1 and to the rotor as I_2, and the commutator maintains δ at the optimum 90° at any speed (Fig. 24.15a).

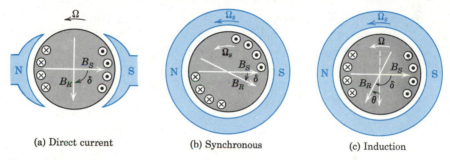

(a) Direct current   (b) Synchronous   (c) Induction

Figure 24.15 Torque production in rotating machines.

In a synchronous machine, polyphase alternating current supplied to the stator produces a rotating magnetic field, and the dc rotor field established by I_2 rotates at synchronous speed to develop a torque proportional to $\sin \delta$ (Fig. 24.15b). In an induction machine, polyphase alternating current supplied to the stator produces a rotating magnetic field (just as in the synchronous machine), and alternating currents are *induced* in the rotor by transformer action. The interaction between the induced rotor currents and the stator field produces torque.

Because the synchronous machine is much better for generating alternating current, the induction generator is not commercially significant and therefore our emphasis is on the three-phase induction motor. This simple, rugged, reliable, inexpensive device (invented by Nikola Tesla in 1887) consumes the major part of all the electric energy generated. It is built in sizes ranging from a fraction of one horsepower to many thousand horsepower. This important electromagnetic device deserves our attention.

Operating Principles

In its most common form, the rotor consists of axial conductors shorted at the ends by circular connectors to form a *squirrel cage* (Fig. 24.16). The stator is identical

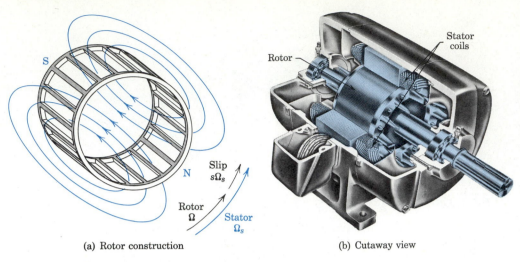

(a) Rotor construction   (b) Cutaway view

Figure 24.16 Induction motor with squirrel-cage rotor.

with that of a synchronous motor with the windings arranged to provide an even number of poles. As the stator field moves past the rotor conductors, emfs are induced by transformer action. The induced emf is a maximum in the rotor conductors under the poles, as shown in Fig. 24.15c. Because of the reactance of the rotor windings, the current and therefore the rotor field lag the emf by angle θ, and the torque angle is greater than 90°.

The stator field rotates at synchronous speed Ω_s determined by the frequency and number of poles (Eq. 24-3). The rotor always turns at a lower speed Ω; if the rotor turned at synchronous speed, there would be no change in flux linkage, no induced current, and no torque. The small difference in speed that produces flux cutting and motor action is called the *slip*. By definition, the slip is

$$s = \frac{\Omega_s - \Omega}{\Omega_s} = \frac{n_s - n}{n_s} \tag{24-12}$$

and is usually expressed in percent. The slip at full load for a squirrel-cage motor is typically 2 to 5%. Note that the rotor field created by the induced rotor currents moves ahead at a speed $s\,\Omega_s$ relative to the rotor structure. The absolute speed of the rotor field is

$$\Omega_{RF} = \Omega + s\,\Omega_s = \Omega_s \tag{24-13}$$

as required for unidirectional torque. In other words, the magnetic pole induced in a portion of the rotor is always of the proper polarity to react with the stator pole passing by. The flexibility of the induction motor, in contrast to the fixed-speed character of the synchronous motor, is due to this principle. (See Example 5.)

Basic Relations

The induction motor is essentially a transformer with a rotating secondary. The force that exists between primary and secondary coils in a transformer appears as useful torque in an induction motor. If the rotor is held stationary (by the inertia of a starting load or by an applied brake), the passing stator field induces in one phase

EXAMPLE 5

A three-phase, 220-V, 60-Hz induction motor runs at 1140 rpm. Determine the number of poles, the slip, and the frequency of the rotor current.

Because the synchronous speed must be slightly higher than the running speed and because at 60 Hz the only possibility is 1200 rpm (Eq. 24-3), we conclude that this is a 6-pole machine. At 1140 rpm the slip is

$$s = \frac{n_s - n}{n_s} = \frac{1200 - 1140}{1200} = 0.05 \text{ or } 5\%$$

The induced emf in the rotor is due to the slip speed and therefore

$$f_2 = sf = 0.05 \times 60 = 3 \text{ Hz}$$

Note that at starting $n = 0$ and $s = 1$; therefore, the frequency of the induced starting current is 60 Hz.

---

of the rotor winding an emf of magnitude E_2 and frequency f. The current that flows at standstill is

$$\mathbf{I}_2 = \frac{\mathbf{E}_2}{R_2 + j2\pi f L_2} = \frac{\mathbf{E}_2}{R_2 + jX_2} \qquad (24\text{-}14)$$

where R_2 and L_2 are the effective resistance and inductance of the rotor winding on a per-phase basis.

Now we are going to take advantage of an ingenious method for representing the power-developing capability of a short-circuited coil. At standstill, the slip $s = 1$ and the rotor frequency is just the stator frequency f. At any other speed, the slip is s, the induced emf is sE_2, the rotor frequency is sf, and the rotor reactance is sX_2. In general, the rotor current at slip s can be expressed as

$$\mathbf{I}_2 = \frac{s\mathbf{E}_2}{R_2 + jsX_2} = \frac{\mathbf{E}_2}{R_2/s + jX_2} = \frac{\mathbf{E}_2}{R_2 + jX_2 + R_2[(1 - s)/s]} \qquad (24\text{-}15)$$

since

$$R_2 + R_2 \frac{1 - s}{s} = R_2 + \frac{R_2}{s} - R_2 = R_2 \qquad (24\text{-}16)$$

Then Eq. 24-15 can be rewritten as

$$\mathbf{E}_2 = \mathbf{I}_2(R_2 + jX_2) + \mathbf{I}_2 R_2 \frac{1 - s}{s} = \mathbf{I}_2 \mathbf{Z}_2 + \mathbf{E}_g \qquad (24\text{-}17)$$

Two circuit models satisfying Eq. 24-17 are shown in Fig. 24.17. The first looks like

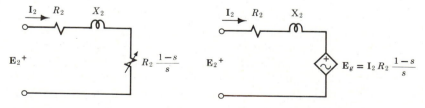

(a) Controlled load resistance   (b) Controlled voltage source

Figure 24.17 Equivalent circuit models of the rotor of an induction motor.

the secondary of a transformer with a variable load resistance; the second represents the load as a reversible energy transformation. In each model, the electrical power transferred to the rotor is $E_2I_2 \cos \theta_2$. The power lost irreversibly is $I_2^2R_2$; the remainder, $I_2^2R_2(1 - s)/s = E_gI_2$, is converted into mechanical power.

In analyzing the energy conversion taking place in an induction motor, the basic relations are

$$E_g = I_2R_2\frac{1 - s}{s} \quad \text{and} \quad T_d = \frac{E_gI_2}{\Omega} \tag{24-18}$$

The first of these relations results from a consideration of electrical energy in the rotor; the second results from the application of the energy-conservation principle to the electromechanical conversion process. To obtain an expression for torque in the form of Eq. 22-30a, consider the total electrical power transferred from the stator at Ω_s to the rotor. From either model of Fig. 24.17, this is $P_2 = E_2I_2 \cos \theta_2$. But E_2 is the rms value of an induced emf directly proportional to B_1, and $\cos \theta_2$ is just equal to $\sin \delta$ (Fig. 24.14c); therefore,

$$T_2 = \frac{P_2}{\Omega_s} = \frac{1}{\Omega_s}E_2I_2 \cos \theta_2 = k_aB_1I_2 \sin \delta \tag{24-19}$$

The developed torque T_d is less than T_2 by an amount corresponding to the $I_2^2R_2$ power loss.

Deriving a Circuit Model

The model for the rotor (Fig. 24.17b) resembles that of the secondary of a transformer. If we use the complete transformer model of Fig. 21.19a, the circuit model of an induction motor is as shown in Fig. 24.18. Since R_2 and X_2 are important

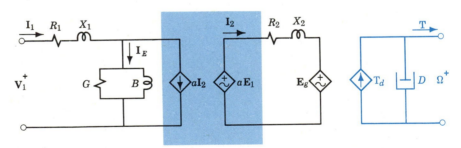

Figure 24.18 Step 1: A general circuit model for an induction motor.

only insofar as they affect input electrical power, it is convenient to assume that the turn ratio a is unity and to define R_2 and X_2 on that basis. If this is done, the ideal transformer consisting of aI_2 and aE_1 is superfluous and can be omitted. The colored line in Fig. 24.19 corresponds to the air gap across which energy is coupled to the rotor. A further simplification results if we transfer the rotational losses Ω^2D to the primary and combine them with the stator core loss represented by G. (The rotor core loss is usually negligibly small because of the low rotor frequency.) The sum of rotational loss and core loss is conveniently determined by measuring the no-load input where I_0 is the no-load current.

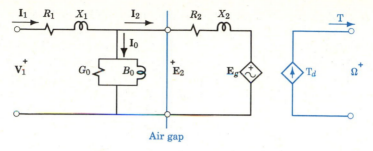

Figure 24.19 Step 2: A circuit model based on no-load measurements.

One other modification reduces the labor of performance calculations and simplifies the general analysis without introducing excessive error. Just as for the transformer (see Fig. 21.18), moving the elements representing no-load loss to the input terminals results in a simpler circuit consisting of two branches in parallel across a fixed voltage. In the transformer the reluctance of the magnetic circuit is small and the exciting current is only a small percentage of rated current. In an induction motor, however, the air gap raises the reluctance and the rotational loss adds to the core loss. As a result, the no-load current is 20 to 40% of rated current, and this produces a voltage drop that is too large to neglect. The effect of the no-load current can be taken into account by reducing the terminal voltage by an amount equal to the $I_0(R_1 + jX_1)$ drop as indicated in Example 6.

EXAMPLE 6

The parameters of a 25-hp, 220-V, 60-Hz induction motor are: $R_1 = 0.05$ Ω, $R_2 = 0.06$ Ω, $X_1 = 0.23$ Ω, and $X_2 = 0.23$ Ω. The no-load power input is 690 W at 220 V and 20 A. Calculate the "adjusted" voltage V_A in Fig. 24.20 and determine the error made by using an algebraic rather than a phasor calculation.

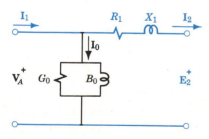

Figure 24.20 Step 3: Using an adjusted terminal voltage.

The stator impedance is

$$R_1 + jX_1 = 0.05 + j0.23 \cong 0.24\underline{/78°}\ \Omega$$

At no load, the power per phase is

or

$$P_0 = V_1 I_0 \cos \theta_0 = 230\ \text{W}$$

and

$$\theta_0 = \cos^{-1}(230/127 \times 20) = -85°$$

$$V_A = V_1 - I_0(R_1 + jX_1)$$

$$= 127\underline{/0°} - 20\underline{/-85°}\ (0.24\underline{/78°})$$

$$= 127\underline{/0°} - 4.8\underline{/-7°} = 127 - 4.8 + j0.6$$

$$= 122.2 + j0.6 \cong 122.2\ \text{V}$$

Algebraically,

$$V_A = V_1 - I_0 \sqrt{R_1^2 + X_1^2} \qquad (24\text{-}20)$$

$$= 127 - 20 \sqrt{(0.05)^2 + (0.23)^2}$$

$$= 127 - 20(0.25) = 122\ \text{V}$$

The error is very small.

As indicated by Example 6, the error made in using an algebraic calculation of V_A is negligibly small and therefore Eq. 24-20 is satisfactory. (Draw the phasor diagram approximately to scale and you will see why.) In many cases R_1 can be neglected compared to X_1 in this calculation, and in some cases use of the unadjusted terminal voltage is satisfactory.

One final simplification is desirable. Combining Eqs. 24-12 and 24-18, we can write

$$T_d = \frac{E_g I_2}{\Omega} = \frac{I_2^2 R_2 (1-s)/s}{\Omega_s(1-s)} = \frac{I_2^2 R_2}{s\Omega_s} \tag{24-21}$$

Since $E_g = I_2 R_2 (1-s)/s$, the total series resistance is

$$R_1 + R_2 + R_2 \frac{1-s}{s} = R_1 + \frac{R_2}{s} \tag{24-22}$$

The reactances X_1 and X_2 are difficult to separate and usually they are assumed equal; the sum is then $X = 2X_1$. The result is the simplified circuit model of Fig. 24.21, where calculations based on V_A are equivalent to calculations based on V_1 in Fig. 24.19.

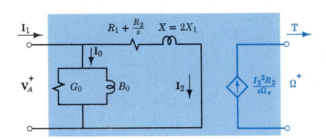

Figure 24.21 Final step: A simplified circuit model for an induction motor.

INDUCTION MOTOR PERFORMANCE

On the basis of the simplified model we can predict the performance of induction motors under varying conditions.

Speed-Torque Characteristics

An important characteristic of any motor is the speed-torque relation. For the induction motor, the total torque T is $3T_d$ or

$$T = \frac{3 I_2^2 R_2}{s\Omega_s} = \frac{3V_A^2}{\Omega_s} \frac{R_2/s}{(R_1 + R_2/s)^2 + X^2} \tag{24-23}$$

The speed-torque curve for a typical squirrel-cage motor is plotted in Fig. 24.22. Under normal operating conditions the slip is small (less than 5%) and the induction motor is essentially a constant-speed machine under load. For s small, R_2/s is large compared to R_1 and to X, and a good approximation is

$$T_{\text{normal}} \cong \frac{3V_A^2 s}{\Omega_s R_2} \tag{24-24}$$

In the normal operating range, torque is directly proportional to slip. Another conclusion from Eq. 24-24 is that at normal speeds the torque is inversely proportional to R_2.

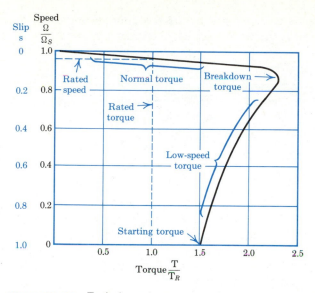

Figure 24.22 Typical speed-torque curve for a squirrel-cage induction motor.

The use of larger rotor conductors permits larger induced currents and greater torque is developed at a given slip.

When the speed is low, $s \cong 1$ and the reactive term in Eq. 24-23 is large compared to the resistive term. An approximation for low-speed torque is

$$T_{\text{low speed}} \cong \frac{3V_A^2 R_2}{s\Omega_s X^2} \qquad (24\text{-}25)$$

and we see that near standstill, torque varies inversely with slip. At standstill, $s = 1$ and Eq. 24-25 becomes

$$T_{\text{starting}} \cong \frac{3V_A^2 R_2}{\Omega_s X^2} \qquad (24\text{-}26)$$

We conclude that starting torque is directly proportional to V_A^2 and to R_2.

Maximum torque occurs at an intermediate speed; how could this maximum be predicted? If we differentiate Eq. 24-23 with respect to slip and set the derivative equal to zero, the slip for maximum torque is found to be

$$s = \frac{R_2}{\sqrt{R_1^2 + X^2}} \qquad (24\text{-}27)$$

At this value of slip, the maximum or *breakdown* torque is

$$T_{\text{max}} = \frac{3V_A^2}{2\Omega_s(R_1 + \sqrt{R_1^2 + X^2})} \qquad (24\text{-}28)$$

An analysis of the effect of R_2 leads to the following conclusions:

- At normal speeds, torque is inversely proportional to R_2.
- At low speeds, torque increases with increasing R_2.
- The slip for maximum torque is directly proportional to R_2.
- The maximum torque is independent of R_2.

These conclusions are supported by the speed-torque curves for similar motors with varying values of R_2 shown in Fig. 24.23. Apparently the design of the squirrel-cage

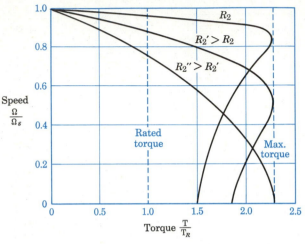

Figure 24.23 The effect of R_2 on speed-torque curves.

rotor is a compromise. Low rotor resistance results in low slip for a given torque and high operating efficiency; high rotor resistance results in high starting torque and relatively low starting current.

EXAMPLE 7

Predict the starting torque and maximum torque for the 6-pole motor of Example 6 ($R_1 = 0.05$ Ω, $R_2 = 0.06$ Ω, $X = 0.46$ Ω, and $V_A = 122$ V) and estimate the rotor resistance for maximum torque at starting.

By Eq. 24-26, the starting torque is

$$T_{start} \cong \frac{3V_A^2 R_2}{\Omega_s X^2} = \frac{3(122)^2 0.06}{(1200 \times 2\pi/60)(0.46)^2} = 101 \text{ N·m}$$

By Eq. 24-28, the maximum torque is

$$T_{max} = \frac{3V_A^2}{2\Omega_s(R_1 + \sqrt{R_1^2 + X^2})}$$

$$\cong \frac{3(122)^2}{2 \times 40\pi(0.05 + 0.46)} \cong 348 \text{ N·m}$$

since R_1^2 is small compared to X_2^2.

To obtain maximum torque at starting, R_2 is given by Eq. 24-27 for $s = 1$ or

$$R_2 = \sqrt{R_1^2 + X^2} = \sqrt{(0.05)^2 + (0.46)^2} \cong 0.46 \text{ Ω}$$

Motors with Variable Rotor Resistance

How can the insertion of additional resistance in the rotor increase the starting torque? At first thought it would seem that insertion of resistance would decrease the induced current and therefore the torque. But at starting, $s = 1$ and $R_2/s = R_2$, which is small compared to X_2, so the magnitude of the current (Eq. 24-15) is affected only slightly by an increase in R_2. In contrast, the power factor (cos θ in Eq. 24-19) is greatly improved and the torque is increased. An alternative viewpoint is that δ is brought closer to 90° and the current I_2 (and therefore B_2; see Fig. 24.15c) bears a more favorable relation to the flux density B_1.

How can the resistance be increased during starting for high torque and reduced to a minimum during normal operation for low slip and high efficiency? One method is to arrange connections to the rotor coils (by using sliprings and brushes) so that additional resistances can be connected in series. In such a *wound-rotor* motor, the rotor winding is similar to that on the stator; an auxiliary *controller* permits the insertion of variable external resistances corresponding to R_2' and R_2'' in Fig. 24.23. The wound-rotor motor provides optimum starting conditions (high torque and low current) and also provides a measure of speed control since, for a given torque, the speed depends on rotor resistance. The disadvantages of the wound-rotor motor include higher first cost, higher maintenance cost, and lower efficiency during variable-speed operation.

The *double-squirrel-cage* motor is a clever design for obtaining some of the advantages of variable resistance without the disadvantages. Each rotor conductor consists of two parts, one in the normal location and a second deeply imbedded in the rotor iron (Fig. 24.24). The leakage reactance and therefore the impedance of the inner

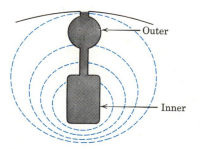

Figure 24.24 Double-rotor conductors and leakage flux.

conductor are higher than those of the outer conductor. At starting, rotor frequency is high and little current flows in the inner conductor; the effective rotor resistance is high. At normal speeds, slip is low, rotor frequency is low, the two conductors are in parallel, and the effective rotor resistance is low. (How would the speed-torque curve of such a motor differ from those in Fig. 24.23?)

Performance of Induction Motors

The similarity between the induction motor and the transformer has been noted; in fact, a wound-rotor motor can be used as a static transformer with a three-phase secondary voltage available at the slip rings. Another similarity is in the method of determining the machine parameters in the circuit model. A *no-load* motor test provides data for the determination of G_0 and B_0 just as an open-circuit transformer test. Corresponding to the short-circuit transformer test is the *blocked-rotor* test; with the rotor held stationary, all the input power is dissipated as losses and R and X can be determined.

The stator reactance X_1 is usually assumed to be half of X, but the division of R must be done carefully because R_2 is a key factor in motor performance. One approach is to measure the dc resistance between terminals which gives an approximate value of $2R_1$, assuming a Y-connected stator. The dc value is then corrected empirically to give an effective ac value of R_1 and the remainder of R is the effective value of R_2 referred to the stator circuit (see Problem 7). When the model parameters are known, the performance of any motor may be predicted from the mathematical relations or from the circuit model itself.

EXAMPLE 8

Predict the input, output, and efficiency of the 1200-rpm motor in Example 6 at a slip of 2.8%. For the model in Fig. 24.25 the parameters are: $V_A = 122$ V, $R_1 = 0.05\ \Omega$, $R_2 = 0.06\ \Omega$, and $X = 0.46\ \Omega$. The no-load power input is 690 W at 20 A and 220 V.

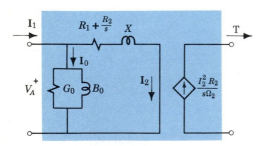

Figure 24.25   Induction motor performance calculations.

At $s = 2.8\%$, the rotor current is

$$\mathbf{I}_2 = \frac{V_A}{R_1 + R_2/s + jX} = \frac{122\,\underline{/0^\circ}}{0.05 + 0.06/0.028 + j0.46}$$

$$= \frac{122\,\underline{/0^\circ}}{2.19 + j0.46} \cong 54.5\,\underline{/-12^\circ}\ \text{A}$$

and

$$\mathbf{I}_1 = \mathbf{I}_0 + \mathbf{I}_2 = 20\,\underline{/-85^\circ} + 54.5\,\underline{/-12^\circ}$$

$$= 63.3\,\underline{/-29.6^\circ}\ \text{A}$$

The total input power is

$$3V_1 I_1 \cos\theta = 3 \times 127 \times 63.3 \times 0.87 \cong 21\ \text{kW}$$

The total torque is (by Eq. 24-23)

$$T = \frac{3I_2^2 R_2}{s\Omega_s} = \frac{3(54.5)^2 \times 0.06}{0.028 \times 40\pi} \cong 152\ \text{N·m}$$

and the output power is

$$P_{\text{hp}} = \frac{T\Omega}{746} = \frac{152 \times (1 - 0.028)40\pi}{746} \cong 25\ \text{hp}$$

The efficiency is

$$\frac{\text{Output}}{\text{Input}} = \frac{25 \times 0.746}{21} = 0.89 \text{ or } 89\%$$

The polyphase squirrel-cage induction motor is a rugged, efficient, easily started, nearly constant-speed machine that is applicable to many purposes. The wound-rotor version provides control of speed and starting torque in return for an increased investment. Induction motors are particularly useful for hoists, mixers, or conveyors that require a smooth start under load. An interesting variation with possible applications for the future is the *linear induction motor* in which the stator windings are arranged on a flat surface to give a linearly moving field.

Single-Phase Induction Motors

To obtain the rotating field necessary for unidirectional torque, a polyphase winding is required. With a single-phase winding, an oscillating field is possible, but the axis of the stator field remains fixed in space. If a squirrel-cage rotor is turning in such an oscillating field, a pulsating torque is developed, which tends to bring the rotor to nearly synchronous speed. If the rotor is stopped and then started in the opposite direction, the developed torque is in the opposite direction and again the rotor is accelerated. The problem in single-phase induction motor design is to get the rotor started. (In the series-wound *universal* motor, the commutator insures unidirectional torque.)

There are several ingenious methods for doing this.[†] In the *shaded-pole* motor (Fig. 24.26a), a heavy copper coil is wound around half of each salient stator pole. Induced currents in the shorted turn delay the buildup of magnetic flux in that region

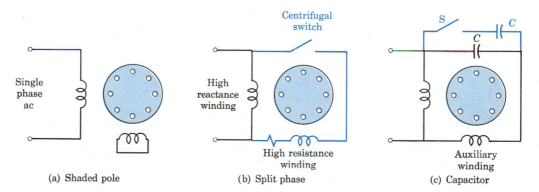

(a) Shaded pole　　　　(b) Split phase　　　　(c) Capacitor

Figure 24.26　Single-phase induction motors.

of the pole. The magnetic flux vector appears to shift as a function of time and the rotor experiences the effect of a partially rotating field.

The *split-phase* motor employs two separate windings having different reactance-resistance ratios. The current reaches its maximum in the high-reactance winding at a later time and the rotor experiences a shift in magnetic field that provides the necessary starting torque. When the motor is nearly up to speed, the high-resistance winding is disconnected by a centrifugal switch.

[†]For further discussion see Chapter 6 of McPherson, *op cit.*

The *capacitor* motor employs a capacitor in series with an auxiliary winding to provide the necessary phase shift. For improved performance two capacitors are used. The larger provides good starting torque and then is switched out by a centrifugal switch. The smaller remains in the circuit to provide good operating efficiency and power factor.

Adjustable Speed Drives

The application of ac motors, synchronous or induction, can be greatly expanded by providing an adjustable frequency power supply. The SCR makes it possible to provide stepless speed control at high efficiency and reasonable cost so that the flexibility of dc machines can be obtained without their inherent maintenance problems and speed limitations.

One approach is to use an SCR *rectifier* to convert ac power to direct current and an SCR *inverter* to provide three-phase, adjustable frequency, adjustable voltage power. In the inverter shown in Fig. 24.27a, six pairs of controlled rectifiers are used to connect the primaries of a transformer to the dc supply. If SCRs 1 and 1′ are fired so that they are ON at the same time, the dc voltage is applied to winding a–a'; when SCRs 4′ and 4 are ON, the dc voltage is applied to the same winding with the reverse polarity. The transformer input is an alternating square wave whose frequency is determined solely by the firing or *gate circuit*.

The magnitude of output voltage is reduced by retarding the conduction period of the second set of SCRs. In Fig. 24.27c, the delay angle is 30° and the simultaneous conduction angle is 150°. The resulting waveforms of phase voltages v_{aa} and v_{bb}, and the corresponding line-to-line voltage, are shown in Fig. 24.27d. The transformer output voltage will approximate a sine wave. To assure adequate torque at all speeds (and maintain optimum flux densities), the output voltage is increased as the frequency increases.

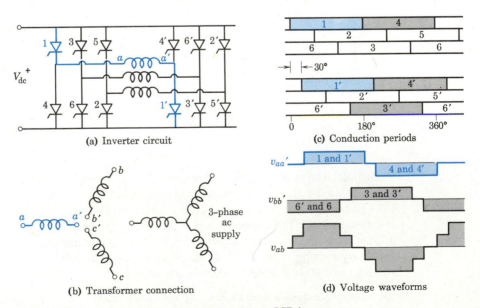

(a) Inverter circuit

(c) Conduction periods

(b) Transformer connection

(d) Voltage waveforms

Figure 24.27 Adjustable frequency supply using an SCR inverter.

With an adjustable frequency drive, speed regulation is precise, starting is smooth, and acceleration or deceleration may be precisely timed. Several motors may be run in synchronism at any selected speed. These characteristics are particularly desirable in the steel and textile industries.

SUMMARY

- In a synchronous machine, the basic relations are

$$E = kn\Phi = k'I_F \qquad \text{and} \qquad T_d = k_T\Phi_R\Phi_S \sin\delta$$

  The speed of the rotating stator field is $n_s = 120\,f/p$.
  Unidirectional torque is developed only when the rotor turns at n_s.
  The direction of power flow depends on torque angle δ.
  Power transformations are explained by phasor diagrams.

- In an alternator, $\mathbf{E} = \mathbf{V} + j\mathbf{I}X_s$ and δ is positive.
  The power factor of the load determines the required E and I_F.
  Mechanical power is absorbed when the rotor is advanced.

$$P_M = 3T_d\Omega = P_E = 3VI\cos\theta = 3\frac{VE}{X_s}\sin\delta$$

- In a synchronous motor, $\mathbf{V} = \mathbf{E} + j\mathbf{I}X_s$ and δ is negative.
  Motor power factor is determined by E and therefore by I_F.
  Electrical power is absorbed when the rotor is retarded.
  Pull-out torque occurs at $\delta = 90°$ with a cylindrical rotor.
  There is no inherent starting torque; usually start by induction motor action.

- In an induction motor, the slip of the rotor coils behind the rotating stator field induces rotor currents that develop torque; $s = (n_s - n)/n_s$.
  The basic energy conversion relations are

$$E_g = I_2R_2\frac{1-s}{s} \qquad \text{and} \qquad T_d = \frac{E_gI_2}{\Omega} = \frac{I_2^2R_2}{s\Omega_s}$$

- Use of an adjusted voltage V_A simplifies the induction motor circuit model. In terms of model parameters the total three-phase torque is

$$T = \frac{3V_A^2}{\Omega_s}\frac{R_2/s}{(R_1 + R_2/s)^2 + X^2}$$

  Useful approximations for normal operation and for starting are

$$T_{normal} \cong \frac{3V_A^2 s}{\Omega_s R_2} \qquad \text{and} \qquad T_{starting} \cong \frac{3V_A^2 R_2}{\Omega_s X^2}$$

  Maximum torque is independent of R_2, but it occurs at a speed determined by R_2. In the wound-rotor motor, R_2 is adjustable.

- Single-phase induction motors depend on a partial rotation of the stator field for starting torque; the running torque is pulsating.

REVIEW QUESTIONS

1. Distinguish between a salient-pole and a cylindrical-rotor machine.
2. Explain the nature of synchronous reactance.
3. Why is the angle between **E** and **V** equal to that between Φ_R and Φ_S?
4. What happens to alternator voltage when the load pf changes?
5. Explain the production of a rotating field by stationary coils.
6. How does a synchronous motor react to an increase in shaft load?
7. What happens if the mechanical input to an alternator is increased continuously?
8. How does reluctance torque affect the pull-out torque of a motor?
9. How can a synchronous motor be used to correct power factor?
10. How would you reverse the rotation of a synchronous motor? An induction motor? A dc motor?
11. Discuss and compare the value of δ in commutator, synchronous, and induction motors.
12. Explain how a short-circuited rotor produces torque.
13. Specify several possible operating speeds for 50-Hz motors.
14. Why is the rotor core loss usually negligibly small?
15. Draw the phasor diagram for Example 6 to scale.
16. Why is induction-motor torque directly proportional to slip at some speeds and inversely proportional at other speeds?
17. Outline an experimental procedure for determining induction-motor parameters.
18. Examine and classify a single-phase motor and explain its operation.
19. Explain how an inverter changes dc to polyphase ac.
20. How is the frequency adjusted in an inverter?

EXERCISES

1. Determine the operating speed of the following 3-phase alternators:
   (a) 12,000 V, 24 pole, 60 Hz.
   (b) 440 V, 4 pole, 50 Hz.
   (c) 220 V, 6 pole, 400 Hz.
2. Determine the number of poles of the following 3-phase alternators:
   (a) 600 V, 600 rpm, 60 Hz.
   (b) 440 V, 750 rpm, 50 Hz.
   (c) 200 V, 12,000 rpm, 400 Hz.
3. A portable, single-phase ac generator is rated at 5 kW at 125 V at 0.8 pf lagging ($R_A = 0.15\ \Omega$, $X_s = 0.6\ \Omega$). Represent the generator by a model and draw phasor diagrams for rated current at: (a) lagging, (b) unity, and (c) leading pf.
4. The generator of Exercise 3 ($E = 150$ V) is connected to a load Z_L. Draw the circuit model, calculate the output current and voltage, and draw the phasor diagram for:
   (a) $Z_L = 4 + j0\ \Omega$
   (b) $Z_L = 3.85 + j2.4\ \Omega$
   (c) $Z_L = 3.85 - j3.6\ \Omega$
5. At a hydroelectric power plant, the 3-phase 600-rpm alternators are rated at 360,000 kW at 18,000 V line-to-line and 0.85 pf lagging. The synchronous reactance per phase[†] is 1 Ω. For rated conditions, determine:
   (a) The number of poles and the current.
   (b) The generated emf and power angle.
   (c) The phasor diagram.
6. A 3-phase, 6-pole synchronous machine is rated at 10 kVA, 220 V, 60 Hz. The magnetization curve (volts rms) is similar to Fig. 24.2b and X_s per phase is 4 Ω.[†]
   (a) Draw an appropriate circuit model for alternator operation.
   (b) Estimate the field current for rated voltage at no load.
   (c) Repeat part (b) for full load at 0.8 pf leading.
   (d) Repeat part (b) for full load at 0.8 pf lagging.
7. The field current of the machine of Exercise 6 is adjusted to give rated voltage at rated current at unity pf. Use a circuit model and phasor diagrams to predict the new output voltage and current if the load is changed to $6 - j10\ \Omega$.
8. A 40-kVA, 220-V, three-phase alternator requires a field current of 2 A for rated voltage at no load. With $I_F = 2$ A and $\mathbf{I}_A = 100\underline{/-70°}$ A,

[†]All 3-phase machines are assumed to be Y-connected.

the terminal voltage is only 133 V. Draw the phasor diagram and evaluate X_s.

9. Continue Fig. 24.6d by drawing the flux vectors corresponding to $\omega t = 180°$ and $210°$. On the stator of Fig. 24.6c, show the location of N and S poles for $\omega t = 180°$.

10. (a) Given a three-phase line with current phasors as shown in Fig. 24.6a, construct a wiring diagram similar to that shown in Fig. 24.6c so that a three-phase synchronous motor will turn in a clockwise direction.
    (b) Assume only two of the phase windings of the motor are connected to the line in part (a). Investigate the possibility of producing a circular rotating stator field by adjusting the relative amplitudes of the two phase voltages.

11. The machine of Exercise 6 is operating as a synchronous motor at rated voltage and current. The field current is increased slightly and it is noticed that the stator current decreases. Draw an appropriate phasor diagram and determine whether the motor is operating with unity, lagging, or leading pf. Explain your reasoning.

12. In deriving a circuit model for predicting the behavior of a synchronous motor, how is the model affected by: (a) neglecting electrical losses, (b) neglecting mechanical losses, (c) assuming a linear magnetization curve, (d) assuming steady-state operation. Draw a circuit model incorporating the above simplifications.

13. A synchronous motor is represented by a simple circuit model. With the motor unloaded, the field current is adjusted until the stator input current I is a minimum $(E \cong V)$. Then the motor is loaded until the torque angle δ is $20°$.
    (a) Draw a labeled phasor diagram for the electrical end of the model, using input voltage as a horizontal reference.
    (b) If the field current is then decreased by 30%, draw the new phasor diagram and predict the new power angle and power factor.

14. A 100-hp, 1200-rpm, 60-Hz synchronous motor $(X_s = 2 \ \Omega/\text{phase})$ is rated at 440 V, 131 A, 0.8 pf leading with $I_F = 6$ A. For rated conditions:
    (a) Draw and label a simple circuit model and construct the phasor diagram.
    (b) Predict generated emf E and angle δ.

15. Repeat Exercise 14 for an 8-pole synchronous motor with per-phase ratings of 200 V and 20 A $(X_s = 5 \ \Omega)$.

16. For the synchronous motor of Exercise 14 operating at rated line current as a "synchronous condenser" without shaft load, draw the phasor diagram and estimate the required field current.

17. The per-phase ratings of a synchronous motor are: 300 V, 15 A, and $X_s = 10 \ \Omega$. Field current for minimum stator current with no shaft load is 2 A. For operation as a synchronous condenser, estimate the field current required for full-rated kVAR. Draw a labeled phasor diagram.

18. When the motor of Exercise 15 is idling with no shaft load, the field current for minimum stator current is 1 A. When the motor is loaded (rated voltage and current at 0.8 leading pf), estimate (stating assumptions):
    (a) Required field current.
    (b) Total developed power (W).
    (c) Total developed torque.

19. Blackouts may occur if heavily loaded alternators are suddenly disconnected from the line to prevent their being pulled out of synchronism. Predict the pull-out power of the alternator of Exercise 5.

20. For the alternator of Exercise 6:
    (a) Draw a simplified model and calculate the rated line current.
    (b) Draw a phasor diagram and estimate the generated emf and the torque angle under rated conditions.
    (c) Estimate the pull-out torque for this machine.
    (d) List all assumptions made in the previous calculations.

21. When the synchronous motor of Exercise 14 is operated at rated conditions:
    (a) What mechanical power (W) is supplied?
    (b) What reactive power can be "supplied"?
    (c) Why is this reactive power significant?

22. A synchronous motor carries a load that results in a power angle $\delta = 30°$; field current is adjusted for minimum stator current.
    (a) Draw a labeled model and phasor diagram for this condition.
    (b) If the load torque is halved, predict the effect on speed, power angle, power factor, and line current.

23. When operating under rated conditions at unity power factor, a synchronous motor draws 30 A from the ac line; the field current is 2 A and $\delta = 20°$. For a constant load (i.e., power out-

put), calculate two additional points and sketch a curve of I versus I_F.

24. A plant load consisting of 850 kW at 0.6 lagging pf is supplied by a 1600-kVA transformer. An additional 400 kW of power is required. If the additional load is carried by synchronous motors at 0.8 pf leading, will a new transformer be required? What will be the new plant pf?

25. The synchronous motor of Fig. 24.28 (built with damping windings) is to be started, brought up to speed, and then adjusted for minimum line current before mechanical load is applied.
   (a) Complete the wiring diagram, including any necessary auxiliary equipment.
   (b) Outline the step-by-step procedure.
   (c) What two functions do the damping windings serve?

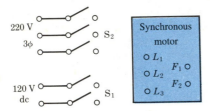

Figure 24.28

26. Determine the number of poles, the slip, and the frequency of the rotor currents at rated load for 3-phase, 10-hp induction motors rated at:
   (a) 220 V, 60 Hz, 1728 rpm.
   (b) 220 V, 50 Hz, 900 rpm.
   (c) 120 V, 400 Hz, 5700 rpm.

27. For a 3-phase, 60-Hz, 15-hp induction motor at rated speed of 1740 rpm, determine:
   (a) The speed of the rotor with respect to the stator (rpm).
   (b) The speed of the rotor field with respect to the rotor.
   (c) The speed of the rotor field with respect to the stator field.
   (d) The frequency of the currents in the rotor.

28. The rotor of the motor in Exercise 27 is 20 cm in diameter and 15 cm long. The maximum air-gap flux density is 1 T. Stating any necessary assumptions, estimate the emf (rms) induced in each rotor conductor at standstill and at full load.

29. The no-load data of Example 6 are used with Fig. 24.19 to calculate the adjusted voltage V_A in the simplified model of Fig. 24.20.
   (a) Draw, to a large scale, a phasor diagram with V_1 as a horizontal reference showing I_0, $I_0 Z_1$, and V_A.

(b) Explain, briefly, why the algebraic calculation of V_A is nearly as accurate as the vector calculation.
   (c) How does use of the model of Fig. 24.20 simplify calculations under load conditions?

30. The per-phase parameters for a 6-pole, 60-Hz, 220-V induction motor are: $R_1 = R_2 = 0.1 \ \Omega$ and $X_1 = X_2 = 0.5 \ \Omega$. The no-load current is $I_0 = 17 \underline{/-78°}$ A. Predict the input current and the horsepower output at 1140 rpm.

31. To obtain motor parameters in the lab, we use the model of Fig. 24.19. With "no load" ($s \cong 0$), per-phase readings are: $V_1 = 254$ V, $I_1 = 13.7$ A, and $W_1 = 900$ W. With the rotor "blocked" ($s = 1$), $V_1 = 91.8$ V, $I_1 = 45$ A, and $W_1 = 810$ W.
   (a) What quantities become negligible under the test conditions?
   (b) Draw simplified models for the "no-load" and "blocked" cases.
   (c) Evaluate the parameters from the lab data.
   (d) Predict the torque developed by this 3-phase, 4-pole, 60-Hz motor at 5% slip.

32. A 100-hp, 3-phase, 440-V, 4-pole, 60-Hz induction motor has the following per-phase parameters: $R_1 = 0.06 \ \Omega$, $R_2 = 0.08 \ \Omega$, $X_1 = X_2 = 0.3 \ \Omega$. The no-load power input is 3420 W at 45 A.
   (a) Draw a simplified circuit model and label it.
   (b) Calculate B_0 and G_0.
   (c) Predict the line current, the developed torque, and the horsepower output at 2% slip.

33. For the motor of Exercise 32, calculate the total torque at 5% slip from the model of Fig. 24.21 and from Eq. 24-24 and draw a conclusion.

34. For the motor of Exercise 32, predict:
   (a) The starting torque.
   (b) The breakdown torque.

35. (a) Estimate the speed of the motor in Exercise 26a at 6 hp output.
   (b) Repeat for the motor in Exercise 26c at 12 hp output.

36. Estimate the starting torque and the breakdown torque of the induction motor of Exercise 30.

37. A three-phase induction motor (440 V, 60 Hz, 6-pole) has the following per-phase parameters: $R_1 = 0.3 \ \Omega, X_1 = 1 \ \Omega, R_2 = 0.2 \ \Omega, X_2 = 1 \ \Omega$, $G_0 = 20$ mS, and $B_0 = -60$ mS.
   (a) Estimate the no-load input current and power.
   (b) Predict the power output at 5% slip.

(c) Predict the motor power factor at 5% slip.

38. A three-phase, 4-pole, 220-V, 60-Hz induction motor has the following per-phase parameters: $R_1 = 0.4\,\Omega$, $X_1 = 1\,\Omega$, $R_2 = 0.6\,\Omega$, $X_2 = 1\,\Omega$, $G_0 = 30$ mS and $B_0 = -40$ mS. Approximately half the no-load input goes to rotational losses.
   (a) Estimate the no-load input current and the no-load speed.
   (b) Estimate the total torque developed at 5% slip.

39. For the induction motor in Exercise 37 estimate:
   (a) The starting torque.
   (b) The starting torque if the line voltage drops to 400 V.

40. Two squirrel-cage induction motors are identical except that the rotor of motor A is made of aluminum of conductivity σ and the rotor of motor C is made of copper of conductivity 1.5 σ. The starting torque of motor A is 120 N·m. Predict the starting torque of motor C.

41. A 100-hp, 440-V, 1200-rpm, 60-Hz synchronous motor is operated from an adjustable frequency supply. For operation at 330 V and 45 Hz with the same stator current and field current, draw the phasor diagrams and estimate the speed, torque, and power output.

42. Specify the type of electric motor that will meet the following requirements best:
   (a) Provide constant speed under all loads.
   (b) Provide an easily adjusted speed which remains nearly constant (within 10%) over a wide range of loads.
   (c) Operate from ordinary "house current" and be cheap. (Starting torque is not important.)
   (d) Operate from a dc line and provide relatively high starting torque.
   (e) Provide rugged, maintenance-free operation and nearly constant speed (within 5%) over a wide range of loads.
   (f) Operate from the single-phase ac line and provide relatively high starting torque.
   (g) Operate on ac, provide high starting torque, and permit some variation of speed under load.
   (h) Draw a leading-pf current while driving a mechanical load.

PROBLEMS

1. The inductive element in Fig. 24.29 might be a transmission line or the air gap of a rotating machine. If $\mathbf{E}_1 = E_1\underline{/0°}$ and $\mathbf{E}_2 = E_2\underline{/\alpha}$ where E_1 and E_2 are constant and α is a variable angle, derive a general expression for the power transferred and determine α for maximum power transfer.

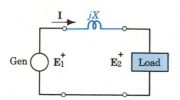

Figure 24.29 Power transfer.

2. A synchronous motor is delivering 1000 hp to a constant load; I_F is adjusted for minimum stator current and $\delta = 30°$. Explaining your reasoning with a large-scale phasor diagram, predict the effect of a 20% increase in I_F on (a) speed, (b) power angle, (c) power factor, (d) line current.

3. A 3-phase synchronous motor is rated at 100 hp, 440 V, 131 A, 1200 rpm at 0.8 pf leading; $X_s = 2\,\Omega$ and $I_F = 6$ A for rated conditions.

(a) What reactive power can be supplied while the motor carries full mechanical load?
(b) If the motor is operated as a synchronous condenser without shaft load, estimate the field current required and the capacitive kVAR available.

4. A transmission line can be modeled (Fig. 24.30 on a per-phase basis) by an inductive reactance $X_T = 20\,\Omega$. The load consists of a resistance $R_L = 100\,\Omega$. For a terminal voltage $V = 3000$ V, the generator voltage is $V_G = 3060$ V.

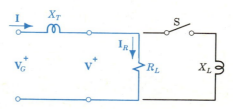

Figure 24.30

(a) Draw a phasor diagram for switch S open.
(b) When an inductive load $X_L = 75\,\Omega$ is connected in parallel with R_L, the terminal voltage drops to approximately 2480 V. Keeping the same V phasor as a reference, modify the phasor diagram to show why under the new

condition V_G must be considerably larger than V.

(c) A synchronous condenser ($X_s = 25 \; \Omega$), whose magnetization curve is described by $E = 200 I_F$, is available for connecting across the terminals at V so that the terminal voltage will be 3000 V with X_L connected. Modify the phasor diagram to include the effect of the synchronous condenser and specify the necessary field current I_F.

(d) Discuss the use of synchronous condensers in voltage regulation.

5. Predict the dynamic behavior of a synchronous motor with rotor inertia J turning at speed Ω_s when a sudden load ΔT is applied to the shaft.

(a) Write a simplified expression for T_d as a function of δ for constant V and Φ_R and small values of δ.

(b) Write an expression for Ω during the time δ is changing to accommodate an increased load.

(c) Neglecting frictional resistance to changes in δ, derive an equation governing the dynamic behavior $\delta(t)$ under a suddenly applied torque.

(d) Determine the form of the variation of δ with time after the torque is applied and draw a conclusion.

6. Determine just enough points to sketch curves showing speed, line current, power factor, and efficiency versus power output for the motor of Exercise 37 in the normal operating region.

7. Laboratory tests of a three-phase, Y-connected, 6-pole, 150-hp, 4000-V induction motor gave the following results:

No-load: 4000 V, 7 A, 7 kW

Blocked rotor: 1340 V, 32 A, 18.2 kW

Effective ac resistance between terminals = 1.4 Ω

(a) Determine the model parameters.

(b) Predict the power output at 1152 rpm.

(c) Predict the efficiency and power factor at rated load.

25

Automatic Control Systems

Control System Characteristics

Dynamic Responses

Feedback Control Systems

One characteristic of life in a technologically developed nation is the use of great amounts of power. A second distinguishing characteristic is the ability to control that power precisely. At one extreme of the power spectrum is the multimegawatt electric power system so precisely controlled that a simple reluctance motor driven by the system becomes an accurate timing device. At the other extreme are the milli-microwatt impulses sent by a cardiac pacemaker to stimulate a defective heart into rhythmic, life-saving contractions. Between these extremes are innumerable applications of control to heating systems, assembly plants, metallurgical processes, and oil refineries and to traffic flow, water supply, medical treatment, aircraft landing, and satellite guidance.

Despite their great variety, control systems can be analyzed into a few basically similar components. In Watt's *flyball governor*, perhaps the first *automatic* control device, a rotating spindle carrying flyweights (Fig. 25.1) is driven by the governed engine. The engine speeds up until the centrifugal force of the flyweights overcomes the force of the speed-adjusting spring and partially closes the throttle valve. An increase in load on the engine momentarily reduces its speed, reduces the centrifugal force

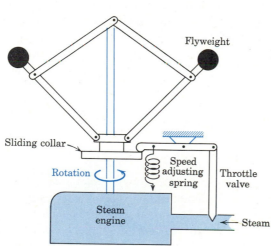

Figure 25.1   A flyball governor.

of the flyweights, allows the throttle valve to open, and accelerates the engine until the set speed is reached. The components of the governor are shown in the block diagram of Fig. 25.2. General terms for the functions performed and the variables existing at various points in the system are shown in italics.

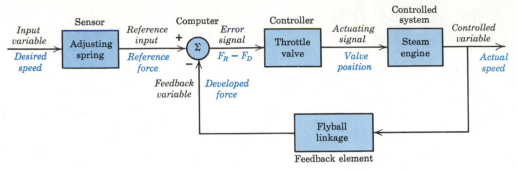

Figure 25.2   Block diagram of a flyball governor.

Most of the elements employed in control systems are familiar to us from our previous study. Transducers are used to measure variables and convert them to electrical form. Computers or comparators are used in determining the difference between reference and feedback variables. Controllers and amplifiers are used to develop signals to actuate the controlled system. Motors are used to provide the "muscle" for system control. Feedback elements are used to facilitate the comparison of output and reference variables. Note that only signals appear in the block diagram of a control system. Power inputs (steam to the engine of Fig. 25.2 or heat from the furnace in Fig. 25.4) are essential to the operation, but ordinarily they do not enter into control calculations.

The study of automatic control represents a continuation of our analysis of linear systems, the central topic of this book, and it provides an opportunity to apply our knowledge of circuit principles, device characteristics, and system techniques. First, we look at the operation of some illustrative examples to discover the essential characteristics of control systems, the benefits obtained by using feedback, and the problems that arise in system design. Then we extend our methods of circuit analysis to permit handling systems incorporating a variety of different elements, each described by a transfer function. Using the new method, we next determine the dynamic behavior of some electromechanical devices. Finally, we consider how to obtain control systems with the necessary accuracy and stability.

CONTROL SYSTEM CHARACTERISTICS

The flyball governor is an example of a *regulator* that maintains a speed, frequency, or voltage constant within specified limits. In a regulator the reference is constant, although it may be adjustable. In a *follower* such as the "self-balancing potentiometer" (Fig. 25.3), the variation of the output displacement duplicates the variation of the input voltage. If the output variable is a translational or angular

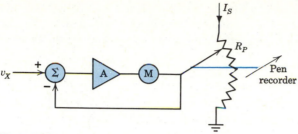

Figure 25.3   A self-balancing potentiometer.

displacement such as the aircraft heading in an autopilot (Fig. 9.1), the follower is called a *servomechanism*. Regulating and following can be performed accurately and rapidly using systems employing feedback.

Open-Loop and Closed-Loop Systems

An ammeter is an example of a useful *open-loop* device. To obtain the desired accuracy, the magnet is carefully formed and patiently aged, the moving coil is exactly designed, and the mechanism is precisely fabricated. After the instrument is calibrated it is assumed that the output deflection is an accurate indication of the input current.

A similar device could be used to control the heating of a home by providing a fuel valve opening directly related to the outside temperature. For comfort, however, the valve controller should also take into account such factors as wind velocity, sun radiation, and the number of people in the room. If sufficient data were available, some of these factors could be taken into account by a computer (Fig. 25.4a), but we

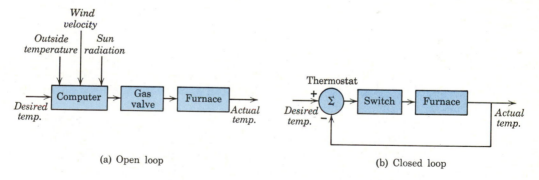

(a) Open loop

(b) Closed loop

Figure 25.4   Temperature control systems.

suspect that there must be a better method of control. In the *closed-loop* system of Fig. 25.4b, a simple thermostat is used to compare the actual temperature with the desired temperature. Whenever the actual temperature is less than the desired temperature, the furnace is ON; otherwise it is OFF.[†] This simple on-off control system illustrates two

---

[†]In practice there is some *dead zone* in which the temperature difference is not sufficient to activate the thermostat.

general benefits of closed-loop systems. It automatically takes into account factors whose effects are known imprecisely or not at all, and it provides accurate overall performance using inexpensive elements of low individual accuracy.

The Control Problem

A characteristic of closed-loop systems which may be disadvantageous is revealed if we assume that the home is heated by means of hot water circulating in pipes buried in a concrete slab. The energy storage capacity of the slab introduces a time lag in the response of the heating system. If the time lag is several hours, as it might well be, the slab would not begin to radiate heat until long after the thermostat turned ON, and it would continue to radiate long after the thermostat turned OFF. The combination of the amplification needed for accurate control and the energy storage inherent in any physical system may result in erratic behavior.

Providing a feedback path in a system containing amplification may create instability. In a *stable* system the response to an impulse disturbance dies away as time increases (Fig. 25.5a). In an *unstable* system a sudden disturbance may give rise to

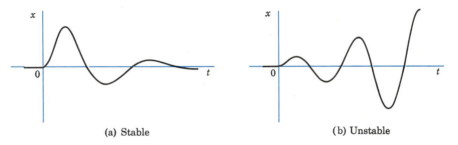

(a) Stable	(b) Unstable

Figure 25.5 Impulse response of different systems.

sustained oscillations or to an uncontrolled increase in a critical variable. By definition, an unstable system is incapable of control. Unfortunately, the requirements of accuracy and stability are incompatible and this creates a difficult design problem for the control engineer.

For high accuracy, only a small error signal (see Fig. 25.2) can be tolerated (in the steady state) and, therefore, the controller must possess high amplification. With high amplification, however, the correction is great and this may lead to instability. Stability can be restored by auxiliary elements that increase the cost and weight and reduce the reliability. The control engineer's problem is to provide the necessary accuracy and speed of response with adequate stability and reliability at a minimum cost and weight.

System Response

One approach to design is to synthesize a possible system, analyze its performance, and compare its performance with the specifications. The performance in terms of accuracy and stability may be deduced from the response of the system to various inputs. In predicting the response, it is convenient to assume that the system is linear (i.e., describable by a linear differential equation) and that the input is an impulse, a step, or a sinusoid.

The basic control system with negative feedback is shown in Fig. 25.6 with

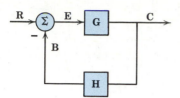

Figure 25.6   The basic feedback control system.

sinusoidal signals represented by phasors and transfer functions shown as complex quantities. The controlled variable **C** is fed back through feedback element **H** and the signal **B = HC** compared to the reference input **R.** In control system nomenclature, the ratio of return signal **B** to error signal **E** is the *open-loop transfer function* **GH.** For this form of the basic feedback circuit, the *closed-loop gain* is

$$\mathbf{G}_F = \frac{\mathbf{G}}{1 + \mathbf{GH}} \qquad (25\text{-}1)$$

A great deal can be learned about a system from its frequency response, i.e., the amplitude and phase of the output signal in response to an input signal of constant amplitude and variable frequency. From our study of feedback we know that if at any frequency the open-loop gain **GH** has a magnitude of unity and a phase angle of ±180°, oscillation buildup is possible. The *Nyquist stability criterion* is based on this principle and establishes the necessary conditions for stability.

The Nyquist criterion is applied to steady-state forced response data. An alternative approach is to consider the natural response of the system as revealed by a pole-zero plot in the complex plane. We know from our study of circuits that the poles of the admittance function of a one-port correspond to natural response current components. In Chapter 9 we extended this concept to include transfer functions of the two-ports employed in control systems. By plotting the possible locations of the poles of the transfer function as system parameters are varied, the so-called *root locus*, we can determine whether a system is stable and, if it is not, what changes can provide stability.

DYNAMIC RESPONSE OF ELECTROMECHANICAL DEVICES

Control systems are useful only insofar as they are able to cope with changing conditions. In evaluating a regulator, we are interested in its *dynamic* behavior, its response to a sudden change in load, for example. In a follower there may be a change in the input variable or there may be a *disturbance* at some point in the system. In general, such changes are random and therefore unpredictable except on a statistical basis. For our purposes, a good approach is to determine the dynamic behavior described by the complete response to a step function. An alternative approach is to consider the frequency response for steady-state sinusoidal inputs. In either approach we rely on the transfer function.

EXAMPLE 1

(a) Write the transfer function for each element of the circuit model of a d'Arsonval movement shown in Fig. 25.7a.

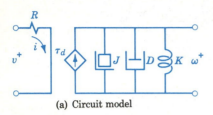

(a) Circuit model

(b) Draw a block diagram for this device.

The governing equations and the corresponding transfer functions are

$$v = Ri \qquad\qquad \frac{I(s)}{V(s)} = \frac{1}{R}$$

$$\tau_d = k_T i \qquad\qquad \frac{T_d(s)}{I(s)} = k_T$$

$$\theta = \int \omega \, dt \qquad\qquad \frac{\Theta(s)}{\Omega(s)} = \frac{1}{s}$$

$$\tau_d = J\frac{d\omega}{dt} + D\omega + \frac{1}{K}\int \omega \, dt \qquad \frac{\Omega(s)}{T_d(s)} = \frac{1}{sJ + D + 1/sK}$$

If the last transfer function is rearranged,

$$\frac{\Omega(s)}{T_d(s)} = \frac{1}{sJ + D + 1/sK} = \frac{sK}{s^2 JK + sDK + 1}$$

and the block diagram is as shown in Fig. 25.7b.

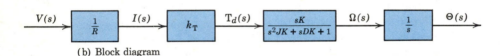

(b) Block diagram

(c) Derive the transfer function for the relation between deflection and applied voltage.

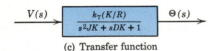

(c) Transfer function

Figure 25.7 Deriving the transfer function of a d'Arsonval movement.

(d) Write the transfer function for sinusoidal excitation.

The transfer functions here correspond to the block "gains" employed in Chapter 9. Following the reasoning displayed in Fig. 9.4, we conclude that the overall transfer function of blocks in cascade is the product of the individual transfer functions or

$$T(s) = \frac{\Theta(s)}{V(s)} = \frac{1}{R} \times k_T \times \frac{sK}{s^2 JK + sDK + 1} \times \frac{1}{s}$$

$$= \frac{k_T(K/R)}{s^2 JK + sDK + 1}$$

For sinusoidal excitation,

$$\mathbf{T} = \mathbf{T}(j\omega) = \frac{\Theta}{\mathbf{V}} = \frac{k_T(K/R)}{1 - \omega^2 JK + j\omega DK}$$

Transfer Functions

The transfer function of a two-port is the ratio of two signals measured at the output and input ports and described by exponentials of the form $a(t) = Ae^{st}$. The two signals may be entirely different in physical nature, but the transfer function is derived and interpreted just as an admittance or impedance function. Block diagram representation is convenient.

Since transfer functions are derived from linear ordinary differential equations, devices characterized by transfer functions must be represented by models made up of lumped, linear elements. The circuit models we have used to represent electrical and mechanical devices satisfy this requirement, so transfer functions can be determined directly as in Example 1.

In some cases further simplification is possible by ignoring factors that have negligible effect on the behavior being investigated. In Example 1, the inductance of the meter coil was ignored for this reason. Once the transfer function is obtained, the complete step response can be determined following the procedure outlined in Chapter 6. The forced response is calculated using the transfer function evaluated at $s = 0$. (How is $s = 0$ related to the step function?) The natural response consists of terms identified with poles of the transfer function. The undetermined coefficients are evaluated from initial conditions. In a control system the initial conditions are usually zero, corresponding to a state of equilibrium prior to the initiation of the change.

First-Order Systems

The order of a system is defined by the highest power of s in the denominator of the transfer function. Frequently encountered first-order transfer functions and their step responses (see Eqs. 9-17 and 9-19) are

$$T(s) = \frac{ks}{s + \alpha} \qquad y(t) = Ye^{-\alpha t} \qquad (25\text{-}2)$$

$$T(s) = \frac{k}{s + \alpha} \qquad y(t) = Y(1 - e^{-\alpha t}) \qquad (25\text{-}3)$$

Example 2 on p. 706 indicates that a shunt motor with constant field current behaves as a first-order system in responding to a step armature voltage input.

Second-Order Systems

In the transfer function of the d'Arsonval movement in Example 1, s appears to the second power in the denominator. By definition, this is a second-order system; the possibility of energy storage in two different forms suggests a more complicated response. However, the transfer function approach is still convenient as illustrated in Example 3 on p. 707.

EXAMPLE 2

A dc shunt motor with constant field excitation is used to position an indicator in a servomechanism. The following simplifying assumptions are reasonable:

Inductance of armature circuit is negligibly small.

Magnetic flux is constant and $K\phi$ is replaced by k_1.

Total friction of motor and load is represented by D.

Total inertia of motor and load is represented by J.

Load torque is negligible compared to D and J.

Derive the transfer function and predict the dynamic speed response.

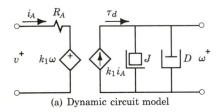

(a) Dynamic circuit model

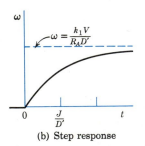

(b) Step response

Figure 25.8   Dynamic response of a first-order system.

On the basis of these assumptions, the circuit model of Fig. 23.9 is modified as shown in Fig. 25.8a. The governing differential equations are

$$J\frac{d\omega}{dt} + D\omega = \tau_d = k_1 i_A \qquad (25\text{-}4)$$

and

$$i_A = \frac{v - k_1\omega}{R_A}$$

Solving for v and collecting terms, we obtain

$$v = \frac{R_A}{k_1}\left(J\frac{d\omega}{dt} + D\omega + \frac{k_1^2}{R_A}\omega\right)$$

Noting that k_1^2/R_A has the same dimensions as the friction coefficient D, let us define an *equivalent friction coefficient D'* where

$$D' = D + \frac{k_1^2}{R_A}$$

Making this substitution and assuming exponential variation of v and ω, we can write the transfer function as

$$T(s) = \frac{\Omega(s)}{V(s)} = \frac{k_1/R_A}{sJ + D'} \qquad (25\text{-}5)$$

This equation is of the same form as Eq. 25-3. The forced response to a step of magnitude V is (for $s = 0$)

$$\omega_f = \frac{k_1 V}{R_A D'}$$

The denominator has a pole at $s = -D'/J$. Therefore, the total step response is

$$\omega = \frac{k_1 V}{R_A D'}(1 - e^{-D't/J}) \qquad (25\text{-}6)$$

as shown in Fig. 25.8b. The speed ω increases until it is limited by equivalent friction torque alone.

Normalized Response of a Second-Order System

The response anticipated in Example 3 is encountered so frequently in control system analysis that it deserves detailed attention. As before (Eq. 4-18b), we define the *undamped natural frequency* as

$$\omega_n = \sqrt{1/JK} \qquad (25\text{-}10)$$

Another useful quantity is the ratio of the actual damping parameter D to D_c, the value for critical damping. Since for critical damping the discriminant is zero, the corre-

EXAMPLE 3

The circuit model and the transfer function of the d'Arsonval movement of Example 1 are shown in Fig. 25.9. With the movement at rest, a dc voltage V is suddenly applied at $t = 0$. Predict the step response of the movement.

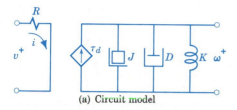

(a) Circuit model

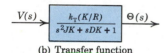

(b) Transfer function

Figure 25.9 Dynamic response of a d'Arsonval meter.

For dc, $s = 0$ in the transfer function and the forced response is

$$\Theta_f = T(0) \times V = \frac{k_T K}{R} V$$

To find the poles of the transfer function, we find the roots of the denominator. Where

$$s^2 JK + sDK + 1 = 0 \qquad (25\text{-}7)$$

the roots are

$$s_1, s_2 = -\frac{D}{2J} \pm \sqrt{\left(\frac{D}{2J}\right)^2 - \frac{1}{JK}} \qquad (25\text{-}8)$$

In general, the complete response is

$$\Theta = \frac{k_T K}{R} V + A_1 e^{s_1 t} + A_2 e^{s_2 t} \qquad (25\text{-}9)$$

Reasoning by analogy from the mathematically similar electrical circuit, we predict that the form of the response is dependent upon the nature of the roots s_1 and s_2. If the roots are real and unequal, the response is overdamped (Fig. 4.8). If the roots are imaginary, the response is underdamped or oscillatory (Fig. 4.10). If the roots are equal, the response is critically damped.

sponding value of D_c is $2\sqrt{J/K}$. The *damping ratio* ζ (zeta) is defined as

$$\zeta = \frac{D}{D_c} = \frac{D}{2\sqrt{J/K}} \qquad (25\text{-}11)$$

The *natural frequency* ω of an underdamped system can be expressed in terms of ω_n and ζ. In this oscillatory case (see Eq. 4-18c), the discriminant is negative and the natural frequency is defined by

$$\omega = \sqrt{\frac{1}{JK} - \frac{D^2}{4J^2}} = \frac{1}{\sqrt{JK}} \sqrt{1 - \frac{D^2}{4J/K}} = \omega_n \sqrt{1 - \zeta^2} \qquad (25\text{-}12)$$

In terms of these definitions, the transfer function of a general second-order system (see Example 3) becomes

$$T(s) = \frac{Y(s)}{X(s)} = \frac{k}{s^2 + 2\omega_n \zeta s + \omega_n^2} \qquad (25\text{-}13)$$

Let us now determine the response $y(t)$ to a step of magnitude X for a system characterized by the transfer function of Eq. 25-13. By inspection, the forced response is

$$y_f = \frac{k}{\omega_n^2} X = Y \qquad (25\text{-}14)$$

The roots of the denominator are determined from

$$s^2 + 2\omega_n\zeta s + \omega_n^2 = 0 \tag{25-15}$$

whence

$$s_1, s_2 = -\omega_n(\zeta \pm \sqrt{\zeta^2 - 1}) \tag{25-16}$$

The complete response is of the form

$$y(t) = \frac{k}{\omega_n^2}X + A_1 e^{s_1 t} + A_2 e^{s_2 t} \tag{25-17}$$

The coefficients A_1 and A_2 can be determined from the initial conditions assuming that at $t = 0^+$, $y = 0$ and $dy/dt = 0$. After several algebraic steps, the normalized response is found to be

$$\frac{y}{Y} = 1 - \frac{\zeta + \sqrt{\zeta^2 - 1}}{2\sqrt{\zeta^2 - 1}}e^{-(\zeta - \sqrt{\zeta^2-1})\omega_n t} + \frac{\zeta - \sqrt{\zeta^2 - 1}}{2\sqrt{\zeta^2 - 1}}e^{-(\zeta + \sqrt{\zeta^2-1})\omega_n t} \tag{25-18}$$

Choosing y/Y and $\omega_n t$ as the normalized variables, the response is as plotted in Fig. 25.10.

These curves, developed by Gordon Brown, are useful in predicting the response of a system whose parameters are known or in estimating the parameters of a system whose response is known. If the damping ratio is small, say, $\zeta = 0.2$, the output variable rises rapidly to the steady-state value, *overshoots* by 50%, and oscillates several times before the natural response component becomes negligible. If the damping ratio is large, say, $\zeta = 1.5$, the output variable approaches the steady-state value very slowly. If $\zeta = 1.0$, the damping is at the critical value and the output variable approaches the steady-state value at the maximum rate possible without overshoot. If rapid response is desired and a small amount of overshoot is tolerable, a design value of $\zeta = 0.8$ might be specified.

EXAMPLE 4

A d'Arsonval movement to be used in a pen recorder has a rotational inertia $J = 2.5 \times 10^{-8}$ kg·m². It is to have a damping ratio $\zeta = 0.8$ and an undamped natural frequency $\omega_n = 200$ rad/s. Specify the spring compliance, determine the developed torque required for a full-scale deflection of 1.2 rad, and estimate the time required for the deflection to reach 0.98 of the steady-state value.

Assuming that the movement is represented by the circuit model of Fig. 25.9a (Example 3), this second-order system is characterized by the transfer function of Eq. 25-13. From Eq. 25-10 the necessary spring compliance is

$$K = \frac{1}{\omega_n^2 J} = \frac{1}{4 \times 10^4 \times 2.5 \times 10^{-8}} = 10^3 \text{ rad/N·m}$$

For a full-scale deflection of 1.2 rad, the torque required is

$$T_d = \frac{\Theta}{K} = \frac{1.2 \text{ rad}}{10^3 \text{ rad/N·m}} = 1.2 \times 10^{-3} \text{ N·m}$$

For $\zeta = 0.8$, Fig. 25.10 indicates that $y/Y = 0.98$ when $\omega_n t \cong 3.6$ rad. Therefore,

$$t \cong \frac{3.6}{\omega_n} = \frac{3.6}{200} = 0.018 \text{ s}$$

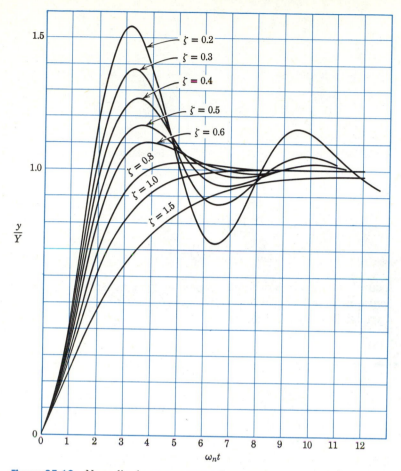

Figure 25.10 Normalized response curves for second-order systems.[†]

System Time Constants

One of the virtues of the transfer function is the variety of useful ways in which it can be interpreted. In several examples we have seen that the form of the step response of a system is determined by the poles of $T(s)$. The *pole* and *zero* locations are explicit when the transfer function is written in the form

$$T(s) = \frac{k(s + \beta)}{(s + \alpha_1)(s + \alpha_2)} \tag{25-19}$$

For the special case of real negative poles and a zero at the origin ($\beta = 0$), the pole-zero diagram is as shown in Fig. 25.11b.

[†]For $\zeta = 1$, equation 25-18 is indeterminate; the response with critical damping is $y/Y = 1 - (1 + \omega_n t)\, e^{-\omega_n t}$. For $\zeta < 1$, it is more convenient to express the response as a damped sinusoid.

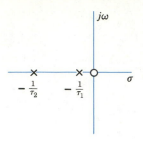

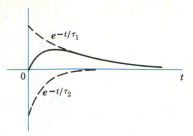

(a) Pole–zero diagram   (b) System time constants

Figure 25.11 Interpretations of the transfer function $T(s) = k's/(s\tau_1 + 1)(s\tau_2 + 1)$.

Rewriting Eq. 25-19 (for $\beta = 0$) as

$$T(s) = \frac{ks/\alpha_1\alpha_2}{(s/\alpha_1 + 1)(s/\alpha_2 + 1)} = \frac{k's}{(s\tau_1 + 1)(s\tau_2 + 1)} \qquad (25\text{-}20)$$

focuses attention on a different set of system characteristics. Here τ_1 and τ_2 are the *time constants* of the natural response terms. The relation of the time constants to the step response is shown in Fig. 25.11b.

EXAMPLE 5

Determine the transfer function of a first-order servomechanism to meet the following specifications:

Maximum displacement: 25°
Error: no greater than 0.5° after 2 s
Static gain: 10

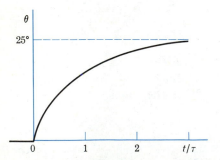

Figure 25.12 Deriving the transfer function of a servomechanism.

The desired behavior is shown in Fig. 25.12 and we conclude that the form of the necessary transfer function is (Eq. 25-3)

$$T(s) = \frac{Y(s)}{X(s)} = \frac{k}{s + \alpha} = \frac{k\tau}{s\tau + 1}$$

to provide a step response

$$\frac{y}{Y} = 1 - e^{-t/\tau}$$

An error of 0.5° corresponds to 2% of the maximum displacement; therefore,

$$1 - 0.98 = 0.02 = e^{-t/\tau}$$

and the time constant should be

$$\tau = \frac{-t}{\ln 0.02} = \frac{-2}{-3.9} \cong 0.51 \text{ s}$$

For a gain of $Y/X = 10$ at $s = 0$, $k\tau = 10$; therefore, the desired system transfer function is

$$T(s) = \frac{10}{0.51s + 1}$$

System Frequency Response

Another useful interpretation of the transfer function is in terms of the steady-state response, i.e., amplitude and phase angle as a function of frequency. In some cases, a laboratory measurement of frequency response is the best method of determining the system characteristics. Given a transfer function, the frequency response is obtained directly by substituting $s = j\omega$. For the system described by $T(s) = 1/(s\tau + 1)$, the frequency response is

$$T(j\omega) = \mathbf{T} = \frac{1}{1 + j\omega\tau} = \frac{1}{\sqrt{1 + (\omega\tau)^2}} \underline{/\tan^{-1} -\omega\tau} \qquad (25\text{-}21a)$$

with a breakpoint at $\omega\tau = 1$. For the system described by $T(s) = s\tau/(s\tau + 1)$, the frequency response is

$$T(j\omega) = \mathbf{T} = \frac{j\omega\tau}{1 + j\omega\tau} = \frac{1}{1 - j\dfrac{1}{\omega\tau}} = \frac{1}{\sqrt{1 + \left(\dfrac{\omega_1}{\omega}\right)^2}} \underline{/\tan^{-1}(\omega_1/\omega)} \qquad (25\text{-}21b)$$

where the breakpoint is defined by $\omega_1\tau_1 = 1$ or $\omega_1 = 1/\tau_1$ in the nomenclature used with amplifiers. If the transfer function of the d'Arsonval movement of Fig. 25.7 has complex roots, the frequency response is a little more complicated but the procedure is the same. (See Problem 2.)

FEEDBACK CONTROL SYSTEMS

With an understanding of the relationship between transfer functions and dynamic responses, we are ready to consider control systems with feedback. Our aim is to learn how they operate, how they can be used to improve performance, and how system stability can be ensured. Because of our experience with circuits and electromechanical devices we choose an electrical system as an illustration.

A Voltage Regulator

The purpose of a voltage regulating system is to hold the controlled voltage within specified limits in spite of changes in load or other operating conditions. The output voltage of the unregulated, separately excited, constant-speed, dc generator in Fig. 25.13a varies with changes in load current because of the armature-resistance voltage drop. The variation can be reduced by "closing the loop" through an operator with an eye on the voltmeter and a hand on the field rheostat (Fig. 25.13b). The

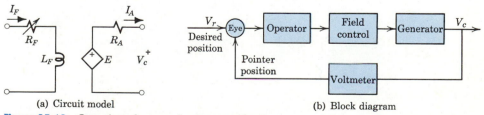

(a) Circuit model   (b) Block diagram

Figure 25.13 Operation of a manual voltage regulator.

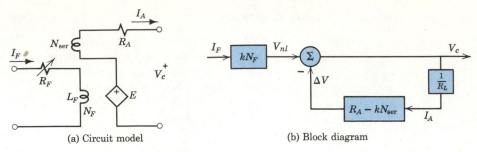

(a) Circuit model (b) Block diagram

Figure 25.14 Steady-state operation of a compound generator.

"controller" notes the "error" between the voltmeter pointer position and the "reference" and takes the necessary "action" to reduce the error to zero.

In the constant-speed compound generator of Fig. 25.14 the effect of a load change is fed back by means of the series winding (see Example 3 in Chapter 23). If operation is on the linear portion of the magnetization curve and changes are slow, the governing equation is

$$V_c = E - I_A R_A = k(N_F I_F + N_{ser} I_A) - I_A R_A$$

$$= V_{nl} - \Delta V = k N_F I_F - I_A(R_A - k N_{ser}) \tag{25-22}$$

where N is the number of turns.

The corresponding block diagram is shown in Fig. 25.14b. For flat compounding, the term $R_A - k N_{ser}$ is made equal to zero, $\Delta V = 0$, and $V_c = V_{nl}$. But there are two inadequacies in this system. First, the feedback loop contains an element R_A that changes with temperature (and therefore with I_A) and factor k that changes with magnetic flux (and therefore with I_A). Second, there is no amplification in the loop and the total error is incorporated in the output.

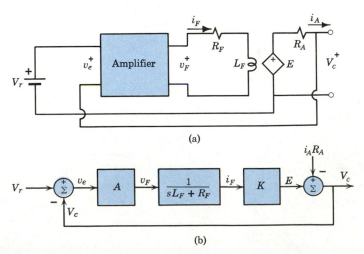

(a)

(b)

Figure 25.15 Wiring and block diagrams for a voltage regulator.

The voltage regulator in Fig. 25.15 represents an improvement. Here the output voltage is compared to a reference, and the difference is amplified and used to decrease the discrepancy. In anticipation of a rapid response, we show the variables as instantaneous values. The operation is: A sudden change in load current reduces output voltage V_c and greatly increases the error voltage $v_e = V_r - V_c$; the error voltage is then amplified and applied to the field circuit, tending to increase i_F and restore the output voltage. For linear operation, the governing equations are

$$v_e = V_r - V_c \qquad\qquad E = Ki_F$$

$$v_F = Av_e \qquad\qquad V_c = E - i_A R_A \qquad\qquad (25\text{-}23)$$

$$v_F = L_F \frac{di_F}{dt} + R_F i_F$$

When each of these relations is represented by an element, the block diagram is as shown in Fig. 25.15b.

In this representation the voltage drop $v_A = i_A R_A$ is introduced as a disturbance to the system. One approach to evaluating the performance of the regulator is to determine its response to a step in load current I_A. In this situation the reference voltage V_r is constant and the important relation is between I_A and V_c. To focus attention on this relation, the block diagram is redrawn as shown in Fig. 25.16. Here the forward

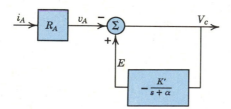

Figure 25.16 A voltage regulator with a disturbance input.

transfer function $G(s) = +1$. With V_r constant, $v_e = -\Delta V_c$ and the feedback transfer function relating ΔE to ΔV_c is

$$H(s) = -\frac{KA}{sL_F + R_F} = -\frac{KA/L_F}{s + R_F/L_F} = -\frac{K'}{s + \alpha} \qquad (25\text{-}24)$$

where K' is a new constant and $\alpha = R_F/L_F$. For this configuration, the overall gain with feedback (Eq. 25-1) is

$$\frac{V_c(s)}{I_A(s)} = -R_A G_F(s) = R_A \frac{-1}{1 - H(s)} = \frac{-R_A}{1 + [K'/(s + \alpha)]} = \frac{-R_A(s + \alpha)}{s + \alpha + K'} \qquad (25\text{-}25)$$

With the transfer function known, the step response of the system is determined from forced and natural components and initial conditions. The forced component of the change in controlled voltage is (for $s = 0$)

$$v_f = -\frac{R_A \alpha}{\alpha + K'} I_A = -\frac{R_A R_F/L_F}{R_F/L_F + KA/L_F} I_A = -\frac{R_A R_F I_A}{R_F + KA} \qquad (25\text{-}26)$$

The natural response is defined by the pole of $G_F(s)$ and

$$v_n = A_n \, e^{-(\alpha + K')t} \tag{25-27}$$

The performance of the regulator is revealed by the variation of the total controlled voltage V_c as a function of time after the introduction of the disturbance I_A. In general,

$$V_c = V_{nl} + \Delta V_c = V_{nl} + v_f + v_n = V_{nl} - \frac{R_A R_F I_A}{R_F + KA} + A_n \, e^{-(\alpha + K')t} \tag{25-28}$$

If speed is constant as assumed, the generated voltage E cannot suddenly change because inductive current i_F cannot suddenly change. Therefore, if the no-load output voltage is V_{nl} and a current I_A is suddenly drawn at $t = 0$, at $t = 0^+$

$$V_c = V_{nl} - I_A R_A = V_{nl} - \frac{R_A R_F I_A}{R_F + KA} + A_n \, e^0 \tag{25-29}$$

Solving Eq. 25-29,

$$A_n = -I_A R_A \left(1 - \frac{R_F}{R_F + KA} \right) \tag{25-30}$$

and, therefore,

$$V_c = V_{nl} - \frac{R_A R_F I_A}{R_F + KA} - I_A R_A \left(1 - \frac{R_F}{R_F + KA} \right) e^{-(R_F + KA)t/L_F} \tag{25-31}$$

The effect of the various parameters is illustrated in Example 6.

Improving Response

An open-loop system is "ignorant of its own output." In contrast, a closed-loop system is continually reminded of what is occurring at the output terminals. The principal virtue of feedback in amplifiers (Chapter 19) is the ability to provide satisfactory operation under widely ranging conditions. In control applications, feedback can greatly improve system performance by reducing errors or by extending the frequency response or by reducing the time constant of a system component.

In the voltage regulator of Example 6, the difference between the controlled voltage and the reference voltage, the *error*, is the input to the amplifier. The purpose of the system is to drive this error to zero. On the other hand, some error must exist to provide the necessary input. As an illustration, consider an aircraft rudder control. The difference between the actual rudder position and that called for by the pilot is sensed and amplified. The output of the amplifier is the input to a motor that provides the torque to actuate the rudder. In a turn, considerable torque is required and therefore there must be some steady-state error. This error is kept small by using high amplification.

It is characteristic of control systems that the input consists of random variations. These may be slow or fast corresponding to low and high frequencies. The ability of the system to respond satisfactorily to random variations can be evaluated in terms of

EXAMPLE 6

The shunt generator in Fig. 25.15 ($R_A = 1\ \Omega$, $R_F = 100\ \Omega$, and $L_F = 20$ H) is rated at 100 V and 10 A. The magnetization curve can be approximated by the relation $E = 100I_F$ V where I_F is in amperes.

Predict the effect of a load current of 10 A on the output voltage of the generator operating without feedback.

Predict the effect of the same load current with the feedback circuit shown incorporating an amplifier with voltage gain $A = 100$.

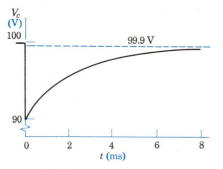

Figure 25.17 Step response of a voltage regulator.

With the field current adjusted for rated voltage at no load but without feedback, the output voltage under load is

$$V = E - I_A R_A = 100 - 10 \times 1 = 90 \text{ V}$$

The application of this load will result in a 10% voltage drop. With feedback, Eq. 25-31 applies. Here

$$\frac{R_F}{R_F + KA} = \frac{100}{100 + 100 \times 100} \cong 0.01$$

and

$$\frac{R_F + KA}{L_F} = \frac{100 + 100 \times 100}{20} \cong 500$$

hence

$$V_c = V_{nl} - 0.01 I_A R_A - I_A R_A (1 - 0.01) e^{-500t}$$

For $v_c = V_{nl} = 100$ V and $I_A R_A = 10 \times 1 = 10$ V introduced at $t = 0$,

$$V_c = 100 - 0.01 \times 10 - 10(0.99) e^{-500t}$$

or

$$V_c = 99.9 - 9.9 e^{-500t}$$

As shown in Fig. 25.17, the voltage will drop abruptly to 90 V, but within a few milliseconds it has been increased to 99.9 V. The application of a 10-A load results in a 0.1% voltage drop compared to the 10% voltage drop in the same generator without feedback.

its steady-state frequency response as shown by a Bode plot (Fig. 9.11c). The frequency response of a mechanical system (such as the d'Arsonval movement in Example 4) can be extended by using feedback. The technique is similar to that employed to extend the frequency response of an amplifier (see Example 11 in Chaper 19). In an amplifier, increased bandwidth is obtained at the expense of gain. In a servo-mechanism, extended frequency response is obtained at the sacrifice of simplicity. Some form of amplification is always required.

The speed of response is another limiting factor in the performance of control systems. It is desirable that the system have a low dynamic error as well as a low steady-state error. How can the time constant of a physical element be reduced using feedback? In Example 6, the response of the basic generator to a change in field voltage v_F was defined by the time constant $\tau = L_F/R_F = 20/100 = 0.2$ s. With feedback, the time constant is reduced to $\tau_F = L_F/(R_F + KA) = 20/(100 + 10,000) \cong 0.002$ s; as expected, the time constant is reduced by a factor that can be interpreted as $(1 + GH)$.

Stability

We see that feedback can improve the performance of a system by making it faster, more accurate, or responsive over a wider range of frequencies. In each case amplification is necessary, and the degree of improvement is related to the amount of amplification. The combination of amplification and feedback may also make the system unstable and therefore ineffective.

The mathematical basis for instability is revealed in the closed-loop gain equation where

$$\mathbf{G}_F = \frac{\mathbf{G}}{1 + \mathbf{GH}} \qquad (25\text{-}32)$$

If for any frequency the denominator is zero, the gain increases without limit and the system is unstable. The denominator is zero if the open-loop transfer function **GH** is just equal to $-1 + j0$. A possible approach is to plot the amplitude and phase of **GH** as functions of frequency as in Fig. 9.11c. If the gain is 1 at a phase shift of $\pm 180°$, the system is unstable.[†]

An alternative statement is that the system is unstable if any poles of the overall transfer function

$$T(s) = \frac{G(s)}{1 + G(s)H(s)} \qquad (25\text{-}33)$$

are on the $j\omega$ axis or in the right half of the complex frequency plane. (Where are the poles for the functions whose responses are given in Fig. 25.5?)

In designing automatic feedback control systems, much of the engineer's work is concerned with stability analysis, and there are many books devoted to the subject. After the system has been designed to perform the desired function, precautions must be taken to ensure stability. Amplifier gains, time constants, gear ratios, and damping ratios must be selected so that the poles of the transfer function are kept out of the right half of the s plane. Sometimes compensating networks are placed in series with system components to cancel out undesirable poles. Auxiliary feedback loops may also be used to provide stability.

The basic question is whether the denominator of $T(s)$ has any roots in the right half of the s plane. If the denominator is known in factored form (see Eq. 4-34), this question is readily answered. In general, however, $G(s)$ and $H(s)$ are polynomials in s and finding the roots is a tedious operation. From the design engineer's viewpoint, the practical problem is to determine the effect on stability of variations in system parameters. To optimize system design, the engineer must provide adequate stability with the specified performance at the least cost (in complexity, reliability, weight, or dollars). Various methods have been developed to provide convenient solutions to this problem in the practical case where the system transfer functions are quite complicated.

One of the most widely used techniques in system analysis is the *root-locus method* developed by W. R. Evans in 1948. By this technique, the poles of the transfer

---

[†]For a discussion of the Nyquist stability criterion see any of the references at the end of this chapter.

function are located graphically as the system parameters are varied. The value of such a graphical method can be illustrated by a specific numerical example. For the system defined by

$$G(s) = \frac{K(s+2)}{s(s+1)(s+4)(s+10)} \qquad H(s) = 1 \qquad (25\text{-}34)$$

the overall transfer function is

$$T(s) = \frac{G(s)}{1 + G(s)} = \frac{K(s+2)}{K(s+2) + s(s+1)(s+4)(s+10)} \qquad (25\text{-}35)$$

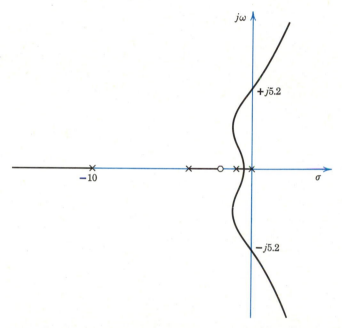

Figure 25.18 All points satisfying $\underline{/G(s)} = \pm 180°, \pm 540°$ (from R. N. Clark, *Introduction Automatic Control Systems,* John Wiley & Sons, New York, 1962, by permission).

The transfer function has a zero at $s = -2$, and for $K = 0$, poles at $s = 0, -1, -4,$ and -10. Using the root-locus technique,[†] the loci of all points satisfying the relation

$$\underline{/G(s)} = \pm 180° + n(360°) \qquad (25\text{-}36)$$

are as shown in Fig. 25.18. The roots move along the loci as K is varied. As K increases from zero, the poles that start at $s = 0$ and $s = -1$ come together and then diverge along the curved locus. For values of K that result in poles close to the $j\omega$ axis, the step response of the system is highly oscillatory. The critical value of K for stability occurs where the poles are on the $j\omega$ axis. On the basis of the root-locus plot, an experienced systems engineer can decide on an appropriate value of K.

[†]See p. 192 of the Clark reference at the end of this chapter.

SUMMARY

■ Automatic feedback control systems employ transducers, error detectors, amplifiers, actuators, and feedback elements to provide rapid and accurate control.
Control system analysis is based on circuit principles, device characteristics, and system techniques.

■ In comparison to an open-loop system, a closed-loop system is capable of greater accuracy over a wider range of conditions using less precise control elements.
The control engineer's problem is to provide the necessary performance with adequate stability at minimum cost.
System performance can be predicted on the basis of step response.

■ The transfer function is the ratio of two exponential functions of time; it is derived and interpreted just as an immittance function.
The dynamic response of a system element is obtained from the transfer function.
System behavior is defined by the overall transfer function.

■ The order of a system is defined by the highest power of s in the denominator of the transfer function. The step response is:

$$y(t) = A_0 + A e^{st} \text{ for a first-order system, and}$$

$$y(t) = A_0 + A_1 e^{s_1 t} + A_2 e^{s_2 t} \text{ for a second-order system.}$$

Normalized curves display the effect on response of damping ratio ζ.

■ Feedback can improve control system performance by reducing errors, by extending the frequency response, or by reducing time constants.

■ The combination of amplification and feedback may result in instability.
Much of the systems engineer's work is concerned with stability analysis.

REFERENCES

1. J. G. Truxal, *Introductory System Engineering*, McGraw-Hill Book Co., New York, 1972.
2. R. N. Clark, *Introduction to Automatic Control Systems*, John Wiley & Sons, New York, 1962.

3. J. J. D'Azzo and C. H. Houpis, *Feedback Control System Analysis and Synthesis*, McGraw-Hill Book Co., New York, 1960.

REVIEW QUESTIONS

1. Why don't power inputs appear in a control system block diagram?
2. Explain in system terminology how highway traffic speed is regulated.
3. Explain in system terminology how body temperature is regulated.
4. What information on system performance is provided by step response?
5. Define "transform," "transfer function," and "step response."
6. Outline the procedure for determining step response from the transfer function.
7. What information is obtained from the poles of $T(s)$? The zeros?
8. What form of step response is indicated by complex poles of $T(s)$?
9. Define ζ. What are its dimensions? Why is it useful?
10. What value of damping ratio results in a 20% overshoot?

EXERCISES

1. In an automobile "Cruise Control" system, the desired speed is manually set into the comparator. The electrical output of the comparator passes through a transistorized amplifier to the solenoid valve of a power unit. The power unit moves the engine throttle which drives the driveshaft. The speedometer is connected to the driveshaft and its reading is introduced into the comparator.
   (a) Draw a block diagram of this automatic control system.
   (b) Identify the controlled output, the feedback element, the error detector, and the reference input.

2. In an antenna-positioning system, the input direction dial and the antenna are connected to similar rotating potentiometers across a common voltage V. The voltage difference is fed into a corrective network, then into a power amplifier that drives a motor that turns the antenna through a set of gears.
   (a) Draw a block diagram of this automatic control system.
   (b) Identify the controlled output, the feedback element, the error detector, and the reference input.

3. For the flywheel of Fig. 25.19:
   (a) Define and derive the transfer function.
   (b) Sketch the pole-zero diagram of $T(s)$.
   (c) Sketch the step response.

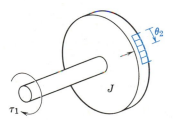

Figure 25.19

4. Given $T(s) = (s + a)/(s + 2a)$:
   (a) Sketch the pole-zero diagram.
   (b) Determine the step response.

5. The reaction time of an automobile driver, $T_R = 0.1$ s, represents a pure time delay in that $x_2(t) = x_1(t - 0.1)$.
   (a) Derive the transfer function $X_2(s)/X_1(s)$.
   (b) Derive the transfer function $T(j\omega)$.

6. Determine the response to a step of amplitude 10 (no initial energy storage) given:
   (a) $T(s) = 2s/(s + 5)$
   (b) $T(s) = 5/(s + 2)$

7. Determine the transfer function given the unit step response:
   (a) $x(t) = 5 - 5e^{-2t}$
   (b) $i(t) = 6e^{-t/50}$

8. Derive the transfer function $T(j\omega)$ relating displacement x to force f for the system of Fig. 9.17. For $M = 0.1$ kg and $K = 0.4$ m/N, sketch the amplitude A and the phase angle ϕ as functions of frequency ω.

9. An amplidyne is a rotating machine in which the voltage generated in one winding provides field current for a second winding. Such a power amplifier can be represented by the circuit in Fig. 25.20.
   (a) Assuming linear magnetic fields, draw a block diagram of the amplidyne.
   (b) Derive the transfer function.
   (c) Determine the forced response to a step input ΔV.
   (d) Determine the *form* of the natural step response.

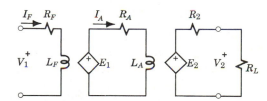

Figure 25.20

10. (a) Explain why the natural response of a system is defined by the poles of the transfer function rather than the zeros.
    (b) Given $T(s) = (s + a)/(s^2 + sb)$, write an expression for the natural response.

11. A shunt motor ($J = 2 \times 10^{-6}$ kg·m² and $D = 8 \times 10^{-5}$ N·m·s/rad) is operated with constant armature current. The inductance of the field circuit is 200 mH and the resistance is 20 Ω. Derive the transfer function relating shaft velocity to field voltage and predict (and sketch) the step response.

12. In Fig. 25.21, a shunt motor with constant field excitation drives a spring-loaded shaft ($K = 3000$ rad/N·m). The equivalent friction torque of the motor is $D' = 3 \times 10^{-5}$ N·m·s/rad and the motor inertia is $J = 6 \times 10^{-6}$ kg·m². Derive the transfer function relating angle θ and input voltage v and sketch the expected step response.

13. Repeat Exercise 12 where K is reduced to 300 rad/N·m.

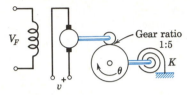

Figure 25.21

14. The frequency response of a transfer function $T(j\omega)$ is given in Fig. 25.22.
    (a) Determine the transfer function.
    (b) Determine the step response, assuming no initial energy storage.

15. In the voltage regulator of Fig. 25.23, the output of the constant-speed generator is $E = 100i_F$ V where i_F is in amperes. The dc amplifier gain is $K = -100$, $R_F = 50$ Ω, $L_F = 10$ H, and $R_A = 1$ Ω.

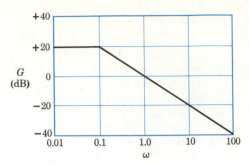

Figure 25.22

(a) Predict V for $R_L = 1$ Ω.
(b) Predict and sketch $V(t)$ if R_L is suddenly changed to 5 Ω.

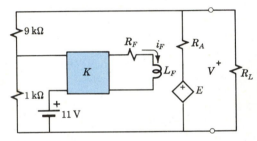

Figure 25.23

PROBLEMS

1. An ac servomotor has a speed-torque curve approximated by a straight line with intercepts $\Omega/\Omega_s = 1.0$ and $T/T_R = 2.0$ (see Fig. 24.22). A handbook states: "The time t_a required for a servomotor to accelerate from rest to 63% of its no-load speed n_o (rpm) is $t_a = (J\ n_o/T_s) \times 1.483 \times 10^{-6}$ s, where J = rotor inertia (g·cm²) and T_s = starting torque (oz·in.)." Stating any necessary assumptions, derive this relationship.

2. The d'Arsonval meter of Fig. 25.7 has a steady deflection θ_o for a dc voltage V. The damping ratio ζ is 0.5.
   (a) Describe the response to a suddenly applied dc voltage V.
   (b) Describe the frequency response, i.e., deflection amplitude as a function of the frequency of the applied voltage.
   (c) Determine and plot the normalized frequency response θ/θ_o vs. ω/ω_o where $\omega_o =$

Appendix

Physical Constants

Conversion Factors

Complex Algebra

Color Code for Resistors

Preferred Resistance Values

Representative TTL IC Units

Device Characteristics

PHYSICAL CONSTANTS[†]

Charge of the electron	e	1.602×10^{-19} C
Mass of the electron (rest)	m	9.109×10^{-31} kg
Velocity of light	c	2.998×10^{8} m/s
Acceleration of gravity	g	9.807 m/s^2
Avogadro's number	N_A	6.023×10^{23} atoms/g $\cdot$ atom
Planck's constant	h	6.626×10^{-34} J $\cdot$ s
Boltzmann's constant	k	8.62×10^{-5} eV/K
Permeability of free space	μ_o	$4\pi \times 10^{-7}$ H/m
Permittivity of free space	ϵ_o	$1/36\pi \times 10^{9}$ F/m

CONVERSION FACTORS

1 inch	$= 2.54$ cm		1 hertz	$= 1$ cycle/s
1 meter	$= 39.37$ in.		1 electron volt	$= 1.60 \times 10^{-19}$ J
1 angstrom unit	$= 10^{-10}$ m		1 joule	$= 10^{7}$ ergs
1 kilogram	$= 2.205$ lb (mass)		1 tesla	$= 1$ Wb/m^2
1 newton	$= 0.2248$ lb (force)			$= 10^{4}$ gauss
1 horsepower	$= 746$ W			

[†]E.A. Mechtly, "The International System of Units: Physical Constants and Conversion Factors," NASA SP-7012, Washington, D.C., 1964.

COMPLEX ALGEBRA

In our numbering system, *positive* numbers correspond to distances measured along a line, starting from an origin. *Negative* numbers enable us to solve equations like $x + 2 = 0$, and they are represented in Fig. A1 by distances to the left of the origin. *Zero* is a relatively recent concept with some peculiar characteristics; for example, division by zero must be handled carefully. Numbers corresponding to distances along the line of Fig. A1 are called *real* numbers. To solve equations like $x^2 + 4 = 0$, so-called *imaginary* numbers were invented; these add a new dimension to our numbering system.

Figure A1 The system of real numbers.

Imaginary Numbers

Imaginary numbers are plotted along the *imaginary axis* of Fig. A2 so that $j2$ lies at a distance of 2 units along an axis at right angles to the *real axis*. The imaginary number $-j2$ lies along the imaginary axis but in the opposite direction from the origin. Note that all numbers are abstractions, and calling one number "real" and another "imaginary" just differentiates between their mathematical properties.

It may be helpful to think of j as an *operator*. An operator like "$\sqrt{\ }$" or "sin" means a particular mathematical operation; $\sqrt{\ }$ operating on 4 yields 2 and sin operating on 30° yields 0.5. In the same way, the minus sign $(-)$ is an operator such that $(-)$ operating on a directed line segment yields a reversal or a 180° rotation. Similarly, we define the operator j to mean a 90° counterclockwise rotation. The operator j taken twice, or $(j)(j)$, indicates a 180° rotation or a reversal, and $(j)(j)$ is equivalent to $(-)$ in effect. This concept is important because it provides an algebraic interpretation for a geometrical operation.

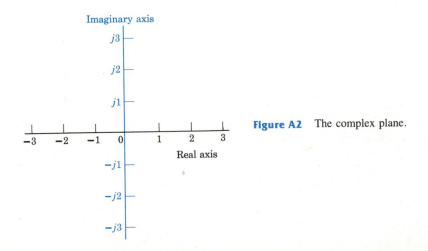

Figure A2 The complex plane.

Just as $\sin^2 \theta$ is defined to mean $(\sin \theta)^2$ and not $\sin (\sin \theta)$, so the properties of jb must be defined. It is understood that

$$j(jb) = j^2 b = -b \qquad \text{or we say} \qquad j^2 = -1$$

$$j(j^2 b) = j^3 b = -jb \qquad \text{or we say} \qquad j^3 = -j$$

$$j(j^3 b) = j^4 b = +b \qquad \text{or we say} \qquad j^4 = +1$$

The equation $j^2 = -1$ is a statement that the two *operations* are equivalent and does *not* imply that j is a number. However, and this is the beauty of the concept, in all algebraic computations imaginary numbers can be handled as if j had a numerical value of $\sqrt{-1}$.

Complex Numbers

The real and imaginary axes define the *complex plane*. The combination of a real number and an imaginary number defines a point in the complex plane and also defines a *complex number*. The complex number may be considered to be the point or the directed line segment to the point; both interpretations are useful.

The complex number $\mathbf{W}$ of magnitude M and direction θ in Fig. 5.3 on p. 125 can be expressed in rectangular form or in polar form where

$$\mathbf{W} = a + jb = M(\cos \theta + j \sin \theta) = Me^{j\theta} = M\underline{/\theta}$$

The conversion from one form to another is facilitated by Euler's theorem which states that $\cos \theta + j \sin \theta = e^{j\theta}$.

The validity of Euler's theorem is evident when series expansions are written as follows:

$$e^\theta = 1 + \theta + \frac{\theta^2}{2!} + \frac{\theta^3}{3!} + \frac{\theta^4}{4!} + \frac{\theta^5}{5!} + \cdots$$

$$\cos \theta = 1 - \frac{\theta^2}{2!} + \frac{\theta^4}{4!} - \cdots \qquad \text{and} \qquad \sin \theta = \theta - \frac{\theta^3}{3!} + \frac{\theta^5}{5!} - \frac{\theta^7}{7!} + \cdots$$

Then

$$\cos \theta + j \sin \theta = 1 + j\theta - \frac{\theta^2}{2!} - j\frac{\theta^3}{3!} + \frac{\theta^4}{4!} + j\frac{\theta^5}{5!} - \cdots = e^{j\theta}$$

Rules of Complex Algebra

Operations such as addition and subtraction, multiplication and division, raising to powers, and extracting roots are performed easily if the complex numbers are in the most convenient form. The following examples indicate the rules of complex algebra as applied to the two complex numbers

and
$$\mathbf{A} = a + jb = A e^{j\alpha} = A\underline{/\alpha} \qquad \text{(read: ``}A\text{ at angle alpha'')}$$

$$\mathbf{C} = c + jd = C e^{j\gamma} = C\underline{/\gamma} \qquad \text{(read: ``}C\text{ at angle gamma'')}$$

Equality

Two complex numbers are equal if and only if the real parts are equal and the imaginary parts are equal.

$$\text{If } \mathbf{A} = \mathbf{C}, \text{ then } a = c \qquad \text{and} \qquad b = d$$

EXAMPLE 1

The voltage phasors across two branches in parallel (Fig. A3) are $V_1 = 2 + jy$ and $V_2 = V\underline{/60°}$. Find y and V.

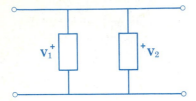

Figure A3   Phasor equality.

For branches in parallel, the voltages are equal and therefore the voltage phasors must be equal; hence,

$$2 + jy = V\underline{/60°}$$

This appears to be a single equation with two unknowns, but applying the rule for equality,

$$2 = V \cos 60° \quad \text{and} \quad y = V \sin 60°$$

Therefore,

$$V = \frac{2}{\cos 60°} = 4$$

and

$$y = 4 \sin 60° = 2\sqrt{3}$$

Addition

Two complex numbers in rectangular form are added by adding the real parts and the imaginary parts separately.

$$A + C = (a + c) + j(b + d)$$

Subtraction can be considered as addition of the negative,

$$A - C = (a - c) + j(b - d)$$

Addition and subtraction are not convenient in the polar form.

EXAMPLE 2

In Fig. A4, $i_2(t) = 2\sqrt{2} \cos (\omega t - 90°)$ and $I_1 = 3 - j4$. Find $i_3(t)$.

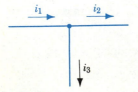

Figure A4   Phasor subtraction.

By inspection we write $I_2 = 2\underline{/-90°}$ but, since the polar form is not convenient, we convert to rectangular form as $I_2 = 0 - j2$. Then

$$I_3 = I_1 - I_2 = (3 - j4) - (0 - j2)$$
$$= 3 - j4 + j2 = 3 - j2$$

In polar form,

$$I_3 = \sqrt{3^2 + 2^2}\underline{/\arctan \tfrac{-2}{+3}} = 3.6\underline{/-33.7°}$$

$$\therefore i_3(t) = 3.6\sqrt{2} \cos(\omega t - 33.7°)$$

Multiplication

The magnitude of the product is the product of the individual magnitudes, and the angle of the product is the sum of the individual angles.

$$(\mathbf{A})(\mathbf{C}) = (A\,e^{j\alpha})(C\,e^{j\gamma}) = AC\,e^{j(\alpha+\gamma)} = AC\underline{/\alpha + \gamma}$$

Multiplication of complex numbers in polar form follows the law of exponents. If the numbers are given in rectangular form and the product is desired in rectangular form, it may be more convenient to perform the multiplication directly, employing the concept that $j = \sqrt{-1}$ and $j^2 = -1$.

$$(\mathbf{A})(\mathbf{C}) = (a + jb)(c + jd)$$

$$= ac + j^2bd + jbc + ajd = (ac - bd) + j(bc + ad)$$

EXAMPLE 3

In a certain alternating-current circuit, $\mathbf{V} = \mathbf{ZI}$ where

$$\mathbf{Z} = 7.07\underline{/-45°} = 5 - j5$$

and

$$\mathbf{I} = 10\underline{/+90°} = 0 + j10.$$

Find $\mathbf{V}$.

In polar form,

$$\mathbf{V} = (7.07\underline{/-45°})(10\underline{/+90°})$$

$$= 7.07 \times 10\underline{/-45° + 90°} = 70.7\underline{/+45°}$$

In rectangular form,

$$\mathbf{V} = (5 - j5)(0 + j10)$$

$$= (5 \times 0 + 5 \times 10) + j(-5 \times 0 + 5 \times 10)$$

$$= 50 + j50$$

Division

The magnitude of the quotient is the quotient of the magnitudes, and the angle of the quotient is the difference of the angles.

$$\frac{\mathbf{A}}{\mathbf{C}} = \frac{A\,e^{j\alpha}}{C\,e^{j\gamma}} = \frac{A\underline{/\alpha}}{C\underline{/\gamma}} = \frac{A}{C}\underline{/\alpha - \gamma}$$

Division of complex numbers in polar form follows the laws of exponents.

Division in rectangular form is inconvenient but possible by using the *complex conjugate* of the denominator. Given a complex number $\mathbf{C} = c + jd$, the complex conjugate of $\mathbf{C}$ is defined as $\mathbf{C}^* = c - jd$; that is, the sign of the imaginary part is reversed. (Here the asterisk means "the complex conjugate of . . . .") Multiplying the denominator by its complex conjugate *rationalizes* the denominator (i.e., converts it to a real number) and simplifies division. To preserve the value of the quotient, the numerator is multiplied by the same factor, as in Example 4.

EXAMPLE 4

In a certain alternating-curent circuit, $\mathbf{V} = \mathbf{ZI}$ where

$$\mathbf{V} = 130\underline{/-67.4^\circ} = 50 - j120$$

and

$$\mathbf{Z} = 5\underline{/53.1^\circ} = 3 + j4$$

Find $\mathbf{I}$.

In polar form,

$$\mathbf{I} = \frac{\mathbf{V}}{\mathbf{Z}} = \frac{130\underline{/-67.4^\circ}}{5\underline{/53.1^\circ}} = 26\underline{/-120.5^\circ}$$

In rectangular form,

$$\mathbf{I} = \frac{\mathbf{V}}{\mathbf{Z}} = \frac{50 - j120}{3 + j4} \cdot \frac{3 - j4}{3 - j4}$$

$$= \frac{(150 - 480) + j(-360 - 200)}{9 + 16}$$

$$= -13.2 - j22.4 = 26\underline{/-120.5^\circ}$$

Powers and Roots. Applying the laws of exponents to a complex number in polar form,

$$\mathbf{A}^n = (A e^{j\alpha})^n = A^n e^{jn\alpha} = A^n\underline{/n\alpha}$$

The same result is obtained from the rule for multiplication by recognizing that raising to the nth power is equal to multiplying a number by itself n times.

Extracting the nth root of a number is equivalent to raising the number to the $1/n$th power and the same rule is applicable:

$$\mathbf{A}^{1/n} = (A e^{j\alpha})^{1/n} = A^{1/n} e^{j\alpha/n} = A^{1/n}\underline{/\alpha/n}$$

The method for finding all n roots is illustrated in Example 5.

EXAMPLE 5

Perform the operations indicated at the right. Recalling that the number of distinct roots is equal to the order of the root (if n is an integer), locate the other roots in (b).

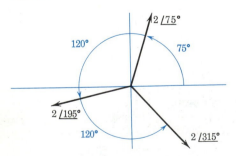

Figure A5   Roots of a complex number.

(a) $(8.66 - j5)^3 = (10\underline{/-30^\circ})^3$

$$= 1000\underline{/-90^\circ} = 0 - j1000$$

(b) $(-5.66 - j5.66)^{1/3} = (8\underline{/+225^\circ})^{1/3} = 2\underline{/+75^\circ}$

Since $8\underline{/225^\circ} = 8\underline{/225^\circ} + N360^\circ$ where N is any integer, three distinct roots are obtained by assigning values of $N = 0, 1, 2$.

For $N = 1$, $(8\underline{/225^\circ} + 360^\circ)^{1/3} = 2\underline{/75^\circ} + 120^\circ$

For $N = 2$, $(8\underline{/225^\circ} + 720^\circ)^{1/3} = 2\underline{/75^\circ} + 240^\circ$

The three cube roots are found to be complex numbers of the same magnitude but differing in angle by $360^\circ/3 = 120^\circ$ (see Fig. A5).

COLOR CODE FOR RESISTORS

Ordinary composition fixed resistors are characterized by power dissipation in watts, resistance in ohms, and tolerance in percent. They vary in size from 0.067 in. (0.1 W) to 0.318 in. (2 W) in diameter. For clarity, the resistance and tolerance values are indicated by color bands according to a standard code. (For example, bands of blue, gray, blue, and black would identify a 68×10^6 or 68 MΩ, 20%-tolerance resistor.)

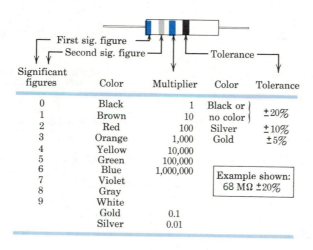

Significant figures	Color	Multiplier	Color	Tolerance
0	Black	1	Black or	
1	Brown	10	no color	±20%
2	Red	100	Silver	±10%
3	Orange	1,000	Gold	±5%
4	Yellow	10,000		
5	Green	100,000		
6	Blue	1,000,000		
7	Violet		Example shown:	
8	Gray		68 MΩ ±20%	
9	White			
	Gold	0.1		
	Silver	0.01		

A similar color scheme is used to indicate the value and tolerance of small capacitors; an extra color band indicates the temperature coefficient.

PREFERRED RESISTANCE VALUES

To provide the necessary variation in resistance while holding the number of resistors to be stocked to a reasonable minimum, the electronics industry has standardized on certain preferred values. The range in significant figures from 1.0 to 10 has been divided into 24 steps, each differing from the next by approximately 10% ($\sqrt[24]{10} = 1.10$). The preferred values, available in all multiples in ±5% tolerance resistors, are:

1.0‡	1.5‡	2.2‡	3.3‡	4.7‡	6.8‡
1.1	1.6	2.4	3.6	5.1	7.5
1.2†	1.8†	2.7†	3.9†	5.6†	8.2†
1.3	2.0	3.0	4.3	6.2	9.1

†Also available in ±10% tolerance resistors.
‡Also available in ±10% and ±20% tolerance resistors.

REPRESENTATIVE TTL IC UNITS

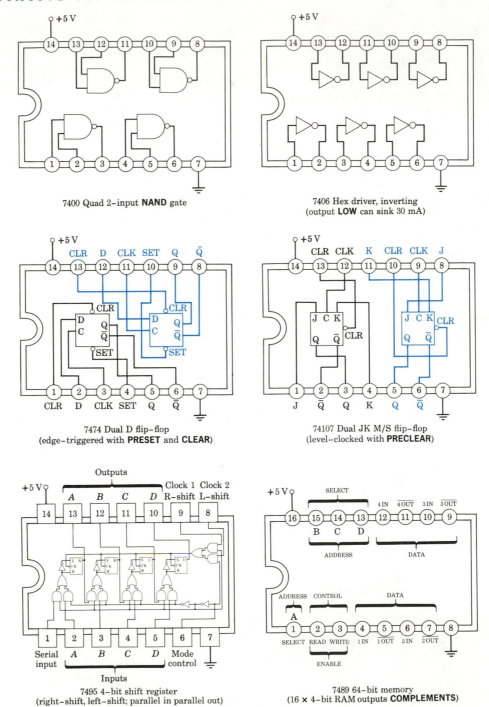

7400 Quad 2–input **NAND** gate

7406 Hex driver, inverting
(output **LOW** can sink 30 mA)

7474 Dual D flip–flop
(edge–triggered with **PRESET** and **CLEAR**)

74107 Dual JK M/S flip–flop
(level–clocked with **PRECLEAR**)

7495 4–bit shift register
(right–shift, left–shift; parallel in parallel out)

7489 64–bit memory
(16 × 4–bit RAM outputs **COMPLEMENTS**)

Figure A6   Some popular TTL integrated circuits.

DEVICE CHARACTERISTICS

2N4416 *n*-Channel Junction Field-Effect Transistor for VHF Amplifiers and Mixers

Maximum Ratings

Gate-Drain or Gate-Source voltage	−30 V
Gate current	10 mA
Total power dissipation	300 mW

Electrical Characteristics

	Min.	Max.
I_{GSS} Gate reverse current		−0.1 nA
I_{GSS} (150°C)		−0.1 μA
BV_{GSS} Gate-Source breakdown voltage	−30 V	
$V_{GS(off)}$ Gate-Source cutoff voltage	−2.5	−6.0 V
I_{DSS} Saturation drain current	5	15 mA

Common-Source Parameters
($V_{DS} = 15$ V, $V_{GS} = 0$ V)

	Min.	Max.
g_{fs} Forward transconductance (= g_{mo})	4500	7500 μS
g_{fs} (400 MHz)		4000 μS
g_{os} Output conductance (= $1/r_d$)		50 μS
C_{rss} Reverse transfer capacitance (= C_{gd})		0.8 pF
C_{iss} Input capacitance (= $C_{gd} + C_{gs}$)		4 pF
C_{oss} Output capacitance (= C_{ds})		2 pF

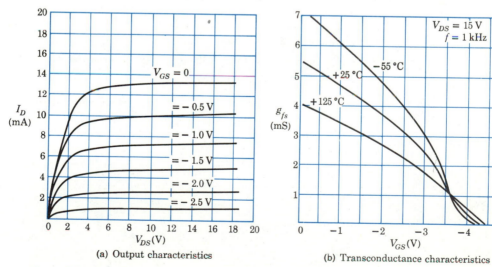

(a) Output characteristics

(b) Transconductance characteristics

Figure A7 Performance curves for 2N4416 JFET.

2N3114 *NPN* High-Voltage, Medium-Power Silicon Planar[†] Transistor

Maximum Ratings			Typical Small-Signal Characteristics	
			($f = 1$ kHz, $I_C = 1$ mA, $V_{CE} = 5$ V)	
Operating junction temperature	200°C	h_{fe}	Small-signal current gain	50
Power dissipation: 25°C case	5 W	h_{ie}	Input resistance	1,500 Ω
25°C free air	0.8 W	h_{oe}	Output conductance	5.3 μS
V_{CBO} Collector-to-Base voltage	150 V	h_{re}	Voltage feedback ratio	1.5×10^{-4}
V_{EBO} Emitter-to-Base voltage	5 V	h_{ib}	Input resistance	27 Ω
		h_{ob}	Output conductance	0.09 μS
		h_{rb}	Voltage feedback ratio	0.25×10^{-4}

Electrical Characteristics (25°C free air)

Symbol	Characteristic	Min.	Typical	Max.	Test conditions
$V_{BE(sat)}$	Base saturation voltage		0.8	0.9 V	$I_C = 50$ mA $I_B = 5.0$ mA
$V_{CE(sat)}$	Collector saturation voltage		0.3	1 V	$I_C = 50$ mA $I_B = 5.0$ mA
I_{CBO}	Collector cutoff current		0.3	10 nA	$I_E = 0$ $V_{CB} = 100$ V
$I_{CBO}(150°C)$	Collector cutoff current		2.7	10 μA	$I_E = 0$ $V_{CB} = 100$ V
$h_{FE}(\beta_{dc})$	DC current gain	15	35		$I_C = 100$ μA $V_{CE} = 10$ V
h_{fe}	20-MHz current gain	2.0	2.7		$I_C = 30$ mA $V_{CE} = 10$ V
C_{ob}	Output capacitance		6	9 pF	$I_E = 0$ $V_{CB} = 20$ V
C_{TE}	Emitter transition capacitance		70	80 pF	$I_C = 0$ $V_{EB} = 0.5$ V

[†]Planar is a patented Fairchild process.

Figure A8 Typical characteristics of the 2N3114 transistor. (Courtesy Fairchild Semiconductor)

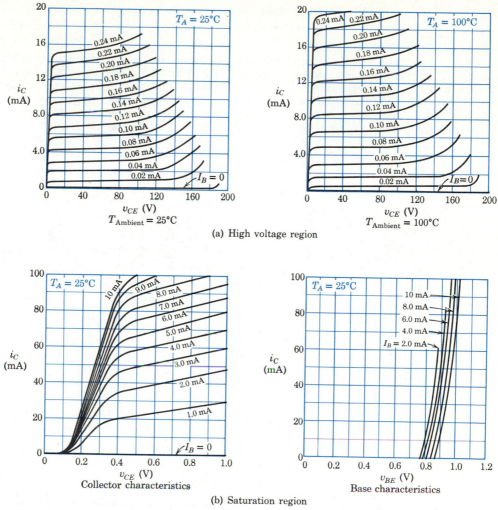

(a) High voltage region

(b) Saturation region

Figure A8   (continued)

μA741 Frequency Compensated Operational Amplifier

General Description

The μA741 is a high performance monolithic operational amplifier constructed on a single silicon chip, using the Fairchild Planar[†] epitaxial process. It is intended for a wide range of analog applications. High common mode voltage range and absence of "latch-up" tendencies make the μA741 ideal for use as a voltage follower. The high gain and wide range of operating voltage provides superior performance in integrator, summing amplifier, and general feedback applications. The μA741 is short-circuit protected, has the same pin configuration as the popular μA709 operation amplifier, but requires no external components for frequency compensation. The internal 6 dB/octave roll-off insures stability in closed loop applications.

[†]Planar is a patented Fairchild process.

Equivalent Circuit

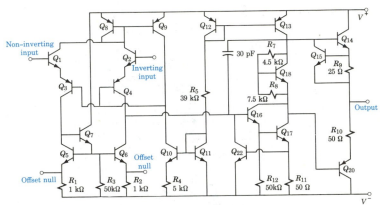

Figure A9 Electrical characteristics and typical performance curves for the μA741. (Courtesy Fairchild Semiconductor.)

Absolute Maximum Ratings

Supply voltage	± 22 V
Internal power dissipation (note 1)	500 mW
Differential input voltage	± 30 V
Input voltage (note 2)	± 15 V
Voltage between offset null and V^-	± 0.5 V
Storage temperature range	$-65°C$ to $+150°C$
Operating temperature range	$-55°C$ to $+125°C$
Lead temperature (soldering, 60 s)	$300°C$
Output short-circuit duration (note 3)	Indefinite

Connection diagram
(Top view)

Note: Pin 4 connected to case

Notes

1. Rating applies for case temperatures to 125°C; derate linearly at 6.5 mW/°C for ambient temperatures above +75°C.
2. For supply voltages less than ± 15 V, the absolute maximum input voltage is equal to the supply voltage.
3. Short circuit may be to ground or either supply. Rating applies to +125°C case temperature or +75°C ambient temperature.

Electrical Characteristics ($V_S = \pm 15$ V, $T_A = 25°C$)

Parameters (see definitions)	Conditions	Min.	Typ.	Max.	Units
Input offset voltage	$R_s \leq 10$ kΩ		1.0	5.0	mV
Input offset current			20	200	nA
Input bias current			80	500	nA
Input resistance		0.3	2.0		MΩ
Input capacitance			1.4		pF
Offset voltage adjustment range			± 15		mV
Large-signal voltage gain	$\begin{cases} R_L \geq 2 \text{ k}\Omega \\ V_{\text{out}} = \pm 10 \text{ V} \end{cases}$	50,000	200,000		
Output resistance			75		Ω
Output short-circuit current			25		mA
Supply current			1.7	2.8	mA
Power consumption			50	85	mW
Transient response (unity gain)	$\begin{cases} V_{\text{in}} = 20 \text{ mV} \\ R_L = 2 \text{ k}\Omega \\ C_L \leq 100 \text{ pF} \end{cases}$				
Rise time			0.3		μs
Overshoot			5.0		%
Slew rate	$R_L \geq 2$ kΩ		0.5		V/μs
CMRR	$R_s \leq 10$ kΩ	70	90.0		dB

Typical
Performance Curves

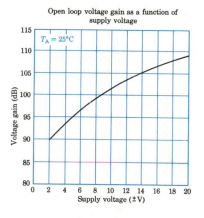

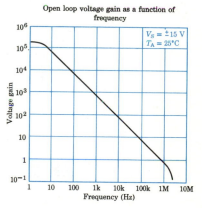

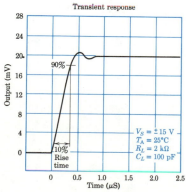

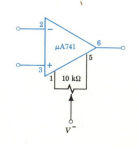

Voltage offset null current

Answers to Selected Exercises

Chapter 1

2. 9.6 mA

4. (a) 200 kV/m
(b) 3.2×10^{14} N, 3.5×10^{16} m/s^2

6. (c) 4.32 ¢ (d) 4.8 Ω

9. 240 W, 77.7%

12. (a) $RI_m \sin \omega t$ (b) $\omega L I_m \cos \omega t$

15. At $t = 1$ ms, $v_L = 25$ V, $v_C = 2.5$ V
At $t = 3$ ms, $v_L = 0$, $v_C = 12.5$ V

18. At $t = 1$ ms, $p = 125$ mW, $w = 562.5$ μJ

21. (b) 20 W (c) 70 W (e) 5 J

Chapter 2

1. 1 A

4. 5 A, 6 A, 10 V

7. (a) $i_L = 4 - 2e^{-2t}$ A (c) $w_C = 9e^{-4t}$ J

10. $v_{bd} = 15$ V, $v_{cd} = 23$ V

13. $v_{cb} = 10$ V

16. 2 A

19. 4 A

22. 500 Ω

25. −3 A

28. 4.33 A

32. 1.5 A

34. (b) $V_T = 0.08$ V, $R_T = 10$ MΩ

37. (b) $9.1 + 0.3 \cos \omega t$ mA

40. (a) 6 mA (b) 20 mA

Chapter 3

2. (a) $v(t) = 25.6e^{-t/2}$ V (c) 0.28 V

5. 0.53 s

7. 16 min 23 s

11. (a) $40 \cos (2000t - \pi/4)$ V,
$40 \cos (2000t + \pi/4)$ V

13. (a) $V_{av} = 15$ V, $V_{eff} = 15.81$ V,
$I_{av} = 4.5$ A, $I_{eff} = 4.9$ A

15. (b) $AM_1 = I_1 = 2.45$ A,
$AM_2 = I_2 = 1.73$ mA

18. (a) Invert. ampl. with $R_1 = 20$ kΩ
(c) Noninvert. ampl. with $R_1 = 76.9$ kΩ

22. (a) $v_o = -2v_1 - 0.4v_2$

25. D$_1$ closed, D$_2$ open, $I = 30$ mA

28. 54 mA, 108 V, 7.2 W

31. (a) 333 μF, 5.66:1 (b) 167 μF, 5.66:1

33. (c) Clips below −2 V

37. Clips above $+2$ V and below -4 V

38. Clamps minimum at zero axis

41. (b) $v_2 = v_1 + 0.5 \, V_m$

(c) $V_m = 35$ V, pk-pk VM

42. Integrator ($RC = 1$) + Summer ($R_F/R_1 = 10$, $R_F/R_2 = 5$)

Chapter 4

2. (b) 40 V, 8 mA (e) $8e^{-2t}$ mA, $T = 0.5$ s

4. (c) $(V/R)e^{-2t/RC}$ (d) $0.5V(1 - e^{-2t/RC})$

7. (a) $i = 2$ A, $i_1 = -2$ A

(b) $i(t) = 2e^{-3t}$ A

11. (b) 50 mA, -50 mA, -50 mA, 60 V,
-30 V, 0, $+30$ V

(c) $di_R/dt = -di_L/dt = -600$ A/s,
$dv_{C1}/dt = 50$ kV/s, $dv_{C2}/dt = 25$ kV/s

13. $R < 20 \, \Omega$

15. (b) -160 A/s, 1.04×10^4 A/s^2

(e) $i(t) = 3e^{-60t} - e^{-20t}$ A

18. (b) $i(t) = 2e^{-2t} - 4te^{-2t}$ A

20. (a) 30 V, 75 mA, 0, -75 mA

(c) $Cs^2 + (1/R)s + (1/L) = 0$

24. (a) $Z(s) = 3(2 + s)/(3 + s)$

(b) 1.33 Ω

(c) For $s = -4$, $i = 0.25 \, e^{-4t}$ A

(d) $8e^{-4t}$ V

27. (b) $Z_{ab}(s) = 2(s^2 + 4s + 5)/(s^2 + 3s + 2)$

(c) $Z(-3) = 2 \, \Omega$, $v_C = 0$

(d) $Z(0) = 5 \, \Omega$, $v = V = 15$ V

30. (a) $i_n = Ae^{-t} \cos (2t + \phi)$,
$v_n = B_1 + B_2 e^{-2.5t}$

(c) C in series with 50 Ω || 20 H

33. $v(t) = 12e^{-t} - 3e^{-2t}$ V

36. (a) $(s^2 LC + sRC + 1)/(sL + R)$

(b) Pole of Y at -2, zeros at $-1 \pm j3$

(c) $v(t) = 16.7e^{-t} \cos (3t + 53.1°)$ V

39. (a) 0, -8 V/s

(b) Pole at -2, zeros at $-2 \pm j1.414$

(c) $v_C(t) = -5.66e^{-2t} \sin 1.414t$ V

Chapter 5

2. $2e^{-2t}$ A, $-8e^{-2t}$ V, $2.8e^{-2t}$ A

5. 8 V

7. (a) 3000 Ω (b) $-500 \, \Omega$ (c) $1179\underline{/-82°} \, \Omega$

10. (a) $10\underline{/83.1°}$ (b) $1.73 + j5$ (e) $17.7 - j11$

(f) $0.69\underline{/-113.1°}$ (g) $4\underline{/60°}$

12. $14.33 \cos (\omega t + 12.37°)$ V

15. $9.24 \sqrt{2} \cos (5000t + 90°)$ V

18. $4 - j3 \, \Omega$

21. (a) $C_s = 2.6 \, \mu$F, $R_s = 76.9 \, \Omega$

24. (a) $1.12\underline{/-63.4°}$ S (b) $2.5\underline{/0°}$ A,
$5.6\underline{/-63.4°}$ A, $5\underline{/-90°}$ A

27. (a) $9.0\underline{/0°}$ A, $225\underline{/-90°}$ V

(b) $16\underline{/-90°}$ A, $18.4\underline{/-60.6°}$ A

(d) $18.4\sqrt{2} \cos (2000t - 60.6°)$ A

30. $20\underline{/0°}$ A, $15\underline{/90°}$ A

32. (b) $f = (1/K) \int_0^t (u_2 - u_1) \, dt$
$= M \, du_1/dt + Du_1$
$v = (1/C) \int_0^t (i_2 - i_1) \, dt$
$= L \, di_1/dt + Ri_1$

35. (b) $v_d = (i_a - i_b)/G_d + L_d \, di_a/dt$
$(i_a - i_b)/G_d = (1/C_d) \int_0^t i_b \, dt$

Chapter 6

2. (a) $V = Ri + L \, di/dt$

4. (a) $(V/R)e^{-t/RC}$, $V(1 - e^{-t/RC})$

(b) $10e^{-100t}$ mA, $20(1 - e^{-100t})$ V

(c) $12.5e^{-100t}$ mA, $20 - 25e^{-100t}$ V

7. (a) $(V/2R) (1 + e^{-2t/RC})$

(b) $(V/2) (1 - e^{-2t/RC})$

10. $t = 5.56$ ms

13. (a) $V_S(1 - e^{-2t/RC})$

(c) $(4V_S/3) \, e^{-3(t-3RC)/2RC} - V_S/3$

16. $i(t) = 0.2(e^{-100t} - e^{-200t})$ A

19. $i_0 = 0$, $i_\infty = V/R_1 \| R_2$, $v_0 = 0$, $v_\infty = V$

21. (b) $8e^{-4t} \sin 3t$ V

24. (a) $9 + Ae^{-2t} \cos (t + \phi)$

26. (c) $2e^{-0.2t}$ A

28. $i(t) = 5 \cos 1000t$ mA

Chapter 7

2. (b) 1800 W, 3000 VA, 2400 VAR, 0.6

5. (a) 100 A, 17.3 kVAR (b) 1 Ω, 1.53 mF,

(c) 4 Ω, 1.15 mF

9. 113 kVA, 0.707

12. (a) $400 + j400 = 566\underline{/45°}$ kVA (c) 14.1 A

(d) 223 kVAR leading

13. Assuming $R = 10$ kΩ, $L = 0.8$ H

17. (a) 3.5 V (b) 0.18 V

20. 99 to 870 pF

23. (b) $V_C = QV$ (c) $V_L = (1 + jQ)V$

(e) $\tan^{-1} Q$

25. (a) 20 pF (b) 400 V (c) 9.52 V

(d) 50 mA

28. (a) 100 rad/s (b) $Q_L = Q_C = 100$, $Q = 50$

(c) 5 kΩ, 4 mA, 200 mA, 200 mA

31. (a) 4.4 kΩ, 0.71 H, 90 pF in series

34. (a) Coil (25 Ω, 3.98 mH) in parallel with
15.9-nF capacitor

37. $I_{ab} = 22\underline{/0°}$ A, $I_a = 38.1\underline{/-30°}$ A

40. 120 V

43. (b) 6.9 A, 6.9 A (d) 2.5 kW

46. 3.8 Ω, 5 kW

49. 10.23 kW

Chapter 8

2. $173\underline{/22.9°}$ V, $77.7\underline{/-120°}$ V
5. $4\underline{/-66°}$ A
8. $8.94\underline{/-26.6°}$ A
11. $\mathbf{Z}_T = 7.07\underline{/-45°}$ Ω, $\mathbf{V}_T = 7.07\underline{/-8.1°}$ V,
    $\mathbf{I}_Z = 10\underline{/-53°}$ A
14. (b) 0.16 nW
17. $5\underline{/-45°}$ A
19. $Z_{31} = Z_{23} = 10$ Ω, $Z_{12} = 5$ Ω
23. (a) $10\underline{/90°}$ Ω, $5\underline{/-90°}$ A (b) $\mathbf{I}_R = 0$
25. $10 - 0.42 \cos 500t - 0.0025 \cos 2000t$ V
29. (a) 1250 V, 2 A (b) 2500 W (c) 625 Ω

Chapter 9

3. AB, $C - 1/B$, $AB/(1 + A - ABC)$
6. (b) $BC/[1 - B(C - A)]$
8. $\mathbf{G} = 1/\mathbf{H} = 200\underline{/0°}$
11. (a) $\mathbf{H} = 0.19\underline{/0°}$ (b) G_F up 65%, unstable
13. (a) $R_F = 0.991$ MΩ
    (b) G_F up 0.002%, stable
16. (a) $T(s) = 1/(KDs + 1)$
    (c) $X_1(1 - e^{-t/KD})$
19. (a) Ae^{-3t} (b) $25(1 - e^{-2t})$
22. (a) 13 dB (b) 6 dB (c) -13 dB
    (d) -20 dB
24. (b) 7500 dB (c) 10^{744} W
27. (a) $1000/(1 + j\omega/\omega_2)$ (b) $200\underline{/1.15°}$
    (c) 5000 rad/s

Chapter 10

3. $-70{,}330$ m/s, 2.25×10^{-21} J, -70.33 μm,
    2.25×10^{-21} J
6. (a) $x = 10$ cm, $y = 0$ cm (b) -100 V
9. (b) 8.53 μm, 0.893 ns
12. $f_e = -1.6 \times 10^{-17}$ N, $f_m = -8 \times 10^{-18}$ N,
    $f_g = 8.9 \times 10^{-30}$ N
15. (a) 1.875×10^7 m/s (b) 80 V
16. (a) 0.02 cm/V (b) 1250 V
    (c) 0.0033 cm/V
19. (a) 125 Hz (b) 2500 Hz

Chapter 11

1. 22.4 A
4. 1.83×10^9 atoms/e-h pair, 0.45 Ω-m
7. (a) $2 \times 10^{20}/\text{m}^3$, $1.1 \times 10^{12}/\text{m}^3$, p-type
    (b) $3.1 \times 10^{21}/\text{m}^3$, $2.9 \times 10^{21}/\text{m}^3$, nearly
    intrinsic
10. (c) $5 \times 10^{21}/\text{m}^3$, $4.5 \times 10^{10}/\text{m}^3$
    (d) 1 mV (e) 245,000
14. $1.56 \times 10^{23}/\text{m}^3$
16. $39{,}330$ A/m²
19. (a) -0.098 V (b) 0.98 mA

23. (a) 8 mA (b) -20 μA (c) 4.85 mA
25. 0.48 A
28. 1.4, 1.4, 0.7, -0.3 V
32. (a) 12 V, 2 Ω (b) 5 Ω (d) 13.5 V

Chapter 12

2. 1.5 V
5. 10
8. (b) 2 mA, 10 V
11. (a) 1.2 V (b) $i_D = (V_{DD} - v_{DS})/(R_D + R_S)$
    (c) 1.2 kΩ, 14.2 V
17. From 0.04 to 2.22 mA
21. (a) 0.7 V (b) 0.7 V, 1.4 V
    (c) 0.1 mA, 3 mA, 2 V
24. 13.5¢
27. $A = 37.5$ mil²
29. 690 Hz
32. (a) 4.77 A, 1608 W (b) 2.36 A, 6.76 W
    (c) 4.77 W

Chapter 13

3. (b) $2 < V < 3$ V, say 2.5 V
5. 2^n
8. **A NOR A** = $\overline{\mathbf{A}}$, $\overline{\mathbf{A}}$ **NOR** $\overline{\mathbf{B}}$ = $\mathbf{A} \cdot \mathbf{B}$
11. (a) 50, 0.98 (c) 0, 0, 15 V
    (d) 0.2 mA, 10 mA, 5 V; 0.42 mA, 14.7 mA,
    0.3 V
    (e) 15 to 0.3 V
14. 6.4 Ω, 480 MΩ; 25 Ω, 14.2 kΩ
17. (a) 1.7 V, < 1 V, 0, 5 V, T_2 **OFF**
    (b) T_2 **ON**, 2.1 V, 1.4 V, 0.5 mA, 0.3 V
20. (a) T_1 **ON**, $V_x = 2.4$ V, $V_y = 1.0$ V,
    $V_F = 0.2$ V
    (b) T_1 **OFF**, $V_x = 1.0$ V, $V_y = 5$ V,
    $V_F = 3.4$ V
23. **ON, ON**, 2.1, 1.4, 0.7, 0.3, >1.4, >1.4 V
    OFF, OFF, 1.0, <1.4, <0.7, $\cong 4.5$, 0.3,
    0.3 V
27. Like 3-input **NOR** gate of Fig. 14.24b
31. (a) 0.3, 5 V (b) 0.3, 5 V (c) 5, 0.3 V
    (d) 5, 0.3 V (e) 0.3, 5 V
34. (c) Divides frequency by four
37. 1000-Hz square wave
38. $V_{\text{out}} = 10$ V, 0 V

Chapter 14

1. (a) **00100, 01100, 10111, 11111**
    (b) **3, 5, 10, 21**
3. (a) **10100101B = 245Q = A5H**
    (b) **205D = 315Q = CDH**
    (c) **10000111B = 135D = 87H**
    (d) **11011000B = 330Q = 216H**

7. 93D = 1011101B = 135Q

10. A

14. $A \cdot B = \overline{\overline{A} + \overline{B}}$

$= \overline{\overline{A} + \overline{A} + \overline{B} + \overline{B}}$

17. (c) Input mode, **Q** follows **DI**; Output disconnected.

20. $f = A\overline{B} + \overline{A}B$

22. $\overline{B}\overline{D}$

27. (b)

CK	Q_D	Q_C	Q_B	Q_A
0	0	0	0	1
1	0	0	1	0
2	0	1	0	0
3	1	0	0	0
4	0	0	0	1

29. (b) **0000** vs. **1010** for binary; ∴ must eliminate two 1s   (c) Make $CK_B = Q_A$ **AND** $\overline{Q}_D$
(d) Make $CK_D = Q_C$ **OR** (Q_A **AND** Q_D)

33. (c) RS = 1, CS_1 = 0, CS_2 = 1
(d) 6 Column Select lines.

35. (c) Let Inputs = Addresses and Outputs = Data for 8 × 1 **ROM**

38. (b) Routes Input to selected Output line

40. (a) **F** = **A** minus **B**, $\overline{CY}$ = Borrow

Chapter 15

4. (a)

0100	**INPUT TO A**
0010	**FROM PORT 2**
1010	**LOAD**
0001	**C FROM A**

(d)

1101	**LOAD**
0010	**D WITH**
0011	**0011**
1110	**LOGIC OP A**
1110	**AND WITH D**

6. (a)

CODE	INSTR
1101	**LOAD A**
0111	**WITH**
0001	**0001**
1000	**ADD TO A**
0000	**B**
1010	**LOAD FROM A**
0000	**B**

9. (a)

0100	**INPUT TO A**
0011	**FROM PORT 3**
1010	**LOAD**
0010	**D FROM A**

(c)

1101	**LOAD**
0000	**B**
0000	**WITH 0000**

12. (a)

1101	**LOAD**
0011	**E**
1111	**WITH 1111**
1011	**LOAD**
0111	**A FROM MEM**
0001	**COMPLEMENT A**
1100	**LOAD MEM**
0111	**FROM A**

15. (a) Perform 5-cycle preparation routine then execute 5-cycle loop 15 times.

(c)

LOCS	CODE
0000	1101, 0111, 1111
0011	1010, 0000
0101	0111, 0000
0111	1111, 1000, 0101

18. Use: $10 = 8 + 2 = 2^3 + 2$

IN A, (1H)	**INPUT N FROM PORT 1**
RLCA	**MULTIPLY BY 2**
LD B, A	**STORE 2N IN B**
RLCA	**MULTIPLY 2N BY 2**
RLCA	**MULTIPLY 4N BY 2**
ADD A, B	**ADD 8N + 2N = 10N**
OUT (2H), A	**OUTPUT 10N TO PORT 2**
HALT	**STOP PROCESSING**

23.

	ORG 0010H
	LD B, FAH
	LD HL, 0100H
LOOP:	**INC L**
	LD A, (HL)
	XOR B
	JPNZ, LOOP
	LDA, L
	OUT (5H), A
	HALT
	END

26. STORE TEST PATTERN IN A
SET HL POINTER TO 1000H MINUS 1
INCR POINTER L
STORE TEST PATTERN IN MEM
COMPARE PATTERN WITH MEM
IF SAME, JUMP TO INCR L INSTR
IF NOT, DECR L POINTER TO LAST WORKING
RAM ADDR

Chapter 16

2. (a) R_1 = 5 kΩ, R_F = 1 MΩ, inverting
(c) for R_1 = 5 kΩ, R_F = 95 kΩ, noninverting

5. For R_1 = 1 kΩ, R = 200 kΩ, C = 795 pF

7. $-10/(1 - j\,100/\omega)$, high-pass filter

12. (a) 10 kΩ   (b) *V*-to-*I* converter, $R_1 = 1$ kΩ,
10 kΩ   (c) 5.7 V, 6.0 V
17. 100 MΩ, 0.1 Ω
18. For $R_i \gg (R_F + R_o)$ and $R_F \gg R_o$,
$R_{iF} = (R_F + R_o)/(A - R_o/R_F) = 0.011$ Ω
20. (b) $H = 0.462$
22. 109 or 40.7 dB
24. $A = 520$, $A_{cm} = 49.5$, CMRR = 10.5
26. (a) $f_2 = 10$ Hz   (b) $f_2 = 10$ kHz
31. (a) $d^2y/dt^2 - ky = 0$
(b) $y(t) = A \cos \sqrt{k}\, t + B \sin \sqrt{k}\, t$

Chapter 17

2. (a) 2 mA, 8 V   (c) 0.55 sin ωt mA   (d) 6.6
5. (b) 5.5 kΩ, 3 kΩ
6. (a) 2 mA, 8 V   (b) 1.8 mA, 9.6 V
10. (a) −6.5 V   (b) 4.7 kΩ
13. (a) 13 V   (b) 1.95 V, 10.2 kΩ   (c) 1.6 mA
14. (c) 18.1 kΩ, 61.8 kΩ, 6 kΩ
18. (a) $R_E = R_C = 6$ kΩ, $R_1 = 18.3$ kΩ,
$R_2 = 34.8$ kΩ
(b) $0.46 < I_C < 0.51$ mA
21. (a) $v_{CE}(\text{sat}) = 0$, $I_{CEO} = 0$, $V_{CE} = 25$ V,
$I_C = 400$ mA
(b) $P_o = 5$ W, $\eta = 25\%$
23. $N_2/N_1 \cong 3.5$, $\eta \cong 50\%$
26. (a) 11.6 W, 6 W, 0.6 W, 5 W, 0
27. (b) 500 Ω   (c) 10 W, 2.5 W, 25%
(d) 7.1, 50%
30. (b) One possibility: $V_{CE} = 25$ V,
$I_C = 250$ mA
(c) 3.125 W, 3.125 W, 25%
(d) $N_2/N_1 \cong 1.25$, 5 W, 5 W, 50%

Chapter 18

3. 2100, 50 krad/s
7. (a) 0, 3, 200, 203, 206 krad/s
(c) If load responds to audio only.
8. $r_{ac} = 1/(1 + 0.02\, v) = 0.83$ Ω
$i(t) = 11 + 1.2 \sin \omega t$
13. $g_m = 1$ mS, $r_d = 60$ kΩ
14. (a) 3000 μS   (b) 2000 μS
17. (b) 11.4
20. 4.5 mA, 2.5 mV
24. (a) npn   (b) yes   (d) $\alpha R_L/r_e$ (e) 39.6
27. (d) 0.1 mA, 4 mA, 12 V
(e) 4 sin ωt mA, 8 sin ωt V
30. (a) Min. 63, Typ. 76, Max. 92
32. $720 < r_\pi < 3030$ Ω; no

Chapter 19

3. (a) 10 kΩ   (b) $-30\underline{/-53°}$
7. (a) $R_D \| R_L \| r_d = R_d$, $R_s + R_1 \| R_2 = R_1 \| R_2$
(b) −10
9. (a) 0.65 mA, 4 kΩ   (b) 6 kΩ   (c) 114
(d) 71
11. (b) 4.4 kΩ   (c) 1.5 kΩ, 15 V
13. 0.8 μF
15. (b) 0.33 μF
17. (a) 16 Hz, 81 Hz   (b) 81 Hz
(c) 1 μF for $R_L = r_\pi \| R_B \cong 1$ kΩ
20. (b) $C_{be} = 100$ pF   (c) 6.4 MHz
23. 64 krad/s
26. −74, 80 kHz
30. (a) 62 dB   (b) 72 dB, 52 dB
34. $f_2 = 0.6$ MHz, 0 dB, 23.5 dB
36. 3
39. $\mathbf{G} = 200\underline{/0°}$
42. $0.0067\underline{/180°}$
44. $dG_F/G_F = (dG/G)(1/1 - GH)$
47. (b) 1%   (c) 30%
50. $H = -0.043$
52. (a) $H = -0.095$
54. 1.9 V, $R_i = 73$ kΩ, $R_o = 53$ Ω
56. (a) 0.32 MHz   (b) 3.2 MHz

Chapter 20

3. $Qu = lANqu = l \cdot dQ/dt = li$
6. $\mathbf{I} \cong 4$ A
8. (a) $H = 796{,}000$ A/m, $M = 16$ A/m
(b) $H = 318$ A/m, $M = 796{,}000$ A/m
11. At $D = 1$ mm, $B = 2$ mT; at $D = 1$ cm,
$B = 0.2$ mT
14. Ni = 320 A·t
17. (a) 300   (b) 1050   (c) 5300   (d) 4200
19. 850 A
21. (a) 11 A   (b) 0.22 A
23. (a) 0.7 A   (b) 4.2 A   (c) impossible
27. 12.3 A

Chapter 21

1. 12,000 V
3. 1.05 mH, 2.1 H
6. (a) 312 t   (b) 0.163 A
10. $N' = 1.18\, N_o$
14. (a) 1250, 15 t   (b) 0.6 A
16. 30 A
19. $a = 4$, $G = 0.5$ mS, $B = 2$ mS, $R = 2$ Ω,
$X = 6$ Ω

21. (a) $253\underline{/1.7°}$ V   (b) $20.1\underline{/-1.4°}$ A
25. (b) $2.4\underline{/-90°}$ A, $80\underline{/-39°}$ A   (c) 97.8%
27. $a = 1/20$, 20 dB
30. (b) 21.8 A, 43.6 A
33. (a) 94.7%   (b) 87.6%
36. (b) 3%
39. 1%, 98.5%

Chapter 22

2. (a) $\mu_o I/2\pi D$, $I/2\pi D$ (b) $eU\mu_o I/2\pi D$
   (c) $\mu_o I\, I_2\, dl/2\pi D$ (d) $\mu_o I\, I_3\, dl/2\pi D$ (e) 0
4. (a) $\beta l\, u^2 t/R$ (b) $\beta^2\, l^2\, u^3 t^2/R$
8. 955 A·t, 30 t
11. 0.03015 Ω, 9994 Ω
13. (a) 1111 Ω, 101 Ω, 10.01 Ω
15. (a) 9.55 A, 10.6 A
17. (a) $(NBlrn\pi/15)\cos(n\pi/30)t$
   (b) $T_m/2NBlr$, $\omega T_m \cos \omega t$
19. (a) 240 V   (b) 57.3 N·m, 12 kW
22. (a) 60 Hz   (b) 251 V
25. 3056 rpm
28. 0.004 N·m
30. (a) $N(\mu_o \mathscr{F}A/2l_a^2)u$ (b) $B^2 A/\mu_o$
33. 10 A

Chapter 23

2. (b) 270 V   (c) 223 V
4. (a) 0.15 Ω   (b) 0.051 N·m/rad/s
   (c) 173 N·m
5. (b) 1.1 A, 1.3 A
8. 275 V, 45 V
11. (a) 250 V   (b) 2.4 A   (c) flat
16. (a) I_A same, $n/2$ (b) $I_A \times 2$, $n/4$
   (c) $I_A/\sqrt{2}$, $n/\sqrt{2}$
18. (b)   0.59 Ω
20. (a) 1300 rpm   (b) 1950 rpm   (c) 83.3 A,
   122 N·m, 680 rpm
22. (a) 1.3 A   (b) −75 A, 152 N·m
26. 89%

Chapter 24

1. (b) 1500 rpm
2. (c) 4
5. (a) 12 poles, 13,600 A   (b) $21,000\underline{/33.4°}$
8. $X_s = 0.5$ Ω
11. I must be lagging V.
13. (b) 29°, 0.66
16. 6.7 A
19. 655 MW
22. (b) $n_2 = n_1$, $\delta_2 \cong \frac{1}{2}\delta_1$, $\theta_2 > \theta_1$, $I_2 \cong \frac{1}{2}I_1$
23. For $I_F = 1.5$ A, $I \cong 39$ A; $I_F = 3$ A,
   $I \cong 55$ A
24. Old transformer ok; new pf = 0.83
28. 2 V, 0.067 V
32. (a) $V_A = 240$ V   (b) $B_0 = -0.2$ S
    $G_0 = 0.018$ S
   (c) 82 A, 218 N·m, 54 hp
36. 33.2, 165 N·m
41. $n_2 = 900$ rpm, $T_2 = 594$ N·m, $P_2 = 75$ hp

Chapter 25

2. (b) Antenna direction, potentiometer, voltage
   difference detector, input direction
4. (a) Pole at $s = -2a$, zero at $s = -a$
   (b) $y(t) = -\frac{1}{2} + \frac{3}{2}e^{-2at}$
7. (a) $10/(s + 2)$ (b) $6s/(s + 0.02)$
10. (a) $T(s) = \infty$ means output with no input.
   (b) $y_n(t) = A_1 + A_2 e^{-bt}$
14. (a) $T(j\omega) = 1/(0.1 + j\omega)$
   (b) $y(t) = 10 - 10e^{-t/10}$
15. (a) 100 V   (b) $V(t) = 104 + 63e^{-5t}$ V

Index

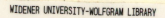

List of Symbols

a acceleration,
turn radio

A amplifier gain,
area

A   amperes

B magnetic flux density,
susceptance

B   binary

BW   bandwidth

C capacitance

C   Celsius
coulombs

C_p thermal capacity

d distance

dB   decibel

D   diffusion coefficient,
electric flux density,
frictional resistance

e charge of an electron,

e, E electromotive force

e natural log base 2.718 . . .

$\mathcal{E}$ electric field strength

f frequency

f, F force

F   farads

$\mathfrak{F}$ magnetomotive force

g thermal generation rate

g_m transconductance

G conductance

G   transfer function

h hybrid parameter,
Planck's constant

h   hours

H magnetic field intensity

H   henrys

H   feedback function

H   hexadecimal

i, I current

IC   integrated circuit

j complex operator $\sqrt{-1}$

J current density,
rotational inertia

J   joules

k Boltzmann's constant

k, K constant of proportionality

K compliance

K   degrees Kelvin

l length

L inductance

m mass of an electron

m   meters

M mass,
mutual inductance

n electron concentration,
rotational speed

n, N number of elements

N   newtons

p hole concentration,
number of poles

p, P power

q, Q electric charge

Q   octal

q_T rate of heat flow

Q quality factor

r radius,
ripple factor,
small-signal resistance

R recombination rate,
resistance

R_T thermal resistance